M. S. Heller

D0881321

PHYSICAL ORGANIC CHEMISTRY

McGRAW-HILL SERIES IN ADVANCED CHEMISTRY

Senior Advisory Board

W. Conard Fernelius   Louis P. Hammett   Harold H. Williams

Editorial Board

David N. Hume   Gilbert Stork
Edward L. King   Dudley R. Herschbach
John A. Pople

BAIR   Introduction to Chemical Instrumentation
BALLHAUSEN   Introduction to Ligand Field Theory
BENSON   The Foundations of Chemical Kinetics
BIEMANN   Mass Spectrometry (Organic Chemical Applications)
DAVIDSON   Statistical Mechanics
DAVYDOV   (*Trans.* Kasha and Oppenheimer)   Theory of Molecular Excitons
DEAN   Flame Photometry
DJERASSI   Optical Rotatory Dispersion
ELIEL   Stereochemistry of Carbon Compounds
FITTS   Nonequilibrium Thermodynamics
FRISTROM AND WESTENBERG   Flame Structure
HELFFERICH   Ion Exchange
HILL   Statistical Mechanics
HINE   Physical Organic Chemistry
KIRKWOOD AND OPPENHEIM   Chemical Thermodynamics
KOSOWER   Molecular Biochemistry
LAITINEN   Chemical Analysis
MANDELKERN   Crystallization of Polymers
McDOWELL   Mass Spectrometry
PITZER AND BREWER   (*Revision of* Lewis and Randall)   Thermodynamics
POPLE, SCHNEIDER, AND BERNSTEIN   High-resolution Nuclear Magnetic
    Resonance
PRYOR   Free Radicals
PRYOR   Mechanisms of Sulfur Reactions
ROBERTS   Nuclear Magnetic Resonance
ROSSOTTI AND ROSSOTTI   The Determination of Stability Constants
SOMMER   Stereochemistry Mechanisms, and Silicon
STREITWIESER   Solvolytic Displacement Reactions
SUNDHEIM   Fused Salts
WIBERG   Laboratory Technique in Organic Chemistry

PHYSICAL ORGANIC CHEMISTRY

JACK HINE

Professor of Chemistry
Georgia Institute of Technology

Second Edition

McGRAW-HILL BOOK COMPANY, INC.   1962

New York   San Francisco   Toronto   London

ALBRIGHT COLLEGE LIBRARY

PHYSICAL ORGANIC CHEMISTRY

Copyright © 1962 by the McGraw-Hill Book Company, Inc. Printed in the United States of America. All rights reserved. This book, or parts thereof, may not be reproduced in any form without permission of the publishers. *Library of Congress Catalog Card Number* 61-18627

IV

28929

THE MAPLE PRESS COMPANY, YORK, PA.

547
H662p2
c.2

218545

PREFACE

A broad definition of the term "physical organic chemistry" might include a major fraction of existing chemical knowledge and theory. As the title for the present book, the term is used in a considerably narrower sense to refer to the mechanisms of organic reactions and the effect of changing reaction variables, particularly reactant structures, on reactivity in these reactions. In order to facilitate consideration of the latter topic, certain aspects of structural theory are discussed in Chap. 1. More than half of the book deals with polar reactions, but space is allotted to free-radical reactions in reasonable accord with their importance and with the extent to which they are presently understood. Multicenter reactions are also treated separately, and because of the rapid advances that have been made in recent years in the chemistry of divalent-carbon intermediates, a chapter on "methylenes" has been added in the second edition. Significant changes have been made in almost every chapter; in particular, the treatment of linear free-energy relationships has been expanded and brought together into one chapter and increased space has been given to consideration of kinetics and its relation to reaction mechanisms. Despite the tremendous practical importance of heterogeneous reactions, their consideration is largely omitted both because of space limitations and because they are in general more poorly understood than homogeneous processes.

Of the vast amount of research in the area thus selected it is possible to mention only a small fraction. The investigations discussed have been chosen because of their relation to the principles the author believes to be most important, but, even so, their choice has often been necessarily arbitrary. Certain topics have been chosen for discussion in some detail, both for their own sake and because it is believed that detailed discussions give the reader viewpoints that could never be gained in any other way. The experimental evidence for reaction mechanisms and the logical methods by which these mechanisms are formulated from experimental data are emphasized throughout.

v

218545

This book is written primarily for graduate students and advanced undergraduates.   Only a good knowledge of the standard undergraduate courses in organic and physical chemistry is assumed.   While no more knowledge of mathematics and physical chemistry is required than may be expected of the student at this level, the knowledge he has is not ignored but used where needed.

I gratefully acknowledge my indebtedness to Dr. P. K. Calaway, Dr. R. L. Sweigert, and Dr. W. M. Spicer for valuable encouragement; to Dr. Erling Grovenstein and Dr. W. H. Eberhardt for many helpful discussions; to Drs. A. I. Turbak, N. W. Burske, L. H. Zalkow, P. E. Robbins, and A. M. Dowell, Jr., who read and criticized parts of the original manuscript; to Drs. O. B. Ramsay and H. E. Harris who similarly assisted with the revised version; and above all to my wife for help at every step of the way.

Jack Hine

CONTENTS

Chapter 1

THE STRUCTURE OF ORGANIC MOLECULES

The representation of electronic structures of organic molecules most commonly used in discussions of reaction mechanisms is that developed by G. N. Lewis,[1] in which only the outer shell of electrons is shown. We shall use a common variation of Lewis's notation, letting the usual line drawn between two atoms represent the bonding electron pair and denoting unshared electron pairs by lines parallel to the sides of the atomic symbols. Unpaired electrons are shown by a dot. Some electronic formulas of this type are shown below.

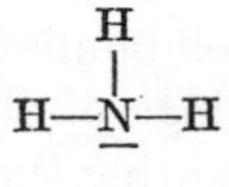

$$
\begin{array}{ccc}
\underset{\text{Ammonia}}{\overset{\displaystyle H}{\underset{\displaystyle H}{H-\underline{N}-H}}}
&
\underset{\text{Acetic acid}}{\overset{\displaystyle H\quad |\overline{O}}{\underset{\displaystyle H}{H-\overset{|}{\underset{|}{C}}-\overset{\|}{C}-\overline{O}-H}}}
&
\underset{\text{Hydrogen atom}}{H\cdot}
\end{array}
$$

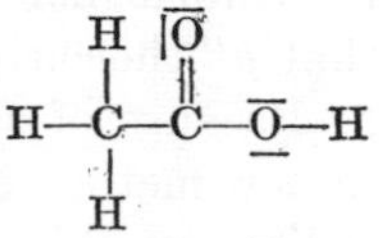

1-1. Resonance. Structural formulas have been of great utility to organic chemists because of the large amount of information they compress into a small space. For example, on seeing the structure

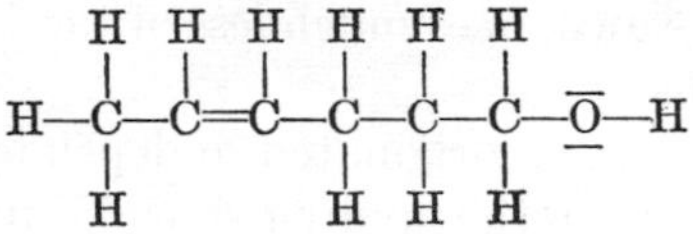

$$H-\overset{H}{\underset{H}{C}}-\overset{H}{C}=\overset{H}{C}-\overset{H}{\underset{H}{C}}-\overset{H}{\underset{H}{C}}-\overset{H}{C}-\overline{O}-H$$

the chemist recognizes a compound that would be expected to add bromine (since most compounds with carbon-carbon double bonds do) and to react with acetic anhydride to yield an ester (a characteristic of compounds with the —CH$_2$OH grouping). The properties of various atomic groupings vary, of course, from compound to compound but often remain within a small enough range for a wide enough variety of compounds to permit the prediction of chemical and physical properties to be made with confidence. Often, however, the properties of compounds containing certain groups differ so widely from those that have

[1] G. N. Lewis, "Valence and the Structure of Atoms and Molecules," Reinhold Publishing Corporation, New York, 1923.

1

become associated with the groups that ordinary structural formulas become of little value. For example, from a knowledge of aliphatic chemistry it would be difficult to understand why the formula

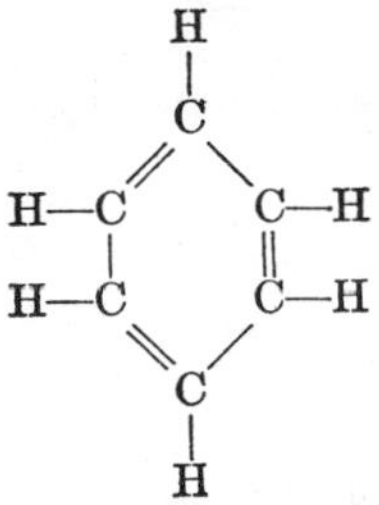

should represent a compound of relatively great stability, which does not add bromine nor reduce permanganate under ordinary conditions. While the reason for this unexpected behavior has been the subject of a great deal of discussion since the Kekulé formula was proposed in 1865, the "aromatic" properties of a system of three alternating double bonds in a six-membered ring were essentially grafted onto organic structure theory as an additional postulate.   Unfortunately, neither the classical method of writing structural formulas nor its improvement by Lewis yields a formula that shows that *all* the carbon-carbon bonds in benzene are equivalent to each other.   Because of shortcomings of this type it became desirable to introduce a new method of representing the structure of molecules, or at least to modify the old method suitably.   The resonance method of describing structures is a modification of the classical method and has the advantage of retaining most of its useful and familiar aspects.   The molecular orbital method (see Sec. 1-2) of describing the structure of molecules is essentially a new method, which, while having many advantages of its own, uses much less of the familiar terminology of classical organic chemistry.

The theory of resonance originated independently from quantum-mechanical calculations on the hydrogen molecule and from the early theory of the English school of physical organic chemists.[2] We shall discuss only *valence-bond resonance, the description of organic molecules in terms of Lewis electronic structures.*   In this method of representation, the Lewis structures in terms of which the molecule is to be described are joined by double-headed arrows, e.g.,

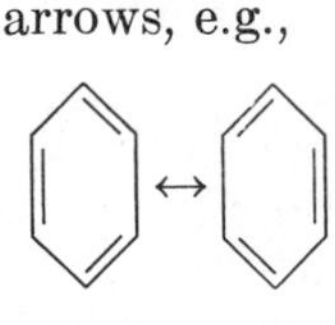

[2] A much more complete discussion of resonance is given in G. W. Wheland, "Resonance in Organic Chemistry," John Wiley & Sons, Inc., New York, 1955.

The two formulas and the arrow between them are taken together as a description of the benzene molecule.   The double-headed arrow should not be confused with the two half arrows ($\rightleftharpoons$) signifying equilibrium, since the benzene molecule does *not* oscillate between the two structures shown but instead has a definite structure of its own.   Certain other methods of notation have been used, such as

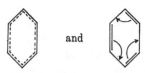

and

which probably make clearer the fact that only one type of molecule is involved, but the use of the "resonance arrow" is probably the most common.

1-1*a*. *Rules for Resonance.*   The resonance description of a molecule in terms of Lewis structures is governed by the following rules:

1. Any compound for which more than one Lewis structure may be written is accurately described by none but is said to be a *resonance hybrid* of them all.   The various structures are called *contributing structures*.   The *extent* to which a contributing structure is said to contribute to the total structure of a resonance hybrid is measured by the extent to which the hybrid has the properties that would be expected from the given structure.   However, the properties of a resonance hybrid are not just an average of the properties of the contributing structures.

2. In addition, *the stability of a resonance hybrid is greater than that which would be expected of any of the contributing structures.*   The energy content of a resonance hybrid is lower than that of the most stable contributing structure by an amount that is referred to as the *resonance energy*.   The hybrid is therefore said to be *stabilized by resonance*.

Note that the actual molecule would be expected to assume the most stable configuration.   From X-ray, electron-diffraction, and spectroscopic determinations of structure in the case of benzene, for example, it is clear that the actual structure is not that which would be expected of a contributing Kekulé structure.   Hence the resonance hybrid is more stable than any contributing structure.

3. The greater the number of important contributing structures, and the more nearly equal their contributions, the greater the resonance energy.

4. The greater the stability, i.e., the lower the energy content, to be expected of a contributing structure, the greater will be its contribution to the total structure of the hybrid.

5. *A structure will not contribute if it has a different number of unpaired electrons from the actual molecule.*   According to quantum mechanics, it

is not possible to specify simultaneously the position and velocity of any electron. However, it is possible in principle to obtain a mathematical function, called the *wave function*, whose square (the *probability-density function*) indicates the probability of finding an electron within any given volume element of the system. For a given position of the atomic nuclei in space and a given energy content the electrons of the molecule will move in harmony with a certain definite probability-density function. In valence-bond resonance we are, in effect, describing this function in terms of certain others (of whose properties we know more) in accord with which the electrons could (in theory only) be caused to move. This is not the same as a description in terms of a distribution function for the electrons when some have been paired or unpaired. Furthermore, the pairing or unpairing would change the energy content of the system. For instance, structure I is not a contributing structure for ethylene (II). It is, rather, a different type of molecule, an excited state of ethylene.

Since the relative contributions of various Lewis structures to the total electronic structure of a system of atomic nuclei in space (a molecule), and hence the resonance stabilization, depend on the stability to be expected for the various structures, it is desirable to have a set of rules for estimating this stability.

A. Other things being equal, the greater the number of covalent bonds, the greater the stability. This follows from the known generalization that covalent bonds stabilize a system. The fact that the union of two hydrogen atoms to form a hydrogen molecule causes the liberation of 103.4 kcal/mole,

$$2H\cdot \rightarrow H_2 \qquad \Delta H = -103.4 \text{ kcal/mole}$$

shows that the system H—H containing a covalent bond is 103.4 kcal more stable than the system $2H\cdot$. Exceptions to this rule are known; e.g., the oxygen molecule has two unpaired electrons that could be written as paired to give an additional bond.

For the rule to hold rigorously, it would be necessary for all covalent bonds to have the same bond energy. This is not the case, but in estimating the relative stability of contributing structures the rule is still very useful. One reason for this is that in comparing two structures A and B with n and $n + x$ covalent bonds, it is found that all, or almost all, of the n covalent bonds of A are also present in B, so that most of the differences in individual bond energies cancel when the total energy content of A is compared with that of B.

B. *Other things being equal, a structure with a negative charge on the most electronegative element will be more stable.* This rule predicts which of several Lewis structures of the type

$$\overset{\oplus}{X} \; |\overset{\ominus}{Y} \quad \text{and} \quad X| \; \overset{\ominus}{} \; \overset{\oplus}{Y}$$

will be the more stable and hence contribute more to the total structure of the molecule. This relative stability is a measure of the electron-attracting power of X and Y and might be expected to be related to their ionization potentials and electron affinities.[3] Thus it is not surprising that in this prediction we shall use *electronegativities,* since they are roughly proportional to the average of the ionization potentials and the electron affinities of the various elements. Electronegativities for some elements of common occurrence in organic molecules, on the most widely used scale, that of Pauling,[4] are shown in Table 1-1.

TABLE 1-1. ELECTRONEGATIVITIES OF VARIOUS ELEMENTS[4]

Fluorine	4.0	Selenium	2.4
Oxygen	3.5	Phosphorus	2.1
Chlorine	3.0	Hydrogen	2.1
Nitrogen	3.0	Arsenic	2.0
Bromine	2.8	Boron	2.0
Iodine	2.5	Silicon	1.8
Sulfur	2.5	Magnesium	1.2
Carbon	2.5	Sodium	0.9

The molecule X—Y *could* be described in terms of the structures

$$\underset{A}{\overset{\oplus}{X} \; |\overset{\ominus}{Y}} \leftrightarrow \underset{B}{X—Y} \leftrightarrow \underset{C}{X| \; \overset{\ominus}{} \; \overset{\oplus}{Y}}$$

but it is more convenient to make the description in terms of a covalent structure (III), which is really a resonance hybrid of B and equal amounts of A and C, and of that ionic structure which contributes more. Thus, by writing a covalent structure, we may *imply* equal contributions of

[3] The ionization potential is the energy required to remove an electron from a neutral atom; the electron affinity is the energy liberated by the combination of a neutral atom with an electron.

$$X\cdot \rightarrow X^+ + \text{electron} \qquad \Delta E = \text{ionization potential of X}$$
$$X\cdot + \text{electron} \rightarrow X^- \qquad \Delta E = -\text{electron affinity of X}$$

[4] L. Pauling, "The Nature of the Chemical Bond," 3d ed., pp. 88–105, Cornell University Press, Ithaca, N.Y., 1960.

ionic structures and reserve the representation

$$X\!-\!Y \leftrightarrow X|\overset{\ominus}{}\overset{\oplus}{Y}$$

III   IV

for cases in which X is more electronegative than Y and IV is therefore the more highly contributing of the two possible ionic structures. Since the contribution of IV will increase with the electronegativity difference between X and Y, the contribution of the two structures will simultaneously become more nearly equal up to the point (when the electronegativities differ by about two or more) where the contribution of IV reaches that of III. An increase in the electronegativity difference would therefore be expected to increase the resonance stabilization (the extent to which the molecule is more stable than would be expected from a purely covalent structure). Since the bonds in X—X and Y—Y *must* be purely covalent, it might be expected that the bond in X—Y should have a strength halfway between X—X and Y—Y if the bond in X—Y is entirely covalent. This is indeed found to be very nearly true for molecules in which the electronegativity difference is very small. However, bonds between elements with considerably different electronegativities are definitely stronger than would be expected, due to the resonance stabilization arising from the contribution of structure IV. In fact, the numerical electronegativity values shown in Table 1-1 were based on bond-energy data, as described in Sec. 1-4c.

C. Other things being equal, the more closely the bond lengths and bond angles to be expected of the contributing structure resemble those of the actual resonance hybrid, the greater the stability.

An example of the application of this rule may be found in the case of ethane.

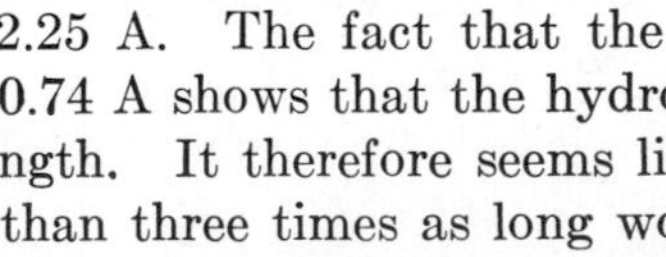

Studies of the actual geometrical configuration of the ethane molecule show that hydrogen atoms on different carbon atoms may not come closer together than about 2.25 A. The fact that the bond distance in the hydrogen molecule is 0.74 A shows that the hydrogen-hydrogen bond is most stable at this length. It therefore seems likely that a hydrogen-hydrogen bond more than three times as long would be very unstable. From similar data it may be shown that bond angles in ethane are very unfavorable for a structure of the type of VI. These and several other points indicate that VI would be very unstable and should contribute negligibly to the structure of ethane.

D. Other things being equal, the greater the separation of like charges, the greater the stability. This follows from the simple physical principle that work is required to bring two like charges closer together. Therefore, any structure with like charges close together, having had work done upon it, will have a high energy content and lower stability.

This rule usually acquires importance only when the two charges are very close together. That is, structures with like charges on adjacent atoms or especially those with a double charge on one atom tend to be unstable. Among the structures that can be written for an alkyl azide are

$$R\text{—}\overset{\oplus}{\overset{\displaystyle\bar{N}}{}}\text{=}N\text{=}\overset{\ominus}{\bar{N}|} \qquad R\text{—}\overset{\ominus}{\bar{N}}\text{—}\overset{\oplus}{N}\text{≡}N| \qquad R\text{—}\overset{\oplus}{N}\text{≡}\overset{\oplus}{N}\text{—}\overset{\ominus\ominus}{\bar{N}|}$$

| VII | VIII | IX |

There is reason to believe that structure IX does not contribute appreciably to the hybrid. This structure would not be expected to be stable because of the double negative charge on one atom and the positive charges on adjacent atoms.

E. Other things being equal, the greater the separation of unlike charges, the less the stability. This follows from the same coulombic principle that was stated under rule *D*. The converse of this rule—the less the separation of unlike charges, the greater the stability—reaches its logical extreme in the case where both of the unlike charges are on the same atom, or, in other words, the charges do not exist.

$$\underset{X}{H\text{—}\overset{\overset{\displaystyle H}{|}}{C}\text{=}\overset{\overset{\displaystyle H}{|}}{C}\text{—}\overset{\overset{\displaystyle CH_3}{|}}{\bar{N}}\text{—}CH_3} \leftrightarrow \underset{XI}{H\text{—}\overset{\overset{\displaystyle H}{|}}{\underset{\ominus}{C}}\text{—}\overset{\overset{\displaystyle H}{|}}{C}\text{=}\overset{\overset{\displaystyle CH_3}{|}}{\underset{\oplus}{N}}\text{—}CH_3}$$

In the above resonance structures for dimethylvinylamine, XI has the destabilization caused by charge separation. It is, however, stabilized by permitting nitrogen to share its electron pairs (see rule *F*).

F. Structures in which basic atoms coordinate their unshared electron pairs with adjacent atoms to form double bonds will often be stable enough to contribute significantly.

Just as basic atoms may be stabilized in intermolecular reactions by coordination through their unshared electron pair with another atom to form a covalent bond, so may contributing structures be similarly stabilized intramolecularly. While structures of this type are *most* important with the more basic elements, they are still of importance with elements whose basicity is often too small to detect in intermolecular reactions. For example, the contribution of structures XIV, XV, and XVI to the total structure of aniline is well evidenced by a number of data. Because of the contributions of these structures, the hybrid has a smaller density of unshared electrons on the nitrogen atom, which therefore is less basic than in aliphatic amines or ammonia. Also, the contributions of these structures increase the electron density in the ortho and para positions of the molecule and thus increase its reactivity (in these positions) toward aromatic substituting reagents (Sec. 16-3*a*).

XII XIII XIV XV XVI

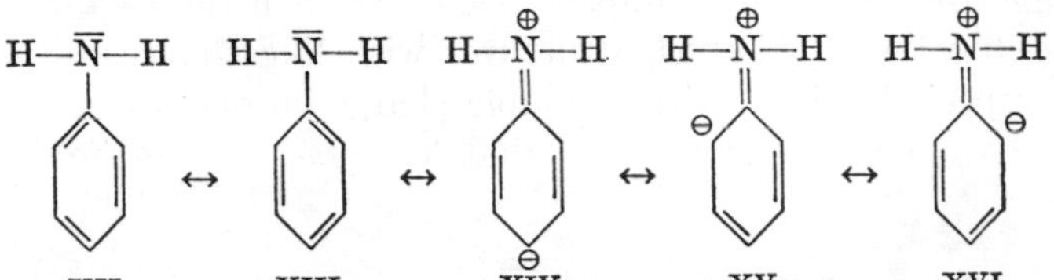

Although the contributions of structures XIX, XX, and XXI to the total structure of chlorobenzene are considerably less than those of the analogous structures of aniline, they are believed to contribute significantly.

XVII XVIII XIX XX XXI

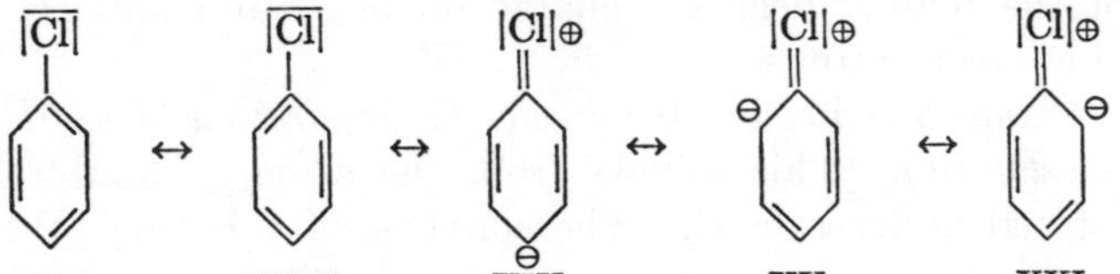

G. Structures with more than two electrons in the outer shell of hydrogen, more than eight in the outer shell of first-row elements, more than twelve in the outer shell of second-row elements, etc., will be unstable. This is a simple extension of the observation that hydrogen never forms a covalent bond with more than one other atom, first-row elements never with more than four, second-row elements never with more than six, etc. For instance, valence-bond structures of the type

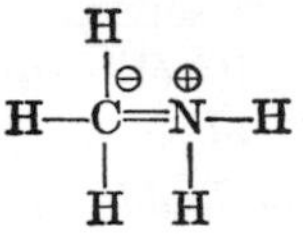

are considered too unstable to contribute significantly.

Unfortunately, in most comparisons of the stabilities of several valence-bond structures "other things" are not equal. Thus, in comparing structures XXII and XXIII

XXII XXIII

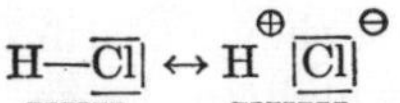

for hydrogen chloride, it may be seen that XXII has the advantage of one more covalent bond (rule *A*) and of less charge separation (rule *E*), while XXIII is stabilized by having a negative charge on the most electronegative element (rule *B*). Since we have no quantitative yard-stick for the operation of these rules, we cannot judge the relative stability, and hence the relative contributions, of the two structures. However, we should predict that, because of the greater electronegativity of fluorine, an ionic structure should contribute more to hydrogen fluoride. Similarly, decreasing contributions of ionic structures would be expected for hydrogen bromide and hydrogen iodide.

This method of describing a bond as a resonance hybrid of a purely covalent and a purely ionic bond is often very useful, but since most bonds have at least some ionic character, the method can become quite tedious for molecules with many bonds. For this reason molecules are ordinarily described not in terms of entirely covalent and ionic bond structures but in terms of the typically partial covalent bond. For example, in describing the resonance stabilization of the propionate anion (responsible for the acidity of propionic acid) we write the structures

$$
\begin{array}{c}
\underset{\overset{|}{H}}{\overset{H}{|}}\ \underset{\overset{|}{H}}{\overset{H}{|}} \\
H{-}C{-}C{-}C \\
\end{array}
\longleftrightarrow
$$

and by each we mean a structure having carbon-hydrogen bonds with their characteristic percentage of ionic character. That is, we *imply* the contributions of the purely covalent and purely ionic structures in the case where we do not specifically state it.

The question is sometimes raised whether the electrons do not move back and forth, assuming sometimes the positions characteristic of one valence-bond structure and sometimes those of another. To this it may be answered that, although the electrons do move, they do not ever assume a position characteristic of one valence-bond structure since *there is no such position*. No particular position of electrons can be said to be characteristic of a carbon-carbon single bond or double bond; only a certain type of probability-density function is characteristic. The probability-density function for the benzene molecule is not like that which would be expected from either Kekulé structure, nor does it change back and forth from one such function to the other.

The atomic nuclei are free to vibrate (although usually with only a relatively small amplitude), and therefore any structural formula showing the exact positions will, in reality, show only the centers around which they vibrate. The electronic probability-density distribution will change somewhat as the nuclei move, but for a given molecule with a given energy content, these centers around which the nuclei vibrate have definite relative positions in space.

1-1b. *A Graphic Treatment of Resonance.*   Among other valence-bond structures for o-dichlorobenzene we may write

XXIV   and   XXV

Since carbon-carbon single bonds that are between double bonds are about 1.48 A in length, we should expect that a molecule that was adequately described by structure XXIV would have C_1—C_2, C_3—C_4, and C_5—C_6 bonds of about this length, with C_2—C_3, C_4—C_5, and C_6—C_1 bonds of about 1.33 A, like most other double bonds.   For XXV these bond lengths should be reversed.   Unlike either of these structures, the actual o-dichlorobenzene molecule is thought to have six carbon-carbon bonds of very nearly equal length (about 1.39 A).   Thus there are actually not two different kinds of molecules corresponding to XXIV and XXV but only one type of molecule, a resonance hybrid of these two structures.

On the other hand, we are well aware not only that we may write valence-bond structures for n-butane and isobutane but also that there are two different types of molecules corresponding to these two different structures.   This raises the question of how we may tell, in general, when, for two or more isomeric valence-bond structures, there will be two or more different kinds of molecules and when there will be only one resonance hybrid.   This question is usually, but not always, an easy one to answer and will be discussed in terms of a graphic treatment of resonance.[5]

Let the left end of the abscissa of Fig. 1-1 represent the geometrical configuration of atoms corresponding to structure A, and let the right end represent the configuration corresponding to valence-bond structure B, where the energy is plotted as the ordinate.   Line a, then, represents the expected energy of the system if it is considered to be in valence-bond structure A, and line b represents the energy of the system considered as valence-bond structure B.   The actual energy of the system, represented by line r, is in all cases lower than the lower of the other two lines (by an amount that is the resonance energy of the system).   C represents a geometrical configuration intermediate between A and B.   This treatment may be applied to benzene, ignoring all but the Kekulé structures.   If the cyclohexatriene (Kekulé) structures accurately represent

[5] This discussion is similar to one by G. E. K. Branch and M. Calvin, "The Theory of Organic Chemistry," p. 74, Prentice-Hall, Inc., Englewood Cliffs, N.J., 1941.

benzene, the compound should have the bond distances shown in formulas A and B.

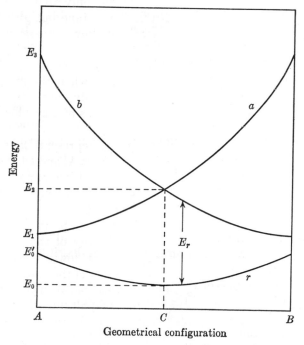

To get the values of the energy corresponding to line a, we must estimate the energy content to be expected of valence-bond structure A for the system whose atoms are in the geometrical configuration represented by a given point on the abscissa. Since geometrical configuration A is the normal one for valence-bond structure A, it must represent an

Fig. 1-1. Graphical representation of resonance between two Kekulé structures for benzene.

energy minimum (E_1 in Fig. 1-1).   On the other hand, if geometrical configuration B were represented as having valence-bond structure A, the stability would be less and the energy content higher, since the single bonds are 0.15 A shorter than their optimum length, and the double

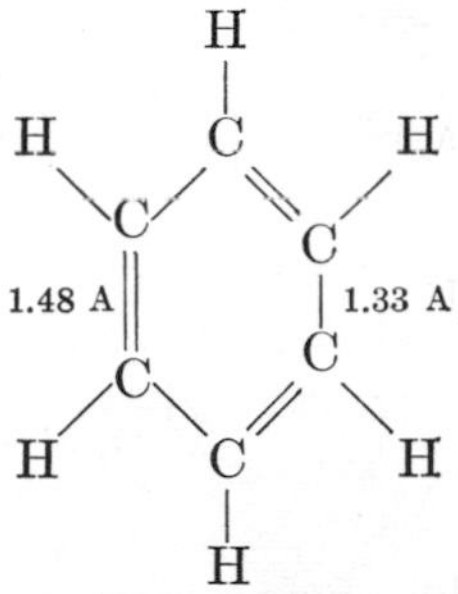

Valence-bond structure A
with geometrical configuration B

bonds are stretched a like amount.   Thus, the energy will have a value E_3, greater than E_1.   The exact shape of line a cannot be predicted, but E_1 will be its lowest value and E_3 its highest.   It seems likely that the line will be concave upward; i.e., a large change in the geometric configuration will cause an increase in the energy content more than proportional to the increase produced by a small change of configuration.   By similar reasoning, line b is readily constructed.   The actual energy of the system in its various geometrical configurations is represented by line r. This will be lower than the lower of lines a and b by an amount equal to the resonance energy of the system.   Since the resonance energy E_r will be greater the more nearly equal the energy content of the contributing structures, it will be at a maximum for geometrical configuration C, where lines a and b cross.   It happens, in the case of benzene, that upon going from configuration A (or B) to C, the resonance energy increases more rapidly than the energy content plotted on line a (or b).   Therefore, the true energy curve r has one minimum at the intermediate configuration C, and in benzene we do not have an equilibrium mixture of forms A and B; we have merely the resonance hybrid C.   This resonance hybrid is halfway between structures A and B in that it is as close to one as it is to the other.

Other specific examples of this general case do not necessarily give the same result as in the case of benzene.   For example, it is known that carboxylic acid dimers exist not only in certain solvents but also in the vapor phase.   In these dimers the two carboxyl groups are joined by hydrogen bonds.   We shall discuss a specific example, formic acid, in the same manner that we have discussed benzene.   The two forms A and B are

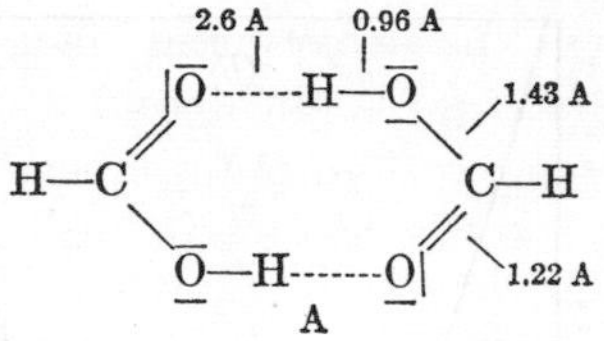

The O—H, C—O, and C=O bond distances are set equal to those found in most compounds with these bonds.   The O--- H distance is set equal to the sum of the van der Waals radii (the distance at which atoms appear to "touch" other atoms to which they are not bonded [see Sec. 1-4a]) of oxygen and hydrogen.   The energy content expected of valence-bond structure A in geometrical configuration B would appear to be relatively high.

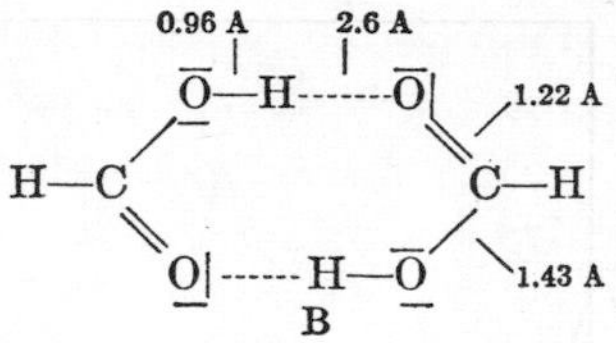

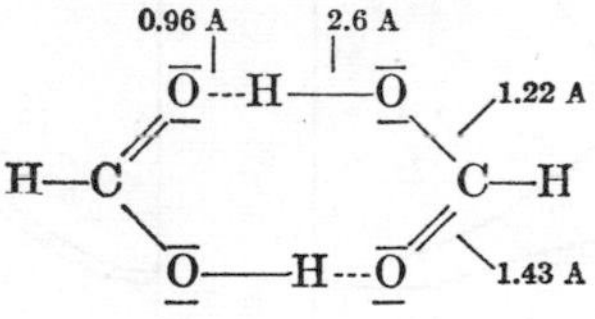

In addition to smaller distortions of the C—O and C=O bonds, the O—H bonds are stretched 1.6 A beyond their optimum length.   Hence in this case the loss of stability occurring in changing valence-bond structure A from geometrical configuration A to geometrical configuration B is considerably greater than it was in benzene; i.e., line a climbs sharply from left to right and line b similarly from right to left.   In this case, too, the resonance energy reaches a maximum at the configuration C, where lines a and b cross (Fig. 1-2a).   However, this increase in resonance energy is not sufficient to offset the increase in energy shown on lines a and b. Therefore it is seen that we have in the present case not a single energy minimum corresponding to one resonance hybrid with two equally contributing structures but two energy minima: D, a resonance hybrid with A the more important contributing structure, and D′, a hybrid toward whose structure B is the more important contributor.

The possibility that formic acid dimer might exist as a resonance hybrid with the hydrogen atoms equidistant between two oxygen atoms and with two equivalent contributing structures could not have been ruled out in

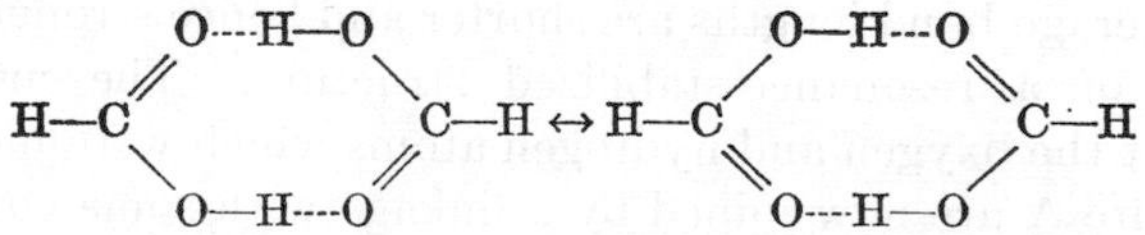

advance.   In fact, it was early thought that this was the case.   However, subsequent work showed that the oxygen atoms are not equidistant from the carbon atoms to which they are attached, as they should be if such a resonance hybrid existed.[6]   Also, the hydrogen atoms are con-

[6] J. Karle and L. O. Brockway, *J. Am. Chem. Soc.*, **66,** 574 (1944).

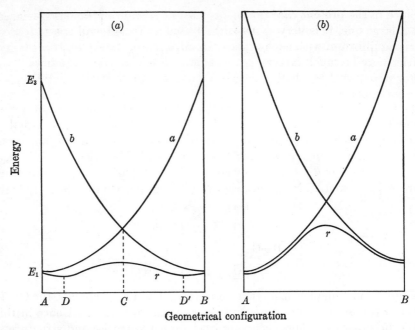

FIG. 1-2. Graphical representation of resonance as in (a) formic acid dimer; (b) cyclo-butene-1,3-butadiene (cis form).

siderably closer to one of the oxygen atoms than to the other.[7]   From these data the bond distances in formic acid dimer are approximately as shown below.

$$\begin{array}{cc}
\text{1.66 A} & \text{1.07 A} \\
\underline{O}\text{------H}\text{---}\underline{O} & \text{1.36 A} \\
\text{H---C} & \text{C---H} \\
\underline{O}\text{---H}\text{------}\underline{O} & \text{1.25 A}
\end{array}$$

That is, the —C=O is longer, since it now has more single-bond character, and the —C—O, with some double-bond character, is shorter.   These changes are due to the contributions of B to D and of A to D′.   On the whole, the average bond lengths are shorter and hence stronger, as would be expected of a resonance-stabilized structure.   The most striking feature is that the oxygen and hydrogen atoms which were merely touching in structure A are now joined by a linkage with some covalent-bond character.   This type of linkage is called a hydrogen bond and will be discussed in more detail later.   It is the opinion of some that most of the stability of the hydrogen bond is better described as an interaction of dipoles than in terms of resonance (see Sec. 1-5).

[7] R. Hofstadter, *J. Chem. Phys.*, **6**, 540 (1938).

Since in the present case the energy barrier separating the two forms is not a large one, it is very easily surmounted and the forms are in tautomeric equilibrium with each other. In many cases, however, this energy barrier may be much larger. An example of such a system follows.

For structure A we shall choose cyclobutene, and for structure B a molecule corresponding to one of the possible conformations of 1,3-butadiene.

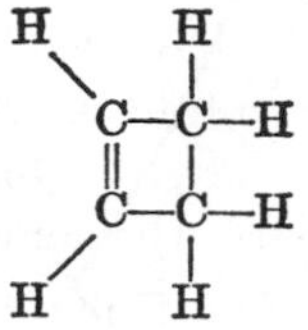

The student can see why valence-bond structure A would be expected to be very unstable in geometrical configuration B. In addition to the differences in bond lengths, all the atoms in B lie in the plane of the paper, whereas the four right-hand hydrogen atoms of A are above and below this plane. Therefore, an energy diagram of the type shown in Fig. 1-2b would be expected. In this case there are seen to be two different types of molecules separated by such a high energy barrier that their interconversion is difficult.

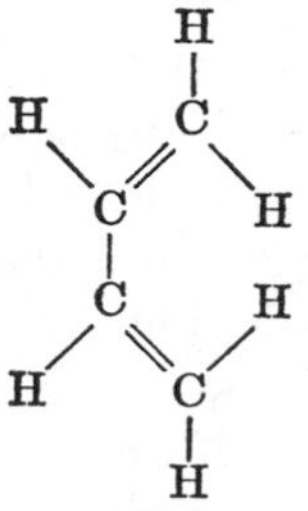

From this discussion it may be observed that the principal factor in determining whether two valence-bond structures will represent one or two different types of molecules is the extent of the difference in the geometrical configuration of the atoms to be expected from the structures. If the positions of the atoms in geometrical configuration A differ from their positions in configuration B by only a few tenths of an angstrom, then only one type of molecule, a resonance hybrid, will exist.[8] How-

---

[8] On the basis of these criteria, such data as those obtained by J. M. Robertson [*J. Chem. Soc.*, 1222 (1951)] on the structure of cupric tropolone (of which one resonance structure is shown below)

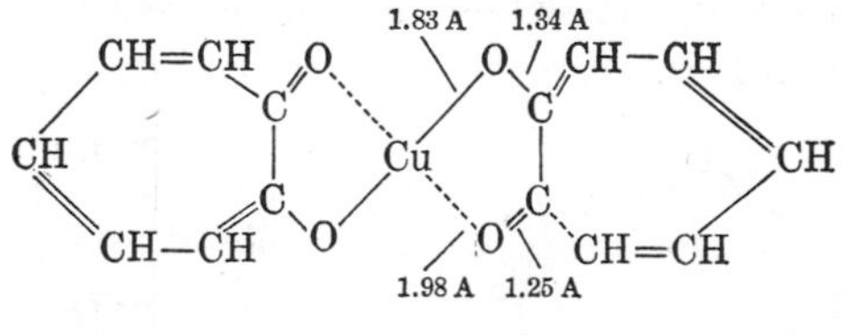

are surprising, since the molecule might be expected to be symmetrical (have all Cu—O bond distances equal). However, the above dimensions were measured in the crystal lattice, where the molecule was subjected to the distorting influence of other molecules in an unsymmetrical manner. This fact, coupled with the experimental error in the structure determination, may account for this lack of symmetry.

ever, if this difference amounts to several angstroms and if there are also considerable differences in bond angles, there will be two types of molecules and valence-bond structure A will contribute only slightly to the total structure of the molecules in geometrical configuration B. We cannot state, however, just where, between several angstroms and a few tenths, a line can be drawn. Fortunately, intermediate cases are not very common. Nevertheless, it is possible to draw any number of r curves varying continuously from the one in Fig. 1-1 to those in Fig. 1-2. Therefore, theoretically at least, there is a gradual and continuous change in going from a resonance hybrid to two forms in tautomeric equilibrium.

Thus whenever two different valence-bond structures (having the same number of unpaired electrons) predict very nearly the same arrangement of atomic nuclei in space, there will be only one type of compound, a resonance hybrid of the two or more structures. In most cases where the resonance itself is not being discussed, it is the most common and simplest procedure to write merely the valence-bond structure thought to contribute most and to assume the reader will realize that the actual molecule will be a resonance hybrid.

1-1c. *Data from Dipole-moment Measurements.* The dipole moment of a molecule is equal to the *distance* between the centers of positive and negative charge multiplied by the size of the charge, and since the distances are commonly on the order of 10^{-8} cm and the charges 10^{-10} esu, it is expressed in Debyes [1 D (Debye) = 10^{-18} esu cm]. Having a positive and a negative end, as well as magnitude, the dipole may be seen to be a vector and is often written with the positive end crossed and the arrow at the negative end.

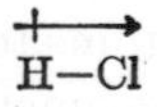

$$\overset{\longrightarrow}{\text{H--Cl}}$$

It is convenient to define a *bond moment* as that part of the dipole moment of a molecule which may be attributed to an individual bond. For a diatomic molecule the bond moment is equal to the dipole moment, but for a polyatomic molecule the dipole moment is equal to the vector sum of the various bond moments. The dipole moment of water, for example, is 1.84 D. Since the angle H—O—H is known to be about 105°, the bond moments must be 1.51 D.

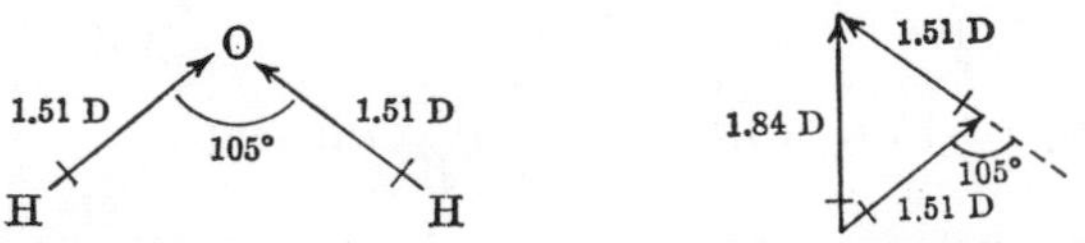

The dipole moment of *p*-chloronitrobenzene is 2.57 D. Since this is much nearer the difference between the values for chlorobenzene (1.56 D)

and nitrobenzene (3.97 D) than it is to their sum, the dipole of the nitro group, like that of the chlorine atom, must be directed *away* from the ring.

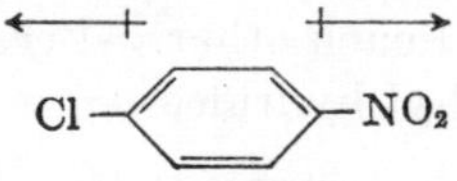

Since the dipole moments of typical saturated aliphatic nitro compounds are around 3.3 D, the larger value for nitrobenzene is evidence for the contributions of structures like XXVIII and XXIX, for which there are no analogs with the aliphatic derivatives.   It should be noted that

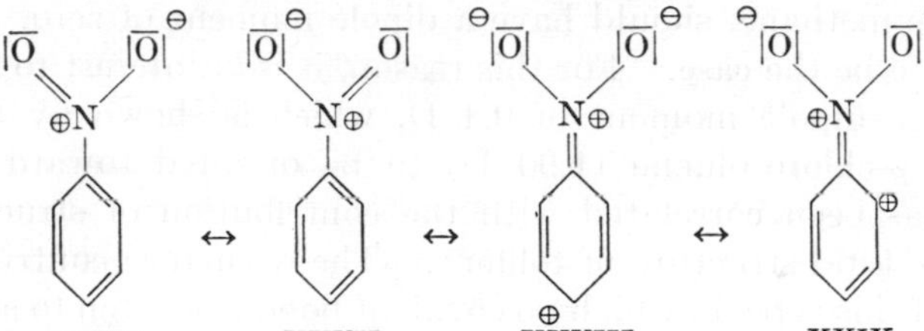

structures XXVIII and XXIX will have their greatest stability when all the atoms are in the same plane, because, in general, when two atoms (of the first row of the periodic table, at least) are joined by a double bond, they, and all the atoms attached directly to them, are preferentially coplanar.   Thus, the fact that nitromesitylene, in which the nitro group is hindered from lying in the plane of the ring by the size of the *o*-methyl groups, has a dipole moment of only 3.65 D (the effects of the three methyl groups should approximately cancel) is added evidence that the exaltation of the dipole moment of nitrobenzene is due to the contribution of structures like XXVIII and XXIX.   Considerable other data have been obtained on the steric inhibition of resonance by dipole-moment

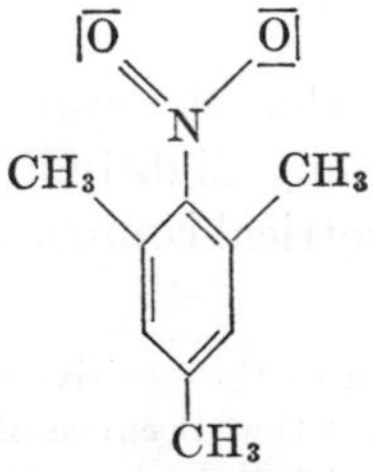

Nitromesitylene

measurements, spectral studies, determinations of the strengths of acids and bases, and studies of the effect of structure on reactivity.[9]

[9] Wheland, *op. cit.*, pp. 232, 314, 367, 508.

1-1d. Hyperconjugation.   Regardless of the direction or magnitude of the C—H bond dipole,[10] the dipole moment of methane would be expected to be zero since, because of the symmetry of the methane molecule, the bond moments would cancel each other.   Because the methane molecule may be represented as methyl hydride,

$$\overset{\longrightarrow}{H_3C}-\overset{\longleftarrow}{H}$$

it may be seen that the moment of the methyl group must exactly equal that of the C—H bond.   Since any saturated aliphatic hydrocarbon may be represented as being formed by the successive replacement of hydrogen atoms with methyl groups, starting with methane, all these hydrocarbons, like methane, should have a dipole moment of zero.   This has been found to be the case.   For this reason, it is of interest to learn that toluene has a dipole moment of 0.4 D, which is shown by the dipole moment of *p*-chlorotoluene (1.90 D) to be oriented toward the ring. This fact has been correlated with the contribution of structures like XXX to the total structure of toluene.   The resonance contributions of structures of this type, in which no covalent bond is written to a hydrogen atom, is called *no-bond resonance*, or *hyperconjugation*.   Much other evidence for resonance of this type has been given.[11,12]   We are aware that if we attempt to describe the molecule in terms of purely covalent and purely ionic bond structures, we must include a definite contribution from a structure like XXXI.

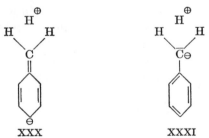

XXX   XXXI

Therefore it is not surprising that structure XXX should contribute significantly, since it should be only slightly less stable than XXXI, having a slightly less favorable geometrical configuration of atoms and somewhat more separation of charge.

[10] The hydrogen has been taken as the positive end because of its smaller electronegativity, but it has been claimed that because of the smaller size of hydrogen the dipole has the opposite direction, although about the same size [W. L. G. Gent, *Quart. Revs. (London)*, **2**, 383 (1948)].

[11] J. W. Baker, "Hyperconjugation," Oxford University Press, London, 1952.

[12] For reports of a conference on hyperconjugation in which most participants supported the concept but some criticized evidence of the type given here and even denied the necessity of the concept, see *Tetrahedron*, **5**, 107–274 (1959).

Hyperconjugation is also important for olefins.

$$
\underset{\overset{\displaystyle |}{H}}{\overset{\displaystyle \overset{H}{|}\ \ \overset{H}{|}\ \ \overset{H}{|}}{H-C-C=C-H}} \qquad \leftrightarrow \qquad \overset{\oplus}{H}\ \ \underset{\overset{\displaystyle |}{H}}{\overset{\displaystyle \overset{H}{|}\ \ \overset{H}{|}}{C=C-\underset{\ominus}{C}-H}}
$$

The stabilization that is due to added hyperconjugative resonance is believed to be responsible for the increased stabilization of nonterminal over terminal olefins; i.e., it explains why the former are favored in equilibria.

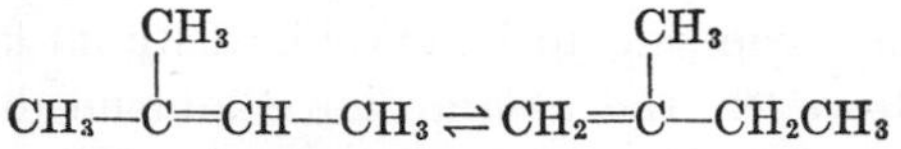

$$
\underset{\overset{\displaystyle |}{CH_3}}{CH_3-C=CH-CH_3} \rightleftharpoons \underset{\overset{\displaystyle |}{CH_3}}{CH_2=C-CH_2CH_3}
$$

Added evidence for hyperconjugation has come from studies by Kistiakowsky and coworkers[13] on the heats of hydrogenation of a number of olefinic hydrocarbons. The heats of hydrogenation were found to decrease as the number of alkyl groups attached to the doubly bound carbons increased (see Table 1-2). The fact that the less-alkylated olefins add hydrogen considerably more exothermically suggests the

TABLE 1-2. HEATS OF HYDROGENATION OF OLEFINS AT 80°[13]

Compound	$-\Delta H$, kcal/mole	Compound	$-\Delta H$, kcal/mole
Ethylene	32.8	Cyclohexene	28.6
Propylene	30.1	Cyclopentene	26.9
1-Heptene	30.1	1,3-Butadiene	57.1
Isobutylene	28.4	1,4-Pentadiene	60.8
cis-2-Butene	28.6	Allene	71.3
trans-2-Butene	27.6	Benzene	49.8
Trimethylethylene	26.9	1,3-Cyclohexadiene	55.4
Tetramethylethylene	26.6	1,3-Cyclopentadiene	50.9

stabilizing influence of alkyl groups on the double bond (although other factors are also at work). The relative stabilities, of course, are not related to the heat contents, or enthalpies, H, but to the free energy, $F = H - TS$. However, the entropies S of these closely related substances differ so little that differences in ΔH of the size observed ensure that ΔF's will vary in the same order. This is also known to be true from the greater stability of more highly alkylated olefins, demonstrated in equilibrium experiments. It should be mentioned, though, that the heats of *bromination* of monoolefins *increase* with the extent of alkylation of the

[13] G. B. Kistiakowsky and coworkers, *J. Am. Chem. Soc.*, **57**, 65, 876 (1935); **58**, 137, 146 (1936); **59**, 831 (1937).

double bond.   This is related to the fact that the stability of alkyl halides decreases in the order: tertiary > secondary > primary.

1-1*e*. *Resonance Energy*.   From thermochemical measurements of the type just described, numerical estimates of the stabilization due to resonance have been made.   Since the heat of hydrogenation of cyclohexene is 28.6 kcal/mole, the heat of hydrogenation of benzene might be expected to be 85.8 kcal/mole if the compound were accurately described by a cyclohexatriene valence-bond structure in which the character of the double bonds is like that of cyclohexene.   Actually, the hydrogenation of benzene to cyclohexane gives off only 49.8 kcal/mole.   For this reason, benzene has been said to be stabilized by 36 kcal/mole of resonance energy.   However, the assumption that the double bonds in a cyclohexatriene structure (A or B, Sec. 1-1*b*) should be identical (in heat of hydrogenation) to those of cyclohexene might very well be challenged. For cyclohexene we might expect stabilization by hyperconjugation (unlikely in a cyclohexatriene structure), and the heat of hydrogenation of this compound is indeed 4.2 kcal/mole lower than that of ethylene. Calculations based on a cyclohexatriene valence-bond structure with ethylenelike double bonds would give a resonance energy of 48.6 kcal/ mole.   Here it might be argued that the mere replacement of two of the hydrogens of ethylene by carbon without hyperconjugation might change the nature of the double bond, and that the double bonds of a cyclohexatriene structure would not be expected to be identical to those of ethylene either.   Upon further thought, it may be seen that there is *no* obvious and logical olefin whose double bond should serve as a standard for those in a cyclohexatriene valence-bond structure.   Therefore, although it is obvious that the "double bonds" in benzene are considerably more resistant to addition than those in most olefins, no quantitative calculation of the resonance energy can have any significance unless the double bonds in terms of which the cyclohexatriene structures are described are clearly defined.   Furthermore, any such definitions will necessarily be arbitrary.

It should be understood that the "resonance energy" discussed above, the stabilization of benzene with respect to structure A (or B) in its most stable geometrical configuration, is not the resonance energy E_r of Fig. 1-1 (which has the value $E_2 - E_0$ for benzene) but rather $E_1 - E_0$. The quantity $E_2 - E_0$, the stabilization of benzene relative to valence-bond structure A (or B) when in the geometrical configuration of benzene, has been estimated (from the energy required to stretch and compress typical double and single bonds) to be about 35 kcal/mole higher than $E_1 - E_0$.[14]

---

[14] R. S. Mulliken, C. A. Rieke, and W. G. Brown, *J. Am. Chem. Soc.*, **63**, 41 (1941). D. F. Hornig [*J. Am. Chem. Soc.*, **72**, 5772 (1950)] has estimated the value $E_2 - E_1$ at 30 kcal and $E_0' - E_0$ at 25 to 27 kcal.

Calculations of resonance energies have also been based on heats of combustion, both directly and by use of bond energies calculated from heat-of-combustion data; they have also been calculated theoretically.[15]

1-2. Atomic and Molecular Orbitals.[16]   From quantum mechanics, it may be shown that the probability-density function for the electron in a hydrogen atom depends upon its energy level (by quantum theory, there are only certain definite amounts of energy possible, so that the energy increases in steps rather than gradually).   The seven lowest types of energy levels, which are the only ones commonly of interest in organic chemistry, are the $1s$, $2s$, $2p$, $3s$, $3p$, $3d$, and $4s$ levels.   Electrons in any of the s levels have probability-density functions that are independent of the direction from the nucleus; i.e., they are spherically symmetrical. For the lowest level, $1s$, the function may be represented by shading the

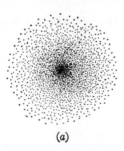   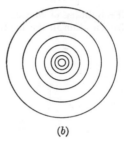

(a)   (b)

FIG. 1-3. Two representations of the probability-density function for the $1s$ level of the hydrogen atom.   (a) Probability proportional to darkness or shading; (b) contour lines connecting points of equal probability.

areas with a darkness proportional to the probability of finding an electron there (Fig. 1-3a) or by a contour map (Fig. 1-3b).   In each case, the three-dimensional representation must be obtained by rotating the two-dimensional representation in Fig. 1-3 about a line through its center. Alternately, we may plot the probability of finding an electron at a given distance r from the nucleus against this distance (Fig. 1-4).

It is interesting to see that, although the probability of finding an electron within a given volume increment increases steadily as the volume increment nears the nucleus, the probability of finding an electron at a given distance is maximum at a distance of about 0.53 A (of course, the number of volume increments at a given distance from the nucleus increases with the distance).   The probability-density function for an electron is called its *orbital* and is most commonly depicted by a line corresponding to a single probability contour (Fig. 1-3b).   An outer

[15] Wheland, *op. cit.*, chap. 3.

[16] A much more complete but still fairly nonmathematical treatment of this topic and much of the other material in this chapter is given in C. A. Coulson, "Valence," Oxford University Press, London, 1952.

contour is chosen to ensure a high probability of finding the electron within the orbital. For the spherically symmetrical s orbitals this contour is simply a sphere (Fig. 1-5a). There are three of each type of p levels. These are referred to as the p_x, p_y, and p_z orbitals and are shown

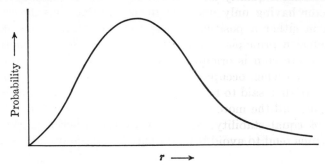

FIG. 1-4. Probability of finding an electron as function of distance r from nucleus (for a hydrogen atom in the 1s state).

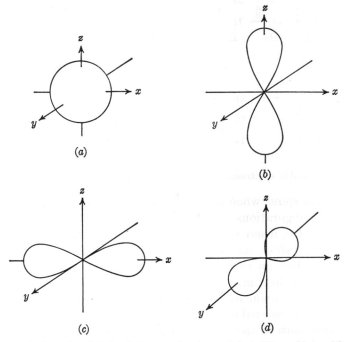

FIG. 1-5. Atomic orbitals of the s and p type: (a) s; (b) p_z; (c) p_x; (d) p_y.

in Figs. 1-5b, c, and d. The boundary surfaces like those shown in Fig. 1-5 have the same shape for all orbitals of a given letter (s or p) and increase in size with the number describing the orbital. By reasonable, but approximate, calculations it may be shown that in atoms containing many electrons these electrons fill the various orbitals described in the order of their decreasing stability in accordance with certain rules.   One of these rules depends upon the fact that, in addition to mass and charge, an electron has a quality known as *spin*.   This spin may be represented as a vector having only one definite magnitude but capable of being oriented in either a positive or negative direction.   According to the Pauli exclusion principle, a given orbital cannot contain more than one electron whose spin is oriented in a given direction, and the orbital is therefore filled when occupied by two electrons with opposite spins (the electrons are then said to be *paired*).   In general, a given orbital will not be filled until all the more stable orbitals have been filled.   When filling orbitals of equal stability, electrons enter according to Hund's rules: (1) Electrons tend to avoid being in the same orbital.   (2) They tend to

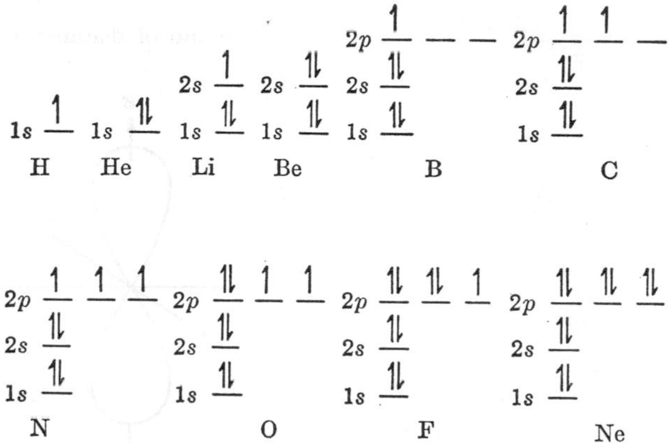

FIG. 1-6. Most stable electronic configurations for the atoms of the first 10 elements.

have identical spins, when possible.   Thus, for atoms of the first 10 elements the configurations of the electrons in the atomic orbitals are as shown in Fig. 1-6, where each electron is represented by an arrow pointed up or down to show the direction of spin.   It may be seen that this formulation offers an explanation for the great stability possible with 2 electrons, as in helium, or 10 as in neon, etc.

In all cases common in organic chemistry, a chemical bond between two atoms is represented as resulting from two electrons' filling an orbital that includes both of the nuclei being bound together.   Orbitals of this type are called *molecular orbitals* in contrast to the *atomic orbitals*, surrounding only one nucleus, which we have been discussing.   The characteristics of that part of such an orbital nearest an individual nucleus are determined largely by that nucleus, and therefore in this region the orbital will resemble the familiar atomic orbital, only somewhat influenced (perturbed) by the presence of the other nucleus.

ALBRIGHT COLLEGE LIBRARY   218545

For this and other reasons, it is useful to regard this bonding orbital as the modified result of the combination or overlap of two atomic orbitals. Thus the bonding orbital in the hydrogen molecule (Fig. 1-7a) may be thought of in terms of the overlap of the two 1s orbitals of the

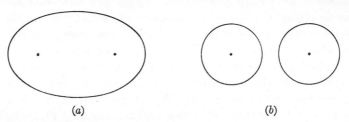

(a)   (b)

FIG. 1-7. (a) Bonding molecular orbital of the hydrogen molecule; (b) 1s orbitals of the two hydrogen atoms.

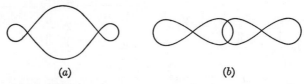

(a)   (b)

FIG. 1-8. (a) Bonding molecular orbital formed by coaxial p orbitals; (b) coaxial p orbitals.

two atoms. Bonding may also occur by the overlap of p orbitals, an s and a p orbital, etc. Bonds formed between p orbitals with a common axis (Fig. 1-8), like those between s orbitals or an s and a p orbital, are called *sigma bonds*. Another type of bond is called the *pi bond* and is produced by the overlap of a p orbital (only) with another p orbital having a parallel axis (Fig. 1-9). Since pi bonds are usually weaker than sigma bonds, they are formed only when a p orbital is forced to be noncoaxial by the presence of a sigma bond formed from other orbitals.

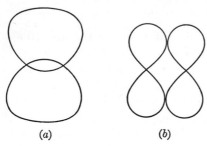

(a)   (b)

FIG. 1-9. (a) Bonding molecular orbital (pi orbital) formed from parallel p orbitals; (b) p orbitals with parallel axes.

Orbital hybridization is an additional important factor, which we shall discuss in its application to the valence of carbon. Instead of forming four bonds with its 2s and three 2p orbitals, tetravalent carbon *mixes* these orbitals to get four equivalent orbitals. The four resultant orbitals are known as sp^3 orbitals, their wave functions being a combination of those for the s and the three p orbitals. They have the shape shown in

Fig. 1-10 and are oriented at angles of 109°28′ from each other (a regular tetrahedral configuration).   The use of orbitals of this sort for bonding is probably related to the fact that there appears to be a definite correlation between the extent to which the two atomic orbitals overlap and the strength of the bond formed thereby.   The carbon atom tends not to form bonds with its *s* orbital, which, because of its concentration near the nucleus, may not overlap extensively with other orbitals, but it may hybridize this *s* orbital with *p*'s to yield orbitals that are very effective at overlapping.   The sp^3 hybridization occurs when carbon is bound to four other atoms.   When it is bound to only three, there may be hybridization of the $2s$ and two of the $2p$ orbitals to give three sp^2 orbitals.   These are coplanar and at 120° angles (Fig. 1-11), while the remaining unhybridized *p* orbital is perpendicular to this plane.   Thus in ethylene each carbon is joined to two hydrogen atoms by sigma bonds formed by

FIG. 1-10. An sp^3 orbital.

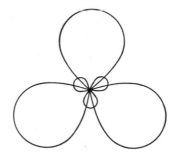

FIG. 1-11. The three sp^2 orbitals.

the overlap of one of the sp^2 orbitals on carbon and the *s* orbital of hydrogen.   In addition, the two carbon atoms are joined by a sigma bond due to sp^2 orbital overlap *and* by a pi bond due to *p* orbital overlap. Carbon atoms attached to two other atoms may hybridize one *s* and one *p* orbital to give two coaxial *sp* orbitals, which form sigma bonds, and to leave two *p* orbitals at right angles to form pi bonds.   Triple bonds may thus be seen to consist of one sigma and two pi bonds.

It is of interest to apply these principles to a few simple molecules. In the water molecule the oxygen atom has available to it one $2s$ and three $2p$ orbitals.   Two orbitals are required for bonding and two for the four unshared electrons of oxygen.   Since electrons in an *s* orbital are more stable than those in a *p* orbital, we should expect that a non-bonding orbital (in which the oxygen has two electrons) will be *s* in character, rather than a bonding orbital (whose electrons spend only part of their time around the oxygen atom).   Furthermore, the *p* orbitals should be preferred for bond formation since they overlap more effectively. Because the three *p* orbitals are perpendicular to each other, we might expect an H—O—H bond angle of 90°. The observed bond angle is about 104°31′. This deviation is believed to be due to repulsion between the partially positive hydrogen atoms and is much smaller with hydrogen sulfide (92° bond angle), where the hydrogen atoms are further apart and less positive. This increase in the bond angle lends *s* character to the bonding orbitals,[17] and hence *p* character to the nonbonding orbitals. The addition of *s* character to the bonding orbitals increases their ability to overlap, and the resultant stronger bonds partially compensate for the energy required to place the unshared electrons in an orbital with more *p* character. Similar considerations may be applied to ammonia (107° valence angle), phosphine (93°), and arsine (92°).

In the molecular orbital description of benzene we should expect each carbon atom to be bonded through sp^2 orbitals to two adjacent carbon atoms and a hydrogen atom. This leaves one electron in a *p* orbital which may overlap to form a pi bond with a similar orbital on an adjacent carbon atom. However, it may be seen that because of the symmetrical arrangement of the atoms, each *p* orbital will overlap as much with the *p* orbital on one adjacent carbon as with that on the other. Thus the atomic orbitals will combine to form molecular orbitals that resemble two doughnuts, one on each side of the plane of the atomic nuclei (Fig. 1-12).

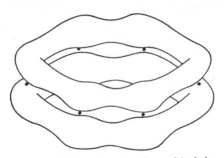

FIG. 1-12. The pi molecular orbital for benzene.

1-3. Aromaticity. From the simple versions of resonance and molecular orbital theory that have been presented we might expect cyclobutadiene and cyclooctatetraene to show aromatic character, as benzene does, since each consists of a ring of sp^2 carbon atoms whose *p* orbitals would seem capable of overlap to give doughnut-shaped molecular orbitals and since, for each, two equivalent valence-bond structures may be written if the ring is planar:

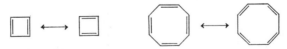

More detailed treatments by either the resonance or molecular-orbital approach, however, lead to the conclusion that a ring of sp^2 atoms will be particularly stable only when the number of electrons present in the

---

[17] Note that *p* orbitals (0 per cent *s*) give 90° bond angles, sp^3 (25 per cent *s*) 109°28′, sp^2 (33 per cent *s*) 120°, and sp (50 per cent *s*) 180°.

various p orbitals of all the atoms of this ring is of the form $4n + 2$, where n is an integer.   The lack of aromatic character found for cyclooctatetraene and the apparent instability of cyclobutadiene have been explained on this basis, but some organic chemists have preferred an alternate explanation based on bond-angle strain.   The tendency of sp^2 atoms to have 120° bond angles could give cyclooctatetraene a nonplanar structure and cause cyclobutadiene to be quite strained (although less so than such known compounds as cyclopropene).   There is a mass of additional evidence, however, that establishes the $4n + 2$ rule as much the more reasonable explanation.   Although attempts to generate the cyclopentadienyl cation as a stable species have all failed, the cycloheptatrienyl cation, containing six pi electrons, is stable in moderately acidic aqueous solution.[18]   In fact, cycloheptatrienyl bromide, unlike practically all other known bromohydrocarbons, appears to be ionized even in the solid form.   These observations cannot be entirely explained by the greater number of resonance structures for the cycloheptatrienyl cation, since the cyclopentadienyl anion, which has six pi electrons, appears to be much more stable than the cycloheptatrienyl anion, which has eight, although seven important contributing structures may be written for the latter anion and only five for the former.

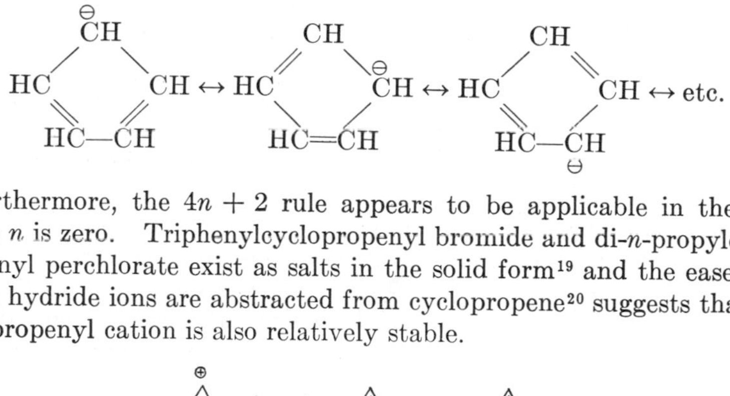

Furthermore, the $4n + 2$ rule appears to be applicable in the case where n is zero.   Triphenylcyclopropenyl bromide and di-n-propylcyclopropenyl perchlorate exist as salts in the solid form[19] and the ease with which hydride ions are abstracted from cyclopropene[20] suggests that the cyclopropenyl cation is also relatively stable.

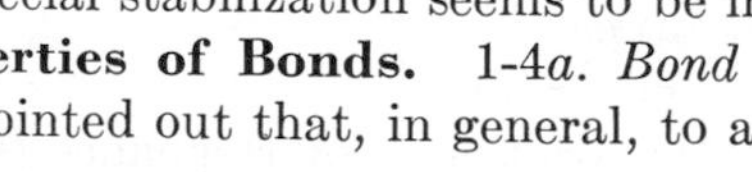

In view of the tremendous strain that there must be in the bond angles of this system, some special stabilization seems to be indicated.

1-4. Some Properties of Bonds.   1-4a. *Bond Distances.*   Pauling and Huggins have pointed out that, in general, to a fair degree of accu-

[18] W. v. E. Doering and L. H. Knox, *J. Am. Chem. Soc.*, **76**, 3203 (1954).

[19] R. Breslow and C. Yuan, *J. Am. Chem. Soc.*, **80**, 5991 (1958); R. Breslow and H. Höver, *J. Am. Chem. Soc.*, **82**, 2644 (1960).

[20] K. B. Wiberg, Abstracts of Papers, ACS meeting, Apr. 7–12, 1957, Miami, Fla., p. 390.

racy, the length of a covalent bond is equal to the sum of what are called the *covalent-bond radii* of the two atoms involved.[21] The numerical values of covalent-bond radii may be simply chosen so as to give the best general agreement with determinations of bond distances by X rays, electron diffraction, and spectroscopic studies. Since multiple bonds are shorter than the corresponding single bonds, it is necessary to have double- and triple- as well as single-bond radii. A number of covalent-bond radii compiled by Pauling are shown in Table 1-3.[22] By use of the values in this table, we may predict the length of various bonds. For example, the carbon-chlorine bond distance in methyl chloride should be equal to the sum of the covalent-bond radii of carbon and chlorine: $0.77 + 0.99 = 1.76$ A. The distance actually found is 1.78 A.

TABLE 1-3. COVALENT-BOND RADII AND VAN DER WAALS RADII, ANGSTROMS[22]

Element	Single-bond radius	Double-bond radius	Van der Waals radius
Hydrogen.........	0.30	.....	1.2
Carbon[a]...........	0.77	0.665	[b]
Nitrogen[c].........	0.70	0.60	1.5
Oxygen...........	0.66	0.55	1.4
Fluorine..........	0.64	0.54	1.35
Phosphorus........	1.10	1.00	1.9
Sulfur............	1.04	0.94	1.85
Chlorine..........	0.99	0.89	1.8
Bromine..........	1.14	1.04	1.95
Iodine............	1.33	1.23	2.15

[a] Triple-bond radius is 0.60.
[b] Radius of a methyl group, 2.0; half thickness of aromatic ring, 1.85.
[c] Triple-bond radius is 0.55.

In addition to the specific case of the hydrogen molecule, whose bond is 0.14 A longer than predicted, there are a number of fairly general deviations in bond lengths. The single-bond radius of sp^2 carbon toward another carbon atom is about 0.74 A and the value for sp carbon is 0.69 A. The bonds that sp and sp^2 carbon form to hydrogen are also shortened relative to the hydrogen-sp^3 carbon bond, but not so much as would be expected from sp and sp^2 carbon-carbon bond lengths.

The contribution of ionic structures to the total structure of a molecule usually leads to a bond length that is shorter than if the bond were purely covalent. Schomaker and Stevenson have suggested an equation for

[21] L. Pauling and M. L. Huggins, *Z. Krist.*, **87A**, 205 (1934).
[22] L. Pauling, "The Nature of the Chemical Bond," 2d ed., Cornell University Press, Ithaca, N.Y., 1945, pp. 164, 189.

predicting bond lengths in which the difference in electronegativity of the atoms involved is used as an added parameter.[23]

Within a distance of a few angstroms, all molecules exert significant attractive forces (van der Waals forces) on each other. If these forces are sufficient to overcome their thermal motion, the molecules are held relatively closely together as a liquid or a solid. From the finite volume of such matter, it is obvious that at sufficiently small distances these van der Waals attractive forces are balanced by repulsive forces due to the interpenetration of the outer electronic orbitals of the atoms involved. As a measure of the equilibrium distances thus possible between atoms in adjacent molecules and between contiguous atoms attached to different parts of the same molecule, the van der Waals radii shown in Table 1-3 are used. Thus, the value 1.8 A listed for chlorine shows that "touching" chlorine atoms in adjacent carbon tetrachloride molecules should have an internuclear distance of about 3.6 A. This would also be expected to be the equilibrium distance between the two chlorine atoms in 2,2'-dichlorobiphenyl (when the molecule is rotated so that those two atoms touch). When attached to the same or adjacent atoms, however, two atoms may approach each other more closely, two chlorine atoms in the same molecule of carbon tetrachloride being less than 3.0 A apart. Van der Waals radii often vary by 0.1 A from the values listed.

1-4b. *Bond Angles.* As described in Sec. 1-2, when carbon is attached to four other atoms, it forms bonds with sp^3 orbitals, which have their maximum stability in a regular tetrahedral configuration with bond angles of 109°28'. In addition to the stability of the bonding orbitals the bond angles are also affected by interactions between the various groups attached. When the four groups attached to carbon are not identical, the angles would not be expected to be exactly those of a regular tetrahedron. In this connection, it is of interest that, despite dipolar repulsions, the F—C—F bond in methylene fluoride is about 108°17',[24] somewhat smaller than the tetrahedral angle. With the larger chlorine atoms, the Cl—C—Cl angle of methylene chloride is about 111°47'.[24] By far the greatest deviations occur with cyclic systems, the bond angles in cyclopropane and cyclobutane being, of course, 60 and 90°, respectively. When a carbon atom with no unshared electrons is attached to three atoms or groups, the bond angles tend toward the 120° characteristic of sp^2 hybridization, whereas attachment to only two atoms or groups ordinarily leads to a linear configuration (180° bond angles). Since nuclear charge usually has only a small effect on molecular geometry, nitrogen has bond angles analogous to those described for carbon when all its outer electrons are involved in bonding,

[23] V. Schomaker and D. P. Stevenson, *J. Am. Chem. Soc.*, **63**, 37 (1941).
[24] D. R. Lide, Jr., *J. Am. Chem. Soc.*, **74**, 3548 (1952).

as in ammonium and nitro compounds. In amines, alcohols, ethers, sulfides, etc., the unshared electrons affect the bond angles as described in Sec. 1-2.

1-4c. *Bond Energies.* In a diatomic molecule the bond energy is customarily defined as the energy required to bring about the fission of the molecule into atoms. For the reaction

$$A—B(g) \rightarrow A\cdot(g) + B\cdot(g)$$

in the vapor phase, we shall define the value of ΔE at 0°K as the bond energy. Frequently, values of ΔH at 298° are used since they are obtainable from the experimental data in cases where the ΔE_0's are not, but the former values relate to further differences in energies of vibration, rotation, and translation and to a pressure-volume term in addition to the energy dependent upon the bond strengths. The errors introduced by using ΔH_{298} values, however, are often no larger than the uncertainty in the ΔH (or analogous ΔE) value.

For polyatomic molecules there are two alternate methods of defining the bond energy. By one definition it is an aliquot part of the energy required to dissociate the molecule completely into atoms. Thus for water, we might define the O—H bond energy as one-half the energy required to dissociate the water molecule into hydrogen and oxygen atoms.

$$H_2O(g) \rightarrow 2H(g) + O(g) \qquad \Delta E_0 = 218.9 \text{ kcal/mole}$$

The O—H bond energy defined in this way (109.4 kcal) is called the *average bond energy.* On the other hand, the O—H bond energy in water may be thought of simply as the energy required to break an O—H bond in water to give a hydroxyl radical and a hydrogen atom. While we know that the ΔE's for the two reactions

$$H_2O(g) \rightarrow H(g) + OH(g) \qquad \Delta E_0 = 118.6 \text{ kcal/mole}$$
$$OH(g) \rightarrow H(g) + O(g) \qquad \Delta E_0 = 100.3 \text{ kcal/mole}$$

must total 218.9 kcal, there is no reason why they should be identical, and indeed, as shown, they are not. The bond energies defined in this manner are called *bond-dissociation energies.* Hereafter average bond energies will usually be referred to simply as bond energies, while the complete name will be used for bond-dissociation energies.

The types of experimental data[25] from which average bond energies may be calculated are illustrated below for methane.

[25] Most of the thermochemical data used herein are taken from Selected Values of Chemical Thermodynamic Properties, *Natl. Bur. Standards Misc. Publ. Ser.* III.

$$CH_4(g) + 2O_2(g) \rightarrow CO_2(g) + 2H_2O(g) \qquad \Delta E_0 = -192.2 \text{ kcal}$$
$$CO_2(g) \rightarrow C \text{ (graphite)} + O_2(g) \qquad\qquad = \quad 94.0$$
$$2H_2O(g) \rightarrow 2H_2(g) + O_2(g) \qquad\qquad = \quad 114.2$$
$$2H_2(g) \rightarrow 4H(g) \qquad\qquad\qquad\qquad = \quad 206.4$$
$$\underline{C \text{ (graphite)} \rightarrow C(g) \qquad\qquad\qquad\qquad = \quad 170.4}$$
$$CH_4(g) \rightarrow C(g) + 4H(g) \qquad\qquad \Delta E_0 = \quad 392.8 \text{ kcal}$$

Therefore the C—H bond energy in methane is $392.8/4 = 98.2$ kcal.

Since methane has four equivalent C—H bonds, it is obvious that its heat of atomization should be divided by 4 in order to get the average C—H bond energy. However, for ethane there is no fundamentally correct way of subdividing the heat of atomization into that due to the C—C and that due to the C—H bonds. For this reason the C—C bond energy in ethane is usually arbitrarily defined as that which is calculated on the assumption that the C—H bonds have the same energy that they do in methane. The value thus obtained is 77.7 kcal. Analogous calculations yield average C—C bond energies approaching 80.5 kcal for higher normal paraffins. This increase in bond energy has no necessary relation to the dissociation energies of the C—C bonds, since it is based on the arbitrary assumption that the C—H bond energy is constant and that only the C—C energy varies. In fact, the thermodynamic data may be explained rather well by assuming that the C—C bond energy has the constant value 84.9 kcal (that in diamond) and that C—H bond energies are 98.2 kcal in methane, and 96.9 kcal for primary, 96.2 kcal for secondary, and 95.6 kcal for tertiary C—H bonds, respectively. Some bond energies of interest are listed in Table 1-4. Most of those in the right-hand column were calculated on the basis of an assumption like that described above for the C—C bond energy. In these cases the value of the bond energy will depend upon the compound on which the calculation is based. Nevertheless such variations are usually small enough for the bond energy calculated from data on one compound to be useful in predicting the properties of another compound with the same type of bond.

Bond-dissociation energies for diatomic molecules are identical to the average bond energies. For polyatomic molecules, however, they may be considerably different, as has been pointed out in the case of water. Spectroscopic and *electron-impact* methods and studies of the kinetics and equilibria of bond-dissociation processes have been used to determine bond-dissociation energies.[26] While these are often more useful in physical organic chemistry than the average bond energies, fewer pertinent data are available.

In general, the energy of a bond between two different atoms A and B is the mean of that for A—A and B—B plus an additional amount due

[26] M. Szwarc, *Quart. Revs. (London)*, **5**, 22 (1951).

to the resonance contribution of ionic structures (see Sec. 1-1a) and dependent upon the electronegativity difference between A and B.   In

TABLE 1-4. SOME AVERAGE BOND ENERGIES OF INTEREST IN ORGANIC CHEMISTRY[a]

Bond	Energy, kcal	Bond	Compound	Energy, kcal
H—H	103.2	O—O	H_2O_2	34
F—F	37	N—N	N_2H_4	37
Cl—Cl	57.1	C—C	C_2H_6	77.7
Br—Br	45.4	C—C	C_3H_8	79.0
I—I	35.6	C—C	$n\text{-}C_4H_{10}$	79.6
N≡N	225.2	C—C	$i\text{-}C_4H_{10}$	80.1
H—F	135	C—C	Diamond	84.9
H—Cl	102.1	C=C	C_2H_4	140.0
H—Br	86.7	C⋯C	C_6H_6	123.8
H—I	70.6	C≡C	C_2H_2	193.3
O—H	109.4	C—O	CH_3OH	78.3[b]
N—H	92.2	C=O	CH_2O	163.3[b]
C—H	98.2	C=O	CH_3CHO	174.4[b]
S—H	81.1[b]	C=O	CH_3COCH_3	185.6[b]
C—F	~102[b]	C—N	CH_3NH_2	66.3[b]
C—Cl	78	C—S	CH_3SH	56.7[b]
C—Br	65			
C—I	57	C≡N	CH_3CN	215.0[b]

[a] Many of these values are from a table in K. S. Pitzer, *J. Am. Chem. Soc.*, **70**, 2140 (1948).

[b] Calculated from ΔH_{298} data.

fact, Pauling's electronegativities (Table 1-1) are numbers chosen to give as good a general fit as possible to the relation

$$x_B - x_A = 0.208 \sqrt{BE_{A-B} - \frac{BE_{A-A} + BE_{B-B}}{2}}$$

where x_A and x_B are electronegativities and BE's are bond energies in kcal/mole.

Wrinch and Harker have pointed out that, in general, for a bond between two given atoms the bond length decreases as the bond energy increases.[27]

1-5. Hydrogen Bonding.[28]   It has long been observed that hydroxy compounds have considerably higher boiling points than their non-hydroxylic isomers.   The boiling point of ethanol, for example, is 103°

[27] D. Wrinch and D. Harker, *J. Chem. Phys.*, **8**, 502 (1940).   See also H. A. Skinner, *Trans. Faraday Soc.*, **41**, 645 (1945) and G. Glockler, *J. Chem. Phys.*, **16**, 842 (1948); **19**, 124 (1951).

[28] For a thorough treatment of this subject, see G. C. Pimentel and A. L. McClellan, "The Hydrogen Bond," W. H. Freeman and Co., San Francisco, 1960.

higher than that of dimethyl ether. This has been explained by the formation of a bond between the hydroxylic hydrogen atom of one molecule and the oxygen atom of another. This bond is called a *hydrogen bond,* and although it is much stronger than the van der Waals attractive forces between molecules, it is much weaker than ordinary covalent bonds, rarely having a bond energy above 9 kcal/mole. Such bonds form between a hydrogen atom and an atom containing an unshared electron pair. The strength of the bond formed is best correlated with the acidity of the hydrogen atom and the basicity of the atom with the unshared electron pair (the *acceptor* atom),[29] although electrostatic interactions are also important. Unless the acceptor atom is at least as basic as fluorine atoms are in uncharged molecules, and unless the hydrogen atoms are much more acidic than those in saturated hydrocarbons, any hydrogen bonds that may form are usually too weak to be of significance. Of course, if the hydrogen atom is too acidic and the acceptor atom too basic, the hydrogen will be transferred as a proton to form a covalent bond with the acceptor atom in a simple acid-base reaction. Hydrogen bonds are usually represented by broken or dotted lines. Thus we shall represent the hydrogen-bonded complex formed between an alcohol and an amine in the following manner:

$$R—\overline{O}—H\text{----}|N—R$$

with R above N and R below N.

The discussion of formic acid dimer in Sec. 1-1b is an explanation of hydrogen bonding in terms of resonance. The objection has been raised that hydrogen bonding cannot be due to resonance because the hydrogen is usually not halfway between the bonded atoms. However, this merely gives the information that the two resonance structures do not contribute equally. Indeed, in the case of the F----H----F⁻ ion, the hydrogen has been shown to be equidistant from the two fluorine atoms,[30] suggesting resonance stabilization by two equally contributing structures. Hydrogen bonding has been depicted as a purely electrostatic interaction, and Coulson has described some of the evidence for electrostatic effects.[31] However, although electrostatic interactions must contribute to

[29] W. Gordy and S. C. Stanford, *J. Chem. Phys.,* **8,** 170 (1940); L. P. Hammett, *J. Chem. Phys.,* **8,** 644 (1940); S. C. Stanford and W. Gordy, *J. Am. Chem. Soc.,* **63,** 1094 (1941); C. Curran, *J. Am. Chem. Soc.,* **67,** 1835 (1945).

[30] E. F. Westrum, Jr., and K. S. Pitzer, *J. Am. Chem. Soc.,* **71,** 1940 (1949); S. W. Peterson and H. A. Levy, *J. Chem. Phys.,* **20,** 704 (1952). For evidence for symmetrical hydrogen bonds in other species, see T. C. Waddington, *J. Chem. Soc.,* 1708 (1958); S. W. Peterson and H. A. Levy, *J. Chem. Phys.,* **29,** 948 (1958); J. C. Speakman, *Proc. Chem. Soc.,* 316 (1959).

[31] Coulson, *op. cit.,* pp. 298–307.

hydrogen bonding, a representation solely on this basis might be expected to yield a correlation of hydrogen-bonding ability with electronegativity. That is, one would expect alkyl fluorides to be better hydrogen-bond acceptors than alcohols and ethers and these latter to be better than amines.   The actual order appears to be the reverse of this, with the more basic atoms leading to stronger hydrogen bonds,[29] as might be expected from a resonance interpretation of hydrogen bonding (in which the hydrogen is covalently bonded to the basic atom in one contributing structure).   The unusually stronger bond in the HF_2^- is no exception to this generalization, since in this case the fairly strongly acidic HF is forming a bond not to an electrically neutral fluoride but to the much more strongly basic fluoride anion.

The formation of intramolecular hydrogen bonds is called *chelation* and is often of interest because of its effect on the properties of compounds involved.   For example, the chelated compound salicylaldehyde

boils at 196° and may be steam-distilled readily, whereas the meta and para isomers, whose hydrogen-bonding tendencies must be satisfied intermolecularly, boil above 240° and are not appreciably volatile in steam.

1-6. Rotation around Carbon–Carbon Single Bonds.   1-6a. *Acyclic Compounds.*[32]   From heat-capacity measurements, X-ray and electron-diffraction determinations of molecular structure, and measurements of dipole moments and of infrared, Raman, ultraviolet, and nuclear-magnetic-resonance spectra has come compelling evidence that there is considerable resistance to rotation around the carbon-carbon single bond of ethane and its derivatives.   This resistance is most simply depicted as arising largely from van der Waals (steric) repulsions between corresponding atoms or groups on adjacent carbons.[33]   Thus, the eclipsed form, in which these atoms are nearest to each other (shown in Fig. 1-13a for the case of ethane), is the least stable, having the highest energy content of any rotational form.   The staggered conformation[34] (Fig. 1-13b), in

---

[32] For a more complete treatment of this subject, see S. Mizushima, "Structure of Molecules and Internal Rotation," Academic Press, Inc., New York, 1954.

[33] J. G. Aston, S. Isserow, G. J. Szasz, and R. M. Kennedy, *J. Chem. Phys.*, **12**, 336 (1944); E. A. Mason and M. M. Kreevoy, *J. Am. Chem. Soc.*, **77**, 5808 (1955); **79**, 4851 (1957).

[34] By *conformation* is meant any one of the arrangements of the atoms of a molecule produced by rotation around a single bond that corresponds to a minimum in a plot of potential energy vs. molecular geometry.

which the atoms on adjacent carbons are farthest apart, is believed to be the most stable.

By a number of independent methods of measurement the difference in energy content between the eclipsed and staggered forms of ethane has

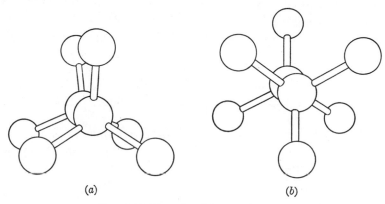

(a)   (b)

FIG. 1-13. Rotational forms of ethane.

been found to be about 2.9 kcal/mole.   This value and those of other barriers to rotation around single bonds of simple organic molecules are listed in Table 1-5.   The relative sizes of the energy barriers for hexachloroethane and hexachlorodisilane are of particular interest.   In the silicon compound, where the chlorines on adjacent atoms are more widely separated, the barrier is much smaller, despite the fact that the

TABLE 1-5. BARRIERS TO ROTATION AROUND SINGLE BONDS[a]

Compound	Barrier, kcal/mole	Compound	Barrier, kcal/mole
CH_3—CH_3	2.9	CH_3—OCH_3	2.8
CH_3—$C(CH_3)_3$	4.3	CH_3—SCH_3	2.0
CH_3—$Si(CH_3)_3$	1.3	CH_3—CH_2CH_3	3.4
CH_3—CF_3	3.3	CH_3—OH	1.1
CH_3—SiF_3	1.2	CH_3—NH_2	1.9
CH_3—$C{\equiv}C$—CH_3	~0	CF_3—CF_3	3.9
Cl_3C—CCl_3	10–15[b]	C_6H_5—CF_3	~0[c]
Cl_3Si—$SiCl_3$	~0	$CH_2{=}CH$—$CH{=}CH_2$	4.9[d]

[a] Unless otherwise stated these data were taken from Ref. 33 and are probably reliable within a few tenths of a kilocalorie per mole.

[b] Y. Morino and M. Iwasaki, *J. Chem. Phys.*, **17**, 216 (1949).

[c] D. W. Scott et al., *J. Am. Chem. Soc.*, **81**, 1015 (1959).   Note that rotation around this single bond should produce not three but six energy maxima per rotation.

[d] J. G. Aston, G. Szasz, H. W. Wooley, and F. G. Brickwedde, *J. Chem. Phys.*, **14**, 67 (1946).   In this case it is the eclipsed forms (all carbons coplanar) that are stable, since in such forms resonance interaction between the double bonds is at a maximum.

Si—Cl bond dipole is probably considerably larger than the C—Cl dipole.

The three conformations of ethane are all identical, but this is not so for a compound like ethylene chloride.[32]   In this case there are two conformations, the *trans* (Fig. 1-14a) and the *gauche*, or skew (Fig. 1-14b). Electron-diffraction studies have shown that the vapor contains about 75 per cent of the trans and 25 per cent of the gauche conformation at

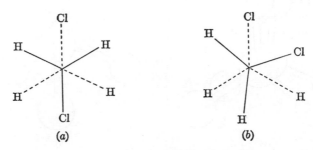

FIG. 1-14. Rotational isomers of ethylene chloride (dashed lines are attached to the carbon atom behind the one to which the solid lines are attached).

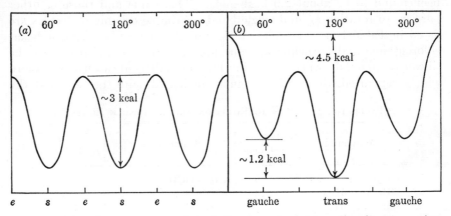

FIG. 1-15. Plot of energy vs. angular rotation from cis and/or eclipsed conformation. (a) Ethane (e, eclipsed; s, staggered); (b) ethylene chloride.

room temperature.   From the manner in which the dipole moment and the infrared spectrum change with changing temperature the trans compound has been shown to be the more stable isomer by about 1.2 kcal/mole in the vapor phase.   As shown in the plot of energy content vs. angular rotation (Fig. 1-15), this is the energy difference between two staggered forms, and the 2.9 kcal/mole value mentioned for ethane is the difference between a staggered and an eclipsed form.   In the pure liquid phase where the "solvent" ethylene chloride, whose dielectric constant is about 10, stabilizes the more polar gauche form to a greater extent than it does the trans form (whose dipole moment is zero), there is little difference in stability between the two forms.   As the liquid mixture is cooled, it is the more symmetrical trans conformation whose solubility is first exceeded.   As it separates, the rapid equilibrium in the liquid phase continually transforms more of the gauche to the trans conformation so that the solid material is pure trans.

1-6b. *Structure of Cyclohexane and Related Compounds.*[35] It was realized for many years that cyclohexane could be written in two conformations, the "boat" form and the "chair" form, but it remained for

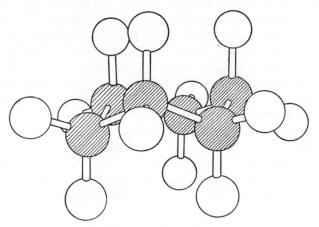

Fig. 1-16. Chair form of cyclohexane.

Hassel to show by electron-diffraction structure determinations that cyclohexane actually has (very predominantly, at least) the chair form[36] (Fig. 1-16).   Pitzer has pointed out that this would be expected, since in this form all the valences attached to adjacent carbon atoms are staggered, whereas in the boat form there are six pairs of opposed valences on adjacent carbon atoms.[37]   A thermochemical study of two stereoisomeric tricyclic compounds, whose only essential difference was that the central six-membered ring of one was in the chair form and that of the other was in the boat form, showed the chair form to be more stable by about

[35] For more complete treatments of this subject, see D. H. R. Barton and R. C. Cookson, *Quart. Revs. (London),* **10,** 44 (1956); W. G. Dauben and K. S. Pitzer, in M. S. Newman (ed.), "Steric Effects in Organic Chemistry," chap. 1, John Wiley & Sons, Inc., New York, 1956.

[36] O. Hassel, *Tidsskr. Kjemi, Bergvesen Met.,* **3,** 32 (1943); *Chem. Abstr.,* **39,** 2244 (1945); and earlier articles.   See also R. S. Rasmussen, *J. Chem. Phys.,* **11,** 249 (1943), where evidence is presented for the chair structure from the Raman spectrum.

[37] K. S. Pitzer, *Science,* **101,** 672 (1945); C. W. Beckett, K. S. Pitzer, and R. Spitzer, *J. Am. Chem. Soc.,* **69,** 2488 (1947).

5.3 kcal/mole.[38]   Hassel has also pointed out that substituents on the cyclohexane ring may be of two different types (see Fig. 1-16).   Three valences are directed straight up from the plane of the ring (actually, two parallel planes, each containing three carbon atoms) and three more directed straight down.   These bonds, which are parallel to the molecule's axis of symmetry, are called *axial* bonds, symbolized *a*.[39]   The other six valence bonds alternate somewhat up and down around the ring but are nearly in the plane of the ring.   These are known as *equatorial*, or *e*, bonds.[39]   It may be noted that each carbon atom contains one axial and one equatorial bond.   Adjacent axial groups of a given size, if significantly larger than hydrogen, are definitely closer to each other than are adjacent equatorial groups.   For this reason it might be expected that steric interactions would tend to keep larger groups out of axial positions, and indeed there is a definite tendency for monoatomic substituents (such as bromine[40]) and a stronger tendency for branched substituents (such as methyl[40]) to be found in equatorial positions.

Two conformers, such as the axial and equatorial forms of cyclohexyl chloride, are in rapid equilibrium.   In the case of cyclohexane itself, nuclear-magnetic-resonance measurements have shown that the average molecule changes its conformation 121 times per second at $-66.5°$.[41] Interactions between nonbonded atoms merely cause cyclohexane to exist largely in the stable, staggered, chair conformation, but they definitely destabilize cyclopentane.   If all the carbon atoms in this molecule were coplanar, the C—C—C bond angles would be 108°, only 1°28′ below the optimum value for tetrahedral carbon.   However, in such a form all the valence bonds would be opposed (eclipsed).   It is therefore not surprising that cyclopentane actually has a nonplanar arrangement of its carbon atoms, decreasing the nonbonding interactions, even though such an arrangement decreases the bond angles even further below their optimum size.[42]

1-7. Solutions of Electrolytes.[43]   In not all compounds are the atoms joined by definite covalent bonds.   On the contrary, the reaction

[38] W. S. Johnson, J. L. Margrave, V. J. Bauer, M. A. Frisch, L. H. Dreger, and W. N. Hubbard, *J. Am. Chem. Soc.*, **82**, 1255 (1960); cf. N. L. Allinger and L. A. Freiberg, *J. Am. Chem. Soc.*, **82**, 2393 (1960).

[39] D. H. R. Barton, O. Hassel, K. S. Pitzer, and V. Prelog, *Science*, **119**, 49 (1954).

[40] Cf. F. R. Jensen and L. H. Gale, *J. Org. Chem.*, **25**, 2075 (1961), and references cited therein.

[41] F. R. Jensen, D. S. Noyce, C. H. Sederholm, and A. J. Berlin, *J. Am. Chem. Soc.*, **82**, 1256 (1960).

[42] J. E. Kilpatrick, K. S. Pitzer, and R. Spitzer, *J. Am. Chem. Soc.*, **69**, 2483 (1947).

[43] For a much more complete treatment, see R. W. Gurney, "Ionic Processes in Solution," McGraw-Hill Book Company, Inc., New York, 1953; H. S. Harned and B. B. Owen, "The Physical Chemistry of Electrolytic Solutions," 3d ed., Reinhold Publishing Corporation, New York, 1958.

of sodium with chlorine consists of the removal of one electron from each sodium atom to give a sodium cation and the addition of one electron to each chlorine atom to give a chloride anion. In the sodium chloride crystal every sodium ion attracts every chloride ion (and repels every other sodium ion) in agreement with Coulomb's law, but no sodium ion is bonded to any particular chloride ion. The crystal lattice consists merely of a systematic array of alternate sodium and chloride ions.

1-7a. Dielectric Constant of Solvents. An important factor permitting the solution of an ionic salt like sodium chloride is the dielectric constant of the solvent (the factor by which the interactions between electrical charges is reduced). That is, in a solution with a dielectric constant of 80, e.g., water, two oppositely charged ions separated by a given distance will have only one-eightieth the attraction for each other that they would in a vacuum and will therefore have a much smaller tendency to recombine to form a crystal. A representative list of dielectric constants for various compounds is given in Table 1-6.

TABLE 1-6. DIELECTRIC CONSTANTS, ϵ, OF VARIOUS COMPOUNDS[a,b]

Compound	ϵ	Compound	ϵ
N-Methylformamide	190.5[c]	Ethanol	25.1
Hydrogen cyanide	115	Acetone	21.2
Sulfuric acid	~110[d]	Acetic anhydride	20
Formamide	109	n-Butyl alcohol	17.8
Water (at 0°C)	88.3	Ammonia	17.3
Water (at 20°C)	80.4	Sulfur dioxide	14.1
Water (at 100°C)	55.1	t-Butyl alcohol (at 30°C)	10.9
Hydrogen fluoride (at 0°C)	84	Ethylene chloride	10.65
Formic acid	57.9	Acetic acid	6.15
Hydrazine	53	Chlorobenzene	5.71
N,N-Dimethylformamide	37.6[c]	Ethyl ether	4.34
Nitromethane	37.5	Benzene	2.28
Acetonitrile	37.5	Carbon tetrachloride	2.24
Nitrobenzene	35.7	Cyclohexane	2.07
Methanol	33.6	n-Hexane	1.89

[a] Largely from A. A. Maryott and E. R. Smith, Table of Dielectric Constants of Pure Liquids, *Natl. Bur. Standards Circ.* 514, 1951.

[b] At 20°C unless otherwise stated.

[c] G. R. Leader and J. F. Gormley, *J. Am. Chem. Soc.*, **73**, 5731 (1951).

[d] J. C. D. Brand, J. C. James, and A. Rutherford, *J. Chem. Phys.*, **20**, 530 (1952).

1-7b. Solvation of Ions. Many facts, including the much more general solubility of ionic substances in water ($\epsilon = 80$) than in hydrogen cyanide ($\epsilon = 115$), show that there is an additional factor operating in solutions of electrolytes that is probably even more important than the dielectric constant. This is a specific interaction between solvent molecules and the dissolved ions called *solvation*.   The attraction of a charge for a dipole must account for part of this interaction.   In the presence of a sodium ion, a dipolar water molecule would be expected to be oriented with its negative (oxygen) end toward the sodium.   Since the negative charge on the oxygen is equal to the sum of the positive charges on the hydrogens and is also closer to the sodium ion, there will be a net attraction.   Thus a sodium ion in an aqueous solution will be surrounded by the oxygen atoms of water molecules held as closely as the balance between the attractive forces and steric repulsive forces permits.

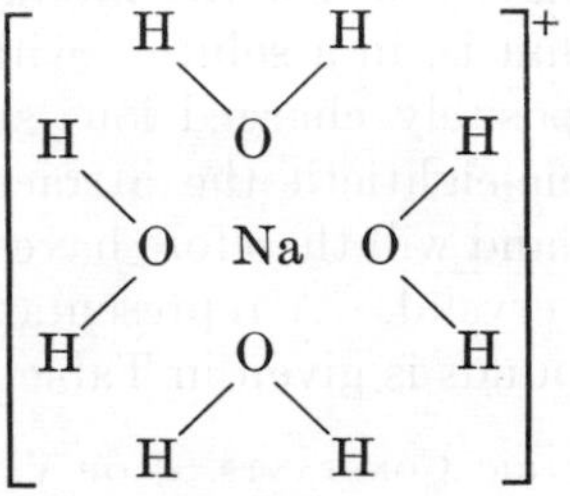

Although four molecules are shown above, the exact number in the primary solvation shell immediately surrounding the sodium ion is not definitely agreed upon.[44]   Also there are secondary, more weakly held solvation shells of water molecules outside the primary one.   An anion would be expected to attract the positive end of a polar solvent molecule,

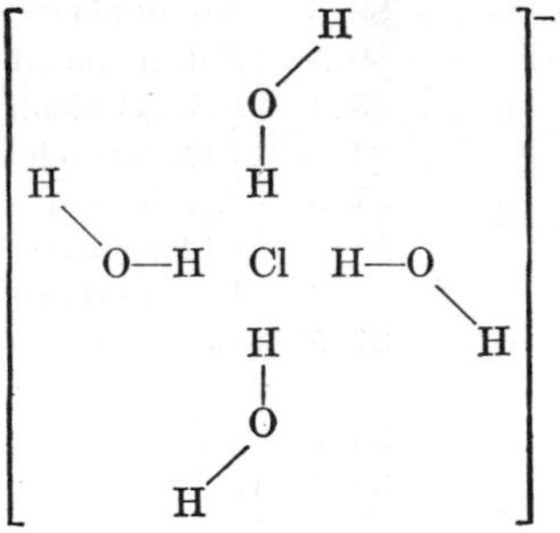

so that for suitable solvents (such as water, above) the solvation interaction is a type of hydrogen bonding.   In this connection, sometimes there is utility in considering resonance contributions of structures with covalent bonds between solvent and ion (with cations as well as anions).

The extent to which an ion tends to be stabilized by solvation, as described above, depends, among other things, on the size of the ion. Since the potential energy associated with a given charge is an inverse function of the volume over which the charge is spread, large ions tend

[44] J. O'M. Bockris, *Quart. Revs. (London)*, **3,** 173 (1949).

to be more stable than small ions.   On the other hand, since the charge on small ions is more concentrated, it interacts more strongly with the solvent.

1-7c. Ion-pair Formation.[45]   In dilute aqueous solutions (<0.01 N) most salts are almost completely dissociated.   In poorer solvating media, however, this is not the case.   In acetone solution, lithium bromide has a dissociation constant of about 4×10^{-4}.   It should not be thought, however, that the undissociated lithium bromide exists as a covalent compound: it consists of a lithium ion and a bromide ion held together by electrostatic attraction, the acetone not having sufficient solvating power to free the ions from each other.   This complex made up of the two ions is called an *ion pair.*   In poorly solvating solvents most salts exist as ion pairs (and triplets and higher aggregates) even at fairly low concentrations.   The phenomenon occurs even in water at higher concentrations.

It is not always easy to distinguish between the association of two ions to form an ion pair and an association to form a covalent compound, and indeed it is quite likely that the types of bonds formed may vary gradually and continuously from highly ionic to highly covalent ones. Nevertheless, it has been found in several cases that the absorption spectrum (in the visible and ultraviolet) is almost the same for an ion whether it is part of an ion pair or completely dissociated.   This is because the formation of an ion pair does not greatly affect the electronic structure of the ion.   For example, it is known that many picrates form ion pairs under certain conditions.   It has been found that the absorption spectrum of these ion pairs is very similar to that of the picrate ion. On the other hand, the formation of a covalent bond to give picric acid gives a colorless compound rather than a yellow one (the yellow color of ordinary picric acid is said to be due to impurities).

Many of the organic reaction mechanisms that we shall discuss in terms of ions are actually reactions of ion pairs, especially in the poor ion-solvating media and high concentrations common in synthetic organic chemistry.   Brady and Jakobovits have demonstrated the importance of this fact in connection with several organic reactions.[46]

PROBLEMS

1. Write Lewis structures for each of the following compounds.   If there is more than one important valence-bond structure, write all of them and discuss their relative contributions to the total structure of the molecule.

[45] For a symposium on ion pairs containing many leading references, see *J. Phys. Chem.,* **60,** 129–190 (1956).

[46] O. L. Brady and J. Jakobovits, *J. Chem. Soc.,* 767 (1950).

(a) CSF_8

(b) Methyl azide

(c) Diazomethane

(d) Bicyclo[2.2.2]-2,5,7-octatriene

(e) Sodium p-acetylphenoxide

(f) o-HOC_6H_4CH=NC_6H_5 (trans-)

(g) Formamide

(h) NH_4BF_4

2. Equilibrium a usually lies to the right and equilibrium b to the left.   Suggest an explanation at least for the *relative* positions of the two equilibria.

(a) R_2CH—$NO \rightleftharpoons R_2C$=$NOH$

(b) $R_2CHNO_2 \rightleftharpoons R_2C$=$NO_2H$

3. There are two kinds of C—N bonds in 2,4,6-trinitroiodobenzene; one is 1.35 A in length and the other 1.45 A.   Tell which is which and why.

4. (a) In terms of x, the C—X bond distance, and data from Chap. 1, derive expressions for the distance between two cis axial X atoms (as in a *cis*-1,3-di-X-cyclohexane) and the distance between two equatorial X atoms attached to adjacent carbons. Assume 109° 28′ bond angles.

(b) Using the relations derived in Prob. 4a, calculate the distances for the cases where X is hydrogen, fluorine, and iodine.

5. Using the bond energies of Table 1-4, estimate the equilibrium constant for the following reaction at 27°:

$$CH_3SH + (CH_3)_2O \rightleftharpoons (CH_3)_2S + CH_3OH$$

6. The vapor pressure of pure methanol is 625 mm at 60°.   Would you expect the partial pressure of methanol over a solution in which the mole fraction of n-octane is 0.99 and that of methanol is 0.01 to be more than, less than, or very nearly equal to 6.25 mm at 60°?   Why?

Chapter 2

ACIDS AND BASES[1]

2-1. Definitions of Acids and Bases. Although the Arrhenius definition of acids (substances that ionize in aqueous solution to produce hydrogen ions) and of bases (those that ionize to produce hydroxide ions) is still widely used, e.g., especially in relation to nomenclature, certain other definitions are more useful in theoretical chemistry.

2-1a. *Lowry-Brønsted Acids and Bases.* According to the definitions of Lowry and Brønsted,[2a] an acid is a *proton donor* and a base is a *proton acceptor.* These definitions may be illustrated by the following equilibria:

(1)	(2)	(3)	(4)
Acid	Base	Acid	Base
H_2SO_4 +	H_2O	$\rightleftharpoons$ H_3O^+	+ HSO_4^-
$(C_6H_5)_3CH$ +	NH_2^-	$\rightleftharpoons$ NH_3	+ $(C_6H_5)_3C^-$
HSO_4^- +	NH_3	$\rightleftharpoons$ NH_4^+	+ SO_4^-

$$HBr + H_3\overset{+}{N}-NH_2 \rightleftharpoons H_3\overset{+}{N}-\overset{+}{N}H_3 + Br^-$$
$$H_3O^+ + OH^- \rightleftharpoons H_2O + H_2O$$

From these equilibria it is seen that acids and bases may have either positive or negative charges or be neutral and that a given molecule or ion may act as an acid in one reaction and as a base in another. It should also be noted that when an acid has donated a proton, it becomes a base. This base is known as the conjugate base of the acid in question. Analogously, each base has its conjugate acid. The acids in columns 1 and 3 above are the conjugate acids of the bases in 4 and 2, respectively.

2-1b. *Lewis Acids and Bases.* According to Lewis's definition,[2b] acids are molecules or ions capable of coordinating with unshared electron pairs, and bases are molecules or ions having unshared electron pairs available for sharing with acids. Since a reagent must have an unshared

---

[1] For a more complete treatment of this subject, see R. P. Bell, "The Proton in Chemistry," Cornell University Press, Ithaca, N.Y., 1959.

[2] (a) T. M. Lowry, *Chem. & Ind. (London),* **42,** 43 (1923); J. N. Brønsted, *Rec. trav. chim.,* **42,** 718 (1923); (b) G. N. Lewis, *J. Franklin Inst.,* **226,** 293 (1938).

electron pair in order to accept a proton, the same reagents may act as bases in either the Lowry-Brønsted or the Lewis sense. All Lowry-Brønsted acids are also Lewis acids, since the proton from any proton donor coordinates with an unshared electron pair. In addition, however, the Lewis definition includes many reagents, such as boron trifluoride, sulfur trioxide, aluminum chloride, etc., that may also neutralize bases.

$$
\underset{\underset{\displaystyle H}{|}}{\overset{\overset{\displaystyle H}{|}}{H\!-\!N|}} + \underset{\underset{\displaystyle |\underline{F}|}{|}}{\overset{\overset{\displaystyle |\overline{F}|}{|}}{B\!-\!\overline{F}|}} \rightarrow \underset{\underset{\displaystyle H}{|}}{\overset{\overset{\displaystyle H}{|\oplus}}{H\!-\!N}}\!-\!\underset{\underset{\displaystyle F}{|}}{\overset{\overset{\displaystyle F}{|\ominus}}{B}}\!-\!F
$$

Although it is therefore true that Lowry-Brønsted acids merely form a special class of Lewis acids, this class is so important and has been so extensively studied that a separate name is useful. For this reason, following common usage, we shall use the terms acid and base in the Lowry-Brønsted sense and call aprotic acids Lewis acids or electrophilic reagents (the respective conjugate terms being Lewis bases and nucleophilic reagents).

2-2. Ionization as an Acid-Base Reaction. *2-2a. Oxonium Ions.* The "ionization" of hydrogen bromide, a covalent compound, that occurs upon solution in water (and many other solvents) involves the action of water as a base in accepting a proton from the acid hydrogen bromide. The hydronium ions (H_3O^+, sometimes called oxonium ions) and bromide ions thus formed are solvated, as are all ions in solution; but since the over-all process also involves the formation of ions from a covalent molecule, it may be seen to differ from the solution of sodium chloride in water, in which the ions that are solvated already existed as ions in the solute. It should be noted that in the solvation of any cation water may be said to act as a base in the Lewis sense in so far as covalent bonds are formed between the cation and solvent. Similarly, the solvation of an anion, involving hydrogen bonding, may be related to solvent acidity.

The basicity of water was not recognized much earlier than it was principally because of the relative weakness of this basicity. Evaporation of a solution of hydrogen bromide in liquid ammonia leaves the solid salt ammonium bromide behind because the equilibrium

$$
NH_3 + HBr \rightleftharpoons NH_4^+ + Br^-
$$

lies so far to the right at room temperature (but not at elevated temperatures). However, the equilibrium constant for the reaction

$$
H_2O + HBr \rightleftharpoons H_3O^+ + Br^-
$$

is considerably smaller, so that at equilibrium appreciable amounts of water and hydrogen bromide are present. When the solution is evaporated, these volatile components are continuously removed and replaced by a shift of equilibrium from the salt, hydronium bromide. However, water has been shown to form stable hydronium salts with sufficiently strong acids. The explosive liquid perchloric acid forms a relatively stable solid salt, hydronium perchlorate, formerly written as $HClO_4 \cdot H_2O$ but shown by X-ray,[3] nuclear-magnetic-resonance,[4] and infrared-spectral[5] studies to have a crystal lattice consisting of H_3O^+ and ClO_4^- ions. The crystalline hydrates of boron trifluoride[6] and nitric acid[4,5] have also been shown to be hydronium salts. Striking further evidence for the formation of the hydronium ion was obtained by Bagster and Cooling in a study in liquid sulfur dioxide solution.[7] In this solvent, water is almost insoluble, whereas hydrogen bromide is soluble but not ionized. A solution of hydrogen bromide in sulfur dioxide, however, dissolves 1 mole of water per mole of hydrogen bromide to give an ionic solution, which upon electrolysis liberates at the cathode 1 mole of water per faraday. All these facts are explained by the equilibrium

$$H_2O + HBr \rightleftharpoons H_3O^+ + Br^-$$

In addition to the simple H_3O^+ ion, substituted oxonium ions are known. Meerwein and coworkers have isolated $(CH_3)_3O^+BF_4^-$ and several other solid oxonium salts.[8] A number of oxonium salts of the general type of the methylpyrone hydrobromide shown below are also known.

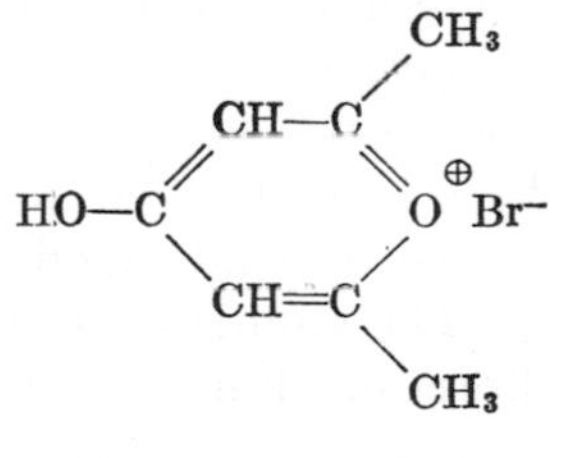

From the similarity of the behavior of acids in alcoholic and in aqueous solutions, there is little doubt that ROH_2^+ ions are present in the former.

2-2b. *Leveling Effects of Solvents.* The compounds commonly referred to as "strong acids" are those for which the equilibrium

$$HA + H_2O \rightleftharpoons H_3O^+ + A^-$$

[3] M. Volmer, *Ann.*, **440**, 200 (1924).

[4] R. E. Richards and J. A. S. Smith, *Trans. Faraday Soc.*, **47**, 1261 (1951); Y. Kakiuchi, H. Shono, K. Komatsu, and K. Kigoshi, *J. Chem. Phys.*, **19**, 1069 (1951).

[5] D. E. Bethell and N. Sheppard, *J. Chem. Phys.*, **21**, 1421 (1953).

[6] L. J. Klinkenberg and J. A. A. Ketelaar, *Rec. trav. chim.*, **54**, 959 (1935).

[7] L. S. Bagster and G. Cooling, *J. Chem. Soc.*, **117**, 693 (1920).

[8] H. Meerwein, G. Hinz, P. Hofmann, E. Kroning, and E. Pfeil, *J. prakt. Chem.*, **147**, 257 (1937).

in dilute aqueous solution is so far to the right that the amount of H_3O^+ formed differs from the amount of HA added by less than the experimental error. The fact that two acids are found to be completely ionized in dilute aqueous solution by present methods of measurement does not, of course, mean they are necessarily equal in strength. For instance, if the ionization constants of HX and HY are 10^4 and 10^2, the hydrogen-ion concentration of their 0.1 N aqueous solutions will be 0.099999 N and 0.0999 N, respectively. Thus, equal concentrations of the two acids will yield equal concentrations of hydrogen ions, within experimental error. In a solvent 10^6 times less basic than water,[9] HX and HY should have ionization constants of 10^{-2} and 10^{-4} and hydrogen-ion concentrations (in 0.1 N solutions) of 0.027 and 0.0031, respectively, and should thus be easily distinguishable in strength. This tendency of a solvent to make all acids whose strength is greater than a certain amount appear equal is called the *leveling effect*.

The strength of bases is also subject to the leveling effect. The basicity of dimethylamine in water is due to the existence of the equilibrium

$$(CH_3)_2NH + H_2O \rightleftharpoons (CH_3)_2NH_2^+ + OH^-$$

The amine is said to be a weak base because not all of it is transformed to its conjugate acid even in fairly dilute aqueous solution. Aniline, which is converted to an even lesser extent into the anilinium ion, is said to be a weaker base. The positions of these equilibria are as much due to the weakness of water as a proton donor as to the proton-accepting ability of the amine. In a more strongly acidic solvent, such as formic acid, both amines would be strong bases and their strengths probably indistinguishable, since both would be transformed entirely (within experimental error) into the corresponding substituted ammonium ions.

2-2c. *Very Weak Acids.* The strengths of acids may be studied only by measurements on equilibria involving at least two acids.

$$HA + B \rightleftharpoons A + HB$$

(In order to make the above equation general we have omitted electrical charges, since in the general case we can say only that the charge on an acid is one unit more positive than that on its conjugate base.) Commonly the solvent acts as one of the acids or bases in equilibria of this type. For fairly strong acids, measurable concentrations of A (and hence a quantitative measurement of the equilibrium constant) are

---

[9] It is not possible in practice to find a solvent that differs from water *only* in basicity, and therefore we have no definite assurance that the relative acidities of HX and HY will remain unchanged. However, experimental data show that the relative acidity of closely related acids of the same electrical charge type does not vary greatly from solvent to solvent (see Sec. 2-3).

obtained when the base B is the solvent. For weaker acids, however, the conjugate base of the solvent must be used as B in order to transform a measurable concentration of HA to A. In this case, as well as in the former case, success depends upon the basicity of B and hence the acidity of HB. In studies of very weak acids it is necessary that HB also be very weak so that B will be very strongly basic. For this reason, the solvent liquid ammonia is often used in studying very weak acids. Even many hydrocarbons, such as diphenylmethane, react as acids toward potassium amide.

$$(C_6H_5)_2CH_2 + NH_2^- \rightleftharpoons (C_6H_5)_2CH^- + NH_3$$

Another technique that has been used in the study of very weak acids involves a direct comparison of two acids by establishment of equilibrium between one and the conjugate base (as the sodium or potassium salt) of the other in a solvent of negligible acidity. Comparisons of this sort have been made by Conant and Wheland and by McEwen using ether and benzene as solvents.[10] By these comparisons, the following order of acidity (in the solvents used) was obtained: t-butyl alcohol $\sim$ acetophenone $>$ phenylacetylene $\sim$ indene $>$ diphenylamine $>$ fluorene $>$ aniline $>$ xanthane $>$ triphenylmethane $>$ diphenylmethane $>$ ammonia $>$ toluene $>$ benzene[11] $>$ pentane.[11]

2-2d. *Very Weak Bases.* Very weak bases may be studied only by their reactions with very strong acids. Sulfuric acid is the most acidic solvent in which it is convenient to make measurements. Many compounds that are too weakly basic to be measured in water are strong bases in this solvent. The ionization of bases in sulfuric acid solution has been studied most usefully by cryoscopic measurements. These measurements were pioneered by Hantzsch,[12] greatly improved by Hammett and coworkers,[13] and further refined by Gillespie, Hughes, and Ingold.[14] Cryoscopic measurements in sulfuric acid are made convenient by its freezing point (10.36°, relatively near room temperature) and fairly high cryoscopic constant (6.12° mole^{-1} kg).[15] The purity of

[10] J. B. Conant and G. W. Wheland, *J. Am. Chem. Soc.*, **54**, 1212 (1932); W. K. McEwen, *J. Am. Chem. Soc.*, **58**, 1124 (1936).

[11] The positions of benzene and pentane followed from the ability of phenylsodium to metalate toluene and of amylsodium to metalate benzene [A. A. Morton and F. Fallwell, Jr., *J. Am. Chem. Soc.*, **60**, 1429, 1924 (1938)].

[12] A. Hantzsch, *Z. physik. Chem.*, **61**, 257 (1907); *Ber.*, **63B**, 1782, 1789 (1930); and intervening papers.

[13] L. P. Hammett and A. J. Deyrup, *J. Am. Chem. Soc.*, **55**, 1900 (1933); H. P. Treffers and L. P. Hammett, *J. Am. Chem. Soc.*, **59**, 1708 (1937).

[14] (a) R. J. Gillespie, E. D. Hughes, and C. K. Ingold, *J. Chem. Soc.*, 2473 (1950); (b) R. J. Gillespie, *J. Chem. Soc.*, 2493, 2537, 2542 (1950).

[15] R. J. Gillespie, *J. Chem. Soc.*, 1851 (1954).

sulfuric acid solutions may be established better by freezing-point measurements than by chemical analysis. In Fig. 2-1, the molality of water and of $H_2S_2O_7$ (the principal form in which a little SO_3 dissolved in H_2SO_4 is believed to exist) is plotted for solutions whose compositions are in the vicinity of that of pure H_2SO_4. From the nature of the curve obtained—straight lines leading to a rounded maximum—it may be seen that after the solution has become about 0.05 molal in H_2O (or $H_2S_2O_7$), the lowering of the freezing point produced by the addition of further H_2O (or $H_2S_2O_7$) is directly proportional to the amount added.

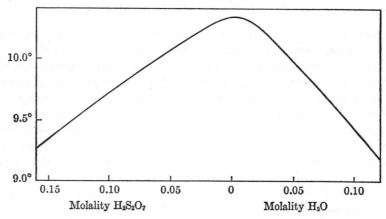

FIG. 2-1. Freezing points of solutions of $H_2S_2O_7$ and H_2O in H_2SO_4.

The addition of small concentrations of H_2O or $H_2S_2O_7$, however, does not lower the freezing point nearly so much as would be expected. This is explained by the considerable self-ionization of sulfuric acid occurring both by autoprotolysis

$$2H_2SO_4 \rightleftharpoons H_3SO_4^+ + HSO_4^- \qquad (2\text{-}1)$$

and by a process called *ionic self-dehydration*

$$2H_2SO_4 \rightleftharpoons H_3O^+ + HS_2O_7^- \qquad (2\text{-}2)$$

It is complicated by such further equilibria as

$$H_3SO_4^+ + HS_2O_7^- \rightleftharpoons H_2SO_4 + H_2S_2O_7 \qquad (2\text{-}3)$$

The addition of water may produce two foreign particles by the reaction

$$H_2O + H_2SO_4 \rightarrow H_3O^+ + HSO_4^- \qquad (2\text{-}4)$$

but the bisulfate ion thus produced may react with some of the approximately 0.013 molal sulfuric acidium ions ($H_3SO_4^+$) present in pure sulfuric acid to reverse equilibrium (2-1). The hydronium ions may similarly reverse equilibrium (2-2).   Thus the net increase in the number of foreign particles will be considerably lower than the number formed from water.   By the time about 0.05 molal water has been added, however, the concentrations of sulfuric acidium ions and hydrogen disulfate ions ($HS_2O_7^-$) left are negligible and so cannot be reduced significantly further.   For this reason it is often more convenient to make cryoscopic measurements not on pure sulfuric acid but on that to which enough water has been added to repress the self-ionization.   One of the most striking features of sulfuric acid as a solvent is the fact that for ionic as well as nonionic solutes, the lowering of the freezing point remains proportional to the concentration added, even to relatively large concentrations. (For most ionic solutes, this ideality continues to higher concentrations only if it is assumed that some of the ions take on sulfuric acid molecules of solvation.)   This fact shows that the activity coefficients of the ions involved remain constant to unusually high concentrations and therefore that the ion-solvating ability of sulfuric acid is probably greater than that of any other known solvent.

Cryoscopic measurement may lead in the present case, as it does in general, to information about the number of moles of foreign (nonsolvent) particles (ions and/or molecules) per formula weight of material added. We shall use the term "ν factor" for this number, in agreement with Gillespie, Hughes, and Ingold,[14] although earlier workers have used the term "i factor."

If water had a ν factor of two, i.e., if it yielded two particles per molecule by acting as a strong base according to Eq. (2-4), and if its solutions in sulfuric acid were ideal, the slope of the straight part of the right branch of the curve in Fig. 2-1 would be equal to twice the cryoscopic constant of sulfuric acid, or 12.24° mole^{-1} kg.   The actual slope is significantly less, being 11.21° mole^{-1} kg, according to Gillespie.[14b]   From his data Gillespie has calculated a value for the basicity constant of water in sulfuric acid (using a somewhat different cryoscopic constant for sulfuric acid from that now accepted).

$$K_b = \frac{[H_3O^+][HSO_4^-]}{[H_2O][H_2SO_4]} = 0.12$$

His data led to a constant value for K only if it was assumed that hydronium bisulfate takes on one molecule of sulfuric acid of solvation.   Deno and Taft[16] have presented a strong argument that K_b is about 50, while a calculation based on the acidity function (Sec. 2-3d) leads to a value of about 10^6.   By determining ν values at various concentrations, Gillespie has determined the basic ionization constants of a number of weak bases such as sulfones, sulfonic acids, sulfates, and nitro compounds.[14b]   By

[16] N. C. Deno and R. W. Taft, Jr., *J. Am. Chem. Soc.*, **76**, 244 (1954).

analogous methods the acidic ionization constants for sulfuric, disulfuric, and perchloric acids have been determined and found to vary in the order $H_2S_2O_7 > HClO_4 > H_2SO_4$, showing disulfuric acid to be considerably stronger than perchloric (sometimes claimed to be the strongest acid known). Higher polysulfuric acids, such as $H_2S_3O_{10}$, are probably even stronger than disulfuric.

The ν factors that have been determined for a number of other substances are also of considerable interest.[12-14] Acetic acid gives a ν factor of two, because of its reaction as a strong base.

$$CH_3CO_2H + H_2SO_4 \rightarrow CH_3C(OH)_2^+ + HSO_4^-$$

Other compounds that appear to be strong bases in sulfuric acid solution, as judged from their ν factors of two, are 2,4-dinitroaniline, azobenzene, benzalacetophenone, anthraquinone, benzophenone, benzoic acid, trimethylacetic acid, acetone, diethyl ether, ethyl acetate, acetaldehyde, benzamide, acetonitrile, and chloroacetic acid. Dichloroacetic acid gives a ν factor between one and two, showing that it is a weak base; trichloroacetic acid is a still weaker base, its ν factor being indistinguishable, by earlier workers, from unity. Certain compounds react in a more complicated manner. Methanol and ethanol give ν factors of three by the reaction

$$ROH + 2H_2SO_4 \rightarrow ROSO_3H + H_3O^+ + HSO_4^-$$

Triphenylcarbinol gives a yellow solution and a ν factor of four. Among the evidence that this is due to the reaction

$$(C_6H_5)_3COH + 2H_2SO_4 \rightarrow (C_6H_5)_3C^+ + H_3O^+ + 2HSO_4^-$$

is the fact that the yellow solution has the same absorption spectrum as the yellow solution of triphenylmethyl chloride in liquid sulfur dioxide (in which triphenylmethyl cations are believed to be present).[17]

Liquid hydrogen fluoride is another solvent in which very weak bases may be studied, and the basicities of aromatic hydrocarbons in this solvent have received considerable attention.[18]

2-3. Effect of Solvents on the Strength of Acids and Bases. The strength of an acid HA in the solvent S is usually defined as being proportional to its *acidity constant*, i.e., the equilibrium constant K_a for the equilibrium

$$HA^n + S \rightleftharpoons A^{n-1} + SH^+$$

$$K_a = \frac{[A^{n-1}][SH^+]}{[HA^n]}$$

[17] A. Hantzsch, *Ber.*, **54B**, 2573 (1921).

[18] D. A. McCaulay and A. P. Lien, *J. Am. Chem. Soc.*, **73**, 2013 (1951); H. C. Brown and J. D. Brady, *J. Am. Chem. Soc.*, **74**, 3570 (1952); M. Kilpatrick and F. E. Luborsky, *J. Am. Chem. Soc.*, **75**, 577 (1953); cf. Sec. 16-1c, Table 16-1.

the constant concentration of the solvent being included in K_a. For an electrically neutral acid ($n = 0$) the acidity constant is the same as the ionization constant, but the acidity constant for an acid with a unit positive charge is not reasonably called an ionization constant, since it is a measure of an equilibrium in which there are as many ions in the reactants as in the products.

2-3a. Effect of the Ion-solvating Powers of the Solvent. The acidity constant of an acid is, as implied in Sec. 2-2b, directly proportional to the basicity of the solvent. However, there are other factors that may affect the acidity constant, and it is not possible to change the solvent basicity without also changing these factors. One of the most important is the ion-solvating power of the solvent. Its influence may be noticed particularly when comparing the effect of solvent changes on the acidity constants of acids of different electrical-charge types. For example, while the ionization constants of carboxylic acids are usually from 10^5 to 10^6 times as large in water as in absolute ethanol, the acidity constants of substituted ammonium ions are, on the average, less than 10 times as large in water as in ethanol. In the latter case,

$$R_3NH^+ + H_2O \rightleftharpoons R_3N + H_3O^+ \tag{2-5}$$
$$R_3NH^+ + EtOH \rightleftharpoons R_3N + EtOH_2^+ \tag{2-6}$$

equilibrium (2-5) usually lies slightly further to the right than (2-6), probably because water is a stronger base than ethanol.[19] This factor has an effect on the equilibria

$$RCO_2H + H_2O \rightleftharpoons RCO_2^- + H_3O^+$$
$$RCO_2H + EtOH \rightleftharpoons RCO_2^- + EtOH_2^+$$

but in addition it must be noted that we are now dealing with equilibria between two neutral molecules and two ions and that the production of ions will be greatly favored by the much better ion-solvating medium, water. Hence the acidity constants for electrically neutral acids should increase more on going from ethanol to water than those for positively charged acids. The principles used here may even be applied to comparisons of acids of the same charge type. For example, the ionization constant of picric acid increases only about 1,500-fold between ethanol and water. Since the charge on the picrate anion is so spread out by resonance, the ion is not so strongly solvated. For this reason its stability does not change so greatly with the ion-solvating power of the solvent as does that of a carboxylate anion, in which the negative charge is largely on two atoms. By analogous reasoning we may explain why the acidity of *p*-nitrobenzamide relative to ethanol increases on going

[19] I. M. Kolthoff and S. Bruckenstein [*J. Am. Chem. Soc.*, **78**, 1, 10 (1956)] have found ethanol to be only about one-fifth as basic as water toward perchloric acid in acetic acid solution.

from ethyl to isopropyl alcohol solution and why that of nitroaniline derivatives increases even more.[20] The ionization of bases may be discussed similarly in terms of the acidity of the solvent and the ion-solvating power of the medium.

2-3b. *Solvent Effects in Terms of Activity Coefficients.* Many useful correlations result from qualitative arguments of the type used above, but further discussion of the subject is made more convenient by the use of activities and activity coefficients. Although the equilibrium constants, expressed in concentrations, that we have used heretofore have definite values only in a particular medium, it is possible by substituting activities for concentrations to obtain equilibrium constants whose values are independent of the medium. These are called *thermodynamic equilibrium constants.* The activity of a molecule or ion X is written a_x and is equal to the concentration of X times its activity coefficient γ_x.[21] It is often useful to ignore the solvent in writing the equilibrium equation for an acid (in a mixed solvent, such as aqueous ethanol, we may not know the relative extent to which the proton coordinates with water and ethanol).

$$HA \rightleftharpoons H^+ + A$$

The thermodynamic equilibrium constant for this equation may be written in terms of activities,

$$K_a = \frac{a_{H^+}a_A}{a_{HA}} \tag{2-7}$$

or in terms of concentrations and activity coefficients,

$$K_a = \frac{[H^+]\gamma_{H^+}[A]\gamma_A}{[HA]\gamma_{HA}} = \frac{[H^+][A]}{[HA]} \frac{\gamma_{H^+}\gamma_A}{\gamma_{HA}}$$

If we define the activity coefficients so that they approach unity at infinite dilution in aqueous solution, the thermodynamic acidity constant K_a is (in dilute aqueous solution) equal to the concentration-acidity constant, which we shall call K_A^W. If, then, γ_x's are activity coefficients in solvent S, referred to dilute aqueous solution,

$$K_a = K_A^S \frac{\gamma_{H^+}\gamma_A}{\gamma_{HA}} = K_A^W \tag{2-8}$$

Analogously, for another acid HB

$$K_a' = K_A^{S'} \frac{\gamma_{H^+}\gamma_B}{\gamma_{HB}} = K_A^{W'} \tag{2-9}$$

[20] J. Hine and M. Hine, *J. Am. Chem. Soc.*, **74**, 5266 (1952).

[21] Discussion of activities, activity coefficients, and their use in equilibrium expressions are given in most beginning physical chemistry texts.

Dividing (2-8) by (2-9),

$$\frac{K_A^S \gamma_A \gamma_{HB}}{K_A^{S'} \gamma_B \gamma_{HA}} = \frac{K_A^W}{K_A^{W'}}$$

or
$$K_A^S = K_A^{S'} \frac{K_A^W}{K_A^{W'}} \frac{\gamma_B \gamma_{HA}}{\gamma_A \gamma_{HB}} \tag{2-10}$$

That is, the acidity constant of an acid HA in any solvent S may be calculated from its acidity constant in water, the acidity constants of some other acid HB in water and S, and the activity-coefficient term shown. For electrically neutral acids HA and HB, this term will have the form $\gamma_B \cdot \gamma_{HA}/\gamma_A \cdot \gamma_{HB}$. The γ_{HA} and γ_{HB} terms may often be evaluated readily from the solubilities, partial pressures, distribution coefficients, etc., of HA and HB in water and solvent S. The ratio of ionic activity coefficients, γ_B-/γ_A-, may be determined from the solubilities of salts, potentiometrically or otherwise. Values of $K_A^{S'}$ calculated in this manner from various types of experimental data have been found in reasonable agreement with those determined experimentally.[22]

2-3c. *Determination of Acidity Constants in Certain Mixed Solvents.* Remembering that $pK = -\log K$, Eq. (2-7) may, for a singly charged cationic acid, be rewritten in the form

$$pK_a = -\log \frac{a_{H^+} a_A}{a_{AH^+}} = \log \frac{a_{AH^+}}{a_A} - \log a_{H^+} \tag{2-11}$$

Since in sufficiently dilute aqueous solution concentrations may be equated to activities,

$$pK_a = \log \frac{[AH^+]}{[A]} + pH$$

the pK_a for AH^+ may be determined from a measurement of the relative concentrations of AH^+ and A in a solution of known pH. In cases where either AH^+ or A absorb at suitable wavelengths, this may often be done conveniently by spectrophotometric measurements. If the base A is too weak, however, it may not be possible to produce from it, in a dilute aqueous solution, an accurately measurable concentration of AH^+. While a relatively strongly acidic solution may produce enough AH^+ to measure, we have no assurance that activities may be equated to concentrations in such a solution. The problem of determining the value of pK_a under such conditions has been treated by Hammett and Deyrup[23] by use of the following method.

[22] N. Bjerrum and E. Larsson, *Z. physik. Chem.*, **127**, 358 (1927); J. O. Halford, *J. Am. Chem. Soc.*, **53**, 2939 (1931); I. M. Kolthoff, J. J. Lingane, and W. D. Larson, *J. Am. Chem. Soc.*, **60**, 2512 (1938).

[23] L. P. Hammett and A. J. Deyrup, *J. Am. Chem. Soc.*, **54**, 2721 (1932); L. P. Hammett, *Chem. Revs.*, **16**, 67 (1935); "Physical Organic Chemistry," chap. 9, McGraw-Hill Book Company, Inc., New York, 1940.

For another acid BH^+ of the same electrical-charge type as AH^+, an equation analogous to (2-11) may be written.

$$pK'_a = \log \frac{a_{BH^+}}{a_B} - \log a_{H^+} \tag{2-12}$$

Subtracting (2-12) from (2-11),

$$pK_a - pK'_a = \log \frac{a_{AH^+}}{a_A} - \log \frac{a_{BH^+}}{a_B}$$

or, in terms of concentrations and activity coefficients,

$$pK_a - pK'_a = \log \frac{[AH^+]}{[A]} - \log \frac{[BH^+]}{[B]} + \log \frac{\gamma_{AH^+}\gamma_B}{\gamma_A\gamma_{BH^+}} \tag{2-13}$$

Since pK_a and pK'_a are defined in terms of activities their values are independent of the medium. Measurements of

$$\log \frac{[AH^+]}{[A]} - \log \frac{[BH^+]}{[B]}$$

were made for a number of pairs of compounds (most of which were aromatic amines) in various mixtures of water and sulfuric acid, and for each pair the values obtained were almost independent of the composition of the medium over the range in which it was possible to make measurements (this range, however, was always a fairly small fraction of the total possible range). In so far as the value of this function for a given pair of compounds is independent of the solvent, we may be sure that the function $\log \gamma_{AH^+}\gamma_B/\gamma_A\gamma_{BH^+}$ is also independent of the solvent. Since this activity-coefficient term has the value zero in aqueous solution and since it is also independent of the nature of the sulfuric acid–water mixture, its value must be zero in all such mixtures. Obviously, if

$$\log \frac{\gamma_{AH^+}\gamma_B}{\gamma_A\gamma_{BH^+}} = \log \frac{\gamma_{AH^+}}{\gamma_A} - \log \frac{\gamma_{BH^+}}{\gamma_B} = 0$$

then

$$\log \frac{\gamma_{AH^+}}{\gamma_A} = \log \frac{\gamma_{BH^+}}{\gamma_B} \tag{2-14}$$

for the compounds originally studied and for a number of others that were studied subsequently. Such compounds are sometimes called *Hammett bases*. For certain other bases, such as water, 1,1-diarylethylenes, etc.,[24,25] that have been studied in aqueous sulfuric acid solutions, the basicities have been found to change with the sulfuric acid content of the solvent in a manner different from that characteristic of

[24] N. C. Deno and C. Perizzolo, *J. Am. Chem. Soc.*, **79**, 1345 (1957); N. C. Deno, P. T. Groves, and G. Saines, *J. Am. Chem. Soc.*, **81**, 5790 (1959).

[25] This topic has been reviewed by M. A. Paul and F. A. Long, *Chem. Revs.*, **57**, 1 (1957).

Hammett bases. It appears that Eq. (2-14) may be extended to mixtures of water with certain other strong acids for at least many Hammett bases. There are fewer data available for most such solvent mixtures, however.

If we choose as the base A one that is barely strong enough to be measured in dilute aqueous solution, the ratio $[AH^+]/[A]$ will also be measurable in certain water–sulfuric acid mixtures. If B is a base that is not quite strong enough to be measured in aqueous solution, $[BH^+]/[B]$ will be measurable in some of the same water–sulfuric acid mixtures, and pK_a', the acidity constant of BH^+, may be determined from Eq. (2-13), which acquires the form

$$pK_a - pK_a' = \log \frac{[AH^+]}{[A]} - \log \frac{[BH^+]}{[B]}$$

when the activity-coefficient term vanishes. From B, the acidity constant of the conjugate acid of a still weaker base C may be determined. By such a stepwise process a value of pK_a can be determined for the conjugate acid of any Hammett base that is strong enough to be measured in a sulfuric acid–water mixture (provided that the base and conjugate acid have different absorption spectra or differ in some other way that permits their concentrations to be determined). Some pK_a values obtained in this way are listed in Table 2-1. Not all the compounds listed were shown clearly to be Hammett bases.

TABLE 2-1. VALUES OF pK_a IN SULFURIC ACID–WATER MIXTURES FOR THE CONJUGATE ACIDS OF SOME WEAK BASES

Base	pK_a	Base	pK_a
p-Nitroaniline.............	$+1.11$	Benzalacetophenone........	-5.61
o-Nitroaniline.............	-0.13	p-Benzoylbiphenyl..........	-6.19
p-Nitrodiphenylamine.......	-2.38	Benzoic acid...............	-7.26
p-Nitroazobenzene..........	-3.35	Anthraquinone.............	-8.15
2,4-Dinitroaniline..........	-4.38	2,4,6-Trinitroaniline........	-9.29

2-3d. *Acidity Functions*.[23,25] Equation (2-11) may be rewritten in the form

$$pK_a = \log \frac{[AH^+]}{[A]} - \log \frac{a_{H^+}\gamma_A}{\gamma_{AH^+}} \qquad (2\text{-}15)$$

Since a_{H^+} and γ_A/γ_{AH^+} both have a definite value in any solvent mixture and since this value is independent of the nature of the particular Hammett base A in solvents for which Eq. (2-14) holds, the term $- \log a_{H^+}\gamma_A/\gamma_{AH^+}$ must have a definite value for any such solvent mixture. It is seen that for a given pK_a the magnitude of this function tells how much of the base is present as its conjugate acid.   Therefore the function is a quantitative measure of the ability of the solvent to donate protons to a Hammett base and is called the *acidity function* H_0.

$$H_0 = -\log \frac{a_{H^+}\gamma_A}{\gamma_{AH^+}} \tag{2-16}$$

By substitution of Eq. (2-16) into (2-15), we get an equation that may be used in the experimental determination of H_0.

$$H_0 = pK_a + \log \frac{[A]}{[AH^+]} \tag{2-17}$$

Using indicators such as those listed in Table 2-1, for which pK_a values are known, values of H_0 have been determined for mixtures of water with sulfuric, hydrochloric, nitric, perchloric, and trichloroacetic acids. Some of the values obtained for sulfuric acid–water mixtures are listed in Table 2-2.   From the value of H_0 for a given medium and the value of $[A]/[AH^+]$ a value of pK_a may be calculated.

TABLE 2-2. H_0 IN VARIOUS SULFURIC ACID–WATER MIXTURES

Wt. H_2SO_4, %	H_0	Wt. H_2SO_4, %	H_0	Wt. H_2SO_4, %	H_0
5	+0.24	40	−2.28	80	−6.82
10	−0.16	50	−3.23	85	−7.62
15	−0.54	60	−4.32	90	−8.17
20	−0.89	70	−5.54	95	−8.74
30	−1.54	75	−6.16	100	−10.60

The antilogarithm of $-H_0$ has been given the symbol h_0 and is often a useful term.

$$\log h_0 = -H_0$$

therefore

$$h_0 = \frac{a_{H^+}\gamma_A}{\gamma_{AH^+}}$$

Deno and Taft[16] have shown that the numerical value of the acidity function H_0 may be calculated in 83 to 99.8 per cent (by weight) sulfuric acid by assuming that the reaction

$$H_2O + H_2SO_4 \rightleftharpoons H_3O^+ + HSO_4^-$$

controls the acidity properties of the solution and that the equilibrium constant for this reaction has a value of about 50 (Sec. 2-2*d*).   These assumptions also permit the calculation of the activity of water in 83 to 95 per cent sulfuric acid.   From these facts it appears likely that the activity coefficients of the species involved in the equilibrium do not change between 83 and 99.8 per cent sulfuric acid.

Gold and Hawes have defined another acidity function J_0 (also called C_0 and H_R) by the equation[26]

$$J_0 = H_0 + \log a_{H_2O}$$

Since a_{H_2O} is defined as unity for pure water, J_0 approaches H_0 and both approach pH as the solvent approaches pure water. Equilibria of the type

$$ROH + H^+ \rightarrow R^+ + H_2O$$

have been found in a number of cases (mostly with triarylmethanol derivatives) to follow J_0 with reasonable agreement.

Gutbezahl and Grunwald have demonstrated that the acidity function H_0 is too crude an approximation to be of much use in mixtures of ethanol and water.[27]

The subscript zero in the acidity function H_0 refers to electrical charge on the base. For mononegatively charged bases in equilibrium with uncharged acids, the acidity function H_- may be defined by equations analogous to (2-16) and (2-17).

$$H_- = pK_{HA} + \log \frac{[A^-]}{[HA]} = \log \frac{\gamma_{HA}}{\gamma_{A^-}} - \log a_{H^+}$$

The acidity function H_- will be generally applicable for solvent changes over which the value of γ_{HA}/γ_{A^-} is essentially independent of the nature of HA. Deno has shown that this is the case for several nitroaniline and nitrotoluene derivatives used as indicators in water-hydrazine mixtures and has calculated values of H_- for these mixtures.[28] The deviation by one indicator and the fact that all the indicators used were rather closely related render it uncertain that the values of H_- calculated may be applied to all electrically neutral acids with a reasonable degree of accuracy. The variations in the relative acidities of electrically neutral acids on going from water to ethanol[27,29] and from ethanol to isopropyl alcohol[20] show that H_- is not a generally useful approximation in these solvent mixtures, and indeed it seems likely that this will prove to be the case for most solvent mixtures.

Schwarzenbach and Sulzberger have described measurements in strong sodium hydroxide and potassium hydroxide solutions, but because of the limited number of types of indicators used and the uncertainty as to the

[26] V. Gold and B. W. V. Hawes, *J. Chem. Soc.*, 2102 (1951); cf. V. Gold, *J. Chem. Soc.*, 1263 (1955); cf. N. C. Deno, H. E. Berkheimer, W. L. Evans, and H. J. Peterson, *J. Am. Chem. Soc.*, **81**, 2344 (1959).

[27] B. Gutbezahl and E. Grunwald, *J. Am. Chem. Soc.*, **75**, 559 (1953).

[28] N. C. Deno, *J. Am. Chem. Soc.*, **74**, 2039 (1952).

[29] E. Grunwald and B. J. Berkowitz, *J. Am. Chem. Soc.*, **73**, 4939 (1951).

exact nature of their color-change reaction, it cannot be stated whether or not H_- is a good approximation in these solutions.[30]

Correlations of the rates of acid- and base-catalyzed reactions with acidity functions are discussed in Sec. 5-3d.

Gutbezahl and Grunwald have suggested a more general rule for the variation of the activity coefficients of species that are closely related to each other and have used this rule, for which considerable evidence is presented, to evaluate the activity coefficients of individual ions.[27]

2-4. Effect of Structure on the Strength of Acids and Bases.
2-4a. *Effect of the Identity of the Acidic or Basic Atom on Acidity and Basicity.* One of the most important factors governing the acidity or basicity of a molecule is the identity of the proton-donating or -accepting atom. Several generalizations may be based on the periodic table.

1. *Within a given period of the periodic table, the acidity of hydrides increases with increasing electronegativity.* Thus, as acids: $HCH_3 <$ $HNH_2 < HOH < HF$. This change is most often attributed to the increasing nuclear charge, but the decrease in the number of electropositive hydrogen atoms attached to the central atom must also be important. This generalization also applies to comparisons of ions of the same charge type, so that the acidity varies thus: $HNH_3^+ < HOH_2^+ <$ HFH^+. These two orders of acidity are, of course, simply another way of stating the following two orders of basicity: $CH_3^- > NH_2^- >$ $OH^- > F^-$ and $NH_3 > H_2O > HF$. There is considerable evidence that *the attachment of a multiple linkage to an atom increases its effective electronegativity.* From this fact and the generalization given above, we should expect acetylene to be a stronger acid than ethylene or benzene and these in turn to be more acidic than ethane. Similarly, pyridine is a much weaker base than saturated amines, and nitriles are still weaker.

2. Another generalization is that *the acidity of hydrides of elements within a given family increases with increasing atomic number.* That is, as acids $HI > HBr > HCl > HF$. This generalization may be correlated with the decrease in bond strengths that occurs with increasing atomic number. The stabilization of the heavier anions due to the spreading of the charge over a larger volume is probably of some importance but is apparently a secondary factor, since HF is the strongest base of the hydrogen halides despite the fact that the addition of a proton gives a relatively small ion. Similarly NH_3 is a stronger base than PH_3.

2-4b. *Resonance Effects on Acidity and Basicity.* Another factor of great importance in determining relative acidities is the degree of resonance stabilization of the acids and conjugate bases involved. For

[30] G. Schwarzenbach and R. Sulzberger, *Helv. Chim. Acta*, **27**, 348 (1944).

instance, in comparing the acidity of an alcohol and a carboxylic acid

$$RCO_2H \rightleftharpoons H^+ + RCO_2^-$$
$$ROH \rightleftharpoons H^+ + RO^-$$

the greater acidity of the carboxylic acid is almost undoubtedly due in part to the resonance stabilization of the carboxylate anion,

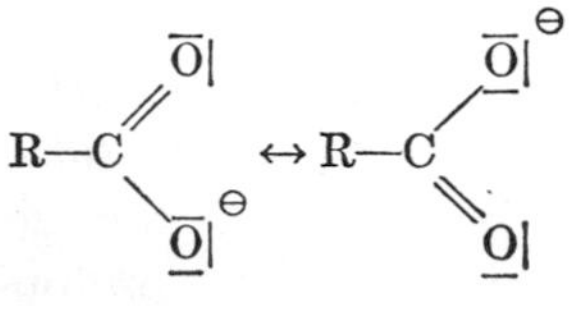

and the resultant spread of the negative charge over a larger volume. Resonance also occurs in the undissociated carboxylic acid, of course, but the resonance stabilization is less, since the principal contributing structures are not of equal energy content.

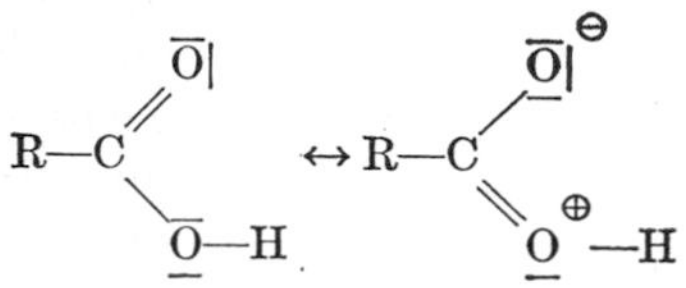

In addition to resonance, inductive effects must also be important in this case, since the double-bonded carbon and oxygen atoms are more electronegative than any atom in the hydrocarbon R group of the alcohol.

The acidity of phenols (relative to alcohols) is similarly attributed to resonance stabilization of the anion,

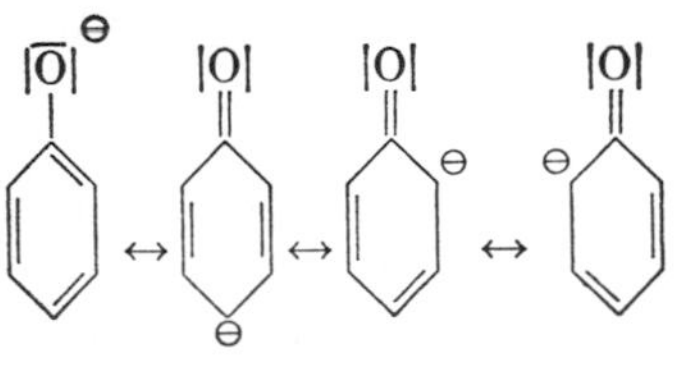

while in aniline the weak basicity (relative to aliphatic amines) is attributed to resonance in the base itself,

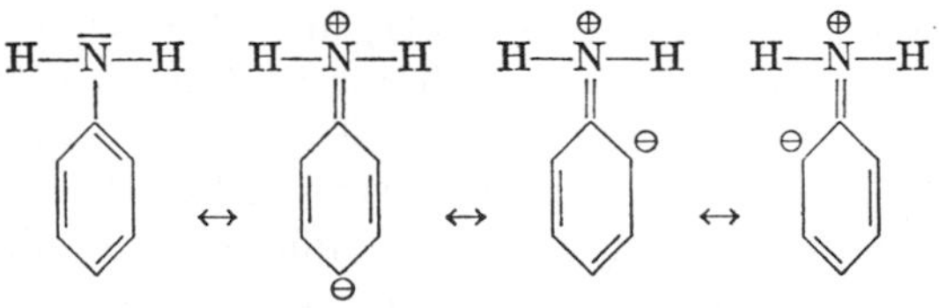

resonance of this type occurring to a negligible extent in the anilinium ion.

In the case of both phenol and aniline the electronegativity of the sp^2 carbon atoms must also be important.

Both effects similarly explain the increased acidity of β-diketones,

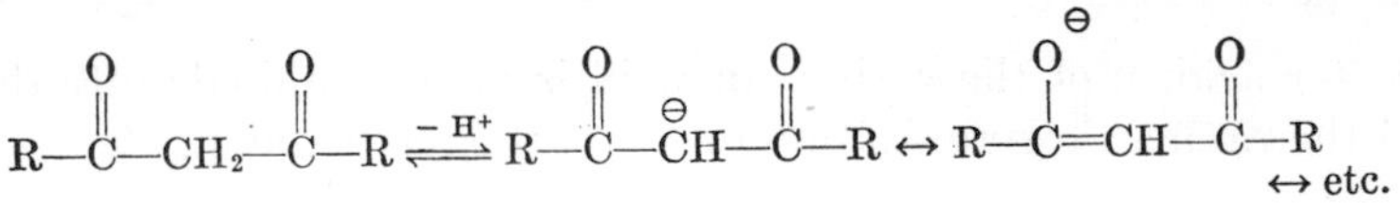

$$\underset{\text{I}}{R-\overset{O}{\overset{\|}{C}}-CH_2-\overset{O}{\overset{\|}{C}}-R} \underset{-H^+}{\rightleftharpoons} R-\overset{O}{\overset{\|}{C}}-\overset{\ominus}{CH}-\overset{O}{\overset{\|}{C}}-R \leftrightarrow R-\overset{\overset{\ominus}{O}}{C}=CH-\overset{O}{\overset{\|}{C}}-R \quad \leftrightarrow \text{etc.}$$

The low basicity of pyrrole can be attributed to the presence of six pi electrons in the aromatic ring (cf. Sec. 1-3) as well as to the inductive effect of the sp^2 carbon atoms.

2-4c. Inductive Effects on Acidity and Basicity.   A number of resonance structures for ethyl chloride involving ionic bonds are written below. Note that a resonance description in terms of ionic structures implies that the structures not written as ionic are purely covalent (see Sec. 1-1a).

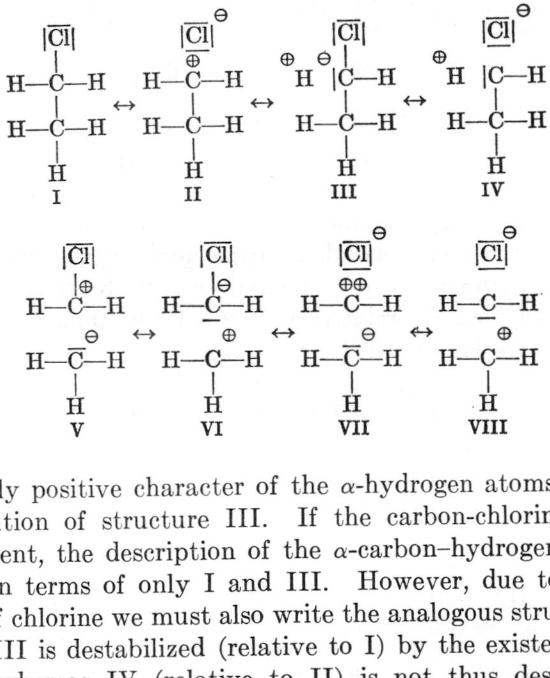

The relatively positive character of the α-hydrogen atoms is shown by the contribution of structure III. If the carbon-chlorine bond were purely covalent, the description of the α-carbon–hydrogen bond might be written in terms of only I and III. However, due to the electronegativity of chlorine we must also write the analogous structures II and IV. Since III is destabilized (relative to I) by the existence of charge separation, whereas IV (relative to II) is not thus destabilized, the contribution of III (relative to I) will be less than that of IV (relative to II). Hence structures involving positively charged hydrogen will contribute more to the total of I, II, III, and IV than they will to the total of I and III alone. In other words, due to the electronegativity of chlorine the α-hydrogen atom is more positively charged than it would otherwise be. Similarly, since VIII will contribute more (relative to

VII) than VI will (relative to V), the β-carbon atom will also be more positive. By analogous arguments it may be seen that the electro-negative character of the chlorine atom would be expected to increase the positive (or decrease the negative) character of every other atom in the molecule, although the magnitude of the effect is expected to decrease with increasing distance. While this effect may be seen to be a result of the resonance contribution of ionic structures, it may be stated by a much less involved argument in terms of the inductive effect. According to this concept, an electron-withdrawing group may, by taking a large share of the bonding electron pair, induce a positive charge on an atom to which it is attached. Since this positive charge will increase the electron-withdrawing power of the atom upon which it resides, the inductive effect will be relayed along a chain of atoms, although with decreasing intensity. The effect is commonly depicted by arrows point-ing in the direction in which the electrons are displaced.

$$\begin{matrix} & H & H & \\ & \downarrow & \downarrow & \\ H \to & C \to & C \to & Cl \\ & \uparrow & \uparrow & \\ & H & H & \end{matrix}$$

Instead of an effect like that described above, in which the influence of an electronegative group is depicted as operating along the atomic chain, it is possible to consider what is called a *field effect*, i.e., the interaction through space of the dipoles or charge of the substituent group with that of the reaction center. A mathematical treatment based on a model of this sort was initiated by Bjerrum[31] and appears to have culminated in the work of Kirkwood and Westheimer.[32,33] Since the positive end of the carbon-chlorine dipole in chloroacetic acid is nearer the acidic proton than is the negative end, there will be a net repulsion of the proton. For this reason less work will be required to remove the proton than if the carbon-chlorine dipole were absent. The difference in the amounts of work required depends on the magnitude and orientation of the dipole, its distance from the acidic proton, and the dielectric constant of the medium through which the interacting lines of force must pass. Kirk-wood and Westheimer's treatment was an improvement over earlier studies (in which this dielectric constant was taken as either a rather high value equal to that of the solvent or a low value characteristic of a hydro-carbon group) in that they used a model consisting of a cavity (spherical or ellipsoidal for mathematical convenience) of low-dielectric constant surrounded by a medium of high dielectric constant. Using reasonable

[31] N. Bjerrum, *Z. physik. Chem.*, **106**, 219 (1923).
[32] J. G. Kirkwood and F. H. Westheimer, *J. Chem. Phys.*, **6**, 506, 513 (1938).
[33] F. H. Westheimer and M. W. Shookhof, *J. Am. Chem. Soc.*, **61**, 555 (1939).

values for their various estimated parameters, they obtained fairly good agreement with experimental data on acids with dipolar *and* with electrically charged substituents.

Roberts and Moreland, however, studied some 4-substituted bicyclo-[2.2.2]octane-1-carboxylic acids,

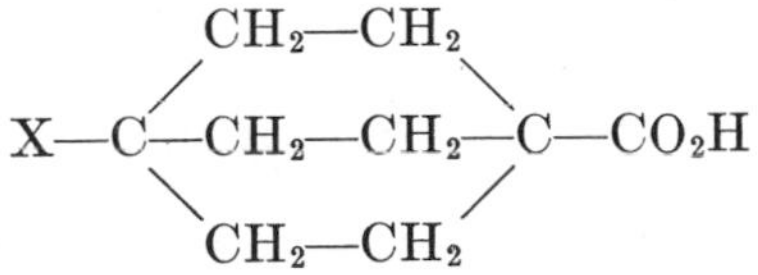

$$X—C \overset{\displaystyle CH_2—CH_2}{\underset{\displaystyle CH_2—CH_2}{—CH_2—CH_2—}} C—CO_2H$$

rigid compounds whose molecular geometry can be estimated with little probable error.[34] They reported that calculations by the Kirkwood-Westheimer method predict substituent effects only about half as large as those actually found. This suggests the necessity of considering the inductive effect and the field, or direct electrostatic, effect as two independent entities. One of the greatest obstacles in evaluating the inductive and field effects separately is that in practically all cases they act in the same direction. The ortho-substituted phenylpropiolic acids provide an exception to this generalization.

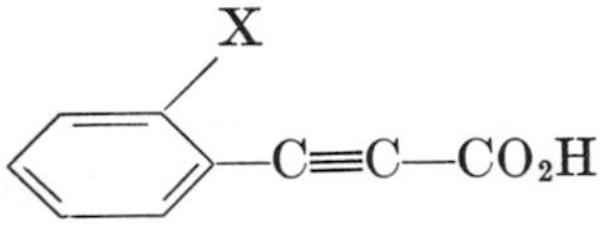

$$\text{(benzene ring)}—C{\equiv}C—CO_2H$$

When X is an electron-withdrawing substituent, such as chlorine, the negative end of the C—X dipole is closer to the acidic proton of the carboxy group than is the positive end. Therefore the field effect should tend to decrease the acidity, in opposition to the inductive effect. Roberts and Carboni found that with both chloro- and nitrophenylpropiolic acids the ortho derivatives were stronger than the unsubstituted compounds but weaker than the meta and para derivatives.[35] Since the inductive effect would traverse a shorter chain of atoms from the ortho position than from the meta or para position, the lesser acidity of *o*-chlorophenylpropiolic acid seems to show that the field effect is operating. On the other hand, the inductive effect appears to be more important since the ortho chloro substituent does increase the acidity of phenylpropiolic acid. It might be suggested that this arises from the fact that a larger fraction of the lines of force from the *negative* end of the dipole must pass through the high-dielectric solvent. However, this argument was met by the observation that the acidity (as measured by the rate of an acid-catalyzed reac-

[34] J. D. Roberts and W. T. Moreland, Jr., *J. Am. Chem. Soc.*, **75**, 2167 (1953).
[35] J. D. Roberts and R. A. Carboni, *J. Am. Chem. Soc.*, **77**, 5554 (1955).

tion) in the low-dielectric solvent dioxane was also increased by the presence of an ortho chloro substituent.

2-4d. Steric Effects on Acidity and Basicity.   Steric effects on acidity and basicity have received considerable attention both in regard to their effect on the action of resonance and in their own right.   A particularly striking case of the former is illustrated by the basicity of benzoquinuclidine.

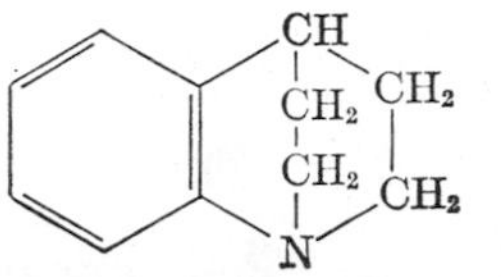

Benzoquinuclidine, $pK = 6.21$

Wepster has shown that this compound is a much stronger base than ordinary dialkylaniline derivatives (dimethylaniline, pK 8.94; diethylaniline, pK 7.44; N-phenylpiperidine, pK 8.80).[36]   This basicity is not due to the bicyclic ring system in itself, since quinuclidine (pK 3.35) is a weaker base than piperidine (pK 2.87).   It seems likely that one important contributing factor is steric inhibition of resonance.   The bicyclic ring system greatly decreases the contributions of structures of the type

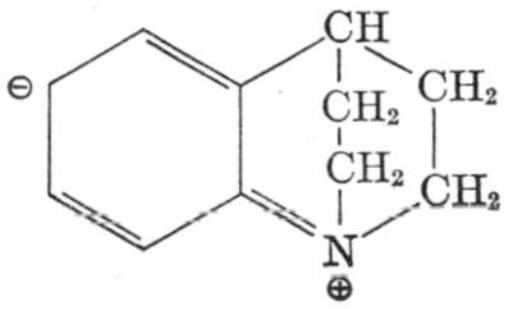

because the two atoms attached directly to the nitrogen atom cannot lie in the plane of the ring.   Since less resonance of this type exists, less can be lost upon formation of the conjugate acid.   The fact that benzoquinuclidine is still a much weaker base than quinuclidine is probably due to the increased electronegativity of the aromatic carbon atoms. Wheland has pointed out one case in which this electronegativity rather than resonance must be important.   Triphenylboron is a much stronger Lewis acid toward ammonia than is trimethylboron, although triphenylboron gains no resonance in forming the addition compound.[37]

An example of a more common type of steric inhibition of resonance has been described by Wheland, Brownell, and Mayo.[38]   These workers

---

[36] B. M. Wepster, *Rec. trav. chim.*, **71**, 1171 (1952).

[37] G. W. Wheland, "Resonance in Organic Chemistry," p. 351, John Wiley & Sons, Inc., New York, 1955.

[38] G. W. Wheland, R. M. Brownell, and E. C. Mayo, *J. Am. Chem. Soc.*, **70**, 2492 (1948).

found that, although 3,5 dimethylation decreases the acidity of phenol by 0.19 and that of *p*-cyanophenol by 0.26 *pK* units, the effect on *p*-nitrophenol is a decrease of 1.09 *pK* units.

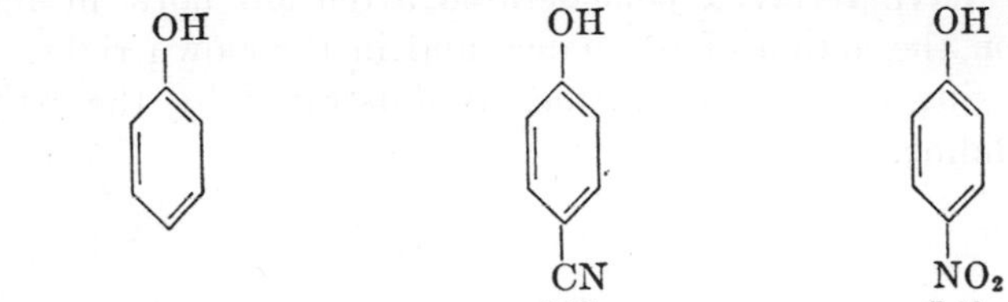

This large difference in behavior is most reasonably explained by the suggestion that a contributing structure of the type of IX, which does much to stabilize the *p*-nitrophenolate, cannot contribute nearly so much

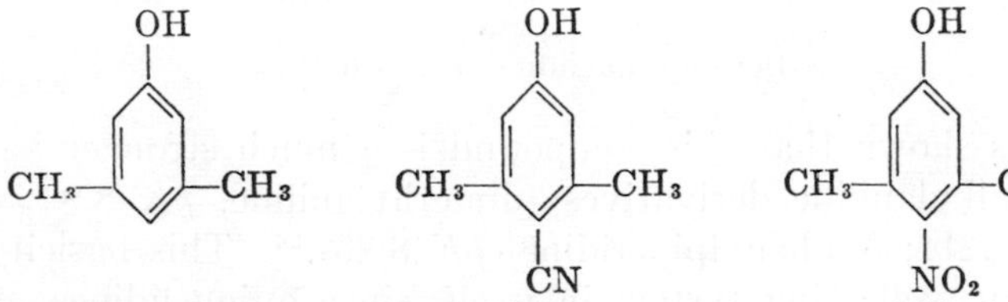

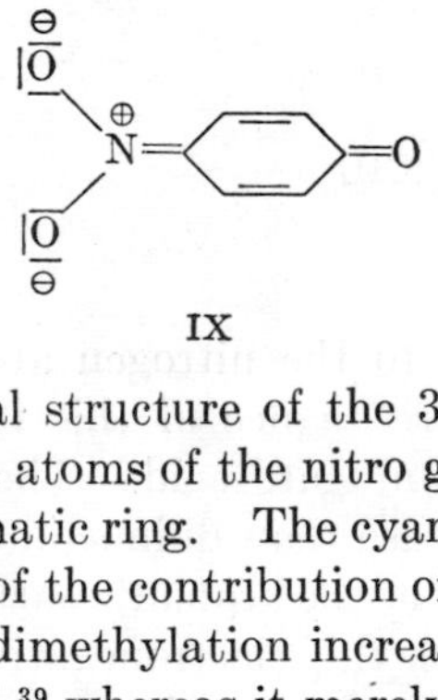

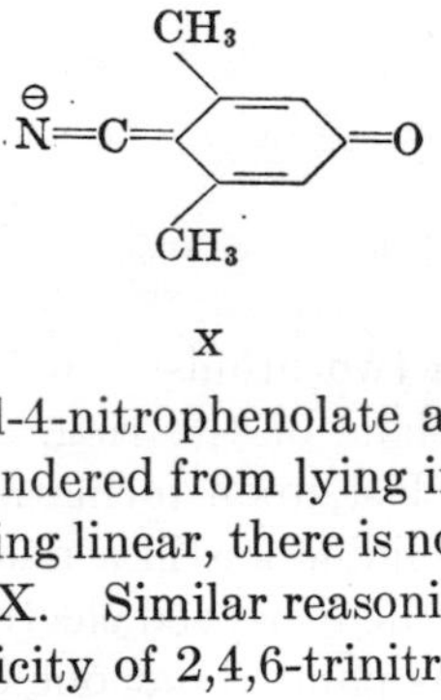

IX   X

to the total structure of the 3,5-dimethyl-4-nitrophenolate anion, since the oxygen atoms of the nitro group are hindered from lying in the plane of the aromatic ring. The cyano group being linear, there is no analogous inhibition of the contribution of structure X. Similar reasoning explains why *N,N* dimethylation increases the basicity of 2,4,6-trinitroaniline by 40,000-fold,[39] whereas it merely triples the basicity of aniline.

Brown and coworkers have accumulated a large amount of excellent evidence that basicity toward reasonably bulky Lewis acids is greatly affected by steric hindrance. For example, the relative stability of addition compounds with trimethylboron varies as follows: $NH_3 <$ $CH_3NH_2 < (CH_3)_2NH > (CH_3)_3N$, the steric effect becoming predominant only when the amine is trimethylated.[40] With tri-*t*-butylboron,

[39] L. P. Hammett and M. A. Paul, *J. Am. Chem. Soc.*, **56**, 827 (1934).

[40] H. C. Brown, H. Bartholomay, Jr., and M. D. Taylor, *J. Am. Chem. Soc.*, **66**, 435 (1944).

however, the stability of the addition compounds with the ethylamines varies as follows: $NH_3 > C_2H_5NH_2 > (C_2H_5)_2NH > (C_2H_5)_3N$, showing that the steric factor is controlling in all cases.[41]   One of the largest F-strain effects that has been observed occurs in the case of 2,6-di-t-butylpyridine, which utterly fails to coordinate with the usually powerful Lewis acid boron trifluoride and even shows diminished basicity toward the hydrogen ion.[42]

PROBLEMS

1. At 25° the autoprotolysis constant of water is 1.0×10^{-14} and that of methanol is 3×10^{-17}.   Describe at least three factors that might contribute to this difference in autoprotolysis constants.

2. A solution of pentamethylbenzotrichloride in sulfuric acid has a ν value of five. Bubbling nitrogen through the solution removes 2 moles of hydrogen chloride and reduces the ν value to three.   When the solution is poured into water, pentamethylbenzoic acid is obtained.   Explain these observations.

3. What percentage of the anthraquinone dissolved in 100 per cent sulfuric acid is in the form of the conjugate acid?   In 90 per cent sulfuric acid?   In 0.1 M hydrochloric acid?

4. From a consideration of inductive effects only, would you expect guanidine to be protonated to a greater extent at its amino or imino nitrogen atoms?   From a consideration of resonance effects only, which would you expect?   How may an argument about the position of protonation be based on the fact that guanidine is much more basic than methylamine or ethylamine?

[41] H. C. Brown, *J. Am. Chem. Soc.*, **67**, 1452 (1945).
[42] H. C. Brown and B. Kanner, *J. Am. Chem. Soc.*, **75**, 3865 (1953).

Chapter 3

KINETICS AND REACTION MECHANISMS

The form of the kinetic equation for a chemical reaction and the effect of reaction conditions, reactant structures, etc., on the rate of reaction are often the strongest evidence a physical organic chemist has that enables him to evaluate the plausibility of the various mechanisms proposed for the reaction. In this chapter we shall discuss certain of the principles of chemical kinetics and also certain generalizations that can be made about organic reaction mechanisms.

3-1. Kinetics.[1] 3-1a. *The Steady-state Treatment.* A fundamental assumption of chemical kinetics is that the kinetic form of a one-step reaction will be identical to its stoichiometric form; e.g., if the reaction

$$A + 2B \rightarrow AB_2$$

is a one-step process, it must be kinetically third-order, first-order in A and second-order in B. Thus all reactions in which the kinetic form and stoichiometry are different (and some in which they are the same) must be multistep processes involving the formation of intermediates. The method to be used in ascertaining the kinetic form of a multistep reaction depends on the relative reactivity of the intermediate(s). The *steady-state treatment,* sometimes called the *method of the reactive intermediate,* is applicable when a given intermediate is so reactive that it is consumed almost as soon as it is formed. When this is the case, its concentration is always negligible in comparison to the changes in concentrations of reactants and products during the reaction. This means, of course, that its rate of formation is essentially equal to its rate of reaction, or that its net rate of formation is zero. If the intermediate is reactive enough, the approximation is quite precise.

---

[1] In this section it is assumed that the reader has a knowledge of the fundamentals of kinetics as presented in most elementary textbooks of physical chemistry. For a more detailed treatment, see A. A. Frost and R. G. Pearson, "Kinetics and Mechanism," 2d ed., John Wiley & Sons, Inc., New York, 1961.

We shall illustrate the principle by applying it to the reaction scheme

$$A + B \underset{k_{-1}}{\overset{k_1}{\rightleftharpoons}} C + D$$

$$C + E \overset{k_2}{\rightarrow} F$$

where C is the reactive intermediate.   By the rate of reaction we mean the rate of formation of F.   Since F is formed only in the second step

$$\frac{dF}{dt} = k_2 CE \qquad (3\text{-}1)$$

According to the steady-state approximation for C,

$$k_1 AB = k_{-1} CD + k_2 CE$$

therefore
$$C = \frac{k_1 AB}{k_{-1}D + k_2 E} \qquad (3\text{-}2)$$

Substituting into Eq. (3-1),

$$\frac{dF}{dt} = \frac{k_1 k_2 ABE}{k_{-1}D + k_2 E}$$

In a given case it may be that $k_{-1}D \gg k_2 E$.   If so,

$$\frac{dF}{dt} = \frac{k_1 k_2}{k_{-1}} \frac{ABE}{D}$$

If $k_2 E \gg k_{-1}D$, then

$$\frac{dF}{dt} = k_1 AB$$

Reaction schemes of this general type are quite common and important in organic chemistry.

3-1b. *Transition-state   Theory*.   According   to   the   transition-state theory,[1] the rate of a one-step reaction depends on the difference in free-energy content between the reactant(s) and the *activated complex*, or *transition state*.   The rates of multistep reactions follow obviously from the rates of their individual steps.   Most of our discussions of the effect of structure on reactivity will be in terms of transition-state theory, and in many of these discussions plots of energy vs. the reaction coordinate (some geometrical measure of the extent to which the reactants have been converted to the products) will be useful.   Several types of such plots will be mentioned here.   For example, if in a reaction of the type

$$A \underset{k_{-1}}{\overset{k_1}{\rightleftharpoons}} B$$

$$B \overset{k_2}{\rightarrow} C$$

B is a reactive intermediate and k_{-1} is much larger than k_2, the reaction plot will have the form shown by the solid curve in Fig. 3-1. The broken curve corresponds to a much slower reaction in which k_2 is larger than k_{-1}. In each case the equilibrium lies well toward the side of the product C. In the reaction plotted with a broken curve the first step is said to be rate-controlling (or rate-determining or rate-limiting), since its rate is essentially equal to the over-all reaction rate and large changes in the rate of the second step can be made without changing the over-all reaction rate. For use in this book we shall define the *rate-controlling step* as the earliest step in the reaction whose rate is equal (within the

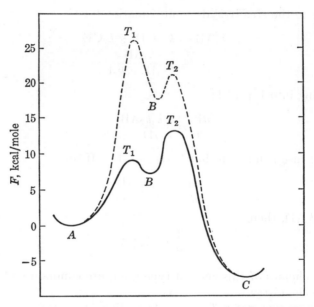

Fig. 3-1. Plots for reactions with first step rate-controlling (broken line) and second step rate-controlling (solid line).

experimental uncertainty) to the over-all reaction rate. The rate-controlling step has also been defined as that step whose transition state has the highest free-energy content, but in some cases this step can be considerably faster than the over-all reaction. Our definition seems more in accord with the words from which the term is made but has the disadvantage of being dependent on the accuracy with which the reaction is studied. In most cases, such as those plotted in Fig. 3-1, both definitions designate the same step as rate-controlling. In the reaction plotted in Fig. 3-2, however, the second step would be rate-controlling according to our definition. According to the alternate definition, the first step, whose rate is almost twice that of the over-all reaction, would be rate-controlling.   It might be noted that according to our definition the rate-controlling step for the forward reaction is not the same as for the reverse reaction in Fig. 3-2.

3-1c. *Kinetic and Thermodynamic Product Control.*   When the reactant A yields two different products B and C,

$$B \underset{k_{-1}}{\overset{k_1}{\rightleftharpoons}} A \underset{k_{-2}}{\overset{k_2}{\rightleftharpoons}} C$$

if k_1 is larger than k_2 but k_2/k_{-2} is larger than k_1/k_{-1}, the reaction plot will be like that shown in Fig. 3-3.   In such a case, if the reaction is

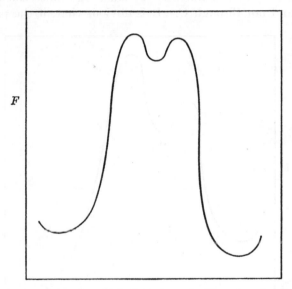

F

FIG. 3-2. Plot for a reaction in which the rate-controlling steps for the forward and reverse processes are different.

stopped before the reversal of B and C to A has had a chance to become important, the principal product will be B since it is formed more rapidly. On the other hand, if the reaction is allowed to continue until the entire system is at equilibrium, the principal product will be C, the most stable of the three species present.   In this reaction B would be called the *kinetically controlled product* and C the *thermodynamically controlled product*.   Such terms are only relative, of course.   There may be some as yet undetected product D, which is less stable but formed faster than B, or a product E, which is more stable but also formed much more slowly than any of the others.

3-1d. *The Principle of Microscopic Reversibility.*   According to the principle of microscopic reversibility, the mechanism of a reversible reaction is the same, in microscopic detail (except for the direction of reaction, of course), for the reaction in one direction as in the other under a given set of conditions. Thus, for example, if the transformation of A to D proceeds only through the intermediate B,

$$A \underset{k_{-1}}{\overset{k_1}{\rightleftharpoons}} B$$
$$k_{-2} \updownarrow k_2 \qquad k_{-3} \updownarrow k_3$$
$$C \underset{k_{-4}}{\overset{k_4}{\rightleftharpoons}} D$$

then, under the same conditions, D must be transformed to A only through B and not through C. It is obvious that, if the path of mini-

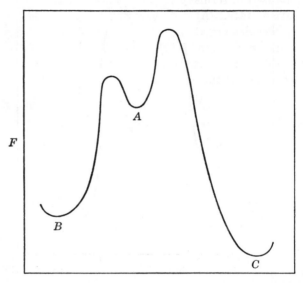

FIG. 3-3. Plot for the reactant A yielding the kinetically controlled product B and the equilibrium-controlled product C.

mum-free-energy expenditure leading from A to D passes through B, the minimum-free-energy path from D to A must also pass through B. Similarly, if the same amount of free energy of activation is required to react via intermediate C as via B, then A will be transformed to D to an equal extent by the two reaction paths and D will revert to A just as frequently via C as via B.

This principle is of importance not only for such qualitative purposes as ruling out possible reaction mechanisms but also in the calculation of rate constants for individual steps of reactions. The principle is ordinarily said not to apply to light-catalyzed reactions. It has alternately been stated that such reactions are usually not really reversible since the reverse reaction is not accompanied by the emission of light (of the same wavelength as that absorbed).

3-1e. *Kinetic Isotope Effects.*[2-4] It has been observed that isotopic substitution can change the rates of chemical reactions by much larger factors than it changes most equilibrium constants. In most reactions that have been studied a bond is broken in the isotopically substituted molecule. When the isotopic change is made at an atom bound by the bond being broken (or formed), the rate change is called a *primary kinetic isotope effect.* The difference in rate between the breaking of an A—B bond and an A—B' bond (perhaps by the attack of some reagent to form a new bond to one of the two atoms), where B and B' are isotopes, can be rationalized in terms of the loss of the A—B stretching frequency in the transition state. At ordinary reaction temperatures most bonds in ordinary molecules are at their lowest vibration level, but even at this level they contain the *zero-point vibrational energy,* which is equal to $\frac{1}{2}h\nu$. The following is a fairly good approximation for the ratio of the stretching frequency of the A—B' bond (ν') to that of the A—B bond (ν):

$$\frac{\nu'}{\nu} = \sqrt{\frac{m_B(m_A + m_{B'})}{m_{B'}(m_A + m_B)}}$$

where the m's are masses. When B' is heavier than B, the ratio ν'/ν is smaller than 1, so that the molecule containing the A—B bond has a larger zero-point energy than the one with the A—B' bond by the amount

$$E_{AB} - E_{AB'} = \frac{1}{2}h\nu \left[1 - \sqrt{\frac{m_B(m_A + m_{B'})}{m_{B'}(m_A + m_B)}} \right] \tag{3-3}$$

This difference in the zero-point energies of the reactants may be compensated by a difference in the zero-point energies of the transition states. In the case of the transfer of the atom B to another atom C, via a linear transition state

$$A\text{----}B\text{----}C$$

there is a "symmetrical" mode of vibration involving the movement of B as well as of A and C, if B is attached to either one of these two atoms more strongly than to the other. To an extent that depends on the difference in the strength of attachment of A to B and C, the frequency of this vibration will depend on the mass of B, and there will be a differ-

[2] L. Melander, "Isotope Effect on Reaction Rates," The Ronald Press Company, New York, 1960.

[3] R. P. Bell, "The Proton in Chemistry," chap. 11, Cornell University Press, Ithaca, N.Y., 1959.

[4] F. H. Westheimer, *Chem. Revs.*, **61**, 265 (1961).

ence in the zero-point energies of the two transition states that may partly or completely balance the difference in zero-point energies of the reactants.   However, when B is attached to A and C with equal strength in the transition state, the symmetrical vibration does not involve any movement of B.   In such a case the symmetrical vibration frequency does not depend on the mass of B and no differences in zero-point energies of the transition state arise from this source.   Neglecting other sources of differences in zero-point energies (which are believed to be less important than those mentioned), the two transition states will have the same zero-point energy, whereas the reactant A—B will have more zero-point energy than A—B', as shown in Eq. (3-3).   It will therefore require less energy for A—B to reach the transition state, and A—B will react faster than A—B' by the following factor:

$$\frac{k_{AB}}{k_{AB'}} = e^{\frac{h\nu}{2kT}\left[1 - \sqrt{\frac{m_B(m_A+m_{B'})}{m_{B'}(m_A+m_B)}}\right]}$$

(3-4)

This is an approximation of the maximum possible kinetic isotope effect, and in Table 3-1 are listed some isotope effects calculated from this equation using representative known values for the stretching frequencies of the various bonds being broken.

TABLE 3-1. APPROXIMATE MAXIMUM VALUES FOR KINETIC ISOTOPE EFFECTS

	Temperature, °C	$k_{AB}/k_{AB'}$
k_{CH}/k_{CD}	0	8.2
k_{CH}/k_{CD}	25	6.9
k_{CH}/k_{CD}	100	4.6
k_{CH}/k_{CD}	500	2.1
k_{NH}/k_{ND}	25	9.2
k_{OH}/k_{OD}	25	11.5
k_{CH}/k_{CT}	25	16
$k_{C^{12}H}/k_{C^{14}H}$	25	1.041
$k_{C^{12}H}/k_{C^{13}H}$	25	1.022
$k_{C^{12}C^{12}}/k_{C^{12}C^{14}}$	25	1.092
$k_{CO^{16}}/k_{CO^{18}}$	25	1.063

In addition to the effect calculated by consideration of the A—B stretching vibration, there may be an effect arising from changes in bending vibrations.   This latter effect is smaller, however, since bending frequencies are usually much smaller than the corresponding stretching frequencies.   In spite of the neglect of the effect of changes in bending vibrations and stretching vibrations other than that of the A—B bond and of other approximations contained in Eq. (3-4), no values of log $(k_{AB}/k_{AB'})$ more than about 25 per cent larger than those listed in Table 3-1 seem to have been established,[5] and, in fact, a number of kinetic isotope effects around the size of those listed have been reported. In other cases, however, much smaller primary kinetic isotope effects have been observed. Some of these were observed for very rapid processes, in which reaction may occur at every collision. Very small kinetic isotope effects would be expected in such reactions, since isotopic substitution has little effect on the collision rates of any but very small molecules. There are other relatively slow reactions for which small primary kinetic isotope effects have been found.[6] For these, the atom B is probably held much more strongly to one of the two atoms A and C than it is to the other. Since some such primary kinetic isotope effects are almost as small as certain *secondary* kinetic isotope effects (in which the isotopic change is made at an atom not involved in any bond breaking during the reaction), it is not possible to distinguish the two types of isotope effects by magnitude alone. Thus, in the case of carbon-bound deuterium a number of secondary kinetic isotope effects ranging up to a maximum of about 30 per cent per deuterium have been reported.[2,3] Hence if a k_H/k_D value of more than about 2.0 were found for a given reaction (near room temperature), one would feel confident that the hydrogen (or deuterium) atom was being transferred during the reaction. With very much smaller k_H/k_D values, however, it would be difficult to tell whether there was a hydrogen transfer in the rate-controlling step of the reaction or not.

3-1f. *Effect of Reaction Medium on Rate.* A number of equations have been proposed for the quantitative correlation of the rates of reactions with the nature of the solvent, but none appears to have anywhere near complete generality.[7] We shall therefore content ourselves here with describing the useful, though still not entirely general, qualitative theory of solvent effects of Hughes and Ingold.[8] According to this theory, *an increase in the ion-solvating power of the medium will accelerate the creation and concentration of charges and inhibit their destruction and diffusion.* Thus the reaction of two like-charged ions involves the concentration of two (or more) charges onto one ion, the transition state. Therefore the rate of this reaction, as well as that of an ionization reaction, will be increased by an increase in ion-solvating power. A reaction between

---

[5] This statement refers to rate constants for individual steps of a reaction. In a few unusual types of reaction mechanisms a small effect on the rate constant for one step can have a much larger effect on the rate of the over-all reaction.

[6] J. Hine et al., *J. Am. Chem. Soc.*, **76**, 827 (1954); **78**, 3337 (1956); **79**, 1406 (1957); **84**, 973 (1962).

[7] Frost and Pearson, *op. cit.*, chap. 7.

[8] E. D. Hughes and C. K. Ingold, *J. Chem. Soc.*, 244 (1935).

oppositely charged ions or a reaction in which the charge is on one atom in the reactant but spread out over several in the transition state will go slower in a better ion-solvating solvent.

Although the rates of most polar reactions are influenced by the ionic strength of the reaction medium, reactions between ions are most strongly influenced.   This is to be expected since activity coefficients of ions are more sensitive to the ionic strength than are those of most neutral molecules.   An equation relating the ionic strength and the rate constant for a reaction between ions may be derived fairly readily from the Debye-Hückel equation and the transition-state theory.   According to the latter, the rate constant k_0 in some standard solution is related to k, the rate constant in another solution, by the equation

$$k = k_0 \frac{\gamma_A \gamma_B \cdots}{\gamma_{\ddagger}} \tag{3-5}$$

where $\gamma_A \gamma_B \cdots$ are the activity coefficients of the reactants and $\gamma_{\ddagger}$ that of the transition state, in the given solution, referred to the standard solution.   According to the Debye-Hückel theory, the activity coefficient of an ion in dilute solution, referred to the pure solvent, follows the equation

$$-\log \gamma = Z^2 \alpha \sqrt{\mu} \tag{3-6}$$

where Z is the charge on the ion, μ is the ionic strength, and α is a constant having the value 0.509 in dilute aqueous solution at 25°.   The combination of Eqs. (3-5) and (3-6) gives[9]

$$\log k = \log k_0 - Z_A^2 \alpha \sqrt{\mu} - Z_B^2 \alpha \sqrt{\mu} + (Z_A + Z_B)^2 \alpha \sqrt{\mu}$$

or
$$\log \frac{k}{k_0} = 2 Z_A Z_B \alpha \sqrt{\mu} \tag{3-7}$$

Thus reactions between like-charged ions proceed more rapidly as the ionic strength increases, whereas those between oppositely charged ions proceed more slowly.   These effects may be rationalized qualitatively in terms of the Hughes-Ingold theory of solvent action when it is realized that ions may be stabilized in a manner analogous to solvation by being surrounded by ions of opposite charge.

Equation (3-7) is applicable to what is known as a primary salt effect. Also of importance is the secondary salt effect.   This pertains to the effect of salts on the rates of acid- and base-catalyzed reactions due to their effects on the ionization constants of the acids and bases involved. Since the ionization of an electrically neutral acid or base, e.g.,

$$HA \rightleftharpoons H^+ + A^-$$

[9] J. N. Brønsted, *Z. physik. Chem.*, **102**, 169 (1922); **115**, 337 (1925); N. Bjerrum, *Z. physik. Chem.*, **108**, 82 (1924); **118**, 251 (1925).

consists of an equilibrium between ions and a neutral molecule, and since the activity coefficient of the neutral molecule is little affected by small changes in the ionic strength, the equilibrium will be shifted to the right by increasing ionic strength in dilute solution.

3-2. Organic Reaction Mechanisms.   The mechanism of a reaction is simply the path the molecules follow in going from reactant to product. A rather crude reaction mechanism may consist of a sequential list of the organic compounds that are intermediates.   The complete elucidation of a reaction mechanism might be expected to include a stereochemically complete description of the movements of every atom throughout the reaction and also a description of the stability, or energy content, of the system in every intermediate configuration, so that transition states, reactive intermediates, stable intermediates, etc., will be distinguishable.   This complete a mechanism may be unattainable, but it can be approached as a limit.

We shall follow the common procedure of judging the correctness of a reaction mechanism, or indeed of any scientific theory, by the extent to which it correlates existing experimental data and predicts the results of new experiments.   Of several mechanisms that are equal by this yardstick we shall prefer the simplest.   The mechanism of a reaction is usually determined, then, by considering all the mechanisms that may be devised in agreement with known data on reaction mechanisms in general and comparing their requirements with the experimental observations on the given reaction.   A proposed mechanism in definite conflict with any experimental observation is ruled out.   New experiments may then be carried out to distinguish between remaining possible mechanisms.   If only one mechanism then remains, it is often said to be the *established* reaction mechanism.   A mechanism "proved" in this way may later be disproved by additional experimental evidence.   We may then modify it or substitute for it a mechanism that is more complicated than the previously existing experimental data had demanded.   We may not, in all cases, be sufficiently ingenious to devise a mechanism in agreement with all the experimental evidence.   The fact that we cannot guarantee that a given proved reaction mechanism will be in complete agreement with all future experiments does not in itself make the reaction mechanism less true than any other class of theories or laws of experimental science, since there are none for which we can make such a guarantee.[10]

In the study of organic reaction mechanisms it has been found that many diverse reactions proceed by a number of simple steps and that

---

[10] For a short discussion of scientific method and references to more complete treatments, see E. B. Wilson, Jr., "An Introduction to Scientific Research," chap. 3, McGraw-Hill Book Company, Inc., New York, 1952.

the number of different types of steps that occur is actually relatively small. In this connection it has been found very useful to divide organic reaction mechanisms into several classes.

3-2a. Polar Reactions. Representatives of one type of organic reaction mechanism are called *polar, ionic,* or *electron-sharing reactions.* These reactions involve Lewis acids (electrophilic reagents), Lewis bases (nucleophilic reagents), and their addition compounds.[11] The driving force for these reactions is the affinity of electrophilic reagents for electron pairs and of nucleophilic reagents for nuclei with which to coordinate. Reactions of this class include:

1. Addition reactions between electrophilic and nucleophilic reagents (and their reversal), e.g.,

$$
\underset{\overset{|}{CH_3}}{\overset{\overset{CH_3}{|}}{CH_3-N|}} + \underset{\overset{|}{CH_3}}{\overset{\overset{CH_3}{|}}{B-CH_3}} \rightleftharpoons \underset{\overset{|}{CH_3}}{\overset{\overset{CH_3}{|}}{CH_3-\overset{\oplus}{N}}} \underset{\overset{|}{CH_3}}{\overset{\overset{CH_3}{|}}{\overset{\ominus}{B}-CH_3}}
$$

2. Displacement reactions in which
   a. One nucleophilic reagent displaces another from its union with an electrophilic reagent, e.g.,

$$(CH_3)_3\overset{\ominus}{B}-\overset{\oplus}{N}(CH_3)_3 + NH_3 \rightleftharpoons (CH_3)_3\overset{\ominus}{B}-\overset{\oplus}{N}H_3 + N(CH_3)_3$$

   b. One electrophilic reagent displaces another from its union with a nucleophilic reagent, e.g.,

$$(CH_3)_3\overset{\oplus}{N}-\overset{\ominus}{B}(CH_3)_3 + BF_3 \rightleftharpoons (CH_3)_3\overset{\oplus}{N}-\overset{\ominus}{B}F_3 + B(CH_3)_3$$

We shall see that a reaction whose over-all result is a nucleophilic displacement (2a) may occur in one simple step or may consist of two steps, each of the type of 1. For example, reaction 2a may be written as proceeding by the mechanism

$$(CH_3)_3\overset{\ominus}{B}-\overset{\oplus}{N}(CH_3)_3 \rightleftharpoons (CH_3)_3B + N(CH_3)_3$$
$$(CH_3)_3B + NH_3 \rightleftharpoons (CH_3)_3\overset{\ominus}{B}-\overset{\oplus}{N}H_3$$

Polar reactions usually involve ions and, at the least, formal charges (see the examples above); in fact, they are frequently called ionic reactions. Many examples involving ions may be given. The fluoride ion, a nucleophilic reagent, will combine with the electrophilic reagent boron trifluoride

$$\overset{\ominus}{F} + BF_3 \rightleftharpoons \overset{\ominus}{BF_4}$$

The hydroxide ion, a nucleophilic reagent, will displace the nucleophilic

[11] For definitions of these terms, see Sec. 2-1b.

reagent ammonia from combination with the electrophilic reagent hydrogen ion (unknown in the uncombined state in solution)

$$\overset{\ominus}{OH} + \overset{\oplus}{NH_4} \rightleftharpoons H_2O + NH_3$$

In fact, all acid-base reactions (in the Lowry-Brønsted sense) involve nucleophilic displacements on hydrogen (the displacement by one nucleophilic reagent of another from combination with the hydrogen ion). The ionization of perchloric acid in methanol solution involves a nucleophilic displacement on hydrogen of perchlorate ion by methanol.

$$CH_3OH + HClO_4 \rightleftharpoons \overset{\oplus}{CH_3OH_2} + \overset{\ominus}{ClO_4}$$

Of great importance in organic chemistry are nucleophilic displacements *on carbon* (Chap. 6); e.g.,

$$\overset{\ominus}{OH} + CH_3I \rightleftharpoons CH_3OH + \overset{\ominus}{I}$$

the displacement by hydroxide ion of iodide ion from its combination with the electrophilic reagent CH_3^+ (unknown in the uncombined state in solution).

Many reactions may be written as combinations of several steps, each of which is a simple polar reaction.

In accordance with a suggestion of Swain and Scott,[12] we shall use the terms *nucleophilicity* and *electrophilicity* in connection with the *rate* of polar reactions, reserving the terms *acidity* and *basicity* for *equilibria*. For instance, the fact that the reaction

$$I^- + C_2H_5OSO_2C_6H_4CH_3\text{-}p \rightarrow C_2H_5I + p\text{-}CH_3C_6H_4SO_3^-$$

has a higher rate constant (in 61 per cent dioxane at 50°, at least) than the reaction[13]

$$OH^- + C_2H_5OSO_2C_6H_4CH_3\text{-}p \rightarrow C_2H_5OH + p\text{-}CH_3C_6H_4SO_3^-$$

will be denoted by the statement that iodide ion is more nucleophilic than hydroxide ion toward ethyl *p*-toluenesulfonate in 61 per cent dioxane at 50°. On the other hand, the fact that the equilibrium is further to the right in the latter reaction will be signified by stating that hydroxide ion is more basic than iodide ion toward ethyl *p*-toluenesulfonate in 61 per cent dioxane at 50°.

3-2b. *Free-radical Reactions.* *A free radical is a molecule or ion that contains one or more unpaired electrons,* and reactions in which electrons are paired or unpaired are called free-radical reactions. Included in this category are:

[12] C. G. Swain and C. B. Scott, *J. Am. Chem. Soc.*, **75**, 141 (1953).
[13] H. R. McCleary and L. P. Hammett, *J. Am. Chem. Soc.*, **63**, 2254 (1941).

1. The formation of free radicals by the dissociation of a nonradical and the converse combination of two free radicals by the pairing of their unpaired electrons

$$(C_6H_5)_3C{-}C(C_6H_5)_3 \rightleftharpoons 2(C_6H_5)_3C\cdot$$

2. Radical-displacement reactions

$$CH_3\cdot + CCl_4 \rightarrow CH_3Cl + \cdot CCl_3$$

3. Radical additions and their reverse

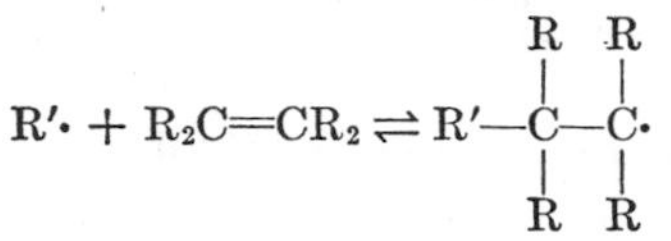

$$R'\cdot + R_2C{=}CR_2 \rightleftharpoons R'{-}\overset{\displaystyle R}{\underset{\displaystyle R}{C}}{-}\overset{\displaystyle R}{\underset{\displaystyle R}{C}}\cdot$$

The classification of bond-cleavage processes as *homolytic*, when the electrons of the bond are shared equally by the resultant fragments (as in the dissociation of hexaphenylethane above), and *heterolytic*, when all the electrons are taken by one of the fragments (as in the dissociation of a boron trifluoride–amine complex), separates single-bond-cleavage reactions into polar and radical categories.   The homolysis of a double bond, however, need not produce a species with unpaired electrons, e.g.,

$$CF_2{=}CF_2 \rightarrow 2|\overline{F}{-}\overline{C}{-}\overline{F}|$$

3-2c. Other Types of Reaction Mechanisms.   Although most organic reactions proceed by polar or free-radical mechanisms, there are two other types of mechanisms recognized.   One is the *simple electron-transfer reaction*, e.g.,

$$Fe^{3+} + Cr^{++} \rightleftharpoons Fe^{++} + Cr^{3+}$$

This type of reaction is quite common in inorganic chemistry, but since it occurs more rarely with organic molecules, it will not be discussed in this book (in at least some cases it might be classed as a free-radical reaction).

Another category is the so-called *multicenter-type reaction*, an inorganic example of which is the reaction of hydrogen iodide to form hydrogen and iodine and its reverse.   It is believed that this reaction involves simply the approach of two molecules of hydrogen iodide to give a transition state in which the hydrogen-iodine bonds are considerably stretched, followed by dissociation in one of the two possible ways.

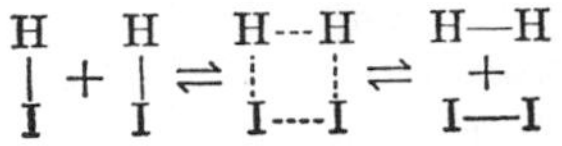

$$\underset{I}{\overset{H}{|}} + \underset{I}{\overset{H}{|}} \rightleftharpoons \underset{I\text{---}I}{\overset{H\text{---}H}{\vdots\ \ \vdots}} \rightleftharpoons \underset{I{-}I}{\overset{H{-}H}{+}}$$

We shall define multicenter-type reactions as those in which the atoms in the reactant(s) simply change their configuration to that of the product(s) without electron pairing or unpairing and without the formation or destruction of ions. There are usually three or four key atoms, each of which, in the transition state, is breaking an old single bond and/or forming a new single bond to an atom to which it was not previously bonded. Thus the most common multicenter-type reactions are *three-center-type reactions* (see Chap. 24) and *four-center-type reactions* (see Chap. 25).

3-2d. *Characteristics of the Various Classes of Organic Reaction Mechanisms.* The extensive investigation of organic reaction mechanisms has made possible some useful generalizations about the various types of mechanisms.

Polar organic reactions, particularly those involving ions, rarely occur in the gas phase. They are very often catalyzed by acids and bases. They are usually not affected by light, traces of oxygen, or peroxides. They usually do not show induction periods or proceed by chain mechanisms. The rate of polar reactions is usually greatly affected by the ion-solvating ability of the reaction medium.

Free-radical reactions frequently occur in the gas phase, often show induction periods, and proceed by a chain mechanism. Many of these reactions are greatly affected by light, traces of oxygen, peroxides, and certain materials known as inhibitors. The reactions are rarely acid- or base-catalyzed, and their rate is not usually affected so greatly by the ion-solvating ability of the medium in which they proceed. Free-radical reactions are often brought about by much higher temperatures than those usually used for polar reactions.

Simple electron-transfer reactions always involve ions and therefore usually take place in solution rather than in the vapor phase. The rate of multicenter-type reactions is usually little affected by a change from the liquid to vapor phase.

In this book our consideration will be limited largely to homogeneous reactions, since these reactions are so much better understood than heterogeneous reactions, although the latter category is also of great practical importance. There is reason to believe that the mechanisms by which many heterogeneous reactions proceed at the "active sites" on catalysts are of essentially the same types as those whose operation in homogeneous reactions is discussed here.

It is not possible to draw a sharp boundary between the various classes of organic reaction mechanisms described. There are instead a few reactions whose mechanisms appear to be rather intermediate between two or more of the categories. In several cases it is possible to imagine, at least, a series of mechanisms, with a gradual and continuous transition from one class to another.

PROBLEMS

1. Making the steady-state assumption for C and E, express the rate of the following reaction in terms of the concentrations of the reactants A and B and the rate constants for the individual steps of the reaction:

$$A + B \underset{k_{-1}}{\overset{k_1}{\rightleftharpoons}} C$$

$$C + A \underset{k_{-2}}{\overset{k_2}{\rightleftharpoons}} E$$

$$E + B \overset{k_3}{\rightarrow} F$$

2. Calculate the values of k_{-1}/k_2, the over-all equilibrium constants, and first-order rate constants for each of the two reactions plotted in Fig. 3-1 (all at 25°).

3. Calculate the approximate maximum kinetic isotope effect due to S^{34} in a reaction that involves breaking a carbon-sulfur bond. In most alkyl sulfides absorption due to the carbon-sulfur stretching vibration occurs at about 15μ.

4. On occasion it has been claimed that since in some reaction of the type

$$A \rightarrow B$$

A_1 gives a better yield of B_1 than A_2 gives of B_2, it follows that A_1 is more reactive than A_2. Is this necessarily true? If it is not, describe, in general terms, as many different types of cases as you can think of in which it might not be true.

Chapter 4

QUANTITATIVE CORRELATIONS OF REACTION RATES AND EQUILIBRIA

Among the most striking developments in the history of physical organic chemistry has been the steady growth and improvement in methods for the quantitative correlation of rate and equilibrium constants. Some of the more general correlations will be considered in this chapter. Others will be discussed in subsequent chapters. Since the validity of quantitative relationships is often the best evidence for the qualitative concepts upon which the relationships were based, a number of important qualitative principles will also be presented here.[1]

4-1. A General Approach. According to the relation

$$\Delta F^\circ = -RT \ln K$$

the equilibrium constant for a reaction depends only on the difference in the standard free energies of the reactants and products. Since, by the transition-state theory (Sec. 3-1b), the activated complex in a rate process is treated as being in equilibrium with the reactants, the rate constant depends similarly on the free-energy difference between the reactants and the activated complex. Thus estimates of rate and equilibrium constants are equivalent to estimates of free-energy differences between various species.

Considerable attention has been given to the problem of estimating the

---

[1] It is difficult to determine specifically the origin of all the concepts presented in this chapter, but many of these concepts have been considered in other discussions of this and related subjects.[2-6]

[2] L. P. Hammett, "Physical Organic Chemistry," chaps. 3, 4, and 7, McGraw-Hill Book Company, Inc., New York, 1940.

[3] G. E. K. Branch and M. Calvin, "The Theory of Organic Chemistry," chap. 6, Prentice-Hall, Inc., Englewood Cliffs, N.J., 1941.

[4] H. H. Jaffé, *Chem. Revs.*, **53**, 191 (1953).

[5] R. W. Taft, Jr., in M. S. Newman (ed.), "Steric Effects in Organic Chemistry," chap. 13, John Wiley & Sons, Inc., New York, 1956.

[6] S. W. Benson and J. H. Buss, *J. Chem. Phys.*, **29**, 546 (1958).

free energies of organic compounds under given sets of conditions.[7]   The methods used serve adequately for many purposes, but the uncertainties are frequently of the order of 4 kcal/mole or more, corresponding to an error of almost 1,000-fold in a rate or equilibrium constant at room temperature.   We shall consider in this section only correlations whose uncertainties are usually much smaller than this.   In many cases such improved correlations are possible because we need, not the absolute free energy of any species, but only some free-energy difference, such as the free energy of reaction or of activation.   Usually, in fact, it is differences between differences, that is, differences in free energies of reaction and activation, that can be correlated most reliably.   Such correlations, which have received the most attention, will be discussed first.

Let us consider the reactions[8]

$$X_1 \!-\! N \!-\! Y_1 + A \overset{K_1}{\rightleftharpoons} X_1 \!-\! N \!-\! Y_2 + B \qquad (4\text{-}1)$$

$$X_2 \!-\! N \!-\! Y_1 + A \overset{K_2}{\rightleftharpoons} X_2 \!-\! N \!-\! Y_2 + B \qquad (4\text{-}2)$$

where X_1 and X_2 are monovalent substituent atoms or groups; Y_1 and Y_2 are monovalent atoms or groups, and may be called the reactant group and product group, respectively; N, that part of the molecule through which the substituent is attached to the reactant or product group, is a divalent atom or group; and A and B are other reactant and product molecules.   In the case, for example, where X_1 is Cl, X_2 is Br, Y_1 is H, Y_2 is I, N is the p-phenylene group, A is I_2, and B is HI, the two equilibria shown are simply the reversible para iodination of chlorobenzene and bromobenzene, respectively.   In consideration of a reaction rate, Y_2 would be the reactant group in its transition-state configuration.   The calculation of K_1/K_2, the ratio of the equilibrium (or rate) constants for the two related reactions, is equivalent to the calculation of the equilibrium constant for the reaction

$$X_1 \!-\! N \!-\! Y_1 + X_2 \!-\! N \!-\! Y_2 \rightleftharpoons X_1 \!-\! N \!-\! Y_2 + X_2 \!-\! N \!-\! Y_1 \quad (4\text{-}3)$$

Therefore

$$2.3RT \log \frac{K_1}{K_2} = F^{\circ}_{X_1NY_2} + F^{\circ}_{X_2NY_1} - F^{\circ}_{X_1NY_1} - F^{\circ}_{X_2NY_2} \qquad (4\text{-}4)$$

where $F^{\circ}_{X_1NY_2}$, etc., are the standard free energies of the various species. The standard free-energy content of a molecule may be expressed as the sum of a number of contributions from the constituent parts of the molecule plus various contributions due to interactions of these groups with each other and with surrounding molecules (e.g., solvent).   Since every

[7] Cf. G. J. Janz, "Estimation of Thermodynamic Properties of Organic Compounds," Academic Press, Inc., New York, 1958; *Quart. Revs. (London)*, **9**, 229 (1955).
[8] Cf. J. Hine, *J. Am. Chem. Soc.*, **82**, 4877 (1960).

molecule has a structurally unique combination of its constituent groups, it will obviously be possible to correlate the standard free energies (or almost any other property) of all known compounds by the use of a sufficient number of contributing structural parameters.   No correlation will be considered as having any great scientific significance, however, unless the existence and relative magnitudes of the parameters used seem to be explainable in terms of other scientific generalizations and unless the parameters are considerably less numerous than the data being correlated.   The free energy of the X_1—N—Y_2 molecule under any given "standard" set of conditions includes terms due to the individual X_1, N, and Y_2 groups (the bond energies of their internal bonds, their energies of interaction with the solvent, etc.), which we shall denote F_{X_1}, F_N, and F_{Y_2}.   There will also be the terms F_{X_1N} and F_{Y_2N} due to the interaction of the X_1 and Y_2 groups with N.   These terms will include the X_1—N and Y_2—N bond energies, energies of steric, dipolar, and resonance interactions between N and the other two groups, and correction factors that serve to modify the F_N term to allow for the fact that N is attached to X_1 and Y_2, etc.   Finally there will be a term $F_{X_1Y_2N}$ due to interactions between the X_1 and Y_2 groups.   Thus the free energy of X_1—N—Y_2 is expressed

$$F^{\circ}_{X_1NY_2} = F_{X_1} + F_N + F_{Y_2} + F_{X_1N} + F_{Y_2N} + F_{X_1Y_2N} \qquad (4\text{-}5)$$

Substitution of this and analogous expressions for $F^{\circ}_{X_1NY_1}$, $F^{\circ}_{X_2NY_1}$, and $F^{\circ}_{X_2NY_2}$ into Eq. (4-4) gives

$$2.3RT \log \frac{K_1}{K_2} = F_{X_1Y_2N} + F_{X_2Y_1N} - F_{X_1Y_1N} - F_{X_2Y_2N} \qquad (4\text{-}6)$$

That is, the ratio of equilibrium constants depends only on the terms for the energy of interaction between X and Y groups.   In general, it is thought that the energy of interaction of groups that are not bonded to each other may be divided into three components: *polar* (including inductive, polarizability, and field, or direct electrostatic, effects); *resonance* (only that resonance involving both groups simultaneously); and *steric* (including interference with internal rotations, etc., as well as steric compressions).   The extent of interaction with the solvent may be influenced by these polar, resonance, and steric factors.   When some of the possible complications are considered, it is not surprising that there appears to be no generally useful method for correlating $F_{XY}N$ terms with satisfactory precision.   We shall, however, discuss several special cases in which useful correlations may be made.

   Let us consider the case in which N is a relatively rigid group [so that the X and Y groups will be the same distance apart in all four species involved in Eq. (4-3), synergistic effects of X and Y on the internal rotations of N will be eliminated, etc.], the groups X and Y are far enough apart that there are no direct steric interactions between them (in the case of groups that interact strongly with the solvent, the closer, more tightly bound solvent molecules may be considered an integral part of the group), and there is no direct resonance interaction between the X and Y groups. Examples of compounds that might meet these specifications are the trans-1,3-substituted cyclobutanes, 1,4-substituted bicyclo-[2.2.2]octanes, meta- and (unless the groups X and Y are capable of resonance interaction with each other) para-substituted benzene derivatives, etc. With such compounds there should be no direct steric and resonance contributions to the $F_{XY}{}^N$ terms. We shall assume that the polar contribution, which we shall call the free energy of polar interaction of substituents ($F_{XY}{}^{Np}$), is proportional to the product of parameters for the substituents that we shall call their *polar substituent constants* ($\sigma_X{}^N$ and $\sigma_Y{}^N$). The proportionality constant, which will be a function of the solvent, the temperature, and the nature of the group (N) through which the two substituents are attached, is a measure of how effectively the influence of one group is *transmitted* to the other and, after separation of the term $2.3RT$ for convenience, will be called τ_N.

$$F_{XY}{}^{Np} = 2.3RT\tau_N\sigma_X{}^N\sigma_Y{}^N \tag{4-7}$$

This equation is somewhat analogous to the Kirkwood-Westheimer treatment (Sec. 2-4c), in which the electrostatic effect is calculated by multiplying the product of the dipole moments (or electric charges) of the substituent groups by a factor that (like τ_N) is a function of the distance between the groups, their orientation with respect to each other, and the dielectric constant of the matter around and between the groups. The substitution of the appropriate specific forms of Eq. (4-7) into Eq. (4-6) gives

$$\log \frac{K_1}{K_2} = \tau_N(\sigma_{X_1}{}^N - \sigma_{X_2}{}^N)(\sigma_{Y_2}{}^N - \sigma_{Y_1}{}^N) \tag{4-8}$$

Within any specific reaction series (by which we mean a given solvent, temperature, Y_1, and Y_2) the value of $\tau_N(\sigma_{Y_2}{}^N - \sigma_{Y_1}{}^N)$ will be a constant, which we shall call the *reaction constant*, $\rho_{Y_1Y_2}{}^N$.

$$\rho_{Y_1Y_2}{}^N = \tau_N(\sigma_{Y_2}{}^N - \sigma_{Y_1}{}^N) \tag{4-9}$$

Since only *differences* between σ's are involved, we can set the absolute value of any one σ arbitrarily without loss of generality. This is done by defining one substituent, usually hydrogen, as the reference substituent, for which σ is defined as zero. Then all the other equilibrium constants are compared to the one (K_0) for the reactant with the reference substituent. Thus, substituting Eq. (4-9) into Eq. (4-8) and then replacing K_2

by K_0 and $\sigma_{X_2}{}^N$ by zero, we obtain

$$\log \frac{K_1}{K_0} = \rho_{Y_1Y_2}{}^N \sigma_{X_1}{}^N \qquad (4\text{-}10)$$

Equations like (4-8), that are based on an assumption that a certain free-energy contribution or free-energy difference is a linear function of some property of a substituent group [cf. Eq. (4-7)], are called *linear free-energy relationships*. Some of the most important quantitative correlations of rate and equilibria are linear free-energy relationships of the form of Eq. (4-10).

4-2. Correlations with meta- and para-substituted Benzene Derivatives. *4-2a. The Hammett Equation.* Hammett found that for a large number of reactions of meta- and para-substituted benzene derivatives a plot of the logarithms of the rate (k) or equilibrium (K) constants

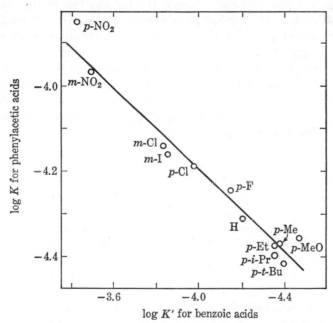

FIG. 4-1. Log-log plot of ionization constants of benzoic and phenylacetic acids in water at 25°.

for one reaction vs. $\log k$ or $\log K$ for another reaction gave reasonably straight lines.[2,4,5,9] Thus in Fig. 4-1 values of $\log K$ for the ionization of benzoic acids have been plotted against those for the ionization of phenylacetic acids. The point labeled p-NO$_2$ has for its abscissa $\log K$ for p-nitrobenzoic acid and for its ordinate $\log K$ for p-nitrophenylacetic

[9] L. P. Hammett, *Chem. Revs.*, **17**, 125 (1935); *Trans. Faraday Soc.*, **34**, 156 (1938).

acid.   These points approximate a line whose equation is

$$\log K = \rho \log K' + C \tag{4-11}$$

where K and K' are the ionization constants for a phenylacetic and a benzoic acid with a given substituent, ρ is the slope of the line, and C is the intercept.   This equation may be used for any substituent including the reference substituent, hydrogen.   Hence,

$$\log K_0 = \rho \log K_0' + C \tag{4-12}$$

where K_0 and K_0' are the ionization constants of phenylacetic and benzoic acids, themselves.   Subtraction of Eq. (4-12) from Eq. (4-11) gives

$$\log \frac{K}{K_0} = \rho \log \frac{K'}{K_0'} \tag{4-13}$$

An equation of this sort could be written for any two of the many reactions for which linear log-log plots can be made.   It is therefore convenient to select some of the reactions as a standard with which to compare all the others.   Because of the large number of accurate data available, the ionization of benzoic acids in aqueous solution at 25° was chosen as the standard reaction, and a new constant, σ, characteristic of a given substituent, was defined as $\log (K'/K_0')$, where K' is the ionization constant of the substituted benzoic acid and K_0' that of benzoic acid. This definition reduces Eq. (4-13) to

$$\log \frac{K}{K_0} = \rho\sigma \tag{4-14}$$

where σ, the substituent constant, is a measure of the electron-donating or electron-withdrawing power of the substituent and ρ, the reaction constant, is a measure of the sensitivity of the equilibrium constant to changes in the σ value of the substituent.   From Eq. (4-13) and the definition of σ it may be seen that ρ has been defined as unity for the standard reaction, the ionization of benzoic acids in water at 25°.   The value of σ may be determined from its definition if the ionization constant of the appropriate benzoic acid has been determined.   Such σ's may be used to calculate ρ constants for other reactions, and from these ρ's new σ values can often be obtained.   A number of σ constants largely based on the ionization of benzoic acids[10] are listed in Table 4-1.   The use of the Hammett equation may be illustrated by a specific example.   Kindler studied the alkaline hydrolysis of the ethyl esters of a number of substituted benzoic acids.[11]   The m-nitro compound was found to react 63.5 times as fast as the unsubstituted compound.   The substitution of this

[10] D. H. McDaniel and H. C. Brown, *J. Org. Chem.*, **23**, 420 (1958).

[11] K. Kindler, *Ann.*, **450**, 1 (1926).

value of k/k_0 and the σ value for the m-nitro group (0.710) into Eq. (4-14) gives a ρ value of 2.54 for the alkaline hydrolysis of ethyl benzoates under the conditions used. From this and from the σ constant for the p-methoxy substituent (-0.268), ethyl p-methoxybenzoate is predicted

TABLE 4-1. HAMMETT SUBSTITUENT CONSTANTS[a]

Substituent	σ		Substituent	σ	
	Meta	Para		Meta	Para
CH_3	-0.069	-0.170	O^{-f}	$-0.708^{c,e}$	$-1.00^{c,h}$
CH_2CH_3	-0.07	-0.151	OH	$+0.121$	-0.37
$CH(CH_3)_2$	$-0.068^{b,c}$	-0.151	OCH_3	$+0.115$	-0.268
$C(CH_3)_3$	-0.10	-0.197	OC_2H_5	$+0.1$	-0.24
C_6H_5	$+0.06$	-0.01	OC_6H_5	$+0.252$	-0.320
$C_6H_4NO_2$-p	....	$+0.26^d$	$OCOCH_3$	$+0.39$	$+0.31$
$C_6H_4OCH_3$-p	....	-0.10^d	F	$+0.337$	$+0.062$
$CH_2Si(CH_3)_3$	-0.16	-0.21	$Si(CH_3)_3$	-0.04	-0.07
$COCH_3$	$+0.376$	$+0.502$	PO_3H^{-f}	$+0.2$	$+0.26$
COC_6H_5	....	$+0.459^{c,e}$	SH	$+0.25$	$+0.15$
CN	$+0.56$	$+0.660$	SCH_3	$+0.15$	0.00
CO_2^{-f}	-0.1	0.0	$SCOCH_3$	$+0.39$	$+0.44$
CO_2H	$+0.35^{b,c}$	$+0.406^{b,c}$	$SOCH_3$	$+0.52$	$+0.49$
CO_2CH_3	$+0.321^{b,c}$	$+0.385^{b,c}$	SO_2CH_3	$+0.60$	$+0.72$
$CO_2C_2H_5$	$+0.37$	$+0.45$	SO_2NH_2	$+0.46$	$+0.57$
CF_3	$+0.43$	$+0.54$	SO_3^{-f}	$+0.05$	$+0.09$
NH_2	-0.16	-0.66	$S(CH_3)_2^{+f}$	$+1.00$	$+0.90$
$N(CH_3)_2$	$-0.211^{c,e}$	-0.83	Cl	$+0.373$	$+0.227$
$NHCOCH_3$	$+0.21$	0.00	Br	$+0.391$	$+0.232$
$N(CH_3)_3^{+f}$	$+0.88$	$+0.82$	I	$+0.352$	$+0.276^{c,d}$
N_2^{+f}	$+1.76^g$	$+1.91^g$	IO_2	$+0.70$	$+0.76$
NO_2	$+0.710$	$+0.778$			

[a] These substituent constants are based on the ionization constants of benzoic acids and are from the compilation of McDaniel and Brown[10] unless noted otherwise.

[b] From van Bekkum, Verkade, and Wepster.[17]

[c] Based on some reaction other than the ionization of benzoic acids.

[d] Calculated from E. Berliner and E. A. Blommers, J. Am. Chem. Soc., **73,** 2479 (1951).

[e] Jaffé.[4]

[f] Values of σ for electrically charged groups may be particularly solvent-dependent.

[g] E. S. Lewis and M. D. Johnson, J. Am. Chem. Soc., **81,** 2070 (1959).

[h] E. Berliner and L. C. Monack, J. Am. Chem. Soc., **74,** 1574 (1952).

to hydrolyze 0.209 times as fast as the unsubstituted compound. Actually it hydrolyzes 0.214 times as fast. Of course, if a least-squares treatment is used to calculate the value of ρ from data on all the compounds studied, instead of just two, a somewhat different ρ value (2.43[4]) is obtained.

The correlation of data on meta- and para-substituted benzene derivatives would be expected from Eq. (4-10) to require two sets of ρ's and two sets of σ's (one for meta and one for para compounds), since two different N groups, m- and p-phenylene, are involved. The Hammett equation uses different σ's for meta and para substituents but only one ρ per reaction (in a given solvent at a given temperature). It may be shown algebraically, without recourse to data on any other reaction, that the substituent constants based on the ionization of benzoic acids could not possibly successfully correlate all equilibrium constants for meta- and para-substituted benzene derivatives if, in each reaction, the same ρ is used for meta- as for para-substituted compounds.[12] Experimentally it is found that although the use of separate ρ's for meta- and para-substituted compounds gives a statistically significant improvement in the correlation of the available data, the improvement is not a very large one.[13] Apparently, for most of the reactions that have been studied, the ratio of $\tau_{\mathrm{para}}(\sigma_{\mathrm{Y_2}}^{\mathrm{para}} - \sigma_{\mathrm{Y_1}}^{\mathrm{para}})$ to $\tau_{\mathrm{meta}}(\sigma_{\mathrm{Y_2}}^{\mathrm{meta}} - \sigma_{\mathrm{Y_1}}^{\mathrm{meta}})$ is about the same as for the reference reaction, in which Y_1 is CO_2H, Y_2 is CO_2^-, and the τ's refer to aqueous solutions at 25° [cf. Eq. (4-9)]. Under these conditions the value of τ_N is around 3.0 to 3.5 for the cases where N is a m- or p-phenylene group.[8,12]

4-2b. Modifications of the Hammett Equation. Probably the most important cause of deviations from the Hammett equation, as Hammett himself pointed out, is direct resonance interaction between the substituent group and the reaction center. Fig. 4-2 shows this point graphically. In the plot of the acidity constants of anilinium ions vs. σ, the points for the meta substituents lie fairly near a straight line, as do the points for all the para substituents except the three that may be classified as resonance-electron-withdrawing groups, whose points lie well above the line. That is, p-nitro-, p-methanesulfonyl-, and p-carbethoxy-anilinium ions are all much more acidic than would be predicted from Eq. (4-14). The equilibria in these three cases are driven to the right by the stabilization of the free amine by the contribution of structures like

$$\overset{\oplus}{H_2N}=\!\!\!\left\langle\underline{}\right\rangle\!\!\!=NO_2^{\ominus}$$

for which there are no counterparts among the contributing structures of the anilinium ions or those of the benzoic acids (and anions) from which the σ constants were obtained. Such deviations from Eq. (4-14) occur regularly with resonance-electron-withdrawing para substituents when such a strongly resonance-electron-donating group as $-NH_2$ (or $-O^-$, $-S^-$, etc.) is present at the reaction center. Similarly, large deviations

[12] J. Hine, *J. Am. Chem. Soc.*, **81**, 1126 (1959).
[13] H. H. Jaffé, *J. Am. Chem. Soc.*, **81**, 3020 (1959).

are found with resonance-electron-donating substituents when the reaction center is capable of strong electron withdrawal. Such deviations due to direct resonance interaction between the substituent group and the reaction center have led to the suggestion of alternate sets of σ constants. One group, known originally as "substituent constants for

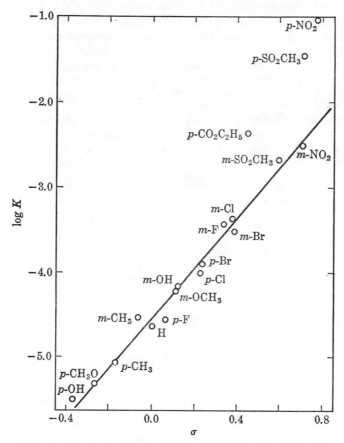

FIG. 4-2. Log K for acidity constants of anilinium ions vs. σ.

use in reactions of aniline and phenol derivatives"[2,4] and also called σ^- (sigma minus) values,[14] contains para substituents capable of resonance-electron withdrawal, whereas σ^+ (sigma plus) values are for resonance-electron-donating para substituents.[14-16] Some values of σ^- and

[14] R. W. Taft, Jr., *J. Am. Chem. Soc.*, **79**, 1045 (1957).

[15] H. C. Brown and Y. Okamoto, *J. Am. Chem. Soc.*, **80**, 4979 (1958).

[16] Although Brown and coworkers have calculated σ^+ values for all types of para substituents, the values obtained are significantly different from the σ values only in the case of substituents capable of resonance-electron donation.

σ^+ are listed in Table 4-2. It has been suggested that all reactions of meta- and para-substituted benzene derivatives may be divided into two classes: those in which the usual σ's (Table 4-1) should be used, and those in which σ^-'s or σ^+'s are required.[2,15] This idea seems intuitively unreasonable, however, since it demands that the amount of direct resonance interaction between substituent group and reaction center be a discontinuous function of the nature of the reaction. It is not surprising that, because of extra resonance, the p-nitro substituent acts as a stronger

TABLE 4-2. VALUES OF σ^-, σ^+, AND σ^n

Substituent	$\sigma^{-\,a}$	Substituent	$\sigma^{+\,b}$	Substituent	$\sigma^{n\,c}$
p-CH=CHC$_6$H$_5$	$+0.619$	p-CH$_3$	-0.311	p-CH$_3$	-0.129
p-CONH$_2$	$+0.627$	p-C(CH$_3$)$_3$	-0.256	m-NH$_2$	-0.038
p-CO$_2$CH$_3$	$+0.636$	p-C$_6$H$_5$	-0.179	m-N(CH$_3$)$_2$	-0.049
p-CO$_2$C$_2$H$_5$	$+0.678$	p-NH$_2$	-1.3	p-NH$_2$	-0.297
p-CO$_2$H	$+0.728$	p-N(CH$_3$)$_2$	-1.7	p-N(CH$_3$)$_2$	-0.219
p-CF$_3$	$+0.74^d$	p-NHCOCH$_3$	-0.6	p-OH	-0.201
p-COCH$_3$	$+0.874$	p-OH	-0.92	p-OCH$_3$	-0.175
p-CN	$+1.000$	p-OCH$_3$	-0.778	p-F	$+0.056$
p-CHO	$+1.126$	p-OC$_6$H$_5$	-0.5	p-Cl	$+0.238$
p-S(CH$_3$)$_3{}^+$	$+1.16^e$	p-F	-0.073	p-Br	$+0.265$
p-SOCH$_3$	$+0.73^e$	p-SCH$_3$	-0.604	p-I	$+0.299$
p-SO$_2$CH$_3$	$+1.049$	p-Cl	$+0.114$		
p-NO$_2$	$+1.27$	p-Br	$+0.150$		
p-N$_2{}^+$	$+3.2^f$	p-I	$+0.135$		

[a] From Ref. 4 unless noted otherwise.
[b] From Ref. 15.
[c] From Ref. 17.
[d] J. D. Roberts, R. A. Clement, and J. J. Drysdale, *J. Am. Chem. Soc.*, **73**, 2181 (1951).
[e] F. G. Bordwell and P. J. Boutan, *J. Am. Chem. Soc.*, **78**, 87 (1956); **79**, 717 (1957).
[f] E. S. Lewis and M. D. Johnson, *J. Am. Chem. Soc.*, **81**, 2070 (1959).

electron withdrawer in the ionization of phenols than in the ionization of benzoic acids and that a σ^- of 1.27 is thus required instead of the usual σ (0.778). However, one would expect that there would be reactions in which the extra resonance between the nitro group and the reaction center would be significant but small enough that a substituent constant between the usual σ and σ^- would be appropriate. Van Bekkum, Verkade, and Wepster[17] have described convincing evidence that not two but a multiplicity of σ values would be required to correlate existing data in cases where there seems to be direct resonance interaction between substituent and reaction center. These workers calculated ρ constants

[17] H. van Bekkum, P. E. Verkade, and B. M. Wepster, *Rec. trav. chim.*, **78**, 815 (1959).

for a number of reactions, using only σ's for substituents for which extra resonance seemed to be impossible. From the ρ constants obtained, σ was then calculated for each substituent in each reaction. If the concept of duality of σ's were correct, one would expect the σ's thus obtained for the p-nitro substituent, for example, to fall in two clusters, one centered around the usual σ (0.778) and the other around σ^- (1.27). Instead, the calculated values varied fairly continuously from 0.766 to 1.378, with 13 of the 34 values being in the range 0.91 to 1.084, about halfway between σ and σ^-. Values of σ^- and σ^+ like those listed in Table 4-2 have been and will continue to be of great utility despite the fact that the amount of extra resonance in a reaction may be more or less than that for which σ^- or σ^+ would be applicable.

Having discussed the complications produced by resonance interactions between the substituent group and reaction center, it would be well to point out that such extra resonance is of importance with certain benzoic acid derivatives and that for this reason some of the σ constants listed in Table 4-1 are considerably more negative than they would have been in the absence of such resonance. For example, the fact that a σ^+ value is required for the carboxy substituent when para to an amino group is good evidence that structures like I contribute significantly to the total structures of such molecules.

$$
\underset{\text{I}}{\text{H}_2\overset{\oplus}{\text{N}}=\hspace{-2pt}\langle\ \rangle\hspace{-2pt}=\text{C}\overset{\displaystyle \overset{\ominus}{\text{O}}}{\underset{\text{OH}}{\Big\backslash}}}
\qquad
\underset{\text{II}}{\text{H}_2\overset{\oplus}{\text{N}}=\hspace{-2pt}\langle\ \rangle\hspace{-2pt}=\text{C}\overset{\displaystyle \overset{\ominus}{\text{O}}}{\underset{\overset{\ominus}{\text{O}}}{\Big\backslash}}}
$$

Structures like II would certainly be expected to contribute less to the total structure of the corresponding benzoate anion. Therefore direct resonance interaction between the amino and carboxy groups contributes to the diminution in the ionization constant of benzoic acid brought about by the p-amino substituent. This point has also been dealt with by van Bekkum, Verkade, and Wepster,[17,18] who calculated substituent constants from data on reactions in which extra resonance seemed quite improbable. Some of these σ^n (sigma, normal) constants are listed in Table 4-2.

4-2c. The Significance of Hammett Substituent Constants. The substituent constants listed in Tables 4-1 and 4-2 provide quantitative substance for many concepts that already rested on a firm qualitative basis. The acid-strengthening substituents, those with positive σ's, are classed

[18] Cf. R. W. Taft, Jr., S. Ehrenson, I. C. Lewis, and R. E. Glick, *J. Am. Chem. Soc.*, **81**, 5352 (1959).

as electron withdrawers, while negative σ's refer to electron-donor substituents. It is useful to divide the cause of substituent electronic effects into two categories: *tautomeric*, or *resonance*, *effects* and *polar effects*. Tautomeric effects act along the pi electron system, involving multiple bonds, unshared electrons, and unfilled outer electronic shells. Polar effects include the *inductive effect*, which acts along single bonds, and the *field*, or *direct electrostatic*, *effect*, which acts through space; sometimes the term "inductive effect" is used so as to include the field effect. Tautomeric and inductive effects are called T and I effects by British workers. Substituents like the nitro group, whose resonance interactions with attached unsaturated systems result in electron withdrawal, are said to have $-T$ effects, while resonance-electron donors, such as the amino group, have $+T$ effects.[19] Groups withdrawing electrons inductively have a $-I$ effect, while those that donate electrons have a $+I$ effect.[20] It appears that inductive effects are transmitted to the reaction center almost as efficiently from the para as from the meta position.[21] Therefore the difference between the meta and para σ constants for a given substituent is largely due to the more efficient operation of the resonance effect from the para position. In fact, the quantity $\sigma_p - \sigma_m$ has been suggested as a measure of the resonance effect of a given substituent. In agreement with this idea, σ_p may be seen (Table 4-1) to be larger than σ_m for all groups, such as —SO$_2$R, —COR, —NO$_2$, —CN, etc., that would be expected to be $-T$ groups; for all $+T$ groups, such as —NR$_2$, —OR, —SR, halogens, and alkyl groups, σ_p is smaller than σ_m. Although it seems clear that resonance effects act more strongly from the para position, it also seems that resonance has some importance for meta substituents. A resonance effect seems to be required to explain the

[19] The T effect is further subdivided into the mesomeric, or M, effect, which is the T effect in normal molecules, and the electromeric, or E, effect, which is the change in the T effect that takes place during a reaction. Most American workers feel that these two extra "effects" are not useful. The usage of plus and minus signs given here in connection with I, T, M, and E effects is that of Ingold; an opposite convention has been used by others. For a much more complete discussion of these terms, see C. K. Ingold, "Structure and Mechanism in Organic Chemistry," sec. 7, Cornell University Press, Ithaca, N.Y., 1953.

[20] This inductive effect in the normal ("static") molecule is sometimes called the I_s effect to distinguish it from the inductomeric, or polarizability, effect (I_d) occurring during a reaction.

[21] One may use the ρ constant for the ionization of cinnamic acids in aqueous solution (0.466)[4] to argue that inductive effects are decreased to 0.466 of their original magnitude by the interposition of two additional sp^2 carbon atoms (the fact that meta substituents, whose effects are mostly inductive, fit the Hammett equation shows that this factor is applicable to inductive effects). Assuming then that each sp^2 carbon diminishes the inductive effect to $\sqrt{0.466}$, and allowing for the two paths of conductance around the benzene ring, it is found that inductive effects should be $1.466/2 \sqrt{0.466}$, or 1.074, times as large in the meta as in the para position.

facts that the meta amino substituent acts as an electron donor in spite of the fact that nitrogen is more electronegative than carbon and that the amino group is found to have a $-I$ effect when attached to saturated carbon. Apparently the carbon atoms adjacent to the one to which the reaction center is attached are made negative by the contribution of structures such as

These negative charges then have an inductive effect on the reaction center. Significant contributions of resonance effects also offer an explanation of why the σ_m's of the halogens do not stand in the order $F > Cl > Br > I$, as would be expected from their electronegativities and the substituent constants when attached to saturated carbon (see Table 4-3). Another measure of the resonance effect may be found in σ^+ and σ^- values. The difference between these and the corresponding σ's (or better, σ^n's) arises from direct resonance interaction with the reaction center. Although σ^+ and σ^- values may be used to classify substituents as $+T$ and $-T$ with the same qualitative results as obtained by the use of $\sigma_p - \sigma_m$ values, the quantitative agreement obtained is not always satisfactory. For example, the resonance effect of the fluorine substituent is 69 per cent of that of the amino group as measured by $\sigma_p - \sigma_m$ but only 13 per cent as measured by $\sigma^+ - \sigma^n$. The quantitative interpretation of substituent constants in terms of resonance and inductive effects has not yet been satisfactorily achieved, but Taft and his coworkers have made major contributions to the problem.[5,22]

The data obtained with substituents, such as amino, hydroxy, alkoxy, and all electrically charged groups, that can interact strongly with the solvent often show significant deviations from the Hammett equation and its usual modifications. These deviations could be minimized by making the σ's for such groups solvent-dependent, but this would complicate the use of the correlation and the data necessary for the reliable determination of a number of such σ's in a variety of solvents does not appear to be available at present.

4-3. Correlations with Relatively Rigid Nonaromatic Compounds. As stated earlier in this chapter, a linear free-energy relationship might be expected to be applicable with any series of relatively rigid compounds in which the substituent group is far enough from the reaction center to avoid direct steric interactions and in which there are no direct

[22] R. W. Taft, Jr., *J. Phys. Chem.*, **64**, 1805 (1960), and earlier references cited therein.

resonance interactions between substituent and reaction center. Roberts and Moreland have studied the 4-substituted bicyclo[2.2.2]octane-1-carboxylic acids and their ethyl esters.[23] For the five substituents, —H, —OH, —CO$_2$C$_2$H$_5$, —Br, and —CN, the strengths of the acids, the rates of reaction of the acids with diphenyldiazomethane, and the rates of the basic hydrolysis of the esters were determined. A log-log plot of the rate or equilibrium constants of any two of these three reactions gives an excellent approximation of a straight line. The available data are quite limited, but it seems probable that compounds of this series should, in general, give reasonably good fits to linear free-energy relationships.

Siegel and Komarmy have studied some *trans*-4-substituted cyclo-hexanecarboxylic acids and their methyl esters, all of which exist largely in the less strained diequatorial forms.[24] Using the same three reactions that had been studied with the bicyclooctyl compounds, but measuring the acidities in three different solvents, an equation of the form

$$\log (k/k_0) = \rho''\sigma''$$

was found to be applicable. The σ'' values obtained (by setting ρ'' equal to 1.0 for acidities measured in water) were proportional to those that could be calculated similarly from data on the bicyclooctyl derivatives, but they were only three-fourths as large. This would be expected since the substituent groups are farther from the reaction center in the case of the 4-substituted cyclohexyl compounds.

Although only polar effects are believed to be important in the case of these saturated compounds, linear free-energy relationships have also been found to be applicable to measurements on nonaromatic systems in which substituent resonance effects are important. Charton and Meis-lich have found that the pK's of *trans*-3-substituted acrylic acids

are linearly related to the Hammett para σ constants for the substituents.[25] Similar results have been found for the rates of reaction of these compounds with diphenyldiazomethane.[26] Apparently the resonance and polar effects of the substituents are mixed in about the same proportion with acrylic acids as with benzoic acids. The ρ constants are about twice as large for the acrylic acid reactions as for the corresponding

[23] J. D. Roberts and W. T. Moreland, Jr., *J. Am. Chem. Soc.*, **75**, 2167 (1953).

[24] S. Siegel and J. M. Komarmy, *J. Am. Chem. Soc.*, **82**, 2547 (1960).

[25] M. Charton and H. Meislich, *J. Am. Chem. Soc.*, **80**, 5940 (1958).

[26] J. Hine and W. C. Bailey, Jr., *J. Am. Chem. Soc.*, **81**, 2075 (1959).

benzoic acid reactions, in which the substituent is separated from the reaction center by twice as many carbon atoms.

4-4. The Taft Equation.   Although linear free-energy relations of the form of Eq. (4-14) are applicable to important but somewhat limited groups of compounds such as meta- and para-substituted benzene derivatives, there are many series of compounds for which such relations do not generally hold.   This is illustrated in Fig. 4-3, where log k for the acid-catalyzed esterification of various carboxylic acids[27] is plotted against log K for the ionization of the acids.   The points are seen to deviate

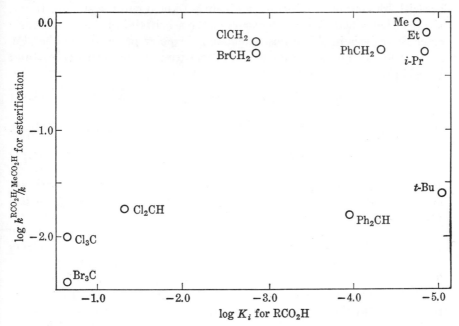

FIG. 4-3. Log-log plot for relative rates of esterification vs. ionization constants of RCO$_2$H's.

markedly from any possible straight line.   It is also seen, though, that the data may be explained, at least qualitatively, by the assumption that the esterification rates are determined practically entirely by steric factors and the ionization constants by polar factors.   In fact, Taft and coworkers have been able to extend the range of linear free-energy relationships vastly by assuming that steric and polar effects may be treated independently.

Taft started by considering the large amount of available data on the acidic and basic hydrolysis of esters of carboxylic acids.[5]   He then utilized Ingold's suggestion that steric effects are the same for the acidic

27 J. J. Sudborough and L. L. Lloyd, *J. Chem. Soc.*, **75**, 467 (1899).

and basic hydrolyses of an ester,[28] and added to it the assumption that resonance effects are also the same.[29]

These assumptions were applied to the rates of hydrolysis of a series of esters, XCO_2R, in which R is held constant and X is varied. They will be applied here in a manner analogous to that used in Sec. 4-1. The rate constant for the basic hydrolysis of a given ester $(k_X{}^B)$ relative to that of the ester defined as the standard reactant $(k_0{}^B)$ will be determined by polar, resonance, and steric factors. The polar factor is measured by $\rho_B\sigma_X$, where the substituent constant σ_X describes the electron-donating or -withdrawing properties of the given X (relative to the standard) and the reaction constant ρ_B is a measure of the sensitivity of the rate of basic ester hydrolysis in this series to changes in polar properties of the X's. The resonance and steric factors are measured by the terms $(E_r{}^X)_B$ and $(E_s{}^X)_B$, giving

$$\log (k_X{}^B/k_0{}^B) = \rho_B\sigma_X + (E_r{}^X)_B + (E_s{}^X)_B \qquad (4\text{-}15)$$

An analogous relation may be written for the rates of acidic hydrolysis.

$$\log (k_X{}^A/k_0{}^A) = \rho_A\sigma_X + (E_r{}^X)_A + (E_s{}^X)_A \qquad (4\text{-}16)$$

If it is now assumed that the steric and resonance factors $(E_r{}^X)_B + (E_s{}^X)_B$ and $(E_r{}^X)_A + (E_s{}^X)_A$ are equal, Eq. (4-16) may be subtracted from Eq. (4-15) to give

$$\log \frac{(k_X{}^B/k_0{}^B)}{(k_X{}^A/k_0{}^A)} = (\rho_B - \rho_A)\sigma_X \qquad (4\text{-}17)$$

Taft has arbitrarily defined the term $(\rho_B - \rho_A)$ for hydrolysis in aqueous alcoholic and aqueous acetone solutions as 2.48 in order to put the resultant substituent constants, which he denotes σ^*'s, on about the same basis as Hammett's substituent constants. (Values of $\rho_B - \rho_A$ for meta- and para-substituted benzoates average around 2.48.) Since ρ_A is considerably smaller than (in fact, almost negligible compared with) ρ_B, the use of Eq. (4-17) gives positive values for those groups that are electron withdrawers relative to the standard substituent, methyl, for which σ^* is, of course, defined as zero.

The validity of Eq. (4-17) is attested by several facts. The resultant values of σ^* are roughly proportional to the corresponding σ' values obtained from data on 4-substituted bicyclooctyl compounds[23] and the σ'' values from data on *trans*-4-substituted cyclohexyl derivatives[24] in the small number of cases where σ' and σ'' values exist. Since significant steric and resonance effects seem unlikely in the case of the bicyclooctyl

---

[28] C. K. Ingold, *J. Chem. Soc.*, 1032 (1930).

[29] The mechanistic significance of this generalization, and also its limitations, will be discussed in Sec. 12-4.

and cyclohexyl compounds, they appear to have been at least largely removed in the treatment by which the σ^* values were obtained. Furthermore, the magnitude of σ^* has been found to vary with the nature of the group in a manner much like that which would be expected on various theoretical grounds.[5] Finally, a number of reactions have been found that obey the simple relation

$$\log (k/k_0) = \rho^*\sigma^* \tag{4-18}$$

These should be reactions in which steric and resonance effects may be neglected, and for most of the reactions it seems obvious *why* these effects

TABLE 4-3. TAFT POLAR SUBSTITUENT CONSTANTS[5]

Substituent	σ^*	Substituent	σ^*
Cl_3C	+2.65	$C_6H_5CH{=}CH$	+0.41
F_2CH	+2.05	$(C_6H_5)_2CH$	+0.40
CH_3OCO	+2.00	$ClCH_2CH_2$	+0.38
Cl_2CH	+1.94	$CH_3CH{=}CH$	+0.36
CH_3CO	+1.65	$F_3CCH_2CH_2$	+0.32
$C_6H_5C{\equiv}C$	+1.35	$C_6H_5CH_2$	+0.22
$CH_3SO_2CH_2$	+1.32	$CH_3CH{=}CHCH_2$	+0.13
$NCCH_2$	+1.30	$F_3C(CH_2)_3$	+0.12
FCH_2	+1.10	$C_6H_5CH_2CH_2$	+0.08
$ClCH_2$	+1.05	CH_3	0.000
$BrCH_2$	+1.00	C_2H_5	-0.10
F_3CCH_2	+0.92	$n\text{-}C_3H_7$	-0.12
ICH_2	+0.85	$n\text{-}C_4H_9$	-0.13
$C_6H_5OCH_2$	+0.85	Cyclohexyl	-0.15
CH_3OCH_2	+0.64[a]	$(CH_3)_3CCH_2$	-0.16
CH_3COCH_2	+0.60	$(CH_3)_2CH$	-0.19
C_6H_5	+0.60	Cyclopentyl	-0.20
$HOCH_2$	+0.56	$(C_2H_5)_2CH$	-0.22
$O_2NCH_2CH_2$	+0.50	$(CH_3)_3SiCH_2$	-0.26
H	+0.49	$(CH_3)_3C$	-0.30

[a] P. Ballinger and F. A. Long, *J. Am. Chem. Soc.*, **82**, 795 (1960).

are negligible. Reaction series will often give good agreement with Eq. (4-18) except for a few compounds, the reasons for whose deviations usually seem obvious. In many cases values of σ^* for additional substituents may be calculated from the data on reactions that follow Eq. (4-18). Some such values along with values calculated from data on ester hydrolysis are listed in Table 4-3.

Among the expected features that may be seen in the relative magnitudes of these polar substituent constants are the facts that

$Cl_3C > Cl_2CH > ClCH_2 > CH_3$; $FCH_2 > ClCH_2 > BrCH_2 > ICH_2$;
$C_6H_5C{\equiv}C > C_6H_5CH{=}CH > C_6H_5CH_2CH_2$

It might also be noted that in a number of cases (but not for the case $X = H$) $\sigma_X^* \sim 2.8\sigma_{XCH_2}^*$.

Taft has also determined polar substituent constants for ortho substituents on a benzene ring by the same process used in determining σ^*'s (except that the standard esters were benzoates instead of acetates). The resultant values (Table 4-4) of σ^* for the ortho substituents are fairly near those already listed for the corresponding para substituents.[30]

TABLE 4-4. TAFT SUBSTITUENT CONSTANTS FOR ORTHO SUBSTITUENTS

Substituent	σ^*	Substituent	σ^*
CH_3O	-0.39	Cl	$+0.20$
C_2H_5O	-0.35	Br	$+0.21$
CH_3	-0.17	I	$+0.21$
H	0.00	NO_2	$+0.80$
F	$+0.24$	$NO_2{}^a$	$+1.22^a$

[a] This value is for use with aniline and phenol reactions and could thus be called a σ^{-*}.

4-5. Quantitative Correlations of Acidity and Basicity. *4-5a. Hammett-equation Correlations with Monosubstituted Acids and Bases.* The utility of the Hammett and Taft equations may be illustrated by application to the equilibrium constants obtained in certain acid-base reactions. In Table 4-5 are listed the Hammett ρ constants obtained for several series of compounds under various conditions. The increase in ρ accompanying a decrease in the ion-solvating power and dielectric constant of the medium that may be seen in comparing reactions 1, 2, and 3 is commonly (but not invariably) found. It is usually explained by pointing out that the electrostatic interaction between the substituents and the reaction center will be stronger in a medium of lower dielectric constant. Perhaps of greater importance, however, is the fact that an ethanol-solvated carboxylate anion group is so much different from the corresponding water-solvated group that the nature of the equilibrium process is different in the two solvents. More specifically, it may be suggested that the carboxylate anion group should distribute its negative charge so much better to the surrounding solvent molecules in water that $\sigma_{CO_2^-}$ should be less negative (or more positive) in water than in ethanol. Since ρ for the ionization of benzoic acids is equal to $\tau(\sigma_{CO_2H} - \sigma_{CO_2^-})$, the increase in ρ on going from water to ethanol as a solvent may be due

[30] For other approaches to the problem of obtaining polar substituent constants for ortho substituents, see D. H. McDaniel and H. C. Brown, *J. Am. Chem. Soc.*, **77**, 3756 (1955); M. S. Newman and S. H. Merrill, *J. Am. Chem. Soc.*, **77**, 5552 (1955); J. D. Roberts and R. A. Carboni, *J. Am. Chem. Soc.*, **77**, 5554 (1955).

largely to the increase in $(\sigma_{CO_2H} - \sigma_{CO_2^-})$. In any case, it seems certain that the σ's for some reaction-center groupings must change with the solvent, since if they did not, any change in solvent would change all ρ's to the same extent (this is not observed), namely, the extent to which τ was changed.

TABLE 4-5. HAMMETT REACTION CONSTANTS FOR ACID-BASE EQUILIBRIA[4]

No.	Acidity constants of	$-\log K_0{}^a$	ρ
1	Benzoic acids in water at 25°	4.20	+1.000
2	Benzoic acids in methanol at 25°	9.40	+1.537
3	Benzoic acids in ethanol at 25°	10.40	+1.957
4	Phenylacetic acids in water at 25°	4.31	+0.489
5	β-Phenylpropionic acids in water at 25°	4.66	+0.212
6	Cinnamic acids in water at 25°	4.44	+0.466
7	p-Phenylbenzoic acids in 50% butyl cellosolve at 25°	5.66	+0.482
8	Phenylboronic acids in 25% ethanol at 25°	9.70	+2.164
9	Benzeneselenic acids in water at 25°	4.79	+0.905
10	Phenols in 95% ethanol at 20–22°	12.80	+2.364
11	Thiophenols in 95% ethanol at 20–22°	9.32	+2.847
12	Anilinium ions in water at 25°	4.63	+2.767

^{a} These are values of $-\log K_0$ for the unsubstituted compounds, not the $-\log K_0$ values required to give the optimum fit to Eq. (4-14).

Comparison of reactions 1, 4, and 5 shows that each additional methylene group interposed between the substituents and reaction center decreases the sensitivity of the equilibrium constants to changes in polar character of the substituents by about 2.2-fold, a factor fairly near that by which an interposed methylene group changes σ^* for a given substituent. From reactions 5, 6, and 7 it may be seen that a chain of sp^2 carbon atoms conducts substituent effects better than a similar chain of saturated carbon atoms. The increased ρ's for phenols, thiophenols, and anilinium ions are at least partly due to the acidic proton being nearer the substituents in these compounds.

In all the cases studied ortho-substituted benzoic acids have been found to be stronger than would be expected from polar factors alone (as measured by Taft's ortho-substituent constants, for example). Thus ortho methyl, ethyl, t-butyl, and hydroxy groups increase the acidity of benzoic acid by 0.29, 0.43, 0.74, and 1.22 pK units, respectively.[31] The effect of the o-alkyl groups is believed due to steric inhibition of resonance.

[31] For these and many other data on acid and base strengths, see H. C. Brown, D. H. McDaniel, and O. Häfliger, in E. A. Braude and F. C. Nachod (eds.), "Determination of Organic Structures by Physical Methods," Chap. 14, Academic Press, Inc., New York, 1955.

Since the carboxy group cannot lie in the same plane as the benzene ring, the substituted benzoic acids are less stabilized by such structures as

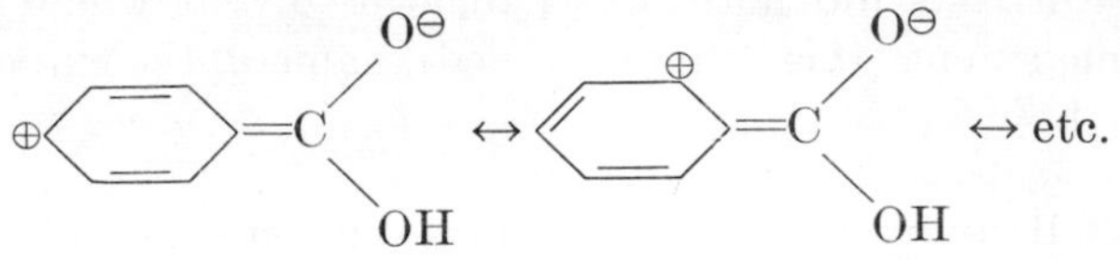

which contribute to the total structure of meta- and para-substituted benzoic acids.   Therefore the loss in this resonance stabilization that accompanies benzoate-anion formation is less for the ortho- than for the meta- and para-substituted compounds.   The larger effect of the smaller hydroxy group appears to be due to stabilization of the anion by hydrogen bonding.

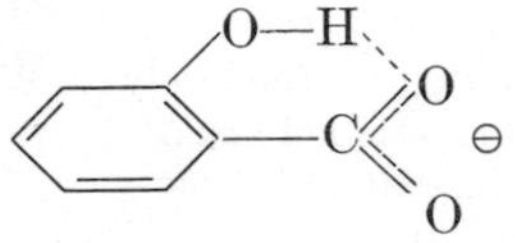

4-5b. Hammett-equation Correlations with Polysubstituted Acids and Bases.   The Hammett equation may be applied to polysubstituted as well as to monosubstituted acids and bases.   This can be done by holding one substituent constant throughout the reaction series.   Thus, for example, ρ is $+0.905$ for the ionization of 4- and 5-substituted 2-nitrobenzoic acids in water at $25°$.[4]   Alternately, the same ρ that is used for the monosubstituted compounds can be used with the *sum* of the σ's for the substituent groups present.   Thus the ionization constant of 3-chloro-5-nitrobenzoic acid in water at $25°$ may be calculated as follows:

$$\log K = \log K_0 + \rho(\sigma_{m-\text{Cl}} + \sigma_{m-\text{NO}_2})$$
$$= -4.20 + 1.08$$
$$= -3.12$$

in good agreement with the observed value, -3.13.   The summation of substituent σ's seems to work well for all 3,5-disubstituted compounds and also for 3,4- and 3,4,5-substituted compounds when the substituents are alkyl groups, halogens, and certain others.   With nitro, dialkylamino, and some other groups, however, deviations attributable to steric inhibition of resonance (cf. Sec. 2-4d) may appear when substituents are ortho to each other or to the reaction center.

The nature of factors that may bring about deviations from the Hammett equation is often quite instructive.   We have already mentioned deviations due to (1) direct resonance interactions between substituent and reaction center, (2) use of the same ρ for meta- as for para-substituted compounds, (3) effects of changes in solvation on the polar character of groups, and (4) steric interactions of substituent groups with each other or with the reaction center.   There is good evidence for two other causes of deviations from the Hammett equation: (5) changes in reaction mechanism within a reaction series and (6) saturation effects. It is obvious how a change in reaction mechanism could complicate the application of the Hammett equation to a rate process.   Specific examples

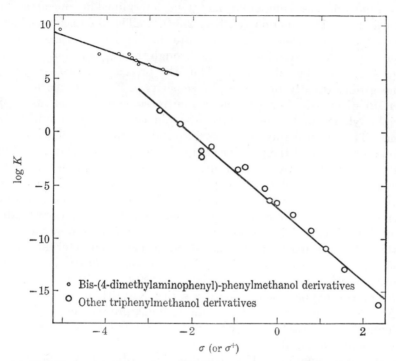

FIG. 4-4. Equilibrium constants for carbonium-ion formation from triarylmethanols vs. σ (or σ^+ for substituents for which σ^+ exists).

of such complications will be described in subsequent discussions of reaction mechanisms (cf. Sec. 11-2d).   A saturation effect is said to exist when two identical substituent groups stabilize a given species less than twice as much as one such group does.   Although some saturation effects are steric in character, there are others for which a polar explanation seems possible.   One such example is found in a Lewis acid-base equilibrium.

$$\text{Ar}_3\text{COH} + \text{H}^+ \rightleftharpoons \text{Ar}_3\text{C}^+ + \text{H}_2\text{O}$$

In Fig. 4-4 values of log K, where

$$K = \frac{[\text{Ar}_3\text{C}^+]}{[\text{H}^+][\text{Ar}_3\text{COH}]}$$

are plotted against the sum of the substituent constants (using σ^+ values for para substituents capable of resonance-electron donation). The small symbols refer to derivatives of malachite green (the carbonium ion derived from 4,4'-bis-dimethylaminotriphenylmethanol), whose pK's were determined in aqueous solution.[32] The large symbols are for triaryl-methanol derivatives that do not contain a p-amino substituent. The pK's of most of these compounds had to be determined in aqueous sulfuric acid using the H_R function (cf. Sec. 2-3d).[33] The points for malachite green derivatives are seen to approximate the line shown, whose slope (ρ) is equal to -1.4. From the best line through the large symbols, however, a ρ of -3.5 may be calculated. Thus the presence of two p-dimethyl-amino groups greatly decreases the sensitivity of pK to the nature of other substituents. For example, a p-nitro substituent changes the pK of triphenylmethanol more than twice as much as it does that of malachite green. This is probably not surprising, since in the malachite green cation a great deal of the positive charge is on the dimethylamino groups, at a considerable distance from any substituent in another ring. This interpretation of the data is complicated by the variation in the extent to which the aromatic rings of different triarylmethyl cations lie in the same plane as the three bonds attached to the central carbon atom. Saturation effects might be observed more often if there were more cases in which K's varying by more than 10^{26}-fold had been determined.

4-5c. *Taft-equation Correlations of Acidity and Basicity.* The Taft equation (4-18) has also been applied successfully to the ionization constants of carboxylic acids. For the series XCO_2H, ρ^* is $+1.721$ and $\log K_0$ is -4.74 in water at 25°. Among the X groups that deviate from the equation are C_6H_5, $C_6H_5CH{=}CH$, and $CH_3CH{=}CH$, whose carboxylic acids are all about 0.6 pK units weaker than would be expected from the Taft equation. This deviation is attributed to stabilization of the acids by resonance interaction between the carboxy group and the carbon-carbon double bond conjugated with it.

Hall has used the Taft equation to correlate the base strengths of amines in which the amino group is not attached directly to an aromatic ring[34] and has thereby helped to explain the anomalous order of base strengths of ordinary aliphatic amines, $NH_3 \ll RNH_2 < R_2NH > R_3N$. In a plot of pK's vs. the sum of the σ^* values for the R groups, the points were seen to approximate three straight lines, one each for primary, secondary, and tertiary amines. The secondary and tertiary amines

[32] M. Gillois and P. Rumpf, *Compt. rend.*, **238,** 591 (1954); O. F. Ginzburg and P. M. Zavlin, *J. Gen. Chem., U.S.S.R. (Eng. Transl.)*, **27,** 747 (1957).

[33] N. C. Deno and A. Schriesheim, *J. Am. Chem. Soc.*, **77,** 3051 (1955); N. C. Deno and W. L. Evans, *J. Am. Chem. Soc.*, **79,** 5804 (1957).

[34] H. K. Hall, Jr., *J. Am. Chem. Soc.*, **79,** 5441 (1957).

were about 1.2 and 2.6 pK units weaker bases than would be expected from the strengths of the primary amines. The fact that the size of the R groups of tertiary amines has no significant effect on their basicity in the cases studied shows that a steric explanation of the order of amine basicities is improbable.[35] It is instead believed that the basicity of ammonia and primary and secondary amines is increased by stabilization of their ammonium ions by hydrogen bonding to the solvent.[36] Trotman-Dickenson has pointed out that such stabilization should increase with an increase in the number of hydrogen atoms attached to nitrogen in the ammonium ion, and he has described other evidence for this point of view. Subsequently, tertiary aliphatic amines have been shown to be stronger bases than the corresponding secondary (and primary) amines toward dinitrophenols in solvents like chlorobenzene, chloroform, and ethylene chloride, which are too weakly basic to hydrogen-bond to ammonium ions appreciably.[37]

PROBLEMS

1. Estimate the value of Hammett's ρ for the reaction

$$ArSCH_3 + CH_3I \rightleftharpoons ArS(CH_3)_2{}^+ + I^-$$

2. For each of the following reaction series tell which para substituent constants for the Hammett equation (σ, σ^+, σ^-, or σ^n) should give the best agreement with the experimental data and why.

(a) $ArCH_3 + NH_2{}^- \rightleftharpoons ArCH_2{}^- + NH_3$
(b) $ArCH_2CO_2H \rightleftharpoons H^+ + ArCH_2CO_2{}^-$
(c) $ArBF_2 + (C_6H_5)_3N \rightleftharpoons ArBF_2{-}N(C_6H_5)_3$

3. Calculate acidity constants for each of the following cases:

(a) m-Nitrobenzoic acid in ethanol at 25°
(b) 4-(4-Nitrophenyl)benzoic acid in 50 per cent butyl cellosolve at 25°
(c) p-Hydroxybenzonitrile in water at 25°
(d) $CH_3SO_2CH_2CO_2H$ in water at 25°

---

[35] Cf. H. C. Brown, H. Bartholomay, Jr., and M. D. Taylor, *J. Am. Chem. Soc.*, **66**, 435 (1944); Ref. 31, pp. 613–614.

[36] A. F. Trotman-Dickenson, *J. Chem. Soc.*, 1293 (1949).

[37] R. P. Bell and J. W. Bayles, *J. Chem. Soc.*, 1518 (1952); R. G. Pearson and D. C. Vogelsong, *J. Am. Chem. Soc.*, **80**, 1038 (1958); J. W. Bayles and A. Chetwyn, *J. Chem. Soc.*, 2328 (1958).

Chapter 5

ACID-BASE CATALYSIS[1]

5-1. Definitions and Examples. Brønsted and coworkers have shown that it is possible to divide acid- (and base-) catalyzed reactions into two categories according to whether the catalysis is kinetically attributable to all the acids (or bases) present in the solution or merely to the conjugate acid (or base) of the solvent. The first type is called a *general* acid- (or base-) catalyzed reaction, and the second, a *specific* acid- (or base-) catalyzed reaction. That is, a reaction in aqueous solution whose rate is merely proportional to the hydronium-ion concentration (and the concentration of reactant, of course) is said to be specific acid-catalyzed (or specific hydronium-ion-catalyzed), whereas in a general acid-catalyzed reaction there will be a term proportional to the concentration of each of the acids in solution.

5-1a. *Hydrolysis of Ethyl Orthoacetate.* An example of a reaction that has been found to be subject to general acid catalysis is the hydrolysis of ethyl orthoacetate

$$CH_3C(OC_2H_5)_3 + H_2O \rightarrow CH_3CO_2C_2H_5 + 2C_2H_5OH$$

which was studied by Brønsted and Wynne-Jones.[2] It is possible to study the reaction under such conditions that the further hydrolysis of the ethyl acetate formed is negligible and the concentrations of the acids and bases present remain essentially constant. Under these conditions it is found the reaction is always of the first order, being pseudounimolecular.

$$v = k \text{ [ethyl orthoacetate]}$$

The extent of acid and/or base catalysis may then be determined by measuring k (dilatometrically) in the presence of varying concentrations of acids and/or bases. Catalysis by bases was shown to be absent by the fact that k had the same value (5.8×10^{-6} sec^{-1}) in 0.1 N and 0.5 N

---

[1] For a much more detailed discussion of acid-base catalysis, see (a) R. P. Bell, "Acid-Base Catalysis," Oxford University Press, London, 1941; (b) R. P. Bell, "The Proton in Chemistry," Cornell University Press, Ithaca, N.Y., 1959.

[2] J. N. Brønsted and W. F. K. Wynne-Jones, *Trans. Faraday Soc.*, **25**, 59 (1929).

aqueous sodium hydroxide solutions at 20°. The rate was then measured in the presence of m-nitrophenol–sodium m-nitrophenolate buffers of various concentrations (enough sodium chloride being present to keep the ionic strength at 0.05 in all cases since the ionization constant of an acid varies with ionic strength). The results obtained are shown in Table 5-1. The rate constants are seen to increase with increasing m-nitrophenol concentration even though the hydronium-ion concentration is kept constant by increasing the sodium m-nitrophenolate concentration to keep the buffer ratio constant. This fact demonstrates

TABLE 5-1. HYDROLYSIS OF ETHYL ORTHOACETATE IN THE PRESENCE OF A
m-NITROPHENOLATE BUFFER

$[NaOC_6H_4NO_2]$	$[HOC_6H_4NO_2]$	$10^9[H_3O^+]$	10^4k
0.00242	0.00242	4.8	1.21
0.00566	0.00566	4.8	1.20
0.0160	0.0160	4.8	1.35
0.0202	0.0202	4.8	1.44
0.00284	0.00384	6.5	1.54
0.00756	0.01025	6.5	1.61
0.0135	0.0183	6.5	1.77
0.00145	0.0031	10.2	2.37
0.00483	0.0103	10.2	2.47
0.0145	0.0309	10.2	2.84

that some of the observed acid catalysis is due to undissociated m-nitrophenol. Plots of k vs. $[HOC_6H_4NO_2\text{-}m]$ for given hydronium-ion concentrations give reasonably straight lines, showing that the part of the reaction due to m-nitrophenol is first-order therein. For hydronium-ion concentrations of 4.8, 6.5, and 10.2×10^{-9} these lines have intercepts of k equal to 1.1, 1.45, and 2.3×10^{-4} sec^{-1}, respectively. A plot of these values of k vs. the hydronium-ion concentration (including the point $k = 0.058 \times 10^{-4}$, where $[H_3O^+] \sim 10^{-13}$, already mentioned) gives another straight line, showing the hydronium-ion-catalyzed portion of the reaction to be first-order in hydronium ion. The rate constant may thus be expressed as the sum of a term for m-nitrophenol, one for hydronium ion, and one for the "uncatalyzed" reaction, probably due to the action of the acid, water.

$$k = k_W[H_2O] + k_h[H_3O^+] + k_n[HOC_6H_4NO_2\text{-}m]$$

At 20°, $k_W = 1 \times 10^{-7}$, $k_h = 2.1 \times 10^4$, and $k_n = 1.7 \times 10^{-3}$ liters mole^{-1} sec^{-1}. The constant k_h is the *catalytic constant* for hydronium ion; k_n, that for m-nitrophenol; and k_W, that for water. In general, for reactions that are subject to general acid catalysis the rate constant is equal to the concentration of every acid present multiplied by its catalytic constant.

$$k = \sum_{i} k_i[A_i] \tag{5-1}$$

5-1b. Other General and Specific Acid- and Base-catalyzed Reactions. Although a specific hydronium-ion-catalyzed reaction is one whose rate constant is simply proportional to the hydronium-ion concentration, this classification depends upon the accuracy of the data used, since for many reactions now classified as specific acid-catalyzed it might be possible to find catalysis by other acids by increasing the accuracy of the kinetic measurements. Brønsted and Wynne-Jones reported that the hydrolysis of ethyl orthopropionate and ethyl orthocarbonate was also general acid-catalyzed, but that the hydrolysis of diethylacetal was specific hydronium-ion-catalyzed. In the latter case the rate constant showed no tendency to increase with the increasing concentration of the weak acid component of the buffer, provided the hydronium-ion concentration was held constant.

Quite analogous to general and specific acid catalysis are general base catalysis, for which

$$k = \sum_{i} k_i[B_i] \tag{5-2}$$

and specific base catalysis, for which

$$k = k_{OH}[OH^-]$$

in aqueous solution. An example of a general base-catalyzed reaction is the decomposition of nitramide.[3] The decomposition of nitrosotriacetonamine appears to be a specific hydroxide-ion-catalyzed reaction.[4]

Many reactions are, of course, catalyzed by both acids and bases. Familiar examples are the hydrolyses of esters, amides, and nitriles. For reactions of this sort either the acidic or the basic catalysis may be general, or both may be. For example, the mutarotation of glucose[5] and the enolization of acetone[6-8] have been found to be general acid- and general base-catalyzed, whereas oxygen exchange between H_2O^{18} and acetone is reported to be general acid-catalyzed but specific base- (hydroxide-ion-) catalyzed.[9]

5-2. Mechanisms of Acid- and Base-catalyzed Reactions. *5-2a. Necessity of Both an Acid and a Base.* If a reaction is truly catalyzed by

[3] J. N. Brønsted and K. Pedersen, *Z. physik. Chem.*, **108**, 185 (1924).

[4] Ref. 1a, p. 77.

[5] J. N. Brønsted and E. A. Guggenheim, *J. Am. Chem. Soc.*, **49**, 2554 (1927).

[6] H. M. Dawson and J. S. Carter, *J. Chem. Soc.*, 2282 (1926).

[7] H. M. Dawson, G. V. Hall, and A. Key, *J. Chem. Soc.*, 2844 (1928).

[8] H. M. Dawson and E. Spivey, *J. Chem. Soc.*, 2180 (1930).

[9] M. Cohn and H. C. Urey, *J. Am. Chem. Soc.*, **60**, 679 (1938).

either an acid or a base, i.e., if the acid or base is not used up as the reaction proceeds, then it follows from a fairly simple argument that both an acid and a base must take a part in the reaction as a whole. Thus the only useful explanation for acid catalysis is to have a proton donated or partially donated to the reactant or some reaction intermediate. However, if the reaction as a whole is not to use up acid, then a proton must also be removed and this removal is the action of a base. Since any reaction mixture that contains an acid capable of donating a proton to the reactant will also contain a base (if only the one resulting from the proton donation) and since a hydroxylic solvent contains a weak acid and base in every solvent molecule, there will always be both an acid and a base present to perform these necessary functions (with the rate depending on their strength).

5-2b. *Mechanisms of Reactions Catalyzed by Acids or by Bases.* Although a reactant may thus require both an acid and a base to assist in its transformation to the product, it does not necessarily follow that both must enter the reaction at or before the rate-controlling step. Under a given set of conditions one can take part in the rate-controlling step and the other enter only into a subsequent rapid step.[10] It is useful to refer to such reactions as being catalyzed by acid *or* base, depending upon which reacts during the rate-controlling portion of the reaction. We shall discuss possible reaction mechanisms for cases of this sort in terms of the transformation of a very weakly basic reactant S to the product P. If the reaction involves the transformation of S to its conjugate acid as its rate-controlling step

$$S + HA \underset{k_{-1}}{\overset{k_1}{\rightleftharpoons}} A^- + SH^+$$

$$SH^+ + A^- \overset{k_2}{\rightarrow} P + HA \qquad k_2 \gg k_{-1} \qquad (5\text{-}3)$$

then the rate equation for the reaction

$$v = k_1[S][HA]$$

will predict general acid catalysis regardless of the nature of the rapid second step of the reaction. A more detailed version of such a rate-controlling protonation of the reactant might suggest that there is a rapid prior equilibrium to form a hydrogen-bonded complex of reactant and catalyzing acid.

$$S + HA \underset{k_{-1}}{\overset{k_1}{\rightleftharpoons}} S\text{---}HA$$

$$S\text{---}HA \underset{k_{-2}}{\overset{k_2}{\rightleftharpoons}} SH^+ + A^- \qquad (5\text{-}4)$$

$$SH^+ + A^- \overset{k_3}{\rightarrow} P + HA \qquad k_3 \gg k_{-2}$$

[10] It might be possible, of course, by changing the conditions, to slow the "subsequent rapid step" sufficiently to make it rate-controlling.

Another possible mechanism involves the rapid reversible transformation of the reactant into its conjugate acid, which then undergoes the first-order rate-controlling step of the reaction. This rate-controlling step may not yield the reaction product(s) directly. It may involve a fission into several fragments, or it may (as shown below) yield a product (PH^+) that is transformed rapidly and inevitably to the final product.

$$S + HA \underset{k_{-1}}{\overset{k_1}{\rightleftharpoons}} SH^+ + A^-$$

$$SH^+ \underset{k_{-2}}{\overset{k_2}{\rightleftharpoons}} PH^+ \qquad (5\text{-}5)$$

$$PH^+ + A^- \overset{k_3}{\rightarrow} P + HA \qquad k_{-1} \gg k_2; \; k_3 \gg k_{-2}$$

For this mechanism the rate equation has the form

$$v = k_2[SH^+] = \frac{k_1 k_2}{k_{-1}} \frac{[S][HA]}{[A^-]}$$

But if the ionization of HA is rapid compared with the reaction in hand, then from the ionization-equilibrium expression it follows that

$$\frac{[HA]}{[A^-]} = \frac{[H_3O^+]}{K_A}$$

By substitution, the rate expression becomes

$$v = \frac{k_1 k_2}{k_{-1} K_A} [S][H_3O^+]$$

showing that the rate depends only on the hydrogen-ion concentration and is not increased by increasing [HA] at constant $[H_3O^+]$. Therefore a mechanism of the type of (5-5) is possible for a specific but not for a general acid-catalyzed reaction.

5-2c. *Mechanisms of Reactions Catalyzed by Acids and Bases.* If both acids and bases take part in the rate-controlling portion of the reaction, certain other mechanistic possibilities appear. For example, the rate-controlling step may be a reaction between the conjugate acid of the reactant and the conjugate base of the acid catalyst.

$$S + HA \underset{k_{-1}}{\overset{k_1}{\rightleftharpoons}} SH^+ + A^-$$

$$SH^+ + A^- \overset{k_2}{\rightarrow} P + HA \qquad k_{-1} \gg k_2 \qquad (5\text{-}6)$$

Here the rate equation is

$$v = k_2[SH^+][A^-] = \frac{k_1 k_2}{k_{-1}} [S][HA]$$

showing that the mechanism may account for general acid catalysis.

Lowry has suggested that the acid and base interact with the reactant simultaneously rather than in two separate steps as shown in (5-6).[11] This does not necessarily require a three-body collision but can instead be accomplished by having one of the catalysts react with a hydrogen-bonded complex of the other and the reactant, e.g.,

$$S + HA \underset{k_{-1}}{\overset{k_1}{\rightleftharpoons}} S\text{----}HA$$

$$B + S\text{----}HA \overset{k_2}{\rightarrow} P + BH^+ + A^- \qquad k_{-1} \gg k_2 \qquad (5\text{-}7)$$

The kinetic equation would be

$$v = \frac{k_1 k_2}{k_{-1}} [S][HA][B] \qquad (5\text{-}8)$$

Now we have seen that the general equation for the pseudo first-order rate constant for general acid-catalyzed reactions (5-1) involves no dependence on base concentration, whereas that for general base-catalyzed reactions (5-2) is independent of acid concentrations. Similarly the pseudo first-order rate constants for reactions that are catalyzed by both acids and bases have almost always been found to give a satisfactory fit to the equation

$$k = \sum^i k_i[A_i] + \sum^i k_i[B_i] \qquad (5\text{-}9)$$

where again no terms involve *both* acid and base. The absence of any termolecular terms in the observed kinetic equations, however, cannot be regarded as compelling evidence that reaction by mechanisms of the type of (5-7) does not occur. These reactions were carried out in the solvent water, which can act as either an acid or a base. Therefore it can be postulated that in all the kinetic-equation terms involving acids water is acting as a base, its constant concentration being absorbed by the catalytic constant, and that in all the terms involving bases water is similarly acting as an acid.

While this argument defends the possibility of mechanism (5-7), the observation of a termolecular term in the kinetic equation of one reaction subject to general acid and base catalysis permits an argument for its probability. Therefore this reaction, the so-called enolization of acetone, will be discussed in some detail.

5-2d. Enolization of Acetone. Lapworth first showed that the bromination of acetone in aqueous solution is a first-order reaction whose rate is proportional to the concentration of acetone but independent of that of bromine.[12] A reasonable interpretation of this fact is the suggestion

---

[11] T. M. Lowry, *J. Chem. Soc.*, 2554 (1927).
[12] A. Lapworth, *J. Chem. Soc.*, **85**, 30 (1904).

that the rate-controlling step of the reaction is the transformation of acetone into the form of its enol or enolate anion, which is then brominated almost instantaneously.

$$CH_3-\overset{O}{\overset{\|}{C}}-CH_3 \rightarrow CH_3-\overset{OH}{\overset{|}{C}}=CH_2 \underset{Br_2}{\overset{fast}{\longrightarrow}} CH_3-\overset{O}{\overset{\|}{C}}-CH_2Br + H^+ + Br^-$$

Lapworth also found the reaction to be accelerated by both acids and bases to an extent proportional to the concentration of the acid and/or base used. The reaction rate, however, remained independent of the bromine concentration so long as any bromine was present. From these facts it would appear reasonable that the formation of the enol is catalyzed by both acids and bases. The suggestion of a rate-controlling enolization has been substantiated by a number of other experimental findings. It has been found that the rate of iodination of acetone not only is similarly independent of the halogen concentration but also proceeds at exactly the same rate as the bromination reaction.[12,13] In Sec. 10-1a it will be seen that for several ketones the rate of the acid-catalyzed or base-catalyzed deuterium exchange is the same as the rate of bromination and also that a similar identity exists between the rates of racemization of optically active ketones of the type RCOCHR′R″ and rates of halogenation.

Let us then consider in more detail the mechanism of the enolization of acetone, as most commonly measured by its halogenation. Dawson and coworkers showed that both the acid and the base catalysis are general and that in an aqueous acetic acid–sodium acetate buffer the pseudo first-order rate constant may be expressed[8]

$$k = 6 \times 10^{-9} + 5.6 \times 10^{-4}[H_3O^+] + 1.3 \times 10^{-6}[HOAc] + 7[OH^-]$$
$$+ 3.3 \times 10^{-6}[AcO^-] + 3.5 \times 10^{-6}[HOAc][AcO^-] \quad (5\text{-}10)$$

The last term, involving both an acid and a base, is of particular interest, of course, in relation to Lowry's hypothesis that the reaction involves the simultaneous action of an acid and a base. It had been argued that this term is much smaller than would be expected from the magnitude of the other terms,[14] but Swain very neatly found the fallacy in this argument.[15] He pointed out that the Lowry termolecular mechanism does not require the part of the reaction rate found proportional to the acetic acid concentration to be due to the action of acetic acid as an acid and water as a base. It may just as well be due to the action of hydronium ion as an acid and acetate ion as a base, since anything found proportional to [HOAc] must also be proportional to the product

[13] P. D. Bartlett, *J. Am. Chem. Soc.*, **56**, 967 (1934).
[14] K. J. Pedersen, *J. Phys. Chem.*, **38**, 590 (1934).
[15] C. G. Swain, *J. Am. Chem. Soc.*, **72**, 4578 (1950).

$[H_3O^+][AcO^-]$. This follows from the ionization equation for acetic acid.

$$K_a = \frac{[H_3O^+][AcO^-]}{[HOAc]}$$

therefore $\qquad K_a[HOAc] = [H_3O^+][AcO^-]$

Similarly the acetate-ion term must contain that part of the rate due to the acid HOAc and the base OH⁻ as well as that due to the acid H_2O and the base AcO⁻, and the first term must be the sum of the $[H_2O][H_2O]$ and the $[H_3O^+][OH^-]$ terms. It seems likely that the part of the reaction due to the [HOAc][AcO⁻] term of Eq. (5-10) proceeds by a Lowry-type termolecular mechanism. Bell and Jones have presented a strong argument, however, that most of the enolization of acetone (and the majority of other acid-base-catalyzed reactions, as well) does not proceed by the termolecular mechanism.[16]

5-2e. *The Acidity of Transition States.* The plausibility of various reaction mechanisms may often be judged by an estimate of the acidity or basicity of various transition states or of various atoms in a given transition state. For example, if a given reaction involves the acid-catalyzed attack of a nucleophilic reagent (B) on an amide to give an intermediate in which the carbonyl carbon has become attached to four atoms, one may wonder whether the acid-catalysis is due to the protonation of the oxygen or the nitrogen atom of the amide. That is, which of the two transition states, I or II, is the more stable?

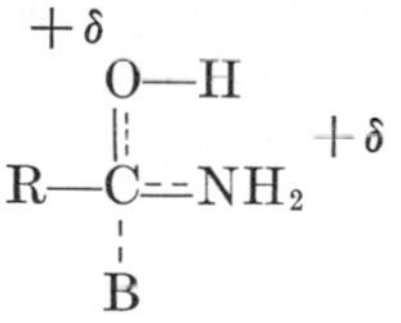

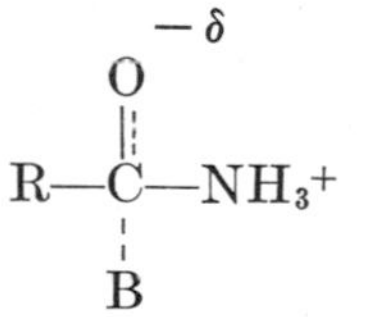

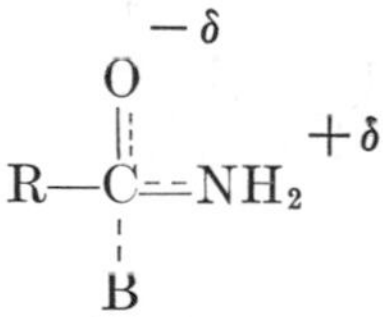

One may state the question in another way by asking whether in III, the transition state for the uncatalyzed attack of the nucleophilic reagent, the oxygen or nitrogen atom is more basic. It seems clear that the oxygen atom in III must be more basic. There is considerable evidence that the oxygen atom in the starting material, an amide, is more basic than the nitrogen.[17] The alkoxide-ion type of oxygen atom in the intermediate IV must be more basic than the amine nitrogen atom. There-

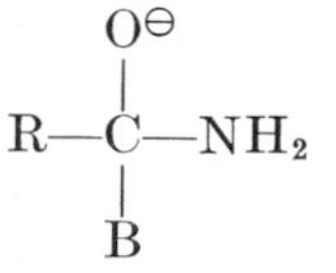

[16] R. P. Bell and P. Jones, *J. Chem. Soc.*, 88 (1953).
[17] Cf. G. Fraenkel and C. Niemann, *Proc. Natl. Acad. Sci. U.S.*, **44**, 688 (1958).

fore the oxygen atom in the intermediate species III must be more basic than the nitrogen, and therefore I must be more stable than II. This type of reasoning may be greatly extended and even used for estimating the rates of acid-catalyzed reactions from the rates of the corresponding base-catalyzed reactions (or vice versa) if it is possible to make satisfactory estimates of the extent to which the various bond cleavages and bond formations that are occurring have progressed in the transition state.

5-3. Rates of Acid- and Base-catalyzed Reactions. *5-3a. Rates of Proton-transfer Reactions.* Since acid- and base-catalyzed reactions involve proton-transfer processes, a survey of what is known about the rates of proton transfers may aid in the mechanistic interpretation of such reactions. Brodskii has pointed out that, in general, proton transfers between atoms with unshared electron pairs are considerably faster than those in which one or both of the atoms involved has no unshared pair.[18] This generalization applies not only to most proton removals from carbon (many are too slow to measure, although a few highly acidic carbon-bound protons are removed very rapidly) but also to silicon-, boron-, and nitrogen-bound protons.[18] Amines and ammonia undergo hydrogen exchange too rapidly to measure by sample-withdrawal-type kinetic studies,[19,20] but the exchange of ammonium ions in acidic solutions may have a half-life of hours.[21-24] An exception has been found in the case of phosphine, whose deuterium exchange with water is quite slow in the absence of an acidic or basic catalyst.[25] Brodskii suggested, quite plausibly, that the ease of hydrogen exchange between atoms with unshared electron pairs indicates a simultaneous four-center-type process.

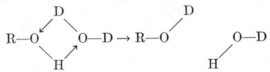

However, subsequent work has shown that hydrogen exchange of hydroxylic and amino compounds proceeds quite rapidly by ionic mechanisms under most conditions. Hydrogen exchange in aqueous methanol,

[18] A. I. Brodskii, *J. Gen. Chem. U.S.S.R. (Eng. Transl.)*, **24**, 421 (1954).

[19] J. Hine and C. H. Thomas, *J. Am. Chem. Soc.*, **76**, 612 (1954).

[20] Cf. R. A. Ogg, Jr., *Discussions Faraday Soc.*, **17**, 215 (1954).

[21] A. I. Brodskii and L. V. Sulima, *Doklady Akad. Nauk S.S.S.R.*, **74**, 513 (1950); *Chem. Abstr.*, **45**, 424a (1951).

[22] L. Kaplan and K. E. Wilzbach, *J. Am. Chem. Soc.*, **76**, 2593 (1954).

[23] C. G. Swain, et al., *J. Am. Chem. Soc.*, **76**, 4243 (1954); **79**, 1084, 1088 (1957).

[24] Cf. S. Meiboom, A. Loewenstein, et al., *J. Chem. Phys.*, **25**, 382 (1956); **27**, 630, 1067 (1957); **29**, 969 (1958).

[25] R. E. Weston, Jr., and J. Bigeleisen, *J. Am. Chem. Soc.*, **76**, 3074 (1954).

ethanol, hydrogen peroxide, and N-methylacetamide has been shown to be very sensitive to acid and base catalysis[26,27]. Hydrogen exchange in liquid ammonia has been found to be powerfully catalyzed by traces of water.[20] Hence most of the hydrogen exchange is due to the reaction of the solvent with a considerably more acidic or basic species. Such information has come from the development of various techniques for measuring the rates of very fast reactions.[28] Some of the rate constants

TABLE 5-2. RATE CONSTANTS FOR PROTON-DONATION REACTIONS

Acid	Base	Solvent	Temperature	k, liters mole^{-1} sec^{-1}
H_3O^+	OH^-	H_2O	25°	1.3×10^{11a}
H_2O	H_2O	H_2O	25	5×10^{-7a}
H_3O^+	$C_6H_5COCO_2^-$	H_2O	25	6×10^{10b}
H_3BO_3	$C_6H_5COCO_2^-$	H_2O	25	6×10^{2b}
H_2O	$C_6H_5COCO_2^-$	H_2O	25	4×10^{-1b}
H_2O	NH_3	H_2O	20	5×10^{5a}
HOAc	H_2O	H_2O	20	8×10^{5a}
HSO_4^-	H_2O	H_2O	20	1.5×10^{9a}
CH_3OH	OH^-	90% CH_3OH	22	3×10^{6c}
CH_3OH	CH_3O^-	90% CH_3OH	22	1×10^{7c}
$CH_3OH_2^+$	CH_3OH	90% CH_3OH	22	9×10^{7c}
H_2O	$EtOH^d$	90% EtOH	22	$8 \times 10^{-1c,d}$
$CH_3NH_3^+$	CH_3NH_2	H_2O	19	3×10^{8e}
$(CH_3)_3NH^+$	H_2O	H_2O	22	6×10^{-2e}
Et_3NH^+	CH_3OH	CH_3OH	0	4×10^{-4f}

[a] M. Eigen et al., *Z. Elektrochem.*, **59**, 483, 986 (1955); *Z. physik. Chem.*, **1**, 176 (1954).
[b] K. Wiesner, M. Wheatley, and J. M. Los, *J. Am. Chem. Soc.*, **76**, 4858 (1954).
[c] Data from Ref. 26.
[d] This is the rate constant for proton *exchange* without the formation of ions, not for simple proton donation.
[e] Data from Ref. 24.
[f] Data from Ref. 23.

that have been reported for proton-transfer reactions between oxygen and nitrogen atoms are listed in Table 5-2. The removal of carbon-bound protons to give carbanions will be discussed in Sec. 10-1c. The reliability of the rate constants in Table 5-2 depends, of course, on the validity of the mechanistic interpretation of the primary data.

Swain and coworkers found the rate of deuterium exchange between

[26] Z. Luz, D. Gill, and S. Meiboom, *J. Chem. Phys.*, **30**, 1540 (1959).
[27] A. Loewenstein, S. Meiboom, et al., *J. Am. Chem. Soc.*, **80**, 2630 (1958); **81**, 62 (1959).
[28] R. P. Bell, *Quart. Revs. (London)*, **13**, 169 (1959).

ammonium ions and alcohols to be proportional to the concentrations of each of the two reactants and, under most conditions, inversely proportional to the acidity of the solution.[23]   Although a reaction mechanism whose rate-controlling step is the attack of alkoxide ion on ammonium ion would give a kinetic equation of the proper form, under many conditions the exchange in acidic solution occurs much more rapidly than alkoxide ions are probably being formed by the autoprotolysis of the solvent.   To explain the observed exchange as a reaction of the alkoxide ions, present in very low concentration, would require an unreasonable rate constant, larger than that for the combination of H^+ and OH^-.   The exchange is more probably initiated by the rapid reversible loss of a proton from a solvating alcohol molecule that is hydrogen-bonded to an acidic hydrogen of the ammonium ion.

$$R_3\overset{\oplus}{N}-H\cdots O-R + ROH \rightleftharpoons R_3N\cdots H-O-R + ROH_2^+$$
$$\underset{H}{|}$$
$$R_3N\cdots H-O-R \rightleftharpoons R_3N + ROH$$

This gives an alcohol-amine complex, held together by a hydrogen bond, which dissociates in the rate-controlling step of the reaction.   Similarly, Meiboom, Loewenstein, and coworkers found that, in addition to the *direct* proton-transfer reaction between methylammonium ions and methylamine, there is a proton transfer that involves a water molecule as well,

$$CH_3\overset{+}{N}H_3 + \overset{H}{\underset{|}{O}}-H + H_2NCH_3 \rightarrow CH_3NH_2 + H-\overset{H}{\underset{|}{O}} + H_3\overset{+}{N}CH_3$$

and, in fact, with trimethylamine, the reaction path involving water is followed to the virtual exclusion of the direct reaction.[24]

5-3b. *The Brønsted Catalysis Equation.*   Since an acid is acting as a proton donor when catalyzing a reaction just as it is when ionizing, it might reasonably be expected that those acids whose ionization constants show them to be the best proton donors at equilibrium should, in general, be the most effective catalysts.   A similar relationship would be expected for base-catalyzed reactions, and in Fig. 5-1 there is a plot of the logarithms of the catalytic constants (k_c) for various bases in the decomposition of nitramide vs. the logarithms of the basicity constants of these bases.[3]

$$H_2NNO_2 \xrightarrow{\text{base}} N_2O + H_2O$$

The data for the monoanions of dicarboxylic acids have been corrected statistically by multiplying the K_b values

$$K_b = \frac{[R(CO_2H)_2][OH^-]}{[HO_2CRCO_2^-]}$$

by 2, since k_c should be related to the basicity of *one* carboxylate anion group whereas K_b is related to the combined proton-donating ability of *two* carboxyl groups. Without this correction the points would be displaced 0.3 log K_b units upward. The K_b values for dicarboxylate dianions have been divided by 2.[29] For a series of related bases, such as carboxylate anions, an excellent approach to a linear relationship is found. The

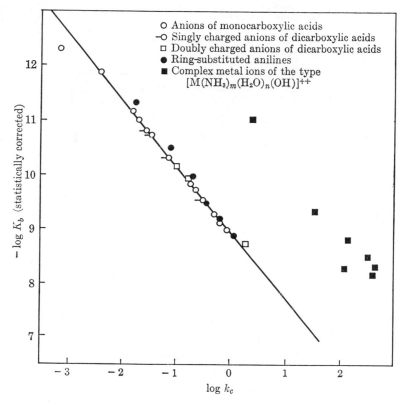

FIG. 5-1. Brønsted plot for basic decomposition of nitramide.

data on ring-substituted anilines give another straight line, which is fairly near the line for carboxylate anions. The deviations shown for certain complex metal ions are unusually large. Fig. 5-1 illustrates what is probably the oldest linear free-energy relationship widely used in organic chemistry, the Brønsted catalysis equation.[3] For a base-catalyzed reaction this relation has the form

$$\log k_c = \beta \log K_b + C \qquad (5\text{-}11)$$

where the constants β and C are the slope and the intercept, respectively.

[29] For a more complete discussion of statistical factors, see S. W. Benson, *J. Am. Chem. Soc.*, **80**, 5151 (1958).

The equation may be expressed exponentially as

$$k_c = G K_b{}^\beta$$

Analogous expressions may be applied to acid-catalyzed reactions.

$$\log k_c = \alpha \log K_a + C \qquad (5\text{-}12)$$

The Brønsted catalysis equation may be discussed in terms of the

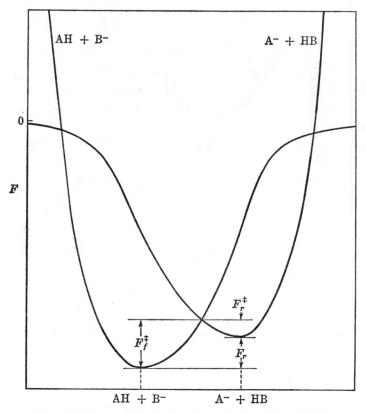

Fig. 5-2. Energy diagram for the reaction of HA with B⁻.

energy curves of Figs. 5-2 and 5-3.[30]   These curves refer to simple acid-base reactions of the type

$$HA + B^- \rightarrow HB + A^-$$

for the cases of several A's and B's.   In all cases the energy content will be referred to that of the completely ionized system $H^+ + A^- + B^-$, whose energy will be defined as zero.   We shall consider only the proton

[30] This treatment is a modification of Bell's (*a*) Ref. 1*a*, chap. 8; (*b*) Ref. 1*b*, chap. 10.

transfer by which the hydrogen-bonded complex A—H····B⁻ gives the complex A⁻····H—B, and we shall neglect any differences in the ease of formation or dissociation of these complexes.   It is convenient to use the position of the proton as a measure of the extent of reaction, but it is

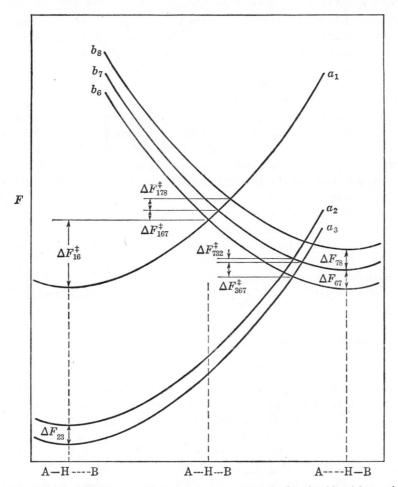

FIG. 5-3. Energy diagrams for the reactions of a series of related acids with a series of related bases.

important to point out that in the transfer of a proton from H—A to B⁻ or in proton removal from H—A, a number of other changes accompany the change in the position of the proton.   Among these are changes in the extent of solvation of A and B, in the distance between A and B, etc. In Figs. 5-2 and 5-3 the abscissa is a measure of the position of the proton. Motion to the right is motion away from A and toward B.   At the point labeled AH + B⁻ the A—H distance is optimum for an A—H bond, and at $A^- + HB$ the B—H distance is optimum. In Fig. 5-2 the curve labeled $AH + B^-$ shows the energy content to be expected of the system if it is considered to consist of HA and B^- and if interactions between these two species are neglected. This curve has a minimum, of course, at the point $AH + B^-$, it rises sharply at smaller A—H distances (to the left), and at large A—H distances the energy approaches zero since the system is becoming $H^+ + A^- + B^-$. Figure 5-2 represents a case in which the base A^- is stronger than the base B^-, and thus the minimum in the $AH + B^-$ curve is lower than that in the $A^- + HB$ curve; i.e., the energy content of the system is lowered more by the combination of the proton with the stronger base. The energy of the system will be affected by interaction of B^- with HA and of A^- with HB, and in the center part of Fig. 5-2, between the two energy minima, the actual system will be more stable (lower in energy) than would be expected from either of the curves shown (cf. Sec. 1-1b). The neglect of such resonance stabilization would give serious errors in the prediction of the activation energy of the forward and reverse reactions ($F_f{}^{\ddagger}$ and $F_r{}^{\ddagger}$) and even the energy of reaction (F_r) from Fig. 5-2. However, it is assumed that neglect of resonance will have about the same effect in several closely related reactions, such as those illustrated in Fig. 5-3, so that there will be no major effect on the *differences* in energies of activation and of reaction.

Figure 5-3 relates to the reaction of the acids HA_1, HA_2, and HA_3 with the bases $B_6{}^-$, $B_7{}^-$, and $B_8{}^-$. Curves a_1, a_2, and a_3 show the energies of the systems HA_1, HA_2, and HA_3 plus any B^-, and curves b_6, b_7, and b_8 show the energies of HB_6, HB_7, and HB_8 plus any A^-. In the specific proton-transfer reaction

$$HA_1 + B_6{}^- \rightarrow A_1{}^- + HB_6$$

the free-energy content of the system will increase (curve a_1) as the A_1—H bond is stretched until the point is reached where curves a_1 and b_6 cross. After this point the system is better described as $A_1{}^- + HB_6$, and the energy content will decrease along curve b_6. Thus the activation energy for the reaction is $F_{16}{}^{\ddagger}$. The minimum in curve b_7 is not so low as that in curve b_6, showing that $B_7{}^-$ is not so basic as $B_6{}^-$. The energy difference ΔF_{67} is a measure of this difference in basicity, and hence of the difference in equilibrium constants for proton removal from HA_1 by $B_6{}^-$ and $B_7{}^-$. The difference in rates for these two processes, governed by the difference in activation energies $\Delta F_{167}{}^{\ddagger}$, will be considerably smaller. For the base $B_8{}^-$, weaker than $B_7{}^-$ by the same factor that $B_7{}^-$ is weaker than $B_6{}^-$ ($\Delta F_{67} = \Delta F_{78}$), the activation energy is higher by an amount $\Delta F_{178}{}^{\ddagger}$, essentially equal to $\Delta F_{167}{}^{\ddagger}$. This proportionality between changes in the energy of reaction and changes in activation energy is equivalent to the Brønsted catalysis equation.

For acids of different types the HA + B⁻ curves need not have the same shapes. This is probably the most common reason for deviations from the Brønsted equation, but even for curves of the same shape certain deviations would be expected when very large changes in the basicities of A and B are made. For example, the rate changes in the reaction of HA_1 with B_6^-, B_7^-, and B_8^- are about half as large as the equilibrium changes ($\Delta F_{167}^{\ddagger} \sim 0.5\ \Delta F_{67}$), so that the Brønsted β [Eq. (5-11)] is about 0.5; but when the B⁻'s are much less basic than A⁻, as in the reaction of HA_3 with B_6^- and B_7^-, the rate changes are almost as large as the equilibrium changes ($\Delta F_{367}^{\ddagger} \sim \Delta F_{23}$), so that β is almost 1.0. On the other hand, in the reaction of HB_7 with the two very strong bases A_2^- and A_3^- the rate change is much smaller than the equilibrium change ($\Delta F_{732}^{\ddagger} \ll \Delta F_{23}$), and thus β is near zero. These qualitative conclusions are in agreement with several experimental observations.[30a,b] For example, the values of α and β in the Brønsted equation are always positive and between 0.0 and 1.0 (when applied to proton transfers); within a series of ketones the β for carbanion formation has been found to decrease with increasing acidity of the ketones; Brønsted plots usually show slight curvature, such that in acid-catalyzed reactions the strongly acidic hydronium ion is less reactive than would be expected and in base-catalyzed reactions the hydroxide ion is less reactive than expected.

Some of the largest deviations from the Brønsted equation are found when the rates of proton transfer from oxygen acids, such as phenol, are compared with the much slower rates of proton transfer from certain carbon acids, such as nitromethane.

Many of the conclusions that we have based on Fig. 5-2 may be rationalized in other terms. Leffler has pointed out that we may rather commonly expect many of the properties of the transition state to be intermediate between the corresponding properties of the reactants and products.[31] For example, if a given change in reaction variables (solvent, substituent, etc.) changes the free energy of the reactants by an amount ΔF_r and that of the products by ΔF_p, the change in the free energy of the transition state $\Delta F_{\ddagger}$ will be intermediate. If the geometry of the transition state is nearer that of the products than that of the reactants (as in the reaction of HA_3 with B_7^-), $\Delta F_{\ddagger}$ will be expected to be nearer ΔF_p than ΔF_r in magnitude. Hammond has discussed this subject in somewhat different terms and has also considered certain cases that are exceptions to the generalizations we have given.[32] Such exceptions, in which reactive intermediates are transformed to less stable products more rapidly than to more stable ones, will be discussed in later chapters (cf. Sec. 10-2c).

[31] J. E. Leffler, *Science*, **117**, 340 (1953).
[32] G. S. Hammond, *J. Am. Chem. Soc.*, **77**, 334 (1955).

5-3c. Acid-Base Catalysis in Heavy Water.[33]   The comparison of the rates of acid- and base-catalyzed reactions in light and heavy water solution is of considerable mechanistic interest, not only because of the possibility of a primary deuterium kinetic isotope effect on the rate of transfer of a hydrogen atom in the rate-controlling step of a reaction, but also because there may be secondary *solvent* isotope effects that are unusually large because the replacement of two protium atoms by deuterium amounts to such a large change in a small molecule like water.   This "solvent" effect may be described briefly by the generalization that the ionic forms of heavy water (D_3O^+ and OD^-) are less stable relative to the neutral molecules than are the analogous species for light water.   Thus the autoprotolysis constant of heavy water is less than one-sixth that of light water.

$$K_{D_2O} = [D_3O^+][OD^-] = 1.54 \times 10^{-15} \qquad \text{at } 25°$$

Similarly, because of the greater difficulty in generating these ionic species, acids and bases have larger ionization constants in light than in heavy water.   The ratio of acidity constants and ionization constants of acids decreases somewhat with increasing acidity, but it is in the range 2 to 5 for most of the acids that have been studied.[34]   Thus, in a reaction whose mechanism is of the general type shown in Eq. (5-5), as in the hydrolysis (specific hydronium-ion-catalyzed) of acetal

$$CH_3CH(OC_2H_5)_2 + H_3O^+ \underset{k_{-1}}{\overset{k_1}{\rightleftharpoons}} CH_3\overset{\overset{H}{\underset{|}{\oplus}}}{C}HOC_2H_5 + H_2O$$
$$\underset{OC_2H_5}{|}$$

$$CH_3\overset{\overset{H}{\underset{|}{\oplus}}}{C}HOC_2H_5 \overset{k_2}{\rightarrow} C_2H_5OH + CH_3\overset{\oplus}{C}HOC_2H_5$$
$$\underset{OC_2H_5}{|} \qquad\qquad\qquad\qquad \overset{H_2O\downarrow fast}{CH_3CHO + \overset{}{C}_2H_5OH + H^+}$$

the reaction is faster in heavy water where the initial equilibrium lies considerably farther to the right.   Just how much faster the reaction goes in heavy water depends on the nature of the reactant, which determines the exact size of $(k_1/k_{-1})^{H_2O}/(k_1/k_{-1})^{D_2O}$ and also the size of the smaller term $k_2^{H_2O}/k_2^{D_2O}$, but a number of reactions believed to proceed by the mechanism given have been found to be from 1.5 to 2.8 times as fast in

[33] For more detailed treatments of this and closely related topics, see F. A. Long and J. Bigeleisen, *Trans. Faraday Soc.*, **55**, 2077 (1959); E. A. Halevi, F. A. Long, and M. A. Paul, *J. Am. Chem. Soc.*, **83**, 305 (1961); C. G. Swain, R. F. W. Bader, and E. R. Thornton, *Tetrahedron*, **10**, 200 (1960); C. A. Bunton and V. J. Shiner, Jr., *J. Am. Chem. Soc.*, **83**, 42, 3207, 3214 (1961).

[34] E. Högfeldt and J. Bigeleisen, *J. Am. Chem. Soc.*, **82**, 15 (1960).

heavy as in light water. In fact, it seems probable that those acid-catalyzed reactions that are considerably faster in light water involve a proton transfer in the rate-controlling step.

Base-catalyzed reactions involving a rapid preequilibrium and no proton transfer in the rate-controlling step are faster, by 45 to 90 per cent, in heavy water. Likewise, deuteroxide ions in heavy water perform nucleophilic attacks on carbon and hydrogen about 20 to 50 per cent faster than do hydroxide ions in light water. On the other hand, base-catalyzed reactions involving a rate-controlling proton transfer in light water but a deuteron transfer in heavy water may be as much as several times as fast in light water. Bases like acetate ion, whose structures are the same in heavy and light water, remove protons about 30 per cent faster in light water.

5-3d. Correlation of Reaction Rates with Acidity Functions.[35] Although there are a number of acid-catalyzed reactions for which the first-order rate constant remains proportional to the concentration of strong acid present, even at acid concentrations of several molar, there are a larger number of reactions for which this is not the case. For some of the first such reactions that were studied carefully the reaction rate was found to be proportional to h_0; that is, a plot of log k vs. H_0 gave a straight line of slope 1. Hammett and Zucker suggested tentatively that those reactions whose rates are proportional to h_0 are examples of specific hydrogen-ion catalysis, having unimolecular reactions of the conjugate acids of the reactants as their rate-controlling steps[36] [as in mechanism (5-5), for example]. The reactions with rates proportional to the *concentration* of catalyzing acid were assumed to be of the general-acid-catalyzed type, with such mechanisms as

$$C_6H_5COCH_3 + H_3O^+ \rightleftharpoons C_6H_5\overset{\overset{OH}{|}}{C}CH_3{}^+ + H_2O$$

$$C_6H_5\overset{\overset{OH}{|}}{C}CH_3{}^+ + H_2O \rightarrow C_6H_5\overset{\overset{OH}{|}}{C}{=}CH_2 + H_3O^+$$

[cf. mechanism (5-3)].

A large number of acid-catalyzed reactions have been studied over a wide enough range of strong acid concentrations to make correlations of rate with h_0 distinguishable from correlations with $[H_3O^+]$. The Hammett-Zucker hypothesis has been quite useful in interpreting the mechanisms of such reactions, particularly when pairs of rather closely related reactions showed different behavior. However, it slowly became obvious

---

[35] This subject has been reviewed in detail by F. A. Long and M. A. Paul, *Chem. Revs.*, **57**, 935 (1957). Cf. L. Melander and P. C. Myhre, *Arkiv Kemi*, **13**, 507 (1959).

[36] L. Zucker and L. P. Hammett, *J. Am. Chem. Soc.*, **61**, 2785, 2791 (1939).

that the hypothesis was not generally reliable. For most of the reactions studied k was proportional to neither $[H_3O^+]$ nor h_0 (although in a number of cases plots of log k vs. H_0 were linear with slopes different from 1). In some cases the deviations seem clearly due to the inapplicability of the steady-state treatment; that is, the reactant is so basic that a significant fraction is present as the conjugate acid. In still more cases, though, the reasons for the deviations seem more fundamental. The H_0 function would not be expected to be any more reliable when applied to rate processes than when applied to equilibrium processes. We have already pointed out (Sec. 2-3c and d) that a number of compounds are known from equilibrium measurements not to be Hammett bases. It seems probable that many of the compounds whose acid-catalyzed reactions have been studied are not Hammett bases.

The theoretical explanation of correlations of reaction rates in strong acid-water mixtures is an important current problem. Bunnett found that plots of log k + H_0 vs. log a_{H_2O}, where a_{H_2O} is the activity of water in the given solvent mixture, are generally linear.[37] The slopes of the lines obtained are certainly useful in providing a method of empirical classification of acid-catalyzed reactions. It also seems clear that Bunnett's approach will be useful in telling whether the mechanisms of various newly studied reactions are the same as those of closely related reactions whose mechanisms have been established by combination of several methods of investigation. This approach may also make it possible to learn the relative number of molecules of water in the reactant and in the transition state, but this point remains to be proved.

PROBLEMS

1. Using appropriate data and equations (e.g., the Hammett equation, the Brønsted catalysis equation), estimate the catalytic constant for the hydrolysis of ethyl orthoacetate by the catalyst phenol in aqueous solution with an ionic strength of 0.05 M at 20°.

2. If the following mechanism for the hydrolysis of ethyl orthocarbonate is correct, what does the observation of general acid catalysis show about the relative magnitudes of any of the rate constants given?

$$(EtO)_4C + HA \underset{k_{-1}}{\overset{k_1}{\rightleftharpoons}} (EtO)_3C\overset{\overset{H}{|}}{\underset{}{O}}\overset{\oplus}{E}t + A^-$$

$$(EtO)_3C\overset{\overset{H}{|}}{\underset{}{O}}\overset{\oplus}{E}t \overset{k_2}{\rightarrow} (EtO)_3C^{\oplus} + EtOH$$

$$(EtO)_3C^{\oplus} + H_2O \xrightarrow[\text{several steps}]{\text{very fast}} (EtO)_2CO + H^+ + EtOH$$

[37] J. F. Bunnett, *J. Am. Chem. Soc.*, **82**, 499 (1960); **83**, 4956, 4968, 4973, 4978 (1961).

Chapter 6

MECHANISMS FOR NUCLEOPHILIC
DISPLACEMENTS ON CARBON

From a number of sources, including particularly a brilliant series of investigations by Hughes, Ingold, and coworkers, there is excellent evidence for the existence of two important mechanisms for nucleophilic substitution reactions at a saturated carbon atom. One, called the S_N2 (substitution, nucleophilic, bimolecular) mechanism,[1] consists of a one-step attack in which one nucleophilic reagent (X) displaces another (Y) from its attachment to a carbon atom. The attacking nucleophilic reagent combines with the carbon atom at the side opposite to that of the nucleophilic group being displaced, thus inverting its steric configuration, as shown in the following equation. No charges are written on X and Y since the mechanism may operate for reactants of a number of different charge types.

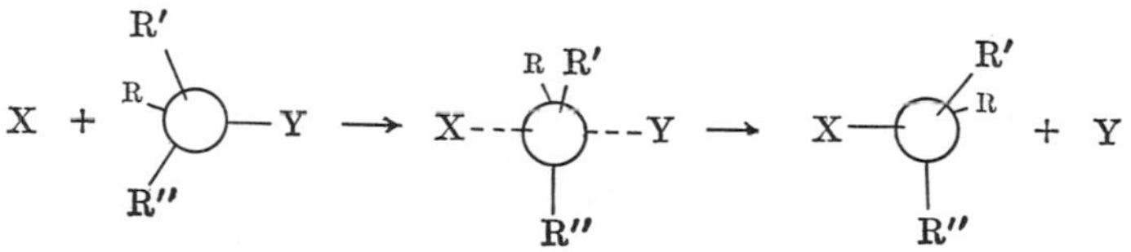

This has also been called the bimolecular-displacement mechanism, the Walden-inversion mechanism, the direct, or one-step, mechanism, etc.

The other mechanism, termed S_N1 (substitution, nucleophilic, unimolecular),[1] consists of a preliminary cleavage to give a carbonium ion, which then reacts (usually quite rapidly) with a nucleophilic reagent.

$$R-Y \rightarrow R^+ + Y$$
$$R^+ + X \rightarrow R-X$$

This is also called the carbonium-ion, or two-step, mechanism.

[1] J. L. Gleave, E. D. Hughes, and C. K. Ingold, *J. Chem. Soc.*, 236 (1935).

It is seen that the S_N2 reaction would be first-order with respect to both [R₃CY] and [X], and the S_N1 reaction as written is first-order with respect to [RY] only. This is the principal difference between the two mechanisms. The rate of reaction of a compound by the S_N2 mechanism is directly proportional to the concentration of the entering nucleophilic reagent. The *true* rate of reaction of RY (this will be different from the *net* rate in those cases where R⁺ recombines with Y) by the S_N1 mechanism is independent of the concentration of the nucleophilic reagent (X) that becomes attached to R in the product (although the extent to which R combines with X as compared with other nucleophilic reagents that are present may depend on the concentration of X). This is because the rate-controlling step precedes the product-controlling step in the S_N1 mechanism, whereas in the S_N2 mechanism the rate-controlling step *is* the product-controlling step. Unfortunately, the dependence of rate on the concentration of the nucleophilic reagent X is not always experimentally detectable. Thus in cases where X is present in considerable excess, e.g., as solvent, the reaction may be pseudounimolecular. For this reason it has not yet been possible to devise, for solvolysis reactions, experimentally applicable definitions of the S_N1 and S_N2 mechanisms that are entirely equivalent to those given above for nonsolvolytic nucleophilic displacements. This problem will be discussed in more detail in Sec. 6-2c.

6-1. The S_N2 Mechanism. The first-order dependence of reaction rate on both X and R₃CY concentrations supports the view that the rate-controlling step of the reaction involves the collision of X and R₃CY. The fact that the attacking group, X, always has an unshared pair of electrons is evidence for the nucleophilic character of the attack. Since the displaced group, Y, also has an unshared pair of electrons, it seems reasonable that the reaction consists of a nucleophilic attack of X on carbon, in which the unshared pair on X forms the new C—X bond and displaces Y with its electron pair. There are good theoretical as well as experimental reasons for believing that this attack occurs on the side of the carbon atom opposite to that to which the displaced group is attached.

6-1a. Racemization of Alkyl Halides by Halide Ions. Some of the best experimental evidence is based on the fact that in acetone solution optically active alkyl iodides are racemized by iodide ion, active alkyl bromides by bromide ion, etc. The rate of racemization is proportional to the concentration of the halide ion,[2] strongly suggesting that the racemization is due to inverting nucleophilic displacements of halogen from the alkyl halide by the halide ions in solution. It is certainly

[2] B. Holmberg, *J. prakt. Chem.*, [2], **88**, 553 (1913).

reasonable to expect nucleophilic substitutions to occur under these conditions, since reactions of the type

$$RBr + I^{\ominus} \xrightarrow{\text{acetone}} RI + Br^{\ominus}$$

$$RCl + I^{\ominus} \xrightarrow{\text{acetone}} RI + Cl^{\ominus}$$

$$RI + Br^{\ominus} \xrightarrow{\text{acetone}} RBr + I^{\ominus}$$

are well known.   In fact, reactions of the type

$$RI + I^{*\ominus} \rightarrow RI^{*} + I^{\ominus}$$

$$RBr + Br^{*\ominus} \rightarrow RBr^{*} + Br^{\ominus}$$

have definitely been shown to occur by the use of radioactive tracers. However, although it might be thus considered proved that inverting nucleophilic displacements are responsible for the observed racemization, it is still possible that only a very small percentage of the nucleophilic displacements occur with inversion, while most give retention of configuration.   Bergmann, Polanyi, and Szabo[3] have shown that this is not likely, since the rate constant for the racemization of *sec*-hexyl iodide by iodide ion in acetone solution is approximately equal to that predicted by extrapolation from the rates with the *sec*-hexyl fluoride, chloride, and bromide.

By studying simultaneously the rate of racemization of 2-octyl iodide by iodide ions in acetone solution and the rate of introduction of the iodide ions (which were labeled with radioactive iodine) into the organic iodide, Hughes and coworkers[4] were able to show that every replacement of organic iodide by an iodide ion produced inversion in the configuration of the compound.   The rate of inversion of 2-octyl iodide molecules may be determined by observing the rate of decrease of the optical rotation of the solution.   The rate of the nucleophilic substitution may be calculated either by determining the rate of increase of radioactivity in the alkyl iodide or the rate of decrease in the radioactivity of the inorganic iodide used.   This calculation is somewhat complicated by the fact that the rate of decay of the radioactive iodine used is comparable to the rate of reaction.   The reaction of optically active α-phenylethyl bromide[5] and

[3] E. Bergmann, M. Polanyi, and A. L. Szabo, *Trans. Faraday Soc.*, **32,** 843 (1936).

[4] E. D. Hughes, F. Juliusberger, S. Masterman, B. Topley, and J. Weiss, *J. Chem. Soc.*, 1525 (1935).

[5] E. D. Hughes, F. Juliusberger, A. D. Scott, B. Topley, and J. Weiss, *J. Chem. Soc.*, 1173 (1936).

α-bromopropionic acid[6] with radioactive bromide ions was studied in an analogous manner. It was again found that the rate constants for inversion and for nucleophilic substitution by bromide ion differed by less than the experimental error. Data on the three reactions described are shown in Table 6-1.

TABLE 6-1. RATE CONSTANTS FOR SUBSTITUTION AND FOR INVERSION OF
ORGANIC HALIDES BY HALIDE IONS IN ACETONE SOLUTION

Compound	Temperature	$10^4 \times$ second-order rate constant for	
		Substitution	Inversion
2-Octyl iodide[4].............	30°	13.6 ± 1.1	13.1 ± 0.1
α-Phenylethyl bromide[5].......	30.2	8.72 ± 0.92	7.95 ± 0.12
α-Bromopropionic acid[6].......	22	5.15 ± 0.50	5.24 ± 0.05

6-1b. *Other Stereochemical Evidence for the Nature of the S_N2 Mechanism.* There is much more stereochemical evidence that points to this same conclusion—that the nucleophilic reagent attacks the carbon from the rear and inverts its configuration while displacing the nucleophilic group originally attached. The experiments with radioactive iodine put this conclusion on a somewhat firmer basis than the other data alone would, since, while we may be very nearly sure that *d*-2-octyl iodide has a configuration enantiomorphic to that of *l*-2-octyl iodide, we are somewhat less certain about the relationship between the configurations of the dextrorotatory isomer of 2-octyl bromide and the dextrorotatory isomer of 2-octyl alcohol (or any other 2-octyl derivative, except the levorotatory bromide). Nevertheless, the data that have been amassed by studying the transformation of RR'CHX compounds to RR'CHY derivatives are rather convincing in their own right and, when combined with the radioisotopic work, make it appear very likely that the mechanism accepted for these specific cases also applies in a very large number of other cases. From data of this type, in fact, Hughes and Ingold have drawn the conclusion that "bimolecular substitutions (S_N2 . . .) are invariably accompanied by steric inversion."[7]

Among the experimental evidence of interest in this regard are the following: The series of reactions below accomplishes the transformation of *d*-benzylmethylcarbinol to its *l* isomer with very little racemization.[8]

[6] W. A. Cowdrey, E. D. Hughes, T. P. Nevell, and C. L. Wilson, *J. Chem. Soc.*, 209 (1938).

[7] W. A. Cowdrey, E. D. Hughes, C. K. Ingold, S. Masterman, and A. D. Scott, *J. Chem. Soc.*, 1252 (1937).

[8] H. Phillips, *J. Chem. Soc.*, **123**, 44 (1923).

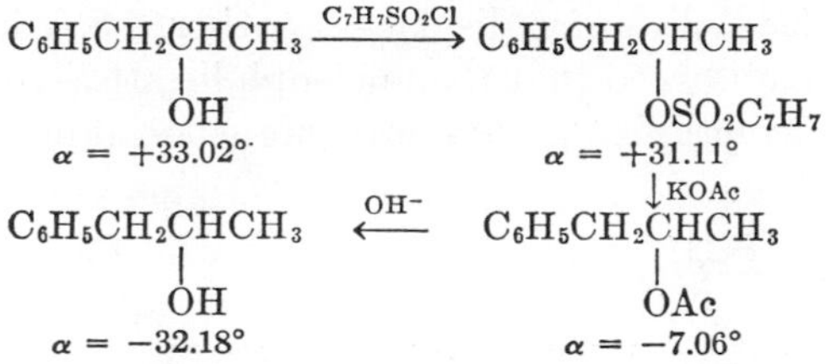

$$C_6H_5CH_2CHCH_3 \xrightarrow{C_7H_7SO_2Cl} C_6H_5CH_2CHCH_3$$

$$\begin{array}{cc}
| & | \\
OH & OSO_2C_7H_7 \\
\alpha = +33.02° & \alpha = +31.11°
\end{array}$$

$$C_6H_5CH_2CHCH_3 \xleftarrow{OH^-} C_6H_5CH_2CHCH_3$$

$$\begin{array}{cc}
| & | \\
OH & OAc \\
\alpha = -32.18° & \alpha = -7.06°
\end{array}$$

The over-all result, inversion, shows merely that during the process an odd number of inverting displacements occurred. However, since the result of the first step is the replacement of the alcoholic hydrogen atom by the p-toluenesulfonyl group ($C_7H_7SO_2$), it seems likely that none of the four valences of the asymmetric carbon atom are changed, and hence its configuration is unaffected. Similarly, it does not seem probable that any change in configuration would accompany the alkaline hydrolysis of the acetate, since (as will be shown in Sec. 12-1a) in the hydrolysis of ordinary esters the cleavage occurs between the oxygen and the *acyl* carbon atoms. Therefore it seems most likely that the inversion occurs during the replacement of the p-toluenesulfonate ion by the acetate ion, a simple example of a nucleophilic substitution reaction.

Similar results have been obtained with *sec*-butyl alcohol, ethyl lactate, lactamide, menthol, *sec*-octyl alcohol, *cis*- and *trans*-2-methylcyclohexanol, α-phenylethyl alcohol, and $C_6H_5CHOHCH_2CO_2H$.[9]

The effect of certain structural features on reactivity, e.g., the inertness of certain halides for which nucleophilic attack from the rear is impossible (see Sec. 7-3a), also constitutes evidence for the S_N2 mechanism.

6-1c. *Nature of the S_N2 Mechanism.* During an S_N2 displacement, the molecule passes through a state (I) in which carbon is pentavalent and has the configuration of a triangular bipyramid.

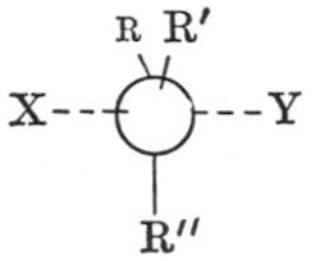

$$X\text{--}\overset{\overset{\textstyle R\ R'}{\diagdown\!/}}{\underset{\underset{\textstyle R''}{|}}{\bigcirc}}\text{--}Y$$

I

It seems reasonable that the partially broken and partially formed C—X and C—Y bonds should be the longer and weaker bonds. The transition state for the S_N2 reaction is usually assumed to be represented by I, but there is no definite assurance that this is not a reactive intermediate preceding or following the transition state.

Substitution with inversion of configuration is not limited to reactions

[9] J. Kenyon, H. Phillips, and coworkers, *J. Chem. Soc.*, 399, 2552 (1925); 2052 (1926); 173 (1933); 1072, 1663 (1935); 303 (1936).

in which one nucleophilic *anion* displaces another.   Snyder and Brewster, for example, have observed that the nucleophilic attack of the acetate ion on the $D(+)$-α-phenethyltrimethylammonium ion produced inversion of configuration.[10]

The attack of the bromide on the α-phenethyldimethylsulfonium ion has also been shown to produce inversion.[11]   If a coulombic effect had predominated, the anions would have been attracted by the positively charged nitrogen and sulfur atoms and attacked from the front side to give retention of configuration.

6-2. The S_N1 Mechanism.   The classical organic chemist proved the existence and structures of various organic species by isolation, elemental analysis, and studies of the properties of pure compounds.   It is therefore natural that he should demand such evidence before accepting such a novel species as a carbonium ion.   In the case of particularly stable carbonium ions, such as the cycloheptatrienyl and the triphenylcyclopropenyl cations (Sec. 1-3) and many triarylmethyl cations, there are pure compounds whose ionic character in the crystalline state seems well established.   Equally convincing evidence for the existence of carbonium ions has been found in other cases where direct measurements were made on carbonium ions even though pure carbonium-ion salts were not isolated.   For example, the dissociation of various organic halides to carbonium ions and halide anions has been studied in various solvents by means of conductivity and spectral measurements.   Even the relatively reactive *t*-butyl cation has been detected directly by spectral measurements on dilute solutions of *t*-butyl alcohol in concentrated sulfuric acid.[12]   The existence of pure crystalline carbonium-ion salts and of direct measurements made on carbonium ions certainly strengthens the evidence that such species are formed as intermediates in many organic reactions in which no such direct evidence can be found.   The fact that the difficulty in isolating pure derivatives and making direct measurements generally increases with the reactivity that would be expected of a carbonium ion is evidence for the validity of our theory of the effects of structure and reaction conditions on the reactivity of various species.   In the case of most S_N1 reactions the evidence for carbonium ions is indirect but still convincing in many instances.

[10] H. R. Snyder and J. H. Brewster, *J. Am. Chem. Soc.*, **71**, 291 (1949).

[11] S. Siegel and A. F. Graefe, *J. Am. Chem. Soc.*, **75**, 4521 (1953).

[12] J. Rosenbaum and M. C. R. Symons, *Proc. Chem. Soc.*, 92 (1959).

6-2a. *Kinetic Evidence.*   Bateman, Hughes, and Ingold have found that in liquid sulfur dioxide solution the initial rate of reaction of benzhydryl chloride is the same with each of the three different nucleophilic reagents: fluoride ion, pyridine, and triethylamine.[13]   It was further shown that the reaction rate was not affected by changes in the concentration of the nucleophilic reagent (to any greater extent than might be attributed to the change in ionic strength).   The most reasonable explanation for these data is that the rate is independent of the nature and concentration of the nucleophilic reagent because the rate-controlling step is the same in all the reactions and does not involve the nucleophilic reagent.   As shown below for reaction with fluoride, the benzhydryl chloride ionizes in the rate-controlling step, and the carbonium ion thus formed then rapidly combines with the fluoride ion.

$$RCl \rightarrow R^+ + Cl^-$$
$$R^+ + F^- \rightarrow RF$$

The independence of the nucleophilic reagent holds, of course, only for the initial stages of the reaction, because as the concentration of chloride ion increases, its ability to compete with the fluoride ion for combination with the carbonium ion (by reversal of the first step) becomes noticeable, and the kinetics become more complicated.   The initial rate of reaction of m-chlorobenzhydryl chloride with fluoride ion in sulfur dioxide was similarly found to be independent of the fluoride-ion concentration[14] and to proceed at the same rate as the reaction with iodide ion and with pyridine.[13]   By supplying chloride ions to compete with the fluoride ions for the intermediate carbonium ions, the addition of tetramethylammonium chloride, it is found, greatly decreases the rate of transformation of the RCl to RF.[14]   Had the reaction mechanism been S_N2, the rate would have been proportional to the fluoride-ion concentration and chloride ions could have exhibited a rate-decreasing effect only by reversing the overall reaction.   This effect could not be very large in the present case since the equilibrium lies well to the right.

In solvolysis reactions,[15] in which the solvent is the nucleophilic reagent, the determination of the reaction mechanism by kinetic measurements is more complicated than in the case described above.   In fact, much of the controversy that has arisen concerning the mechanisms of nucleophilic substitution reactions has centered around solvolyses.

[13] L. C. Bateman, E. D. Hughes, and C. K. Ingold, *J. Chem. Soc.*, 1011 (1940); cf. M. L. Bird, E. D. Hughes, and C. K. Ingold, *J. Chem. Soc.*, 634 (1954); C. A. Bunton, C. H. Greenstreet, E. D. Hughes, and C. K. Ingold, *J. Chem. Soc.*, 642, 647 (1954).

[14] L. C. Bateman, E. D. Hughes, and C. K. Ingold, *J. Chem. Soc.*, 1017 (1940).

[15] For a detailed discussion of the mechanisms of solvolysis reactions, see A. Streitwieser, Jr., *Chem. Revs.*, **56**, 571 (1956).

Despite continued uncertainty about the mechanisms of many such reactions, their practical importance and the volume of work that has been expended in their study make it desirable for us to devote considerable attention to them.

The hydrolysis of t-butyl bromide in 90 per cent acetone[16] would be expected to be a first-order reaction (if studied in dilute enough solution to avoid interference by the hydrogen bromide formed as a by-product) whether the reaction mechanism was S_N1 or S_N2, since even in the latter case the nucleophilic reagent water would be present in such large excess (5.5 M) over the t-butyl bromide that its concentration would not be changed significantly by the reaction of all the bromide. Thus if the second-order rate equation

$$v = k_2[H_2O][t\text{-BuBr}]$$

is followed, then the first-order equation

$$v = k_1[t\text{-BuBr}]$$

where k_1 is equal to $k_2[H_2O]$, must also be followed, since the term $[H_2O]$ is essentially constant. One might consider changing the concentration of water in the belief that there would be no change in the first-order rate constant if the reaction mechanism is S_N1 but that the rate constant would be proportional to the water concentration if the S_N2 mechanism is operating. The rate of solvolysis in 70 per cent acetone shows that tripling the water concentration increases the rate by about fortyfold.[17] This result may easily be explained qualitatively on the basis of the theory of solvent effects described earlier (Sec. 3-1f), since either an ionization to a t-butyl cation and a bromide anion or a nucleophilic attack by water to give a t-butyloxonium ion and a bromide ion would be expected to be accelerated by an increase in the ion-solvating power of the reaction medium. In the absence of a quantitative theory of solvent effects, however, we cannot tell whether the first-order ionization rate constant is being increased by fortyfold or the second-order rate constant for nucleophilic attack by three times as much water is being increased by $13\frac{1}{3}$-fold. Quantitative theories of solvent effects will be discussed in the next chapter, but we can state in advance that these theories seem never to provide as powerful evidence for the nature of the reaction mechanism as certain other types of evidence described in this chapter do in favorable cases. It might seem possible to minimize the solvent effect and still vary reagent concentrations considerably by studying the

---

[16] Among workers who study solvolyses, n per cent acetone (or some other water-soluble solvent) usually means a solvent mixture containing n per cent acetone by volume (measured at a stated temperature) and $100 - n$ per cent water.

[17] L. C. Bateman, E. D. Hughes, and C. K. Ingold, $J.$ $Chem.$ $Soc.$, 960 (1940).

reaction of an alkyl halide with small amounts of water or some other nucleophilic reagent in an inert solvent. A number of such studies have been made and strong disagreement about their interpretation (and even about some of the experimental observations) still exists.[18]

One of the earliest criteria for the mechanism of solvolysis reactions is whether or not the reaction is accelerated by the conjugate base of the solvent. Hughes, for example, found that the solvolysis of t-butyl chloride in aqueous ethanol was not noticeably speeded by the presence of 0.05 M alkali (a rate increase would have had to amount to several per cent to have been noticeable).[19] It was therefore stated that the solvolysis mechanism must have been S_N1 because if the solvent molecules had been attacking the halide by an S_N2 mechanism, the conjugate base of the solvent should attack much more rapidly.[20] It was further pointed out that the solvolyses of methyl and most other primary aliphatic halides, which are believed to react by the S_N2 mechanism, are markedly accelerated by small concentrations of the conjugate base of the solvent. However, it might alternatively be suggested that although hydroxide ions are more reactive than water toward t-butyl chloride, just as they are toward primary halides, with the tertiary halides the difference in reactivity is not so great. Even if hydroxide ions are 20 times as reactive as water molecules, 0.05 M aqueous sodium hydroxide will be only 2 per cent more reactive than pure (55 M) water. Although this criterion for reaction mechanism is thus not completely convincing in its own right, it is usually found to give results in general agreement with those obtained from more rigorous criteria.

Some of the most powerful evidence that some organic halides solvolyze by an S_N1 mechanism has come from studies of the effect of added salts on solvolysis reactions. The hydrolysis of t-butyl bromide in "90 per cent aqueous acetone" (90 per cent acetone by volume) is typical of that of a number of alkyl halides, since the hydrolysis of dilute solutions (less than 0.02 M) follows good first-order kinetics, whereas the kinetic study of the hydrolysis of more concentrated solutions, e.g., 0.1 M, shows that first-order rate "constants" increase as the reaction proceeds.[17] This result is attributed to the increase in ionic strength (from 0.00 to 0.10) resulting from the reaction and to the fact that a reaction of this type would be expected to have a positive ionic strength effect (see Sec. 3-1f). However, in the same solvent the rate "constants" for the hydrolysis of p,p'-dimethylbenzhydryl chloride not only do not increase, they actually

[18] Cf. E. D. Hughes, C. K. Ingold, Y. Pocker, and coworkers, *J. Chem. Soc.*, 1206, 1220, 1230, 1238, 1256, 1265, 1279 (1957); C. G. Swain and E. E. Pegues, *J. Am. Chem. Soc.*, **80,** 812 (1958); and earlier references cited therein.

[19] E. D. Hughes, *J. Chem. Soc.*, 255 (1935).

[20] E. D. Hughes and C. K. Ingold, *J. Chem. Soc.*, 244 (1935).

decrease (from $8.68 \times 10^{-5} \sec^{-1}$ at 8 per cent reaction to $6.7 \times 10^{-5} \sec^{-1}$ at 79.7 per cent reaction).[21] This decrease is ascribed to the recombination of chloride ions with the carbonium ions by the reversal of the first step of the reaction. In the early part of the reaction no appreciable number of carbonium ions combine with chloride ion (step 2 in the scheme below), because the concentration of chloride ion is too small. However, later in the reaction this combination becomes important, and although the alkyl halide is ionizing even more rapidly than at the start (due to the increased ionic strength), so many of the carbonium ions are reverted to RCl that the rate of formation of ROH actually decreases. This decrease is the result of the mass-law effect.

With regard to some of the quantitative aspects of the reaction of carbonium ions with nucleophilic reagents other than solvent, the investigation of the hydrolysis of p,p'-dimethylbenzhydryl chloride in 85 per cent aqueous acetone[21] is of interest. In this work, the effect of the addition of four salts was studied. (1) A "common-ion" salt, lithium chloride. This would be expected to transform many carbonium ions back to RCl and thus slow the rate of disappearance of RCl and of formation of ROH. There would also be an ionic-strength effect working in the opposite direction. (2) Lithium bromide. Although bromide ions would be expected to combine with the carbonium ion, the product RBr is known to be much more reactive than RCl and would rapidly re-form the carbonium ion. Hence the total effect might be much the same as if the RBr had not been formed; that is, the principal effect may be simply an ionic-strength effect. (3) Tetramethylammonium nitrate. Nitrate ions are so weakly nucleophilic (cf. Sec. 7-2) that, at the concentrations used, they should not capture any significant fraction of the intermediate carbonium ions. Hence this salt should exhibit only an ionic-strength effect. (4) Sodium azide. The combination of azide ions with the carbonium ions would yield the stable compound RN_3. Thus, sodium azide would have only an ionic-strength effect on the rate of reaction of RCl but should give a final reaction product containing RN_3 as well as ROH. We then arrive at the reaction scheme

$$RCl \underset{(2)}{\overset{(1)}{\rightleftharpoons}} R^+ + Cl^-$$

$$R^+ + H_2O \overset{(3)}{\rightarrow} ROH + H^+$$

$$R^+ + N_3{}^- \overset{(4)}{\rightarrow} RN_3$$
$$R^+ + Br^- \rightleftharpoons RBr$$

All salts were used in the same concentration, $0.051\ N$. When lithium bromide, sodium azide, and tetramethylammonium nitrate were used, it

[21] L. C. Bateman, E. D. Hughes, and C. K. Ingold, *J. Chem. Soc.*, 974 (1940).

was found that the initial rate of reaction of RCl was, respectively, 1.46, 1.50, and 1.53 times the initial rate in the absence of any added salt. Thus it seems that this concentration of salt increases the initial reaction rate by about 50 per cent.   When 0.051 N lithium chloride was used, however, it was found that the initial reaction rate was only 0.49 times that in the absence of added salt (and hence only about one-third of the reaction rate to be expected at this ionic strength).   The explanation of Bateman, Hughes, and Ingold is that the alkyl chloride ionizes at the expected rate under these conditions but that two out of three of the carbonium ions combine with chloride ion to regenerate the starting material.   Thus ROH is formed only one-third as fast as it would have been if no chloride ions were present.

As evidence for the combination of nucleophilic ions with the carbonium ions it is pointed out that although the RCl disappears at the expected rate in the presence of NaN_3, the product is not all alcohol but is 64 per cent RN_3, indicating that almost two-thirds of the carbonium ions combine with azide ions in this case.   The explanation that this RN_3 is formed by an S_N2 attack of azide ion on RCl seems unlikely, since a reaction of this kind would not interfere with the ionization already occurring in the absence of NaN_3.   Therefore the existence of this additional mode of reaction for RCl should cause a corresponding increase in the *rate* of reaction of RCl.   Actually the rate in the presence of sodium azide does not differ beyond the experimental error from the rate in the presence of equal concentrations of lithium bromide or tetramethylammonium nitrate.

When the alkyl chloride is hydrolyzing in the absence of added anions (other than chloride), the size of the mass effect is governed only by the concentration of chloride ions and by the *relative* magnitudes of k_2 and k_3, the rate constants for the combination of the carbonium ion with chloride ion and water, respectively.   This may be seen from the following analysis.[22]   If we assume that the carbonium ion is so highly reactive an intermediate that its concentration at any time is negligible compared with that of the alkyl chloride or alcohol, then the rate of reaction as measured by formation of chloride or hydrogen ions will simply be equal to the rate of step 3 in the reaction scheme above.   Now the rate of step 3 will be the rate at which carbonium ions are formed—the rate of step 1— multiplied by the fraction of those carbonium ions that react with water.

$$\text{Rate} = k_1[\text{RCl}]F$$

where F is the fraction of carbonium ions reacting with water.   This fraction will be equal to the rate of reaction of carbonium ions with

[22] L. C. Bateman, M. G. Church, E. D. Hughes, C. K. Ingold, and N. A. Taher, *J. Chem. Soc.*, 979 (1940).

water divided by their total rate of reaction. Since water is present in essentially constant excess, step 3 may be treated as a first-order reaction.

$$F = \frac{k_3[R^+]}{k_2[R^+][Cl^-] + k_3[R^+]} = \frac{1}{(k_2/k_3)[Cl^-] + 1}$$

Letting $\alpha = k_2/k_3$,

$$\text{Rate} = \frac{k_1[RCl]}{\alpha[Cl^-] + 1} \tag{6-1}$$

Values for α, the mass-law constant, have been obtained using a form of the above equation (modified in a somewhat complicated way to allow for ionic-strength effects) for several alkyl halides and are listed in Table 6-2.[22]

TABLE 6-2. VALUES OF THE MASS-LAW CONSTANT, α, FOR VARIOUS ALKYL HALIDES IN 85 PER CENT ACETONE[22]

Alkyl halide	α
Triphenylmethyl chloride	~400[a]
p,p'-Dimethylbenzhydryl chloride	68–69
p-Methylbenzhydryl chloride	28–35
p-t-Butylbenzhydryl chloride	20–43
Benzhydryl chloride	10–16
Benzhydryl bromide	50–70[b]
t-Butyl bromide	1–2[b,c]

[a] C. G. Swain, C. B. Scott, and K. H. Lohmann, *J. Am. Chem. Soc.*, **75**, 136 (1953).

[b] With these two alkyl bromides, α is the rate constant for the reaction of the carbonium ion with bromide (rather than chloride) ion divided by the rate constant for its reaction with water. The fact that the value of α for benzhydryl bromide is about five times as large as for benzhydryl chloride shows that the bromide ion is about five times as adept at combining with the benzhydryl carbonium ion as the chloride ion is. This is in agreement with the generalization that bromide ion is usually more nucleophilic than chloride ion (see Sec. 7-2).

[c] This value is estimated from rather small deviations in reaction rate, which may be specific salt effects rather than mass effects. Hence α might even be zero in this case.

It may be noted that the values of α increase with the expected increasing stability of the carbonium ions. This agrees with the suggestion that the less stable and more reactive carbonium ions have a tendency to combine with one of the first molecules they meet, very probably a solvent molecule, whereas the more stable carbonium ions have a long enough average life to be able to show a preference for reacting with a more nucleophilic reagent.

In relation to an investigation of the mechanism of the hydrolysis of t-butyl nitrate, Lucas and Hammett suggested that the effect of dissolved salts may depend in part on their relative affinity for water molecules.[23] The retardation of the hydrolysis of an alkyl chloride produced by lithium chloride might thus be explained by the assumption that lithium chloride requires so much water for solvation that the solvent is rendered essentially less aqueous.   Since it is well known that these hydrolysis reactions proceed more slowly in a solvent mixture containing less water, it appears possible that such an effect might override the positive ionic-strength effect also present.   Benfey, Hughes, and Ingold have presented additional evidence to show that this explanation does not suffice for the data on benzhydryl halides, at least.[24]   It may be seen that the explanation supposes the effect of lithium chloride to be a general action on the solvent, whereas the mass-law explanation attributes it to a specific action on the compound being solvolyzed.   According to the former explanation, if a certain alkyl chloride solvolyzes more slowly in the presence of lithium chloride than in the presence of the same concentration of lithium bromide, then so should the analogous alkyl bromide. It was found, however, that although 0.1 N lithium bromide speeded and 0.1 N lithium chloride slowed the solvolysis of benzhydryl chloride in "80 per cent aqueous acetone," exactly the opposite result occurred with benzhydryl bromide.   Here lithium bromide decreased and lithium chloride increased the rate.   This, of course, is in agreement with the mass-law explanation, which requires that the decrease be the property of a common-ion salt.

Actually the observed effects of added salts on the solvolysis of benzhydryl halides primarily give evidence of the following:

1. The reaction proceeds in at least two steps, with the rate-controlling step preceding the product-controlling step.

2. Any new reaction paths made possible by the addition of non-common-ion salts come after the rate-controlling step.

3. Some step other than the last produces halide ion and is reversible. It might be argued that the S_N2 mechanism

$$RCl + H_2O \rightleftharpoons ROH_2^+ + Cl^-$$
$$ROH_2^+ + H_2O \rightleftharpoons ROH + H_3O^+$$
$$ROH_2^+ + N_3^- \rightarrow RN_3$$

could explain the data.   However, a more detailed analysis shows that such a mechanistic explanation would require the proton-transfer reactions shown to be relatively slow, whereas all the direct observations that have been made show that such proton transfers between oxygen atoms in hydroxylic solvents are very fast (cf. Sec. 5-3a).

[23] G. R. Lucas and L. P. Hammett, *J. Am. Chem. Soc.*, **64**, 1928 (1942).
[24] O. T. Benfey, E. D. Hughes, and C. K. Ingold, *J. Chem. Soc.*, 2488 (1952).

The mass-law effect constitutes some of the most convincing evidence we have for the S_N1 mechanism, but it is large enough to be detected only in the case of rather stable carbonium ions. There appears to be no direct mass-law evidence that simple aliphatic tertiary halides react by the S_N1 mechanism in a hydroxylic solvent, although there is considerable other evidence that they do. Some such evidence will be discussed in the following sections.

6-2b. *Stereochemical Evidence.* The evidence that bimolecular nucleophilic displacement reactions by halide ions in acetone solution proceed by an inverting mechanism has been given in Sec. 6-1a. In order to add to the evidence that this result is general in reactions by the bimolecular mechanism and to learn more about the stereochemical consequences of the S_N1 reaction, Hughes, Ingold, and coworkers carried out an interesting series of investigations. They found that treatment of D-2-octyl bromide with 1 M potassium hydroxide in 60 per cent ethanol (both hydroxide and ethoxide ions would be present in such a solution) gave L-2-octyl alcohol and L-2-octyl ethyl ether with essentially 100 per cent inversion of configuration, as would be expected for an S_N2 reaction.[25] When the active 2-octyl bromide was heated with 60 per cent ethanol in the absence of added base, the solvolysis products obtained were found to be about 62 per cent racemized. There are three possible ways in which this racemization could have occurred. Some of the octyl bromide could have racemized (e.g. by S_N2 attack by the bromide ions produced in the reaction) *before* it solvolyzed. The products could have racemized *after* they were formed. The racemization could have occurred *during* the replacement of bromide. It was found that the octyl bromide did racemize somewhat before it reacted. Unfortunately, however, the possibility of racemization of the products after their formation does not appear to have been considered. Somewhat more convincing results were obtained in a study of the solvolysis of optically active α-phenylethyl chloride.[26] The second-order reactions with sodium methoxide and sodium ethoxide gave ethers with practically pure inversion of configuration, but the first-order solvolyses gave highly (up to 90 per cent) racemized products. Racemization of the reacting α-phenylethyl chloride was shown to be negligible. Racemization of the products after their formation was not investigated, but in at least one case, in which the solution was kept slightly basic throughout the reaction by an amount of alkali too small to bring about the contribution of any significant amount of second-order reaction, it seems unlikely that the product would have racemized.

It has thus been reasonably well demonstrated in at least one case

[25] E. D. Hughes, C. K. Ingold, and S. Masterman, *J. Chem. Soc.*, 1196 (1937).
[26] E. D. Hughes, C. K. Ingold, and A. D. Scott, *J. Chem. Soc.*, 1201 (1937).

(and shown to be quite probable in a number of other cases) that racemization can occur as an integral part of a nucleophilic substitution on carbon. It is therefore clear that some reaction mechanism other than the S_N2 is operative. Since a carbonium ion has sp^2 hybridization at its positively charged carbon, it should be subject to nucleophilic attack from either side. The fact that significant amounts of inversion of configuration accompany the racemization observed in practically all reported studies of the stereochemistry of solvolysis shows that these solvolyses can not be proceeding *entirely* via free planar carbonium ions. It has been suggested that the part of a solvolysis that gives inversion proceeds by the S_N2 mechanism and the part that gives racemization is an S_N1 reaction. Although some of the inversion that is found in a number of solvolyses is no doubt due to reaction by the S_N2 mechanism, it so happens that partial inversion can also be accommodated in a more detailed picture of the S_N1 mechanism. For reasons to be described in Sec. 6-5c, it is believed that those alkyl halides that react by the ionization mechanism do so to give an ion pair as the initial reaction product. In this ion pair the halide ion may still be bonded weakly to the positive carbon atom through its vacant p orbital.[27] In any case, it should be, for a time, close enough to one side of the positive carbon atom to shield it preferentially from nucleophilic attack by the solvent.[28] With unstable, highly reactive carbonium ions, combination with solvent will often occur before the departing ion is very far away; therefore inversion of configuration because of the shielding effect should be, and is generally found to be, more marked than in the case of relatively stable carbonium ions.

6-2c. *Transition between the S_N1 and S_N2 Mechanisms.* Although there are a number of nucleophilic substitution reactions at saturated carbon atoms whose mechanisms are generally thought to be fairly well approximated by our concept of the S_N2 mechanism and a number of others for which the S_N1 mechanism is believed to be a reasonably good representation, there is considerably less agreement about the mechanisms of certain intermediate cases. There are several reasons for this disagreement. There are no widely accepted definitions of the two mechanisms in terms of the operations one carries out in studying such reactions, and it is not clear that such definitions would provide any significant improvement in our understanding. The existing criteria for the nature of the reaction mechanism are not mutually exclusive. An S_N2 reaction must give inversion, but the observation of inversion is not usually accepted as proof of the S_N2 mechanism. There is no demonstrable dependence

[27] W. v. E. Doering and H. H. Zeiss, *J. Am. Chem. Soc.*, **75**, 4733 (1953).
[28] W. A. Cowdrey, E. D. Hughes, C. K. Ingold, S. Masterman, and A. D. Scott, *J. Chem. Soc.*, 1252 (1937).

of the reaction rate on the concentration of nucleophilic reagent in an S_N2 solvolysis.   Conversely, one can imagine a reaction whose rate was proportional to the concentration of the nucleophilic reagent nevertheless proceeding via an intermediate carbonium ion, as would be the case, for example, if RX ionized rapidly and reversibly to an ion pair that underwent nucleophilic attack as the rate-controlling step of the reaction.

Thus the question whether the intermediate cases involve the simultaneous operation of the S_N1 and S_N2 mechanisms or the operation of a single reaction mechanism of intermediate character is to some extent a question of definition.   The author's view of the matter, which follows, differs from some views held by other workers more in method of presentation than in actual content.[29]   In almost all nucleophilic substitution reactions of alkyl halides carried out in ordinary solvents there is an ion or a molecule (the *nucleophile*) containing an atom with an unshared electron pair "behind" the alkyl halide molecule as it begins to react. As the carbon-halogen bond is stretched and the halogen atom acquires more halide-ion character and more solvation, the unshared electron pair of the nucleophile interacts with the orbital on carbon whose electrons are being removed by the departure of the halide ion.   When this interaction is weak, it is called *solvation;* when it is strong, it is called *covalent-bond formation.*   There is no distinct line between solvation and covalent-bond formation, however.   The amount of interaction varies continuously with the extent of reaction and, even at constant extent of reaction, with the nature of the solvent, nucleophile, alkyl halide, and temperature. In S_N2 reactions the amount of such interaction is relatively large, whereas in some S_N1 reactions, such as those of bridgehead bicyclic halides and those involving very stable triarylmethyl cations, there may be, in the transition state, practically no direct back-side interaction at the carbon atom at which substitution occurs.   The gradation of possibilities may be discussed in terms of the following representation of the system at its half-reaction point, where the three R groups have become coplanar with the carbon atom at which substitution is taking place.

$$\text{X---}\underset{\text{R}''}{\overset{\text{R R}'}{\bigcirc}}\text{---Y}$$

In an extreme S_N1 reaction this will represent a relatively stable intermediate carbonium ion whose positive charge is well distributed over the

[29] For other views on this subject, with varying amounts of agreement, see S. Winstein, E. Grunwald, and H. W. Jones, *J. Am. Chem. Soc.*, **73**, 2700 (1951); M. L. Bird, E. D. Hughes, and C. K. Ingold, *J. Chem. Soc.*, 634 (1954); V. Gold, *J. Chem. Soc.*, 4633 (1956).

three R groups as well as the α-carbon atom.   In this case the departing halide ion Y and the nucleophile X are so loosely attached to the α-carbon atom that they are easily replaced by other molecules and ions.   This rate of exchange of solvating species is so fast compared with the rate at which the carbonium ion combines with X that it is unlikely that the combination will occur before X and Y have been replaced.   (Recombination with Y before its replacement is usually not experimentally detectable.)   Under such conditions it seems likely that the species with which the carbonium ion finally does combine will be as likely to be on one side as the other, so that a racemic product would be expected.   However, if the R groups are less capable of supplying electrons to the carbon atom, the "half-reaction species" will be a less stable intermediate in which X and Y are more firmly attached, and X may often form a covalent bond to carbon before Y has been replaced by any other species, so that a partly inverted product is obtained.   With a greater amount of bond formation at the half-reaction point, and therefore presumably in the transition state, the stability and hence the reaction rate will show a greater dependence on the ability of X to use its unshared electron pair to form a bond.   In a typical S_N2 reaction the half-reaction species is believed to be a transition state instead of a reactive intermediate; at least X and Y are so tightly held that they are never exchanged, and thus a completely inverted product is obtained.   Factors affecting the relation between the extent of bond formation to X and Y and the nature of X, Y, the R groups, the solvent, etc., will be discussed in Chap. 7.

6-3. The S_Ni Mechanism.   The reactions of optically active alcohols with nonmetal halides and oxyhalides usually result in inversion of configuration accompanied by varying amounts of racemization.[30]   This is the result often obtained on reaction with thionyl chloride in the presence of pyridine,[31] but under some other conditions predominant retention of configuration is observed.[32]   Inversion and racemization can be rationalized on the basis of S_N2 and S_N1 mechanisms, but a somewhat different reaction path, termed the S_Ni mechanism (substitution, nucleophilic, internal) by Hughes, Ingold, and coworkers,[33] is needed to explain retention.   The reaction of an alcohol with thionyl chloride is apparently initiated by the nucleophilic attack of oxygen on sulfur and the loss of hydrogen chloride to give an intermediate chlorosulfite.   A base like

[30] A. McKenzie and G. W. Clough, *J. Chem. Soc.*, **97,** 1016, 2564 (1910); R. H. Pickard and J. Kenyon, *J. Chem. Soc.*, **99,** 45 (1911).

[31] J. Kenyon, H. Phillips, and F. M. H. Taylor, *J. Chem. Soc.*, 382 (1931).

[32] A. McKenzie and G. W. Clough, *J. Chem. Soc.*, **103,** 687 (1913).

[33] W. A. Cowdrey, E. D. Hughes, C. K. Ingold, S. Masterman, and A. D. Scott, *J. Chem. Soc.*, 1252 (1937); E. D. Hughes, C. K. Ingold, and I. C. Whitfield, *Nature*, **147,** 206 (1941).

pyridine transforms the hydrogen chloride into a salt whose chloride ions may attack the chlorosulfite by an S_N2 process.

$$Cl^- + \overset{\diagdown}{\underset{\diagup}{-}}C\!-\!OSOCl \rightarrow Cl\!-\!C\overset{\diagup}{\underset{\diagdown}{-}} + SO_2 + Cl^-$$

The mechanism for producing chloride with retention, that is, the S_Ni mechanism, as Cram has pointed out, seems to be closely related to the S_N1 mechanism, since (1) the S_Ni mechanism best competes with other reaction paths when a relatively stable carbonium ion can be formed and (2) the reaction proceeds more rapidly in better ion-solvating media.[34] In fact, Cram makes the plausible postulate that the reaction begins, like a typical S_N1 reaction, with the formation of an ion pair. However, in the case of a chlorosulfite the anion formed is SO_2Cl^- which may rapidly decompose to sulfur dioxide and a chloride ion. Recombination of the ion pair then gives the chloride, with the same configuration as the original alcohol.

$$\overset{\diagdown}{\underset{\diagup}{-}}C\!-\!OSOCl \rightarrow \overset{\diagdown\;\diagup}{\underset{|}{C^+SO_2Cl^-}} \rightarrow SO_2 + \overset{\diagdown\;\diagup}{\underset{|}{C^+Cl^-}} \rightarrow \overset{\diagdown}{\underset{\diagup}{-}}C\!-\!Cl$$

A particularly careful study of the reaction has been carried out by Lewis and Boozer. They were the first to prove the intermediacy of alkyl chlorosulfites in the S_Ni by actually decomposing compounds of this type and observing retention of configuration.[35] They also showed that the reaction is kinetically first-order, as demanded by the S_Ni mechanism. The decomposition in dioxane solution was found to give almost complete retention of configuration. This may result from back-side displacement of sulfur dioxide from the ion pair by dioxane, followed by a second back-side displacement of dioxane from the dioxane-solvated carbonium ion by chloride ion. The variation of the reaction rate and steric result with the nature of the solvent are also discussed.[36]

Phosgene reacts with alcohols in a manner somewhat analogous to that of thionyl chloride. Alkyl chloroformates are formed, which may decompose alone to give chlorides with retention of configuration or in the presence of pyridine to give inverted chlorides.[37]

The formation of a bromide with retained configuration in the reaction of hydrogen bromide with several phenylalkyl carbinols is another example of the S_Ni reaction,[38] probably involving a hydrogen-bonded complex between the acid and alcohol and an intermediate ion pair.

[34] D. J. Cram, *J. Am. Chem. Soc.*, **75**, 332 (1953).

[35] E. S. Lewis and C. E. Boozer, *J. Am. Chem. Soc.*, **74**, 308 (1952).

[36] C. E. Boozer and E. S. Lewis, *J. Am. Chem. Soc.*, **75**, 3182 (1953).

[37] M. B. Harford, J. Kenyon, and H. Phillips, *J. Chem. Soc.*, 179 (1933); K. B. Wiberg and T. M. Shryne, *J. Am. Chem. Soc.*, **77**, 2774 (1955).

[38] P. A. Levene and A. Rothen, *J. Biol. Chem.*, **127**, 237 (1939).

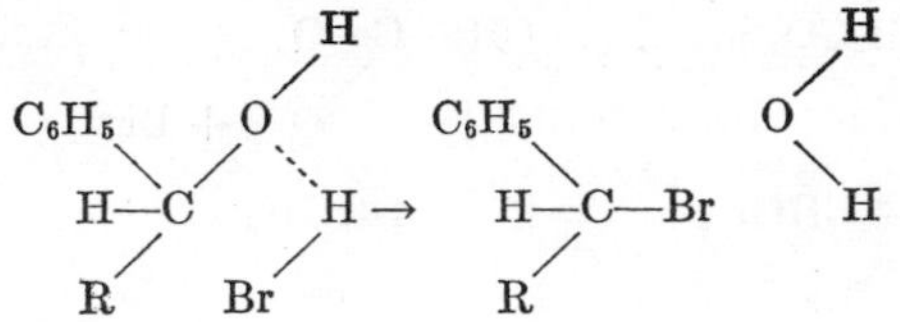

In addition to examples of the type described, there appears to be a somewhat different mechanism for substitution with retention of configuration due to the front-side attack. It bears the same relation to those $S_N i$ reactions proceeding by ionization to an ion pair that the $S_N 2$ mechanism bears to the $S_N 1$. Cases of this type involve a one-step nucleophilic displacement by a group that is attached to the carbon atom attacked only through the atom displaced. The Stevens rearrangement is of this type. In one example, phenacyl-α-phenethyldimethylammonium (II) salts rearrange to α-dimethylamino-β-phenylbutyrophenone (III) in the presence of alkali. The following mechanism[39]

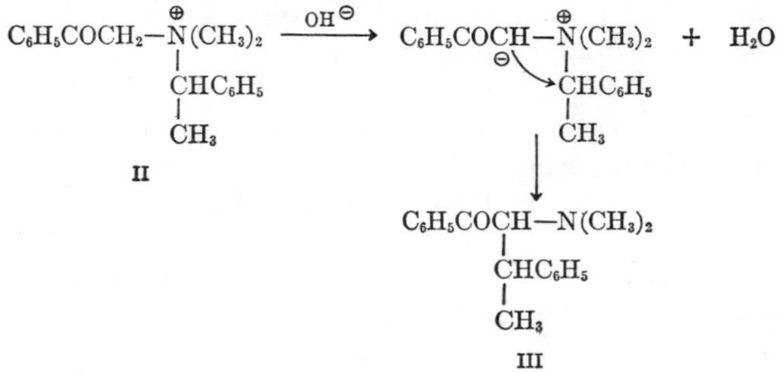

is in agreement with the evidence of Stevens that the reaction is kinetically first-order in quaternary salt and intramolecular,[40] and the observation of Brewster and Kline that the α-phenethyl group migrates with retention of configuration.[41] The mechanism of reactions of this type has been discussed in some detail by Hauser and Kantor.[39]

6-4. Participation of Neighboring Groups. We have discussed under $S_N i$ reactions those nucleophilic displacements by groups that are connected to the carbon atom attacked only through the atom displaced. There are probably a larger number of intramolecular displacements in which the nucleophilic group is not connected through the atom displaced. Some examples of reactions of this type are

[39] H. B. Watson, "Modern Theories of Organic Chemistry," 2d ed., p. 205, Oxford University Press, London, 1941; C. R. Hauser and S. W. Kantor, *J. Am. Chem. Soc.*, **73**, 1437 (1951).

[40] T. Thomson and T. S. Stevens, *J. Chem. Soc.*, 55 (1932); T. S. Stevens, *J. Chem. Soc.*, 2107 (1930).

[41] J. H. Brewster and M. W. Kline, *J. Am. Chem. Soc.*, **74**, 5179 (1952).

$$
\begin{array}{c}
CH_2CO_2{}^- \\
\diagup \\
NH \qquad\qquad\qquad NH \qquad\qquad O \quad +\ Br^- \\
\diagdown \qquad\qquad\qquad \diagdown \qquad \diagup \\
O{=}C{-}CHCH_3 \qquad O{=}C{-}{-}CHCH_3 \\
\ | \\
Br
\end{array}
$$

$$
\begin{array}{c}
OH \qquad\qquad O^- \\
| \qquad\qquad \overset{OH^-}{\ } \ | \\
CH_2{-}CH_2Cl \xrightarrow{\ OH^-\ } CH_2{-}CH_2Cl \rightarrow CH_2{-}CH_2 + Cl^-
\end{array}
$$

Intramolecular substitution reactions of this type usually occur much more rapidly than the analogous intermolecular reactions, since a nucleophilic reagent located in the same molecule within five or six atoms of the carbon atom at which displacement occurs will necessarily spend more of its time in a position suitable for nucleophilic attack than if it were a part of a different molecule. Like the two examples shown, many "neighboring-group" displacement reactions are simply intramolecular instances of reactions well known intermolecularly. However, because of the great facility of cyclizations of this type, many neighboring-group reactions have few, if any, intermolecular analogs. Some of the immediate products of neighboring-group displacements are stable compounds, but others may be very reactive intermediates because of strain present in their small rings. Even ethylene oxide, although capable of being isolated, is a vastly more reactive compound than a typical ether.

6-4a. *The Neighboring Carboxylate Anion.* The neighboring-group participation of the carboxylate anion has been investigated considerably, and some of the most interesting data have been obtained for cases in which the carboxylate group is attached directly to the reaction center.

Optically active α-bromopropionic acid and its derivatives have been studied by several investigators. Cowdrey, Hughes, and Ingold[42] found that the reactions of the methyl ester are normal, both the second-order reaction with methoxide ion and the first-order methanolysis giving inversion of configuration. In fact, when corrections are made for the racemization of the unreacted ester and the product by methoxide ion (probably due to carbanion formation) and for the racemization of the unreacted ester by inverting displacements by bromide ion, etc., the inversion appears to be complete. In concentrated methanolic sodium methoxide solution, the α-bromopropionate anion undergoes a second-order reaction with methoxide ion to give an α-methoxypropionate anion of inverted configuration. However, in dilute alkaline solution in methanol the decomposition of the α-bromopropionate anion is first-order, and the α-methoxypropionate anion formed has the same configuration as the starting material. These results are shown in the reaction scheme below. The letters D and L refer to stereochemical configuration.

[42] W. A. Cowdrey, E. D. Hughes, and C. K. Ingold, *J. Chem. Soc.,* 1208 (1937).

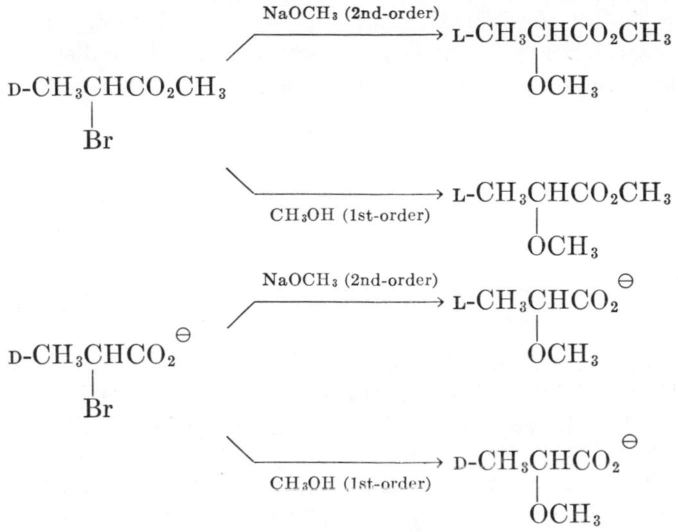

$$\text{D-CH}_3\text{CHCO}_2\text{CH}_3 \quad\begin{array}{c}\xrightarrow{\text{NaOCH}_3\ \text{(2nd-order)}}\ \text{L-CH}_3\text{CHCO}_2\text{CH}_3\\ |\\ \text{OCH}_3\end{array}$$

with Br substituent:

$$\xrightarrow[\text{CH}_3\text{OH (1st-order)}]{}\ \text{L-CH}_3\text{CHCO}_2\text{CH}_3\ \ |\ \text{OCH}_3$$

$$\text{D-CH}_3\text{CHCO}_2^{\ominus} \quad \xrightarrow{\text{NaOCH}_3\ \text{(2nd-order)}}\ \text{L-CH}_3\text{CHCO}_2^{\ominus}\ \ |\ \text{OCH}_3$$

$$\xrightarrow[\text{CH}_3\text{OH (1st-order)}]{}\ \text{D-CH}_3\text{CHCO}_2^{\ominus}\ \ |\ \text{OCH}_3$$

In the case of hydroxylation the results found are like those described for methoxylation above. The hydrolysis of the acid (in 0.5 N aqueous H_2SO_4 to suppress the ionization) gives inversion, as does the second-order reaction of the anion with hydroxide ion. However, the first-order hydrolysis of the anion gives retention of configuration.

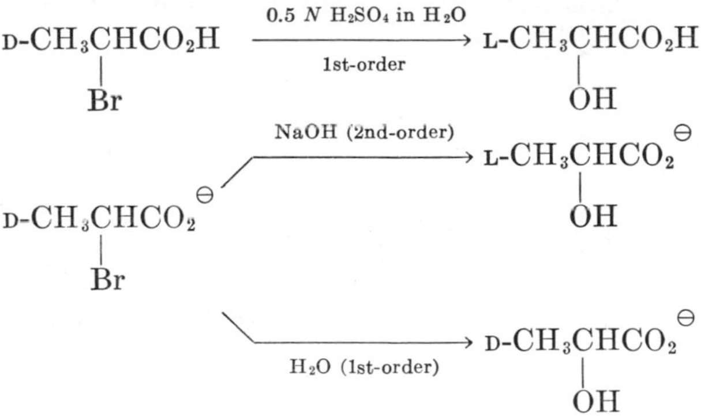

$$\text{D-CH}_3\text{CHCO}_2\text{H}\ (\text{Br}) \quad \xrightarrow[\text{1st-order}]{0.5\ N\ \text{H}_2\text{SO}_4\ \text{in}\ \text{H}_2\text{O}}\ \text{L-CH}_3\text{CHCO}_2\text{H}\ \ |\ \text{OH}$$

$$\text{D-CH}_3\text{CHCO}_2^{\ominus}\ (\text{Br}) \quad \xrightarrow{\text{NaOH (2nd-order)}}\ \text{L-CH}_3\text{CHCO}_2^{\ominus}\ \ |\ \text{OH}$$

$$\xrightarrow[\text{H}_2\text{O (1st-order)}]{}\ \text{D-CH}_3\text{CHCO}_2^{\ominus}\ \ |\ \text{OH}$$

It appears that the most probable explanation for these results is that all the reactions proceed by the S_N2 mechanism except the first-order decompositions of the anions. The rate-controlling step in these cases appears to be the nucleophilic displacement of the bromide ion by the carboxylate anion of the same molecule.[43] Thus there is formed,

[43] S. Winstein and H. J. Lucas, *J. Am. Chem. Soc.*, **61**, 1576 (1939); S. Winstein, *J. Am. Chem. Soc.*, **61**, 1635 (1939); the same mechanism was proposed for a different compound by C. M. Bean, J. Kenyon, and H. Phillips, *J. Chem. Soc.*, 303 (1936). Cowdrey, Hughes, and Ingold prefer to consider the reaction an ionization to a carbonium ion that is stabilized in a tetrahedral configuration by the negative charge of the carboxylate group.

with inversion of configuration, an α-lactone.   This very reactive intermediate may then rapidly undergo a second nucleophilic displacement by solvent to give the final product with retention of configuration.

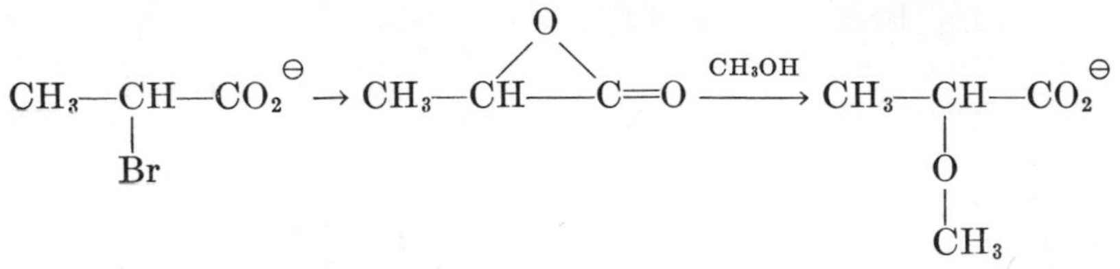

$$CH_3\!-\!CH\!-\!CO_2^{\ominus} \rightarrow CH_3\!-\!CH\!-\!\!\overset{\displaystyle O}{\overset{\displaystyle /\ \ \backslash}{}}\!\!C\!=\!O \xrightarrow{\ CH_3OH\ } CH_3\!-\!CH\!-\!CO_2^{\ominus}$$

with Br below the left CH and $O\!-\!CH_3$ below the right CH.

The formation of an intermediate of some sort is also indicated by Grunwald and Winstein's discovery of a mass-law effect in the reaction. That is, although the reaction is speeded by an inert salt such as perchlorate, it is slowed by a bromide, since bromide ions may react with the α-lactone reconverting it to α-bromopropionate ion.[44]   The formation of reactive intermediate α-lactones from α-halo acids is quite credible in view of the reported isolations of β- and γ-lactones from reactions of β- and γ-halo acids.

6-4b. *The Neighboring Hydroxyl Group.*   The neighboring hydroxyl group as such has not been studied greatly, but reactions in alkaline solution due to its equilibrium transformation to an alkoxide anion have received considerable attention.   Bartlett has shown that the action of alkali on *trans*-2-chlorocyclohexanol yields cyclohexene oxide, apparently by the mechanism[45]

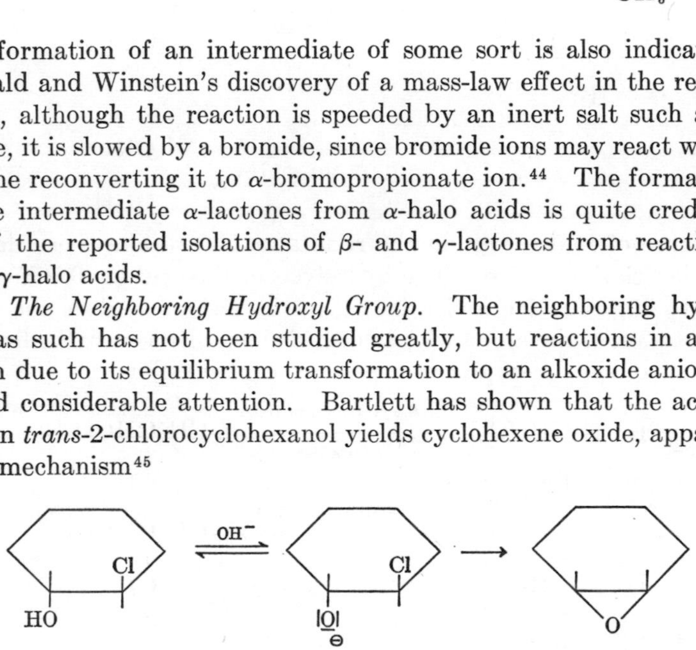

The cis compound reacts less than one-hundredth as rapidly with alkali,[46] although there is no reason to believe that its alkoxide anion should be formed significantly less readily.   The lack of reactivity is almost certainly due to the inability of the alkoxide ion to attack the *rear* of the adjacent carbon atom without subjecting the molecule to considerable strain.   Under the somewhat more vigorous conditions required for reaction *cis*-2-chlorocyclohexanol yields cyclohexanone.

6-4c. *The Neighboring RS Group.*   One compound with a neighboring RS group that has received considerable study is mustard gas (β,β'-

[44] E. Grunwald and S. Winstein, *J. Am. Chem. Soc.*, **70**, 841 (1948).

[45] P. D. Bartlett, *J. Am. Chem. Soc.*, **57**, 224 (1935).

[46] T. Bergkvist, *Svensk Kem. Tidskr.*, **59**, 215 (1947).

dichlorodiethyl sulfide).[47]   The hydrolysis of this compound in its initial stages is purely first-order in mustard gas and independent of the concentration of added alkali.   The rate constants decrease with time because the reaction is slowed by the chloride ions formed.   When thiosulfate ions or certain other nucleophilic reagents are added to the reaction mixture, they react quantitatively but without changing the rate of reaction of the mustard gas.   These data, of course, imply that the mustard gas reacts slowly to give a reactive intermediate that may react with (1) water, (2) chloride ion to regenerate the starting material, or (3) any other nucleophilic reagent present.   The fact that the intermediate has the ability to react with small amounts of certain nucleophilic reagents in preference to the water, present in a much larger concentration, shows that the intermediate has some stability (see Sec. 6-2a).   There is no reason to expect the primary carbonium ion, $ClCH_2CH_2SCH_2CH_2^{\oplus}$, to have this much stability or to be formed in a first-order solvolysis reaction at the rate found for mustard gas (*very* much faster than that for typical primary chlorides).   However, an intramolecular attack by the sulfur atom (whose nucleophilicity in intermolecular reactions is well known) displacing the chloride ion seems a reasonable rate-controlling step, especially since the three-membered-ring sulfonium ion might well be expected to be a reactive intermediate.[47,48]

$$ClCH_2CH_2\overset{\frown}{S}CH_2CH_2Cl \rightleftharpoons ClCH_2CH_2\overset{\oplus}{S}\underset{CH_2}{\overset{CH_2}{<}} \;+\; Cl^-$$

$$\swarrow 2H_2O \qquad\qquad \downarrow S_2O_3^{=}$$

$$H_3O^+ \;+\; ClCH_2CH_2SCH_2CH_2OH \qquad\qquad ClCH_2CH_2SCH_2CH_2SSO_3^-$$

Intermediate substituted ethylene sulfonium ions seem to offer the most probable explanation for a number of rearrangements observed with hydroxy and halo sulfides.   For example, Fuson, Price, and Burness[49] have found that the reaction of ethyl 1-hydroxy-2-propyl sulfide with hydrochloric acid yields a rearranged chloride.

$$\underset{\displaystyle C_2H_5SCHCH_2OH}{\overset{\displaystyle CH_3}{\overset{|}{}}} \xrightarrow{HCl} \underset{\displaystyle C_2H_5SCHCH_2OH_2^{\oplus}}{\overset{\displaystyle CH_3}{\overset{|}{}}}$$

$$\downarrow$$

$$\underset{\displaystyle \underset{Cl}{\overset{|}{C_2H_5SCH_2CHCH_3}}}{} \xleftarrow{Cl^-} C_2H_5-\overset{\oplus}{S}\underset{CH_2}{\overset{CHCH_3}{<}}$$

[47] R. A. Peters and E. Walker, *Biochem. J.* (*London*), **17**, 260 (1923); A. G. Ogston, E. R. Holiday, J. St. L. Philpot, and L. A. Stocken, *Trans. Faraday Soc.*, **44**, 45 (1948); P. D. Bartlett and C. G. Swain, *J. Am. Chem. Soc.*, **71**, 1406 (1949).

[48] Cf. F. E. Ray and I. Levine, *J. Org. Chem.*, **2**, 267 (1937).

[49] R. C. Fuson, C. C. Price, and D. M. Burness, *J. Org. Chem.*, **11**, 475 (1946).

6-4d. *Neighboring Amino Groups.*   In view of the fact that the nucleophilicity of the amino group is even more familiar than that of the sulfide group, it is not surprising that similar rearrangements have been found in the preparation and reactions of β-haloamines.   Several workers[50] have reported the rearrangement of 1-chloro-2-dialkylaminopropanes to 1-dialkylamino-2-chloropropanes.   Fuson and Zirkle have studied the following ring expansion, which probably proceeds by the mechanism shown.[51]

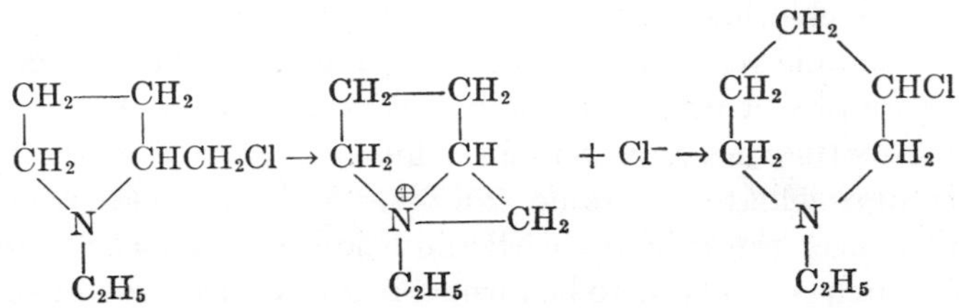

In all the cases mentioned a primary chloride rearranged to a secondary chloride.   Since primary positions are usually more reactive by the S_N2 mechanism and secondary positions almost invariably more reactive by the S_N1 mechanism, it has been suggested that the "ethylene immonium" ring opens by forming a carbonium ion.[52]   However, the isomerization of an amino primary chloride to an amino secondary chloride cannot, in itself, give any information about the mechanism of ring opening.   It merely shows that the secondary chloride is more stable than the primary. In the reaction scheme below it seems assured that the primary chloride is transformed to the cyclic intermediate by an internal nucleophilic attack by the amino group.

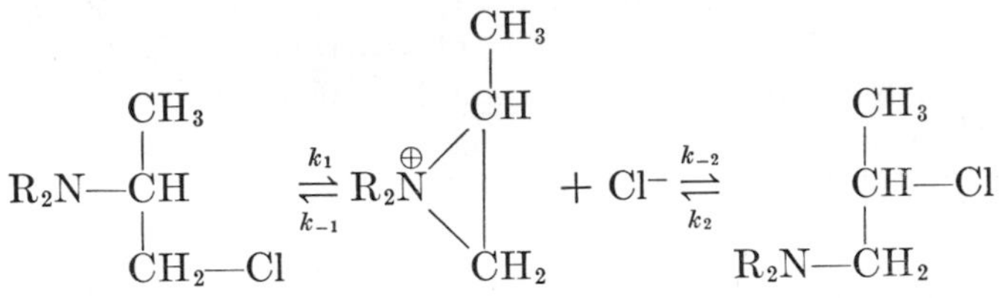

This follows from the fact that β-amino primary chlorides solvolyze *much* more rapidly in aqueous solution than do the corresponding unsubstituted primary chlorides, a fact that seems to have no reasonable explanation other than participation of the amino group in the rate-controlling step of

[50] (a) J. F. Kerwin, G. E. Ullyot, R. C. Fuson, and C. L. Zirkle, *J. Am. Chem. Soc.,* **69,** 2961 (1947); (b) E. M. Schultz and J. M. Sprague, *J. Am. Chem. Soc.,* **70,** 48 (1948).

[51] R. C. Fuson and C. L. Zirkle, *J. Am. Chem. Soc.,* **70,** 2760 (1948).

[52] E. R. Alexander, "Principles of Ionic Organic Reactions," p. 99, John Wiley & Sons, Inc., New York, 1950.

the reaction.   From the principle of microscopic reversibility (Sec. 3-1d) then, it is clear that the ring-opening reaction that gives primary chloride must be S_N2 rather than S_N1 in character.   Reaction by the S_N1 mechanism would be more probable at a secondary than at a primary position, but even so, in the case where R is hydrogen in the reaction scheme above, ring opening to give the secondary chloride appears to proceed by an S_N2 mechanism in aqueous solution.   This follows from the fact that the cyclization of 1-amino-2-chloropropane[53] proceeds more than one hundred times as fast as the estimated solvolysis rate of 2-chloropropane[54] in water at 25°.   The reaction mechanism could probably be made S_N1 by the use of sufficiently strongly electron-withdrawing R groups to decrease the rate of internal nucleophilic attack.

Kinetic investigations of certain reactions of β-haloamines also support the concept of the intermediacy of ethylene immonium ions.   Bartlett, Ross, and Swain have studied the cyclic dimerization of methyl-bis-β-chloroethylamine in aqueous acetone solution.[55]   Since the reaction involves several steps with comparable rate constants, and since the substituted ethylene immonium ion is stable enough under most conditions to accumulate in the solution to a considerable extent (rendering the steady-state approximation inapplicable), the kinetics of the reaction were complicated and certain rate constants were obtained by use of a mechanical differential analyzer.   The mechanism of the dimerization is as follows:

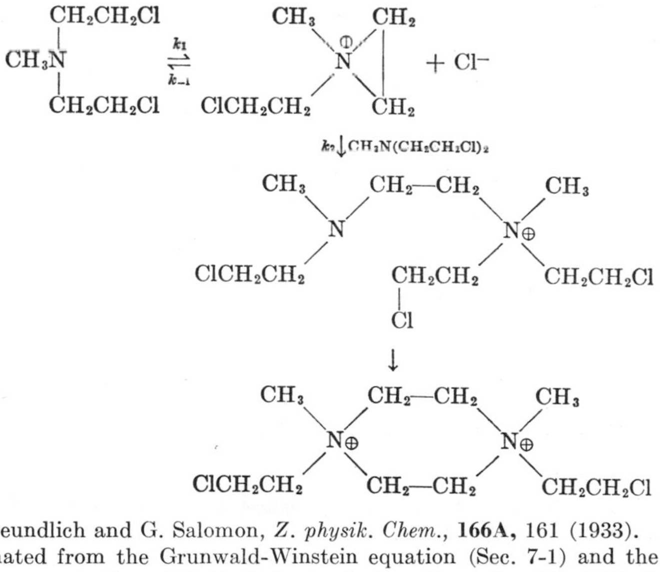

---

[53] H. Freundlich and G. Salomon, *Z. physik. Chem.*, **166A,** 161 (1933).

[54] Estimated from the Grunwald-Winstein equation (Sec. 7-1) and the solvolysis rate in 50 per cent ethanol [J. D. Roberts, *J. Am. Chem. Soc.*, **71,** 1880 (1949)].

[55] P. D. Bartlett, S. D. Ross, and C. G. Swain, *J. Am. Chem. Soc.*, **69,** 2971 (1947).

In the presence of excess hydroxide ion, thiosulfate ion, or triethylamine, the over-all course of the reaction becomes largely a replacement of chlorine by these nucleophilic reagents. As the reactivity and concentration of the nucleophilic reagent is increased, the reaction rate approaches but does not exceed $k_1[CH_3N(C_2H_4Cl)_2]$, the rate of ethylene immonium–ion formation. This shows that these reactions, too, are nucleophilic attacks not on the haloamine but on the cyclic immonium ion. As expected, the reactions may be slowed by chloride ion because of a mass-law effect. The reactions of other β-haloamines with nucleophilic reagents have also been studied kinetically.[56]

6-4e. *Neighboring Halogen Atoms.* Bromonium ions were first suggested as intermediates in the addition of bromine to olefins (Sec. 9-1a). Although analogous to no stable compounds known at the time,[57] the bromonium ion was the first intermediate investigated extensively in connection with the effect of neighboring groups on nucleophilic substitutions on carbon. Most of the pioneering work in this field was done by Winstein, Lucas, and colleagues. These workers found that upon reaction with fuming hydrobromic acid, optically active *erythro*-3-bromo-2-butanol (IV)[58] yielded *meso*-2,3-dibromobutane, and *threo*-3-bromo-2-butanol (VIII)[58] gave *dl*-2,3-dibromobutane.[59] The workers suggest that

the alcohol coordinates with a proton to give the oxonium ion, V. A water molecule is displaced from this intermediate by the nucleophilic attack of the neighboring bromine atom to give the bromonium ion, VI. The opening of this three-membered ring by attack of a bromide ion on either carbon atom will yield the observed meso product, VII. With *threo*-3-bromo-2-butanol, the mechanism is analogous.

[56] P. D. Bartlett, J. W. Davis, S. D. Ross, and C. G. Swain, *J. Am. Chem. Soc.*, **69**, 2977 (1947); B. Cohen, E. R. Van Artsdalen, and J. Harris, *J. Am. Chem. Soc.*, **70**, 281 (1948); P. D. Bartlett, S. D. Ross, and C. G. Swain, *J. Am. Chem. Soc.*, **71**, 1415 (1949).

[57] Although iodonium compounds had been long known, the first stable bromonium and chloronium salts appear to have been prepared by R. B. Sandin and A. S. Hay [*J. Am. Chem. Soc.*, **74**, 274 (1952)].

[58] The prefix *erythro*- means "having a configuration analogous to that of the aldotetrose erythrose"; *threo*- relates to the diastereomeric threose.

[59] S. Winstein and H. J. Lucas, *J. Am. Chem. Soc.*, **61**, 1576, 2845 (1939).

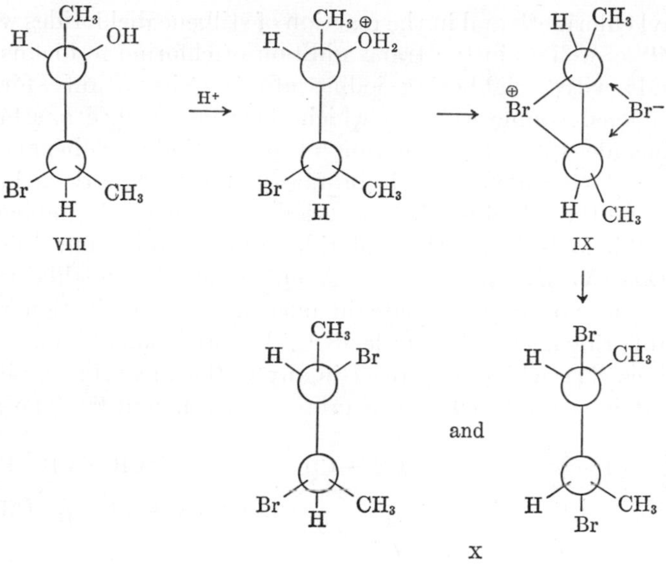

$$\text{VIII} \xrightarrow{\ H^+\ } \longrightarrow \text{IX}$$

IX $\downarrow$

X (and)

In this case the bromonium ion (IX) is symmetrical and therefore inactive; it will be attacked equally at carbon atoms 2 and 3 to yield a *dl* mixture (X). There appears to be no other reasonable mechanism to explain the experimental data. The reaction cannot be a simple S_N2 replacement of the protonated hydroxyl group, because such a mechanism would yield meso dibromide from threo bromohydrin rather than from erythro, as found. The mechanism cannot be of the $S_N i$ type, since the threo bromohydrin yields *dl* rather than optically active dibromide. An S_N1 reaction to give a planar carbonium ion capable of reacting on either side would produce a mixture of meso and active dibromides from either starting material. The objections to an S_N1 mechanism with inversion due to the shielding effect of the departing water molecule or with retention due to some effect of the adjacent bromine atom are the same as those to the S_N2 and $S_N i$ mechanisms, respectively. The preparation of optically active 2,3-dibromobutane with a significant specific rotation[60] shows that optical activity in the reaction would have been easily detectable.

Although no experimental test was made, it is improbable that the isolation of a racemic product is due either to the racemization of the bromohydrin before reaction or racemization of the dibromide after its formation.

There is evidence for the intermediacy of an analogous chloronium ion in the transformation of 3-chloro-2-butanol to 2,3-dichlorobutane

[60] H. J. Lucas and C. W. Gould, Jr., *J. Am. Chem. Soc.*, **64,** 601 (1942).

by thionyl chloride[61a] and in the reaction of stilbene dichlorides with silver acetate,[61b] as well as in the trans addition of chlorine to olefins. Many compounds with neighboring iodine atoms quite readily form cyclic iodonium ions. Some cases in which their importance has been mentioned include the transformation of *trans*-2-iodocyclohexanol to the bromide and chloride by hydrobromic and hydrochloric acid, respectively,[62a] and the solvolysis of *trans*-2-iodocyclohexyl *p*-bromobenzenesulfonate,[62b] 2-iodo-1-phenylethyl chloride,[62c] and β-iodo-*t*-butyl chloride.[62c]

6-4f. Other Neighboring Groups. A number of other neighboring groups have been found to participate in nucleophilic substitution reactions. The neighboring acetoxy group leads to the formation of an acetoxonium ion (such as XI), which may react in any of the ways shown below, provided that it is produced in the proper environment.[63] The action of

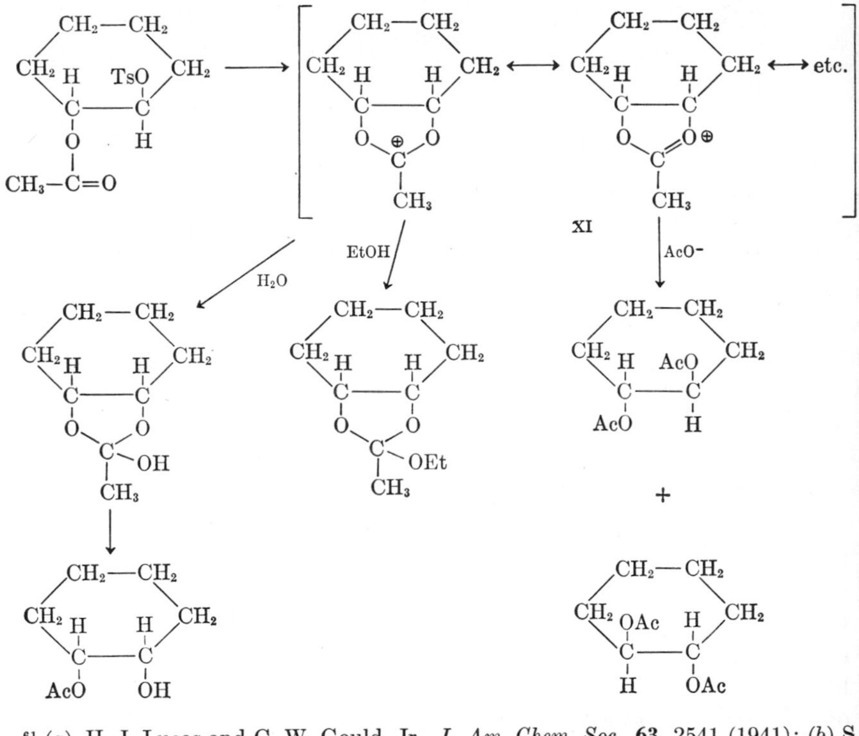

[61] (a) H. J. Lucas and C. W. Gould, Jr., *J. Am. Chem. Soc.*, **63**, 2541 (1941); (b) S. Winstein and D. Seymour, *J. Am. Chem. Soc.*, **68**, 119 (1946).

[62] (a) S. Winstein, E. Grunwald, R. E. Buckles, and C. Hanson, *J. Am. Chem. Soc.*, **70**, 816 (1948); (b) S. Winstein, E. Grunwald, and L. L. Ingraham, *J. Am. Chem. Soc.*, **70**, 821 (1948); (c) S. Winstein and E. Grunwald, *J. Am. Chem. Soc.*, **70**, 828 (1948).

[63] S. Winstein, H. V. Hess, and R. E. Buckles, *J. Am. Chem. Soc.*, **64**, 2796 (1942); S. Winstein and R. E. Buckles, *J. Am. Chem. Soc.*, **64**, 2780, 2787 (1942); **65**, 613 (1943).

hydrocarbon radicals as neighboring groups is described under carbon-skeleton rearrangements in Chap. 14. Carbonium ions with the structure that would result from the neighboring-group participation of hydrogen are discussed in Sec. 9-2a. Neighboring-group participation also occurs with the methoxy[64] and the acylamino and related groups.[65]

6-5. Reactions of Allylic Halides.[66]   *6-5a. The S_N1 and S_N2 Mechanisms.* Nucleophilic substitution reactions of allylic halides may in some cases occur by mechanisms that are new or that at least represent a considerable modification of any of the mechanisms discussed so far. The S_N2 mechanism may proceed in essentially its normal manner, but in the case of the S_N1 mechanism, if a really free carbonium ion is formed, there will be two electrophilic carbon atoms and, if the allyl group is unsymmetrically substituted, two products may be formed.

$$R\text{---}CH\text{=}CH\text{---}CH_2X \rightarrow \quad \begin{matrix} R\text{---}CH\text{=}CH\text{---}\overset{\oplus}{CH_2} \\ \updownarrow \\ R\text{---}\overset{\oplus}{CH}\text{---}CH\text{=}CH_2 \end{matrix} \quad \overset{Y^-}{\rightarrow} \quad \begin{matrix} R\text{---}CH\text{=}CH\text{---}CH_2Y \\ \text{and} \\ R\text{---}CH\text{---}CH\text{=}CH_2 \\ \quad\quad | \\ \quad\quad Y \end{matrix}$$

The formation of a rearranged product is, however, not regarded as sufficient evidence in itself for the intermediacy of a carbonium ion, since rearrangement could occur as a result of the so-called S_N2' mechanism, in which a bimolecular nucleophilic attack with allylic rearrangement occurs.

$$\underset{\underset{Y^-}{\uparrow}}{R\text{---}\overset{\overset{H}{|}}{C}\text{=}\overset{\overset{H}{|}}{C}\text{---}\overset{\overset{H}{|}}{\underset{\underset{H}{|}}{C}}\text{---}X} \rightarrow R\text{---}\overset{\overset{H}{|}}{C}\text{=}\overset{\overset{H}{|}}{\underset{\underset{Y}{|}}{C}}\text{=}\overset{\overset{H}{|}}{\underset{\underset{H}{|}}{C}}\text{---}X \rightarrow R\text{---}\overset{\overset{H}{|}}{\underset{\underset{Y}{|}}{C}}\text{---}\overset{\overset{H}{|}}{C}\text{=}\overset{\overset{H}{|}}{\underset{\underset{H}{|}}{C}} + X^-$$

In a number of instances the product of the solvolysis of an allylic halide has been found to be a mixture of the two possible allylic isomers. It seems reasonable that this is due to reaction by the S_N1 mechanism. Catchpole and Hughes found the first-order ethanolysis of α-methylallyl chloride to yield 82 per cent ethyl crotyl ether.[67] It does not seem likely that the nucleophilic attack of the ethanol solvent would have occurred 82 per cent by the S_N2' mechanism when it was demonstrated

---

[64] S. Winstein and L. L. Ingraham, *J. Am. Chem. Soc.*, **74**, 1160 (1952); S. Winstein, C. R. Lindegren, and L. L. Ingraham, *J. Am. Chem. Soc.*, **75**, 155 (1953).

[65] G. E. McCasland, R. K. Clark, Jr., and H. E. Carter, *J. Am. Chem. Soc.*, **71**, 637 (1949); S. Winstein and R. Boschan, *J. Am. Chem. Soc.*, **72**, 4669 (1950).

[66] For a comprehensive review of this subject, see R. H. DeWolfe and W. G. Young, *Chem. Revs.*, **56**, 753 (1956).

[67] A. G. Catchpole and E. D. Hughes, *J. Chem. Soc.*, 4 (1948).

that the kinetically proved nucleophilic attack by ethoxide ion in the same solvent proceeded 100 per cent (within experimental error) by the normal S_N2 mechanism. Apparently, the solvolysis of most pairs of isomeric allylic halides does not proceed entirely through a free carbonium ion, since in most such cases different product mixtures are obtained from the two different reactants. This result can be explained by the incursion of the S_N2 mechanism or in other ways.[66–68] This interpretation is supported by the observation that as the structures of the reactants and the reaction conditions are changed so as to favor the formation of more stable, longer-lived carbonium ions, the compositions of the reaction-product mixtures obtained from the two isomers become more nearly identical.[66]

It appears that many allylic rearrangements are simply S_N1 reactions in which the carbonium ion recombines (at a different carbon atom) with the anion to which it was originally attached. Catchpole and Hughes have suggested the following mechanism for the rearrangement of α-phenylallyl p-nitrobenzoate to cinnamyl p-nitrobenzoate.[69]

$$C_6H_5-\overset{\displaystyle |}{\underset{\displaystyle X}{C}}HCH=CH_2 \rightleftharpoons C_6H_5-\overset{\oplus}{C}HCH=CH_2 \quad \updownarrow \quad + X^-$$

$$C_6H_5-CH=CHCH_2X \rightleftharpoons C_6H_5-CH=CH\overset{\oplus}{C}H_2$$

where $X = p\text{-}O_2NC_6H_4CO_2^-$. The cinnamyl ester, in which the double bond is conjugated with the benzene ring, is the more stable of the two isomers. Compounds with terminal double bonds usually seem to be less stable, even if in the other isomer the double bond is merely stabilized by additional hyperconjugation.

6-5b. *The S_N2' Mechanism.* The first good evidence for the S_N2' (substitution, nucleophilic, bimolecular, with allylic rearrangement) mechanism appears to be due to Kepner, Winstein, and Young,[70] who found that in the reaction of α-ethylallyl chloride with sodiomalonic ester 23 per cent of the product was the rearranged compound XII.

$$C_2H_5\overset{\displaystyle |}{\underset{\displaystyle Cl}{C}}HCH=CH_2 + \overset{\ominus}{C}H(CO_2Et)_2 \rightarrow C_2H_5CH=CHCH_2CH(CO_2Et)_2$$

$$\text{XII}$$

Most of this compound must have been formed by the nucleophilic attack of the dicarbethoxymethyl anion, since kinetic studies show that at least

[68] W. G. Young and L. J. Andrews, *J. Am. Chem. Soc.*, **66**, 421 (1944).
[69] A. G. Catchpole and E. D. Hughes, *J. Chem. Soc.*, 1 (1948).
[70] R. E. Kepner, S. Winstein, and W. G. Young, *J. Am. Chem. Soc.*, **71**, 115 (1949).

99 per cent of the reaction was of a second-order character, first-order in α-ethylallyl chloride and first-order in sodium dicarbethoxymethide. This attack must have been on α-ethylallyl chloride, since this chloride rearranges to γ-ethylallyl chloride much too slowly to explain these results.   It was also shown that the isolation of the rearranged product could not have been due to a rearrangement of the normal product after its formation.

Young, Webb, and Goering have similarly shown that the reaction of diethylamine with α-methylallyl chloride in benzene, which yields at least 85 per cent crotyldiethylamine, also proceeds by the S_N2' mechanism,[71] as does the reaction of trimethylamine with α-methylallyl chloride.[72]

De la Mare and Vernon found the product of the S_N2' reaction of sodium thiophenoxide with 3,3-dichloropropene to consist of approximately equal amounts of the cis and trans isomers of phenyl 3-chloroallyl sulfide.[73]   Apparently each of two different types of rotational conformers reacted to about the same extent.   Stork and White, studying the S_N2' reactions of the 2,6-dichlorobenzoates of several *trans*-6-alkyl-2-cyclo-hexen-1-ols with piperidine, showed that the entering nucleophilic reagent attacks the six-membered ring from the side cis to the 2,6-dichlorobenzo-ate group being displaced [to give N-(*trans*-4-alkyl-2-cyclohexenyl) piperidines].[74]   This observed cis attack in the S_N2' reaction had been predicted by Young, Webb, and Goering.[71]

In the case of many allylic halides the S_N2 reaction occurs so rapidly that it is not possible to detect any reaction by the S_N2' mechanism. England and Hughes have ingeniously surmounted this difficulty by studying the reactions of crotyl bromide and α-methylallyl bromide with lithium bromide in acetone.[75]   The S_N2' mechanism may be studied in this case since reaction by the S_N2 mechanism leaves the allylic bromide chemically unchanged and hence does not render it unavailable for future S_N2' attack (as would an S_N2 attack by a foreign anion such as ethoxide). The lithium bromide is found to transform either allylic bromide into the equilibrium mixture of the two by a purely second-order process at a rate that is the rate of S_N2' attack.   The rate of S_N2 attack was measured simultaneously by use of radioactive bromide ions.   The rate constants shown in the accompanying table were obtained.

[71] W. G. Young, I. D. Webb, and H. L. Goering, *J. Am. Chem. Soc.*, **73**, 1076 (1951).
[72] W. G. Young, R. A. Clement, and C. H. Shih, *J. Am. Chem. Soc.*, **77**, 3061 (1955).
[73] P. B. D. de la Mare and C. A. Vernon, *J. Chem. Soc.*, 3331 (1952); 3679 (1954).
[74] G. Stork and W. N. White, *J. Am. Chem. Soc.*, **75**, 4119 (1953); **78**, 4609 (1956).
[75] B. D. England and E. D. Hughes, *Nature*, **168**, 1002 (1951); B. D. England, *J. Chem. Soc.*, 1615 (1955).

	Mechanism	$10^6 k_2$, 25°
Crotyl bromide..............	S_N2	141,000
	S_N2'	5
α-Methylallyl bromide.........	S_N2	879
	S_N2'	14.9

Bulky α substituents can increase the extent of S_N2' reaction largely because they so markedly reduce the rate of the competing S_N2 reaction. Thus, for example, 1-t-butylallyl chloride reacts with sodium ethoxide by a second-order process to give essentially pure 3-t-butylallyl ethyl ether.[76] Polar factors can also be important. For example, the ready occurrence of bimolecular nucleophilic substitution with allylic rearrangement in the reactions of highly fluorinated allylic halides, established by Miller and coworkers,[77] is undoubtedly related to the tendency of fluoro substituents to facilitate nucleophilic attack at olefinic double bonds (cf. Sec. 9-4c). In some such compounds it is possible that reaction occurs by a two-step mechanism involving an intermediate carbanion

$$X^- + CF_2{=}CF{-}CF_2Y \rightarrow XCF_2{-}\overset{\ominus}{C}F{-}CF_2Y \rightarrow XCF_2{-}CF{=}CF_2 \\ + Y^-$$

With most allylic halides, however, the S_N2' mechanism appears to be a concerted process.

6-5c. *Ion Pairs in the Solvolysis of Allyl Halides.* Young, Winstein, and Goering have shown that the acetolysis of α,α-dimethylallyl chloride involves simultaneous solvolysis and rearrangement to γ,γ-dimethylallyl chloride.[78] This was shown by the fact that (1) the solvolysis rate constants fell from the initial high values that might be expected for the α,α-dimethyl isomer to the much lower values characteristic of γ,γ-dimethylallyl chloride, and (2) when the solvolysis of originally pure α,α-dimethylallyl chloride was interrupted after about 35 per cent completion, the unreacted halide was found to consist almost entirely of γ,γ-dimethylallyl chloride. It was shown that the rearrangement is not due to a mass-law effect in which free carbonium ions combine with chloride ions to form γ,γ-dimethylallyl chloride. Similar results have been obtained in the acetolysis of other compounds (Sec. 14-1). The reaction could be considered as an internal rearrangement of the multi-

[76] P. B. D. de la Mare, E. D. Hughes, P. C. Merriman, L. Pichat, and C. A. Vernon, *J. Chem. Soc.*, 2563 (1958).

[77] W. T. Miller, Jr., and A. H. Fainberg, *J. Am. Chem. Soc.*, **79**, 4164, 4170 (1957); J. H. Fried and W. T. Miller, Jr., *J. Am. Chem. Soc.*, **81**, 2078 (1959).

[78] W. G. Young, S. Winstein, and H. L. Goering, *J. Am. Chem. Soc.*, **73**, 1958 (1951).

center type (Sec. 25-1), but if it is, we shall have to point out that the transition state is much more polar than the reacting molecule since the reaction increases with the ion-solvating power of the medium in much the same way that rates of carbonium-ion formation do. For this reason, in fact, it is probably more useful to consider the reaction as an ionization to an ion pair (XIII) that is transformed to a covalent chloride at a rate comparable with its dissociation to a carbonium ion able to combine with solvent.[78]

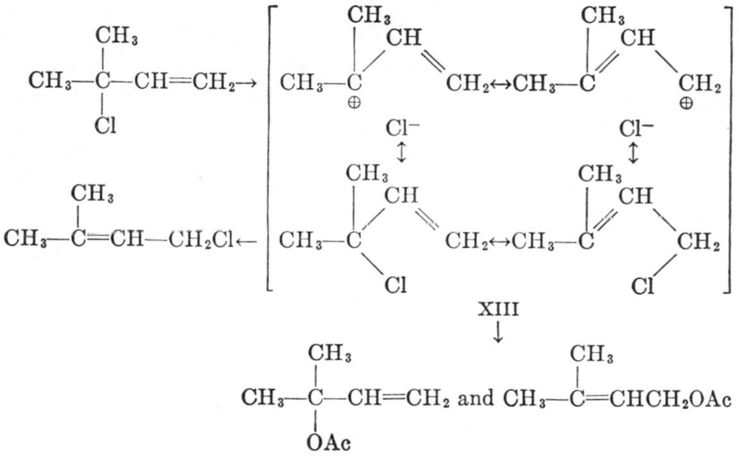

The discovery of this solvolysis through an ion-pair intermediate in the case of several different compounds makes it appear likely that ion pairs are intermediate in many, perhaps most, solvolyses, especially in poorly ion-solvating media. This problem has been studied in detail by Winstein and coworkers.[79]

PROBLEMS

1. Without using data on any specific compound, prove that in the optically active compound

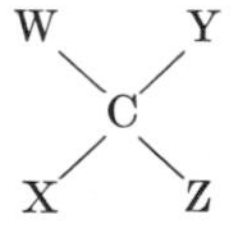

the replacement of the group X by another group, without change in configuration, could not possibly always result in retention (or always result in inversion) of the direction of rotation of polarized light.

2. Compare the solvolysis rates of triphenylmethyl chloride in 85 per cent acetone in the presence of 0.06 M sodium perchlorate and 0.06 M sodium chloride.

[79] S. Winstein et al., *J. Am. Chem. Soc.*, **76**, 2597 (1954); **78**, 328, 2763, 2767, 2777, 2780, 2784 (1956); *Helv. Chim. Acta*, **41**, 807 (1958).

3. The reaction of 2-hydroxy-2-methylpropyl chloride with sodium ethoxide in ethanol is found to be a second-order process when measured either by the rate of appearance of sodium chloride or by the rate of disappearance of ethoxide ions. When the reaction mixture containing an excess of sodium ethoxide is allowed to stand a long time to assure the reaction of all the organic chloride, the reaction product, ethyl 2-hydroxy-2-methylpropyl ether, may be isolated in more than 90 per cent yield. Tell what has been established about the reaction mechanism. Suggest experiments to distinguish between any mechanisms that still seem possible in the face of the experimental evidence described.

4. In the Stevens rearrangement (Sec. 6-3) the nucleophilic atom is separated from the atom undergoing nucleophilic attack by only one other atom. Suggest one-step S_Ni reactions that might be studied in which the nucleophilic atom is separated from the atom undergoing nucleophilic attack by 2, 3, 4, 5, 6, 7, 8, 9, and 10 atoms (including the atom displaced by the nucleophilic attack). In which cases would you expect retention and in which inversion of configuration?

5. Describe experiments designed to give several types of evidence whether the ethanolysis of $CH_3SeCH_2CH_2Cl$ involves neighboring-group participation or not.

Chapter 7

REACTIVITY IN NUCLEOPHILIC
DISPLACEMENTS ON CARBON

**7-1. Effect of Solvent on Reactivity in Nucleophilic Displace-
ments.** The Hughes-Ingold theory of solvent effects was described in
Sec. 3-1f. According to this theory, the rate of solvolysis of electrically
neutral molecules by either the S_N1 or S_N2 mechanism would be expected
to increase with the ion-solvating power of the medium, since the reaction,
by either mechanism, involves charge formation in the transition state.
From the empirical observation that the solvolysis rate increases much
faster with compounds thought for other reasons to react by the S_N1
mechanism than with those thought to react by the S_N2 mechanism, it
may be rationalized that a larger or more concentrated charge is formed
in the S_N1 than in the S_N2 transition state. The magnitudes of solvolysis
effects, however, are not separated into two sharply defined categories
but vary continuously.

In addition to this qualitative theory, several quantitative correlations
of the rates of solvolysis reactions have been suggested. One of these
is due to Grunwald and Winstein, who assigned to each of several solvents
a number, Y, which is a quantitative measure of its "ionizing" power, as
manifested in its effect on the rate of solvolyses by the S_N1 mechanism.[1]
Y was defined by the relation

$$Y = \log \frac{k^{\text{BuCl}}}{k_0{}^{\text{BuCl}}} \tag{7-1}$$

where k^{BuCl} and $k_0{}^{\text{BuCl}}$ are the rate constants for the solvolysis of t-butyl
chloride in the given solvent and in the reference solvent (80 per cent
ethanol), respectively. The following equation was then suggested to
correlate the rates of solvolysis of all compounds that react by the S_N1
mechanism:

$$\log \frac{k}{k_0} = mY \tag{7-2}$$

[1] E. Grunwald and S. Winstein, *J. Am. Chem. Soc.*, **70**, 846 (1948); S. Winstein,
E. Grunwald, and H. W. Jones, *J. Am. Chem. Soc.*, **73**, 2700 (1951).

where k and k_0 are the solvolysis rates of a certain compound in a given solvent and in 80 per cent ethanol, respectively, Y is characteristic of the given solvent, and m is a constant characteristic of the compound being solvolyzed.   A number of values of Y are given in Table 7-1.[2]

TABLE 7-1. Y VALUES FOR VARIOUS SOLVENTS[2]

Solvent[a]	Y	Solvent[a]	Y
100% EtOH	−2.033	i-PrOH	−2.73
95% EtOH	−1.287	t-BuOH	−3.26
90% EtOH	−0.747	100% HCO_2H	2.054
80% EtOH	0.000	50% HCO_2H	2.644
70% EtOH	0.595	100% AcOH	−1.639
50% EtOH	1.655	50% AcOH	1.938
30% EtOH	2.721	50% HCO_2H–50% AcOH	0.757
Water	3.493	90% Dioxane	−2.030
100% MeOH′	−1.090	50% Dioxane	1.361
90% MeOH	−0.301	90% Acetone	−1.856
70% MeOH	0.961	50% Acetone	1.398
50% MeOH	1.972	100% $HCONH_2$	0.604
30% MeOH	2.753	n-$C_3F_7CO_2H$	1.7

[a] By "$x\%$ ROH" is meant a solution made from x volumes of ROH and $100 - x$ volumes of water.

Subsequent work showed that Eq. (7-2) was not as widely applicable as had been originally hoped.   Plots of log k for solvolyses vs. Y gave marked scattering in many cases.   However, in each case the scattered points could be resolved into a number of straight lines, one for each binary solvent mixture.[3]   Hence we may use the modified equation

$$\log \frac{k}{k_0{}^s} = m_s Y \qquad (7\text{-}3)$$

where k and Y have the definitions given previously, but m_s is a function not only of the nature of the compound undergoing solvolysis and of the reaction temperature but also of the nature of the pair of solvents in the mixture of which the solvolysis is being studied; log $k_0{}^s$ is also a disposable parameter whose value varies with the reactant, temperature, and solvent pair.   Values of log $k_0{}^s$ and m_s are listed in Table 7-2.

Although Eq. (7-3) uses more disposable parameters than does (7-2), it is capable of correlating the rates of solvolysis of most compounds, including many that react by the S_N2 mechanism (in the solvents listed in Table 7-1) with an average deviation from the observed k values of less than 10 per cent.[4]

[2] A. H. Fainberg and S. Winstein, *J. Am. Chem. Soc.*, **78**, 2770 (1956).

[3] A. H. Fainberg and S. Winstein, *J. Am. Chem. Soc.*, **79**, 1597, 1602, 1608 (1957).

[4] S. Winstein, A. H. Fainberg, and E. Grunwald, *J. Am. Chem. Soc.*, **79**, 4146 (1957).

Other equations for the correlation of solvolysis rates have been suggested by Swain and coworkers.[5]

TABLE 7-2. VALUES OF LOG $k_0{}^s$ AND m_s FOR EQ. (7-3)[2-4]

Reactant	Solvent pair	Temperature	Log $k_0{}^s$*	m_s
t-BuCl	All	25°	−5.033	1.000
t-BuBr	EtOH-H$_2$O	0	−4.967	1.017
t-BuBr	EtOH-H$_2$O	25	−3.455	0.941
t-BuBr	AcOH-HCO$_2$H	25	−3.972	0.946
HOCH$_2$CH$_2$Br	EtOH-H$_2$O	70	−6.245	0.226
C$_6$H$_5$CH$_2$OSO$_2$C$_6$H$_4$CH$_3$-p	Acetone-H$_2$O	25.3	−4.711	0.650
Ac$_2$O	Acetone-H$_2$O	25	−4.636	0.579
CH$_3$SO$_2$Cl	Dioxane-H$_2$O	25	−4.568	0.465
C$_6$H$_5$CHClCH$_3$	EtOH-H$_2$O	25	−4.939	0.966
C$_6$H$_5$CHClCH$_3$	AcOH-H$_2$O	25	−5.180	1.136
C$_6$H$_5$CHClCH$_3$	Dioxane-H$_2$O	25	−5.493	1.136
C$_6$H$_5$C(CH$_3$)$_2$CH$_2$Cl	EtOH-H$_2$O	50	−8.535	0.833
(C$_6$H$_5$)$_2$CHCl	EtOH-H$_2$O	25	−2.763	0.740
(C$_6$H$_5$)$_2$CHCl	AcOH-H$_2$O	25	−2.660	1.561
(C$_6$H$_5$)$_3$CF	EtOH-H$_2$O	25	−3.530	0.890
(C$_6$H$_5$)$_3$CF	Acetone-H$_2$O	25	−5.02	1.58

* In sec^{-1}.

7-2. Effect of the Nucleophilic Reagent on the Rates of Nucleophilic Displacements.[6]

It has long been recognized that for a given nucleophilic atom nucleophilicity is fairly well correlated with the basicity of the nucleophilic reagent. For example, Smith has pointed out that in the reaction of the chloroacetate ion with 32 nucleophilic anions whose nucleophilic atom is oxygen the variation in the logarithm of the rate constants is very nearly proportional to the differences in the logarithms of the basicity constants.[7] The sulfite and thiosulfate anions, which form carbon-sulfur bonds rather than carbon-oxygen bonds, were much more nucleophilic than would be expected from their basicity. By restricting the variation in structure to meta and para substituents on the aromatic ring, a still better correlation may be obtained, at least partially because of the elimination of steric effects, which can destroy the correlation, since nucleophilic attack on carbon is usually more subject to steric hindrance than is coordination with a proton. Reaction constants, ρ, have been calculated for several series of reactions involving aniline derivatives and substituted phenolate anions as nucleophilic

[5] C. G. Swain et al., *J. Am. Chem. Soc.*, **75**, 4627 (1953); **77**, 3727, 3731, 3737 (1955); C. G. Swain and C. B. Scott, *J. Am. Chem. Soc.*, **75**, 141 (1953).

[6] Cf. J. O. Edwards and R. G. Pearson, *J. Am. Chem. Soc.*, **84**, 16 (1962).

[7] G. F. Smith, *J. Chem. Soc.*, 521 (1943).

reagents[8] (see Table 7-3). The negative values of ρ show the nucleo-philicity to be increased by electron-donating groups in all cases. Similar results are found in the nucleophilic attack of anilines on halo-nitro aromatic compounds and on acid halides.[8]

TABLE 7-3. REACTION CONSTANTS, ρ, FOR REACTIONS OF PHENOXIDE IONS
AND ANILINE DERIVATIVES AS NUCLEOPHILIC REAGENTS[8]

Reaction	$k_0{}^a$	ρ
Phenolate ions with ethylene oxide in 98% ethanol at 70.4°.	-4.254	-0.947
Phenolate ions with propylene oxide in 98% ethanol at 70.4°.	-4.698	-0.771
Phenolate ions with ethyl iodide in ethanol at 42.5°.	-3.955	-0.994
Phenolate ions with $CH_3OSO_3{}^-$ in H_2O at 100°.	-4.121	-0.813
Dimethylanilines with methyl iodide in aqueous acetone at 35°.	-3.366	-3.303
Dimethylanilines with methyl picrate in acetone at 35°.	-4.505	-2.378
Dimethylanilines with trinitrocresol methyl ether in acetone at 35°.	-4.916	-2.776
Phenyldiethylphosphines with ethyl iodide in acetone at 35°.	-3.290	-1.094

^{a} All k's in liters mole^{-1} sec^{-1}.

Another generalization is that nucleophilicity increases, within a given group of the periodic table, with the atomic number of the atom forming the new bond to carbon. That is, $I^- > Br^- > Cl^- > F^-$; $RS^- > RO^-$; $R_2S > R_2O$, etc. It is noted that this variation, which is the one usually but not invariably observed, is in the direction opposite to that which would be expected from the basicities of the various nucleophilic reagents. The increased nucleophilicity is most commonly attributed to the increase in polarizability that accompanies the increase in the distance of the outer electronic shell from the nucleus. The greater ease of distortion of the outer shell permits an easier adjustment to the requirements of a stable transition state. The nucleophilicity of some of the anions, particularly F^- and RO^- is decreased considerably by their strong solvation and is markedly increased in nonhydroxylic solvents.

The relative reactivity of various nucleophilic reagents has been treated more quantitatively by Swain and Scott. For one-step nucleophilic attack in aqueous solution they suggested the equation

$$\log \frac{k}{k_0} = sn \qquad (7\text{-}4)$$

where k is the second-order rate constant for a nucleophilic displacement by a reagent whose nucleophilicity is n on a compound whose sensitivity

[8] H. H. Jaffé, *Chem. Revs.*, **53**, 191 (1953).

to change in nucleophilicity is s, and k_0 is the second-order rate constant for nucleophilic attack by water, the standard nucleophile. As a standard for the sensitivity values, s was set equal to 1.0 for methyl bromide. From 47 values of k/k_0, determined in aqueous acetone and aqueous dioxane solution ranging from 39 to 100 per cent in water content and at temperatures from 0 to 50°, optimum values of n's and s's were determined. Some of these, with values determined later by other workers, are listed in Tables 7-4 and 7-5. The values shown can be used to predict

TABLE 7-4. NUCLEOPHILIC CONSTANTS, n, OF VARIOUS NUCLEOPHILIC REAGENTS[a]

Reagent	n	Reagent	n
ClO_3^-, ClO_4^-, BrO_3^-, IO_3^-	<0[b]	Br^-	3.89
H_2O	0.00	N_3^-	4.00
$p\text{-}CH_3C_6H_4SO_3^-$	<1.0	$(NH_2)_2CS$	4.1
NO_3^-	1.03[b]	OH^-	4.20
Picrate anion	1.9	$C_6H_5NH_2$	4.49
F^-	2.0	SCN^-	4.77
SO_4^-	2.5	I^-	5.04
$CH_3CO_2^-$	2.72	CN^-	5.1[c]
Cl^-	3.04	SH^-	5.1
C_5H_5N	3.6	SO_3^-	5.1
HCO_3^-	3.8	$S_2O_3^-$	6.36
HPO_4^-	3.8	$HPSO_3^-$	6.6

[a] From Ref. 5 unless otherwise stated.
[b] W. L. Petty and P. L. Nichols, Jr., *J. Am. Chem. Soc.*, **76**, 4385 (1954).
[c] M. F. Hawthorne, G. S. Hammond, and B. M. Graybill, *J. Am. Chem. Soc.*, **77**, 486 (1955).

the k/k_0 values with an average deviation of about 60 per cent. A somewhat smaller deviation might have resulted if the data had all referred to exactly the same solvent and temperature, but many large deviations would not be materially diminished by such a modification. It may be noted that Eq. (7-4) requires that the relative order of nucleophilicities of nucleophilic reagents be invariable in a given solvent. That is, since the iodide ion is more reactive (by almost 10-fold) than the hydroxide ion toward 2,3-epoxypropanol in aqueous solution,[5] it should be more reactive than hydroxide ion in all nucleophilic displacements on carbon taking place in aqueous solution. It has been found, however, that toward the ethylene-β-chloroethylsulfonium ion the hydroxide ion is about 12 times as reactive as the iodide ion.[5,9] There is thus a change in the relative reactivities of the two ions of more than 100-fold. Similarly, triethylamine is 26 times as reactive as pyridine toward methyl iodide, whereas toward isopropyl iodide pyridine is 6 times as reactive as triethyl-

[9] A. G. Ogston, E. R. Holiday, J. St. L. Philpot, and L. A. Stocken, *Trans. Faraday Soc.*, **44**, 45 (1948).

amine. This 150-fold reversal in the relative reactivity of the two amines is reasonably explained on the basis of steric hindrance.[10]   Despite its greater steric requirements, the more basic triethylamine is more reactive than pyridine toward methyl iodide. Since any attack on isopropyl iodide is much more hindered, the steric effects are too large to

TABLE 7-5. RELATIVE SUSCEPTIBILITIES, s, OF VARIOUS REACTANTS TO CHANGES IN NUCLEOPHILICITY OF ATTACKING REAGENT[5]

Reactant	s
$C_2H_5OSO_2C_6H_4CH_3\text{-}p$	0.66
$C_6H_5CH_2Cl$	0.87
β-Propiolactone[a]	0.77
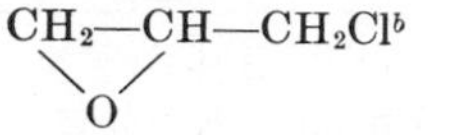 $CH_2\!-\!CH\!-\!CH_2Cl$[b] $\diagdown\!\!O\!\!\diagup$	0.93
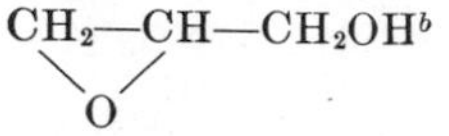 $CH_2\!-\!CH\!-\!CH_2OH$[b] $\diagdown\!\!O\!\!\diagup$	1.00
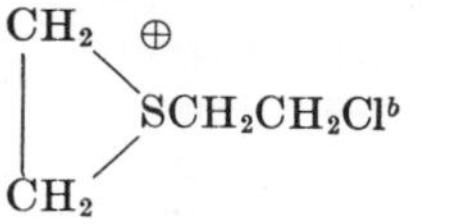 CH_2 $\oplus$ $\quad\diagdown SCH_2CH_2Cl$[b] $CH_2\diagup$	0.95
CH_3Br	1.00
$C_6H_5SO_2Cl$	1.25
C_6H_5COCl	1.43

[a] Refers to a nucleophilic attack on the β-carbon atom.
[b] Refers to a ring-opening attack.

overcome in this case and pyridine is the more reactive. Some other explanation must be given for the increased reactivity of the hydroxide ion toward the ethylene-β-chloroethylsulfonium ion, however.[5]

Edwards has correlated the rates of nucleophilic substitution reactions by use of the relation

$$\log \frac{k}{k_0} = AP + BH \tag{7-5}$$

where k and k_0 are rate constants for reaction with the given nucleophilic reagent and with water, H is a measure of the basicity of the nucleophilic reagent (H $= pK_a + 1.74$), P is a measure of its polarizability [P $= \log (R/R_{H_2O})$, where the R's are molar refractions], and A and B are measures of the sensitivity of the reactant to changes in the polarizability and basicity, respectively, of the nucleophilic reagent.[11]   In addition to correlating data on nucleophilic displacements, Eq. (7-5) gives good correlations for rates of displacements on oxygen, hydrogen, and sulfur, and for equilibrium constants in complex-ion formation.

[10] H. C. Brown and N. R. Eldred, *J. Am. Chem. Soc.*, **71**, 445 (1949).   Cf. H. C. Brown et al., *J. Am. Chem. Soc.*, **78**, 5387 (1956).

[11] J. O. Edwards, *J. Am. Chem. Soc.*, **76**, 1540 (1954); **78**, 1819 (1956).

7-3. Effect of Structure of R on the Rates of Nucleophilic Displacements of RY.

7-3a. Reactivity in S_N1 Reactions. In most of the S_N1 reactions that have been studied kinetically the carbonium ion is a very reactive intermediate that is rapidly transformed into the final product, only small amounts usually reverting to reactant. Therefore we should discuss S_N1 reactivity in terms of the energy required to transform the reactant into the transition state leading to the intermediate carbonium ion. However, since this transition state usually differs from the reactant only in its greater similarity to the carbonium ion, the same structural changes that stabilize a carbonium ion will usually decrease the free-energy content of a transition state leading to a carbonium ion. For this reason and for the sake of the resultant simplification in discussion, we shall follow the common practice of often discussing S_N1 reactivity in terms of the stability of the intermediate carbonium ion. However, it will be well to keep in mind the fact that it is the transition-state stability that is of fundamental importance.

The S_N1 transition state (and the intermediate carbonium ion) differs from the reactant in several important ways. First, there is a much larger partial positive charge on the carbon atom at which displacement occurs. Second, the three groups attached to this carbon atom, but not directly involved in the displacement, become much more nearly coplanar with it. Other differences include the fact that the bond to the group being replaced is considerably extended.

In many studies of S_N1 reactivity the rate of reaction of a halide with silver (or occasionally mercuric) ions is measured. This reaction is used because the metal ion, by coordinating with the halide ion, greatly facilitates its removal from carbon without simultaneously strengthening the nucleophilic attack on carbon. This greatly increases the probability that the reaction mechanism will be S_N1 but does not make it certain. The reaction with silver ion may be complicated by catalysis by the precipitated silver halide, and the reaction with mercuric ion, by the existence of several mercury-halide complex ions.

In a discussion of electronic effects it is useful to consider cases in which steric factors are held constant. An example of this type appears in the solvolysis of a series of substituted benzhydryl chlorides. We have described excellent evidence that benzhydryl chloride and its p,p'-dimethyl derivative react by the S_N1 mechanism in aqueous acetone (Sec. 6-2a), and it seems very likely that the ethanolysis also proceeds by this mechanism.[12] When the Hammett equation is applied to the

[12] J. F. Norris et al., *J. Am. Chem. Soc.*, **50**, 1795, 1804, 1808 (1928); E. D. Hughes, C. K. Ingold, and N. A. Taher, *J. Chem. Soc.*, 949 (1940); J. Packer, J. Vaughan, and A. F. Wilson, *J. Org. Chem.*, **23**, 1215 (1958); G. Kohnstam, *J. Chem. Soc.*, 2066, (1960).

resultant data, compounds substituted with para substituents capable of resonance-electron donation are seen to react much faster than would be expected from ordinary σ constants. A satisfactory correlation ($\rho = -4.03$) may be obtained by the use of σ^+ values, however. In another S_N1 solvolysis reaction σ^+ values may be used to treat the rates of hydrolysis of t-cumyl chlorides [ArC(CH$_3$)$_2$Cl] to give a ρ of -4.54.[13] These observations show that in the transition states of these reactions the carbon atom at which substitution is occurring has become highly electron-deficient; that is, the transition states have considerable carbonium-ion character. When the reaction center is located farther from the aromatic ring, as in the solvolysis of ArCH$_2$C(CH$_3$)$_2$Cl's,[14] the value of ρ is considerably smaller (-1.11) and the use of σ's gives better agreement than does the use of σ^+'s, since the electron-deficient carbon atom is now insulated from direct resonance interaction with the ring.

The effect of ortho substituents on the S_N1 reactivity of the benzhydryl and other benzyl-type halides that have been studied may be rationalized by assuming a resonance effect about like that in the para position, a larger inductive effect (since the substituent is nearer the reaction center), and, in many cases, a steric effect that diminishes the resonance interaction of the positively charged carbon atom with the aromatic ring.

The replacement of the α-hydrogen atoms of an alkyl halide by saturated alkyl groups causes a large increase in reactivity by the S_N1 mechanism, since the alkyl groups are much more effective than the hydrogen atoms at supplying electrons to the electron-deficient carbon atom. For this reason, tertiary alkyl halides solvolyze 10^3 to 10^4 times as rapidly as secondary halides in most solvents. The secondary halides are not always more reactive than primary halides in solvolyses, however, because of the incursion of the S_N2 mechanism in the latter case. Streitwieser found that the rates of solvolysis of several series of compounds fit the Taft equation.[15] For example, in the solvolysis of tertiary chlorides, R$_3$CCl, in 80 per cent ethanol, using the sums of the σ^*'s for the R groups, ρ^* is -3.29. Examples of neighboring-group participation were revealed by their significantly greater reaction rates. With very bulky R groups, such as neopentyl, the compounds also solvolyzed much faster than would be expected from the Taft equation.

Brown and coworkers pointed out that interference between such large alkyl groups may be relieved upon transformation to a carbonium ion, since an increase (from about 109.5 to about 120°) in the size of the angle between the bonds attaching the three alkyl groups to the

[13] H. C. Brown et al., *J. Am. Chem. Soc.*, **79**, 1913 (1957); **80**, 4964, 4969, 4972, 4976, 4979 (1958).

[14] A. Landis and C. A. VanderWerf, *J. Am. Chem. Soc.*, **80**, 5277 (1958).

[15] A. Streitwieser, Jr., *J. Am. Chem. Soc.*, **78**, 4935 (1956).

α-carbon accompanies this transformation.[16] Hughes and coworkers also reported evidence for this type of *steric acceleration* in the solvolysis of tertiary halides.[17]

The S_N1 reactivities of halogen atoms attached to alicyclic rings vary in an interesting manner, depending on the ring size to a considerable extent. Cyclopropyl chloride is very unreactive, and cyclopropyl *p*-toluenesulfonate has been found to undergo acetolysis at only one-fifty-thousandth the rate for cyclohexyl *p*-toluenesulfonate and to yield the rearranged product allyl acetate.[18] Roberts and Chambers have pointed out that the unreactivity of these compounds is probably related to that of vinyl and phenyl halides. Like carbon atoms joined by a double bond (a two-membered ring), but to a somewhat smaller extent, the carbon atoms in cyclopropane (and probably also cyclobutane) rings

TABLE 7-6. RATES OF SOLVOLYSIS OF 1-METHYL-1-CHLOROCYCLOALKANES IN "80 PER CENT ETHANOL" AT 25°[19]

Cycloalkane	10^6k, sec^{-1}	Cycloalkane	10^6k, sec^{-1}
Cyclobutane..............	0.62	Cyclotridecane..............	8.4
Cyclopentane..............	367	Cyclopentadecane...........	5.3
Cyclohexane..............	2.94	Cycloheptadecane...........	5.6
Cycloheptane..............	320		
Cyclooctane..............	842	Analogous aliphatic chlorides:	
Cyclononane..............	129	*t*-Butyl....................	8.9
Cyclodecane..............	52.3	3-Methylpentyl............	23.9
Cyclohendecane...........	35.3	6-Methylhendecyl..........	13.3

appear somewhat more electronegative than ordinary saturated aliphatic carbon atoms. Also, the carbon-halogen bonds in cyclopropyl halides, like those in vinyl halides, may have partial double-bond character. Brown and coworkers have explained the unreactivity of cyclopropyl halides in terms of another factor, which they call *I strain* (*internal strain*).[19] The internal angles of a cyclopropyl compound, being about 60°, are strained by 49.5° from the optimum value (109.5°) for tetrahedral carbon. The formation of a carbonium ion with optimum bond angles of 120° would increase this strain and is hence more difficult. Similar arguments would lead to the prediction of successively smaller

[16] H. C. Brown and R. S. Fletcher, *J. Am. Chem. Soc.*, **71**, 1845 (1949); H. C. Brown and A. Stern, *J. Am. Chem. Soc.*, **72**, 5068 (1950); H. C. Brown and H. L. Berneis, *J. Am. Chem. Soc.*, **75**, 10 (1953).

[17] F. Brown, T. D. Davies, I. Dostrovsky, O. J. Evans, and E. D. Hughes, *Nature*, **167**, 987 (1951).

[18] J. D. Roberts and V. C. Chambers, *J. Am. Chem. Soc.*, **73**, 5034 (1951).

[19] H. C. Brown, R. S. Fletcher, and R. B. Johannesen, *J. Am. Chem. Soc.*, **73**, 212 (1951); H. C. Brown and M. Borkowski, *J. Am. Chem. Soc.*, **74**, 1894 (1952).

diminutions in the S_N1 reactivities of cyclobutyl and cyclopentyl halides. As shown in Table 7-6, the decrease in reactivity of a cyclobutyl halide has been observed in the case of 1-chloro-1-methylcyclobutane, which undergoes solvolysis in 80 per cent ethanol at less than one-tenth the rate for t-butyl chloride. The solvolysis of cyclobutyl chloride in 50 per cent ethanol is about 15 times as rapid as that of sec-butyl chloride,[20] but some of the driving force in the solvolysis of cyclobutyl chloride may be related to the carbon-skeleton rearrangement that accompanies the reaction.[18] On the other hand, evidently I strain is not the most important factor with cyclopentyl halides, since these compounds are more reactive than their aliphatic analogs.[18-20] Roberts and Chambers suggested that this is because carbonium-ion formation involves a partial relief of the strain caused by opposed valences which gives the cyclopentane ring a nonplanar configuration (see Sec. 1-6). Brown and coworkers give a similar explanation and add that the slightly decreased S_N1 reactivity of cyclohexyl halides (see Table 7-6) results from the fact that the valences in the cyclohexyl transition state do not have the stable, perfectly staggered configuration of the unreacted molecule. Halogen atoms attached to rings with from seven to about ten members are said to be S_N1 reactive because more strain is required to stagger the valences in the reacting molecule than in the transition state.[19] The reactivity of the larger ring halides is seen to approach that of typical aliphatic compounds.

A group of alicyclic halides whose reactivity is of particular theoretical interest is composed of those bicyclic compounds having a halogen atom at a bridgehead. Bartlett and Knox reported the first study on a compound of this type, apocamphyl chloride.[21] This halide gave no reaction upon being refluxed with alcoholic silver nitrate for 48 hr or with 30 per

cent potassium hydroxide in aqueous ethanol for 21 hr; it is thus seen to be remarkably unreactive, since, for example, typical tertiary chlorides give almost immediate precipitates with ethanolic silver nitrate at room temperature. The low S_N1 reactivity of this compound is attributed to the fact that if a carbonium ion were formed at the bridgehead, the bicyclic ring system would prevent its assuming a planar configuration.

[20] J. D. Roberts, *J. Am. Chem. Soc.*, **71**, 1880 (1949).

[21] P. D. Bartlett and L. H. Knox, *J. Am. Chem. Soc.*, **61**, 3184 (1939).

Since the electron-deficient carbon atom of a carbonium ion should form bonds by sp^2 hybridization, the most stable configuration for these bonds would be a coplanar one (the carbon-boron bonds in trialkylboron compounds are known to be coplanar), and therefore apocamphyl chloride could form only a very highly strained carbonium ion.   Reaction by the S_N2 mechanism is prevented by the impossibility of rearward attack and of inversion without enormous distortion of bond angles and distances.   Using more vigorous reaction conditions, Doering and coworkers found 1-bromobicyclo[2.2.1]heptane (I) to yield the corresponding alcohol upon treatment with aqueous silver nitrate at 150° for 2 days.[22]   The corresponding bicyclo[2.2.2]octane derivative (II) reacted analogously at

room temperature in 4 hr.   Although II is still far less reactive than a typical tertiary alkyl bromide, it is much more reactive than I, a fact that is attributed to the possibility of spreading over a larger number of bonds the strain required to approach planarity.

Because of resonance stabilization of the intermediate carbonium ion, the replacement of α-hydrogen atoms by phenyl radicals increases the S_N1 reactivity even more than replacement by alkyl groups does.   Thus it may be reasonably estimated that in "80 per cent acetone" triphenylmethyl chloride hydrolyzes about 10^6 times as rapidly as benzhydryl chloride, which in turn reacts about 10^3 times as rapidly as α-phenylethyl chloride.[23]

Unsaturation can have profound effects on S_N1 reactivity, the nature of these effects depending upon the location of the multiple bond.   Several reasons may be suggested for the familiar inertness of vinyl and phenyl halides (except for some negatively substituted derivatives whose reactions are discussed in Sec. 17-2).   The multiply bound carbon atoms have a greater electronegativity (see Sec. 2-4a) and hence decrease the ease with which a halide ion may be removed with its bonding electron pair.   Furthermore, the resonance stabilization resulting from the considerable contribution of structures like III is lost upon carbonium-

[22] W. von E. Doering, M. Levitz, A. Sayigh, M. Sprecher, and W. P. Whelan, Jr., *J. Am. Chem. Soc.*, **75,** 1008 (1953).

[23] E. D. Hughes, C. K. Ingold, and A. D. Scott, *J. Chem. Soc.*, 1201 (1937); M. G. Church, E. D. Hughes, and C. K. Ingold, *J. Chem. Soc.*, 966 (1940); C. G. Swain, C. B. Scott, and K. H. Lohmann, *J. Am. Chem. Soc.*, **75,** 136 (1953).

ion formation. On the other hand, allyl halides, like benzyl halides,

$$-\overset{|}{C}=\overset{|}{C}-\underline{\overline{X}}| \leftrightarrow -\overset{|}{C}-\overset{|}{\underset{\ominus}{C}}=\overset{\oplus}{\underline{\overline{X}}}$$

III

display considerably increased S_N1 reactivity, owing to resonance stabilization of the intermediate carbonium ions. More distantly located double bonds usually have little effect except in cases where neighboring-group participation (Sec. 14-1c) may occur.

The great solvolytic reactivity of α-halo ethers shows that the powerful resonance-electron-donating ability of the α-alkoxy substituent is much more important than its electron-withdrawing inductive effect.[24,25] The α-RS— group also appears to increase S_N1 reactivity by a large factor, although not so much as that found for the RO— group.[24,26]

With α-halogen substituents resonance and inductive effects are more closely balanced. Studies of benzyl halides have shown that α-chlorine substituents increase the S_N1 reactivity considerably, α-bromines do so to a lesser extent, and α-fluorines produce little, if any, activation (compared with α-hydrogens).[27,28] The larger effect of α-chlorine compared with α-bromine suggests that its greater resonance-electron-donating ability is more important than its greater negativity (3.0, compared to 2.8 for bromine). The sharp increase in electronegativity (to 4.0) on going to the α-fluorine substituent decreases the S_N1 reactivity in spite of fluorine's greater ability to donate electrons by resonance. This is thought to result from the manifestation of fluorine's great electronegativity in the form of resonance stabilization of the reactant α-fluorobenzyl halide by the contribution of such no-bond structures as

The contribution of such structures is supported by the particular stability of compounds with several fluorines on the same carbon atom[29]

[24] H. Böhme, *Ber.*, **74B**, 248 (1941).

[25] R. Leimu and P. Salomaa, *Acta Chem. Scand.*, **1**, 353 (1947).

[26] H. Böhme, H. Fischer, and R. Frank, *Ann.*, **563**, 54 (1949); F. G. Bordwell, G. D. Cooper, and H. Morita, *J. Am. Chem. Soc.*, **79**, 376 (1957).

[27] S. C. J. Olivier and A. P. Weber, *Rec. trav. chim.*, **53**, 869 (1934).

[28] J. Hine and D. E. Lee, *J. Am. Chem. Soc.*, **73**, 22 (1951); **74**, 3182 (1952).

[29] L. O. Brockway, *J. Phys. Chem.*, **41**, 185, 747 (1937); D. E. Petersen and K. S. Pitzer, *J. Phys. Chem.*, **61**, 1252 (1957).

and by the fact that $\sigma_{p\text{-}CF_3}$ is larger than $\sigma_{m\text{-}CF_3}$.[30] They have no counterparts among the contributing structures for the carbonium ion. It should not be surprising that if the presence of a highly electronegative fluorine atom on the same carbon as a phenyl group could bring about such resonance stabilization of the reactant, such an effect might also be noted when a chlorine atom is on the same carbon as some group that is a stronger resonance-electron donor than phenyl. Indeed, α-chlorine substituents have been found to decrease the S_N1 reactivity of chloromethyl methyl ether, presumably due to resonance stabilization of the reactants[31]

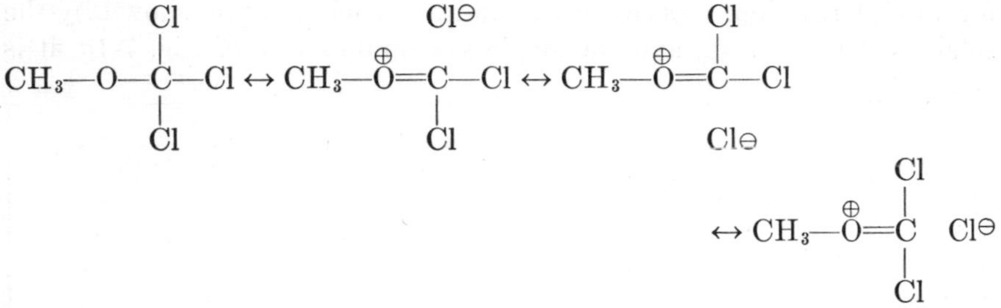

$$\begin{array}{c} \text{Cl} \\ | \\ CH_3—O—\overset{|}{\underset{|}{C}}—Cl \\ | \\ \text{Cl} \end{array} \leftrightarrow \begin{array}{c} \text{Cl}^\ominus \\ | \\ CH_3—\overset{\oplus}{O}{=}C—Cl \\ | \\ \text{Cl} \end{array} \leftrightarrow \begin{array}{c} \text{Cl} \\ | \\ CH_3—\overset{\oplus}{O}{=}C—Cl \\ \\ \text{Cl}^\ominus \end{array}$$

$$\leftrightarrow \begin{array}{c} \text{Cl} \\ | \\ CH_3—\overset{\oplus}{O}{=}C \quad Cl^\ominus \\ | \\ \text{Cl} \end{array}$$

As β substituents, where the electron-withdrawing inductive effect may operate but the electron-donating resonance effect may not, such groups as halogen, —OH, —OR, —OAc, —SR, —NR$_2$, etc., all decrease the ease of carbonium-ion formation, although the unimolecular reactivity may increase because of neighboring-group participation. The magnitude of both these effects is discussed in Sec. 7-3c (see Table 7-10).

Carbonyl groups appear to diminish S_N1 reactivity because of the partial positive charge placed on carbon by the contribution of resonance structures of the type of IV.

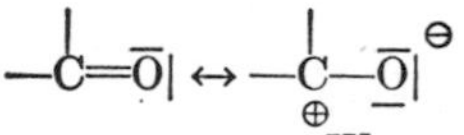

$$\begin{array}{c} | \\ —C{=}\bar{O}| \end{array} \leftrightarrow \begin{array}{c} | \\ —\underset{\oplus}{C}—\bar{O}|^\ominus \end{array}$$
$$\text{IV}$$

The effect of —CN, —SO$_2$R, —NO$_2$, and related groups is similar to that of carbonyl groups.

7-3b. Reactivity in S_N2 Reactions. In this section we shall usually make the common assumption that in the S_N2 transition state the three atoms covalently bound to the α-carbon atom are essentially coplanar with it and that the entering nucleophilic group (X) and the group being displaced (Y) are attached to the α-carbon atom to a nearly equal extent and are approximately collinear with it, as shown below (if X = Y, the α-carbon atom is exactly coplanar with R, R', and R'' and exactly

[30] J. D. Roberts, R. L. Webb, and E. A. McElhill, *J. Am. Chem. Soc.*, **72**, 408 (1950).
[31] J. Hine and R. J. Rosscup, *J. Am. Chem. Soc.*, **82**, 6115 (1960).

equally attached to X and Y; if R = R' = R'', X, Y, and C_α are exactly collinear).

$$\begin{array}{ccc} R & & R' \\ & \diagdown \quad \diagup & \\ X \text{------} & C & \text{------} Y \\ & \big| & \\ & R'' & \end{array}$$

This assumption implies that the plot of free energy vs. extent of reaction contains but a single maximum, like that shown in Fig. 7-1a.   By contrast, the proof of the existence of an intermediate shows that there is an intermediate minimum and hence two maxima in the energy curve for an S_N1 reaction.   Between the limits of such a curve, shown by the solid line in Fig. 7-1b, and the single maximum curve of Fig. 7-1a, it is

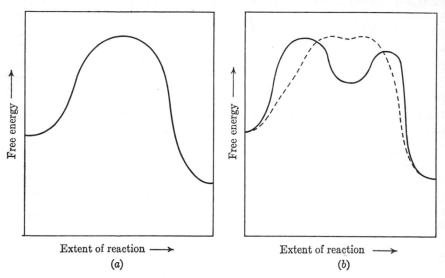

FIG. 7-1. Energy diagrams for reactions with and without an intermediate.

possible to construct any number of intermediate curves, such as that shown by dashed lines in Fig. 7-1b.   Although we know that there is no very stable intermediate in the S_N2 mechanism, it is possible that the energy diagram does contain a small minimum.[32]

In any event, in the transition state of a nucleophilic displacement on carbon, the nature of the C—X and C—Y bonds may vary considerably. They may be relatively long and ionic, as in the S_N1 transition state. Here X need not even be the nucleophilic reagent to which the α-carbon atom later forms a covalent bond but may instead be a molecule of solvent.   In this case the alkyl group (R_3C) will be positively charged.

---

[32] For a suggestion that there is an intermediate in the S_N2 reaction, see S. Winstein, D. Darwish, and N. J. Holness, *J. Am. Chem. Soc.*, **78**, 2915 (1956).

On the other hand, if the bonds are shorter and fairly covalent, as they appear to be in the S_N2 transition state, then the alkyl group may be negatively charged, or at least less positive than it was in the reacting molecule. This idea has been expressed in a somewhat different manner by Baker and Nathan,[33] and more recently by Swain and Langsdorf,[34] who point out that as an over-all result of reaction a new bond must be formed and the old bond broken and that the relative extent to which each process has occurred in the transition state may vary. If, in the transition state, the new bond has been formed to a greater extent than the old one has been broken, the alkyl group will be more negative than it was in the reacting molecule, and the reactivity will be increased by electron-withdrawing substituents, since substituents of this type are stabilized by being made negative. On the other hand, if the old bond has been broken to a greater extent than the new one has been formed, the alkyl group will be more positive in the transition state, and the reactivity will be increased by electron-donating groups, since these groups tend to be positive. These workers also suggest that, since electron-donating groups aid bond breaking and electron-withdrawing groups aid bond making, the reactivity of a compound for which bond making and bond breaking have proceeded to an approximately equal extent in the transition state may be increased by either type of group, since electron-donating groups may lead to a transition state in which the alkyl group is relatively positive and electron-withdrawing groups to one in which the alkyl group is more negative. In a plot of reactivity vs. the electron-withdrawing power of substituents (as in a Hammett-equation plot, for example) this would lead to a curve with a minimum at the point at which the charge on the alkyl group was the same in the transition state as in the reacting molecule, and indeed minima in Hammett-equation plots have been reported.[33,34]

Bond breaking is always vastly more extensive than bond making in the S_N1 transition state, but either may predominate in the S_N2 transition state, so that S_N2 reactions are sometimes aided by electron withdrawal and sometimes by electron supply. Furthermore, electronic effects are usually smaller for S_N2 displacements at a saturated carbon atom than for S_N1 reactions (although with acyl, vinyl, and phenyl halides, where the new bond may be very largely formed before the old one is broken, electron-withdrawing groups may have very large effects). However, since the α-carbon atom is bound to X and Y by relatively short partial covalent bonds in the S_N2 transition state, the reaction rate is quite sensitive to steric hindrance.

In discussing electronic effects alone, it is useful to consider the S_N2

[33] J. W. Baker and W. S. Nathan, *J. Chem. Soc.*, 1840 (1935); cf. E. D. Hughes, C. K. Ingold, and U. G. Shapiro, *J. Chem. Soc.*, 228 (1936).

[34] C. G. Swain and W. P. Langsdorf, Jr., *J. Am. Chem. Soc.*, **73**, 2813 (1951).

reactions of some meta- and para-substituted benzyl halides and related compounds. In reactions of benzyl bromides,[35] chlorides,[36] and fluorides[37] with nucleophilic anions, such as bromide, iodide, nitrate, and ethoxide, electron-withdrawing groups increased the reactivity, and so did electron-donating groups in most of the cases studied. Tertiary amines react with benzyl halides to form quaternary ammonium salts in which the benzyl group is relatively positive-charged due to the inductive effect of the attached nitrogen atom. Thus, it seems possible that the charge on the benzyl group in the transition state would be considerably greater than in S_N2 reactions involving attack by anions. Since in cases of the latter type the charge in the transition state appears to be in the vicinity of that in the halide molecule for the *unsubstituted* benzyl halides, the unsubstituted benzyl group should be more positive in the transition state of an S_N2 attack by a tertiary amine, so that the reactivity would be increased by electron-donating groups and decreased by electron-withdrawing groups (until the electron withdrawal becomes strong enough to produce a rate minimum). This seems to be the case. Baker and Nathan found the reactivities of benzyl bromides toward pyridine in acetone to be increased by electron-donating and decreased by electron-withdrawing groups until a rate minimum is reached in the vicinity of the *p*-nitro compound.[33] The data seem to be approaching a minimum in this vicinity in the reactions of trimethylamine with benzyl halides.[34]

The effect of ortho substituents on the reactivity of benzyl halides is fairly predictable qualitatively when steric hindrance (usually not large for one ortho substituent) is added to electronic effects of the type described.

Because of the complications produced by the possibility of steric interference with the attack of the nucleophilic reagent, of steric acceleration of the departure of the group displaced, and of entropy effects due to interference with rotation around single bonds and changes in the extent of solvation, it is very difficult to learn what influences electronic effects have on the S_N2 reactivity of alkyl halides and even more difficult to predict the reactivity that will result from the addition of an electronic effect to the other effects described. Nevertheless, some arguments will be presented leading to some generalizations that are in agreement with many of the known data and that permit predictions in untested cases.

[35] J. W. Baker and W. S. Nathan, *J. Chem. Soc.*, 236 (1936); S. Sugden and J. B. Willis, *J. Chem. Soc.*, 1360 (1951).

[36] H. Franzen, *J. prakt. Chem.*, **97**, 82 (1918); H. Franzen and I. Rosenberg, *J. prakt. Chem.*, **101**, 333 (1921); G. M. Bennett and B. Jones, *J. Chem. Soc.*, 1815 (1935).

[37] W. T. Miller, Jr., and J. Bernstein, *J. Am. Chem. Soc.*, **70**, 3600 (1948).

In view of the fact that the S_N2 reactivity (in the cases studied) of methyl halides is decreased by the introduction of either an α-methyl group (electron donating) or an α-halogen atom (electron withdrawing), it seems likely that the decrease is at least partially due to steric hindrance. Since it appears that steric effects are often the dominant factor in determining S_N2 reactivity, the discussion of this subject by Dostrovsky, Hughes, and Ingold[38] is of particular interest. This discussion is in terms of an S_N2 transition state of the type shown in Fig. 7-2. Either Fig. 7-2a or b may be considered as a model for the transition state of an S_N2 reaction of a methyl halide, where Cα represents the carbon atom,

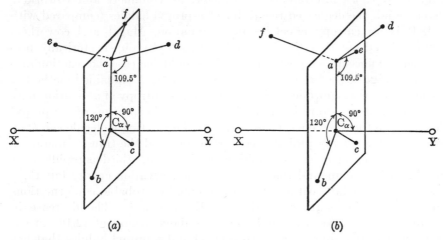

(a)   (b)

FIG. 7-2. Transition states for reaction by the S_N2 mechanism. [*Reproduced with permission from Dostrovsky, Hughes, and Ingold, J. Chem. Soc., 173 (1946).*]

a, b, and c the hydrogen atoms, and X and Y the entering and departing nucleophilic groups. We shall use this model as a basis for comparison. In the transition state for ethyl halides a represents the β-carbon atom, d, e, and f being β-hydrogens. Fig. 7-2a very probably represents a more stable ethyl transition state than Fig. 7-2b, in which one hydrogen is placed as closely as possible to the entering nucleophilic group. Nevertheless, from calculations based on certain reasonable assumptions, it seems that the interaction of the β-hydrogen atoms with X and Y, as shown in Fig. 7-2a, will produce steric strains.[39] This agrees with the

[38] I. Dostrovsky, E. D. Hughes, and C. K. Ingold, *J. Chem. Soc.*, 173 (1946).

[39] For calculations of steric hindrance in S_N2 reactions other than those of Ref. 38, see E. C. Baughan and M. Polanyi, *Trans. Faraday Soc.*, **37**, 648 (1941); A. G. Evans and M. Polanyi, *Nature*, **149**, 608 (1942); A. G. Evans, *Trans. Faraday Soc.*, **42**, 719 (1946); P. B. D. de la Mare, L. Fowden, E. D. Hughes, C. K. Ingold, and J. D. H. Mackie, *J. Chem. Soc.*, 3200 (1955).

previous suggestion that the greater reactivity of methyl compared with ethyl halides, which has been observed in all nucleophilic displacements proved to be S_N2 in character, is at least partly due to steric hindrance. As would be expected, continued replacement of α-hydrogen atoms by methyl groups produces continued deactivation due to steric hindrance. For example, the following relative reactivities of alkyl bromides toward iodide ions in acetone have been reported: Me, 10,000; Et, 65; i-Pr, 0.50.[40]

The introduction of a β-methyl group into an ethyl halide to get the n-propyl compound need not produce added steric hindrance, since the carbon of the β-methyl group may be at f in Fig. 7-2a, too distant from X and Y to cause interference. However, as Hammett and coworkers have found, the decreased reactivity of n-propyl halides (compared with ethyl) is an entropy effect.[41] In explanation, Magat and coworkers have pointed out that the fact that the methyl group is sterically prevented from occupying positions d and e should decrease the reaction rate by changing the probability or entropy factor.[42] A simplified method of estimating the magnitude of this change in the entropy of activation may be based on the assumption that the three rotational forms of a n-propyl halide have the same energy content. If the reacting molecule can then exist in three equally probable conformations and the transition state is limited to one, the reaction should proceed only one-third as rapidly as it would with a compound that has the same activation energy but that, like an ethyl halide, can exist in three equally probable conformations in the transition state as well as in the reactant. The decrease in entropy of activation expected from the above argument (2.18 e.u.) is not far from the difference between ethyl and n-propyl halides that has been observed in several cases (2.3 e.u. for reaction with sodium thiophenolate in methanol,[43] 3.0 e.u. for reaction with sodium thiosulfate in 50 per cent ethanol[41]). When two β-methyl groups are present, however, as in an isobutyl halide, there will be increased steric repulsions even in the more favorable transition state shown in Fig. 7-2b (methyl groups at d and e), and indeed isobutyl halides show a decrease in reactivity in which increased activation energy is an important factor.[38,41] The presence of three β-methyl groups, as in a neopentyl halide, produces greatly increased strain in the transition state (Fig. 7-2a).

In accord with these generalizations, in S_N2 reactions methyl halides

[40] L. Fowden, E. D. Hughes, and C. K. Ingold, *J. Chem. Soc.*, 3187 (1955); the relative reactivity reported by these workers for t-butyl bromide has been shown to be unreliable by S. Winstein, S. Smith, and D. Darwish, *Tetrahedron Letters*, **16**, 24 (1959).

[41] T. I. Crowell and L. P. Hammett, *J. Am. Chem. Soc.*, **70**, 3444 (1948); P. M. Dunbar and L. P. Hammett, *J. Am. Chem. Soc.*, **72**, 109 (1950).

[42] N. Ivanoff and M. Magat, *J. chim. phys.*, **47**, 914 (1950); E. Bauer and M. Magat, *J. chim. phys.*, **47**, 922 (1950).

[43] J. Hine and W. H. Brader, Jr., *J. Am. Chem. Soc.*, **75**, 3964 (1953).

are usually found to be from 4 to 150 times as reactive as the corresponding ethyl halides, which in turn are usually from 1.5 to 5 times as reactive as the n-propyl halides. The reactivity of higher straight-chain primary halides, including n-propyl, does not usually differ from the average value by more than 50 per cent. If there is branching on the β-carbon (or considerable branching on the γ-carbon), the reactivity will be less; e.g., toward sodium ethoxide in ethanol at 95° the relative S_N2 reactivities of ethyl, isobutyl, and neopentyl bromides are 1, 0.04, and 10^{-5}, respectively.[44]

No completely successful general theory of electronic effects on the S_N2 reactivity of aliphatic compounds appears to have been suggested, but several useful generalizations can be made. With saturated halides, electron-withdrawing substituents usually decrease the S_N2 reactivity, and electron-donating groups increase it. Some examples may be seen in Table 7-7, where data on the reactivity of alkyl bromides toward

TABLE 7-7. REACTIVITIES OF ALKYL BROMIDES TOWARD SODIUM THIOPHENOLATE
IN METHANOL AT 20°[43]

Alkyl group	k, liters mole^{-1} sec^{-1}, $\times 10^4$	$\Delta H^{\ddagger}$,[a] kcal/mole	$\Delta S^{\ddagger}$,[b] e.u./mole
HCH_2CH_2	39.1	18.1	-7.7
$CH_3CH_2CH_2$	25.6	17.7	-10.0
$C_2H_5CH_2CH_2$	26.9	17.7	-10.0
FCH_2CH_2	4.95	19.4	-7.3
$ClCH_2CH_2$	5.61	19.1	-8.3
$BrCH_2CH_2$	4.99[c]	18.7	-9.7

[a] ± 0.3 kcal.
[b] ± 0.7 e.u.
[c] Rate constant per bromine (one-half the total rate constant for ethylene bromide).

sodium thiophenoxide in methanol are listed.[43] The replacement of β-hydrogen atoms by fluorine, chlorine, or bromine atoms was found to decrease the reactivity. The decrease in entropy of activation that is seen to accompany the increasing size of the β substituent may be due to increasing interference in rotation around the C_α—C_β bond in the transition state. The increase in heat of activation that accompanies the increased electron-withdrawing power of the β substituent is believed to be an electronic effect. The introduction of a halogen atom on the γ-, or a more distant, carbon atom of an alkyl chain usually has little effect on the S_N2 reactivity.[45] Some of the data of Table 7-8, relating to the reactivities of organic chlorides toward potassium iodide in acetone,

[44] I. Dostrovsky and E. D. Hughes, *J. Chem. Soc.*, 157 (1946).
[45] J. Hine, S. J. Ehrenson, and W. H. Brader, Jr., *J. Am. Chem. Soc.*, **78**, 2282 (1956).

may also be interpreted in this light. The α-acetoxy and α-methoxy groups cause an increase in reactivity, since their net effect is to donate electrons by contributions of structures like

$$CH_3\overset{\oplus}{-}\overset{H}{\underset{|}{\overset{|}{O}}}=\overset{|}{\underset{H}{\overset{|}{C}}}\quad Cl^{\ominus}$$

In the β position, however, an alkoxy or phenoxy group can only withdraw electrons and thus bring about deactivation. To just what mixture of

TABLE 7-8. RELATIVE REACTIVITIES OF ALKYL CHLORIDES TOWARD POTASSIUM IODIDE IN ACETONE AT 50°[46]

Alkyl group	k^a	Alkyl group	k^a
CH_3	200^b	$NCCH_2{}^c$	3,000
C_2H_5	2.5	$C_6H_5COCH_2{}^c$	100,000
$n\text{-}C_3H_7$	1.1	$C_6H_5COCH_2CH_2$	87
$n\text{-}C_4H_9$	1.0	$C_6H_5CO(CH_2)_3{}^c$	370
$n\text{-}C_5H_{11}$	1.4	$CH_3COCH_2{}^c$	36,000
$n\text{-}C_6H_{13}$	1.3	$EtO_2CCH_2{}^c$	1,700
$n\text{-}C_{12}H_{25}$	1.2	$EtO_2CCH_2CH_2$	1.6
$n\text{-}C_{30}H_{61}$	0.9	$EtO_2C(CH_2)_3$	1.6
$i\text{-}C_3H_7{}^c$	0.02	$CH_3CO_2CH_2{}^c$	270
$sec\text{-}C_4H_9{}^c$	0.02	$CH_3CO_2CH_2CH_2$	0.4
$CH_2{=}CHCH_2{}^c$	79	$CH_3CO_2(CH_2)_3{}^c$	40
$C_6H_5CH_2{}^c$	200	$CH_3OCH_2{}^c$	920
$C_6H_5CH_2CH_2$	1.2	$C_6H_5OCH_2CH_2$	0.3
Cyclohexyl	$<10^{-3}$	H_2NCOCH_2	99
C_6H_5	$<10^{-3}$	$CH_3SCH_2CH_2$	1.5
$ClCH_2{}^d$	0.2^e	$HOCH_2CH_2$	1.7
$ICH_2{}^d$	0.1	$HOCH_2CH_2CH_2{}^c$	2.1^c

^{a} Relative to $n\text{-}C_4H_9Cl$.

^{b} On the basis of the assumption that the reactivity of CH_3Br relative to $n\text{-}C_3H_7Br$[40] is the same as that of CH_3Cl relative to $n\text{-}C_3H_7Cl$.

^{c} Calculated from data at other temperatures.

^{d} J. Hine, C. H. Thomas, and S. J. Ehrenson, *J. Am. Chem. Soc.*, **77**, 3886 (1955).

^{e} A statistical factor of two has been applied in this case to obtain the reactivity per chlorine.

electronic and steric effects the deactivating influence of α-halogens is due is not known, but both factors are probably important.

Multiple bonds near the reaction center have a strong effect on S_N2 reactivity. Nucleophilic substitution reactions of phenyl and vinyl

[46] J. B. Conant et al., *J. Am. Chem. Soc.*, **46**, 232 (1924); **47**, 476, 488 (1925); W. R. Kirner, *J. Am. Chem. Soc.*, **48**, 2745 (1926); **50**, 2446 (1928).

halides are extremely slow (unless strongly electron-withdrawing substituents are present), but according to a narrow definition of S_N1 and S_N2
(as referring to substitution at saturated carbon), these compounds
cannot react by the S_N2 mechanism; they will be discussed in Chap. 17.
When the carbon atom attached directly to the reaction center is unsaturated, a marked increase in reactivity is observed.   This may be seen in
Table 7-8 for the cases of allyl, benzyl, and phenacyl chlorides and the
α-chloro derivatives of acetonitrile, acetone, acetamide, and ethyl
acetate.   The increased reactivity of these compounds compared with
typical primary chlorides may be due in part to a steric effect.   Thus, in
the S_N2 transition state shown in Fig. 7-2 an sp^2 carbon atom at a could
have all three of its bonds in the plane perpendicular to the line described
by X, C, and Y.   There should then be less interference with X and Y
than if this carbon were sp^3.   From the relative reactivities listed it also
seems clear that a steric effect is not the only factor of importance.
Stabilization by overlap of the orbital by which X and Y are bound
with the p orbital on the adjacent sp^2 (or sp) carbon atom may also be
important.

This explanation is supported by Bartlett and Trachtenberg's evidence
that the great S_N2 reactivity characteristic of phenacyl compounds is
not found in the case of 5,7-dinitro-3-coumaranone

where orbital overlap of the type suggested is made impossible by the
fact that the CH_2—O bond cannot become perpendicular to the plane of
the carbonyl group.[47]   Other, somewhat related explanations are not
ruled out, however.

The reactivity of compounds with sp^2 atoms further from the reaction
center sometimes seems rather anomalous.   For example, the reactivity

[47] P. D. Bartlett and E. N. Trachtenberg, *J. Am. Chem. Soc.*, **80**, 5808 (1958).

of β-phenylethyl chloride toward potassium iodide in acetone is increased by p-halogen and p-nitro substituents[48] despite the fact that such β-electron-withdrawing substituents as halogens usually decrease S_N2 reactivity.   Other rather surprising data are seen in Table 7-8.   The compounds $C_6H_5COCH_2CH_2Cl$ and $C_6H_5COCH_2CH_2CH_2Cl$, among others, are unexpectedly reactive.   It is possible that not all the reactions really proceed by the S_N2 mechanism.   Furthermore, there are complications due to ion-pair formation.[49]   Even allowing for these factors, though, some anomalies still seem to be present.

The decreased S_N2 reactivities of cyclopropyl and cyclobutyl halides (see Table 7-9) have been attributed to the same factors that explained

TABLE 7-9.   REACTIVITY OF CYCLOALKYL BROMIDES TOWARD IODIDE ION
IN ACETONE[50]

Alkyl group	10^7k at 70°, liters mole^{-1} sec^{-1}
Cyclopropyl	<0.01[a]
Cyclobutyl	0.98
Cyclopentyl	208
Cyclohexyl	1.29
Cycloheptyl	127
Isopropyl	130

[a] Based on the report of "no reaction" at 100° in J. D. Roberts and V. C. Chambers, *J. Am. Chem. Soc.*, **73**, 5034 (1951).

their S_N1 inertness, since a more electronegative carbon atom should be less susceptible to nucleophilic attack, a carbon-halogen bond with double-bond character should be more difficult to break, and I strain should still be present, the S_N2 transition state, too, having optimum bond angles of 120°.[18,19,50]   Cyclohexyl bromide is also seen to be rather unreactive, and, indeed, from an inspection of molecular models it is clear that there will be considerable steric hindrance in the S_N2 transition state for a cyclohexyl halide.   Cyclopentyl and cycloheptyl bromides are found to be about as reactive as the isopropyl compounds, probably because of a compensation of effects.

7-3c. Reactivity in Neighboring-group Displacements.   Winstein, Grunwald, and coworkers have given an excellent discussion of the effect of structure on the rate of reactions involving neighboring groups.[51]   Using solvolysis data for reactions in which S_N2 attack by solvent seems very unlikely, these workers point out that the reaction of a compound con-

---

[48] G. Baddeley and G. M. Bennett, *J. Chem. Soc.*, 1819 (1935).

[49] S. Winstein, L. G. Savedoff, S. Smith, I. D. R. Stevens, and J. S. Gall, *Tetrahedron Letters*, **9**, 24 (1960).

[50] P. J. C. Fierens and P. Verschelden, *Bull. soc. chim. Belges*, **61**, 427 (1952).

[51] S. Winstein, E. Grunwald, et al., *J. Am. Chem. Soc.*, **70**, 812, 816, 821, 828 (1948).

taining the β neighboring group —SA may have as its rate-controlling step either a carbonium-ion formation

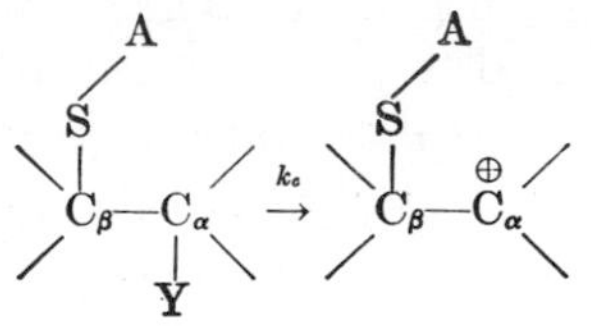

or a nucleophilic displacement by the neighboring group.

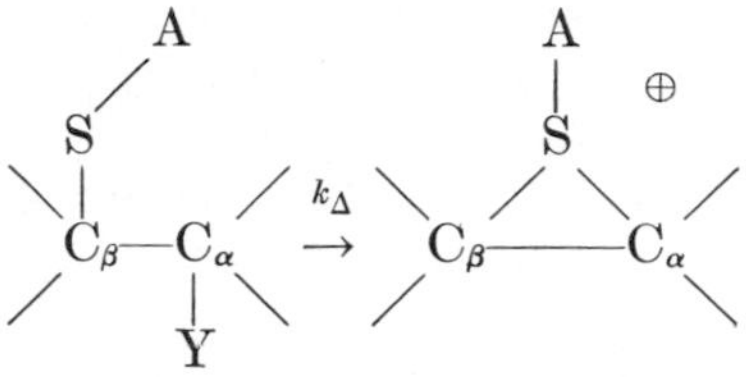

Before considering the effect of the nature of —SA on the reactivity, it is useful to discuss the effect of the nature of C_α and C_β. As previously stated, methyl groups on C_α increase k_c, the rate constant for carbonium-ion formation, by a factor of about 10^4. Methyl groups on C_β usually have little effect. The initial product of neighboring-group participation will be a resonance hybrid of the structures

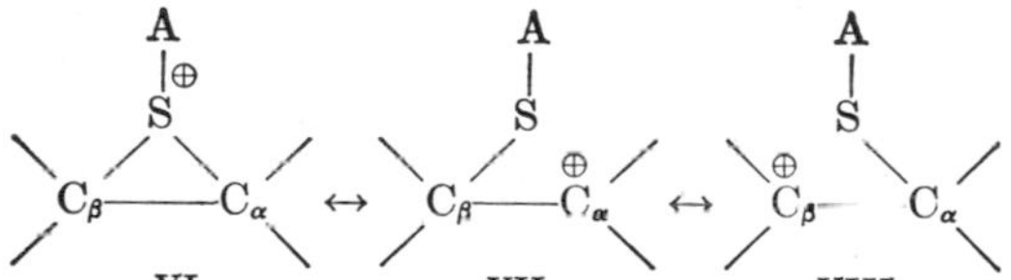

to which VII and VIII will contribute equally if C_α and C_β are identically substituted. In the transition state leading to this intermediate, however, a structure similar to VII should contribute more than one analogous to VIII. Therefore it is reasonable that the rate should be increased by methylation of either carbon atom but more so by methylation of C_α. Winstein and Grunwald state that a methyl group on C_α increases k_Δ, the rate constant for neighboring-group displacement, by about 200-fold and on C_β by about 20-fold. This stabilizing effect of both α- and β-methyl groups on the three-membered ring of the transition state may be related to their stabilization of double bonds, or two-membered rings (see Sec. 1-1d).

The nucleophilic driving force that the participation of the neighboring group adds to the solvolysis rate may be estimated from a determination of k, the solvolysis rate constant, and an estimate of k_c, since

$$k = k_c + k_\Delta$$

Since β-hydrogen does not appear to participate in the cases studied, Winstein and coworkers approximated k_c by estimating the effect that the transformation of β-hydrogen to the substituent in question would have on the energy required to reach the transition state for carbonium-ion formation. Then k_c was determined by the amount by which this change in energy would affect k_H, the solvolysis rate constant for the unsubstituted compound. The estimation involved a calculation of the interaction between the dipole due to —SA and the one being formed by ionization of the C—Y bond. It was improved by treating the effective dielectric constant as a disposable parameter, which was given the value necessary to make k equal k_c for the substituent β-chlorine, for which there is good evidence for nonparticipation in the cases studied. The nucleophilic driving force of the β substituent may be expressed either in terms of k_Δ/k_c, the ratio of the rate of the neighboring-group displacement to that of the carbonium-ion formation, or in terms of k/k_H, the factor by which the presence of the β substituent changes the solvolysis rate. Estimates of these factors and of k_c/k_H, the factor by which the β substituent changes the rate of carbonium-ion formation, are given for ethyl compounds in Table 7-10. To apply these generalizations to other than ethyl compounds, note should be taken of the effects that substitutions on C_α and C_β have on k_Δ and k_c. A correlation may be noted between the neighboring-group nucleophilicities of various groups shown in Table 7-10 and the nucleophilic constants of related species listed in Table 7-4.

TABLE 7-10. EFFECTS OF β SUBSTITUENTS ON RATES OF NEIGHBORING-GROUP DISPLACEMENTS AND CARBONIUM-ION FORMATION OF ETHYL COMPOUNDS[51]

β group	k/k_H	k_Δ/k_c	k_c/k_H
O^{-a}	10^{10}	10^4	10^6
SCH$_2$CH$_2$OH	10^7	10^9	10^{-2}
NH$_2$	10^4	10^5	10^{-1}
I	1.6×10^3	2×10^6	7×10^{-4}
Br	0.4	2×10^3	2×10^{-4}
OH, OCH$_3$	0.1	10	10^{-2}

[a] The values for this substituent were obtained from data on base-catalyzed transformations of halohydrins to epoxides and involve estimates of the acidities of the halohydrins.

A factor of considerable importance in determining the reactivity of a neighboring group is its distance from the carbon atom it attacks. Although nearness of a neighboring group should in itself facilitate reaction, the nearer neighboring groups will produce small, and hence strained, rings. To avoid strain a ring of five members or more should be formed. In all cases that have been investigated there have been other factors operating in addition to ring size.   Freundlich, Salomon, and coworkers studied the cyclization of compounds of the type $H_2N(CH_2)_nBr$ when n was varied from two to six with the results shown in Table 7-11.[52]   Here it

TABLE 7-11. RATES OF CYCLIZATION OF ω-AMINOALKYL BROMIDES IN WATER AT 25°[52]

Compound	10^4k, sec^{-1}
$H_2N(CH_2)_2Br$	6
$H_2N(CH_2)_3Br$	0.08
$H_2N(CH_2)_4Br$	5,000[a]
$H_2N(CH_2)_5Br$	80
$H_2N(CH_2)_6Br$	0.1

[a] Estimated from k for the chloride (80×10^{-4} sec^{-1}).

seems that an increase in ring size from three to four does not offset the effect of separating the two reactive centers, but the almost total removal of ring strain in the five-membered case does.   Thereafter, with no further large change in ring strain, the distance factor controls.   In addition to these factors, however, it might also be noted that the amino group in aminoethyl bromide is probably the least basic of those listed, so that its nucleophilicity (toward external carbon atoms anyway) should be the least.   Similarly the ease of replacement of the bromine in this molecule will be affected by the presence, on the β-carbon atom, of an amino group (becoming an ammonium group as the reaction proceeds).

TABLE 7-12. RATES OF LACTONIZATION OF THE ANIONS OF BROMO ACIDS IN WATER AT 25°[53]

Acid	10^6k, sec^{-1}
α-Bromopropionic.........	0.42
α-Bromohexanoic..........	0.55
β-Bromopropionic.........	3.5
β-Bromohexanoic..........	35
γ-Bromopentanoic.........	5,500
ϵ-Bromohexanoic..........	1.7
ζ-Bromoheptanoic.........	0.43[a]

[a] This reaction may just be hydrolysis, and if so, the k is a maximum for lactonization.

The substitution of alkyl groups alpha to the bromine atom in the series above could change the order of reactivities considerably since, as mentioned earlier, this would increase the ease of formation of the three-membered ring, but, as Freundlich and Salomon have found, it decreases the rate of reaction to give a five-membered ring (just as it decreases intermolecular S_N2 reactivity).

[52] This work is summarized and discussed by G. Salomon, *Helv. Chim. Acta*, **16**, 1361 (1933); **17**, 851 (1934).

The first-order decomposition of the anions of a number of bromo acids, believed to involve the intermediate formation of lactones in all cases except those noted, was studied by Lane, Heine, and coworkers.[53]   The data (Table 7-12) show that in this case the four-membered ring is formed faster than the three.   This is probably due to the fact that, with the lactones but not the cyclic amines, the three-membered ring has added destabilization (I-strain) due to the presence of an sp^2 carbon atom.

7-4. Effect of the Nature of Y on the Reactivity of RY.   Since the group Y, in being displaced, is acquiring sole possession of an electron pair that it previously merely shared, it is reasonable that its reactivity should increase as the basicity of Y| decreases.   This correlation is only very approximate unless one restricts the comparison to Y groups in which the nature of the atom forming the bond to carbon is the same, and even in this case it is not at all perfect.   Fairly good Hammett-equation correlations have been obtained in reactions of ethyl esters of various benzenesulfonic acids.   For the solvolysis in 30 per cent ethanol at 25°, log k_0 (sec^{-1}) = -5.254, and $\rho = 1.168$, whereas for reaction with sodium ethoxide in ethanol at 35°, log k_0 (liters mole^{-1} sec^{-1}) = -2.881 and $\rho = 1.372$.[8]   A considerably broader linear free-energy relationship has been found by Hammett and Pfluger to exist for the reaction of trimethylamine with methyl esters.[54]

$$RCO_2CH_3 + (CH_3)_3N \rightarrow RCO_2^{\ominus} + (CH_3)_4N^{\oplus}$$

These workers found that a logarithmic plot of the rate constants for these reactions against the ionization constants of the corresponding acids, RCO_2H, approximates a straight line even when data on aliphatic acids and ortho- as well as meta- and para-substituted benzoic acids are used.   They state that the relation does not extend to the reaction of trimethylamine with the methyl "esters" of phenols, however.

Since the coordination of a proton with Y| will always decrease its basicity, the group YH will always be more readily displaced than Y itself.   This fact probably finds its greatest utility in the reactions of alcohols and ethers, since there are very few instances of nucleophilic displacements of OH and OR groups attached to saturated carbon atoms. These groups are, however, sufficiently basic to exist to a considerable extent as OH_2^+ and OHR^+ under acidic conditions, and it is almost always in these forms that they react.   The solvolysis of fluorides has also been shown to be acid-catalyzed.[37,55]   While the solvolysis of chlorides has been found to be insensitive to acid catalysis in aqueous

[53] J. F. Lane and H. W. Heine, *J. Am. Chem. Soc.*, **73**, 1348 (1951); H. W. Heine, E. Becker, and J. F. Lane, *J. Am. Chem. Soc.*, **75**, 4514 (1953).

[54] L. P. Hammett and H. L. Pfluger, *J. Am. Chem. Soc.*, **55**, 4079 (1933).

[55] N. B. Chapman and J. L. Levy, *J. Chem. Soc.*, 1677 (1952).

solution,[56] certain reactions of chlorides are subject to acid catalysis in more weakly basic solvents, such as nitrobenzene.[57]   The Friedel-Crafts reaction is an example of a Lewis acid–catalyzed nucleophilic displacement reaction that proceeds with bromides and iodides as well as chlorides and fluorides.   In general, the rate of displacement of Y cannot be increased by its coordination with a proton (or other acid) by a factor any larger than the ratio $[HY]/[Y^-]$ that is characteristic of Y^- under the conditions of acidity being employed.

Another factor that may have an enormous effect on reactivity is the release during a nucleophilic displacement of the strain associated with a three- or four-membered ring.   Although unprotonated OR groups are ordinarily displaced only with the greatest difficulty, ethylene oxide is so reactive that it will render an aqueous solution of potassium chloride basic to phenolphthalein quickly at room temperature due to the reaction[58]

$$Cl^- + CH_2CH_2 \rightarrow ClCH_2CH_2O^\ominus + H_2O \rightarrow ClCH_2CH_2OH + OH^-$$

Brønsted, Kilpatrick, and Kilpatrick made a careful kinetic study of the analogous reactions of glycidol (2,3-epoxypropanol) and epichlorohydrin (2,3-epoxypropyl chloride) and showed that they involve a bimolecular attack by the chloride ion in the rate-controlling (first) step.   Similar reactions have been found for a number of other nucleophilic anions, including hydroxide ion.   In addition, at pH's lower than about 4, acid-catalyzed reactions become noticeable.   These include third-order reactions such as one whose rate may be expressed

$$v = k[C_2H_4O][H^+][Br^-]$$

and which probably proceeds by the mechanism

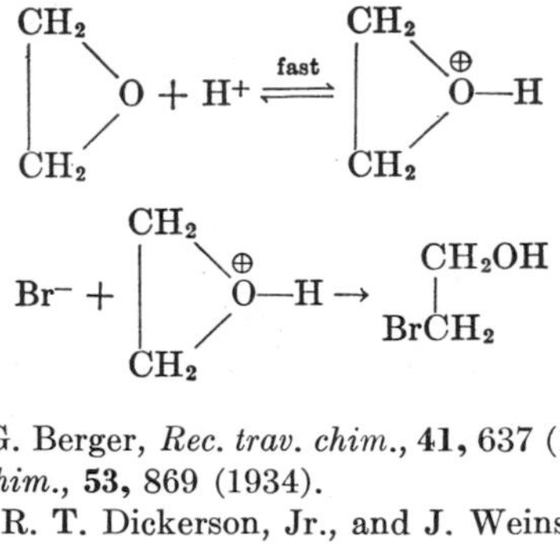

[56] S. C. J. Olivier and G. Berger, *Rec. trav. chim.*, **41,** 637 (1922); S. C. J. Olivier and A. P. Weber, *Rec. trav. chim.*, **53,** 869 (1934).

[57] H. F. Herbrandson, R. T. Dickerson, Jr., and J. Weinstein, *J. Am. Chem. Soc.*, **76,** 4046 (1954).

[58] J. N. Brønsted, M. Kilpatrick, and M. Kilpatrick, *J. Am. Chem. Soc.*, **51,** 428 (1929).

and also a second-order acid-catalyzed hydration, which may be either an S_N2 attack of water on the conjugate acid of the epoxide or the spontaneous formation of a carbonium ion from this conjugate acid followed by combination with water, depending on the structure of the epoxide.[59] There is also a first-order uncatalyzed hydration reaction that, in many cases at least, probably involves an S_N2 attack by solvent molecules. A mechanism involving an S_N2 attack by hydroxide ions on the conjugate acid of the epoxide is in agreement with the kinetics, since the product, $[H^+][OH^-]$, is a constant. But in the cases of ethylene oxide and glycidol it does not seem likely that this mode of reaction comprises a very large fraction of the total in view of the magnitude of the rate constants for reaction of hydroxide ions with the epoxides themselves[6,60] and the extent to which acid is found to catalyze the reactions with other nucleophilic reagents.[58]

Searles has shown that a change in ring size has the expected effect, the reaction of thiosulfate ion with ethylene oxide proceeding more than 15 times as rapidly as the reaction with trimethylene oxide.[61] The acid-catalyzed reaction, however, appears to proceed at about the same rate for both compounds. This is probably due to the fact that the greater basicity of trimethylene oxide increases the equilibrium concentration of its conjugate acid. The difficulty of cleavage of the five-membered ring of tetrahydrofuran is comparable to that of an aliphatic ether.

In comparisons of the ease of displacement of various Y groups when there are variations in the nature of the atom whose bond to carbon is broken, there appears to be no over-all theory to correlate the existing data, although the basicity of the group displaced is still important. Some of the more common groups, in the order of their decreasing ease of displacement, are as follows: $-\overset{+}{N}{\equiv}N > OSO_2R > I \sim Br > NO_3 \sim Cl > OH_2^+ \sim SMe_2^+ > F > OSO_3^- > NR_3^+ > OR > NR_2$. This list is by no means invariable. There will be a large change in the relative reactivities of charged groups compared with uncharged ones as the solvent is changed (the list above was largely estimated from data in aqueous solution). These changes may be predicted qualitatively from the theory of solvent effects of Sec. 3-1f. There will also be changes in the series with variations in the nature of the alkyl group to which Y is attached and with variations in the nucleophilic reagent. Most of the data on variations of Y concern alkyl halides. Glew and Moelwyn-Hughes found the relative rates of solvolysis in water at 100° to be

[59] The mechanisms of this and related reactions are discussed by S. Winstein and R. B. Henderson, in R. C. Elderfield, "Heterocyclic Compounds," vol. I, chap. 1. John Wiley & Sons, Inc., New York, 1950.

[60] H. J. Lichtenstein and G. H. Twigg, *Trans. Faraday Soc.*, **44**, 905 (1948).

[61] S. Searles, *J. Am. Chem. Soc.*, **73**, 4515 (1951).

$CH_3F:CH_3Cl:CH_3Br:CH_3I::1:25:300:100.$[62] More often the iodide is found to be the most reactive of the halides, as in the example of Tronov and Krüger, who report the relative reactivities of isoamyl halides toward piperidine to be $RF:RCl:RBr:RI::1:68:17,800:50,500.$[63] The reactivity of fluorides relative to chlorides is often much lower than in the two preceding cases, k_{RCl}/k_{RF} being about 10^6 in the solvolysis of triphenylmethyl halides in 85 per cent acetone.[64]

PROBLEMS

1. Calculate rate constants for each of the following reactions:

(a) The solvolysis of t-butyl chloride in pure water at 25°

(b) The solvolysis of triphenylmethyl fluoride in 50 per cent ethanol at 25°

(c) The reaction of p-nitrophenoxide ions with ethylene oxide in 98 per cent ethanol at 70.4°

2. With which compound would solvolysis be more likely to involve neighboring-group participation by the dimethylamino substituent: p-$CH_3OC_6H_4CHClCH_2$-$N(CH_3)_2$ or p-$O_2NC_6H_4CHClCH_2N(CH_3)_2$? Why?

3. Suggest reasonable structural formulas for each of the following compounds:

(a) Compound A (C_2H_5OCl) solvolyzes, liberating HCl, with a half-life of less than 10 min in 50 per cent ethanol–50 per cent ether at 0°.

(b) Compound B ($C_5H_{11}Cl$) reacts with sodium ethoxide in ethanol at 80° less than one-thousandth as fast as ethyl chloride.

(c) Compound C (C_3H_8OCl) reacts with sodium ethoxide in ethanol more slowly than ethyl chloride does.

4. Number the compounds in the following series in the order of their decreasing rates of solvolysis in ethanol: C_6H_5Cl, $(p$-$CH_3OC_6H_4)_2CHCl$, $(C_6H_5)_2CHCl$, n-$C_5H_{11}Cl$, p-$ClC_6H_4CH_2Cl$, $C_6H_5CHClCH_3$. Give your reasons.

[62] D. N. Glew and E. A. Moelwyn-Hughes, *Proc. Roy. Soc. (London)*, **211A,** 254 (1952).

[63] B. V. Tronov and E. A. Krüger, *Zhur. Russ. Fiz.-Khim. Obshchestva,* **58,** 1270 (1926); *Chem. Abstr.,* **21,** 3887[9] (1927).

[64] C. G. Swain and C. B. Scott, *J. Am. Chem. Soc.,* **75,** 246 (1953).

Chapter 8

POLAR ELIMINATION REACTIONS

Elimination reactions are processes in which two atoms or groups are removed from a molecule without being replaced by other atoms or groups. Usually these groups are on adjacent atoms, so that a multiple bond is formed, e.g.,

$$X-C_\beta-C_\alpha-Y \xrightarrow{-X \text{ and } Y} C=C$$

Elimination reactions of this type are called beta eliminations, and since they are the most common, the "beta" is often omitted. Such reactions as

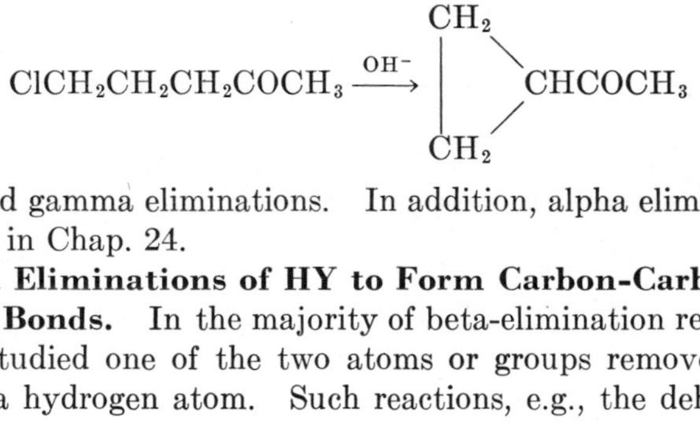

$$ClCH_2CH_2CH_2COCH_3 \xrightarrow{OH^-} \begin{array}{c} CH_2 \\ | \\ CH_2 \end{array}\!\!\!\!\!\!CHCOCH_3$$

may be called gamma eliminations. In addition, alpha eliminations will be discussed in Chap. 24.

8-1. Beta Eliminations of HY to Form Carbon-Carbon Double and Triple Bonds. In the majority of beta-elimination reactions that have been studied one of the two atoms or groups removed from the molecule is a hydrogen atom. Such reactions, e.g., the dehydration of alcohols,

$$H-C-C-OH \xrightarrow{H_2SO_4} C=C + H_2O$$

the Hofmann degradation of quaternary ammonium hydroxides,

$$H-C-C-NR_3{}^+ \xrightarrow{OH^-} C=C + R_3N$$

and the dehydrohalogenation of alkyl halides

$$H-C-C-Y + KOH \rightarrow C=C + KX + H_2O$$

186

may proceed by any of three different polar reaction mechanisms, depending on whether the H is removed *before* the Y is, *as* the Y is, or *after* the Y is. The first mechanism ($E1cB$, unimolecular elimination in the conjugate base of the substrate[1]) is a two-step process involving the intermediate formation of a carbanion that subsequently loses Y to give the olefin. The second mechanism ($E2$, bimolecular elimination[1]) is a concerted reaction in which a base removes H, a double bond forms, and Y is ejected, all in a single act. In the third mechanism ($E1$, unimolecular elimination[1]), Y is first removed to give a carbonium ion, which then loses a β-hydrogen to form the olefin.

Although there appear to be well-authenticated examples of the operation of each of the three mechanisms, it does not seem possible to draw sharp, operationally significant lines between them. There are thus a number of cases in which the mechanism has not yet been clearly determined.

8-1a. The Carbanion ($E1cB$) Mechanism for Elimination. From the nature of the carbanion mechanism for beta elimination

$$H{-}\overset{|}{\underset{|}{C}}{-}\overset{|}{\underset{|}{C}}{-}Y + B^- \rightleftharpoons BH + {}^{\ominus}|\overset{|}{\underset{|}{C}}{-}\overset{|}{\underset{|}{C}}{-}Y$$

$$ {}^{\ominus}|\overset{|}{\underset{|}{C}}{-}\overset{|}{\underset{|}{C}}{-}Y \rightarrow \overset{|}{C}{=}\overset{|}{C} + Y^-$$

it seems probable that base-induced elimination reactions will be most likely to proceed by this mechanism rather than by the concerted $E2$ mechanism when (1) Y is a group that is not easily lost with its bonding electron pair, (2) the intermediate carbanion is reasonably stable relative to the reactants and products, and (3) the unsaturated product is comparatively unstable relative to the reactants. Various of these factors may be seen to be important in the elimination reactions that have been shown to proceed by the carbanion mechanism.

Hughes, Ingold, and Patel suggested that the decomposition of 2-(p-nitrophenyl)ethyltrimethylammonium ion to p-nitrostyrene, which proceeds much faster in neutral solution than in 0.5 M hydrochloric acid, may proceed by the carbanion mechanism, as follows:[2]

$$ArCH_2CH_2\overset{\oplus}{N}Me_3 + H_2O \rightleftharpoons Ar\overset{\ominus}{C}HCH_2\overset{\oplus}{N}Me_3 + H_3O^+$$

$$Ar\overset{\ominus}{C}HCH_2\overset{\oplus}{N}Me_3 \rightarrow ArCH{=}CH_2 + Me_3N$$

[1] The terms $E1cB$, $E2$, and $E1$ were originated by C. K. Ingold ("Structure and Mechanism in Organic Chemistry," chap. 8, Cornell University Press, Ithaca, N.Y., 1953).

[2] E. D. Hughes and C. K. Ingold, *J. Chem. Soc.*, 523 (1933); E. D. Hughes, C. K. Ingold, and C. S. Patel, *J. Chem. Soc.*, 526 (1933).

with the 0.5 M acid slowing the reaction by shifting the first equilibrium to the left.   Hodnett and Flynn, however, observed that the formation of p-nitrostyrene from p-$O_2NC_6H_4CHTCH_2NMe_3^+$ is not accompanied by tritium exchange of the reactant at pH 7.0, and that the tritium kinetic isotope effect ($k_H/k_T \sim 2.2$) seems too large to be a secondary kinetic isotope effect.   This shows that there was no reversible carbanion formation at pH 7.0 but gives no good evidence concerning reversible carbanion formation at pH 1.0 and below.   It would be interesting to learn the relationship between rate and pH over the range 1.0 to 7.0 and to learn whether there is hydrogen exchange in strongly acidic solution.

Stronger evidence for the carbanion mechanism may be found in the observations that trichloroethylene[4] and several of the dihaloethylenes[5] undergo base-catalyzed deuterium exchange at a rate considerably faster than their dehydrohalogenation.

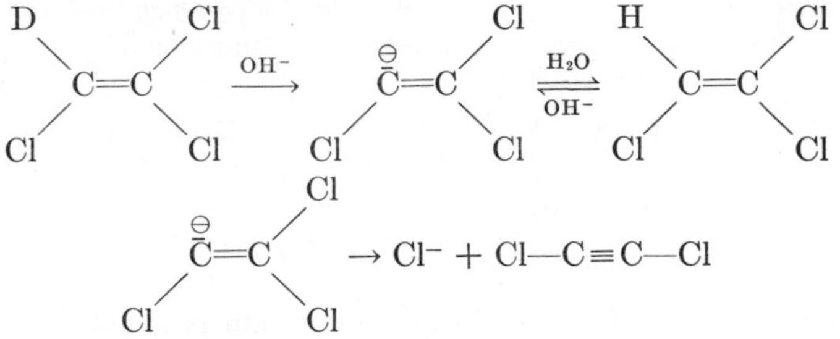

Assuming that carbanion formation is proved by the occurrence of deuterium exchange (see Sec. 10-1a), it hardly seems likely that the carbanion formation would merely be an irrelevant side reaction taking place at the same time as a concerted $E2$ reaction.   Even in the $E2$ mechanism the function of the removal by base of β-hydrogen without its bonding electron pair must be to liberate this pair so that it can form the extra carbon-carbon bond being produced.   If the *complete* freeing of the electron pair, as in carbanion formation, is often not enough to bring about the ejection of the adjacent halogen, it seems improbable that *partial* freeing, as in an $E2$ reaction, would be effective.   Other observations of base-catalyzed deuterium exchange that was rapid compared with elimination have been found in the transformation of malic to fumaric acid[6] and in the dehydrofluorination of various 2,2-dihalo-1,1,1-

[3] E. M. Hodnett and J. J. Flynn, Jr., *J. Am. Chem. Soc.*, **79**, 2300 (1957).

[4] L. C. Leitch and H. J. Bernstein, *Can. J. Research*, **28B**, 35 (1950); T. J. Houser, R. B. Bernstein, R. G. Miekka, and J. C. Angus, *J. Am. Chem. Soc.*, **77**, 6201 (1955).

[5] S. I. Miller and W. G. Lee, *J. Am. Chem. Soc.*, **81**, 6313 (1959).

[6] L. E. Erickson and R. A. Alberty, *J. Phys. Chem.*, **63**, 705 (1959).

trifluoroethanes[7] (in the latter case the olefinic product added alcohol too fast to be isolated). The occurrence of the carbanion mechanism in these cases instead of the more common $E2$ mechanism may be rationalized in terms of the three factors listed at the beginning of this subsection.

8-1b. *The E2 Mechanism for Elimination.* Many elimination reactions carried out in the presence of base are kinetically second-order, first-order in base, and first-order in substrate. Such kinetic evidence rules out the unimolecular $E1$ mechanism, but it does not distinguish between the carbanion and $E2$ mechanisms. Evidence on this point has been obtained by Skell and Hauser, who studied the dehydrohalogenation of β-phenylethyl bromide by sodium ethoxide in C_2H_5OD solution.[8] They found that the unreacted organic bromide, isolated after the reaction had gone about halfway to completion, contained no significant amount of deuterium. This evidence for the concerted nature of the reaction is necessary but not sufficient, since the same results would have been obtained if the reaction had consisted of the formation of a carbanion whose rate of reprotonation was quite slow compared with the rate of loss of bromide ion to give styrene. Supplementary and probably stronger evidence for the $E2$ mechanism is the fact that many second-order elimination reactions take place under conditions that are quite inadequate for the formation of carbanions with no better stabilizing substituents than those present.

Additional strong evidence for the concerted character of $E2$ reactions is the striking stereospecificity observed in many olefin-forming eliminations. Because of the ability of vinyl carbanions to maintain their configurations (see Sec. 10-1b) the preferential trans elimination observed in the formation of many acetylene derivatives[9] cannot be considered evidence for a concerted mechanism, and, in fact, some and perhaps all of these reactions occur by the carbanion mechanism.[4,5]

Results obtained in the formation of 2- and 3-menthene from menthyl and neomenthyl derivatives are shown in the following reaction scheme.[10]

---

[7] J. Hine, R. Wiesboeck, and R. G. Ghirardelli, *J. Am. Chem. Soc.*, **83**, 1219 (1961); J. Hine, R. Wiesboeck, and O. B. Ramsay, *J. Am. Chem. Soc.*, **83**, 1222 (1961).

[8] P. S. Skell and C. R. Hauser, *J. Am. Chem. Soc.*, **67**, 1661 (1945).

[9] A. Michael, *J. prakt. Chem.*, **52**, 308 (1895); S. I. Miller and R. M. Noyes, *J. Am. Chem. Soc.*, **74**, 629 (1952); S. J. Cristol and A. Begoon, *J. Am. Chem. Soc.*, **74**, 5025 (1952).

[10] The data on the trimethylammonium ions are from N. L. McNiven and J. Read [*J. Chem. Soc.*, 153 (1952)] and those on the chlorides from W. Hückel, W. Tappe, and G. Legutke [*Ann.*, **543**, 191 (1940)]. The reactions of the chlorides and of the trimethylneomenthyl ion were shown to be kinetically second-order by E. D. Hughes, C. K. Ingold, and J. B. Rose [*J. Chem. Soc.*, 3839 (1953)] and E. D. Hughes and J. Wilby [*J. Chem. Soc.*, 4094 (1960)], respectively.

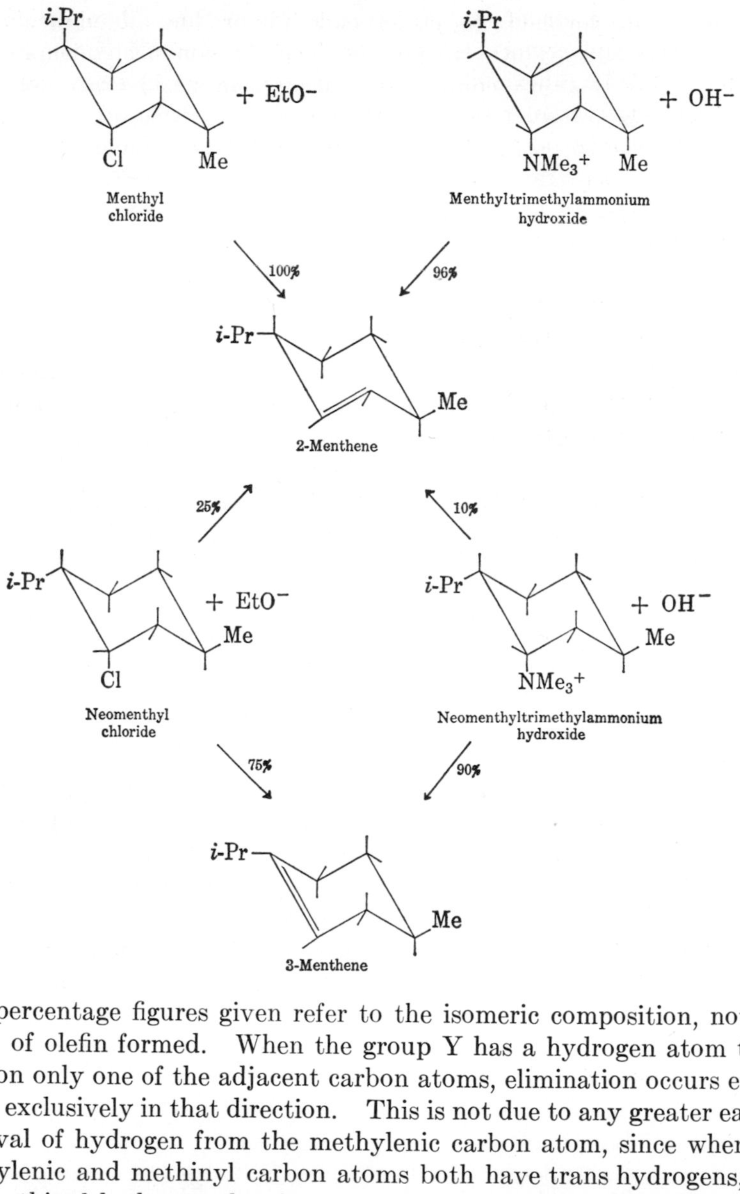

i-Pr

Cl   Me   + EtO⁻

Menthyl
chloride

i-Pr

NMe₃⁺   Me   + OH⁻

Menthyltrimethylammonium
hydroxide

100%   96%

i-Pr

Me

2-Menthene

25%   10%

i-Pr   + EtO⁻

Me

Cl

Neomenthyl
chloride

i-Pr   + OH⁻

Me

NMe₃⁺

Neomenthyltrimethylammonium
hydroxide

75%   90%

i-Pr

Me

3-Menthene

The percentage figures given refer to the isomeric composition, not the yield, of olefin formed. When the group Y has a hydrogen atom trans to it on only one of the adjacent carbon atoms, elimination occurs essentially exclusively in that direction. This is not due to any greater ease of removal of hydrogen from the methylenic carbon atom, since when the methylenic and methinyl carbon atoms both have trans hydrogens, it is the methinyl hydrogen that is removed more rapidly. Thus the difference in behavior is due to a strong tendency toward trans rather than cis elimination. Although the conformations shown in the reaction scheme are not in all cases the most stable, they are the ones through which the elimination reactions shown proceed, since they are the only ones in which the C—Y bond lies in the same plane with a β-C—H bond trans to it.

Another illustration of the stereospecificity of the $E2$ mechanism is due to Cristol and coworkers, who studied the isomeric benzene hexachlorides (1,2,3,4,5,6-hexachlorocyclohexanes), of which there are eight possible geometric isomers. The kinetics of the dehydrochlorination of five of these isomers was studied, and one isomer was found to react at less than 10^{-4} the rate of any of the others (whose rates differed from the average by no more than twofold).[11] This one, the so-called β isomer, had already been shown independently to be 1,3,5-*cis*-2,4,6-*trans*-hexachlorocyclohexane, and hence the only isomer that contains no hydrogen atom trans to, and on a carbon adjacent to, a chlorine atom.

More precise information as to the nature of the transition state in the $E2$ reaction was obtained by Cristol and Hause, who investigated the dehydrochlorination of the cis (I) and trans (II) isomers of 11,12-dichloro-9,10-dihydro-9,10-ethanoanthracene.

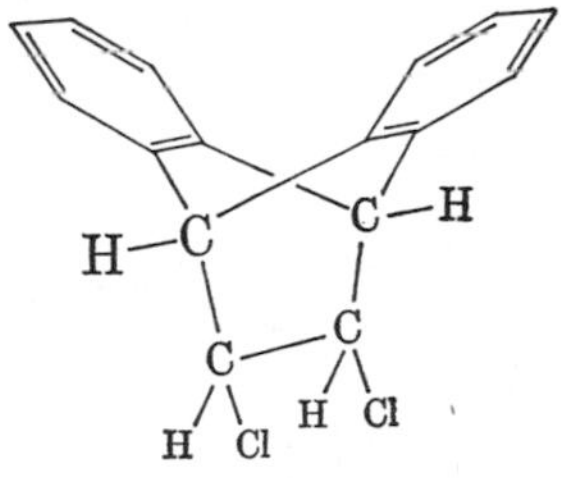

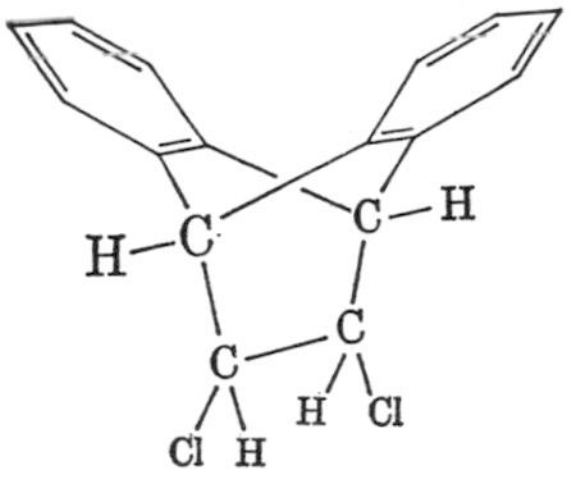

The markedly greater rate usually found for trans eliminations was not observed in this case. In fact, the trans isomer (whose dehydrohalogenation is cis) reacts somewhat faster than the cis isomer (which gives trans elimination), but both compounds are quite unreactive.[12] Apparently, in the concerted $E2$ mechanism, as the electrons of the β-C— H bond are freed by removal of the β-hydrogen, they attack the α-carbon atom from the rear (as in the S_N2 mechanism), displacing Y and forming the double bond. This type of interlocking of the removal of H, formation of a double bond, and displacement of Y is relatively (perhaps completely) unimportant unless the β-hydrogen is almost exactly trans, so that Y, C_α, C_β, and H are very nearly coplanar in the transition state. Such a coplanar orientation would introduce considerable strain into molecule I whose C—Cl bonds are forced to be very nearly eclipsed. For this reason I does not undergo dehydrochlorination with the facility characteristic of most vicinal dichlorides.

Although it seems clear that the base-induced dehydrohalogenation of most organic halides is a one-step stereospecific trans process, it is of

---

[11] S. J. Cristol, *J. Am. Chem. Soc.*, **69**, 338 (1947); **71**, 1894 (1949); S. J. Cristol, N. L. Hause, and J. S. Meek, *J. Am. Chem. Soc.*, **73**, 674 (1951).

[12] S. J. Cristol and N. L. Hause, *J. Am. Chem. Soc.*, **74**, 2193 (1952).

interest to inquire as to the reaction mechanism in cases where trans elimination is not possible. This point was considered by Cristol and Fix, who studied the reaction of β-benzene hexachloride (β-BHC) with sodium ethoxide in C_2H_5OD solution.[13] If the attack of base brings about the formation of a carbanion, this carbanion may have three possible fates, as shown in the following reaction scheme.

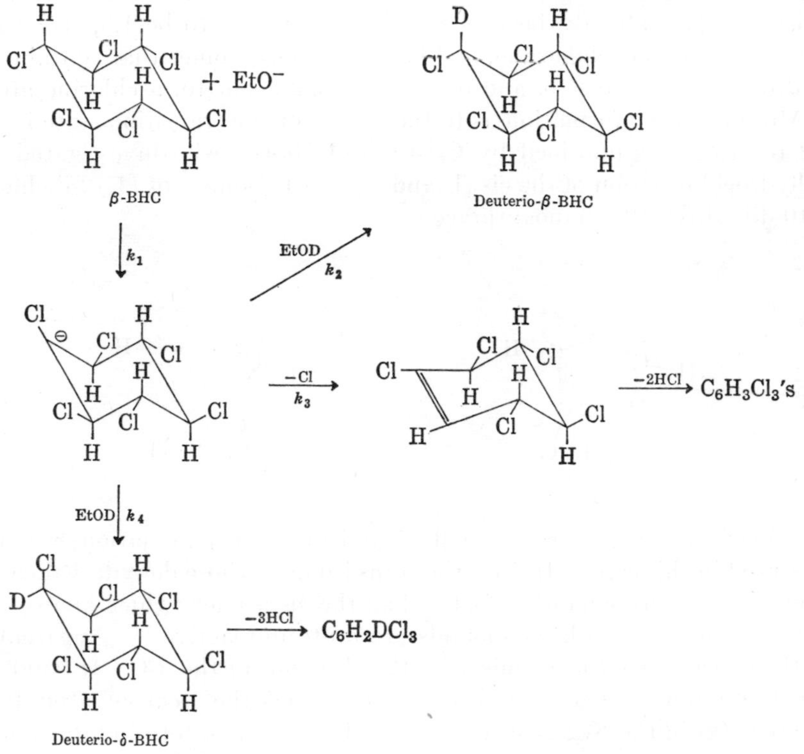

It may lose chloride ion to give a $C_6H_5Cl_5$ isomer that will be dehydrochlorinated rapidly (compared with the rate for β-BHC) to trichlorobenzenes. It may be deuteriated with retention of configuration to give deuterio-β-BHC, or with inversion of configuration to give a deuteriated form of δ-BHC, which is known to undergo dehydrochlorination much faster than β-BHC. When unreacted β-BHC was isolated after the dehydrohalogenation had gone about 50 per cent to completion, it was found to contain 0.08 atom per cent excess (over that normally present in hydrogen compounds) deuterium. This observation seems to be best

---

[13] S. J. Cristol and D. D. Fix, *J. Am. Chem. Soc.*, **75**, 2647 (1953); consideration of the $\sim$30 per cent C_2H_5OH also present in their solvent would complicate the ensuing discussion in a quantitative but not a qualitative sense.

explained on the basis of intermediate carbanion formation. The low extent of deuteriation found could mean that most of the reaction did not go through a carbanion, or it could mean that the intermediate carbanion was reprotonated to β-BHC only a small percentage of the time [that is, that $(k_3 + k_4)$ is much larger than k_2]. If the relatively large magnitude of $(k_3 + k_4)$ is due to k_4, the trichlorobenzene produced in the reaction should contain considerable deuterium. Unfortunately, its deuterium content was not investigated.

A more detailed investigation of cis eliminations was carried out by Bordwell, Pearson, and their coworkers, who studied the base-induced elimination reactions of some cycloalkyl p-toluenesulfonates with β-p-toluenesulfonyl substituents. Compound III, in which a trans elimination is possible, reacts with hydroxide ions in 50 per cent dioxane at 25° 435 times as fast as does its geometric isomer IV, which gives the same unsaturated product but via a cis elimination.[14]

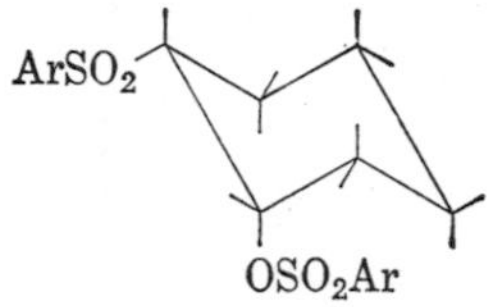

III 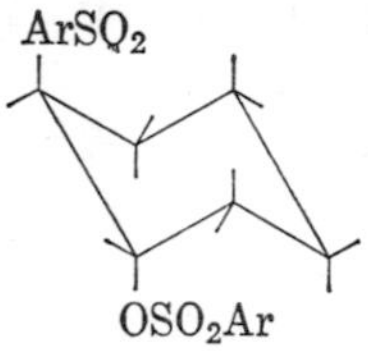   IV

In the analogous cyclopentane derivatives, the trans elimination is 20 times as fast as the cis. In consideration of the possibility of a carbanion mechanism for the cis eliminations, the kinetics of the amine-induced reaction were studied. The rate was found to be unaffected by large changes in the concentration of the conjugate acid of the amine.[14] This shows that no carbanion was formed reversibly since, as shown in the reaction scheme

$$\text{TsCHRCHROTs} + \text{R}_3\text{N} \rightleftharpoons \text{Ts}\overset{\ominus}{\text{C}}\text{RCHROTs} + \text{R}_3\text{NH}^+$$

$$\text{Ts}\overset{\ominus}{\text{C}}\text{RCHROTs} \rightarrow \text{TsCR}\!\!=\!\!\text{CHR} + \text{TsO}^-$$

the addition of R_3NH^+ would shift the initial equilibrium to the left and thus slow the reaction. A mechanism involving the rate-controlling formation of a carbanion is not excluded, however. The strongest present evidence on this matter deals with whether the cis-elimination reaction rate is a reasonable one for the formation of such a carbanion under the given conditions. The rate of carbanion formation, as measured by deuterium exchange, of 1-deuteriocyclohexyl p-tolyl sulfone was found to be only about 10^{-5} as large as the rate of the corresponding cis

[14] J. Weinstock, R. G. Pearson, and F. G. Bordwell, *J. Am. Chem. Soc.*, **78**, 3468, 3473 (1956).

elimination.[15]   At first it seemed improbable that a β-p-toluenesulfonoxy (tosyloxy) group's polar effects would increase the rate of carbanion formation by a factor so large as 10^5, but subsequent work, using other β-substituted compounds and applying the Taft equation, showed that the polar effect of a β-tosyloxy group should increase the carbanion-formation rate by somewhat more than 10^5-fold.[16]   Hence it is believed that the cis eliminations observed with these sulfones involved the rate-controlling formation of carbanions, a process that was slowed significantly in the cyclohexyl series by the steric effect of the *cis*-tosyloxy group.

8-1c. *The E1 Mechanism for Elimination.*   In the $E1$ mechanism the reactant first loses Y to yield a carbonium ion, which may then donate a β proton to some base to give an olefin.   For this reason the possibility of elimination by this mechanism accompanies any S_N1 reaction whose intermediate carbonium ion bears a β-hydrogen atom.

$$H{-}\underset{|}{\overset{|}{C}}{-}\underset{|}{\overset{|}{C}}{-}Y \rightarrow H{-}\underset{|}{\overset{|}{C}}{-}\underset{|}{\overset{|}{C}}{\overset{\oplus}{}} \overset{X}{} \rightarrow H{-}\underset{|}{\overset{|}{C}}{-}\underset{|}{\overset{|}{C}}{-}X \qquad (S_N1)$$

$$\downarrow -H^+$$

$$\underset{|}{\overset{|}{C}}{=}\underset{|}{\overset{|}{C}} \qquad (E1)$$

Although elimination by the $E1$ mechanism will usually be a first-order process (no examples involving mass-law effects having yet been investigated), this kinetic order will not serve to distinguish it from an $E2$ attack by solvent, which would be pseudounimolecular.   There is, however, evidence from several sources that at least some elimination reactions do not involve a bimolecular attack by solvent in the rate-controlling step but do proceed through an intermediate carbonium ion. For one thing, it may be seen that to a first approximation the fraction of RY solvolyzing which gives elimination and that which gives substitution should be independent of the nature of Y, since these fractions are determined by reactions of the *carbonium ion*.   A bimolecular nucleophilic attack should also produce a mixture of substitution and elimination by the simultaneous occurrence of the S_N2 and $E2$ mechanisms. But since both are reactions of RY, there is no assurance that the ratio of elimination to substitution will remain constant as the nature of Y is changed.   Cooper, Hughes, Ingold, and MacNulty have found the ratio of elimination to substitution to be fairly nearly constant in the solvolysis reactions of various t-butyl and t-amyl derivatives.[17]   The

[15] J. Weinstock, J. L. Bernardi, and R. G. Pearson, *J. Am. Chem. Soc.*, **80**, 4961 (1958).

[16] J. Hine and O. B. Ramsay, *J. Am. Chem. Soc.*, **83**, 973 (1961).

[17] K. A. Cooper, E. D. Hughes, C. K. Ingold, and B. J. MacNulty, *J. Chem. Soc.*, 2038 (1948).

constancy of ratio that accompanies the change of Y from Cl to Br to I is, of course, required by the $E1$ mechanism but can hardly be considered good evidence against a bimolecular attack by solvent, since in known bimolecular nucleophilic attacks on alkyl halides the change in the ratio of elimination to substitution with a change in Y from Cl to Br to I is not always great. It is striking, however, that the solvolysis of t-butyl- and t-amyldimethylsulfonium ions gives about the same fraction of elimination as that of a halide (Table 8-1), since in the known bimolecular nucleophilic attacks that have been studied the sulfonium ions tend to give a far larger fraction of elimination.[18] To explain the small differences in the fraction of elimination found in solvolyses, it has been suggested that the group displaced may have a small influence if the carbonium ion is so reactive that it reacts before this group is very far away.

TABLE 8-1. FRACTION OF ELIMINATION OCCURRING IN THE SOLVOLYSIS OF SOME t-ALKYL CHLORIDES AND DIMETHYLSULFONIUM IONS IN "80 PER CENT ETHANOL"[17]

Compound	Temperature, °C	Fraction elimination
t-BuCl	65.3	0.36
t-BuSMe$_2^+$	65.3	0.36
t-AmCl	50	0.40
t-AmSMe$_2^+$	50	0.48

A somewhat different type of product-distribution evidence can be presented in reactions in which more than one olefin is formed. The reaction of sodium ethoxide with t-amyl bromide yields 71 per cent 2-methyl-2-butene and 29 per cent 2-methyl-1-butene,[19] whereas its reaction with the t-amyldimethylsulfonium ion at the same temperature yields 14 per cent 2-methyl-2-butene and 86 per cent 2-methyl-1-butene.[18] In view of this pronounced effect of the change from bromine to the dimethylsulfonium group, we may consider as evidence for the $E1$ mechanism the fact that the olefin obtained from the ethanolysis of t-amyl bromide (82 per cent 2-methyl-2-butene) and the one obtained from the t-amyldimethylsulfonium ion (87 per cent 2-methyl-2-butene) have compositions that differ by little more than the experimental error.[17]

Further strong evidence for the $E1$ mechanism is found in certain examples of nonstereospecific elimination. For example, Hückel, Tappe, and Legutke found that the solvolysis of menthyl chloride or p-toluenesulfonate gives about 70 per cent 3-menthene and 30 per cent 2-menthene,[10] despite the fact that the 3-menthene could not have been formed by a trans elimination (see the reaction scheme in Sec. 8-1b).

[18] E. D. Hughes, C. K. Ingold, and L. I. Woolf, *J. Chem. Soc.*, 2084 (1948).
[19] M. L. Dhar, E. D. Hughes, and C. K. Ingold, *J. Chem. Soc.*, 2065 (1948).

The mechanism of a number of vapor-phase elimination reactions varies rather continuously from four-center-type reactions to $E1$ reactions. Although all these related reactions will be discussed in Sec. 25-3, it should be pointed out here that some, such as the unimolecular thermal dehydrohalogenation of secondary and tertiary chlorides and bromides, usually appear to be essentially $E1$ in character. Similarly, there is a continual gradation between the concerted $E2$ mechanism and the stepwise $E1$ mechanism, on one hand, and the $E1cB$ mechanism, on the other.

8-1d. Reactivity in E2 Reactions. Since the formation of two olefins from a given starting material may be considered as the operation of two different elimination reactions, the question of predicting the orientation, i.e., of what relative amounts of the two are formed, is simply one of predicting the relative rates of the two reactions. If the establishment of equilibrium between the various conformers of the reactant is rapid compared with the subsequent reaction, the factors that influence $E2$ reactivity can be discussed in terms of the following generalized $E2$ transition state.

$$\begin{array}{c} \text{B} \\ \ddots \\ \text{H} \qquad\qquad \text{R}_3 \\ \ddots \qquad\qquad \diagup \\ \text{R}_1{\text{-}}\text{C}{=}\text{C}{-}\text{R}_4 \\ \diagup \qquad \ddots \\ \text{R}_2 \qquad \text{Y} \end{array}$$

In the case of acyclic compounds in which B, Y, and the R's are all relatively small, the reaction rate seems to be controlled largely by the solvent, temperature, basicity of B, ease of displacement of Y with its bonding electron pairs, and the stability of the olefin being formed. In addition, if the β-C—H bond has been broken in the transition state to a greater extent than the C—C double bond has been formed, then the transition state will have a certain amount of "carbanion character" and the ability of the R's (especially R_1 and R_2) to stabilize a partial negative charge will be important. Conversely, if the C—Y bond is broken to a greater extent than the C—C double bond has been formed, the ability to stabilize a partial positive charge will be important. In agreement with the *Zaitsev* (Saytzeff) *rule*, according to which "dehydrohalogenation of secondary and tertiary halides proceeds by the preferential removal of β-hydrogen from the carbon that has the smaller number of hydrogens," olefin stability has been shown to be the most important factor in a number of cases.[20] Thus, the reaction of t-amyl bromide with sodium ethoxide at 25° gives 28 per cent 2-methyl-1-butene and 72 per cent 2-methyl-2-butene.[21] Since the reactants are the same in each of these two competing $E2$ reactions, the difference in rates depends only on the

[20] Ingold, *op. cit.*, sec. 31.
[21] Dhar, Hughes, and Ingold, *loc. cit.*

free-energy contents of the respective transition states. Since there are six β-hydrogens whose removal gives the 1-olefin, and only two whose removal gives the 2-olefin, the formation of the 1-olefin will be favored by an entropy increment corresponding to a threefold faster rate. Although the over-all free energy of formation of the trialkylated ethylene, 2-methyl-2-butene, is about 1.3 kcal/mole more favorable than that of the dialkylated ethylene, 2-methyl-1-butene[22] (cf. Table 1-2), we would not expect a rate difference of a magnitude (about ninefold at 25°) that would completely reflect this difference in stabilities, since in the transition state, where the double bond is only partly formed, the difference in stabilities should be less.   The smaller difference in rates observed may be due in part to this, but judging from the importance of steric effects in reactions where B and/or Y are somewhat bulkier, it seems probable that steric factors are of some importance even in the present case.

As the attacking base is made progressively larger, steric hindrance, particularly between the base and R_1 and R_2, becomes more significant. Brown and coworkers obtained evidence for this point.[23]   As shown in Table 8-2, the larger alkoxide ions have a greater tendency to attack the

TABLE 8-2. ORIENTATION IN DEHYDROHALOGENATIONS USING VARIOUS
POTASSIUM ALKOXIDES[23a]

Alkyl halide	1-Olefin, %			
	EtOK	t-BuOK	t-AmOK	Et₃COK
$C_2H_5CHBrCH_3$	19	53		
$C_3H_7CHBrCH_3$	31	66		
$C_2H_5CBr(CH_3)_2$	30	72	78	89
$(CH_3)_2CHCBr(CH_3)_2$	21	73	81	92
$(CH_3)_3CCH_2CBr(CH_3)_2$	86	98	...	97

less hindered primary hydrogens, so that in many cases the less stable 1-olefin is the principal product.   The greater tendency to attack the more acidic primary hydrogens may be due in part to increased carbanion character in the transition state brought about by the increased basicity of the larger alkoxide ions.   However, in dehydrohalogenations brought about by pyridine bases the yield of 1-olefin increases with the effective size of the base but not with its basicity.[23b]

The last compound in Table 8-2 is interesting in that steric effects

[22] F. D. Rossini, K. S. Pitzer, R. L. Arnett, R. M. Braun, and G. C. Pimentel, "Selected Values of Physical and Thermodynamic Properties of Hydrocarbons and Related Compounds," p. 475, Carnegie Press, Pittsburgh, Pa., 1953.

[23] (a) H. C. Brown and I. Moritani, *J. Am. Chem. Soc.*, **75**, 4112 (1953); **78**, 2203 (1956); (b) H. C. Brown and M. Nakagawa, *J. Am. Chem. Soc.*, **78**, 2197 (1956).

appear to have made the 1-olefin more stable than the 2-olefin. The relative instability of the latter is due to interaction between the t-butyl group attached to the double bond and the methyl group cis to it.[24]

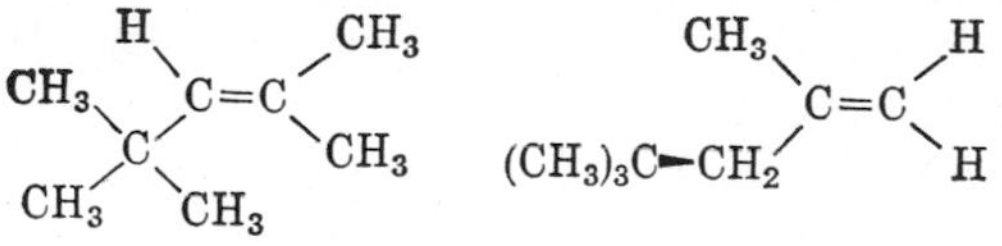

2,4,4-Trimethyl-2-pentene   2,4,4-Trimethyl-1-pentene

There are no such interactions in the 1-olefin, where the t-butyl **group** need not even lie in the same plane as the four atoms attached to the carbons of the double bond.

Steric effects are also important when the group being displaced (Y) is large.[23] This may account for the *Hofmann rule*, according to which "in the decomposition of a quaternary ammonium hydroxide the alkyl group with the largest number of β-hydrogens will be preferentially eliminated as an olefin."[25] Thus, for example, the decomposition of n-propylethyldimethylammonium hydroxide yields far more ethylene than propylene. It seems unreasonable to attribute the more rapid formation of ethylene, in any large measure, to any greater ease of displacement of n-propyldimethylamine compared with ethyldimethylamine, since both amines are of about the same basicity and effective size. In terms of the transition state shown at the beginning of this subsection, the formation of propylene will be hampered by steric interaction between the methyl group at R_1 (or R_2) and the bulky Y group, whereas there will be no such interaction in the formation of ethylene. In less rigorous terms we may point out that by far the most stable conformation of the reactant will be one in which there is on the propyl group no β-hydrogen trans to the nitrogen atom.

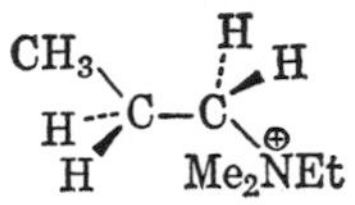

Hence, propylene can be formed by the concerted trans $E2$ mechanism only when the molecule is in a considerably less stable conformation,

[24] H. C. Brown and H. L. Berneis, *J. Am. Chem. Soc.*, **75**, 10 (1953); H. C. Brown and I. Moritani, *J. Am. Chem. Soc.*, **77**, 3607 (1955).

[25] The Hofmann rule was originally stated somewhat differently but, as given above, is applicable to most $E2$ reactions of quaternary ammonium ions, sulfonium ions, and sulfones, where the alkyl groups present are acyclic and saturated and contain no elements other than carbon and hydrogen.

whereas ethylene can be formed by an $E2$ process from any conformation. In addition to this steric effect, there should be a polar effect due to the greater carbanion character[20] of the transition state (compared with the transition state in dehydrohalogenation), which will also favor ethylene formation. Hofmann-type orientation appears rather general in the $E2$ reactions of quaternary ammonium ions, sulfonium ions, and sulfones, where there is double or triple branching at the atom by which the Y group is attached. There is one exception to this generalization that indicates that the *major* cause of Hofmann orientation is steric. This is the decomposition of neomenthyltrimethylammonium hydroxide, shown in the first reaction scheme of Sec. 8-1b. In this quaternary ammonium salt the only conformation in which the less highly alkylated β-carbon bears a hydrogen trans to Y is also one in which the more highly alkylated β-carbon has a trans hydrogen, too. In this case, there is no steric effect due to Y, and olefin stability controls the orientation, 3-menthene being the predominant product.

Hammett-equation correlations give information as to the polar character of the $E2$ transition state (relative to the reactant molecule). Unfortunately the only available data (Table 8-3) are for a number of substituted β-phenylethyl compounds.[26-29] All the ρ values are rather large and positive, showing that the β-C—H bond has been broken to a considerably greater extent than the C—C double bond has been formed in the transition state. The considerable carbanion character of the transition state is further shown by the fact that for the p-nitro and p-acetyl substituents σ^- values give a better correlation than σ values.[29] It may be seen that ρ tends to increase with decreasing ease of displacement of Y, as measured by the reaction rate (log k_0).[27,28] The only significant exception to this generalization is the dimethylsulfonium-ion reaction, where ρ is probably a poorer measure of the relative amount of carbanion character of the transition state [cf. Eq. (4-9)]. The expulsion of Y from the molecule is, to a considerable extent, due to the creation of a partial negative charge on the β-carbon atom. The more difficult it is to expel Y, the greater is the negative charge required (other things being equal). The large amount of carbanion character of these transition states must be due in part to the ability of any β-aryl substituent to stabilize a carbanion. Both α- and β-phenyl substituents increase $E2$ reactivity, since each will stabilize the double bond being formed in the transition state.[20] In reactions with bases as strong as ethoxide ions, at

[26] S. J. Cristol, N. L. Hause, A. J. Quant, H. W. Miller, K. R. Eilar, and J. S. Meek, *J. Am. Chem. Soc.*, **74**, 3333 (1952).

[27] C. H. DePuy and D. H. Froemsdorf, *J. Am. Chem. Soc.*, **79**, 3710 (1957).

[28] C. H. DePuy and C. A. Bishop, *J. Am. Chem. Soc.*, **82**, 2532, 2535 (1960).

[29] W. H. Saunders, Jr., and R. A. Williams, *J. Am. Chem. Soc.*, **79**, 3712 (1957).

least, the β-phenyl substituent has considerably more effect because of its ability to stabilize a negative charge on the β-carbon.

As α substituents, the influence of bromine and chlorine on $E2$ reactivity is fairly small. Thus α-chlorine substituents approximately triple the $E2$ reactivity (per chlorine) of $(p\text{-ClC}_6\text{H}_4)_2\text{CHCH}_2\text{Cl}$,[26] and increase the $E2$ reactivity of cyclohexyl bromide and chloride by a smaller factor.[30] An α-bromo substituent decreases the $E2$ reactivity of cyclohexyl bromide slightly.[30] On the other hand, β-bromine and -chlorine have a strong activating influence, as shown by Olivier and Weber's observation that ethylene bromide is transformed by alkali to vinyl bromide about 200

TABLE 8-3. HAMMETT-EQUATION CORRELATIONS IN $E2$ REACTIONS WITH
SODIUM ETHOXIDE IN ETHANOL AT 30°

Reactants	Log $k_0{}^a$	ρ
$\text{Ar}_2\text{CHCCl}_3{}^b$	-2.82	2.73
$\text{Ar}_2\text{CHCHCl}_2{}^b$	-3.43	2.46
$\text{ArCH}_2\text{CH}_2\text{I}^c$	-2.50	2.07
$\text{ArCH}_2\text{CH}_2\text{Br}^d$	-3.27	2.34
$\text{ArCH}_2\text{CH}_2\text{OSO}_2\text{C}_6\text{H}_4\text{CH}_3\text{-}p^c$	-4.34	2.27
$\text{ArCH}_2\text{CH}_2\text{Cl}^c$	-4.8	2.61
$\text{ArCH}_2\text{CH}_2\text{SMe}_2{}^{+d}$	-2.36	2.64
$\text{ArCH}_2\text{CH}_2\text{F}^{c,e}$	-7.0	3.12

^{a} In liters mole^{-1} sec^{-1}.
^{b} Data from Ref. 26 for the reaction in 92.6 per cent ethanol at 20°.
^{c} From Refs. 27 and 28.
^{d} From Ref. 29.
^{e} Calculated from data at higher temperatures.

times as fast as ethylidene bromide is.[31] It is interesting that the free energy of activation is lower for ethylene bromide, although the free energy of reaction is lower for ethylidene bromide (this follows from the fact that ethylene bromide is more stable than ethylidene bromide[32]). The effect of a β-fluoro substituent is considerably different from that of β-bromine or -chlorine, the relative $E2$ reactivities (per bromine) of $\text{BrCH}_2\text{CH}_2\text{Br}$, $\text{ClCH}_2\text{CH}_2\text{Br}$, and $\text{FCH}_2\text{CH}_2\text{Br}$ toward sodium hydroxide in aqueous dioxane at 70° being 190:130:1.[33] The effect of β-bromine and -chlorine on $E2$ reactivity is probably not due to stabilization of the olefin being formed, since if it were, α-bromine and -chlorine should have the same effect. Instead, their activating influence is believed to be due to their ability to stabilize a negative charge on the carbon atom to which

[30] H. L. Goering and H. H. Espy, *J. Am. Chem. Soc.*, **78**, 1454 (1956).
[31] S. C. J. Olivier and A. P. Weber, *Rec. trav. chim.*, **53**, 1087 (1934).
[32] F. R. Mayo and A. A. Dolnick, *J. Am. Chem. Soc.*, **66**, 985 (1944).
[33] J. Hine and P. B. Langford, *J. Am. Chem. Soc.*, **78**, 5002 (1956).

they are attached. Indeed, the qualitative order observed, Br > Cl > F, is the same as that found for the ability of α-halogen substituents to stabilize carbanions.[34] It is possible, however, that the effect of fluorine is due in part to its tendency to destabilize olefins.[35] Goering and Espy found β-bromine and -chlorine to increase $E2$ reactivity strongly in the *cis*-1,2-dihalocyclohexanes, where an activated trans elimination may take place, but not with the *trans*-1,2-dihalocyclohexanes, where the only possible trans dehydrohalogenation does not remove a hydrogen atom attached to the same carbon as a halogen.[30] With the more strongly electron-withdrawing *p*-toluenesulfonyl group as a β substituent, however, an activated cis elimination (presumably by the carbanion mechanism[16]) has been found to take place in preference to a nonactivated trans elimination.[14,36,37] The effect of other carbanion-stabilizing groups, such as —CN, —NO$_2$, —COR, etc., appears to be qualitatively similar to that of —SO$_2$R groups.

The effect of α-alkoxy and -alkylthio groups on $E2$ reactivity does not appear to have been determined, probably because α-halo ethers and sulfides are so reactive in substitution processes that the $E2$ mechanism is difficult to isolate. The effect of β-alkoxy groups on $E2$ reactivity appears to be deactivation. McElvain, Clarke, and Jones found that the dehydrobromination of the diethylacetal of α-bromoisovaleraldehyde gives the acetal of an α,β-unsaturated aldehyde rather than the ketene acetal (which is known to be stable under the reaction conditions).[38]

$$
\begin{array}{c}
\quad\;\; CH_3 \qquad OEt \\
\quad\;\; | \qquad\quad\;\; | \\
CH_3\!-\!C\!-\!CH\!-\!C\!-\!OEt \xrightarrow{\;t\text{-}BuOK\;} CH_3\!-\!C\!=\!CH\!-\!C\!-\!OEt \\
\quad\;\; | \quad\;\; | \quad\; | \qquad\qquad\qquad\qquad\quad | \\
\quad\;\; H \;\; Br \;\; H \qquad\qquad\qquad\qquad\qquad\; H
\end{array}
$$

The reason for this behavior is not known. The β-alkylthio substituent does appear to increase $E2$ reactivity,[39] and in explanation alkylthio groups are known to stabilize carbanions[40,41] and double bonds.[42]

[34] J. Hine, N. W. Burske, M. Hine, and P. B. Langford, *J. Am. Chem. Soc.*, **79,** 1406 (1957).

[35] J. R. Lacher, J. D. Park, et al., *J. Phys. Chem.*, **60,** 608 (1956); **61,** 584, 1125 (1957).

[36] J. Weinstock, R. G. Pearson, and F. G. Bordwell, *J. Am. Chem. Soc.*, **76,** 4748 (1954).

[37] F. G. Bordwell and R. J. Kern, *J. Am. Chem. Soc.,* **77,** 1141 (1955).

[38] S. M. McElvain, R. L. Clarke, and G. D. Jones, *J. Am. Chem. Soc.*, **64,** 1966 (1942).

[39] E. Rothstein, *J. Chem. Soc.*, 1550, 1553, 1558, 1560 (1940).

[40] H. Gilman and F. J. Webb, *J. Am. Chem. Soc.*, **62,** 987 (1940).

[41] R. B. Woodward and R. H. Eastman, *J. Am. Chem. Soc.*, **68,** 2229 (1946).

[42] D. S. Tarbell and M. A. McCall, *J. Am. Chem. Soc.*, **74,** 48 (1952); D. S. Tarbell and W. E. Lovett, *J. Am. Chem. Soc.*, **78,** 2259 (1956).

The rates of $E2$ reactions can be influenced considerably by the steric compressions and interferences that arise (in the transition state) in the formation of most cis olefins. Cristol and Bly, for example, found that the dehydrochlorination of dl-stilbene dichloride, which gives chloro-$trans$-stilbene, is about eight times as rapid as the dehydrochlorination of $meso$-stilbene dichloride to chloro-cis-stilbene.[43] Even in the trans isomer, of course, there is interaction between the chlorine atom and the phenyl group cis to it. When this partially compensating factor is absent, a still larger effect is expected, and, indeed, Curtin and Kellom observed that the ratio of $trans$- to cis-stilbene was 130 in the reaction of the 2,4,6-triethylbenzoate of benzylphenyl carbinol with potassium t-butoxide.[44]

$$C_6H_5CH_2-\underset{\underset{ArCO_2}{|}}{C}HC_6H_5 \xrightarrow{t\text{-BuOK}} C_6H_5CH{=}CHC_6H_5$$

In their work on the $E2$ reactions of 1,2-diphenyl-1-propyl derivatives, Cram, Greene, and DePuy pointed out that the extent to which the instability of the product affects the reaction rate depends upon how closely the geometry of the transition state approaches that of the product.[45] $E2$ reactions with considerable carbanion or carbonium-ion character will thus show smaller "cis effects" than those in which the C—H and C—Y bonds have been broken to about the same extent in the transition state.

8-1e. *Olefin Yield in E2 Reactions.* Since the attack of a nucleophilic reagent on RY may result in either an $E2$ or an S_N2 reaction, the yield of olefin is determined by the relative rates of the reactions (except in cases where carbonium-ion formation or some other competing reaction occurs). Although the previous discussions of the effect of structure on the rates of $E2$ and S_N2 reactions might make possible a comparison of the relative yields of olefins obtainable from various compounds, they do not tell which reactions are really useful for the preparation of olefins (or substitution products), a purpose for which some numerical "orienting" data on specific compounds are desirable. Table 8-4 lists some of the data of Hughes, Ingold, and coworkers on the reaction of alkyl bromides with sodium ethoxide.[46] It is seen that olefin is obtained from ethyl bromide rather slowly and in very poor yield. The *rate* of olefin

[43] S. J. Cristol and R. S. Bly, Jr., *J. Am. Chem. Soc.*, **82**, 142 (1960).

[44] D. Y. Curtin and D. B. Kellom, *J. Am. Chem. Soc.*, **75**, 6011 (1953).

[45] D. J. Cram, F. D. Greene and C. H. DePuy, *J. Am. Chem. Soc.*, **78**, 790 (1956).

[46] M. L. Dhar, E. D. Hughes, C. K. Ingold, and S. Masterman, *J. Chem. Soc.*, 2055, 2058, 2065 (1948).

formation is increased by either α- or β-methyl groups, but the *yield* of olefin is increased more by α-methyl groups, largely because the α-methyl group slows the competing S_N2 reaction more.

The yield of olefin does not change greatly with the nature of the halogen atom, but the order appears to be RI > RBr > RCl from the fact that the reaction of alkali at 80° gives 58 per cent and 61 per cent olefin from i-PrCl and i-PrBr in "80 per cent ethanol" and 58 per cent and 74 per cent olefin from i-PrBr and i-PrI in "60 per cent ethanol."[47]

TABLE 8-4. RATES OF ELIMINATION AND SUBSTITUTION IN THE REACTION OF ALKYL BROMIDES WITH SODIUM ETHOXIDE[46]

Alkyl bromide	Temp.,[a] °C	k_2, liters mole^{-1} sec^{-1}, $\times$ 10^5			Olefin, %
		Total	S_N2	E2	
CH_3CH_2Br	55	174	172	1.6	0.9
$CH_3CH_2CH_2Br$	55	60	54.7	5.3	8.9
$CH_3(CH_2)_3Br$	55	43.9	39.6	4.3	9.8
$CH_3(CH_2)_4Br$	55	39.2	35.7	3.5	8.9
$(CH_3)_2CHCH_2Br$	55	14.3	5.8	8.5	59.5
$C_6H_5CH_2CH_2Br$	55	593	32	561	94.6
$CH_3CHBrCH_3$	25	0.295	0.058	0.237	80.3
$C_2H_5CHBrCH_3$	25	0.422	0.075	0.347	82.2[b]
n-$C_3H_7CHBrCH_3$	25	0.343	0.067	0.276	80.7[c]
$C_2H_5CHBrC_2H_5$	25	0.455	0.054	0.401	88.1
$CH_3CBr(CH_3)_2$[d]	25	4.17	<0.1	4.17	>97
$C_2H_5CBr(CH_3)_2$[d]	25	9.44	<0.2	9.44	>97[e]

[a] Between 25 and 55° the k's should change by factors of from 20 to 40.

[b] Olefin: 81 per cent 2-butene, 19 per cent 1-butene.

[c] Olefin: 71 per cent 2-pentene, 29 per cent 1-pentene.

[d] The data for this compound had to be corrected for a considerable amount of solvolysis.

[e] Olefin: 72 per cent 2-methyl-2-butene, 28 per cent 2-methyl-1-butene.

When the structure of Y is so changed as to make RY an onium ion, there is a large increase in the relative extent to which attacks on R give elimination rather than substitution. For example, the reaction of ethyl bromide with alkali in ethanol yields only 0.9 per cent ethylene (Table 8-4) and, judging from solvent effects found with other bromides, probably yields even less olefin in aqueous ethanol, but the reaction of the triethylsulfonium ion with alkali in either 60 or 80 per cent ethanol yields essentially 100 per cent ethylene.[48] Since β-methyl groups increase the

[47] E. D. Hughes and U. G. Shapiro, *J. Chem. Soc.*, 1177 (1937).

[48] J. L. Gleave, E. D. Hughes, and C. K. Ingold, *J. Chem. Soc.*, 236 (1935).

*E*2 rates for halides and decrease them for onium ions, it would be of interest to compare the 59.5 per cent of elimination found with isobutyl bromide (Table 8-4) with that for triisobutylsulfonium ion. Unfortunately, the latter data are not available. Hughes, Ingold, and Maw found that the isobutyldimethylsulfonium ion reacts with sodium ethoxide to give 2.4 per cent isobutylene and 97.6 per cent substitution.[49] Almost undoubtedly, however, the substitution occurs largely at a methyl rather than an isobutyl group. If these attacks on methyl occur between 32 and 6,200 times as fast as on isobutyl, as has been found for some other nucleophilic displacements,[50] we may estimate that the attack on the isobutyl group in this case gives between 61 and 99.7 per cent elimination.

Several other factors may influence the ratio of substitution to elimination. An increase in the basicity of the nucleophilic reagent used is usually accompanied by an increase in the percentage of elimination. The greater sensitivity of S_N2 reactions to steric effects has made it worthwhile to use relatively large bases, such as potassium *t*-butoxide and 2,6-dialkylpyridines, to obtain better yields of elimination products. It appears that the ratio of elimination to substitution obtained with alkali–metal amide derivatives decreases in the order $KNR_2 > NaNR_2 > LiNR_2$.[51] Bishop and DePuy have found that *p*-toluenesulfonates give a smaller fraction of elimination than do the corresponding halides.[28,52]

Hughes, Ingold, and coworkers have examined the available data on *E*1 and *E*2 eliminations of alkyl chlorides, bromides, iodides, and sulfonium ions and have pointed out that in all cases the elimination reactions have the higher activation energy, i.e., that the proportion of olefin formed increases with the temperature.[53] A typical example is that of isopropyl bromide, which upon reaction with alkali in "60 per cent ethanol" gives 53.2 per cent olefin at 45° and 63.6 per cent at 100°. According to Chapman and Levy, however, the bimolecular reaction of sodium ethoxide with alkyl fluorides yields less olefin at higher temperatures.[54]

Since the *E*2 attack of an anion on a neutral molecule involves charge dispersal in the transition state,

[49] E. D. Hughes, C. K. Ingold, and G. A. Maw, *J. Chem. Soc.*, 2072 (1948).

[50] S. S. Woolf, *J. Chem. Soc.*, 1172 (1937); D. Segaller, *J. Chem. Soc.*, **105,** 106 (1914); I. Dostrovsky and E. D. Hughes, *J. Chem. Soc.*, 157 (1946); S. F. Acree and G. H. Shadinger, *Am. Chem. J.*, **39,** 226 (1908); P. M. Dunbar and L. P. Hammett, *J. Am. Chem. Soc.*, **72,** 109 (1950).

[51] W. H. Puterbaugh and C. R. Hauser, *J. Org. Chem.*, **24,** 416 (1959), and references cited therein.

[52] C. A. Bishop and C. H. DePuy, *Chem. & Ind.* (*London*), 297 (1959).

[53] E. D. Hughès, C. K. Ingold, et al., *J. Chem. Soc.*, 2043, 2049 (1948).

[54] N. B. Chapman and J. L. Levy, *J. Chem. Soc.*, 1673 (1952).

the Hughes-Ingold theory of solvent effects (Sec. 3-1f) would predict an increase in rate with the decreasing ion-solvating power of the medium, just as for the related S_N2 reaction. Hughes, Ingold, and coworkers state that the charge is *more* dispersed in the $E2$ transition state, so that the $E2$ rate changes faster with solvent.[53] Therefore the yield of olefin increases with the alcohol content of aqueous ethanol solvents in the base-catalyzed second-order reaction of alkali with alkyl halides.[53] This is not an entirely satisfactory test of the theory, however, since the nature of the nucleophilic reagent changes (from hydroxide to ethoxide ion) with the composition of the solvent in this case.[55] Elimination reactions of other charge types are also discussed, and in all cases for which data are given the olefin yield increases with the alcohol content of the solvent.

8-1f. *Reactivity and Olefin Yield in $E1$ Reactions.* The rate of an $E1$ reaction is simply the rate of formation of carbonium ions multiplied by the fraction of the carbonium ions that gives olefins. The rate of formation of carbonium ions was discussed in Sec. 7-3a. The yields of olefins obtained from a carbonium ion appear to depend largely on the relative stabilities of the various olefins and of the substitution product. In most cases the larger changes in stability occur with the olefin and may often be attributed to hyperconjugation. Thus in ethanol at 25°, the t-butyl carbonium ion gives 19 per cent olefin, and the t-amyl carbonium ion, which can form the more stable 2-methyl-2-butene, gives 36.3 per cent olefin.[19] However, in 80 per cent ethanol, where the t-butyl and t-amyl carbonium ions yield 16 and 34 per cent olefin, respectively, the solvolyses of dimethyl-t-butylcarbinyl chloride and dimethylneopentylcarbinyl chloride yield 61 and 65 per cent olefin.[56] Brown and Fletcher proposed a steric explanation for this large increase in olefin yield. They pointed out that the substitution product may contain steric strains not present in the olefin, so that the increased yield need not be due to an increased rate of transformation of carbonium ion to olefin but may be due to a decrease in its rate of transformation to substitution product.

Praill and Saville found an exceptional case in the dehydration of 2-methyl-2,4-pentanediol, which gives mostly the 1-olefin.[57] It is suggested that this comes about because the carbonium ion loses its β-hydrogen internally, transferring it to the secondary hydroxyl group via a transition state containing a six-membered ring.

[55] Cf. R. G. Burns and B. D. England, *Tetrahedron Letters*, **24**, 1 (1960).
[56] H. C. Brown and R. S. Fletcher, *J. Am. Chem. Soc.*, **72**, 1223 (1950).
[57] P. F. G. Praill and B. Saville, *Chem. & Ind. (London)*, 495 (1960).

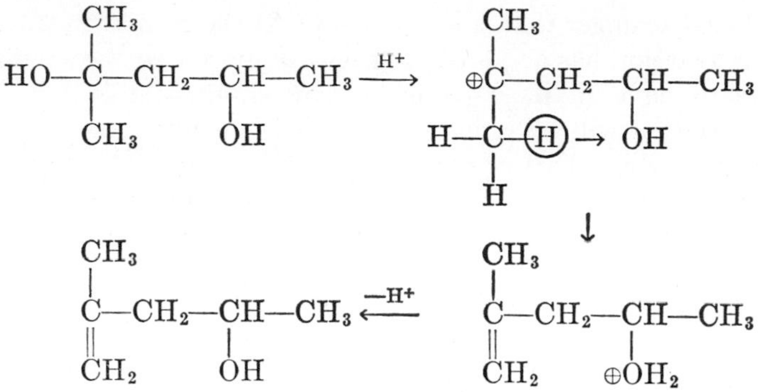

$$\underset{\substack{\text{CH}_3 \\ | \\ \text{CH}_3}}{\overset{\text{CH}_3}{\text{HO}-\text{C}-\text{CH}_2-\text{CH}-\text{CH}_3}} \overset{\text{H}^+}{\longrightarrow}$$

(see structures)

8-2. Other Types of Elimination Reactions. *8-2a. Eliminations to Form Multiple Bonds to Nitrogen, Oxygen, and Sulfur.* While elimination reactions to form $C{=}N$, $C{\equiv}N$, $N{=}N$, $C{=}O$, $C{=}S$, etc., bonds share many of the characteristics of the elimination reactions that yield olefins, they also have certain characteristics of their own that warrant discussion. The elimination of HX from compounds of the type

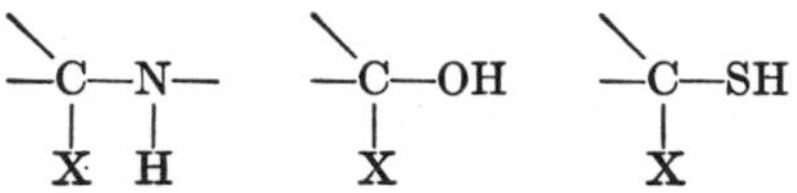

$$\underset{\substack{| \\ X\ H}}{\overset{\diagdown}{-}\text{C}-\text{N}-} \qquad \underset{\substack{| \\ X}}{\overset{\diagdown}{-}\text{C}-\text{OH}} \qquad \underset{\substack{| \\ X}}{\overset{\diagdown}{-}\text{C}-\text{SH}} \qquad \text{etc.}$$

occurs with particular facility, but since these reactions are simply the reverse of carbonyl addition–type reactions, they are considered in Chap. 11.

In comparing the ease of removal of HX from compounds of the type

$$\underset{\substack{| \quad | \\ H\ \ X}}{-\text{C}{=}\text{N}-} \qquad \underset{\substack{| \quad | \\ H\ \ X}}{-\text{C}-\text{N}-} \qquad \underset{\substack{| \quad | \\ H\ \ X}}{-\text{C}-\text{O}} \qquad \underset{\substack{| \quad | \\ H\ \ X}}{-\text{C}-\text{S}}$$

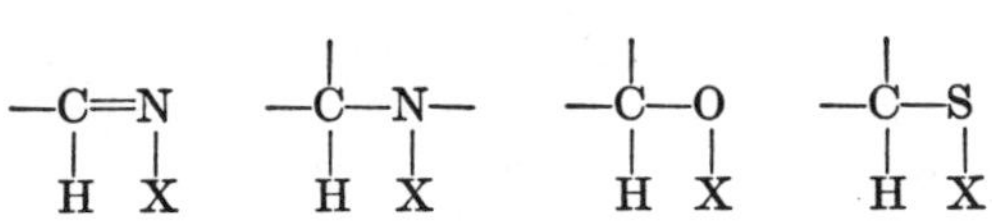

with the ease of analogous eliminations of HX to form olefins, note that X is practically always either a halogen atom or a group with a relatively electronegative atom, such as oxygen or nitrogen, at its point of attachment. In the *olefin-forming* eliminations the bond to X derives considerable strength from the difference in electronegativity between carbon and the atom by which X is attached (see Sec. 1-4c). When X is attached to oxygen, nitrogen, or sulfur, much of this extra stability is usually absent. Furthermore, it may be seen from Table 1-4 that a carbon-carbon double bond is about 60 kcal/mole stronger than the corresponding single bond, whereas a carbon-oxygen double bond is about 94 kcal stronger than the single bond. Similarly, a carbon-carbon triple bond is about 113 kcal stronger than a single bond, whereas a carbon-nitrogen triple bond is about 143 kcal stronger than a single bond. If the transition state in the elimination reaction has acquired sufficient similarity to the products (as it appears to have in most olefin-forming eliminations), the lowered energy of reaction will be reflected in a lower energy of activation and hence a faster rate. This provides a probable explanation for the fact that the bimolecular removal of HX from compounds of the type shown above by the action of basic reagents usually proceeds much more rapidly than the analogous acetylene- or olefin-forming eliminations (the simple removal of X with its bonding electron pair often causes rearrangement [Chap. 15] rather than elimination by a mechanism of the $E1$ type). Increased acidity of the hydrogen being removed (because of the adjacent oxygen, nitrogen, or sulfur atom) is probably also a factor in producing this increase in reactivity.

Hauser, LeMaistre, and Rainsford studied the alkaline dehydrochlorination of some aldochlorimines of substituted benzaldehydes in 92.5 per cent ethanol at $0°$.[58] The reaction is speeded by electron-withdrawing

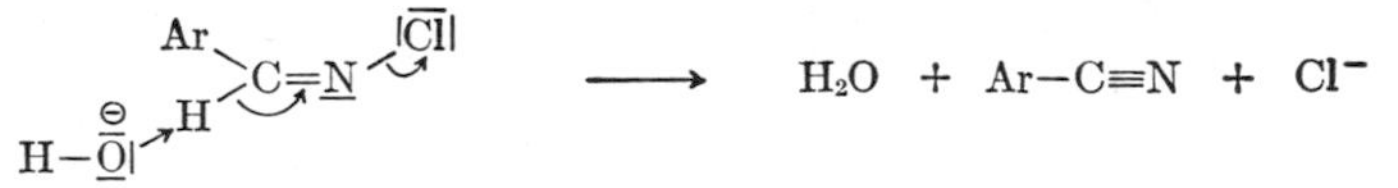

$$\underset{H-\overset{\ominus}{\underline{O}}|}{\overset{Ar}{\underset{H}{>}}} C{=}\underline{N} \overset{|\overline{C}l|}{} \longrightarrow H_2O + Ar-C{\equiv}N + Cl^-$$

groups, having $\rho = +2.240$ and log k_0 (liters mole^{-1} sec^{-1}) $= -1.796$.[59] The positive character of ρ and the fact that the p-nitro compound is somewhat more reactive than predicted from the Hammett equation suggest that the benzyl carbon atom may have acquired considerable carbanion character in the transition state, although it does not demand the intermediacy of a free carbanion. The ease of formation of a carbon-carbon triple bond is much less. Log k for the dehydrohalogenation of cis-β-bromostyrene by sodium hydroxide in isopropyl alcohol at $0°$ is about -4.7,[60] and the chloro compound would be much less reactive. A number of other studies of elimination reactions that form carbon-nitrogen bonds have been reported.[61]

Kornblum and DeLaMare[62] described the base-catalyzed transformation of α-phenylethyl t-butyl peroxide to acetophenone and t-butyl alcohol and pointed out that the mechanism is probably of the $E2$ type.

[58] C. R. Hauser, J. W. LeMaistre, and A. E. Rainsford, *J. Am. Chem. Soc.*, **57**, 1056 (1935).

[59] L. P. Hammett, "Physical Organic Chemistry," p. 190, McGraw-Hill Book Company, Inc., New York, 1940.

[60] S. J. Cristol and W. P. Norris, *J. Am. Chem. Soc.*, **76**, 3005 (1954).

[61] Among these are W. E. Jordan, H. E. Dyas, and D. G. Hill, *J. Am. Chem. Soc.*, **63**, 2383 (1941); O. L. Brady et al., *J. Chem. Soc.*, 1221, 1227, 1232, 1234, 1243 (1950); W. R. Bamford and T. S. Stevens, *J. Chem. Soc.*, 4735 (1952); R. H. Wiley, H. L. Davis, D. E. Gensheimer, and N. R. Smith, *J. Am. Chem. Soc.*, **74**, 936 (1952).

[62] N. Kornblum and H. E. DeLaMare, *J. Am. Chem. Soc.*, **73**, 880 (1951).

$$\begin{matrix} & CH_3 & & CH_3 \\ & | & & | \\ C_6H_5-C-\overline{O}-\overline{O}-CMe_3 & \longrightarrow & C_6H_5-C=\overline{O}| & + & {}^{\ominus}|\overline{O}-CMe_3 \\ & | & & \\ & H & & \\ Et-\overline{O}|^{\ominus} & & EtOH \end{matrix}$$

In olefin-forming eliminations alkoxide ions are displaced only under drastic conditions or with particularly favorable compounds.

Baker and Easty have studied in more detail the bimolecular transformation of alkyl nitrates to carbonyl compounds.[63] This reaction was

$$\begin{matrix} & | & & | \\ & -C-\overline{O}-NO_2 & \longrightarrow & -C=\overline{O}| & + & NO_2^- \\ & | & & \\ & H & & \\ Et-\overline{O}|^{\ominus} & & EtOH \end{matrix}$$

complicated, although in a fairly predictable manner, by the simultaneous occurrence of hydrolysis and olefin-forming elimination reactions. The rate of the carbonyl-forming elimination reaction was found to increase on going from methyl to ethyl nitrate but then to decrease on going to isopropyl nitrate. The first increase may be attributable to the stabilizing influence of a methyl radical on a carbonyl group, and the decrease can be rationalized as a steric effect.

Teich and Curtin have suggested a similar mechanism for the alkaline cleavage of S-desylthioglycolic acid, which yields desoxybenzoin.[64]

8-2b. *Elimination of Groups Other than HY.* In $E2$ reactions the incipient removal of a β-hydrogen atom without its bonding electron pair can make the β-carbon atom so electron-rich that a nucleophilic group attached to the α-carbon atom is displaced to permit the formation of a carbon-carbon double bond. It therefore seems reasonable that any other β-group X that can be removed readily without its bonding electron pair may give an elimination reaction similarly. Examples of this type are numerous and include the reaction of β-haloalkylsilanes with alkali.[65]

$$(C_2H_5)_3Si-CH_2-CH_2-Cl \longrightarrow (C_2H_5)_3SiOH + C_2H_4 + Cl^- \\ \uparrow \\ OH^-$$

Olefin formation in the reaction of metals with 1,2-dihalides and β-halo ethers, sulfides, amines, etc., probably proceeds by a similar mechanism,

[63] J. W. Baker and D. M. Easty, *Nature*, **166**, 156 (1950); *J. Chem. Soc.*, 1193, 1208 (1952); cf. G. R. Lucas and L. P. Hammett, *J. Am. Chem. Soc.*, **64**, 1928 (1942); S. J. Cristol, B. Franzus, and A. Shadan, *J. Am. Chem. Soc.*, **77**, 2512 (1955).

[64] S. Teich and D. Y. Curtin, *J. Am. Chem. Soc.*, **72**, 2481 (1950).

[65] L. H. Sommer, L. J. Tyler, and F. C. Whitmore, *J. Am. Chem. Soc.*, **70**, 2872 (1948).

in which an organometallic compound may be formed, but at the same time as, or after, the metal—C_β bond is formed, the low electronegativity of the metal produces such a high electron density on the β-carbon atom that Y is eliminated and a double bond formed.[66]

$$X-\underset{/\backslash}{C}-\underset{/\backslash}{C}-Y \ + \ M \longrightarrow \ M \overset{\frown}{\underset{/\backslash}{C}} \overset{\curvearrowright}{\underset{/\backslash}{C}} Y \longrightarrow \ M^+ \ + \ \underset{}{>}C{=}C\underset{}{<} \ + \ Y$$

Winstein, Pressman, and Young have found that the second-order (first-order in each reactant) dehalogenation of the 2,3-dibromobutanes by iodide ion transforms the meso isomer to *trans*-2-butene and the *dl* compound to *cis*-2-butene, showing that the elimination goes trans.[67] They point out that one bromine atom undergoes nucleophilic attack by iodide ion while simultaneously the double bond is formed and the other bromine is displaced as an anion. It has been suggested that the reaction involves the formation of a vicinal diiodide, which then loses iodine, and it may indeed be seen from the larger rates of dehalogenation of ethylene iodide and ethylene bromoiodide compared to ethylene bromide[68] that the replacement of bromine by iodine increases the reactivity. Nevertheless, the nucleophilic displacement of bromide by iodide ions does not appear to be important in the dehalogenation of 2,3-dibromobutane. A mechanism involving a rate-controlling *displacement* followed by rapid elimination of iodine bromide would disagree with the observed stereochemistry. A mechanism involving two consecutive displacements would demand the second to be faster than the first (since the over-all reaction is cleanly second-order) and thus require that a β-iodobromide have much greater S_N2 reactivity than the corresponding β-bromobromide. Studies on the effect of halogen substituents on S_N2 reactivity show that such a situation is very improbable.[69] A reaction mechanism involving rate-controlling nucleophilic displacement on carbon seems very plausible for the iodide-ion dehalogenation of many vicinal dibromides (such as 1,2-dibromides) with more S_N2 reactivity than 2,3-dibromobutane.[70] Such a mechanism appears to offer a more reasonable explanation of why the terminal olefins, ethylene, propylene, and 1-butene, are formed from their dibromides faster than are the more stable 2-butenes[71] (Table 8-5). This mechanism is further supported by the observation that the rate constant for dehalogenation of ethylene

[66] E. D. Amstutz, *J. Org. Chem.*, **9**, 310 (1944).

[67] S. Winstein, D. Pressman, and W. G. Young, *J. Am. Chem. Soc.*, **61**, 1645 (1939).

[68] A. Slator, *J. Chem. Soc.*, **85**, 1697 (1904); C. F. van Duin, *Rec. trav. chim.*, **45**, 345 (1926).

[69] J. Hine and W. H. Brader, Jr., *J. Am. Chem. Soc.*, **75**, 3964 (1953); J. Hine, S. J. Ehrenson, and W. H. Brader, Jr., *J. Am. Chem. Soc.*, **78**, 2282 (1956).

[70] J. Hine and W. H. Brader, Jr., *J. Am. Chem. Soc.*, **77**, 361 (1955).

[71] R. T. Dillon, *J. Am. Chem. Soc.*, **54**, 952 (1932).

TABLE 8-5. RATES OF DEHALOGENATION OF VICINAL DIBROMIDES BY IODIDE ION IN 99 PER CENT METHANOL AT $59.72°$[71]

Dibromide	$10^6 k$, liters mole^{-1} sec^{-1}
$BrCH_2CH_2Br$	83.2
$CH_3CHBrCH_2Br$	3.1
$C_2H_5CHBrCH_2Br$	4.1
meso-$CH_3CHBrCHBrCH_3$	2.5
dl-$CH_3CHBrCHBrCH_3$	1.3

bromide has just the value expected for the nucleophilic displacement of bromide by iodide ions under the given conditions.[70] The mechanism may be regarded as established in view of Schubert, Steadly, and Rabinovitch's observation that meso-1,2-dibromo-1,2-dideuterioethane reacts with iodide ions to give cis-1,2-dideuterioethylene.[72]

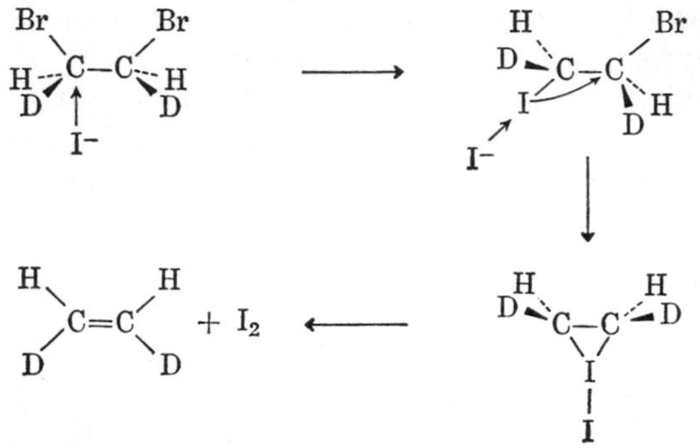

It is probably the greater ease with which iodine expands its outer octet of electrons that makes it more susceptible to nucleophilic attack than bromine. Reasons for suggesting the intermediacy of a species in which one, but only one, halogen atom is bound directly to carbon will be described in Sec. 9-1a.

Weinstock, Lewis, and Bordwell found that trans-1,2-dibromocycloalkanes are dehalogenated by iodide ion at the following relative rates: $C_7 > C_8 \sim C_5 > C_6 \gg C_4$.[73] The unreactivity of cyclobutene dibromide probably arises in part from the difficulty in achieving a completely trans orientation for the two bromine atoms, but the increased I strain that would come from putting two sp^2 carbon atoms in a four-membered ring must also tend to lower the reaction rate. The reactivity of the other cycloalkene dibromides was in the general vicinity of that of the 2,3-dibromobutanes and is probably the result of several opposed influences.

[72] W. M. Schubert, H. Steadly, and B. S. Rabinovitch, *J. Am. Chem. Soc.*, **77**, 5755 (1955).

[73] J. Weinstock, S. N. Lewis, and F. G. Bordwell, *J. Am. Chem. Soc.*, **78**, 6072 (1956).

8-2c. 1,4 *Eliminations.*   In addition to the β- or 1,2-elimination reactions, a number of 1,4 eliminations are known.   An example of one kind of 1,4 elimination is the formation of anthracene derivatives from related 9,10-dihydroanthracene compounds.   Cristol, Barasch, and Tieman found that base eliminated benzoic acid from *trans*-1,5-dichloro-9,10-dihydro-9,10-anthrandiol dibenzoate more than 1,000 times as fast as from the corresponding cis compound.[74]

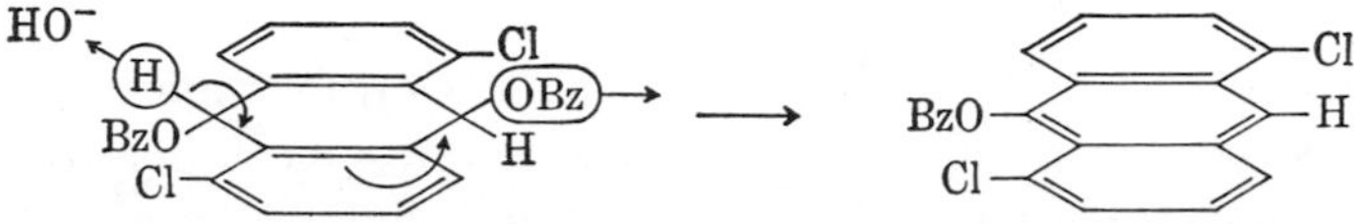

This shows that elimination reactions are preferentially cis in this series. One might point out that just as trans 1,2 eliminations are analogous to trans S_N2 reactions, so cis 1,4 eliminations are analogous to cis S_N2' reactions (Sec. 6-5b).

Another type of reaction that may be classified as a 1,4 elimination involves the formation of two new multiple bonds and the cleavage of a single bond.   The basic decomposition of 3-bromo-2,2-dimethyl-1-propanol[75] is an example of this type.

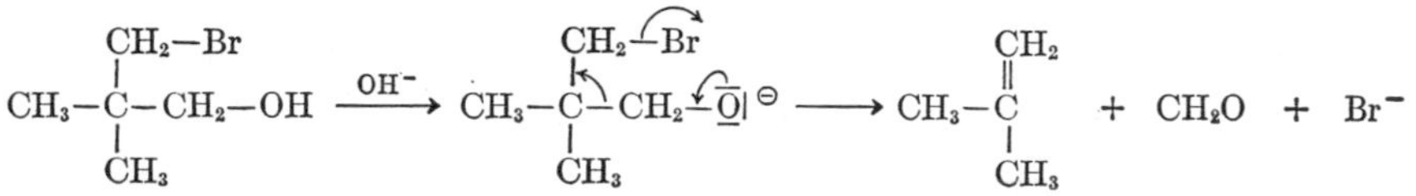

Eliminations of the 1,4 as well as the 1,2 type may proceed by $E1$ mechanisms too, and, of course, this carbonium-ion mechanism and the $E2$ type of mechanism are not separated sharply but blend gradually into each other.

Grovenstein and Lee and also Cristol and Norris have studied a case of this sort, the first-order transformation of the anion of cinnamic acid dibromide to β-bromostyrene.[76]   In aqueous solution the reaction of the dibromide prepared from *trans*-cinnamic acid yields β-bromostyrene containing about 78 per cent of the more stable trans isomer and 22 per cent cis.   The investigators have pointed out that this suggests the following mechanism:

$$C_6H_5CHBrCHBrCO_2^- \rightarrow Br^- + C_6H_5\overset{\oplus}{C}HCHBrCO_2^-$$
$$\downarrow$$
$$C_6H_5CH{=}CHBr + CO_2$$

[74] S. J. Cristol, W. Barasch, and C. H. Tieman, *J. Am. Chem. Soc.*, **77**, 583 (1955).

[75] S. Searles and M. J. Gortatowski, *J. Am. Chem. Soc.*, **75**, 3030 (1953).

[76] E. Grovenstein, Jr., and D. E. Lee, *J. Am. Chem. Soc.*, **75**, 2639 (1953); S. J. Cristol and W. P. Norris, *J. Am. Chem. Soc.*, **75**, 2645 (1953).

since rotation around the C_α—C_β bond of the carbonium ion prior to decarboxylation may account for the lack of stereospecificity. If carbonium-ion formation is discouraged by decreasing the ion-solvating power of the medium, as by running the reaction in ethanol or acetone, the percentage of cis isomer in the β-bromostyrene produced increases, reaching essentially 100 per cent in acetone. Here, apparently, a trans elimination of the $E2$ type is occurring.

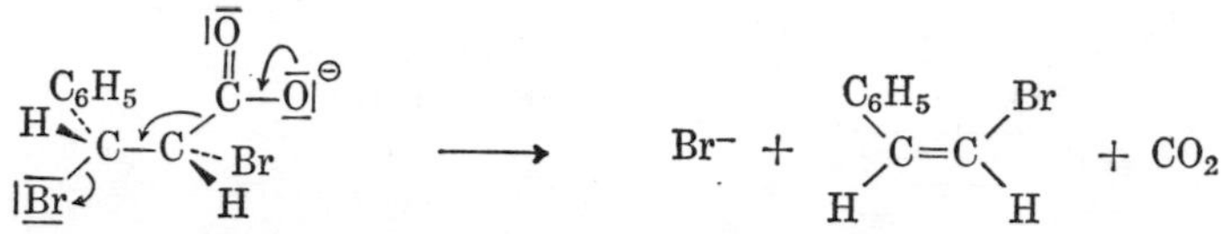

Our knowledge of the mechanism of 1,4 eliminations with fragmentation has been reviewed and extended by Grob and coworkers.[77]

PROBLEMS

1. Write a structural formula for the principal olefinic product of each of the following reactions and give the reasons for your choice:

a) $ClCH_2CHCl_2 + NaOEt \longrightarrow$

b)

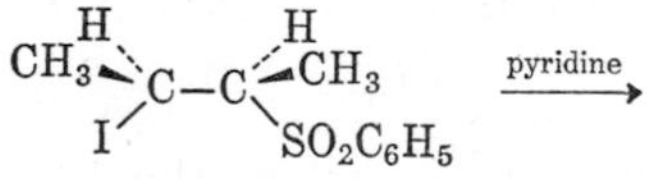

c) $(CH_3)_2CHC(CH_3)_2 \xrightarrow{H_2SO_4}$
         |
        OH

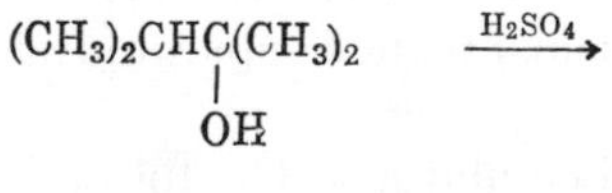

2. Suggest experiments designed to show where, in several series of related elimination reactions, the mechanism changes from $E1$ to $E2$.

3. The reactions of both the cis and trans isomers of 4-t-butylcyclohexyltrimethylammonium chloride with potassium t-butoxide have been studied. One isomer yields 10 per cent 4-t-butylcyclohexyldimethylamine and 90 per cent 4-t-butylcyclohexene, and the other gives only 4-t-butylcyclohexyldimethylamine. Identify the isomers and explain their behavior.

4. The dehalogenation of $trans$-1,2-dibromocyclohexane by iodide ion is about 12 times as fast as that of the cis isomer. The iodide-ion dehalogenation of $trans$-1-bromo-2-chlorocyclohexane proceeds at about the same rate as that of its cis isomer. Tell why the reactivity of the trans compound relative to the cis is greater with the dibromides than with the chlorobromides. Estimate the relative reactivities of the cis dibromide and the cis chlorobromide.

[77] C. A. Grob et al., *Helv. Chim. Acta*, **38**, 594 (1955); **42**, 872 (1959); *Experientia*, **13**, 126 (1957); *Chem. & Ind.* (*London*), 757 (1958).

5. The compounds A, B, and C

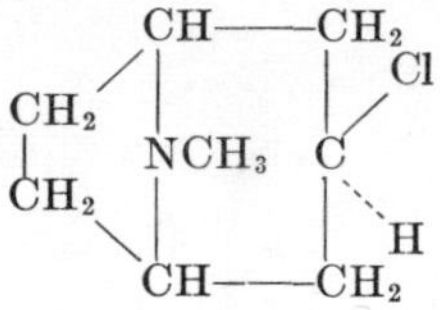

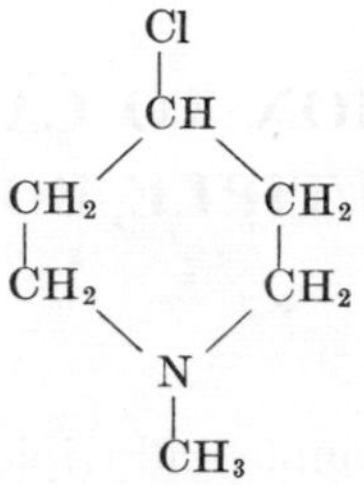

undergo solvolysis in 80 per cent ethanol at a rate that is unaffected by the addition of dilute alkali. At 118° the compounds solvolyzed 11,500, 780, and 140 times as fast as cyclohexyl chloride, respectively. Compounds B and C each gives the alcohol (and ethyl ether) related to the chloride shown. Compound A gives an unsaturated monocyclic product. Explain the relative rates of solvolysis and give the structural formula of the product obtained from A.

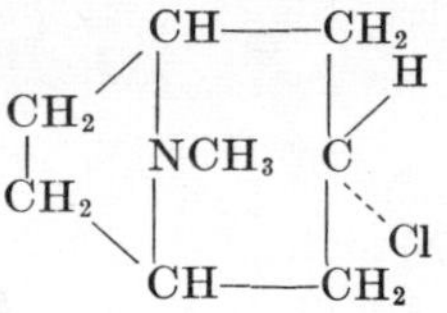

Chapter 9

POLAR ADDITION TO CARBON-CARBON
MULTIPLE BONDS

Although carbon-carbon double and triple bonds may react with either nucleophilic or electrophilic reagents, and the relative tendencies to do so depend on the exact structure of the unsaturated compound, the most common reactions are those with electrophilic reagents. These will be discussed first.

9-1. Addition of Halogens.[1] *9-1a. Mechanism of Halogen Additions.* The addition of halogen to olefins often occurs by a free-radical mechanism (Sec. 20-2a), particularly in light-catalyzed and/or vapor-phase reactions. However, there is also a polar mechanism for addition, whose operation is encouraged by ion-solvating media and the absence of light and peroxides. To explain the fact that electron-donating substituents increase the reactivity in the polar addition of halogens to olefins,[2] we must assume that the halogen is behaving as an electrophilic reagent and that the olefin is nucleophilic. Since it has also been found that the addition of bromine in the presence of nucleophilic reagents often yields products in which the nucleophilic reagent and a bromine atom have added, it was reasonable to suggest that the reaction consists of the addition of a bromine *cation* in the rate-controlling step to yield an intermediate carbonium ion, which may then rapidly react with any available nucleophilic reagent. For instance, this explains the observations of Francis, who found that the reaction of ethylene with bromine in aqueous solution yields, in addition to ethylene bromide, some β-bromoethyl chloride in the presence of sodium chloride and β-bromoethyl nitrate in the presence of sodium nitrate.[3]

---

[1] This subject has been reviewed through 1949 in P. B. D. de la Mare, *Quart. Revs. (London)*, **3**, 126 (1949).

[2] C. K. Ingold and E. H. Ingold, *J. Chem. Soc.*, 2354 (1931); S. V. Anantakrishnan and C. K. Ingold, *J. Chem. Soc.*, 984, 1396 (1935).

[3] A. W. Francis, *J. Am. Chem. Soc.*, **47**, 2340 (1925).

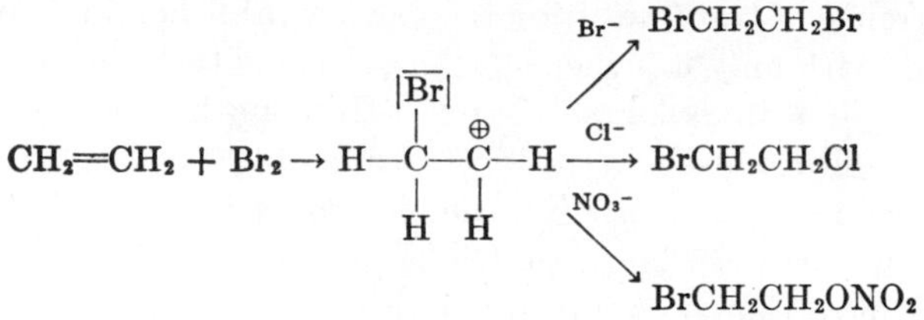

$$CH_2{=}CH_2 + Br_2 \rightarrow H{-}\overset{\overset{\displaystyle |\overline{Br}|}{|}}{\underset{\underset{\displaystyle H}{|}}{C}}{-}\overset{\oplus}{\underset{\underset{\displaystyle H}{|}}{C}}{-}H$$

$$\xrightarrow{Br^-} BrCH_2CH_2Br$$

$$\xrightarrow{Cl^-} BrCH_2CH_2Cl$$

$$\xrightarrow{NO_3^-} BrCH_2CH_2ONO_2$$

Combination of such a carbonium ion with solvent could explain the formation of halohydrins in aqueous solution and their ethers in alcoholic solvents. Roberts and Kimball pointed out that the electron-deficient carbon atom in such an intermediate might very well satisfy its deficiency by coordination with the unshared pairs of the bromine atom to yield an intermediate with a three-membered ring, called a "bromonium ion."[4] Since the opening of this ring by the attack of a nucleophilic reagent would result in a Walden inversion at the carbon atom attacked, this mechanism also has the great advantage of explaining trans addition. Thus, maleic acid yields the racemic dibromide, and fumaric acid gives the meso compound.[5]

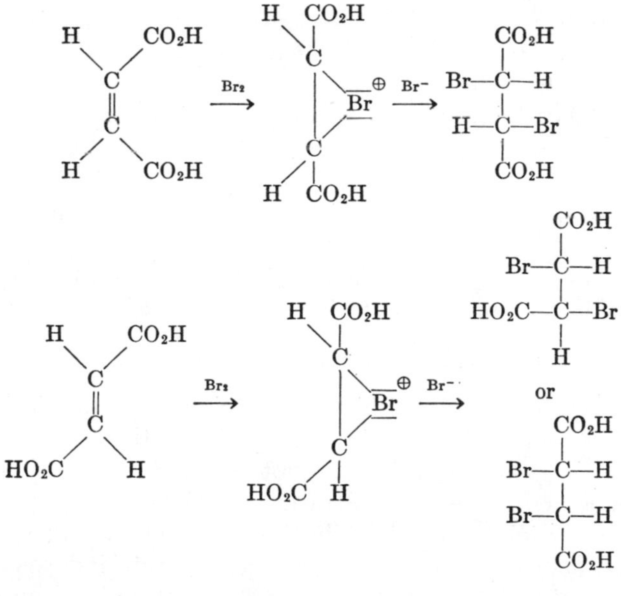

[4] I. Roberts and G. E. Kimball, *J. Am. Chem. Soc.*, **59**, 947 (1937).

[5] A. McKenzie, *Proc. Chem. Soc.*, **27**, 150 (1911); *J. Chem. Soc.*, **101**, 1196 (1912); B. Holmberg, *Chem. Abstr.*, **6**, 2072 (1912); P. F. Frankland, *J. Chem. Soc.*, **101**, 673 (1912).

The occurrence of trans addition has been established in a large number of other cases with only a few exceptions, such as the addition to maleate and fumarate ions in polar solvents.[6] Here both ions yield the same product, *meso*-dibromosuccinate ion, presumably because in the bromonium ion formed from maleate ion the electrostatic repulsion between the carboxylate groups is so large that it ruptures the three-membered ring with isomerism to the trans intermediate.[5] It has more recently become possible to obtain much support for the bromonium ion from investigations of neighboring-group participation (Secs. 6-4e and 7-3c). An intermediate like I would also explain trans addition and would be the bromine

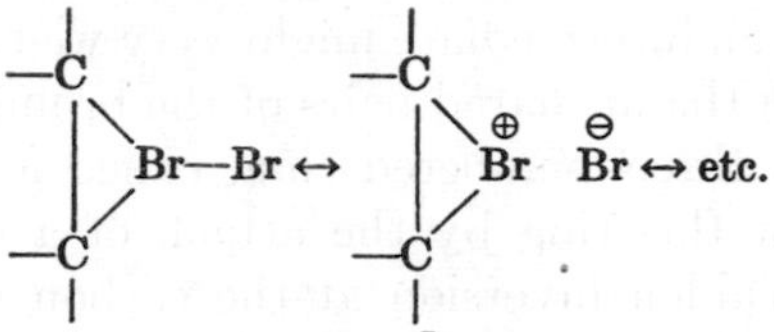

analog of an intermediate that seems from studies of deiodination (Sec. 8-2b) to be a very likely one in the addition of iodine to double bonds.

Although these observations, particularly that of trans addition, show the improbability of a simple "broadside" mechanism in which the halogen molecule attacks the olefin to add both halogen atoms simultaneously, the argument has been further strengthened by demonstrating that bromochlorides, bromonitrates, bromohydrins, etc., are not necessarily formed from BrCl, $BrONO_2$, BrOH, etc. Bartlett and Tarbell have presented some of the most convincing evidence of this sort.[7] These workers studied the reaction of stilbene with bromine in methanol to yield mostly stilbene methoxybromide but some stilbene dibromide. If the major product were due to the reaction of stilbene with methyl hypobromite, the latter would have to have been formed by the reaction

$$Br_2 + MeOH \rightleftharpoons MeOBr + H^+ + Br^- \qquad (9\text{-}1)$$

Although it was found that within the experimental error all the bromine remained in the molecular form in methanolic solution even at concentrations as low as 0.004 M, this does not rule out the possibility of the presence of a very small concentration of MeOBr through which the entire reaction might proceed. The formation of MeOBr could not be rate-controlling, because this would render the rate independent of the concentration of stilbene, when actually the reaction is found to be first-order in stilbene. It is obvious from Eq. (9-1) that the addition of acid

[6] R. Kuhn and T. Wagner-Jauregg, *Ber.*, **61**, 519 (1928).

[7] P. D. Bartlett and D. S. Tarbell, *J. Am. Chem. Soc.*, **58**, 466 (1936).

will decrease the concentration of hypobromite present at equilibrium. The fact, then, that the reaction rate is not slowed by acid shows that the reaction cannot be simply the addition of MeOBr to stilbene.

Further evidence that the reaction does not consist of the direct one-step addition of halogen or a hypohalite comes from the observation of Tarbell and Bartlett that the addition of chlorine to the dimethyl-maleate and dimethylfumarate ions yields β-lactones, presumably by the mechanism

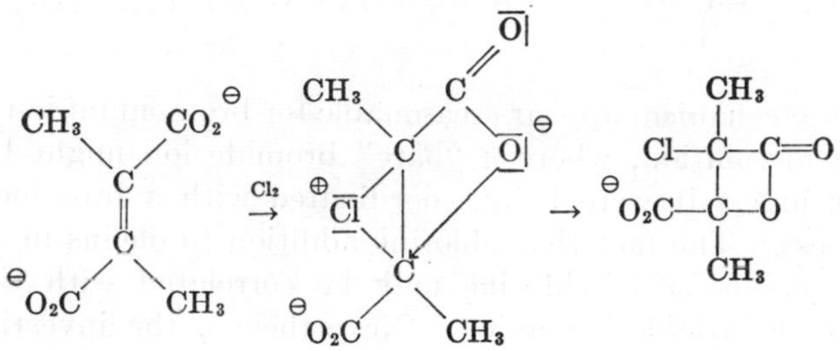

since neither of the β-lactones can be made from either the one dichloro acid or the one chlorohydrin known under the conditions used.[8]

Although there appear to be no other studies of the kinetics and product composition of the addition of halogen to an olefinic double bond in aqueous or alcoholic solutions comparable in thoroughness to the work of Bartlett and Tarbell on stilbene,[7] the data that have been obtained in these solvents are of the same type as theirs.   For example, the additions of bromine to maleic and fumaric acids and of iodine to allyl alcohol are all first-order in halogen and first-order in olefin,[9] as are the additions of bromine to cis-cinnamic and acrylic acid in aqueous acetic acid and water.[10]

Many studies have been carried out in acetic acid solution, particularly by Robertson and coworkers.   The reaction kinetics are often more complicated in this solvent, with third-order terms (first-order in olefin and second-order in halogen) appearing and overshadowing the second-order reaction even in fairly dilute solutions in the reactions of bromine, bromine chloride, iodine, and iodine chloride with a variety of olefins.[10,11] White and Robertson suggested that this may indicate that the olefin and halogen form a complex that reacts with another molecule of halogen, as in a scheme of the following type:

[8] D. S. Tarbell and P. D. Bartlett, *J. Am. Chem. Soc.*, **59**, 407 (1937).

[9] A. Berthoud and M. Mosset, *J. chim. phys.*, **33**, 272 (1936).

[10] P. W. Robertson, N. T. Clare, K. J. McNaught, and G. W. Paul, *J. Chem. Soc.*, 335 (1937).

[11] E. P. White and P. W. Robertson, *J. Chem. Soc.*, 1509 (1939).

$$\begin{array}{c} -\overset{|}{\underset{|}{C}} \\ \parallel \\ -\overset{|}{\underset{|}{C}} \end{array} + Br_2 \quad \overset{\text{fast}}{\rightleftharpoons} \quad \begin{array}{c} \overset{|}{\underset{}{C}} \\ \diagdown \\ \diagup \quad Br\!-\!Br \\ -\overset{|}{\underset{|}{C}} \end{array}$$

$$\begin{array}{c} -\overset{|}{\underset{}{C}} \\ \diagdown \\ \diagup \quad Br\!-\!Br + Br_2 \\ -\overset{|}{\underset{|}{C}} \end{array} \quad \overset{\text{rate}}{\underset{\text{controlling}}{\longrightarrow}} \quad \begin{array}{c} -\overset{|}{\underset{}{C}} \\ \diagdown \\ \diagup \quad \overset{\oplus}{Br} + Br_3^- \\ -\overset{|}{\underset{|}{C}} \end{array}$$

This sort of mechanism appears reasonable for bromonium-ion formation in acetic acid solution, where a "bare" bromide ion might be difficult to displace unless it were being coordinated with a bromine molecule in the process. The fact that chlorine addition to olefins in acetic acid solution is *first-order* in chlorine may be correlated with the lack of stability of the trichloride anion. Nevertheless, the investigations of these third-order halogenations that have been carried out to date do not seem to have yielded enough information to distinguish between this mechanism and several other possibilities.

In nonpolar solvents, such as carbon tetrachloride and hydrocarbons, halogen addition is much slower and is complicated by catalysis by the surface of the reaction vessel and traces of water and other ion-solvating molecules.[10]

Some unsaturated compounds, particularly vinyl and allyl halides, when undergoing halogen addition in the presence of halide ions follow a kinetic equation with a third-order term containing the concentrations of olefin, halogen, and halide ion.[12] This may indicate a nucleophilic attack by the halide ion on a reversibly formed olefin-halogen complex, but other interpretations are also available.

For many compounds like α,β-unsaturated carbonyl compounds, nitriles, etc., the situation is further complicated by acid catalysis.[13] This may involve the formation of the conjugate acid of the unsaturated compound, toward which the halogen acts as a nucleophilic reagent.

In the absence of the free halogens it is usually possible to study reactions of olefins with hypohalous acids. The situation here is rather analogous to that found in aromatic halogenation reactions and therefore will not be discussed in detail here.

With conjugated dienes the possibility of the formation of more than one product provides complications. In the addition of chlorine to 1,3-butadiene, a relatively strainless chloronium ion

[12] K. Nozaki and R. A. Ogg, Jr., *J. Am. Chem. Soc.*, **64**, 697, 704, 709 (1942).
[13] P. B. D. de la Mare and P. W. Robertson, *J. Chem. Soc.*, 888 (1945).

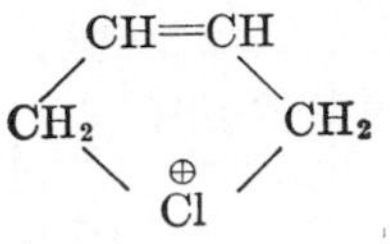

$$CH=CH$$

$$CH_2 \qquad CH_2$$

$$Cl^{\oplus}$$

may be written as an intermediate, but this possibility has been ruled out by Mislow and Hellman, who found that the 1,4-dichloro-2-butene produced in the addition of chlorine to butadiene is entirely trans rather than cis, as would be expected from the intermediate above.[14]   It therefore appears that in the reactive intermediate, chlorine interacts with only one of the double bonds, giving an intermediate that may be a chloronium ion

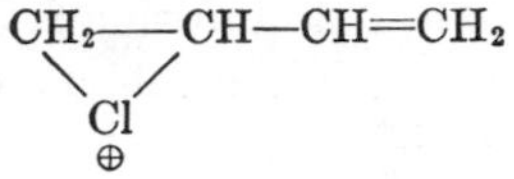

$$CH_2 \text{---} CH \text{---} CH=CH_2$$

$$Cl \atop \oplus$$

or may have the carbonium-ion structure

$$Cl\text{---}CH_2\text{---}\overset{\oplus}{C}H\text{---}CH=CH_2 \leftrightarrow Cl\text{---}CH_2\text{---}CH=CH\text{---}\overset{\oplus}{C}H_2$$

since chlorine is not a very effective neighboring group and since this carbonium ion is resonance-stabilized.   The intermediate may combine with a chloride ion at either of two carbon atoms regardless of whether it is a chloronium or carbonium ion (in the former case Cl^- may attack by either the S_N2 or S_N2' mechanism).   In general, in cases of this sort the product isolated need not be the one formed more rapidly from the reactive intermediate, since in all cases the products are allyl halides and may ionize to the reactive intermediate cation with some degree of ease.   If the two possible products are interconvertible under the reaction conditions, then the more stable, or *thermodynamically controlled*, product will be formed preferentially.   If, on the other hand, the possible products are not interconvertible under the reaction conditions, the more rapidly formed, or *kinetically controlled*, product will be isolated in larger yield. The fact that Muskat and Northrup obtained about twice as much 1,2-dichloro-3-butene as 1,4-dichloro-2-butene under conditions where the two compounds are not interconvertible[15] shows that the former is the kinetically controlled product, and the observation of Pudovik that zinc chloride will isomerize either dichloride to a mixture containing about 70 per cent of the 1,4 isomer[16] shows that the latter is the thermodynamically controlled product.

[14] K. Mislow and H. M. Hellman, *J. Am. Chem. Soc.*, **73**, 244 (1951).

[15] I. E. Muskat and H. E. Northrup, *J. Am. Chem. Soc.*, **52**, 4043 (1930).

[16] A. N. Pudovik, *Zhur. Obshchei Khim.*, **19**, 1179 (1949); *Chem. Abstr.*, **44**, 1005 (1950).

As implied in the preceding paragraph, although it seems probable that many products of the addition of halogens to olefins are formed by the nucleophilic attack of a halide ion on an intermediate cyclic halonium cation, it is also quite likely that in some cases the intermediate is a simple β-halocarbonium ion

$$X-\underset{|}{\overset{|}{C}}-\underset{|}{\overset{|}{C}}\oplus$$

It seems clear that the probability and extent of the β-halocarbonium-ion reaction mechanism will increase with increasing stabilization of the carbonium ion and with decreasing ability of X to act as a neighboring group, but there seems to be little information on just which reactions proceed by each of the two possible mechanisms. In at least one case, however, the preferentially cis addition of chlorine to phenanthrene in acetic acid, a case in which the halonium-ion mechanism appears particularly improbable, the simple carbonium-ion mechanism seems required.[17] This reaction may be initiated by the donation of Cl+ to give an ion pair that is sufficiently reactive to combine with chloride ion, frequently before the chloride ion can diffuse to the other side of the aromatic ring system.

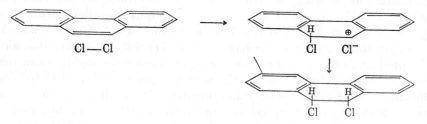

9-1b. Reactivity in Halogen Additions.   Since the double bond behaves as a nucleophilic reagent in the addition of halogens, hypohalous acids, etc., it is reasonable to expect electron-donating groups to increase the reactivity of olefins toward such reagents.   In the addition of chlorine to compounds of the type $RC_6H_4CH{=}CHCOC_6H_5$ in acetic acid solution, for example, ρ appears to have a value of about -4.5.[18] Relative reactivities toward bromine in acetic acid are listed in Table 9-1 for a number of olefins. In general, the reactivity is seen to increase with the electron-donating power of the substituents. The greater reactivity of either *cis-* or *trans-*cinnamic acid compared with phenylpropiolic acid seems to be common for olefins compared with their corresponding acetylenes.   Although the free energy of reaction is probably more

[17] P. B. D. de la Mare and N. V. Klassen, *Chem. & Ind. (London)*, 498 (1960).
[18] From data of P. W. Robertson, quoted by de la Mare in Ref. 1.

TABLE 9-1. RELATIVE REACTIVITIES TOWARD BROMINE IN ACETIC ACID
AT 24°[18]

Compound	Relative reactivity
$C_6H_5CH=CH_2$	Very fast
$C_6H_5CH=CHC_6H_5$	18
$CH_2=CHCH_2Cl$	1.6
$CH_2=CHCH_2Br$	1.0
$C_6H_5CH=CHOC_6H_5$	0.33
$C_6H_5CH=CHBr$	0.11
cis-$C_6H_5CH=CHCO_2H$	0.063
trans-$C_6H_5CH=CHCO_2H$	0.017
$C_6H_5C\equiv CCO_2H$	0.0053
$CH_2=CHBr$	0.0011

favorable for the acetylenes, the rate is slower, probably because of the instability of either a vinyl carbonium ion or an unsaturated three-membered ring.

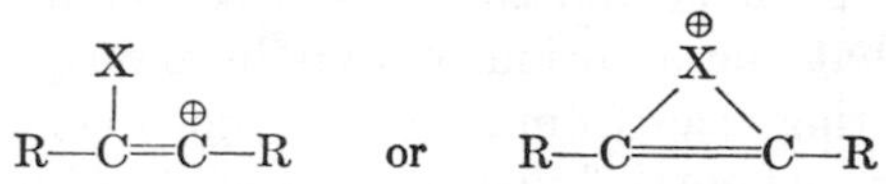

$$\overset{X}{\underset{|}{}}\qquad\qquad \overset{\oplus}{X}$$

R—C=$\overset{\oplus}{C}$—R   or   R—C$\diagup\diagdown$C—R

The difference in reactivity between allyl chloride and bromide, though not large, is not in the direction that would be expected.

Another type of reactivity problem arises when the addition of such unsymmetrical reagents as ICl, HOCl, etc., to unsymmetrical olefins is considered. Often in the major product of such reactions the more electronegative half of the adding reagent has become attached to the carbon that is more capable of supporting a positive charge. In some cases, such as the addition of hypochlorous acid to isobutylene, this orientation may be attributed to the preferential formation of the more stable of the two possible acyclic β-halocarbonium ions.[19]

$$(CH_3)_2C=CH_2 \xrightarrow{HOCl} (CH_3)_2\overset{\oplus}{C}CH_2Cl \xrightarrow{H_2O} (CH_3)_2\underset{\underset{OH}{|}}{C}CH_2Cl + H^+$$

In other cases, however, such as the addition of iodine chloride to crotonic acid,[20] a cyclic halonium-ion intermediate seems much more probable.

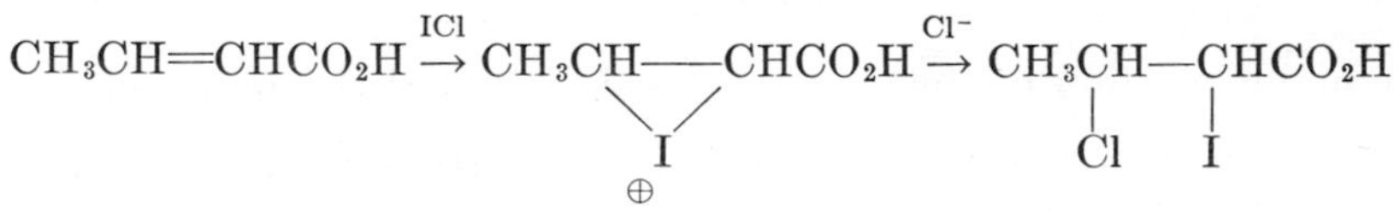

$$CH_3CH=CHCO_2H \xrightarrow{ICl} CH_3CH\underset{\underset{\oplus}{I}}{\diagdown\diagup}CHCO_2H \xrightarrow{Cl^-} CH_3CH\underset{Cl}{|}—\underset{I}{|}CHCO_2H$$

[19] P. B. D. de la Mare and A. Salama, J. Chem. Soc., 3337 (1956).
[20] C. K. Ingold and H. G. Smith, J. Chem. Soc., 2742 (1931).

Apparently bond breaking is more important than bond making in the nucleophilic attack of chloride on this iodonium ion.

There are a number of addition reactions known in which the *less* electronegative half of the adding reagent becomes attached largely to the more highly alkylated carbon atom of the double bond. For example, in the addition of bromine in methanol to t-butylethylene the only methoxybromide isolated was the secondary bromide.[21] Puterbaugh and Newman pointed out that nucleophilic attack at the neopentyl-like secondary position of the intermediate bromonium ion in this case must be highly sterically hindered.

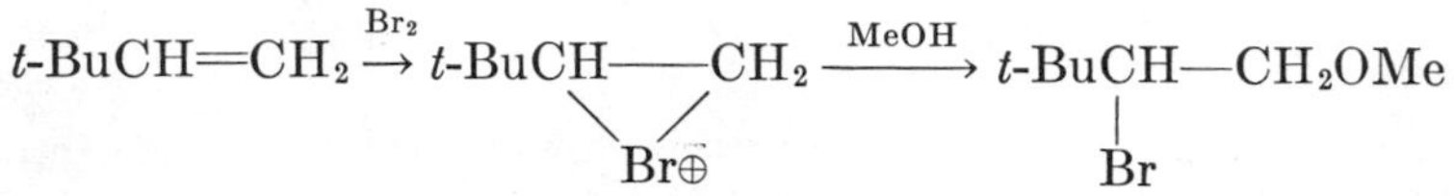

$$t\text{-BuCH}{=}\text{CH}_2 \xrightarrow{\text{Br}_2} t\text{-BuCH}\!\!-\!\!\!\overset{}{\underset{\text{Br}\oplus}{\diagdown\diagup}}\!\!\!-\!\!\text{CH}_2 \xrightarrow{\text{MeOH}} t\text{-BuCH}\!\!\underset{\text{Br}}{\overset{|}{-}}\!\!\text{CH}_2\text{OMe}$$

Under the same conditions 1-hexene gives a methoxybromide mixture that is 80 per cent primary and 20 per cent secondary. Traynham and Pascual showed that the addition of hypobromous acid to methylenecycloalkanes with ring sizes from four through seven carbons gave only primary bromides, whereas hypochlorous-acid addition gave largely primary chlorides only with the four- and six-membered-ring reactants.[22] The presence of halogen and certain other substituent groups in the olefinic reactant can produce complications. Ballinger and de la Mare found that in the addition of hypochlorous acid to Cl^{36}-labeled methallyl chloride, the 2,3-dichloro-2-methylpropanol obtained (in about 5 per cent yield) has 19 per cent of its Cl^{36} attached to the central carbon atom.[23] This suggests that the intermediate chloronium ion rearranges, perhaps by way of a carbonium ion.

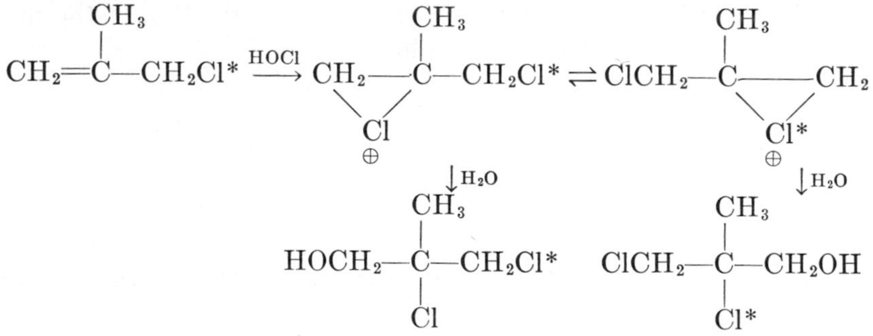

The major reaction product was 1,3-dichloro-2-methyl-2-propanol.

[21] W. H. Puterbaugh and M. S. Newman, *J. Am. Chem. Soc.*, **79**, 3469 (1957).
[22] J. G. Traynham and O. S. Pascual, *Tetrahedron*, **7**, 165 (1959).
[23] P. Ballinger and P. B. D. de la Mare, *J. Chem. Soc.*, 1481 (1957).

The reactivity may be affected as strongly by the nature of the attacking halogen or halogen derivative as by that of the olefin.   The reactivity of X—Y (where at least X is halogen) is increased by increasing the electron-withdrawing power of either X or Y.   From data on third-order halogenation in acetic acid solution, White and Robertson have estimated the following relative reactivities:[11]

I_2	IBr	ICl	Br_2	BrCl
1	3×10^3	10^5	10^4	4×10^6

If it is X that is forming the bond to carbon in the rate-controlling step, then an increase in its electronegativity will increase the strength of this bond.   An increase in the electron-withdrawing power of Y will make X even more electrophilic toward the nucleophilic olefin.   That is, in these nucleophilic displacements on X, bond making is always a more important factor than bond breaking.   This might be expected since X, not being a first-row element, may easily expand its octet to form the new bond before the old bond is broken.

9-2. Additions of Strong Acids.   *9-2a. Mechanism of Additions of Acids.*   As nucleophilic reagents olefins would be expected to react with acids, and it is reasonable that the reaction should be initiated by the donation of a proton to the carbon-carbon multiple bond.   This may indeed happen, but it is also possible for some acids to add by a free-radical mechanism.   This is true for hydrogen bromide, the acid whose addition to olefins has been studied most often.   For this reason, studies of the addition of hydrogen bromide must be viewed with caution if they were carried out before 1933, when Kharasch and Mayo showed that minute amounts of peroxides formed by the exposure of olefins to air may almost entirely reverse the direction of addition.[24]   The mechanism of this free-radical reaction and the explanation of the resulting orientation of addition will be discussed in Sec. 20-2b, but here it will suffice to say that the radical reaction may be excluded by working with highly purified reactants in the dark or, more easily, by the use of suitable inhibitors.

The simplest plausible mechanism for addition to olefins under acidic conditions involves an initial protonation to give a carbonium ion, followed by combination of the carbonium ion with a nucleophilic reagent.   For the acid-catalyzed hydration of 2-methyl-2-butene this mechanism may be written

[24] M. S. Kharasch and F. R. Mayo, *J. Am. Chem. Soc.*, **55**, 2468 (1933); F. R. Mayo and C. Walling, *Chem. Revs.*, **27**, 351 (1940).

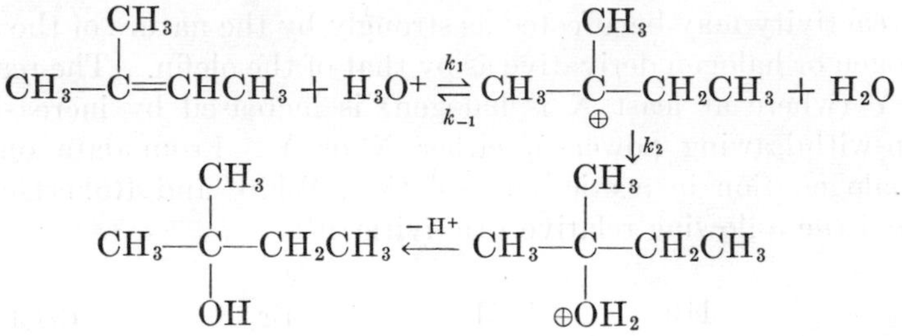

$$CH_3-\underset{\underset{CH_3}{|}}{C}=CHCH_3 + H_3O^+ \underset{k_{-1}}{\overset{k_1}{\rightleftharpoons}} CH_3-\underset{\underset{CH_3}{|}}{\overset{\oplus}{C}}-CH_2CH_3 + H_2O$$

$$CH_3-\underset{\underset{OH}{|}}{\overset{\overset{CH_3}{|}}{C}}-CH_2CH_3 \overset{-H^+}{\longleftarrow} CH_3-\underset{\underset{\oplus OH_2}{|}}{\overset{\overset{CH_3}{|}}{C}}-CH_2CH_3$$

It appears that the first step of the reaction is rate-controlling (that is, $k_2 \gg k_{-1}$) in aqueous solution. This follows from Purlee and Taft's observation that the unreacted olefin recovered after 50 per cent hydration in deuteriated water was essentially free of deuterium.[25] Kinetic-isotope-effect studies[25] and h_0-rate correlations[26] have been interpreted in terms of a more complicated mechanism, in which the first reaction intermediate is a pi complex, wherein a proton has coordinated with the pi electrons of the olefin as shown below,

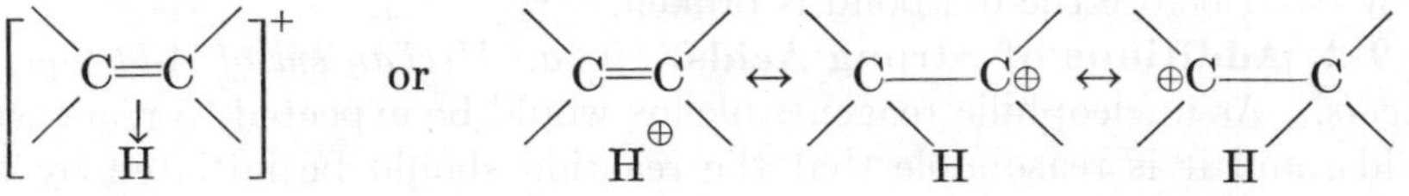

$$\left[\underset{\underset{H}{\downarrow}}{\overset{}{C}=C} \right]^+ \quad or \quad \underset{\underset{H}{\oplus}}{C=C} \leftrightarrow \underset{\underset{H}{}}{C-C\oplus} \leftrightarrow \underset{\underset{H}{}}{\oplus C-C}$$

but this point does not appear to have been settled yet.[27,28]

As might be expected from a reaction mechanism involving an intermediate carbonium ion, the acidic hydration of 1,2-dimethylcyclohexene results in considerable amounts of both cis and trans addition under conditions where the two products are not appreciably interconvertible.[29] On the other hand, Hammond and Nevitt found that the 1,2-dimethylcyclohexyl bromides formed by the addition of hydrogen bromide in acetic acid to 1,2-dimethylcyclohexene, 1-methylene-2-methylcyclohexane, and 2,3-dimethylcyclohexene contained 0, 35, and 13 per cent of the cis isomer, respectively.[30] This difference in results rules out any mechanism in which the same 1,2-dimethylcyclohexyl cation is an intermediate in all three reactions. The purely trans addition observed with 1,2-dimethylcyclohexene could be explained by trans nucleophilic attack of bromide ion on a symmetrically protonated olefin molecule.

[25] E. L. Purlee and R. W. Taft, Jr., *J. Am. Chem. Soc.*, **78**, 5807 (1956).

[26] R. W. Taft, Jr., *J. Am. Chem. Soc.*, **74**, 5372 (1952).

[27] I. Dostrovsky and F. S. Klein, *J. Chem. Soc.*, 791, 4401 (1955).

[28] R. H. Boyd, R. W. Taft, Jr., A. P. Wolf, and D. Christman, *J. Am. Chem. Soc.*, **82**, 4729 (1960).

[29] C. H. Collins and G. S. Hammond, *J. Org. Chem.*, **25**, 911 (1960).

[30] G. S. Hammond and T. D. Nevitt, *J. Am. Chem. Soc.*, **76**, 4121 (1954).

$$
\begin{array}{ccc}
\underset{\displaystyle CH_2}{\overset{\displaystyle CH_2-CH_2}{\diagup\quad\diagdown}}\underset{\displaystyle CH_2}{}\\
CH_2 \qquad\qquad CH_2 \\
C = C \\
CH_3 \qquad\qquad CH_3
\end{array}
\rightarrow
\begin{array}{ccc}
CH_2-CH_2 \\
CH_2 \quad H\oplus \quad CH_2 \\
\uparrow \\
C = C \\
CH_3 \qquad CH_3 \\
Br^-
\end{array}
$$

$$\downarrow$$

$$
\begin{array}{c}
CH_2 - CH_2 \\
CH_2 \quad H \quad CH_3 \quad CH_2 \\
\big| \qquad \big| \\
C - C \\
CH_3 \quad Br
\end{array}
$$

The intermediacy and appreciable lifetime of an ordinary carbonium ion seems to be favored by the use of an effective ion-solvating medium, such as water.

9-2b. *Reactivity in Additions of Acids.* A discussion of the effect of the structure of the olefin on reactivity in additions by acids may be divided into two parts. One aspect to be considered is the relative reactivity of various olefins, and the other is the reactivity of the two carbon atoms joined by a given multiple bond. We shall consider first the latter problem, the orientation of addition of acids to unsymmetrically substituted multiple bonds, which has been studied more and is probably more important to the synthetic organic chemist.

Markownikoff was the first to formulate a generalization applicable to this problem. His rule for the addition of hydrogen halides to olefins states that the addition occurs so as to place the halogen atom on the carbon bearing the least number of hydrogen atoms.[31] This rule, given for additions to hydrocarbons only, appears to be entirely general for monoolefins of this type when the addition proceeds by the polar mechanism. It is also applicable to a majority of olefins in general, but it is the nature of the numerous exceptions that has probably contributed the most to the development of current theory. According to this view, the reaction is initiated by a proton donation to the olefin (this may involve the formation of a pi complex) to yield largely the more stable possible carbonium ion, which may then combine with a nucleophilic reagent to give the observed final product. Since a halogen substituent appears to yield a more stable carbonium ion when in the α position than in the β (Sec. 7-3a), it is not surprising that the addition of hydrogen

[31] W. Markownikoff, *Ann.*, **153**, 256 (1870).

bromide to vinyl bromide and vinyl chloride has been found to give ethylidene halides,

$$CH_2=CHCl + HBr \longrightarrow CH_3-\overset{\overset{\displaystyle H}{|}}{\underset{\oplus}{C}}\overset{\frown}{\,}\overline{Cl} \longrightarrow CH_3CHClBr$$

although it is interesting to note that in the β position cyclic halonium ions may be written as intermediates. Although such strongly electron-withdrawing groups as CN, COR, CO_2H, NO_2, SO_2R, etc., when attached directly to one of the two carbon atoms of the double bond, cause the halogen atom of HX being added to become attached to the other carbon atom, this may be due to addition by a 1,4 mechanism followed by tautomerism of the resultant enol,

$$CH_2=CH-CH=O \overset{HX}{\rightleftharpoons} \begin{bmatrix} CH_2=CH-CH=\overset{\oplus}{O}H \\ CH_2=CH-\overset{\oplus}{C}H-OH \\ \overset{\oplus}{C}H_2-CH=CH-OH \end{bmatrix} \to \begin{matrix} \overset{\overset{\displaystyle X}{|}}{C}H_2-CH=CHOH \\ \downarrow \\ CH_2CH_2CHO \\ | \\ X \end{matrix}$$

This explanation, however, cannot be applied to the similar orientation in additions to the vinyltrimethylammonium ion[32]

$$(CH_3)_3\overset{\oplus}{N}CH=CH_2 + HI \to (CH_3)_3\overset{\oplus}{N}CH_2CH_2I$$

or to 1,1,1-trifluoro-2-propene[33]

$$CH_2=CHCF_3 + HBr \overset{AlBr_3}{\longrightarrow} BrCH_2CH_2CF_3$$

In these cases the strongly electron-withdrawing character of the substituent would cause the secondary carbonium ion, in which there is a positive charge on the adjacent carbon atom, to be less stable than the primary carbonium ion, in which the charge is one atom further removed (if indeed the reaction involves a carbonium-ion intermediate).

Perhaps the best generalization that can be made about the relative ease with which various olefins add strong acids is that the ease of additions increases with the stability of the intermediate carbonium ion, e.g.,

$$(CH_3)_2C=CH_2 \sim (CH_3)_2C=CHCH_3 > CH_3CH=CH_2 > CH_2=CH_2$$

This generalization fails in some cases where the double bond of the olefin is particularly stabilized by resonance, as in certain stilbene derivatives.

The reactivity of the halogen acids toward olefins increases with their acidity in the order HF < HCl < HBr < HI. It may be seen that HI should not only be best at donating a proton to a multiple bond but also at furnishing the most nucleophilic anion to combine with the carbonium

[32] E. Schmidt, *Ann.*, **267**, 300 (1892).
[33] A. L. Henne and S. Kaye, *J. Am. Chem. Soc.*, **72**, 3369 (1950).

ion formed.   The reactivity increases with the ion-solvating power of the medium, and it is often useful to employ Friedel-Crafts catalysts, such as $AlBr_3$, BF_3, $SnCl_4$, etc., which are specifically active in coordinating with halide ions or solvating them.   Often opposing this ion-solvating effect is a decrease in reactivity with increasing basicity of the solvent.   The enormous reactivity of sulfuric acid toward olefins would be expected in view of its high acidity and ion-solvating power and its low basicity.

9-3. Additions of Other Electrophilic Reagents. 9-3a. *Additions of Carbonium Ions.*   Whitmore was one of the first to use the carbonium-ion concept widely in explaining the mechanisms of organic reactions and was apparently the first to suggest that the acid-catalyzed polymerization of olefins may involve the formation of a carbonium ion and its combination with a molecule of olefin to yield a new carbonium ion.[34]

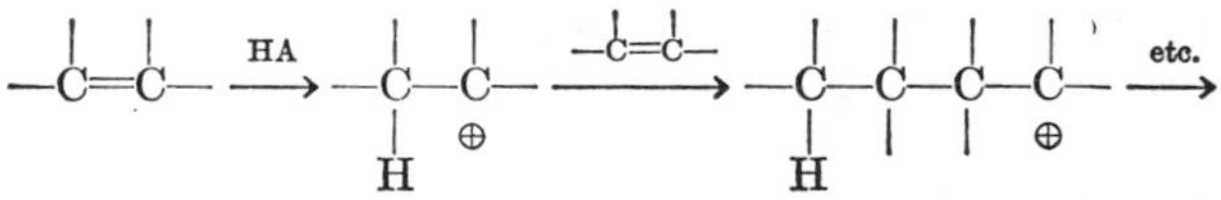

The polymerization may be terminated by the combination of the carbonium ion with some nucleophilic reagent other than an olefin or by loss of a β-proton to give an olefin.   The structure of polymers thus formed may be predicted by the same rules used in predicting the direction of addition of other unsymmetrical reagents to unsymmetrical olefins, as shown in the following mechanism for the dimerization of isobutylene.

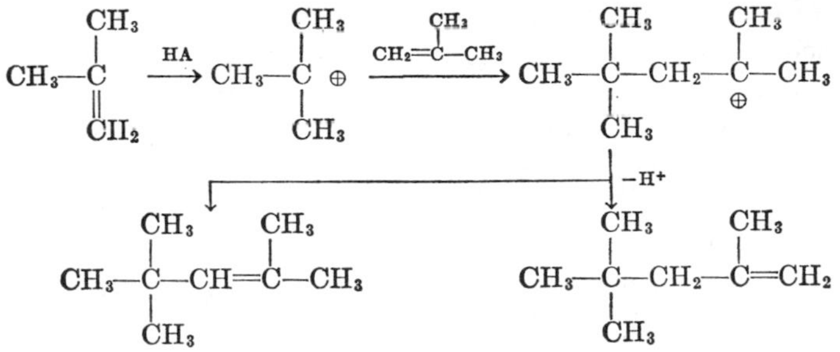

Thus a proton adds to isobutylene to give the tertiary rather than a primary carbonium ion, and the intermediate tertiary butyl cation combines with another isobutylene molecule so as to give a new tertiary carbonium ion, which may (among other possible fates) lose a β-proton to become an olefin in either of two possible ways.

In many polymerizations catalyzed by Friedel-Crafts catalysts, especially in hydrocarbon solvents, the presence of small amounts of certain "promoters" (especially water) seems to be required.   The

[34] F. C. Whitmore, *Ind. Eng. Chem.*, **26**, 94 (1934).

function of the promoter may be to solvate ions and/or furnish protons for donation to olefin molecules, but this subject does not appear to have been studied in detail.

A number of studies have been made of the relative reactivity of various olefins in Friedel-Crafts polymerizations, and especially copolymerizations, by several methods, including the determination of monomer reactivity ratios[35] (see Sec. 20-1c).

The addition of alkyl halides to olefins probably proceeds by an analogous mechanism, e.g.,

$$CHCl_3 + AlCl_3 \rightleftharpoons \overset{\oplus}{CHCl_2} + AlCl_4^-$$

$$\overset{\oplus}{CHCl_2} + Cl_2C{=}CCl_2 \rightleftharpoons Cl_2\overset{\oplus}{C}{-}CCl_2{-}CHCl_2$$

$$Cl_2\overset{\oplus}{C}{-}CCl_2{-}CHCl_2 + AlCl_4^- \rightleftharpoons Cl_3C{-}CCl_2{-}CHCl_2 + AlCl_3$$

It is often possible to add such saturated hydrocarbons as isobutane to such olefins as isobutylene. The unmodified Whitmore mechanism gives no hint as to how the isobutane becomes involved in the reaction, but this was provided by a brilliant experiment of Bartlett, Condon, and Schneider. These workers mixed two rapidly flowing isopentane streams, one containing dissolved aluminum bromide and the other t-butyl chloride, in a nozzle from which the mixture was forced into a vigorously stirred water bath.[36] Only about 0.001 sec was thus allowed for reaction before the reaction was quenched by combination of the aluminum bromide with water. Nevertheless, during this time the reaction of t-butyl chloride was essentially complete, and t-amyl bromide was formed in about 60 per cent yield. The most reasonable mechanism for this reaction involves the abstraction of a hydride anion from isopentane by the t-butyl carbonium ion to yield a t-amyl carbonium ion.

$$(CH_3)_3CCl + AlBr_3 \rightarrow (CH_3)_3\overset{\oplus}{C} + ClAlBr_3^-$$

$$(CH_3)_3\overset{\oplus}{C} + C_2H_5CH(CH_3)_2 \rightarrow (CH_3)_3CH + C_2H_5\overset{\oplus}{C}(CH_3)_2$$

$$C_2H_5\overset{\oplus}{C}(CH_3)_2 + ClAlBr_3^- \rightarrow C_2H_5C(CH_3)_2 + ClAlBr_2$$
$$\underset{Br}{|}$$

The possibility of hydride-ion transfers to carbonium ions has since found wide application in many reaction mechanisms and, as Bartlett, Condon, and Schneider pointed out, permits a suitable mechanism for the addition of isobutane to isobutene, the isobutane being changed to a t-butyl carbonium ion by reaction with an "isooctyl" carbonium ion.

[35] R. E. Florin, *J. Am. Chem. Soc.*, **71**, 1867 (1949); **73**, 4468 (1951); C. G. Overberger et al., *J. Am. Chem. Soc.*, **73**, 5541 (1951); **74**, 4848 (1952); **75**, 6349 (1953).

[36] P. D. Bartlett, F. E. Condon, and A. Schneider, *J. Am. Chem. Soc.*, **66**, 1531 (1944).

9-3b. *Additions of Metal Cations.* The ability of certain metallic cations to coordinate with carbon-carbon double and triple bonds has long been recognized. In many cases solid addition compounds are formed. By measuring the effect of silver nitrate on increasing the solubility of olefins in water, Winstein and Lucas determined equilibrium constants for the coordination of silver ions with several olefins.[37] They suggested the structure

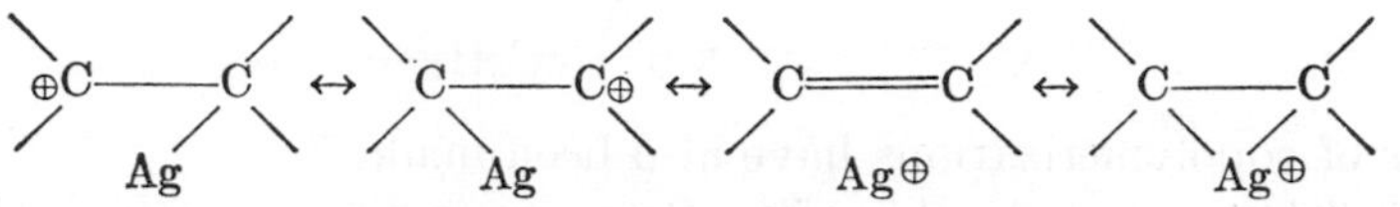

for the complex, and the particular aptness of this formulation for metal ion–olefin complexes makes the formulation appear more plausible for related but less stable species such as the proton-olefin pi complex.

9-4. Additions of Nucleophilic Reagents. 9-4a. *Additions to α,β-unsaturated Carbonyl Compounds, Nitriles, etc.* Just as the addition of electrophilic reagents has been found to occur more readily when a relatively stable intermediate carbonium ion may be formed, so might it be expected that nucleophilic reagents would add to carbon-carbon multiple bonds if a sufficiently stable intermediate carbanion may be formed.

In the Michael reaction the nucleophilic reagent is a carbanion. A kinetic study of the Michael addition of nitroform to β-nitrostyrene in methanol (and the reverse reaction) gave results in agreement with the following mechanism:[38]

$$HC(NO_2)_3 + MeOH \rightleftharpoons C(NO_2)_3^- + MeOH_2^+$$

$$C_6H_5CH{=}CHNO_2 + C(NO_2)_3^- \rightleftharpoons C_6H_5CH\overset{\ominus}{C}HNO_2$$
$$\underset{|}{\phantom{C_6H_5CH\overset{\ominus}{C}HNO_2}}$$
$$C(NO_2)_3$$

$$C_6H_5CH\overset{\ominus}{C}HNO_2 + MeOH_2^+ \rightleftharpoons C_6H_5CHCH_2NO_2 + MeOH$$
$$\underset{C(NO_2)_3}{|} \qquad\qquad \underset{C(NO_2)_3}{|}$$

Most Michael reactions consist in the addition of much less acidic molecules than nitroform. In these cases a basic catalyst is usually added to generate the required intermediate carbanion.

9-4b. *Carbanion-catalyzed Polymerizations.* When a carbanion adds to an olefin to form a new carbanion capable of addition to another molecule of olefin, it may become possible to form quite long chains

---

[37] S. Winstein and H. J. Lucas, *J. Am. Chem. Soc.*, **60**, 836 (1938); H. J. Lucas, R. S. Moore, and D. Pressman, *J. Am. Chem. Soc.*, **65**, 227 (1943).

[38] J. Hine and L. Kaplan, *J. Am. Chem. Soc.*, **82**, 2915 (1960).

by the continuation of the reaction. Numerous cases of this sort are known;[39] e.g., styrene may be polymerized by the action of potassium amide in liquid ammonia.[40]

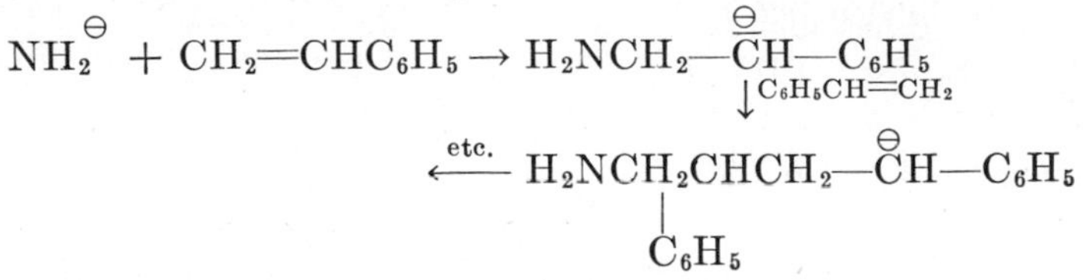

$$NH_2^{\ominus} + CH_2{=}CHC_6H_5 \rightarrow H_2NCH_2{-}\overset{\ominus}{C}H{-}C_6H_5$$
$$\downarrow C_6H_5CH{=}CH_2$$

$$\overset{\text{etc.}}{\longleftarrow} H_2NCH_2CHCH_2{-}\overset{\ominus}{C}H{-}C_6H_5$$
$$\underset{C_6H_5}{|}$$

Studies of copolymerizations have also been made.[39,41]

9-4c. Additions of Nucleophilic Reagents to Fluoroolefins. Although fluorine atoms and fluorinated alkyl groups do not have an ability to stabilize carbanions comparable to that of carbonyl, cyano, nitro, etc., groups, they are able to make possible the addition of nucleophilic reagents to carbon-carbon multiple bonds. The addition of ethanol in the presence of sodium ethoxide occurs readily with tetrafluoroethylene, trifluorochloroethylene, 1,1-difluoro-2,2-dichloroethylene, and 1,1-difluoro-2-chloroethylene, where the intermediate carbanion may be stabilized by both α- and β-halogen atoms.[42]

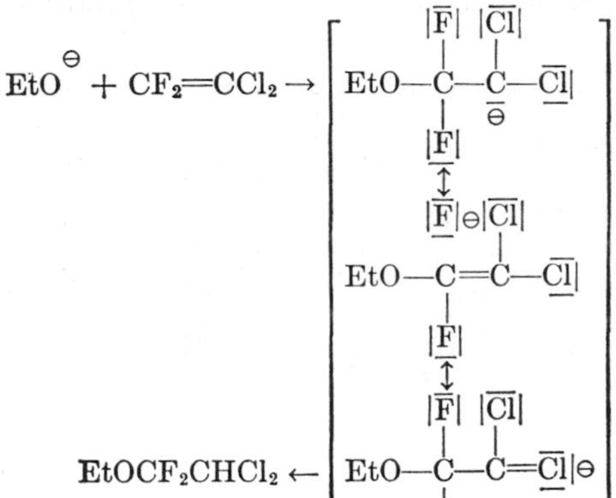

$$EtO^{\ominus} + CF_2{=}CCl_2 \rightarrow \left[\begin{array}{c} |\overline{F}| \ |\overline{Cl}| \\ | \quad | \\ EtO{-}C{-}\underset{\ominus}{C}{-}\overline{Cl}| \\ | \\ |\overline{F}| \\ \updownarrow \\ |\overline{F}|\ominus|\overline{Cl}| \\ | \\ EtO{-}C{=}C{-}\overline{Cl}| \\ | \\ |\overline{F}| \\ \updownarrow \\ |\overline{F}| \ |\overline{Cl}| \\ | \quad | \\ EtO{-}C{-}C{=}\overline{Cl}|\ominus \\ | \\ |\overline{F}| \end{array} \right]$$

$$EtOCF_2CHCl_2 \leftarrow$$

In the case above, the carbanion is formed at the chlorine-bearing carbon atom. This is probably because in the β position the carbanion stabilization is solely due to electronegativity, whereas in the α position the ease

[39] F. R. Mayo and C. Walling, *Chem. Revs.*, **46**, 191 (1950); C. Walling, E. R. Briggs, W. Cummings, and F. R. Mayo, *J. Am. Chem. Soc.*, **72**, 48 (1950).

[40] W. C. E. Higginson and N. S. Wooding, *J. Chem. Soc.*, 760, 1178 (1952).

[41] F. C. Foster, *J. Am. Chem. Soc.*, **74**, 2299 (1952).

[42] W. E. Hanford and G. W. Rigby, U.S. Patent 2,409,274, Oct. 15, 1946; *Chem. Abstr.*, **41**, 982b (1947); W. T. Miller, Jr., E. W. Fager, and P. H. Griswold, *J. Am. Chem. Soc.*, **70**, 431 (1948).

of expanding the outer shell to accommodate 10 electrons may also be important.   Thus fluorine is best as a β substituent, and chlorine is better in the α position.   Addition may also occur when the intermediate contains only β-halogen substituents, but this apparently occurs somewhat less readily.   Ethanol may be added to $CH_2{=}CH{-}CF_3$ and $CH{\equiv}C{-}CF_3$ in the presence of sodium ethoxide.[43]

The cyanide ion also adds readily to polyfluoroolefins,[44] and Miller and coworkers showed that the relatively weakly basic fluoride ion is particularly effective in nucleophilic attack on $CF_2{=}\overset{|}{C}{-}$ groupings.[45]   The unusually high nucleophilicity of fluoride ions toward fluorinated carbon atoms may stem from the stabilization of the $-CF_3$ group by resonance of the type

$$\underset{\underset{\oplus|\underline{F}}{\|}}{\overset{|\overline{F}|}{\underset{|}{-C}}}\;|\underline{F}|^{\ominus}\;\leftrightarrow\;\underset{\underset{|\underline{F}|}{|}}{\overset{|\overline{F}|^{\ominus}}{-C{=}\overline{F}|}}^{\oplus}\;\leftrightarrow\;\text{etc.}$$

which has also been used to explain the unusually short C—F bond distances[46] and particularly great stability[47] of molecules like CF_4 and CF_3Cl.

PROBLEMS

1. Describe experiments designed to distinguish between the following two possible mechanisms for the reaction of stilbene with bromine in methanol.

(a) The reaction involves the irreversible rate-controlling formation of an intermediate bromonium ion (or species of the type of I) by two parallel paths, one involving attack of bromine and the other of tribromide ion on stilbene.   This intermediate then reacts rapidly, either with bromide ion to give stilbene dibromide or with the solvent to give stilbene methoxybromide.

(b) The reaction involves the initial equilibrium

$$C_6H_5CH{=}CHC_6H_5 + Br_2 \rightleftarrows \underset{\underset{\underset{\text{I}}{Br}}{\overset{|}{Br}}}{C_6H_5CH{-}CHC_6H_5}$$

[43] A. L. Henne, M. A. Smook, and R. L. Pelley, *J. Am. Chem. Soc.*, **72**, 4756 (1950); A. L. Henne and M. Nager, *J. Am. Chem. Soc.*, **74**, 650 (1952).

[44] D. C. England, R. V. Lindsey, Jr., and L. R. Melby, *J. Am. Chem. Soc.*, **80**, 6442 (1958).

[45] J. H. Fried and W. T. Miller, Jr., *J. Am. Chem. Soc.*, **81**, 2078 (1959); W. T. Miller, Jr., J. H. Fried, and H. Goldwhite, *J. Am. Chem. Soc.*, **82**, 3091 (1960).

[46] L. O. Brockway, *J. Phys. Chem.*, **41**, 185, 747 (1937).

[47] D. E. Petersen and K. S. Pitzer, *J. Phys. Chem.*, **61**, 1252 (1957).

followed by the rate-controlling reaction of the intermediate with bromide ion or methanol to give the dibromide or methoxybromide.

2. The addition of hydrogen iodide to tiglic acid gives a different 2-methyl-3-iodobutyric acid than does the addition to angelic acid.

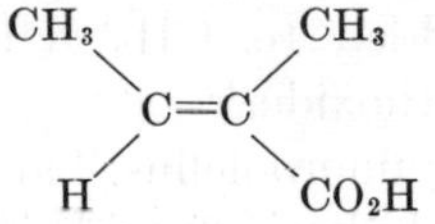

Tiglic acid 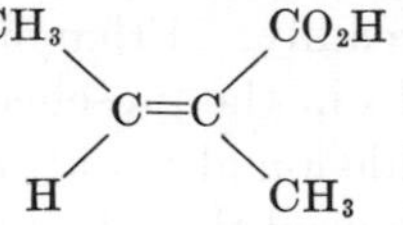   Angelic acid

When a solution of the sodium salt of the tiglic acid hydroiodide is warmed, *trans*-2-butene is obtained; analogous treatment of angelic acid hydroiodide yields *cis*-2-butene. Write reasonable, stereochemically complete mechanisms for each of these reactions.

3. In the presence of acidic catalysts, certain 1,1-diarylethylenes undergo the following dimerization:

$$2Ar_2C=CH_2 \rightarrow Ar_2C=CH\overset{\overset{\textstyle CH_3}{\textstyle |}}{C}Ar_2$$

In the presence of a given amount of acid the reaction is second-order in olefin. Suggest a reasonable reaction mechanism. Would you expect the Hammett ρ for this reaction to be positive or negative? Which type of σ value do you think would give the best Hammett correlation?

4. Suggest an alternate interpretation of the experimental data[13] that led to the proposal that halogens may act as nucleophilic reagents toward the conjugate acids of α,β-unsaturated carbonyl compounds.

Chapter 10

CARBANIONS AND ENOLIZATION

10-1. Mechanism and Reactivity in Carbanion Formation.
10-1a. Enolization and Halogenation, Racemization, and Deuterium Exchange. The kinetics of halogenation of one ketone, acetone, and the identity of the rates of bromination and iodination were described in Sec. 5-2*d*, along with the statement that these processes involve as the rate-controlling step the reaction with an acid and/or base. The reaction with acid yields an enol and that with base probably an enolate anion, either of which may react almost instantaneously with halogen. Further evidence for such a mechanism comes from several observations of the identity of rates of halogenation, deuterium exchange, and racemization of suitable compounds.[1] The largest amount of work appears to have been done with phenyl *sec*-butyl ketone. Bartlett and Stauffer have found that the rate constant for acid-catalyzed iodination is 0.0298 under conditions where the rate constant for racemization is 0.0296, the two figures being well within experimental error.[2] It is reasonable that both reactions involve the rate-controlling formation of the enol, which, having a plane of symmetry, is as likely upon reketonization to give the *d* isomer as the *l*.

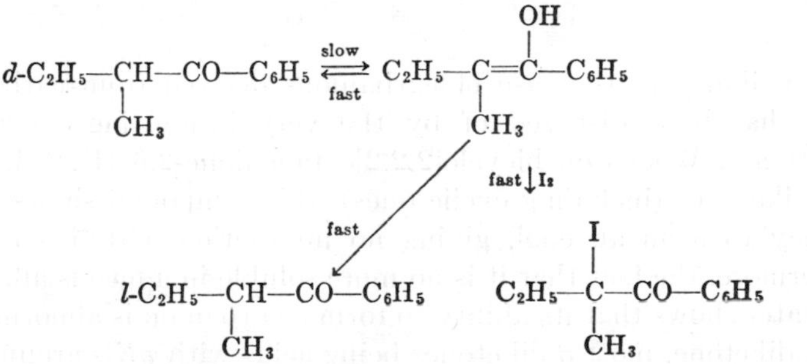

$$d\text{-}C_2H_5\text{---}CH\text{---}CO\text{---}C_6H_5 \underset{\text{fast}}{\overset{\text{slow}}{\rightleftharpoons}} C_2H_5\text{---}\overset{\overset{\displaystyle OH}{|}}{C}\text{=}\overset{}{C}\text{---}C_6H_5$$

$$\underset{\overset{|}{CH_3}}{} \qquad \underset{\overset{|}{CH_3}}{}$$

$$\text{fast} \downarrow I_2$$

$$l\text{-}C_2H_5\text{---}CH\text{---}CO\text{---}C_6H_5 \qquad C_2H_5\text{---}\overset{\overset{\displaystyle I}{|}}{C}\text{---}CO\text{---}C_6H_5$$

$$\underset{\overset{|}{CH_3}}{} \qquad \underset{\overset{|}{CH_3}}{}$$

[1] L. Ramberg and A. Mellander, *Arkiv Kemi, Mineral. Geol.*, **11B**(31) (1934); L. Ramberg and I. Hedlund, *Arkiv Kemi, Mineral. Geol.*, **11B**(41) (1934); C. K. Ingold and C. L. Wilson, *J. Chem. Soc.*, 773 (1934).

[2] P. D. Bartlett and C. H. Stauffer, *J. Am. Chem. Soc.*, **57**, 2580 (1935).

Using the same ketone, Hsü and Wilson have found the rate of the base-catalyzed bromination (acetate ion was the base) to be equal to the rate of racemization under the same conditions.[3]   Furthermore it has been demonstrated that the rates of the deuteroxide-catalyzed racemization and deuterium exchange are identical.[4]   In the base-catalyzed reactions it is probable that it is the planar enolate anion that is being formed in the rate-controlling step.

10-1b. *Stereochemistry of Carbanions.*   In order for there to be any considerable contribution of structures like that shown on the right

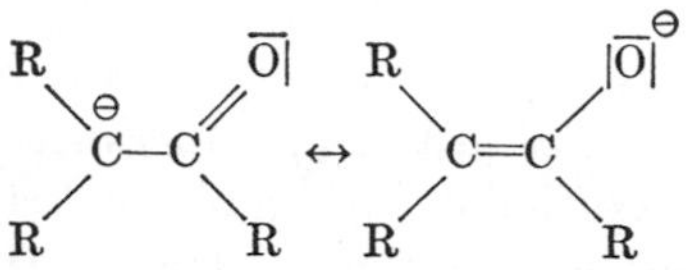

$$
\begin{array}{c}
\underset{R}{\overset{R}{\diagdown}} \overset{\ominus}{\underset{}{C}} - \overset{\overline{O}}{\underset{R}{\diagdown}} C \quad \leftrightarrow \quad \underset{R}{\overset{R}{\diagdown}} C = \overset{\overline{|O|}^{\ominus}}{\underset{R}{\diagup}} C
\end{array}
$$

to the total structure of an enolate anion, it is necessary that the
$\diagup\diagdown C = \overset{|}{C} - \overline{O|}^{\ominus}$ system and the three atoms attached directly thereto lie at least very nearly in the same plane.   The fact that proton removal from an asymmetric carbon atom to form a carbanion results in racemization, mentioned in the previous section, does not demonstrate this point unambiguously, since another explanation is possible.   That is, since carbanions are isoelectronic with amines, it is possible that their normal structure is pyramidal and that, like amines, their optical inactivity is due to a rapid equilibrium between two enantiomorphic structures.

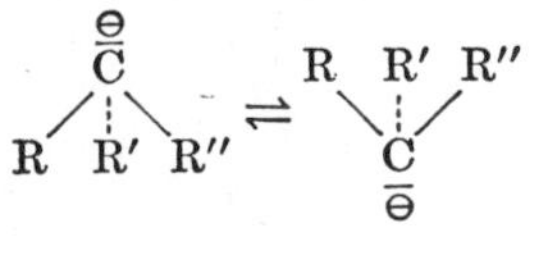

$$
\overset{\ominus}{\underset{R\ \ R'\ \ R''}{\overset{C}{\diagup}\!|\!\diagdown}} \quad \rightleftarrows \quad \underset{\overset{C}{\underset{\ominus}{}}}{\overset{R\ \ R'\ \ R''}{\diagdown\!|\!\diagup}}
$$

This possibility in the case of carbanions derived from carbonyl compounds has been disposed of by the very interesting observation of Bartlett and Woods on bicyclo[2.2.2]octanedione-2,6 (I).[5]   Unlike typical β-diketones (including cyclic ones), this compound shows very little tendency to form an enol, giving no interaction with $FeCl_3$ or Cu^{++}. Furthermore, the fact that it is no more soluble in aqueous alkali than in pure water shows that its ability to form a carbanion is abnormally small for a β-diketone, most β-diketones being acids with pK's around 9.   The logical explanation is that resonance contributions of structures like Ia

[3] S. K. Hsü and C. L. Wilson, *J. Chem. Soc.*, 623 (1936).
[4] S. K. Hsü, C. K. Ingold, and C. L. Wilson, *J. Chem. Soc.*, 78 (1938).
[5] P. D. Bartlett and G. F. Woods, *J. Am. Chem. Soc.*, **62**, 2933 (1940).

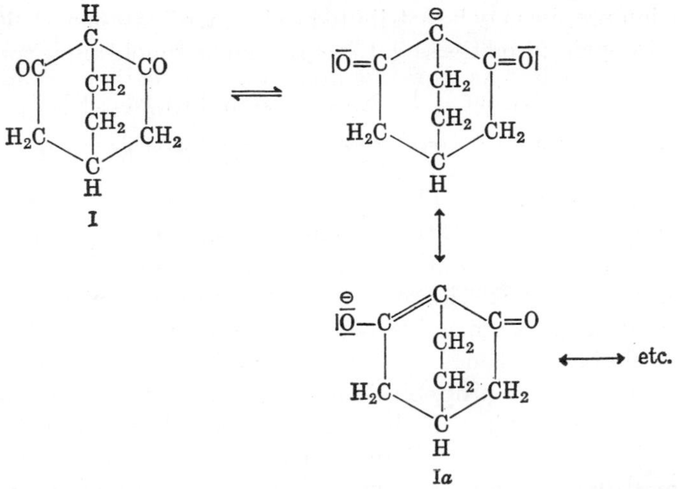

are made negligible by the fact that the bicyclic ring system prevents the coplanarity of the four atoms attached to the double bond (cf. Bredt's rule).   For similar reasons, tri-o-tolylmethane (II) and triptycene (III) are much weaker acids than triphenylmethane.[6]

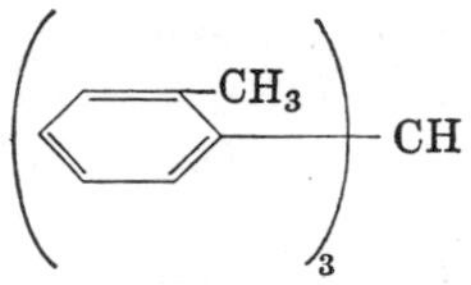

It thus seems established that carbanions in which the negative carbon is attached directly to three other atoms will be most stable when coplanar if they are stabilized by the contribution of structures involving double bonds between the negative carbon and some other first-row atom.   For carbanions that lack such stabilization, however, a pyramidal structure would be expected.   The evidence for this latter point is often rather indirect, since most of the stereochemical studies believed to involve such carbanions show that they racemize readily but do not distinguish between coplanar and rapidly racemizing pyramidal structures.   For example, Letsinger found that in petroleum ether–diethyl ether 2-octyllithium racemized slowly at $-70°$ but within minutes at $0°$,[7] whereas Curtin and Koehl reported that the compound required hours to racemize at $-6°$ in pentane.[8]   The catalytic action of diethyl ether on this

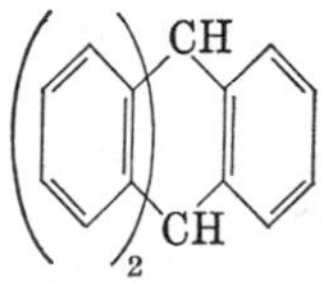

[6] P. D. Bartlett and J. E. Jones, *J. Am. Chem. Soc.*, **64,** 1837 (1942); P. D. Bartlett, M. J. Ryan, and S. G. Cohen, *J. Am. Chem. Soc.*, **64,** 2649 (1942).

[7] R. L. Letsinger, *J. Am. Chem. Soc.*, **72,** 4842 (1950).

[8] D. Y. Curtin and W. J. Koehl, *Chem. & Ind.* (*London*), 262 (1960).

racemization process may be attributed to its coordination with lithium to facilitate ion-pair formation by the covalent organolithium compound.

$$d\text{-}C_8H_{17}Li\text{----}(OEt_2)_n \rightleftharpoons d\text{-}C_8H_{17}^-Li(OEt_2)_n^+$$
$$l\text{-}C_8H_{17}Li\text{----}(OEt_2)_n \leftrightharpoons l\text{-}C_8H_{17}^-Li(OEt_2)_n^+$$

Vinyl carbanions have considerably more configurational stability than their saturated analogs. Either *cis-* or *trans*-propenyllithium maintains most of its stereochemical purity even after more than an hour in refluxing diethyl ether.[9,10] The cyclopropyl compounds appear, as they often do, to have properties intermediate between those of their vinyl and their saturated aliphatic analogs. Walborsky and Impastato reported 2,2-diphenyl-1-methylcyclopropyllithium to be about 20 per cent racemized after a half-hour at 5° in a diethyl ether–hydrocarbon solvent.[11]

Cram and coworkers[13] and Corey and Kaiser[12] described evidence for a nonplanar structure for carbanions stabilized by α-sulfonyl substituents. The rate constant for the base-catalyzed deuterium exchange of phenyl 2-octyl sulfone in aqueous ethanol is about 40 times as large as that for base-catalyzed racemization under the same conditions.[12] These observations are most reasonably explained by the following mechanism:

$$d\text{-}C_6H_{13}\underset{\underset{SO_2C_6H_5}{|}}{C}HCH_3 + OH^- \rightleftharpoons d\text{-}C_6H_{13}\underset{\underset{SO_2C_6H_5}{|}}{\overset{\ominus}{C}}CH_3 + H_2O$$

$$l\text{-}C_6H_{13}\underset{\underset{SO_2C_6H_5}{|}}{C}HCH_3 + OH^- \rightleftharpoons l\text{-}C_6H_{13}\underset{\underset{SO_2C_6H_5}{|}}{\overset{\ominus}{C}}CH_3 + H_2O$$

The stereochemical results of reactions involving the intermediate formation of carbanions depend on more factors than just the structure of the carbanion. Cram and coworkers described ingenious experiments showing how a given reaction can yield any result from predominant inversion through almost complete retention of configuration simply by changing the solvent.[13] These results were said to arise from reactions of two different types of ion pairs, one with and one without a molecule of solvent between the cation and anion.

10-1c. *Relative Ease of Carbanion Formation.* Pearson and Dillon have tabulated many of the existing data on rate and equilibrium constants

[9] E. A. Braude and J. A. Coles, *J. Chem. Soc.*, 2078, 2085 (1951).

[10] D. Y. Curtin and J. W. Crump, *J. Am. Chem. Soc.*, **80**, 1922 (1958).

[11] H. M. Walborsky and F. J. Impastato, *J. Am. Chem. Soc.*, **81**, 5835 (1959).

[12] E. J. Corey and E. T. Kaiser, *J. Am. Chem. Soc.*, **83**, 490 (1961).

[13] D. J. Cram et al., *J. Am. Chem. Soc.*, **81**, 5740, 5750, 5754, 5760, 5767, 5774, 5835 (1959); **82**, 6415 (1960); **83**, 3696 (1961).

for the reaction

$$HA + H_2O \overset{k_1}{\rightleftharpoons} H_3O^+ + A^-$$

where the hydrogen atom removed from HA was attached to carbon.[14] Some of their data are listed in Table 10-1.

TABLE 10-1. RATE AND EQUILIBRIUM CONSTANTS FOR THE IONIZATION OF CARBON-BOUND HYDROGEN ATOMS IN WATER AT 25°[14]

Compound	K_a	k_1
CH_3NO_2	6.1×10^{-11}	2.6×10^{-6}
$CH_3CH_2NO_2$	2.5×10^{-9}	2.2×10^{-6}
$CH_2(NO_2)_2$	2.7×10^{-4}	~ 50
CH_3COCH_3	$\sim 10^{-20}$	2.8×10^{-8}
$CH_2(COCH_3)_2$	1.0×10^{-9}	1.0
$CH(COCH_3)_3$	1.4×10^{-6}	
$CH_2(SO_2CH_3)_2$	1×10^{-14}	
$CH(SO_2CH_3)_3$	Strong	
$CH_2(CN)_2$	6.5×10^{-12}	9.0×10^{-1}
$CH_2(CO_2Et)_2$	5×10^{-14}	1.5×10^{-3}

Taft correlated the rates of several such carbanion-formation reactions in terms of resonance and inductive substituent constants,[15] but it is clear from the existing data that no very simple linear free-energy relation could correlate all the observed rate and equilibrium constants for carbanion formation. Steric factors are very probably of importance in certain cases. Thus, $(CH_3CO)_2CH_2$ is a much stronger acid than $(CH_3SO_2)_2CH_2$, but $(CH_3CO)_3CH$ is much weaker than $(CH_3SO_2)_3CH$. The acidity of $(CH_3CO)_3CH$ is probably diminished by steric inhibition of resonance in the anion, in which it is difficult for all the carbon atoms to lie in the same plane.

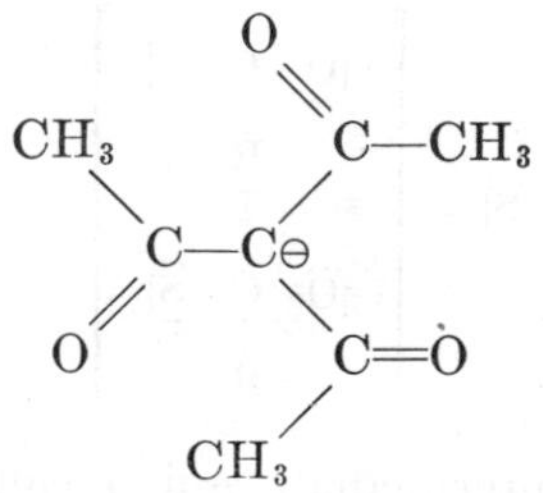

There is no need for all the carbon atoms in the $(CH_3SO_2)_3C^-$ anion to be coplanar, and even if there were, less crowding would result than in the

[14] R. G. Pearson and R. L. Dillon, *J. Am. Chem. Soc.*, **75**, 2439 (1953).
[15] R. W. Taft, Jr., *J. Am. Chem. Soc.*, **79**, 5075 (1957).

case of the triacetylmethide ion.   Bell[16] and Pearson and Dillon discussed the very important problem of why these deviations from linear free-energy relationships occur and have referred to other discussions, but the problem is still worth additional study.

Bonhoeffer, Geib, and Reitz have tabulated data on the rate of removal of carbon-bound hydrogen by deuteroxide ions in deuterium oxide solution.[17]   In addition to groups such as —NO$_2$, —COR, etc., it might be mentioned that α-halogen atoms also increase the ease of carbanion formation, chloroform undergoing base-catalyzed deuterium exchange about as readily as acetone (cf. Table 24-1).[18]

10-2. Carbanions and Tautomerism.   *10-2a. The Stepwise Mechanism for Tautomerism.*   Isomerization reactions having the form

$$\text{H—X—Y=Z} \rightleftharpoons \text{X=Y—Z—H}$$

have been found to occur very commonly with organic compounds. Such an isomerism is an example of tautomerism, and the atomic system shown is called a triad system.   Two types of mechanisms, concerted and stepwise, may be suggested for these reactions, depending on whether the removal of a hydrogen from X occurs simultaneously with, or in a separate step from, the addition of hydrogen to Z.   We shall deal here only with prototropic tautomerism (overwhelmingly the most common kind), in which the hydrogen atoms are added and removed as protons. There is evidence for the occurrence of both concerted and stepwise mechanisms in reactions of this type.

Tautomerism occurs most rapidly when both X and Z are oxygen or nitrogen atoms.   In many cases of this sort the evidence for the stepwise mechanism[19] appears clear, since the (usually) ionic intermediate may be stable enough to be studied independently without having to rely on indirect arguments for its formation.   In the tautomerism of a thio acid, for example,

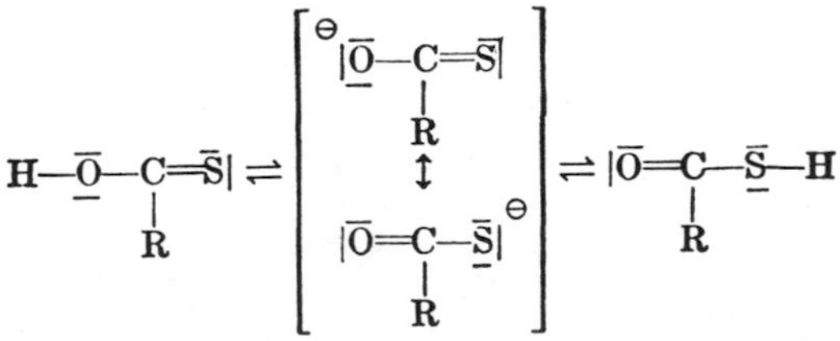

the formation of the intermediate anion can easily be demonstrated.

[16] R. P. Bell, "The Proton in Chemistry," chap. 10, Cornell University Press, Ithaca, N.Y., 1959.

[17] K. F. Bonhoeffer, K. H. Geib, and O. Reitz, *J. Chem. Phys.*, **7**, 664 (1939).

[18] J. Hine, R. C. Peek, Jr., and B. D. Oakes, *J. Am. Chem. Soc.*, **76**, 827 (1954).

[19] C. K. Ingold, C. W. Shoppee, and J. F. Thorpe, *J. Chem. Soc.*, 1477 (1926).

Although one proton is removed before the other is added in the reaction above, there are many reactions, such as the tautomerism of amidines, in which first one proton is added and then the other removed.

10-2*b*. *The Concerted Mechanism for Tautomerism.*   In an investigation of the range of applicability of the two-step mechanism for tautomerism, Ingold and Wilson studied the establishment of equilibrium between the methyleneazomethines IV and V.

$$
\begin{array}{ccc}
\underset{\text{CH}_3}{\overset{\text{C}_6\text{H}_5}{\diagdown}}\text{CH}-\text{N}=\text{C}\underset{\text{C}_6\text{H}_5}{\overset{\text{C}_6\text{H}_4\text{Cl-}p}{\diagup}} & \rightleftarrows & \underset{\text{CH}_3}{\overset{\text{C}_6\text{H}_5}{\diagup}}\text{C}=\text{N}-\text{CH}\underset{\text{C}_6\text{H}_5}{\overset{\text{C}_6\text{H}_4\text{Cl-}p}{\diagdown}} \\
\text{IV} & & \text{V}
\end{array}
$$

The approach to equilibrium in ethanol solution is a second-order process, first-order in the basic catalyst sodium ethoxide and first-order in the methyleneazomethine. When optically active IV is isomerized, the V formed might be expected to be partly racemic and partly optically active, inasmuch as the entering proton may add to the right-hand carbon in two stereochemically different ways but, because of the influence of the original asymmetric carbon atom, not necessarily to the same extent. Apparently, however, the proton addition does occur to the same extent in each of the two different ways, for the V formed is completely racemic even early in the reaction, when the unreacted IV is still almost optically pure. The first mechanism considered for the reaction involved the intermediate formation of a resonance-stabilized anion.

$$
\text{IV} + \text{EtO}^- \underset{k_{-1}}{\overset{k_1}{\rightleftarrows}} \left[\underset{\text{CH}_3}{\overset{\text{C}_6\text{H}_5}{\diagdown}}\text{C}\text{---}\text{N}\text{---}\text{C}\underset{\text{C}_6\text{H}_5}{\overset{\text{C}_6\text{H}_4\text{Cl-}p}{\diagdown}} \right]^- \underset{k_{-2}}{\overset{k_2}{\rightleftarrows}} \text{V} + \text{EtO}^-
$$

$$\text{VI}$$

If this two-step process is the correct mechanism, the intermediate anion VI should add protons to give either racemic IV or racemic V. Thus, if VI does return to IV an appreciable fraction of the time (that is, if k_2 is not too much larger than k_{-1}), the isomerization of IV will be accompanied by its racemization even from the first of the reaction, before the concentration of V has become sufficient for the formation of racemic IV from racemic V to be a significant process. Experimentally, no such accompanying racemization was observed; all the racemization of IV involved the intermediate formation of V.[20] Therefore, either VI is not a reaction intermediate, or k_2 happens in this case to be much larger than

[20] C. K. Ingold and C. L. Wilson, *J. Chem. Soc.*, 1493 (1933); 93 (1934); S. K. Hsü, C. K. Ingold, and C. L. Wilson, *J. Chem. Soc.*, 1778 (1935).

k_{-1}; that is, the diarylated carbon is protonated faster than the mono-arylated one. The obvious approach at this point would be to study the isomerization of optically active V where accompanying racemization would be fast if $k_2 \gg k_{-1}$. Perhaps because of experimental difficulties, this was not done. Instead, investigations were made on two other methyleneazomethines VII and VIII. In neither case was the isomeriza-

$$p\text{-}C_6H_5C_6H_4 \diagdown \qquad\qquad C_6H_5 \qquad\qquad C_6H_5 \diagdown \qquad\qquad C_6H_5$$

$$\text{CH---N}{=}\text{C} \qquad\qquad\qquad \text{CH---N}{=}\text{C}$$

$$C_6H_5 \diagup \qquad\qquad H \qquad\qquad CH_3 \diagup \qquad\qquad C_6H_5$$

$$\text{VII} \qquad\qquad\qquad\qquad \text{VIII}$$

tion accompanied by any racemization except that involving the inter-mediate formation of the tautomer.[20] Rather than to assume that all three methyleneazomethines studied just happened to yield anions that are protonated to form product much faster than reactant, it seems preferable to regard the intermediacy of free ions like VI as disproved. The reaction may proceed by a concerted process in which an ethoxide ion removes one proton as an ethanol molecule donates the other, as Ingold and Wilson have suggested.

$$\begin{array}{ccc} \text{EtO}^- \ \text{H} & \text{H---OEt} & \quad \text{EtO---H} \quad \text{H}^-\text{OEt} \\ \mid \quad \mid & & \mid \quad \mid \\ \text{RR'C---N}{=}\text{CR''R'''} & \longrightarrow & \text{RR'C}{=}\text{N---CR''R'''} \end{array}$$

It seems possible, though, that there is formed in the reaction an asym-metric ion pair rather like those suggested by Cram and coworkers in other reactions.[13]

As Hsü, Ingold, and Wilson pointed out, the stepwise mechanism for tautomerism involving an ionic intermediate appears to be favored when this ion is relatively stable.

10-2c. *Tautomerism of α,β- and β,γ-unsaturated Carbonyl and Related Compounds.* The isomerization of IX to X in alkaline "deuteriated" ethanol solution was studied by Ingold, de Salas, and Wilson, who found that IX undergoes a deuterium exchange to equilibrate two of its hydro-gen atoms with the solvent at a rate vastly in excess of the rate at which it isomerizes to X or the rate at which X exchanges.[21] The establishment of the equilibrium, which lies almost entirely to the right, is catalyzed by alkali. These results are explained by a mechanism in which the basic catalyst transforms IX into a resonance-stabilized anion that may com-bine with a hydrogen ion to yield either IX or X. It is of particular interest to note, though, that the rapidity of deuterium exchange requires that the intermediate anion accept a proton more rapidly to form IX than

[21] C. K. Ingold, E. de Salas, and C. L. Wilson, *J. Chem. Soc.*, 1328 (1936).

the much more stable X.   This is in agreement with Ingold's rule that

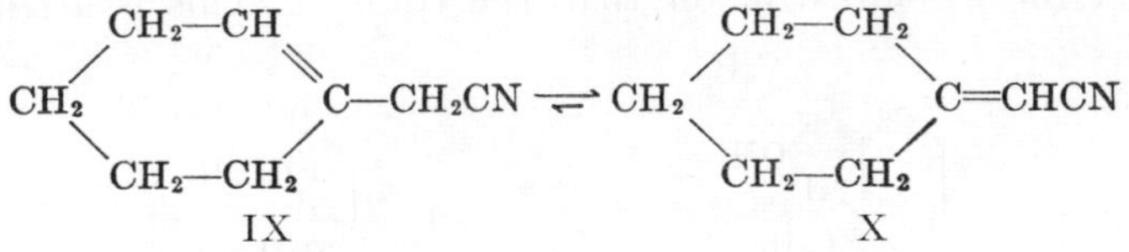

$$
\underset{\text{IX}}{
\begin{array}{c}
\text{CH}_2\!-\!\text{CH} \\
\diagup \qquad \diagdown \\
\text{CH}_2 \qquad\qquad \text{C}\!-\!\text{CH}_2\text{CN} \\
\diagdown \qquad \diagup \\
\text{CH}_2\!-\!\text{CH}_2
\end{array}}
\;\rightleftharpoons\;
\underset{\text{X}}{
\begin{array}{c}
\text{CH}_2\!-\!\text{CH}_2 \\
\diagup \qquad \diagdown \\
\text{CH}_2 \qquad\qquad \text{C}\!=\!\text{CHCN} \\
\diagdown \qquad \diagup \\
\text{CH}_2\!-\!\text{CH}_2
\end{array}}
$$

"when a proton is supplied by acids to the mesomeric anion of weakly ionizing tautomers of markedly unequal stability, then the tautomer which is most quickly formed is the thermodynamically least stable . . . ."[22]   Although it is not clear how "weakly ionizing" (weakly acidic) tautomers must be, or how unequal their stability, to fall within the dominion of the rule, there appear to be exceptions.   Russell, for example, studied the protonation of the potassium salt of cumene, a very weak acid that is much more stable than its tautomers, the isopropylidene-cyclohexadienes.[23]   Protonation by D_2O, DOAc, and DCl gives, respectively, 0, 1.4, and 20 per cent ring deuteriation, showing that the α,α-dimethylbenzyl (cumyl) anion was in all cases protonated preferentially on the benzyl carbon atom.   As the acidity of the protonating agent increased, the protonation became, in general, more nearly random.

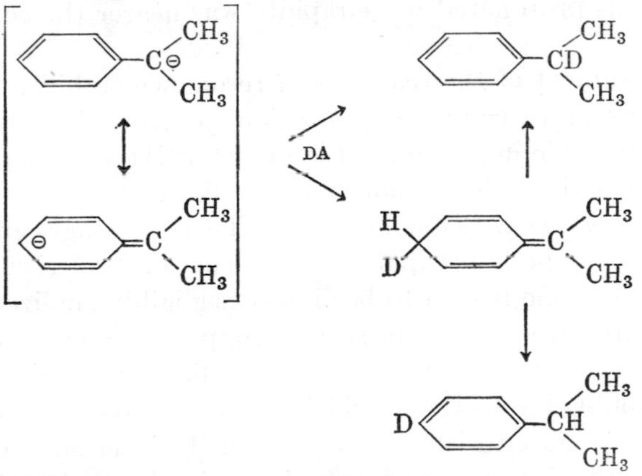

This may be explained in terms of energy diagrams analogous to those in Fig. 5-3.   Although Ingold's generalization is thus not completely general, there are other cases in which it predicts the correct results.   In some of these cases the results seem explicable in more commonly accepted terms.   Zimmerman, for example, showed that in the formation of

[22] C. K. Ingold, "Structure and Mechanism in Organic Chemistry," p. 565, Cornell University Press, Ithaca, N.Y., 1953; cf. A. G. Catchpole, E. D. Hughes, and C. K. Ingold, *J. Chem. Soc.*, 11 (1948).

[23] G. A. Russell, *J. Am. Chem. Soc.*, **81**, 2017 (1959).

1-benzoyl-2-phenylcyclohexane by ketonization of its enol, the less stable cis isomer formed much faster than the trans.[24]   This is attributed to

greater ease of protonation on the less hindered side of the cyclohexane ring, and other examples are quoted to show that this factor is often of importance.   The isomerization of vinylacetic acid, like that of cyclohexenylacetonitrile, does not seem capable of rationalization by any such steric effect, however.   In strongly basic solution at 100° the vinylacetate anion undergoes deuterium exchange much faster than it is isomerized to crotonate, whose exchange rate is relatively slow.[25]   Hammond suggested that the protonation of resonance-stabilized anions may occur preferentially as near as possible to the center of negative charge, or at the atom with the largest negative charge in those cases in which the transition states come so early in the reaction that the relative stability of the various products is not the controlling factor.[26]   In the cases of the anions derived from cyclohexenylacetonitrile and vinylacetate anion the carbon that was protonated more rapidly was nearer the center of negative charge.

The entire subject of the reactions of resonance-stabilized carbanions, carbonium ions, and free radicals at various positions is worth additional study.   The possible importance of ion pairs as intermediates in the first two types of reactions should not be overlooked.

10-2d. *Equilibrium in Tautomerism.*   The percentages of enol found in a number of compounds capable of keto-enol tautomerism are listed in Table 10-2.   Acetone is seen to be almost negligibly enolized.   Replacement of a hydrogen atom by an acetyl group gives a compound (acetylacetone), however, that is 80 per cent enolized at equilibrium.   This result is no doubt due to the stabilizing influence of the unenolized carbonyl group conjugated with the enol and of the internal hydrogen bond. The effect of the internal hydrogen bond is shown by the fact that in water, where the keto form may also be hydrogen-bonded (by the solvent), only 15 per cent enol is present at equilibrium, whereas in hexane or in the vapor phase about 92 per cent enol is present at equilibrium.[27]   The carbethoxy group is considerably less effective than the acetyl group at promoting enolization, and the replacement of an active hydrogen atom of acetoacetic ester by a methyl group is interestingly

[24] H. E. Zimmerman, *J. Org. Chem.*, **20**, 549 (1955).

[25] D. J. G. Ives and H. N. Rydon, *J. Chem. Soc.*, 1735 (1935).

[26] G. S. Hammond, *J. Am. Chem. Soc.*, **77**, 334 (1955).

[27] J. B. Conant and A. F. Thompson, Jr., *J. Am. Chem. Soc.*, **54**, 4039 (1932).

TABLE 10-2. PERCENTAGE OF ENOL IN KETO-ENOL TAUTOMERS AT EQUILIBRIUM

Compound	Enol, %	
	Pure	In water
CH_3COCH_3	0.00025[a]	
$CH_3COCH_2COCH_3$	80[b]	15[c]
$CH_3COCH_2CO_2Et$	7.5[b]	
$CH_3COCHMeCO_2Et$	4.1[b]	
$CH_3COCH(C_6H_5)CO_2Et$	30[b]	
Cyclopentanone	0.0048[a]	
Cyclohexanone	0.020[a]	
$CH_3COCOCH_3$	0.0056[a]	
1,2-Cyclohexanedione	~100[a]	40[a]
5,5-Dimethyl-1,3-cyclohexanedione	....	95[c]

[a] G. Schwarzenbach and C. Wittwer, *Helv. Chim. Acta*, **30**, 656, 659, 663, 669 (1947).
[b] J. B. Conant and A. F. Thompson, Jr., *J. Am. Chem. Soc.*, **54**, 4039 (1932).
[c] G. Schwarzenbach and E. Felder, *Helv. Chim. Acta*, **27**, 1044 (1944).

seen to have a small but negative influence.   The phenyl group shows its expected favorable effect, and there appears to be more enolization in six- than in five-membered rings.[28]   Aliphatic α-diketones are but little enolized and probably exist in a trans conformation

because of the large repulsion between the dipoles of the carbonyl groups. In cyclic α-diketones, where such stabilization of the keto form is not possible, extensive enolization occurs.   In these cases a hydrogen-bonded ring would contain only four atoms other than hydrogen and would thus probably not be very stable.   It is probably for this reason and because of hydrate formation of the keto form that the enol is relatively less stable in water.   Part of the stabilizing influence of the six-membered ring on the enol form is probably unrelated to its being an α-diketone, since 5,5-dimethyl-1,3-cyclohexanedione is also largely enolic.

Kon, Linstead, and coworkers studied the effect of structure on the equilibrium between α,β- and β,γ-unsaturated acids, esters, nitriles, etc.[29]

[28] Cf. H. C. Brown, J. H. Brewster, and H. Shechter, *J. Am. Chem. Soc.*, **76**, 467 (1954).
[29] For summaries of this work, see J. W. Baker, "Tautomerism," chap. 9, Routledge & Kegan Paul, Ltd., London, 1934; R. P. Linstead, in H. Gilman, "Organic Chemistry," p. 819, John Wiley & Sons, Inc., New York, 1938.

$$-\overset{|}{\underset{|}{C}}-\overset{|}{C}=\overset{|}{C}-CO_2{}^- \rightleftarrows -\overset{|}{C}=\overset{|}{C}-\overset{|}{\underset{\cdot|}{C}}-CO_2{}^-$$

Most of their results may be explained qualitatively on the basis of the effect of alkyl groups on stabilizing the double bonds by hyperconjugation and the more powerful effect of $-CO_2{}^-$, $-CO_2H$, $-CO_2R$, $-COR$, $-CN$, phenyl, etc., groups in stabilizing them by resonance.

The fact that 2,5-dihydrofuran rearranges to 2,3-dihydrofuran in the presence of alkali-metal alkoxides[30] shows that carbon-carbon double bonds may be stabilized by groups strongly supplying electrons by resonance as well as by strongly electron-withdrawing groups.

10-3. Carbanions in Displacement Reactions. *10-3a. Alkylation of β-Diketones and Related Compounds.* The alkylation of malonic ester, acetoacetic ester, β-diketones, etc., carried out most commonly by the reaction of a sodium salt with an alkyl halide, is best interpreted as a nucleophilic substitution of a carbanion for a halide anion. The reaction usually proceeds satisfactorily with halides with sufficient S_N2 reactivity, but with tertiary halides the basicity of the carbanion usually causes elimination reactions to predominate greatly over substitution processes. Evidence for the S_N2 character of the reaction has been obtained in several cases, as in the demonstration that the anion from malonic ester causes a Walden inversion in its reaction with cyclopentene oxide[31] and the observation that the reaction of this anion with ethyl bromide is kinetically first-order in each reactant.[32]

Kornblum and coworkers pointed out that in the alkylation of *ambident* anions, resonance-stabilized species such as

$$\left[CH_3-\overset{\overset{\textstyle O}{\|}}{C}\text{---}CH\text{---}\overset{\overset{\textstyle O}{\|}}{C}H-CH_3 \right]^- \quad \text{and} \quad \left[CH_2\text{---}N\overset{\nearrow O}{\underset{\searrow O}{}} \right]^-$$

which may react at any of several atoms, the tendency for alkylation at the most electronegative atom usually increases with the S_N1 character of the reaction.[33] Heterogeneity of the reaction mixture can change the character of the reaction drastically. Benzyl and allyl halides react with sodium and potassium salts of phenol to give almost exclusive oxygen

[30] R. Paul, M. Fluchaire, and G. Collardeau, *Bull. soc. chim. France,* 668 (1950).

[31] W. E. Grigsby, J. Hind, J. Chanley, and F. H. Westheimer, *J. Am. Chem. Soc.,* **64,** 2606 (1942).

[32] R. G. Pearson, *J. Am. Chem. Soc.,* **71,** 2212 (1949).

[33] N. Kornblum, R. A. Smiley, R. K. Blackwood, and D. C. Iffland, *J. Am. Chem. Soc.,* **77,** 6269 (1955).

alkylation in solution and almost exclusive carbon alkylation when the salts are present only in the solid form.[34]

10-3*b*. *The Wurtz Reaction.* Two plausible mechanisms have been suggested for the Wurtz reaction. In one, the alkyl halide molecules react with sodium to give free radicals, which may then couple or disproportionate. In the second, one molecule of alkyl halide is changed to an alkylsodium molecule, which then reacts with the other molecule of alkyl halide. Although a radical mechanism seems probable for the reaction of alkyl halides with sodium in the vapor phase at elevated temperatures (see Sec. 19-2*b*), the reaction as ordinarily carried out in solution may be satisfactorily explained on the basis of an alkylsodium intermediate. By vigorously stirring the reaction mixture so that the formation of alkylsodium could successfully compete with its subsequent reaction with alkyl halide, Morton and coworkers were able to transform alkyl halides to alkylsodium compounds in high yield, showing that, although the sodium surface may intermediately transform the alkyl halide into a free radical, this free radical is probably then transformed to an alkylsodium compound before it can escape.[35] Whitmore and Zook pointed out that the disproportionation products (alkane and alkene of the same carbon content as the alkyl halide used) whose presence had been cited as evidence for a free-radical mechanism could be explained as well by the simultaneous occurrence of elimination and substitution in the reaction of alkylsodium with alkyl halide, e.g.,

$$n\text{-}C_4H_9Cl + 2Na \rightarrow NaCl + n\text{-}C_4H_9Na$$

$$n\text{-}C_4H_9Na + n\text{-}C_4H_9Cl \nearrow \begin{matrix} C_2H_5CH{=}CH_2 + n\text{-}C_4H_{10} + NaCl \\ \\ n\text{-}C_8H_{18} + NaCl \end{matrix}$$

and that these disproportionation-type products were obtained in the reaction of preformed alkylsodium compound with alkyl halide.[36] Additional evidence against the free-radical mechanism for the Wurtz reaction as ordinarily carried out is that several optically active halides have been found to give active products with predominant inversion of configuration.[37–39] In this connection it is of interest that ethylsodium

[34] N. Kornblum and A. P. Lurie, *J. Am. Chem. Soc.*, **81**, 2705 (1959).

[35] A. A. Morton, J. B. Davidson, and H. A. Newey, *J. Am. Chem. Soc.*, **64**, 2240 (1942).

[36] F. C. Whitmore and H. D. Zook, *J. Am. Chem. Soc.*, **64**, 1783 (1942); cf. A. A. Morton, J. B. Davidson, and B. L. Hakan, *J. Am. Chem. Soc.*, **64**, 2242 (1942).

[37] E. LeGoff, S. E. Ulrich, and D. B. Denney, *J. Am. Chem. Soc.*, **80**, 622 (1958).

[38] J. F. Lane and S. E. Ulrich, *J. Am. Chem. Soc.*, **72**, 5132 (1950).

[39] For evidence that an intermediate free radical would have been expected to racemize, see Sec. 23-1*a*.

reacts with 2-chlorooctane to give 3-methylnonane with but 20 per cent racemization,[40] whereas with 2-bromooctane 97 per cent racemization occurs.[41]   This racemization is probably due to the formation, by metal-halogen interchange, of the 2-octylsodium compound

$$C_2H_5Na + n\text{-}C_6H_{13}\underset{\underset{\displaystyle Br}{|}}{C}HCH_3 \rightarrow C_2H_5Br + n\text{-}C_6H_{13}\underset{\underset{\displaystyle Na}{|}}{C}HCH_3$$

which has too much carbanion character to maintain its configuration. Since this metal-halogen interchange is essentially a nucleophilic displacement on halogen, it occurs *much* more rapidly on bromine, which can more easily expand its outer valence shell.   In fact, in the Wurtz reaction with active *sec*-butyl bromide, where no bromide as S_N2-reactive as ethyl bromide is present to compete with reversible metal-halogen interchange, racemic 3,4-dimethylhexane is formed,[42] with analogous results being obtained with 1,2-diphenylethyl bromide[42] and 2-bromooctane.[38]   Lack of racemization in the reaction of benzylsodium with 2-bromobutane[43] may be attributed to inhibition of metal-halogen interchange, in this case by the much greater acidity of toluene than butane.

10-3c. *The Favorskii Rearrangement.*[44]   The Favorskii rearrangement is a reaction of α-halo ketones with hydroxide (or alkoxide) ions to give salts (or esters) of acids with a somewhat rearranged carbon skeleton but the same number of carbon atoms.   When the α-halo ketone has available an α'-hydrogen atom, the reaction proceeds through an intermediate containing a cyclopropanone ring.   This mechanism predicts that α-chlorocyclohexanone labeled with C^{14} at the chlorine-bearing carbon atom, when treated with alkoxide ions, will yield an ester of cyclopentane-carboxylic acid containing half the C^{14} at the α-carbon atom and half at the β-carbon.   Loftfield has shown that this prediction is in agreement

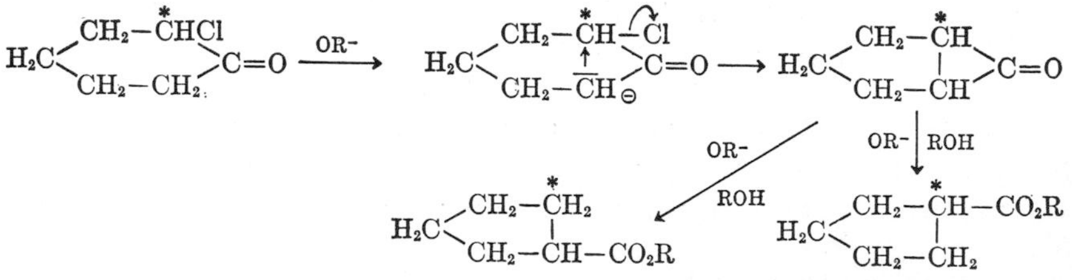

with the experimental facts, an important observation since most of the

[40] S. E. Ulrich, F. H. Gentes, J. F. Lane, and E. S. Wallis, *J. Am. Chem. Soc.*, **72**, 5127 (1950).

[41] N. G. Brink, J. F. Lane, and E. S. Wallis, *J. Am. Chem. Soc.*, **65**, 943 (1943).

[42] E. S. Wallis and F. H. Adams, *J. Am. Chem. Soc.*, **55**, 3838 (1933).

[43] R. L. Letsinger et al., *J. Am. Chem. Soc.*, **70**, 406 (1948); **73**, 2373 (1951).

[44] Cf. A. S. Kende, *Org. Reactions*, **11**, 261 (1960).

other reasonable mechanisms predict that the C^{14} should be entirely in the α position.[45] Furthermore, Stork and Borowitz have shown that with the 1-acetyl-1-chloro-2-methylcyclohexanes there is inversion of configuration at the carbon atom to which chlorine was initially attached.[46]

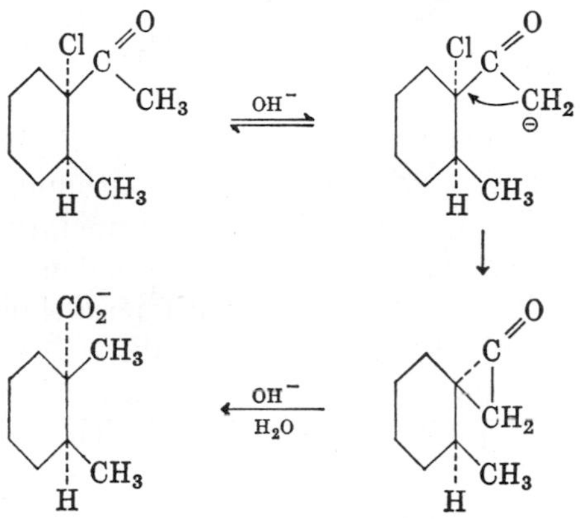

Nevertheless, in those cases where the absence of α'-hydrogen atoms makes the cyclopropanone mechanism impossible (and perhaps in some other cases also), an alternate mechanism must be operating.   In the rearrangement of α-chlorocyclohexyl phenyl ketone, the mechanism is very probably closely related to that of the benzilic-acid rearrangement[47] (cf. Sec. 14-3).   With an α,α'-dibromo ketone, the opening of the cyclopropanone ring may be accompanied by a loss of halide ion to yield an α,β-unsaturated acid, e.g.,[48]

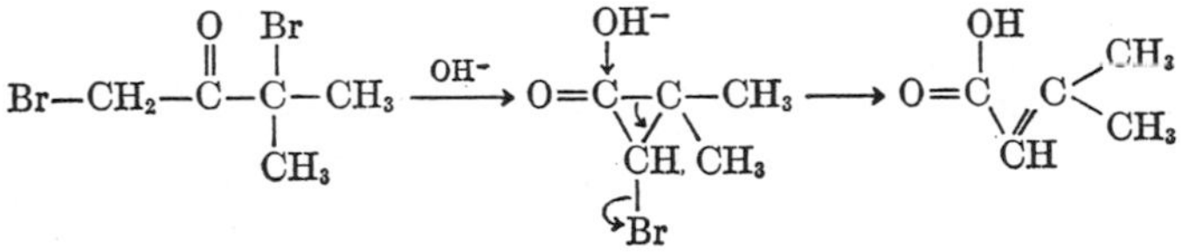

10-3d. Other Carbanion Rearrangements.   Although rearrangements of purely hydrocarbon anions are not nearly so common as those of carbonium ions, a reaction apparently involving such a rearrangement was discovered by Grovenstein[49] and confirmed by Zimmerman and Smentowski.[50]   The reaction of 2,2,2-triphenylethyl chloride with sodium in the

[45] R. B. Loftfield, *J. Am. Chem. Soc.*, **72**, 632 (1950); **73**, 4707 (1951); **76**, 35 (1954).
[46] G. Stork and I. J. Borowitz, *J. Am. Chem. Soc.*, **82**, 4307 (1960).
[47] B. Tchoubar and O. Sackur, *Compt. rend.*, **208**, 1020 (1939); C. L. Stevens and E. Farkas, *J. Am. Chem. Soc.*, **74**, 5352 (1952).
[48] A. E. Favorskii, *J. prakt. Chem.*, [2], **88**, 658 (1913).
[49] E. Grovenstein, Jr., *J. Am. Chem. Soc.*, **79**, 4985 (1957).
[50] H. E. Zimmerman and F. J. Smentowski, *J. Am. Chem. Soc.*, **79**, 5455 (1957).

presence of t-amyl alcohol or ammonia gives 1,1,1,-triphenylethane, but in an ether solvent the product is 1,1,2-triphenylethylsodium.

$$(C_6H_5)_3CCH_2Cl \xrightarrow{\text{Na}} (C_6H_5)_3CCH_2Na \xrightarrow{\text{ROH}} (C_6H_5)_3CCH_3$$

$$\downarrow$$

$$(C_6H_5)_2\overset{\ominus}{C}CH_2C_6H_5$$
$$Na^{\oplus}$$

Lithium was found to react with 2,2,2-triphenylethyl chloride in tetrahydrofuran at $-60°$ to give 2,2,2-triphenylethyllithium, which rearranged readily to 1,1,2-triphenylethyllithium at temperatures around $0°$.[51] Apparently the lithium compound would rearrange much more rapidly if it had more carbanion character, for with potassium the rearranged product is the only one observed even at $-50°$.

Other examples of rearrangements involving carbanions include the Stevens rearrangement and related reactions, which were discussed under S_{Ni} reactions (Sec. 6-3).

PROBLEMS

1. Which pair of geometric isomers would be interconverted more rapidly, cis- and $trans$-2-p-chlorophenyl-1,2-diphenylvinyllithium or cis- and $trans$-propenyllithium?[10] Give your reasons.

2. Draw energy diagrams to show why the reaction of cumylpotassium with deuterium chloride should give more p-deuteriocumene than the reaction with heavy water does.

3. Describe experiments designed to show whether the deuterium exchange and isomerization of cyclohexenylacetonitrile follow the reaction mechanism given in Sec. 10-2c and to give a reasonable explanation for the relative rates of the various reactions of any intermediate(s) involved.

4. As an alternative to the cyclopropanone intermediate described in Sec. 10-3c, the following resonance-stabilized species has been suggested:

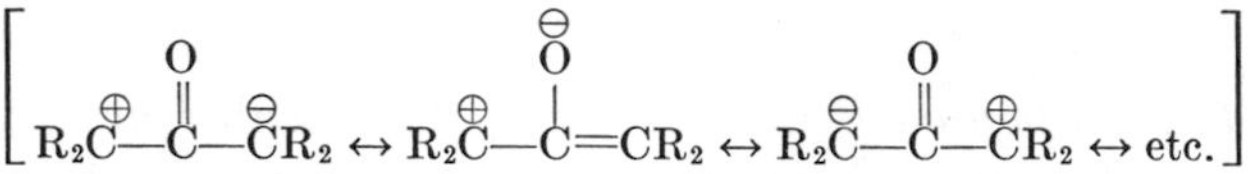

$$\left[R_2\overset{\oplus}{C}-\overset{O}{\underset{\|}{C}}-\overset{\ominus}{C}R_2 \leftrightarrow R_2\overset{\oplus}{C}-\overset{\overset{\ominus}{O}}{\underset{|}{C}}=CR_2 \leftrightarrow R_2\overset{\ominus}{C}-\overset{O}{\underset{\|}{C}}-\overset{\oplus}{C}R_2 \leftrightarrow \text{etc.} \right]$$

Tell whether you think this intermediate is probably formed in the cases referred to, and give your reasons. What conditions would you expect to favor the formation of such an intermediate?

5. Suggest a reasonable mechanism for the formation of each of the products shown in the following reaction:

$$(CH_2=CHCH_2)_2S \xrightarrow{\text{NaOEt}} (CH_3CH=CH)_2S \text{ and } CH_2=CHCH_2S\overset{SH}{\underset{|}{C}}H_2CH=CH_2$$

[51] E. Grovenstein, Jr., and L. P. Williams, Jr., *J. Am. Chem. Soc.*, **83**, 412 (1961).

Chapter 11

ADDITION TO ALDEHYDES AND KETONES

Because of the considerable contribution of structures of the type of II to the total structure of carbonyl compounds,

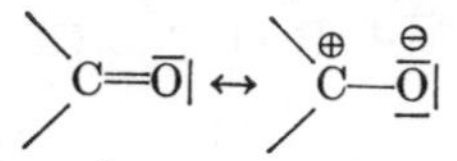

$$\begin{array}{ccc} \diagup \diagdown & & \diagup \diagdown \\ C=\overline{O}| & \leftrightarrow & \overset{\oplus}{C}-\overset{\ominus}{\underline{O}}| \\ \diagup & & \diagup \end{array}$$

$$\text{I} \qquad\qquad \text{II}$$

the carbon atom has marked electrophilic character. Therefore, addition reactions to carbonyl compounds almost invariably involve the attack of a nucleophilic reagent on the carbonyl carbon atom. Under acidic conditions the carbonyl compound may be transformed partially into its conjugate acid, in which the carbon atom is much more strongly electrophilic.

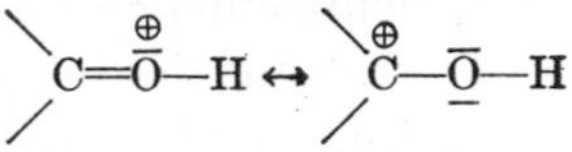

$$\begin{array}{ccc} \diagup \diagdown & & \diagup \diagdown \\ C=\overset{\oplus}{O}-H & \leftrightarrow & \overset{\oplus}{C}-\underline{O}-H \\ \diagup & & \diagup \end{array}$$

$$\text{III} \qquad\qquad \text{IV}$$

That is, IV contributes more to the total structure of the conjugate acid than II (in which there is charge separation) does to the structure of the carbonyl compound itself. Thus carbonyl addition reactions may be catalyzed by acids, since the acids convert the carbonyl compound to a form that is more reactive toward nucleophilic reagents. Even the partial donation of a proton to the carbonyl group, as in a hydrogen-bonded complex with an acid, would be expected to increase its electrophilicity.

11-1. Addition of Water and Alcohols. *11-1a. Hydrates of Aldehydes and Ketones.* Any electron-donating group attached to a carbonyl group will decrease its electrophilicity and stabilize it by distributing the positive charge. This stabilizing influence, and the converse destabilizing influence of electron-withdrawing groups, is reflected in the extent to which aldehydes and ketones form hydrates. In dilute aqueous solution

$$\begin{array}{ccc} \diagup \diagdown & & \diagup \diagdown & OH \\ C=O + H_2O & \rightleftharpoons & C & \diagup \\ \diagup & & \diagup \diagdown & OH \end{array}$$

at 20° formaldehyde exists about 99.99 per cent in the hydrated form,[1] acetaldehyde is about 58 per cent hydrated,[2] and acetone is almost negligibly so (all at equilibrium). Chloral and a number of compounds with the —COCOCO— grouping form stable crystalline hydrates.

11-1b. *Mechanism of the Hydration of Acetaldehyde.* Bell and coworkers studied the kinetics of the hydration of acetaldehyde.[3] The reaction was found to be subject to both general acid and general base catalysis. For the acid-catalyzed reaction the following mechanism was suggested:

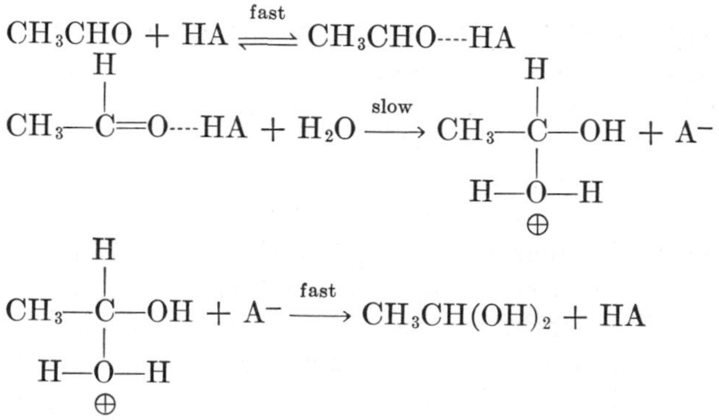

$$CH_3CHO + HA \underset{}{\overset{fast}{\rightleftharpoons}} CH_3CHO\text{----}HA$$

$$CH_3\text{—}\underset{H}{\overset{H}{C}}\text{=}O\text{----}HA + H_2O \xrightarrow{slow} CH_3\text{—}\underset{\underset{\oplus}{H\text{—}O\text{—}H}}{\overset{H}{C}}\text{—}OH + A^- \qquad (11\text{-}1)$$

$$CH_3\text{—}\underset{\underset{\oplus}{H\text{—}O\text{—}H}}{\overset{H}{C}}\text{—}OH + A^- \xrightarrow{fast} CH_3CH(OH)_2 + HA$$

A mechanism involving a rate-controlling proton donation to the carbonyl oxygen atom could also be suggested if the possibility were admitted that such a reaction could occur slowly enough to measure. For the base-catalyzed reaction the mechanism

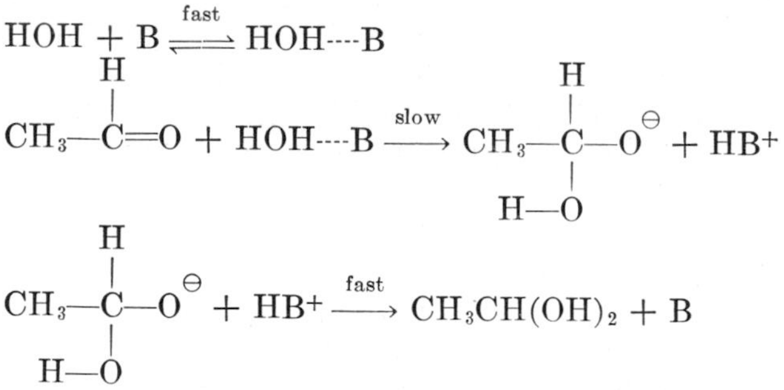

$$HOH + B \underset{}{\overset{fast}{\rightleftharpoons}} HOH\text{----}B$$

$$CH_3\text{—}\underset{H}{\overset{H}{C}}\text{=}O + HOH\text{----}B \xrightarrow{slow} CH_3\text{—}\underset{\underset{H\text{—}O}{}}{\overset{H}{C}}\text{—}O^{\ominus} + HB^+ \qquad (11\text{-}2)$$

$$CH_3\text{—}\underset{\underset{H\text{—}O}{}}{\overset{H}{C}}\text{—}O^{\ominus} + HB^+ \xrightarrow{fast} CH_3CH(OH)_2 + B$$

was proposed. The possibility that the reaction proceeds by a Lowry-type termolecular mechanism (see Sec. 5-2c) was considered. From the rate of the spontaneous (water-catalyzed) reaction and the catalytic

[1] R. Bieber and G. Trümpler, *Helv. Chim. Acta*, **30**, 1860 (1947).
[2] R. P. Bell and J. C. Clunie, *Trans. Faraday Soc.*, **48**, 439 (1952).
[3] R. P. Bell and B. deB. Darwent, *Trans. Faraday Soc.*, **46**, 34 (1950); R. P. Bell and J. C. Clunie, *Proc. Roy. Soc.* (*London*), **212A**, 33 (1952); R. P. Bell, M. H. Rand, and K. M. A. Wynne-Jones, *Trans. Faraday Soc.*, **52**, 1093 (1956).

constants for hydronium, hydroxide, and acetate ions and for acetic acid, the method of Swain[4] was used to estimate the magnitude of the termolecular rate term in which acetic acid acted as the acid and acetate ion as the base. Although the predicted magnitude of the termolecular term was far too large for it to have been overlooked, none was observed experimentally, and it was concluded that the termolecular mechanism is probably unimportant in this instance. It might be argued that this is one reaction for which the method of estimation would not work. The hydroxide ion, like any other base, can remove a proton from a water molecule as the water molecule attacks the carbonyl carbon atom [mechanism (11-2)], and in addition, being itself a fragment of water, hydroxide ion is unique among bases in being able to initiate hydrate formation by direct attack on the carbonyl carbon atom. However, this objection is not relevant in the present case, since it so happens that the value of the catalytic constant for hydroxide ion used in this estimation can be reduced as much as desired without significantly changing the magnitude of the termolecular term predicted.

11-1c. O^{18} *Exchange of Acetone.* To study the rate of formation of a hydrate so unstable that its equilibrium concentration is too small to measure, it is necessary to use a method that tells not how much hydrate *is present* but how much *has been formed.* Exchange with water containing excess O^{18} provides a method of this type, because the hydrate, when formed, will lose H_2O^{16} and H_2O^{18} with almost equal likelihood.

$$\begin{array}{ccc} CH_3 & CH_3 \quad OH & CH_3 \\ \diagdown & \diagdown \diagup & \diagdown \\ C=O + H_2O^{18} \rightleftharpoons & C & \rightleftharpoons \quad C=O^{18} + H_2O \\ \diagup & \diagup \diagdown & \diagup \\ CH_3 & CH_3 \quad O^{18}H & CH_3 \end{array}$$

The O^{18} exchange of acetone was studied by Cohn and Urey, who found the reaction in the absence of catalysts to proceed very slowly even at 100°, but in the presence of 10^{-4} N hydrochloric acid or 10^{-3} N sodium hydroxide to occur too rapidly to measure conveniently even at 25°.[5] The acid catalysis was shown to be general, and although *general* base catalysis was not detected (in the hydration of acetaldehyde, catalysis by bases other than hydroxide ion is quite small), the reaction mechanism is very probably the same as that of the hydration of acetaldehyde.

11-1d. *Mutarotation of Glucose.* Alcohols add to carbonyl groups with a facility comparable to that of water. There have been a number of studies of the equilibria in such reactions,[6] but only a few studies of the

[4] C. G. Swain, *J. Am. Chem. Soc.*, **72**, 4578 (1950); see also Sec. 5-2d.

[5] M. Cohn and H. C. Urey, *J. Am. Chem. Soc.*, **60**, 679 (1938).

[6] See for example K. L. Wolf and K. Merkel, *Z. physik. Chem.*, **187A**, 61 (1940); I. L. Gauditz, *Z. physik. Chem.*, **48B**, 228 (1941); F. E. McKenna, H. V. Tartar, and E. C. Lingafelter, *J. Am. Chem. Soc.*, **75**, 604 (1953).

reaction kinetics have been made.[7]   The mutarotation of sugars, particularly glucose, has received a great deal of attention, and since these reactions involve the establishment of equilibrium between diastereomeric cyclic hemiacetals through the open-chain hydroxyaldehyde (or ketone) as a reactive intermediate, their rate-controlling step is the reverse of the addition of an alcohol to an aldehyde (or ketone).   Thus in the case of glucose the reaction path is

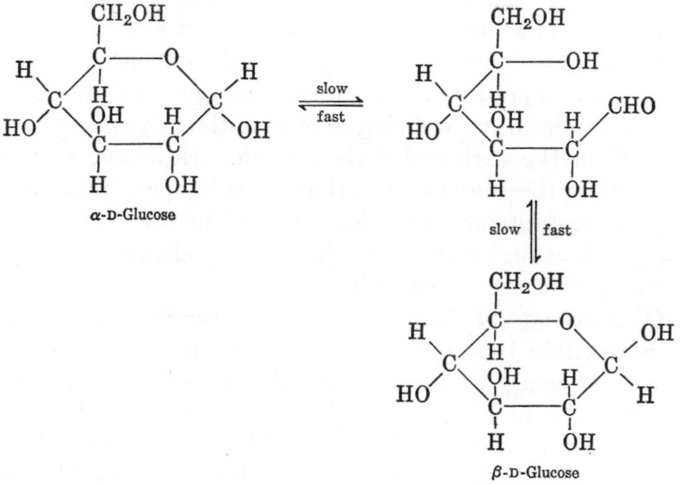

ignoring the smaller concentrations of furanose and other forms present at equilibrium.   As ordinarily studied, the equilibrium is approached from the side of the pure α isomer and the extent of reaction followed polarimetrically.   Brønsted and Guggenheim found the mutarotation of glucose to be both general acid- and general base-catalyzed.[8]   Lowry and Faulkner carried out some striking experiments whose results are in support of the concerted termolecular mechanism.[9]   Using 2,3,4,6-tetramethylglucose (the nonparticipating hydroxyl groups were methylated to make the compound soluble in organic solvents), these workers found mutarotation to proceed very slowly in dry pyridine, a definitely basic but almost negligibly acidic solvent.   Similar results were obtained in the acidic but very weakly basic solvent cresol.   In a mixture of cresol and pyridine, however, the mutarotation proceeded quite rapidly.   In water, which is both basic and acidic, the reaction proceeds fairly rapidly. These facts were interpreted as indicating that the mechanism involves the simultaneous removal of a proton from the hydroxy group and dona-

---

[7] See, however, (a) I. Lauder, *Trans. Faraday Soc.*, **44,** 734 (1948); (b) G. W. Meadows and B. deB. Darwent, *Trans. Faraday Soc.*, **48,** 1015 (1952).

[8] J. N. Brønsted and E. A. Guggenheim, *J. Am. Chem. Soc.*, **49,** 2554 (1927).

[9] T. M. Lowry and I. J. Faulkner, *J. Chem. Soc.*, **127,** 2883 (1925).

tion of one to the ethereal oxygen atom of the hemiacetal group.[10]   Swain
and Brown studied the kinetics of the amine- and phenol-catalyzed
mutarotation of tetramethylglucose in benzene solution and observed
third-order kinetics, first-order in amine, phenol, and tetramethylglucose.[11]   In further striking agreement with the concerted mechanism,
these workers found that 2-hydroxypyridine, which may donate a proton
to, and accept one from, the hemiacetal group, is a powerful specific
catalyst for the mutarotation.   At a concentration of 0.001 M this compound is 7,000 times as effective a catalyst as a mixture of 0.001 M
pyridine and 0.001 M phenol, although it is only one ten-thousandth as
strong a base as pyridine and only one-hundredth as strong an acid as
phenol.   Also, in its presence the reaction becomes second-order, first-
order in 2-hydroxypyridine and first-order in tetramethylglucose.   Other
evidence showed that this catalyst forms a complex with the reactant,
and so there is excellent evidence for a mechanism of the type

As Swain and Brown point out, the action of enzymes may be due to
polyfunctional catalysis of this sort.   For 2-hydroxypyridine the concerted mechanism might be favored over a stepwise mechanism in an
organic solvent as compared with water, since the stepwise mechanism
involves the intermediate formation of ions.   Indeed, the concerted
mechanism has not been proved (or disproved) for the reaction in aqueous
solution.   Although no third-order term is required to explain the reaction kinetics, the predicted size of the third-order term is so small that it
could not have been detected if present.[4]

**11-2. Addition of Nucleophilic Reagents Containing Sulfur and
Nitrogen.** 11-2a. *Additions of Hydrogen Sulfide and Mercaptans.*
Hydrogen sulfide and mercaptans have a considerably greater tendency
to add to carbonyl groups than do water and alcohols.   The basicity of

[10] T. M. Lowry, *J. Chem. Soc.*, 2554 (1927).
[11] C. G. Swain and J. F. Brown, Jr., *J. Am. Chem. Soc.*, **74**, 2534, 2538 (1952).

sulfur compounds toward carbon compared with their basicity toward hydrogen is usually greater than that of the analogous oxygen compounds. This fact is, no doubt, related to the considerable nucleophilicity of certain sulfur compounds (Sec. 7-2) and might be predicted from the bond energies given in Table 1-4. For example, the reaction of hydrogen sulfide with aldehydes and ketones to yield 1,1-dithiols

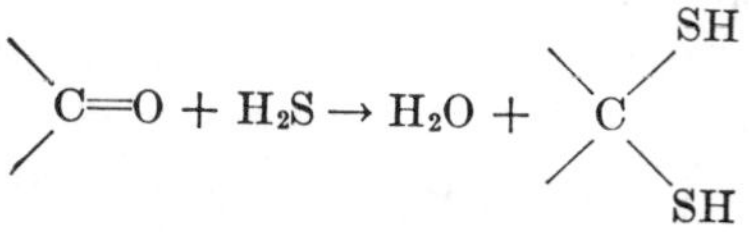

$$\ce{\chemfig{C=O}} + H_2S \rightarrow H_2O + \ce{\chemfig{C(-[1]SH)(-[7]SH)}}$$

appears to be rather general.[12] In addition to forming thioacetals more easily than alcohols form acetals, mercaptans form thioketals quite generally, whereas ketal formation from alcohols is uncommon.

11-2b. Addition of Sodium Bisulfite. Sodium bisulfite adds to many aldehydes and ketones. The products have been shown to be the salts of α-hydroxysulfonic acids.[13] From the manner in which the rate changes with changing pH it appears that the more nucleophilic sulfite ion is the effective attacking reagent even under conditions where its concentration is much smaller than that of bisulfite.[14]

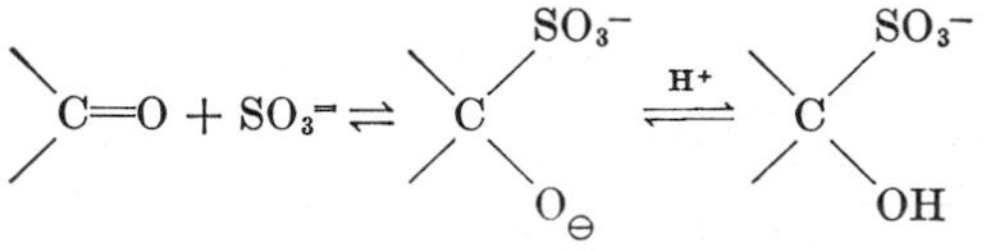

$$\ce{\chemfig{C=O}} + SO_3^- \rightleftharpoons \ce{\chemfig{C(-[1]SO_3^-)(-[7]O^\ominus)}} \overset{H^+}{\rightleftharpoons} \ce{\chemfig{C(-[1]SO_3^-)(-[7]OH)}}$$

Gubareva[15] measured the equilibrium constants for the formation of the bisulfite addition compounds from a number of aldehydes and ketones and found them to decrease in the order $CH_3CHO > C_2H_5CHO > C_6H_5CHO > CH_3COCH_3 > C_2H_5COCH_3 > (CH_3)_3CCOCH_3$. This order is probably a result of both steric and electronic factors.

11-2c. Mechanism of Formation of Hydrazone Derivatives and Oximes. Simple imines in which the carbon-nitrogen double bond has no particular stabilization tend to polymerize or react in some other way to give compounds with carbon-nitrogen single bonds. This tendency is decreased when the double bond is conjugated with an aromatic ring, as in anils and benzalimines. Conjugation with two aromatic rings (benzalaniline derivatives) or with the unshared pairs of oxygen or nitrogen atoms gives

[12] T. L. Cairns, G. L. Evans, A. W. Larchar, and B. C. McKusick, *J. Am. Chem. Soc.*, **74**, 3982 (1952).

[13] F. Raschig and W. Prahl, *Ann.*, **448**, 265 (1926).

[14] T. D. Stewart and L. H. Donnally, *J. Am. Chem. Soc.*, **54**, 3559 (1932).

[15] M. A. Gubareva, *Zhur. Obshchei Khim.*, **17**, 2259 (1947); *Chem. Abstr.*, **42**, 4820a (1948).

quite stable carbon-nitrogen double bonds. For the latter reason it is nitrogen compounds like hydroxylamine, phenylhydrazine, semicarbazide, etc., that are useful in preparing derivatives of aldehydes and ketones. The rate and equilibrium constants in reactions of this sort have been studied considerably.

In a number of cases the reactions have been found to be first-order in carbonyl compound, first-order in nitrogen compound, and under some conditions, at least, subject to general acid catalysis.[16-18] Plots of second-order rate constants vs. pH have been found to give maxima in the weakly acidic range.[16-18] Under the conditions of acid catalysis it is obvious that the carbonyl compound, nitrogen compound, and catalyzing acid (in some form) must all appear in the reaction mechanism at or before the rate-controlling step. Rather than require all three reagents to collide simultaneously, it seems more reasonable to assume the rapid reversible association of two of the reagents to give a complex that then reacts with the third. It is often difficult to decide which two reagents form the initial complex, particularly in cases where the complex is never formed in sufficient amounts to detect. This difficulty does not arise in the present case, however. Jencks showed that in the presence of fairly high concentrations of hydroxylamine, semicarbazide, etc., many aldehydes and ketones are transformed largely to adducts.[19] The presence of 0.5 M hydroxylamine reduces the absorbance of a furfural solution at 2,750 A to about one-third its original value within 1 min. This decrease in absorbance cannot be due to the formation of furfuraldoxime, which absorbs more strongly than furfural at 2,750 A.

Apparently there is an initial rapid reversible addition to the carbonyl compound to give an intermediate that undergoes acid-catalyzed dehydration in the rate-controlling step.

$$RCH{=}O + H_2NOH \underset{}{\overset{fast}{\rightleftharpoons}} RCH{-}NHOH$$
$$\underset{OH}{|}$$

$$RCH{-}NHOH + HA \overset{slow}{\longrightarrow} RCH{=}NOH + H_2O + HA \qquad (11\text{-}3)$$
$$\underset{OH}{|}$$

This mechanism will lead to third-order kinetics only if the equilibrium concentration of the intermediate adduct is much smaller than either the aldehyde or hydroxylamine concentration.

A decrease in pH from 7 to 5 increases the reaction rate because it

[16] E. Barrett and A. Lapworth, *J. Chem. Soc.*, **93,** 85 (1908).

[17] A. Olander, *Z. physik. Chem.*, **129,** 1 (1927).

[18] J. B. Conant and P. D. Bartlett, *J. Am. Chem. Soc.*, **54,** 2881 (1932).

[19] W. P. Jencks, *J. Am. Chem. Soc.*, **81,** 475 (1959).

increases the rate of the dehydration step. With further increases in acidity this effect begins to be countered by the transformation of an increasing fraction of the hydroxylamine into its inert conjugate acid. Still further increases in acidity change the rate-controlling step and decrease the reaction rate. The dehydration in strongly acidic solution is so fast that essentially every molecule of adduct formed is dehydrated, but the concentration of free oxime has become so low that its addition to the carbonyl compound is the rate-controlling step of the reaction. There may be an increase in reaction rate in basic solution[16] due to basic catalysis of the dehydration step.[20]

11-2d. Reactivity in Semicarbazone Formation. Anderson and Jencks studied the kinetics of the formation of semicarbazones by a number of substituted benzaldehydes.[20] Satisfactory Hammett-equation correlations were obtained both for the equilibrium constants for adduct formation ($\rho = 1.81$) and the rate constants for the acid-catalyzed dehydration

TABLE 11-1. REACTION RATES AND HEATS AND ENTROPIES OF ACTIVATION
FOR SEMICARBAZONE FORMATION[21]

Compound	10^3k, at $0.03°$	$\Delta H,^{\ddagger}$ kcal	$\Delta S^{\ddagger} - \Delta S^{\ddagger}_{acetone},$ e.u.
Acetone..................	63.5	2.0	0.0
Diethyl ketone...........	6.88	1.4	−6.5
Pinacolone...............	0.771	1.8	−9.7
Cyclopentanone..........	8.26	4.0	3.3
Cyclohexanone...........	437	1.1	0.4
Furfural.................	6.56	4.1	3.2
Acetophenone...........	0.204	4.6	−1.9

of the adducts ($\rho = -1.74$). The over-all rate constant in neutral solution, which is the product of the equilibrium constant for the first step and the rate constant for the second step, is therefore little affected by meta and para substituents ($\rho = 0.07$). In acid solution, however, where the first step is rate-controlling, ρ is 0.91. At certain intermediate pH's, where the rate-controlling step is not the same for all the benzaldehydes, there is a definite break in the Hammett-equation plot.

When structural changes are made nearer the reaction center, the rate and equilibrium constants for the over-all process of semicarbazone formation do not seem to follow any simple linear free-energy relationship. A particularly careful discussion has been given by Price and Hammett, whose data are listed in Table 11-1.[21] These data were determined at pH 7.0, and it is therefore probable that the dehydration

[20] B. M. Anderson and W. P. Jencks, *J. Am. Chem. Soc.*, **82**, 1773 (1960).
[21] F. P. Price, Jr., and L. P. Hammett, *J. Am. Chem. Soc.*, **63**, 2387 (1941).

of the intermediate complex was rate-controlling in all cases. As Price and Hammett note, the eightyfold difference in rate between acetone and pinacolone (methyl *t*-butyl ketone) is almost entirely due to entropy factors, suggesting that the *t*-butyl group has a number of modes of internal motion in pinacolone that are not possible in the more crowded transition state. On the other hand, the increased reactivity of cyclohexanone (compared with acetone) is almost entirely due to a lower heat of activation. Brown, Fletcher, and Johannesen have pointed out that a cyclohexane ring all of whose carbon atoms are tetragonal may exist in the particularly stable chair form, in which all the valences are staggered (see Sec. 1-6), but that when one of the carbon atoms is trigonal, as in cyclohexanone, this stable configuration is impossible.[22] Since the carbonyl carbon atom has acquired considerable tetragonal character in the transition state of the rate-controlling step for semicarbazone formation, the reaction occurs particularly easily with cyclohexanone. In the case of cyclopentanone the valences are more easily staggered in the ketone than in the reactive intermediate, and a decrease in reactivity is observed. The differences in reactivity on this basis would be expected to be, and indeed are found to be, unrelated to the stabilities of the semicarbazones as measured by the equilibrium constants for their formation. The equilibrium constant for the formation of the derivative from cyclopentanone is about twice that for acetone, and cyclohexanone has an intermediate value.[21]

In addition to the small polar effect on reactivity that is reflected by the small ρ constant found with benzaldehydes,[20] there is another important electronic effect in operation. Because of conjugation with the C—O or C—N double bonds, benzaldehyde derivatives and other α,β-unsaturated aldehydes and ketones and their semicarbazones, oximes, etc., are stabilized relative to the transition state, in which there is only partial double-bond character. For this reason, such aldehydes and ketones usually react much more slowly than analogous saturated species, even though their reactions may be just as complete (or more so) at equilibrium. Conant and Bartlett, for example, found that acetaldehyde reacted more than 100 times as fast as benzaldehyde under their conditions, although the equilibrium constant for the formation of benzaldehyde semicarbazone is about 7 times as large as that for the acetaldehyde derivative.[18] In view of the fact that the phenyl group is an electron withdrawer relative to methyl, whereas —CH=O must be more strongly electron-withdrawing than —CH=NNHCONH$_2$, the equilibrium would be expected (cf. Sec. 4-1) to be more favorable for the benzaldehyde derivative.

[22] H. C. Brown, R. S. Fletcher, and R. B. Johannesen, *J. Am. Chem. Soc.*, **73**, 212 (1951).

11-3. Carbon-Carbon Condensations. 11-3a. *Cyanohydrin Formation.* Lapworth demonstrated that the addition of HCN to aldehydes and ketones is a base-catalyzed reaction; he therefore concluded that the rate-controlling step of the reaction is the combination of the carbonyl compound with a cyanide ion.[23] Svirbely and Roth found the additions of cyanide ion to acetone, acetaldehyde, and propionaldehyde to be slightly subject to general acid catalysis, the effect being beyond the experimental error only for the latter compound.[24]

The equilibrium constants for cyanohydrin formation have been determined for a number of substituted benzaldehydes.[25] The cyanohydrins are found to be stabilized by electron-withdrawing groups and destabilized by electron-donating groups, as would be expected. The fact that pinacolone forms as stable a cyanohydrin as acetone and that the cyanohydrin of phenyl *t*-butyl ketone is more stable than that of acetophenone[26] suggests that the increased ability of the methyl group to stabilize the carbonyl group by hyperconjugative electron donation may equal or surpass the destabilization of the cyanohydrin by steric interactions with the *t*-butyl group.

Prelog and Kobelt measured the dissociation constants for the cyanohydrins of a number of cyclic ketones with the result shown in Table 11-2.[27] These workers discussed the theoretical basis for these results, as have Brown, Fletcher, and Johannesen.[22]

11-3b. *Aldol Condensation of Acetaldehyde.* It is well known that aldol condensations may be base-catalyzed. The most likely function to attribute to the base is the formation of a carbanion, which may then add to a carbonyl group; e.g., in the case of acetaldehyde

$$CH_3CHO + OH^- \underset{k_{-1}}{\overset{k_1}{\rightleftharpoons}} \overset{\ominus}{C}H_2CHO + H_2O$$

$$CH_3CHO + \overset{\ominus}{C}H_2CHO \underset{k_{-2}}{\overset{k_2}{\rightleftharpoons}} CH_3\underset{\underset{O\ominus}{|}}{C}HCH_2CHO \tag{11-4}$$

$$CH_3\underset{\underset{O\ominus}{|}}{C}HCH_2CHO + H_2O \underset{k_{-3}}{\overset{k_3}{\rightleftharpoons}} CH_3\underset{\underset{OH}{|}}{C}HCH_2CHO + OH^-$$

Bonhoeffer and Walters found that when this reaction was carried out with 10 M acetaldehyde in heavy-water solution, the aldol isolated, as soon as an appreciable amount was formed, contained only a negligible

[23] A. Lapworth, *J. Chem. Soc.,* **83,** 995 (1903); **85,** 1206 (1904).

[24] W. J. Svirbely and J. F. Roth, *J. Am. Chem. Soc.,* **75,** 3106 (1953).

[25] A. Lapworth and R. H. F. Manske, *J. Chem. Soc.,* 2533 (1928); J. W. Baker et al., *J. Chem. Soc.,* 191 (1942); 1089 (1949); 2831 (1952); 404 (1956).

[26] A. Lapworth and R. H. F. Manske, *J. Chem. Soc.,* 1976 (1930).

[27] V. Prelog and M. Kobelt, *Helv. Chim. Acta,* **32,** 1187 (1949).

amount of carbon-bound deuterium.[28]   This showed that essentially every carbanion formed added to another molecule of acetaldehyde, since if a significant fraction reacted with water, the acetaldehyde would acquire deuterium in its methyl group and hence would yield carbon-deuteriated aldol.   Broche and Gibert, however, showed that in more dilute solution the reaction was neither cleanly first-order in acetaldehyde, as it should be if $k_2[CH_3CHO]$ is much larger than k_{-1}, nor cleanly second-order, as when k_{-1} is much larger than $k_2[CH_3CHO]$.   The observed kinetics instead satisfactorily fit the expected, more complicated kinetic equation applicable to the case where k_{-1} and $k_2[CH_3CHO]$

TABLE 11-2. DISSOCIATION CONSTANTS OF CYANOHYDRINS OF UNSUBSTITUTED CYCLIC KETONES[27]

Ring size	$100K$	Ring size	$100K$	Ring size	$100K$
5	2.1	11	112	16	9
6	0.1	12	31	17	12
7	13	13	26	18	10
8	86	14	6	19	10
9	170	15	11	20	7

are comparable.[29]   Broche and Gibert's conclusions were confirmed by Bell and Smith, who found that at acetaldehyde concentrations up to 1.4 M the aldolization in heavy water is accompanied by considerable deuterium exchange of the acetaldehyde.[30]   Bell and McTigue showed that the reaction is complicated by certain other factors, including diminution of the hydroxide-ion concentration by the action of the weak acid $CH_3CH(OH)_2$, whose conjugate base is also probably capable of transforming acetaldehyde to a carbanion.[31]   The existence of about half the acetaldehyde as the hydrate does not seriously change the form of the kinetic equation, since the establishment of the $CH_3CHO-CH_3CH(OH)_2$ equilibrium is quite rapid compared with the aldolization reaction.[2,3]

11-3c. *Aldol Condensation of Acetone*.   Because of the stabilizing influence of the extra methyl radical on the carbonyl group of acetone, the equilibrium constant for the aldol condensation of this compound is much less favorable than that for acetaldehyde.   Although the condensation does not proceed to a sufficient extent to make kinetic study convenient, the reverse reaction, the dealdolization of diacetone alcohol, has received considerable study.   This reaction is base-catalyzed, as expected, and

[28] K. F. Bonhoeffer and W. D. Walters, *Z. physik. Chem.*, **181A**, 441 (1938).
[29] A. Broche and R. Gibert, *Bull. Soc. chim. France*, 131 (1955).
[30] R. P. Bell and M. J. Smith, *J. Chem. Soc.*, 1691 (1958).
[31] R. P. Bell and P. T. McTigue, *J. Chem. Soc.*, 2983 (1960).

the basic catalysis is specific,[32] the rate equation being[33]

$$v = k[\text{diacetone alcohol}][\text{OH}^-]$$

The mechanism of the aldolization of acetone is very probably

$$CH_3COCH_3 + OH^- \underset{k_{-1}}{\overset{k_1}{\rightleftharpoons}} \overset{\ominus}{C}H_2COCH_3 + H_2O$$

$$CH_3COCH_3 + \overset{\ominus}{C}H_2COCH_3 \underset{k_{-2}}{\overset{k_2}{\rightleftharpoons}} \underset{\underset{O}{|\ominus}}{(CH_3)_2C}\text{—}CH_2COCH_3 \tag{11-5}$$

$$\underset{\underset{O}{|\ominus}}{(CH_3)_2C}\text{—}CH_2COCH_3 + H_2O \underset{k_{-3}}{\overset{k_3}{\rightleftharpoons}} \underset{\underset{OH}{|}}{(CH_3)_2C}\text{—}CH_2COCH_3 + OH^-$$

Since the rate of dealdolization is independent of the acetone concentration, the rate-controlling step cannot be the one governed by k_{-1}. The step governed by k_{-3} is probably not rate-controlling because k_3, the rate constant for a proton transfer from an alkoxide ion to water, should be much larger than k_{-2}, the rate constant for a C—C bond cleavage. Therefore the second step of the reaction must be rate-controlling. This must mean that k_{-1} is much larger than k_2, and, in fact, Walters and Bonhoeffer calculated (from deuterium-exchange data on acetone) that it is more than a thousand times as large.[34] The striking difference in the ratio k_{-1}/k_2 between the two mechanisms (11-4) and (11-5) is in the direction that might have been predicted, since the more reactive carbonyl group of acetaldehyde should coordinate with a carbanion more rapidly than that of acetone.

Although the reaction is catalyzed by certain amines,[35] Westheimer and Cohen have shown that this is a specific effect of primary and secondary (but not tertiary) amines rather than a general base catalysis.[32] It therefore appears that this specific effect is due to the reaction of the amine with the carbonyl group to give the intermediate[36]

$$\underset{\underset{O}{\ominus|}}{(CH_3)_2C}\text{—}CH_2\text{—}\underset{\underset{\underset{R \qquad R}{}}{\overset{||\oplus}{N}}}{C}\text{—}CH_3$$

which should cleave much more rapidly than the conjugate base of diacetone alcohol since the positively charged nitrogen atom is a more strongly electron-withdrawing group than neutral oxygen.

[32] F. H. Westheimer and H. Cohen, *J. Am. Chem. Soc.*, **60**, 90 (1938).

[33] K. Koelichen, *Z. physik. Chem.*, **33**, 129 (1900); V. K. La Mer and M. L. Miller, *J. Am. Chem. Soc.*, **57**, 2674 (1935).

[34] W. D. Walters and K. F. Bonhoeffer, *Z. physik. Chem.*, **182A**, 265 (1938).

[35] J. G. Miller and M. Kilpatrick, *J. Am. Chem. Soc.*, **53**, 3217 (1931).

[36] L. P. Hammett, "Physical Organic Chemistry," p. 345, McGraw-Hill Book Company, Inc., New York, 1940.

11-3d. Other Additions of Carbanions to Aldehydes and Ketones. There are a number of other base-catalyzed carbon-carbon condensations that have been found to be kinetically first-order in base, active hydrogen compound, and carbonyl compound, i.e., third-order over-all. Coombs and Evans found the condensation of benzaldehyde with acetophenone in ethanol to give benzalacetophenone to obey the rate equation[37]

$$v = k[C_6H_5CHO][C_6H_5COCH_3][OEt^-]$$

The following mechanism seems plausible for such a reaction:

$$C_6H_5COCH_3 + EtO^- \underset{k_{-1}}{\overset{k_1}{\rightleftharpoons}} C_6H_5COCH_2^{\ominus} + EtOH$$

$$C_6H_5COCH_2^{\ominus} + C_6H_5CHO \underset{k_{-2}}{\overset{k_2}{\rightleftharpoons}} \overset{\overset{\displaystyle O^{\ominus}}{|}}{C_6H_5COCH_2CHC_6H_5}$$

$$\overset{\overset{\displaystyle O^{\ominus}}{|}}{C_6H_5COCH_2CHC_6H_5} + EtOH \underset{k_{-3}}{\overset{k_3}{\rightleftharpoons}} \overset{\overset{\displaystyle OH}{|}}{C_6H_5COCH_2CHC_6H_5} + EtO^-$$

$$\overset{\overset{\displaystyle OH}{|}}{C_6H_5COCH_2CHC_6H_5} + EtO^- \underset{k_{-4}}{\overset{k_4}{\rightleftharpoons}} \overset{\overset{\displaystyle OH}{|}}{C_6H_5CO\overset{\ominus}{C}HCHC_6H_5} + EtOH$$

$$\overset{\overset{\displaystyle OH}{|}}{C_6H_5CO\overset{\ominus}{C}HCHC_6H_5} \underset{k_{-5}}{\overset{k_5}{\rightleftharpoons}} C_6H_5COCH=CHC_6H_5 + OH^-$$

(11-6)

The observed kinetic order will be found if any step except the first is rate-controlling, but it is not likely that the third step, a proton transfer between an alcohol and an alkoxide ion, would be slow enough to be rate-controlling. Noyce, Pryor, and Bottini's observation that alkali transforms the intermediate β-hydroxy ketone to benzaldehyde and acetophenone more rapidly than it dehydrates it shows that the second step of the reaction is not rate-controlling.[38] Although the equilibrium in the formation of the β-hydroxy ketone may not be particularly favorable because of the loss of conjugation in the benzaldehyde part of the molecule, the reaction is probably driven to the right by the subsequent dehydration reaction. This dehydration is written as proceeding by the carbanion mechanism in view of the relative stability of the intermediate carbanion and the difficulty of displacing hydroxide ions from organic molecules. A *p*-methoxy substituent in the benzaldehyde ring decreases the rate (to one-seventh of that of the unsubstituted compound) as

[37] E. Coombs and D. P. Evans, *J. Chem. Soc.*, 1295 (1940).
[38] D. S. Noyce, W. A. Pryor, and A. H. Bottini, *J. Am. Chem. Soc.*, **77**, 1402 (1955).

expected. The effect of a p-methoxy substituent in the acetophenone ring might seem more difficult to predict, but when it is noted that the acetophenone moiety has some carbanion character in the transition state of the rate-controlling step of the reaction (regardless of whether this is the fourth or fifth step), it is clear that a decrease in reactivity would be expected. Acetophenone is found to react four times as fast as its p-methoxy derivative.[37]

Other third-order carbanion condensations include the reaction of benzaldehyde with phenacyl chloride, in which the initial addition is followed by epoxide formation.[39]

$$C_6H_5COCH_2Cl + OH^- \overset{fast}{\rightleftharpoons} C_6H_5CO\overset{\ominus}{C}HCl + H_2O$$

$$C_6H_5CHO + C_6H_5CO\overset{\ominus}{C}HCl \overset{slow}{\longrightarrow} \underset{\underset{O}{\overset{\ominus}{|}}\quad\underset{Cl}{|}}{C_6H_5CH-CHCOC_6H_5}$$

$$\underset{\underset{O}{\overset{\ominus}{|}}\quad\underset{Cl}{|}}{C_6H_5CH-CHCOC_6H_5} \longrightarrow \underset{\overset{\diagdown\,\diagup}{O}}{C_6H_5CH-CHCOC_6H_5}$$

Ballester and Pérez-Blanco secured additional evidence for this mechanism by isolating two diastereomeric chlorohydrins from the reaction of m-nitrobenzaldehyde and 2,4,6-trimethoxyphenacyl chloride.[40]

Earlier in the history of organic chemistry there was considerable debate as to which component of the Perkin reaction mixture was the reactant and which the catalyst. However, as Breslow and Hauser noted, from the standpoint of current theory it may be seen that the α-hydrogen atoms of acetic anhydride must be vastly easier to remove by base than those of sodium acetate, and therefore it seems most likely that acetic anhydride is the active hydrogen compound and that the sodium acetate acts as a basic catalyst.[41] This view is supported by Kalnin's observation[42] that sodium acetate may be replaced by other bases, such as sodium carbonate, sodium phosphate, quinoline, pyridine, and triethylamine, whose effectiveness appears to increase with their basicity; it also agrees with the kinetic study of Buckles and Bremer.[43]

The reaction

$$CH_2O \rightarrow HOCH_2CHO \rightarrow HOCH_2(CHOH)_nCHO$$

occurs in the presence of thallium hydroxide, calcium hydroxide, or certain other bases and bears a superficial resemblance to an ordinary

[39] M. Ballester and P. D. Bartlett, *J. Am. Chem. Soc.*, **75**, 2042 (1953).
[40] M. Ballester and D. Pérez-Blanco, *J. Org. Chem.*, **23**, 652 (1958).
[41] D. S. Breslow and C. R. Hauser, *J. Am. Chem. Soc.*, **61**, 786, 793 (1939).
[42] P. Kalnin, *Helv. Chim. Acta*, **11**, 977 (1928).
[43] R. E. Buckles and K. G. Bremer, *J. Am. Chem. Soc.*, **75**, 1487 (1953).

aldol condensation. It does not appear to be of the ordinary aldol type, however, because formaldehyde would not be expected to form a very stable carbanion, and the reaction is much less effectively catalyzed by sodium, lithium, or barium hydroxide (a Cannizzaro reaction takes place instead). Breslow pointed out that a polar chain reaction is probably involved. The initiation mechanism is not known, but the following chain propagation steps are quite plausible:[44]

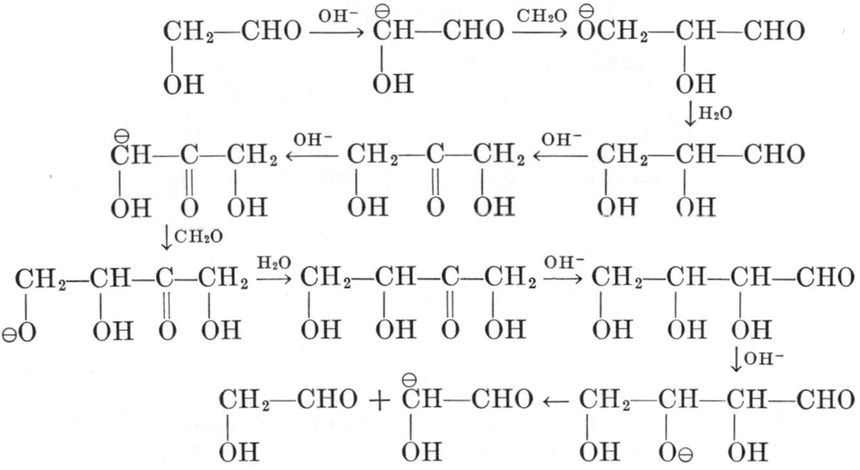

$$CH_2\!-\!CHO \xrightarrow{OH^-} \overset{\ominus}{C}H\!-\!CHO \xrightarrow{CH_2O} \overset{\ominus}{O}CH_2\!-\!CH\!-\!CHO$$

$$\overset{\ominus}{C}H\!-\!C\!-\!CH_2 \xleftarrow{OH^-} CH_2\!-\!C\!-\!CH_2 \xleftarrow{OH^-} CH_2\!-\!CH\!-\!CHO$$

$$CH_2\!-\!CH\!-\!C\!-\!CH_2 \xrightarrow{H_2O} CH_2\!-\!CH\!-\!C\!-\!CH_2 \xrightarrow{OH^-} CH_2\!-\!CH\!-\!CH\!-\!CHO$$

$$CH_2\!-\!CHO + \overset{\ominus}{C}H\!-\!CHO \leftarrow CH_2\!-\!CH\!-\!CH\!-\!CHO$$

11-3e. The Mannich Reaction. In the Mannich reaction an active hydrogen compound, formaldehyde, and ammonia (or a primary or secondary amine) react as shown below.

$$-\!\overset{|}{\underset{|}{C}}\!-\!H + CH_2O + HNR_2 \rightarrow -\!\overset{|}{\underset{|}{C}}\!-\!CH_2NR_2 + H_2O$$

Alexander and Underhill found the Mannich reaction involving ethylmalonic acid and dimethylamine to be a third-order reaction, first-order in each reactant.[45] They obtained good evidence that the reaction involves an initial addition of dimethylamine to formaldehyde to give dimethylaminomethanol. It therefore seems reasonable that the rate-controlling step of the reaction is of the following type:[46]

$$(CH_3)_2\overset{\oplus}{N}\!=\!CH_2 + \overset{\ominus}{C}(CO_2H)_2 \rightarrow (CH_3)_2N\!-\!CH_2\!-\!\underset{\underset{C_2H_5}{|}}{C}(CO_2H)_2 \tag{11-7}$$

The reaction was found to have a maximum rate at a pH of about 3.8.

[44] R. Breslow, *Tetrahedron Letters*, **21**, 22 (1959).
[45] E. R. Alexander and E. J. Underhill, *J. Am. Chem. Soc.*, **71**, 4014 (1949).
[46] S. V. Lieberman and E. C. Wagner, *J. Org. Chem.*, **14**, 1001 (1949).

This is not surprising, since at equilibrium the concentration of the intermediate $C_2H_5\overset{\ominus}{C}(CO_2H)_2$ must be directly proportional to that of the monoanion of ethylmalonic acid. This anion will be changed largely to the acid in strongly acidic solutions and to the dianion in basic solutions, and it will have a maximum concentration at an intermediate pH. Alexander and Underhill suggested a somewhat different mechanism. It is possible that Mannich reactions on other compounds or under other conditions proceed by other mechanisms, such as

$$R\text{—}COCH_3 + CH_2O \rightleftharpoons R\text{—}COCH_2CH_2OH$$
$$R\text{—}COCH_2CH_2OH \rightleftharpoons R\text{—}COCH{=}CH_2 + H_2O$$
$$R\text{—}COCH{=}CH_2 + Me_2NH \rightleftharpoons R\text{—}CO\overset{\ominus}{C}HCH_2\overset{\oplus}{N}HMe_2$$
$$R\text{—}CO\overset{\ominus}{C}HCH_2\overset{\oplus}{N}HMe_2 \rightleftharpoons R\text{—}COCH_2CH_2NMe_2$$

Mechanisms involving the nucleophilic displacement of hydroxide ions from carbon under mild conditions seem quite improbable, however, in view of the stability of acetals toward bases and many other relevant observations.

There are a few reports that are anomalous in view of the mechanisms described for the Mannich reaction.[47]

11-3f. *The Benzoin Condensation.* The benzoin condensation has the over-all appearance of a simple addition of an active hydrogen compound to an aldehyde. However, it would not be expected that the hydrogen atom attached to the carbonyl group would be active enough to be removed easily in aqueous solution. This fact—together with the fact that the reaction is not catalyzed by hydroxide ion, or by bases in general, but is specifically catalyzed by cyanide—makes a mechanism of the type outlined by Lapworth[48] plausible. In this mechanism the addition of the cyanide ion to the carbonyl group places the attached hydrogen atom in the alpha position of a nitrile and hence makes it active.

$$C_6H_5\text{—}CHO + CN^- \rightleftharpoons C_6H_5\text{—}\underset{\underset{O}{|}\ominus}{CH}\text{—}CN \rightleftharpoons C_6H_5\text{—}\overset{\ominus}{\underset{\underset{OH}{|}}{C}}\text{—}CN$$

$$C_6H_5\text{—}\overset{\ominus}{\underset{\underset{OH}{|}}{C}}\text{—}CN + C_6H_5\text{—}CHO \rightleftharpoons C_6H_5\text{—}\underset{\underset{O}{\ominus|}}{CH}\text{—}\underset{\underset{OH}{|}}{\overset{\overset{CN}{|}}{C}}\text{—}C_6H_5$$

$$C_6H_5\text{—}\underset{\underset{O}{\ominus|}}{CH}\text{—}\underset{\underset{OH}{|}}{\overset{\overset{CN}{|}}{C}}\text{—}C_6H_5 \rightleftharpoons C_6H_5\text{—}\underset{\underset{OH}{|}}{CH}\text{—}\underset{\underset{O}{||}}{C}\text{—}C_6H_5 + CN^-$$

[47] K. Bodendorf and G. Koralewski, *Arch. Pharm.*, **271**, 101 (1933); G. F. Grillot and R. I. Bashford, Jr., *J. Am. Chem. Soc.*, **73**, 5598 (1951).

[48] A. Lapworth, *J. Chem. Soc.*, **83**, 995 (1903); **85**, 1206 (1904).

This mechanism, with either the last or the next to the last step rate-controlling and all previous steps rapid and reversible, agrees with the observation that the reaction is first-order in cyanide ion and second-order in benzaldehyde[49] but does not agree with Wiberg's report that the deuterium exchange of benzaldehyde that accompanies the benzoin condensation run in a deuteriated solvent occurs at a rate comparable to that of the condensation reaction.[50] A more thorough investigation was promised.[50]

11-3g. *Acid-catalyzed Aldol Condensations.* The mechanism of acid-catalyzed aldol condensations has been studied extensively by Noyce and coworkers.[51] The reaction of benzaldehyde with acetophenone in acetic acid containing various amounts of water and of sulfuric acid as a catalyst was found to be first-order in each of the organic reactants, with the observed second-order rate constants being proportional to the Hammett acidity function h_0. These observations are consistent with the following reaction mechanism, with either the dehydration or the carbon-carbon bond-forming step rate-controlling:

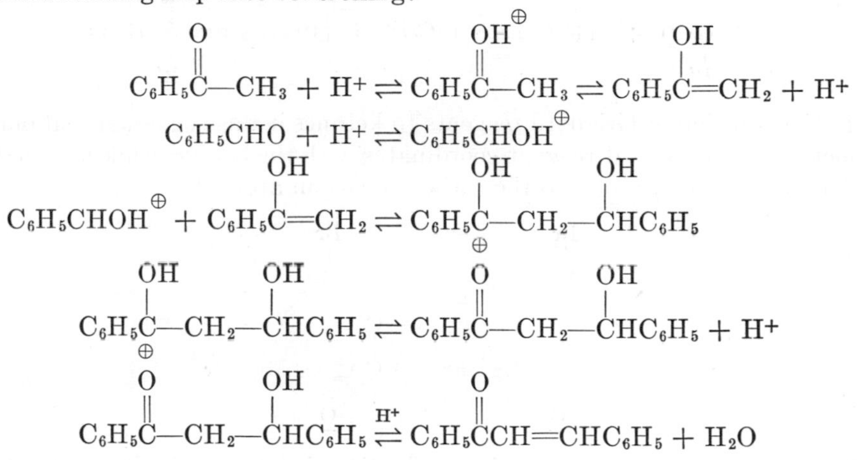

$$\underset{\substack{\| \\ O}}{C_6H_5C}-CH_3 + H^+ \rightleftharpoons \underset{\substack{\| \\ \overset{\oplus}{O}H}}{C_6H_5C}-CH_3 \rightleftharpoons \underset{\substack{| \\ OH}}{C_6H_5C}=CH_2 + H^+$$

$$C_6H_5CHO + H^+ \rightleftharpoons C_6H_5\overset{\oplus}{C}HOH$$

$$C_6H_5\overset{\oplus}{C}HOH + \underset{\substack{| \\ OH}}{C_6H_5C}=CH_2 \rightleftharpoons \underset{\substack{| \\ OH}}{C_6H_5C}-\underset{\oplus}{CH_2}-\underset{\substack{| \\ OH}}{CH}C_6H_5$$

$$\underset{\substack{| \\ \overset{OH}{\underset{\oplus}{}}}}{C_6H_5C}-CH_2-\underset{\substack{| \\ OH}}{CH}C_6H_5 \rightleftharpoons \underset{\substack{\| \\ O}}{C_6H_5C}-CH_2-\underset{\substack{| \\ OH}}{CH}C_6H_5 + H^+$$

$$\underset{\substack{\| \\ O}}{C_6H_5C}-CH_2-\underset{\substack{| \\ OH}}{CH}C_6H_5 \overset{H^+}{\rightleftharpoons} \underset{\substack{\| \\ O}}{C_6H_5C}CH=CHC_6H_5 + H_2O$$

Studies on the intermediate β-hydroxy ketone in this and related reactions show that the reaction mechanisms are rather more complicated than that written above. In acetic acid solution most of the β-hydroxy ketone may be transformed to its acetate before the α,β-unsaturated ketone is formed. The fraction of the β-hydroxy ketone that reverts to reactants varies with structure and reaction conditions, so that in some cases the dehydration step is rate-controlling and in other cases the carbon-carbon bond-forming step is.

[49] G. Bredig and E. Stern, *Z. Elektrochem.*, **10**, 582 (1904); E. Stern, *Z. physik. Chem.*, **50**, 513 (1905).

[50] K. B. Wiberg, *J. Am. Chem. Soc.*, **76**, 5371 (1954).

[51] D. S. Noyce et al., *J. Am. Chem. Soc.*, **77**, 1397, 1402 (1955); **80**, 4033, 4324, 5539 (1958); **81**, 618, 620, 624 (1959).

11-3h. *Addition of Organometallic Compounds to Aldehydes and Ketones.* Because of the rapidity of the additions of many organometallic compounds to aldehydes and ketones, kinetic studies are often not feasible. For this and other reasons there have been few thorough studies of the reaction mechanisms involved. One of the most important characteristics of organometallic compounds is the polar nature of the carbon-metal bond. Although this does give the alkyl group considerable carbanion character, the bond also has definite covalent character. This fact and the low ion-solvating power of the typical reaction media make the reactions much more complicated than a simple combination of carbanions with a carbonyl group. The action of the metal as a Lewis acid by coordinating with the carbonyl oxygen atom is usually an important factor in the reaction.

Swain and Kent obtained evidence that in the reaction of organolithium compounds with ketones a reversibly formed complex decomposes to give the final product.[52]

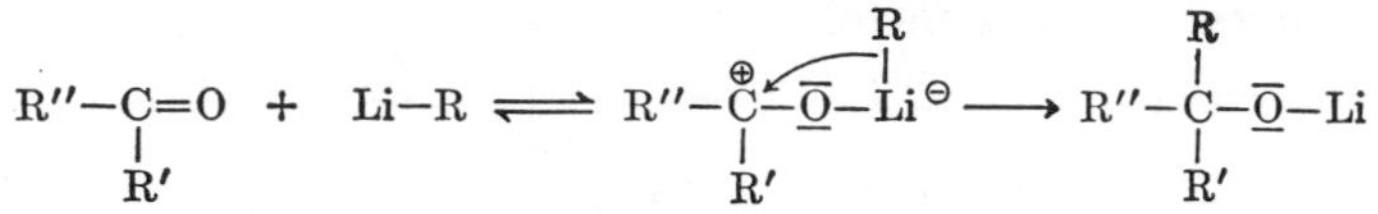

$$ R''-\underset{\underset{R'}{|}}{C}=O \;+\; Li-R \;\rightleftharpoons\; R''-\overset{\oplus}{\underset{\underset{R'}{|}}{C}}\!\!-\overset{R}{\overset{|}{\underline{O}}}-Li^{\ominus} \;\longrightarrow\; R''-\underset{\underset{R'}{|}}{\overset{\overset{R}{|}}{C}}-\underline{O}-Li $$

In the addition of Grignard reagents to ketones it was suggested that one molecule of Grignard reagent coordinates with the ketone, while a second donates the alkyl group to the carbonyl carbon atom.[53]

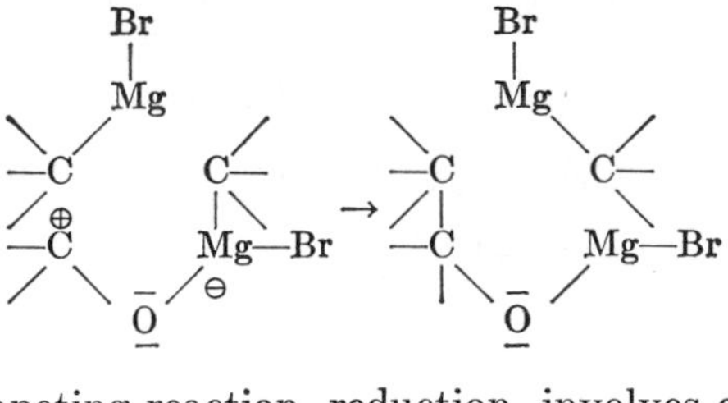

The common competing reaction, reduction, involves only one molecule of Grignard reagent.

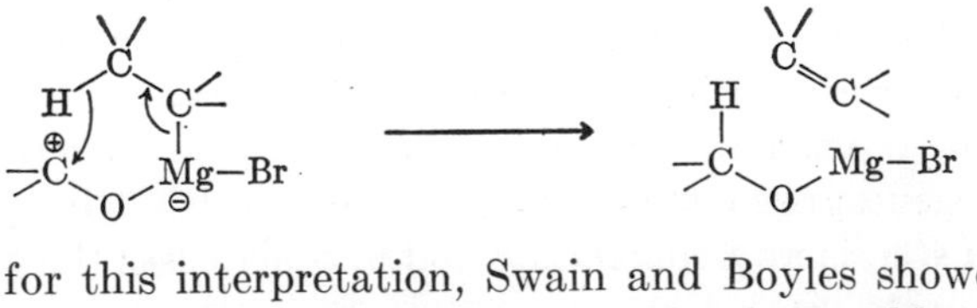

As evidence for this interpretation, Swain and Boyles showed that the addition of magnesium bromide (which should aid the addition reaction, since it should coordinate with a carbonyl group better than a Grignard

---

[52] C. G. Swain and L. Kent, *J. Am. Chem. Soc.*, **72**, 518 (1950).
[53] C. G. Swain and H. B. Boyles, *J. Am. Chem. Soc.*, **73**, 870 (1951).

reagent) approximately doubles the yield of addition product in the reaction of the *n*-propyl Grignard reagent with diisopropyl ketone.[53]

11-4. Hydride-transfer Reactions. 11-4*a. The Cannizzaro Reaction.* The transformation of an aldehyde into an equimolar mixture of the corresponding alcohol and acid (or its salt) is usually known as the Cannizzaro reaction. The reaction may be carried out in several ways: enzymatically; by the use of metal catalysts, such as nickel and platinum; in a two-phase system (an organic phase and a strongly alkaline aqueous phase); and in homogeneous alkaline solution. We shall discuss mainly the homogeneous reaction.

When benzaldehyde or formaldehyde undergoes the Cannizzaro reaction in heavy water solution, the alcohol produced contains no carbon-bound deuterium, showing that the hydrogen is transferred directly from one molecule of aldehyde to the other.[54] The Cannizzaro reactions of furfural,[55] formaldehyde,[56] and sodium benzaldehyde-*m*-sulfonate,[57] under certain conditions, have been found to be fourth-order reactions, second-order in aldehyde and second-order in alkali. From data of this sort, Hammett suggested the following reaction mechanism,[58]

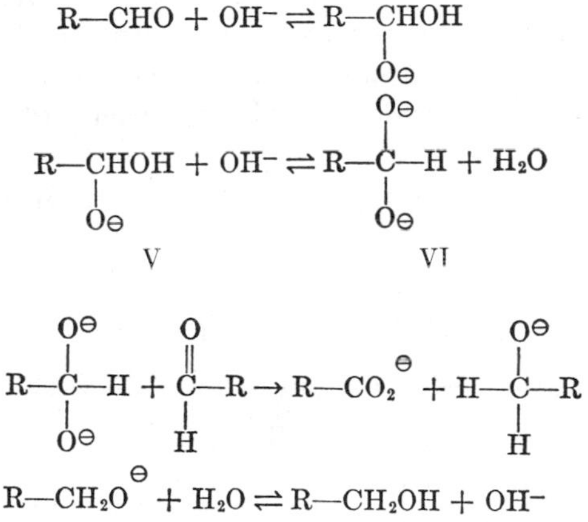

$$R-CHO + OH^- \rightleftharpoons R-\underset{\underset{O\ominus}{|}}{C}HOH$$

$$R-\underset{\underset{O\ominus}{|}}{C}HOH + OH^- \rightleftharpoons R-\underset{\underset{O\ominus}{|}}{\overset{\overset{O\ominus}{|}}{C}}-H + H_2O$$

$$\underset{V}{\qquad} \qquad \underset{VI}{\qquad}$$

$$R-\underset{\underset{O\ominus}{|}}{\overset{\overset{O\ominus}{|}}{C}}-H + \underset{\underset{H}{|}}{\overset{\overset{O}{\|}}{C}}-R \rightarrow R-CO_2{}^{\ominus} + H-\underset{\underset{H}{|}}{\overset{\overset{O\ominus}{|}}{C}}-R$$

$$R-CH_2O{}^{\ominus} + H_2O \rightleftharpoons R-CH_2OH + OH^-$$

in which the rate-controlling step consists of the donation of a hydride ion from the reactive intermediate VI to the carbonyl carbon atom of an aldehyde molecule. It appears that the *α*-hydrogens on all alkoxide ions may be easily removed as hydride ions, due to the tendency of the negatively charged oxygen atom to form a double bond with carbon. The intermediate VI has a particularly strong tendency to donate hydride

---

[54] H. Fredenhagen and K. F. Bonhoeffer, *Z. physik. Chem.*, **181A**, 379 (1938).

[55] K. H. Geib, *Z. physik. Chem.*, **169A**, 41 (1934).

[56] H. v. Euler and T. Lövgren, *Z. anorg. Chem.*, **147**, 123 (1925).

[57] E. A. Shilov and G. I. Kudryavtsev, *Doklady Akad. Nauk S.S.S.R.*, **63**, 681 (1948).

[58] Hammett, *op. cit.*, pp. 350–352.

ions because of the presence of two negatively charged oxygen atoms on the same carbon (the product being a resonance-stabilized carboxylate anion). The intermediate V should also be capable of acting as a hydride-ion donor, and since it should be only very weakly acidic, it must be present in a much higher concentration than VI at equilibrium. It must be the greater reactivity of VI, then, that makes it the principal reacting species in those cases for which the Cannizzaro reaction follows fourth-order kinetics. The Cannizzaro reaction of benzaldehyde and several derivatives has been found to be third-order (second-order in aldehyde and first-order in base);[59] and under some conditions, at least, this is true for formaldehyde and furfural.[60] Evidently it is the intermediate V that is the principal hydride-ion donor in these cases.

It is reported that electron-withdrawing substituents increase the reactivity of aldehydes in the Cannizzaro reaction.[59] This would be expected, since both molecules of the original aldehyde have acquired a certain amount of alkoxide-ion character in the transition state.

The Cannizzaro reaction may occur intramolecularly with dialdehydes[61] and α-keto aldehydes. The base-catalyzed rearrangement of phenyl-glyoxal to the salt of mandelic acid has been studied carefully. Alexander found the reaction to be second-order, first-order in hydroxide ion and first-order in phenylglyoxal, and suggested the following mechanism by analogy with the intermolecular Cannizzaro reaction:[62]

$$C_6H_5COCHO + OH^- \underset{}{\overset{\text{fast}}{\rightleftharpoons}} C_6H_5COCHOH$$
$$\underset{O^\ominus}{\big|}$$

Further support for this mechanism may be found in the following: when the reaction is run in deuterium oxide solution, the product contains no carbon-bound deuterium;[63] it is the hydrogen atom and not the phenyl

[59] E. L. Molt, *Rec. trav. chim.*, **56**, 233 (1937); A. Eitel and G. Lock, *Monatsh.*, **72**, 392 (1939); E. Tommila, *Ann. Acad. Sci. Fennicae*, **59A**(8) (1942).

[60] I. I. Paul, *Zhur. Obshchei Khim.*, **11**, 1121 (1941); V. Pajunen, *Suomen Kemistilehti*, **21B**, 21 (1948); *Ann. Acad. Sci. Fennicae*, ser. A, II, **37**, 7 (1950); A. Eitel, *Monatsh.*, **74**, 124 (1942).

[61] J. Thiele and O. Günther, *Ann.*, **347**, 106 (1906); E. M. Fry, E. J. Wilson, Jr., and C. S. Hudson, *J. Am. Chem. Soc.*, **64**, 872 (1942).

[62] E. R. Alexander, *J. Am. Chem. Soc.*, **69**, 289 (1947).

[63] W. von E. Doering, T. I. Taylor, and E. F. Schoenewaldt, *J. Am. Chem. Soc.*, **70**, 455 (1948).

group that migrates;[63,64] and 2,4,6-trimethylphenylglyoxal undergoes the rearrangement readily.[65]

Franzen showed that β-diethylaminoethyl mercaptan and related compounds are effective catalysts for the transformation of phenylglyoxal to mandelic acid derivatives.[66] The activity of these catalysts must be related to the great ability of SH groups to add to carbonyl groups and to the basicity of the amino group, as suggested in the following mechanism:

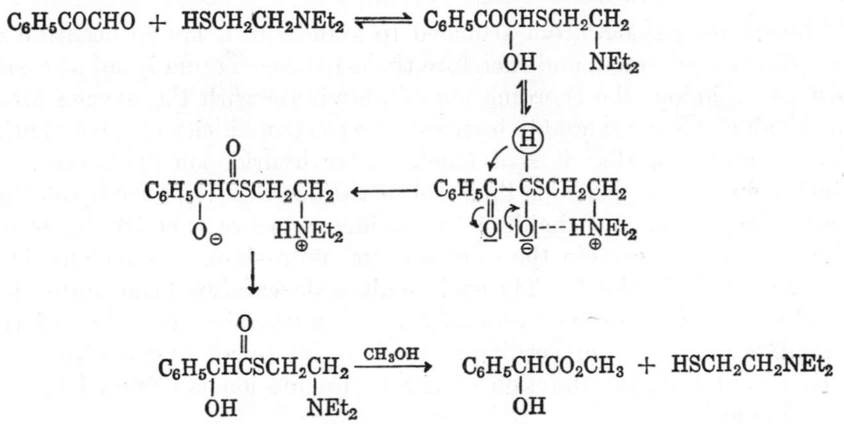

$$C_6H_5COCHO + HSCH_2CH_2NEt_2 \rightleftharpoons C_6H_5COCHSCH_2CH_2$$

It was suggested that the reaction mechanism may resemble that by which the enzyme glyoxalase acts.

11-4b. *The Tishchenko Reaction.* The hydride-ion transfer characteristic of the Cannizzaro reaction has a more than formal resemblance to the transfer of a hydride ion from a hydrocarbon to a carbonium ion, described in Sec. 9-3a, since in so far as the polar resonance structure contributes to the total structure of an aldehyde, the aldehyde may be thought of as a carbonium ion. Although the aldehyde is not nearly so electrophilic as a carbonium ion, the intermediate V (or VI) must be a vastly better hydride-ion donor than is a hydrocarbon. Viewed in these terms, the Tishchenko reaction (the aluminum alkoxide–catalyzed transformation of an aldehyde into the ester of the corresponding acid and alcohol) is explained readily by a mechanism analogous to that first written for the Meerwein-Ponndorf-Oppenauer equilibration by Woodward, Wendler, and Brutschy.[67]

[64] O. K. Neville, *J. Am. Chem. Soc.*, **70**, 3499 (1948).

[65] A. R. Gray and R. C. Fuson, *J. Am. Chem. Soc.*, **56**, 739 (1934).

[66] V. Franzen, *Chem. Ber.*, **88**, 1361 (1955); **89**, 1020 (1956); **90**, 623, 2036 (1957).

[67] R. B. Woodward, N. L Wendler, and F. J. Brutschy, *J. Am. Chem. Soc.*, **67**, 1425 (1945).

$$C_6H_5-CHO + Al(OCH_2C_6H_5)_3 \rightleftharpoons C_6H_5-\underset{\underset{OCH_2C_6H_5}{|}}{CH}OAl(OCH_2C_6H_5)_2$$

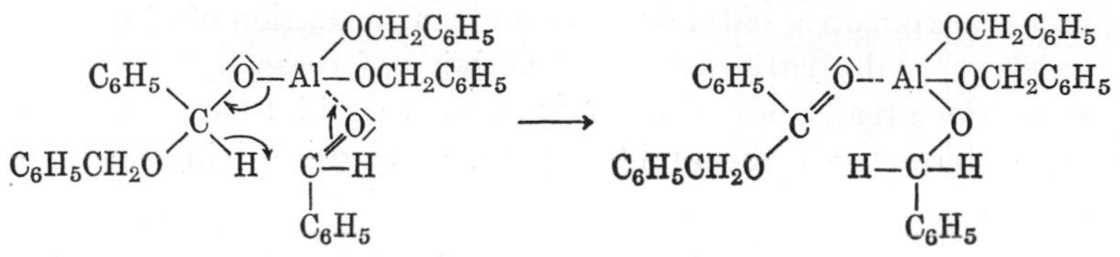

Although the oxygen atom attached to aluminum is not so negative as in a dissociated anion, and therefore the α-hydrogen atom is not so easily lost as an anion, the coordination of aluminum with the oxygen atom of the aldehyde considerably increases the electrophilicity of the carbonyl carbon atom, so that it is a much better hydride-ion abstractor. A similar mechanism explains Pfeil's observations that in aqueous solution barium, calcium, and thallium hydroxide are more effective bases for the Cannizzaro reaction than are sodium, potassium, and tetramethylammonium hydroxide.[68]   The sodium alkoxide–catalyzed transformation of aldehydes into esters[69] probably has a mechanism like that of the Cannizzaro reaction, proceeding through an intermediate like V (but not VI), except that the function of the hydroxide ion is assumed by the alkoxide ion.

Kharasch and Snyder showed that in the case of the Cannizzaro reaction carried out with an organic phase and an aqueous phase, most of the reaction takes place in the organic phase if the aqueous phase is sufficiently strongly alkaline.[70]   They further showed that this reaction in the organic phase is powerfully catalyzed by the benzyl alcohol initially produced by the slower reaction in the aqueous layer. Evidently the sodium or potassium benzylate formed from the alcohol and alkali dissolves in the organic phase to catalyze a Tishchenko reaction to yield benzyl benzoate, which may be hydrolyzed by the alkali. Indeed Lachman isolated benzyl benzoate from the reaction by avoiding overheating and excess alkali.[71]   There is evidence that in at least some cases, however, the heterogeneous Cannizzaro reaction may proceed by a free-radical mechanism.[72]

Just as the reaction has been found to be catalyzed by the strongly basic alkali-metal alkoxides and the amphoteric aluminum alkoxides,

---

[68] E. Pfeil, *Chem. Ber.*, **84**, 229 (1951).

[69] L. Claisen, *Ber.*, **20**, 649 (1887); O. Kamm and W. F. Kamm, "Organic Syntheses," 2d ed., vol. I, p. 104, John Wiley & Sons, Inc., New York, 1941.

[70] M. S. Kharasch and R. H. Snyder, *J. Org. Chem.*, **14**, 819 (1949).

[71] A. Lachman, *J. Am. Chem. Soc.*, **45**, 2356 (1923).

[72] M. S. Kharasch and M. Foy, *J. Am. Chem. Soc.*, **57**, 1510 (1935).

it may also be catalyzed by strong acids. Nemtsov and Trenke studied such a case in the aqueous sulfuric acid–catalyzed transformation of formaldehyde to methanol, formic acid, and products of their further reaction.[73] The mechanism is probably of the form

$$H_2C(OH)_2 \overset{H_2SO_4}{\rightleftharpoons} \overset{\oplus}{HOCH_2OH_2} \rightleftharpoons \overset{\oplus}{HOCH_2}$$

$$\overset{\oplus}{HOCH_2} + H_2C(OH)_2 \rightarrow HOCH_3 + \overset{\oplus}{HC(OH)_2}$$

$$\overset{\oplus}{HC(OH)_2} \rightleftharpoons \overset{\oplus}{H} + HCO_2H$$

11-4c. The Meerwein-Ponndorf-Oppenauer Equilibrium. The Meerwein-Ponndorf reduction and the Oppenauer oxidation are two applications of the fact that aluminum and other alkoxides catalyze the establishment of equilibrium between **primary** or **secondary** alcohols and the corresponding aldehydes or ketones.

$$RCHOHR' + R''COR''' \overset{Al(OR)_3}{\rightleftharpoons} RCOR' + R''CHOHR'''$$

The aluminum alkoxide–catalyzed reaction probably proceeds by a mechanism of the type written for the Tishchenko reaction and suggested by Woodward, Wendler, and Brutschy.[67] A mechanism for the alkali-metal alkoxide–catalyzed reaction can easily be written by analogy with the Cannizzaro reaction.

The Lobry de Bruyn–Alberda van Ekenstein rearrangement, in which glucose, fructose, and mannose are interconverted by use of an alkaline catalyst,[74] is an interesting reaction whose over-all result is the intramolecular oxidation-reduction characteristic of the Meerwein-Ponndorf-Oppenauer equilibrium.

$$\begin{array}{ccc} CHO & CH_2OH & CHO \\ | & | & | \\ H-C-OH \rightleftharpoons & C=O \rightleftharpoons & HO-C-H \\ | & | & | \\ R & R & R \end{array}$$

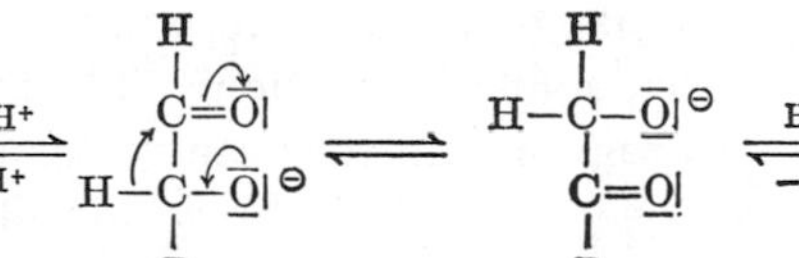

Two reasonable mechanisms may be suggested for this reaction. One is an internal oxidation-reduction involving a hydride-ion shift

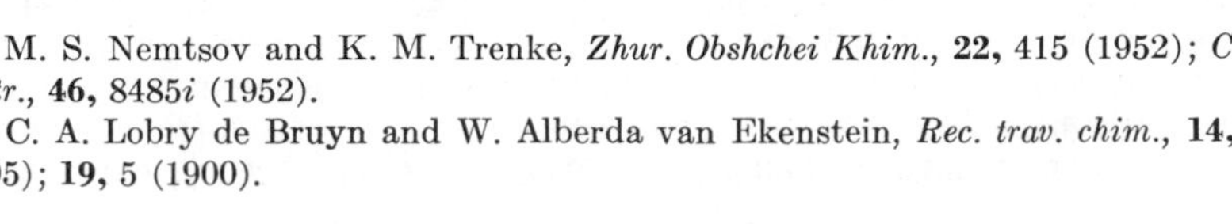

[73] M. S. Nemtsov and K. M. Trenke, *Zhur. Obshchei Khim.*, **22**, 415 (1952); *Chem. Abstr.*, **46**, 8485*i* (1952).
[74] C. A. Lobry de Bruyn and W. Alberda van Ekenstein, *Rec. trav. chim.*, **14**, 203 (1895); **19**, 5 (1900).

and the other is a base-catalyzed enolization to give an enediol that may either revert to starting aldose or be converted to the corresponding ketose or the epimeric aldose.

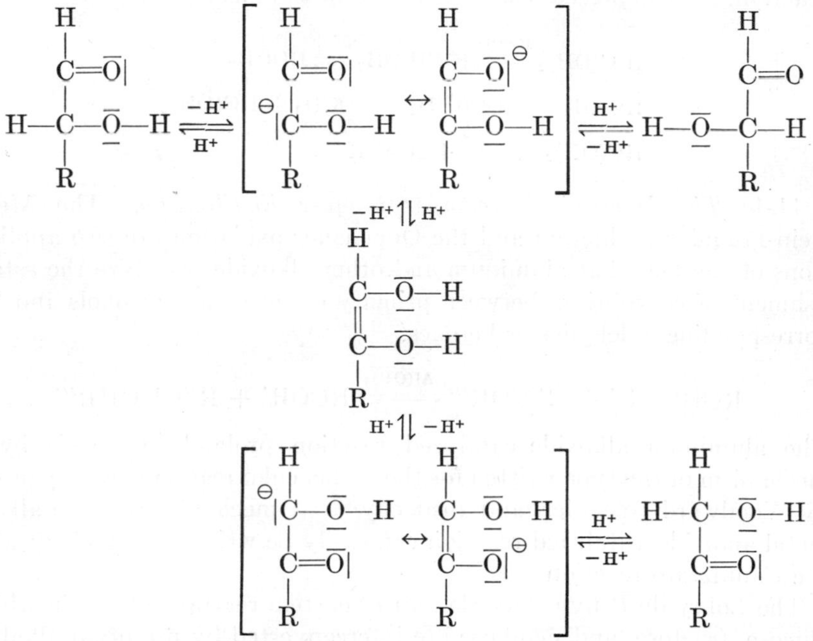

In various investigations of the reaction mechanism it has been reported that in heavy water glucose yields fructose containing no carbon-bound deuterium;[75] in heavy water glucose yields fructose containing considerable amounts of carbon-bound deuterium;[76,77] mannose is formed from glucose only via fructose;[76] and mannose is formed directly from glucose.[77,78] It seems probable that the reaction can proceed by each of the two mechanisms at comparable rates and that the relative extent to which the two mechanisms operate depends on the nature of the basic catalysts, the temperature, the structure of the reactant, etc.

11-4d. *The Leuckart Reaction.* The reductive amination of an aldehyde or a ketone by the use of formamide, ammonium formate, or an amine formate is called the Leuckart reaction. The fact that the reaction may be carried out with ammonium formate under conditions at which formamide will not work suggests that the action of the latter may be

[75] H. Fredenhagen and K. F. Bonhoeffer, *Z. physik. Chem.*, **181A**, 392 (1938); K. Goto, *J. Chem. Soc. Japan*, **63**, 217 (1942); *Chem. Abstr.*, **41**, 3062g (1947).

[76] Y. J. Topper and D. Stetten, Jr., *J. Biol. Chem.*, **189**, 191 (1951).

[77] J. C. Sowden and R. Schaffer, *J. Am. Chem. Soc.*, **74**, 499, 505 (1952).

[78] C. H. Bamford and J. R. Collins, *Proc. Roy. Soc. (London)*, **228A**, 100 (1955).

due to its hydrolysis to ammonium formate.[79]   The reaction probably involves a hydride transfer from a formate ion to an immonium ion.[80]

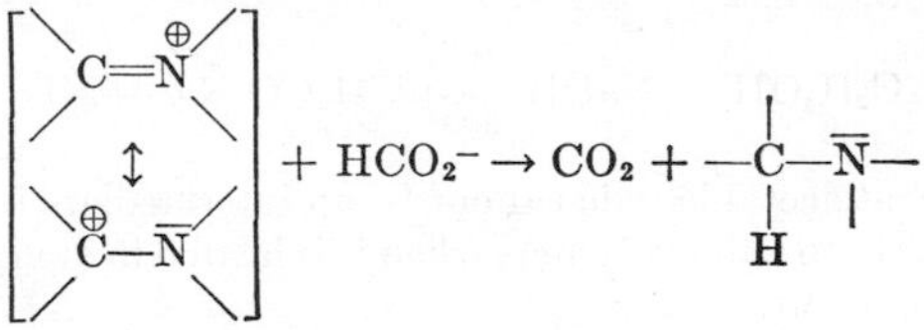

$$
\left[
\begin{array}{c}
\diagup \\[-2pt]
\mathrm{C}\!=\!\overset{\oplus}{\mathrm{N}} \\
\diagup \qquad \diagdown \\
\updownarrow \\
\diagdown \qquad \diagup \\
\overset{\oplus}{\mathrm{C}}\!-\!\overline{\mathrm{N}} \\
\diagup \qquad \diagdown
\end{array}
\right]
+ \mathrm{HCO_2^-} \rightarrow \mathrm{CO_2} + \overset{\displaystyle |}{\underset{\underset{\mathrm{H}}{\displaystyle |}}{\mathrm{C}}}\!-\!\overset{|}{\underset{|}{\overline{\mathrm{N}}}}\!-
$$

This mechanism is strongly supported by the observation that in the closely related formic acid reduction of an enamine the hydrogen atom that becomes attached to the carbon adjacent to nitrogen came from the formate anion.[81]   The formate ion would not be expected to be a very strong hydride donor, since it is considerably stabilized by resonance and since its oxygen atoms bear only a fractional negative charge.   However, the Leuckart reaction is aided by the fact that the hydride donor is present in high concentration, and the immonium ion is a better hydride-ion acceptor than is an aldehyde or ketone.

A number of other hydride-ion transfer reactions have been discussed in the review by Deno, Peterson, and Saines.[82]

PROBLEMS

1. Write a detailed mechanism for the acid-catalyzed transformation of acetaldehyde to its diethyl acetal in ethanol solution.   Tell which step would probably be rate-controlling and why.

2. Write reasonable mechanisms for the following reactions, which do not occur in the absence of cyanide:

(a) $2\mathrm{C_6H_5COCHO} \xrightarrow[\mathrm{KHCO_3}]{\mathrm{KCN}} \mathrm{C_6H_5COCH_2OH} + \mathrm{C_6H_5COCO_2K}$

(b) $\mathrm{Cl_3CCHO} + \mathrm{H_2O} \xrightarrow{\mathrm{KCN}} \mathrm{Cl_2CHCO_2K}$

3. Suggest a mechanism for the reaction

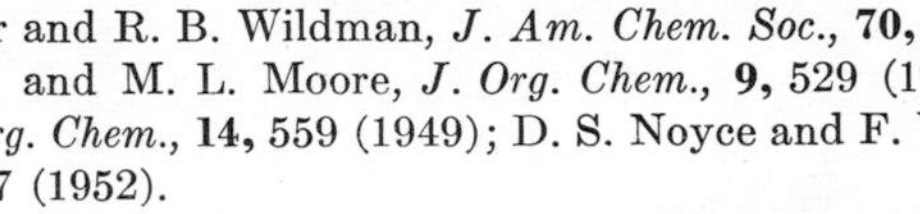

$$\text{(cyclic structure)}\!-\!\mathrm{CH_2OH} \xrightarrow[\mathrm{H_2O}]{\mathrm{H^+}} \mathrm{CH_3COCH_2CH_2CO_2H}$$

4. Two plausible mechanisms may be written for the acid-catalyzed dehydration of $\mathrm{ArCHOHCH_2COCH_3}$'s to $\mathrm{ArCH}\!=\!\mathrm{CHCOCH_3}$'s.   One mechanism appears to operate for the p-methoxyphenyl compound and the other for the phenyl and p-nitrophenyl

[79] E. R. Alexander and R. B. Wildman, *J. Am. Chem. Soc.*, **70**, 1187 (1948).

[80] F. S. Crossley and M. L. Moore, *J. Org. Chem.*, **9**, 529 (1944); E. Staple and E. C. Wagner, *J. Org. Chem.*, **14**, 559 (1949); D. S. Noyce and F. W. Bachelor, *J. Am. Chem. Soc.*, **74**, 4577 (1952).

[81] N. J. Leonard and R. R. Sauers, *J. Am. Chem. Soc.*, **79**, 6210 (1957).

[82] N. C. Deno, H. J. Peterson, and G. S. Saines, *Chem. Revs.*, **60**, 7 (1960).

compounds.   What are these two mechanisms and for which compound(s) does each operate?   Give reasons for your conclusions.

5. Suggest a reasonable mechanism for the following reaction, which proceeds in high yield:

$$C_2H_5OH + NaOH \xrightarrow[300°]{H_2O} CH_3CO_2Na + 2H_2$$

It has been argued that acetaldehyde cannot be an intermediate in this reaction, since it is transformed largely to aldol polymers when it is heated in aqueous alkali.   Discuss the merits of this argument.

6. Although *o*-fluorobenzaldehyde undergoes the Cannizzaro reaction upon treatment with potassium hydroxide, 2,6-difluorobenzaldehyde gives potassium formate and *m*-difluorobenzene.   Write a reasonable mechanism for each reaction and suggest a reason for the difference in behavior.

7. Explain, using a reaction mechanism, how *N*-ethylthiazolium ions act as catalysts in the benzoin condensation.

Chapter 12

FORMATION AND REACTIONS OF ESTERS
AND RELATED COMPOUNDS

Since esterification and hydrolysis may be part of the same equilibrium, information about the mechanism of both may be obtained from a study of one. The majority of the studies that have been made in this case have been on ester hydrolysis. A considerable number of mechanisms may be visualized for ester hydrolysis. Of these, only a few are common, and some are hardly more than mechanistic curiosities. These mechanisms may be subdivided on three different bases depending upon (1) whether the cleavage takes place between the ethereal oxygen and the acyl carbon atom or the alkyl carbon atom, (2) whether or not the reaction is acid- or base-catalyzed, and (3) whether certain parts of the reaction occur by a concerted or a stepwise mechanism. Since most of the ester hydrolysis reactions carried out synthetically probably occur by acyl-oxygen fission mechanisms, these will be discussed first.

12-1. Acyl-Oxygen Fission Mechanisms for Esterification and Hydrolysis. *12-1a. Evidence for Acyl-Oxygen Fission.* Holmberg showed that either the acid or alkaline hydrolysis of acetoxysuccinic acid gave malic acid with retention of configuration at the asymmetric carbon atom.[1]

$$HO_2C-CH_2\overset{*}{C}H-CO_2H \xrightarrow[H_2O]{H^+ \text{ or } OH^-} HO_2C-CH_2\overset{*}{C}H-CO_2H$$
$$\underset{OAc}{|} \qquad\qquad\qquad \underset{OH}{|}$$

Polanyi and Szabo[2] reported that the alkaline hydrolysis of n-amyl acetate in water containing excess O^{18} yielded n-amyl alcohol that contained no excess O^{18}. Polanyi and Szabo's experimental evidence, but not their conclusion, was criticized by Kursanov and Kudryavtsev,[3] who reported that ethyl propionate containing excess O^{18} in the ethereal oxygen underwent alkaline hydrolysis to give propionic acid without and ethyl alcohol with excess O^{18}.

$$C_2H_5\overset{\overset{\displaystyle O}{\|}}{C}O^{18}-C_2H_5 \xrightarrow[H_2O]{OH^-} C_2H_5CO_2^- + C_2H_5O^{18}-H$$

[1] B. Holmberg, *Ber.*, **45**, 2997 (1912).

[2] M. Polanyi and A. L. Szabo, *Trans. Faraday Soc.*, **30**, 508 (1934).

[3] D. N. Kursanov and R. V. Kudryavtsev, *J. Gen. Chem. U.S.S.R.* (*Eng. Transl.*), **26**, 1183 (1956).

Similar evidence for acyl-oxygen cleavage was reported by Datta, Day, and Ingold for the acid hydrolysis of methyl hydrogen succinate,[4] by Long and Friedman for both the acid and basic hydrolysis of γ-butyrolactone,[5] and by Roberts and Urey for the esterification of benzoic acid by methanol.[6] Prévost[7] and Ingold and Ingold[8] showed that both the alkaline and acid hydrolysis of crotyl acetate and of α-methylallyl acetate yielded the corresponding alcohol. If the reaction had proceeded by a carbonium-ion mechanism, both reactants should have yielded the same product, since the carbonium ions formed should be identical.

In addition to the rather direct evidence described above, it might be mentioned that ester hydrolysis often occurs at a rate unreasonably fast for an alternate mechanism, such as an S_N2 reaction at the alkyl carbon atom. For these reasons it seems likely that most common esterification and hydrolysis reactions involve acyl-oxygen fission.

12-1b. *Evidence for the Carbonyl-addition Mechanism of Esterification and Hydrolysis.* The alkaline hydrolysis of most esters is a kinetically second-order reaction, first-order in ester and first-order in hydroxide ion. To agree with this and the fact of acyl-oxygen fission, two reaction mechanisms have been suggested. One of these mechanisms consists of the addition of hydroxide ion to the carbonyl group to give the intermediate I, which may revert to reactants, decompose to alkoxide ion and carboxylic acid, or be protonated to give II, the monoester of an ortho acid. This mechanism is shown below for the case of an ester labeled in the carbonyl group with O^{18} to distinguish two of the oxygen atoms of II (for reasons that will become apparent shortly).

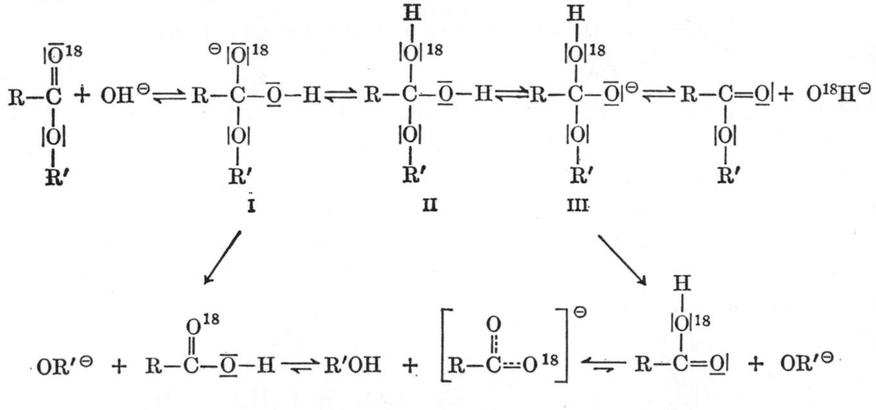

[4] S. C. Datta, J. N. E. Day, and C. K. Ingold, *J. Chem. Soc.*, 838 (1939).

[5] F. A. Long and L. Friedman, *J. Am. Chem. Soc.*, **72**, 3692 (1950).

[6] I. Roberts and H. C. Urey, *J. Am. Chem. Soc.*, **60**, 2391 (1938).

[7] C. Prévost, *Ann. chim. (Paris)*, **10**, 147 (1928).

[8] E. H. Ingold and C. K. Ingold, *J. Chem. Soc.*, 756 (1932).

In the alternate mechanism the hydroxide ion simply displaces the alkoxide ion in a concerted one-step reaction.

$$OH^- + \underset{\underset{R}{|}}{\overset{\overset{O}{\|}}{C}}-OR \rightleftharpoons \left[HO\text{---}\underset{\underset{R}{|}}{\overset{\overset{O}{\|}}{C}}\text{---}OR \right]^{\ominus} \rightleftharpoons HO-\underset{\underset{R}{|}}{\overset{\overset{O}{\|}}{C}} + OR'^{\ominus}$$

$$\rightarrow RCO_2^- + R'OH$$

IV

Bender solved the problem of deciding between these two mechanisms very neatly.[9] The mechanisms differ in their description of the ester–hydroxide-ion adduct.   In the concerted mechanism this (IV) is a transition state, lying at the top of a potential-energy curve, whereas in the carbonyl-addition mechanism it (I) is an intermediate, lying at an energy minimum.   Since the intermediate I is an alkoxide anion, the equilibrium between it and its conjugate acid, II, should be very rapidly established. Upon reionization the intermediate II would be as likely to form III as I. Since hydroxide ions and alkoxide ions are of comparable basicity, the intermediates I and III would be expected to lose these two ions at comparable rates (although the greater size of the alkoxide anions might somewhat favor their loss).   Thus the carbonyl-addition mechanism leads to the prediction that, if the protonation of 1 is comparable in rate to its decomposition (or faster), the hydrolysis of ester labeled in the carbonyl group with O^{18} should be accompanied by O^{18} loss from the unhydrolyzed ester (due to re-formation of the ester from III).   Bender found this decrease in O^{18} content to occur during the alkaline hydrolysis of three different esters: ethyl, isopropyl, and t-butyl benzoate.   This establishes the carbonyl-addition mechanism, since the concerted mechanism gives no explanation for how this loss of O^{18} could occur.[9]

For acid hydrolysis the carbonyl-addition mechanism has the form outlined below.

$$R-\overset{\overset{O}{\|}}{C}-OR' \underset{}{\overset{H^+}{\rightleftharpoons}} \left[R-\overset{\overset{OH}{\|}}{C}\text{---}OR' \right]^{\oplus} \overset{H_2O}{\rightleftharpoons} R-\underset{\underset{OH_2}{\underset{\oplus}{|}}}{\overset{\overset{OH}{|}}{C}}-OR' \rightleftharpoons R-\underset{\underset{OH}{|}}{\overset{\overset{OH}{|}}{C}}-OR'$$

$$H^+ \updownarrow$$

$$RCO_2H \rightleftharpoons RC(OH)_2^+ + R'OH \rightleftharpoons R-\underset{\underset{OH}{|}}{\overset{\overset{OH}{|}}{C}}-OH\overset{\oplus}{R}'$$

[9] M. L. Bender, *J. Am. Chem. Soc.*, **73**, 1626 (1951).

The concerted mechanism may be written

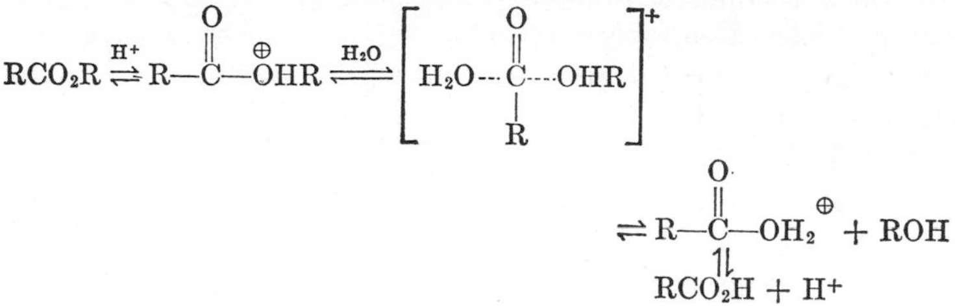

$$RCO_2R \overset{H^+}{\rightleftharpoons} R\!-\!\overset{\overset{O}{\|}}{C}\!-\!\overset{\oplus}{O}HR \overset{H_2O}{\rightleftharpoons} \left[H_2O \cdots \overset{\overset{O}{\|}}{\underset{R}{C}} \cdots OHR \right]^+$$

$$\rightleftharpoons R\!-\!\overset{\overset{O\cdot}{\|}}{C}\!-\!\overset{\oplus}{O}H_2 + ROH$$

$$\updownarrow$$

$$RCO_2H + H^+$$

Here too the carbonyl-addition mechanism may be distinguished by the loss of O^{18} during reaction, and again Bender observed such a loss (in the acid hydrolysis of ethyl benzoate).[9]

Swarts described evidence for the formation of an addition compound between ethyl trifluoroacetate and sodium ethoxide.[10] Bender confirmed and extended this observation by infrared measurements and

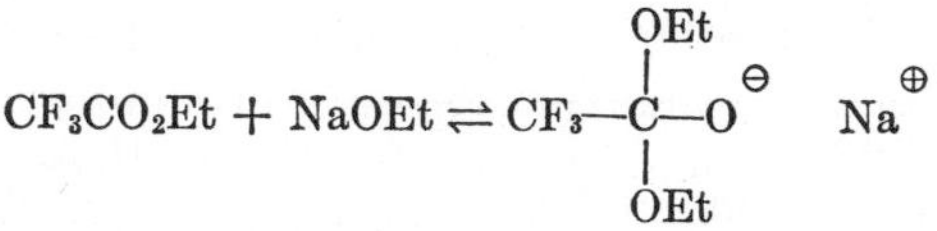

$$CF_3CO_2Et + NaOEt \rightleftharpoons CF_3\!-\!\overset{\overset{OEt}{|}}{\underset{OEt}{C}}\!-\!O^{\ominus} \quad Na^{\oplus}$$

pointed out its support for the carbonyl-addition mechanism.[11]

12-1c. *The Oxocarbonium-ion Mechanism for Esterification and Hydrolysis.* Another method by which esters can be formed and hydrolyzed with acyl-oxygen fission involves the intermediate formation of an oxocarbonium ion (RCO^+), also called an acylium ion. Treffers and Hammett obtained good evidence for the existence of such an ion by observing that 2,4,6-trimethylbenzoic acid dissolves in sulfuric acid to form four particles.[12]

$$RCO_2H + 2H_2SO_4 \rightarrow RCO^+ + H_3O^+ + 2HSO_4^-$$

This interpretation is further supported by the following facts: the methyl ester of the acid may be hydrolyzed easily by dissolving it in concentrated sulfuric acid and pouring the solution into ice water;[12] the methyl ester may be easily prepared by pouring a sulfuric acid solution of the acid into cold methanol.[13] These facts are significant because analogous treatment fails to hydrolyze methyl benzoate or to esterify benzoic acid. Furthermore, with dilute acid catalysts, 2,4,6-trimethylbenzoic acid is tremendously more difficult to esterify than benzoic acid, and its esters are much more difficult to hydrolyze than benzoates. The behavior of the trimethylbenzoic acid does not appear to be due to steric factors

[10] F. Swarts, *Bull. soc. chim. Belges*, **35**, 414 (1926).
[11] M. L. Bender, *J. Am. Chem. Soc.*, **75**, 5986 (1953).
[12] H. P. Treffers and L. P. Hammett, *J. Am. Chem. Soc.*, **59**, 1708 (1937).
[13] M. S. Newman, *J. Am. Chem. Soc.*, **63**, 2431 (1941).

alone, since it is not shared by 2,4,6-tribromobenzoic acid, nor is it entirely electronic, since 2,6-dimethylbenzoic acid yields about three and one-half particles in sulfuric acid, whereas 2,4-dimethylbenzoic acid yields only two. As Newman pointed out, both factors appear to be important. It is probable that the carbonyl oxygen atom of benzoic acid is more basic than the hydroxylic oxygen because the donation of a proton to the former gives a resonance-stabilized cation (V). However, for the maximum contribution of structures of the type shown on the right the oxygen atoms must lie in the same plane as the ring. Two *o*-methyl groups must

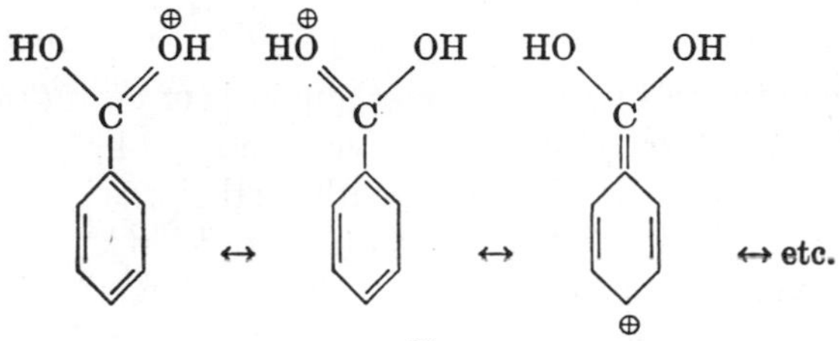

V

make this very difficult. On the other hand, *o*- and *p*-methyl groups may stabilize the oxocarbonium ion (VI) by the contribution of resonance structures of the type shown.

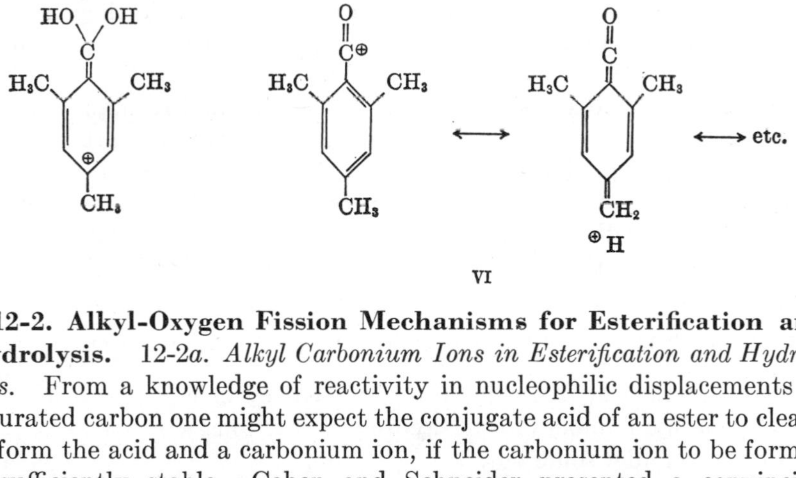

VI

12-2. Alkyl-Oxygen Fission Mechanisms for Esterification and Hydrolysis. 12-2*a*. *Alkyl Carbonium Ions in Esterification and Hydrolysis.* From a knowledge of reactivity in nucleophilic displacements at saturated carbon one might expect the conjugate acid of an ester to cleave to form the acid and a carbonium ion, if the carbonium ion to be formed is sufficiently stable. Cohen and Schneider presented a convincing argument that a mechanism of this sort occurs in the acid hydrolysis of *t*-butyl benzoate and probably most other esters of tertiary alcohols.[14]

$$R{-}\overset{\overset{\displaystyle O}{\|}}{C}{-}OR' \overset{H^+}{\rightleftharpoons} R{-}\overset{\overset{\displaystyle \oplus\, OH}{\|}}{C}{-}OR' \rightleftharpoons RCO_2H + R'^{\oplus}$$

$$R'^{\oplus} + H_2O \rightleftharpoons R'OH_2^{\oplus} \rightleftharpoons R'OH + H^{\oplus}$$

[14] S. G. Cohen and A. Schneider, *J. Am. Chem. Soc.*, **63**, 3382 (1941).

Although the chemical nature of the products of ester *hydrolysis* does not tell which bond to oxygen has been broken, this is not true of *alcoholysis*, where acyl-oxygen fission gives ester interchange and alkyl-oxygen fission produces ether and acid.

$$R{-}\overset{\displaystyle O}{\overset{\|}{C}}{\vdots}O{-}R' + R''OH \rightarrow R{-}\overset{\displaystyle O}{\overset{\|}{C}}{-}O{-}R'' + R'OH$$

$$R{-}\overset{\displaystyle O}{\overset{\|}{C}}{-}O{\vdots}R' + R''OH \rightarrow R{-}\overset{\displaystyle O}{\overset{\|}{C}}{-}O{-}H + R'OR''$$

It is therefore noteworthy that the methanolysis of both *t*-butyl benzoate and *t*-butyl 2,4,6-trimethylbenzoate gives methyl *t*-butyl ether and the acid in good yield in acid solution.   With methanolic sodium methoxide *t*-butyl benzoate gives methyl benzoate and *t*-butyl alcohol by ester interchange involving nucleophilic attack of the methoxide ion on the carbonyl group.   Under the same conditions *t*-butyl 2,4,6-trimethylbenzoate gives no reaction, presumably because of hindrance at the carbonyl group.   Cohen and Schneider also showed that although *t*-butyl 2,4,6-trimethylbenzoate is easily cleaved by hydrochloric acid, the methyl ester is unaffected.   From their data it appears that the esters of most tertiary alcohols may be cleaved to carbonium ions if they are first protonated.   In order for this cleavage to occur without acid catalysis, the carbonium ion to be formed should be more stable or the anion to be formed should be less basic, or both.   Hammond and Rudesill found an example of such an uncatalyzed ionization of an ester to a carbonium ion and a carboxylate anion in the solvolysis of triphenylmethyl benzoate.[15] They found that the ethanolysis of this ester yields benzoic acid and ethyl triphenylmethyl ether and the addition of sodium ethoxide had only a "salt effect" on the reaction rate.   Here the reaction involves only a simple ionization of the ester.

$$C_6H_5CO_2C(C_6H_5)_3 \rightarrow C_6H_5CO_2^- + (C_6H_5)_3C^{\oplus} \xrightarrow{\text{EtOH}} (C_6H_5)_3COEt + H^+$$

Stereochemical evidence for the intermediacy of carbonium ions has been found in a number of ester-hydrolysis reactions, some of which were acid-catalyzed and some not.[16]

12-2*b*. *The S_N2 Mechanism for Ester Hydrolysis.*   The alkaline hydrolysis of esters ordinarily occurs by the acyl-oxygen fission mechanism, because the carbonyl carbon atom is so much more susceptible to nucleophilic attack.   There are several methods by which nucleophilic attack at the alkyl carbon atom may be observed.

[15] G. S. Hammond and J. T. Rudesill, *J. Am. Chem. Soc.*, **72**, 2769 (1950).
[16] A. G. Davies and J. Kenyon, *Quart. Revs. (London)*, **9**, 203 (1955).

The first method was not designed for the purpose described but was discovered accidentally. Hughes, Ingold, and coworkers pointed out that the hydrolysis of malolactone in neutral solution gives inversion at the asymmetric carbon atom,[17] and Olson and Miller made the same observation for β-butyrolactone.[18] These facts seem rational in consideration of the strain present in the four-membered rings of these compounds. According to the I-strain concept of Brown,[19] the sp^2 carbon atom of a carbonyl group, having optimum bond angles of 120°, would be strained by about 30°. The hydration of this carbonyl group (the first step of ordinary ester hydrolysis) would make this carbon atom sp^3 and thus reduce the strain to about 19.5°. This 10.5° decrease in bond strain at this atom is said to be responsible for the observed high reactivity of carbonyl groups in four-membered rings.[19] The S_N2 attack of a solvent molecule on the alkyl carbon atom, however, is accompanied by the release of 10 to 30° of strain at each of four different atoms. Thus it might have been expected that a β-lactone would show much more increased reactivity (compared with an open-chain ester) at its alkyl carbon atom than at the acyl carbon atom. That this larger increase would have been sufficient to cause the mechanism of hydrolysis in neutral solution to involve an S_N2 attack by water molecules could not have been predicted, however. In fact, the nucleophilic attack of hydroxide ions on β-lactones does occur at the acyl carbon atom, the hydrolysis reactions in basic solution giving retention of configuration.[18] The explanation for this is the same, no doubt, as for the fact that hydroxide ions are about 10^4 times as reactive as water molecules in most substitution reactions at a saturated carbon atom but more than 10^{10} times as reactive toward the carbonyl group of an ester.[20] Attack on the carbonyl group involves transformation of the reactants to a relatively unstable intermediate, and therefore the transition state lies considerably nearer the products than the reactants (cf. Fig. 5-3). In the S_N2 reaction, however, the reactants are transformed directly to products of comparable stability, and the transition state occurs earlier in the reaction. In attack on the carbonyl group, where the new bond has been formed to a larger extent in the transition state, the greater ability of hydroxide ion to form new bonds should be reflected by a larger difference in reaction rate.

[17] W. A. Cowdrey, E. D. Hughes, C. K. Ingold, S. Masterman, and A. D. Scott, *J. Chem. Soc.*, 1264 (1937).

[18] A. R. Olson and R. J. Miller, *J. Am. Chem. Soc.*, **60**, 2687 (1938); cf. F. A. Long and M. Purchase, *J. Am. Chem. Soc.*, **72**, 3267 (1950); P. D. Bartlett and P. N. Rylander, *J. Am. Chem. Soc.*, **73**, 4273 (1951).

[19] H. C. Brown, R. S. Fletcher, and R. B. Johannesen, *J. Am. Chem. Soc.*, **73**, 212 (1951).

[20] C. G. Swain and C. B. Scott, *J. Am. Chem. Soc.*, **75**, 141 (1953).

Long and coworkers obtained evidence from correlations of rates with h_0[21] and the concentration of catalyzing acid[22] that β-propiolactone and β-butyrolactone hydrolyze by the acyl–carbonium-ion mechanism,[21] whereas γ-butyrolactone hydrolyzes by the carbonyl-addition mechanism.[22,23] The acyl–carbonium-ion mechanism, like nucleophilic attack on alkyl carbon, has the advantage of simultaneously releasing strain all over the four-membered ring. In the case of β-butyrolactone this mechanism is supported by the observation that acid hydrolysis proceeds with retention of configuration.[17]

Other methods of observing the S_N2 reactivity at the alkyl carbon atom of an ester include using an alkoxide anion whose nucleophilic attack at acyl carbon will merely yield the starting materials. Bunnett, Robison, and Pennington isolated dimethyl ether in good yield from the reaction of anhydrous methanolic sodium methoxide with methyl benzoate and several of its derivatives.[24] Hammett and Pfluger studied the reaction of trimethylamine, a reagent not capable of displacing the strongly basic alkoxide ions to a great extent at equilibrium because it offers no way of later supplying a proton to them.[25] Tani and Fudo studied the similar reagent sodium thiomethylate.[26] With methyl 2,4,6-trimethylbenzoate the S_N2 hydrolysis should be greatly encouraged with respect to hydrolysis via acyl-oxygen fission because of the large amount of steric hindrance to nucleophilic attack on the carbonyl group. The alkaline hydrolysis of this ester is indeed slow,[27] but Bender and Dewey have shown by use of H_2O^{18} that the reaction involves largely acyl-oxygen fission.[28]

12-3. Catalysis of Ester Hydrolysis.[29] The mechanism of hydrolysis of esters by the action of strong acids and bases has already been considered in Secs. 12-1 and 12-2. These are the reagents most often used in the laboratory, but there are a number of other catalysts for ester hydrolysis whose mode of action has been studied. The investigation of such catalysts should be of great importance in understanding the manner in

[21] F. A. Long and M. Purchase, *J. Am. Chem. Soc.*, **72**, 3267 (1950).

[22] F. A. Long, F. B. Dunkle, and W. F. McDevit, *J. Phys. Chem.*, **55**, 829 (1951).

[23] This work is reviewed by A. A. Frost and R. G. Pearson, "Kinetics and Mechanism," 2d ed., chap. 12C, John Wiley & Sons, Inc., New York, 1961.

[24] J. F. Bunnett, M. M. Robison, and F. C. Pennington, *J. Am. Chem. Soc.*, **72**, 2378 (1950).

[25] L. P. Hammett and H. L. Pfluger, *J. Am. Chem. Soc.*, **55**, 4079 (1933).

[26] H. Tani and K. Fudo, *Chem. Abstr.*, **45**, 10198c (1951).

[27] Harvey L. Goering, T. Rubin, and M. S. Newman, *J. Am. Chem. Soc.*, **76**, 787 (1954).

[28] M. L. Bender and R. S. Dewey, *J. Am. Chem. Soc.*, **78**, 317 (1956).

[29] This subject has been reviewed comprehensively by M. Bender, *Chem. Revs.*, **60**, 53 (1960).

which enzymes act, and it may lead to the development of new procedures for the synthesis and determination of the structure of organic compounds.

Ester hydrolysis may be catalyzed by Lewis acids as well as Brønsted acids. Bender and Turnquest showed that the hydrolyses of glycine methyl ester and phenylalanine methyl ester in a glycine buffer at pH 7.3 is powerfully catalyzed by cupric ions.[30] This catalysis is believed due to reaction by a mechanism of the type

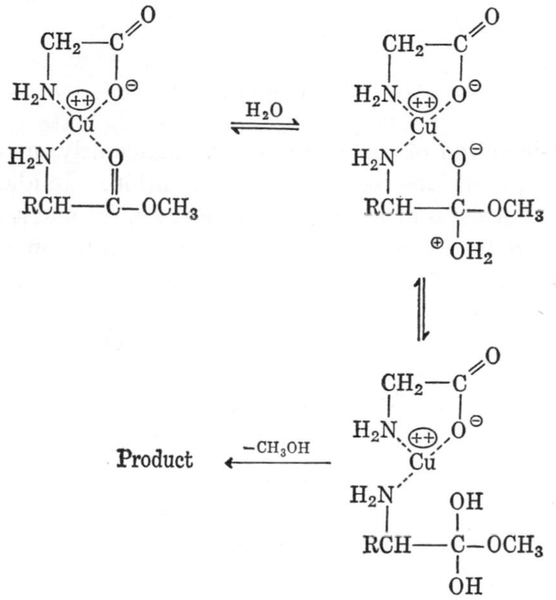

It is necessary to postulate the orthoester-type intermediate, in which the two hydroxy groups are equivalent, in order to explain the observation that the hydrolysis of the phenylalanine ester is accompanied by oxygen exchange of the reactant.

Nucleophilic catalysis of ester hydrolysis may be observed when a nucleophilic reagent attacks an ester (at an observable rate) to give an intermediate that hydrolyzes more rapidly than the original ester. Such a mode of action provides a reasonable explanation for the acetate-ion catalysis of the hydrolysis of 2,4-dinitrophenyl acetate,[31] p-nitrophenyl acetate,[32,33] phenyl acetate,[32] and other esters of phenols. Although there is no direct evidence for the formation of an intermediate acid anhydride in such reactions, Bender and Neveu found that the benzoic

[30] M. L. Bender and B. W. Turnquest, *J. Am. Chem. Soc.*, **79**, 1889 (1957).
[31] M. L. Bender and B. W. Turnquest, *J. Am. Chem. Soc.*, **79**, 1652, 1656 (1957).
[32] M. L. Bender and M. C. Neveu, *J. Am. Chem. Soc.*, **80**, 5388 (1958).
[33] T. C. Bruice and R. Lapinski, *J. Am. Chem. Soc.*, **80**, 2265 (1958).

acid produced in the hydrolysis of 2,4-dinitrophenyl benzoate catalyzed by acetate ions containing excess O^{18} had in it a considerable fraction of the O^{18} from the catalyst.[32]

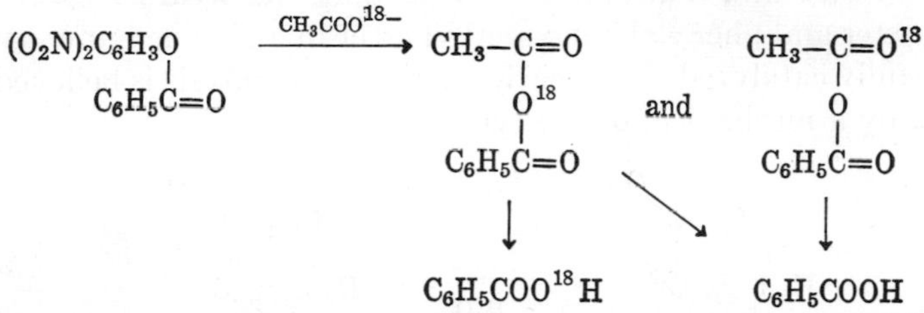

$$(O_2N)_2C_6H_3O \overset{CH_3COO^{18-}}{\longrightarrow} \quad CH_3-C=O \qquad CH_3-C=O^{18}$$

$$C_6H_5C=O \qquad\qquad O^{18} \qquad and \qquad O$$

$$C_6H_5C=O \qquad\qquad C_6H_5C=O$$

$$\downarrow \qquad\qquad\searrow \qquad\qquad \downarrow$$

$$C_6H_5COO^{18}H \qquad\qquad C_6H_5COOH$$

The catalytic action of tertiary amines, particularly imidazole and its derivatives, has received considerable attention. Imidazole strongly catalyzes the hydrolysis of p-nitrophenyl acetate,[31,34] and acetylimidazole has been detected spectrophotometrically as a reaction intermediate[31] and even isolated from the reaction mixture.[35]

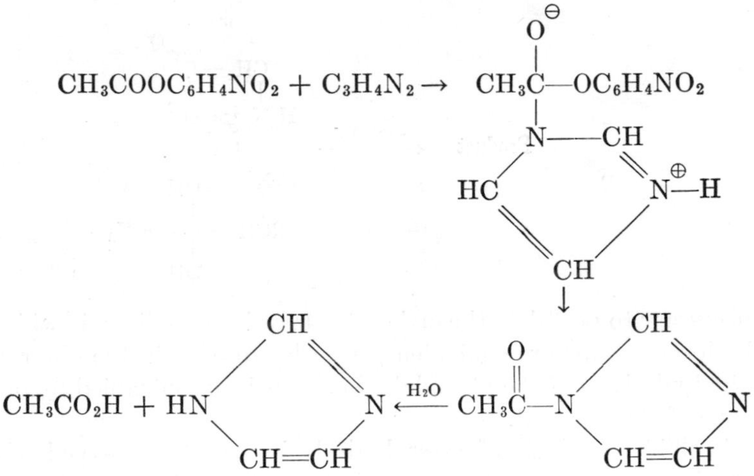

$$CH_3COOC_6H_4NO_2 + C_3H_4N_2 \rightarrow \quad \overset{O^{\ominus}}{CH_3C}-OC_6H_4NO_2$$

$$N\!-\!-CH$$

$$HC \qquad\qquad N\!-\!H^{\oplus}$$

$$CH$$

$$\downarrow$$

$$CH_3CO_2H + HN \qquad\qquad N \overset{H_2O}{\longleftarrow} CH_3\overset{O}{C}-N \qquad\qquad N$$

$$CH\!=\!CH \qquad\qquad CH\!=\!CH$$

Although imidazole is an effective catalyst in the hydrolysis of esters of relatively acidic alcohols, such as phenols and thiols, it is almost without effect on the esters of ordinary alcohols, perhaps because of the greater difficulty of expelling the more basic alkoxide ions.

As mentioned previously (Sec. 6-4), the presence of two groups in the same molecule can tremendously encourage their interaction. Zimmering, Westhead, and Morawetz showed that the hydrolysis of mono-p-nitrophenyl glutarate, for example, was about one million times as fast

---

[34] T. C. Bruice and G. L. Schmir, *J. Am. Chem. Soc.*, **79**, 1663 (1957).

[35] W. Langenbeck and R. Mahrwald, *Chem. Ber.*, **90**, 2423 (1957).

at pH 5.0 as that of p-nitrophenyl pivalate.[36]   A copolymer of 9 per cent p-nitrophenyl methacrylate and 91 per cent acrylic acid, which, like the pivalate, has a trisubstituted α-carbon atom, hydrolyzed even faster than the glutarate.   Apparently there is neighboring-group attack by the carboxylate-anion groups to give a highly reactive anhydride intermediate.

$$\begin{array}{ccccccccc}
 & & & & \mathrm{CH_3} & & & & \\
 & & & & | & & & & \\
-\mathrm{CH_2} & - & \mathrm{CH} & - & \mathrm{CH_2} - \mathrm{C} - \mathrm{CH_2} & - & \mathrm{CH} & - & \mathrm{CH_2} - \\
 & & | & & | & & | & & \\
 & & \mathrm{CO_2^-} & \longrightarrow & \mathrm{C}{=}\mathrm{O} & & \mathrm{CO_2^-} & & \\
 & & & & | & & & & \\
 & & & & \mathrm{O} & & & & \\
 & & & & | & & & & \\
 & & & & \mathrm{C_6H_4NO_2} & & & &
\end{array}$$

Neighboring-group facilitation of ester hydrolysis is not limited to esters of highly acidic alcohols.   The hydrolysis of monomethyl phthalate appears to involve the intermediate formation of phthalic anhydride.[37]

In the hydrolysis of certain esters bifunctional catalysis by both an acid and a nucleophilic reagent appears to be important.[38]

12-4. Reactivity in the Hydrolysis and Formation of Esters. Our discussion of reactivity in esterification and hydrolysis will be limited almost entirely to reactions proceeding by the common mechanism, the carbonyl-addition mechanism with acyl-oxygen fission.   Ideas on reactivity by carbonium-ion mechanisms and the S_N2 mechanism may be obtained from the discussion in Chap. 7.

The rate of basic hydrolysis by the carbonyl-addition mechanism is equal to the rate of attack of hydroxide ions on the ester multiplied by the fraction of these attacks that lead to hydrolysis.   It appears that this fraction is usually large enough that the over-all reaction rate depends mainly on the rate of the initial attack by hydroxide ions.   There appear to be three ways in which the structure of the reacting ester can influence the rate of attack by hydroxide ions.   These three ways relate to (1) the electrophilic character of the carbonyl carbon atom, (2) steric hindrance, and (3) stabilization of the carbonyl group by conjugation.

Obviously, electron-withdrawing groups in either the acyl or alkyl part of the ester would be expected to facilitate attack by hydroxide ions.   Accordingly, all the ρ's listed by Jaffé for the alkaline hydrolysis of esters of meta- and para-substituted benzoic acids are positive (ranging between 1.82 and 2.85).[39]

[36] P. E. Zimmering, E. W. Westhead, Jr., and H. Morawetz, *Biochim. et Biophys. Acta*, **25**, 376 (1957).
[37] M. L. Bender, F. Chloupek, and M. C. Neveu, *J. Am. Chem. Soc.*, **80**, 5384 (1958).
[38] Cf. H. Morawetz and I. Oreskes, *J. Am. Chem. Soc.*, **80**, 2591 (1958).
[39] H. H. Jaffé, *Chem. Revs.*, **53**, 191 (1953).

As stated in Sec. 4-4, the hypothesis that steric and resonance effects (we refer here only to resonance involving the carbalkoxy group) are the same in corresponding basic and acidic ester hydrolyses has permitted the quantitative separation of polar and steric-plus-resonance factors in these reactions. The excellent internal agreement obtained in the treatment of ester hydrolysis data and the applicability of the resultant Taft polar substituent constants (Table 4-3) to a wide variety of organic reactions provides evidence that the claimed separation of factors has in large measure, at least, been achieved. It is therefore probable that the attacking nucleophilic reagent has formed a new bond to about the same extent in the transition state for acidic (VII) and basic (VIII) ester hydrolysis.

$$
\begin{array}{cc}
\mathrm{H} & \\
| & \\
\mathrm{O} +\delta & \mathrm{O} -\delta \\
\| & \| \\
\mathrm{R{-}C{-}OR'} & \mathrm{R{-}C{-}OR'} \\
| & | \\
+\delta\,\mathrm{OH_2} & -\delta\,\mathrm{OH} \\
\mathrm{VII} & \mathrm{VIII}
\end{array}
$$

Since the hydrogen atoms are relatively small and need not be oriented toward a bulky R or R′ group, transition states VII and VIII should be about equally crowded, and since the carbonyl group has about the same amount of double-bond character in each case, resonance effects should be about equal. The Taft equation for the acidic hydrolysis of esters may be written in the form

$$\log\,(k/k_0)_\mathrm{A} = \rho_\mathrm{A}{}^{*}\sigma^{*} + E_r + E_s \tag{12-1}$$

where k and k_0 are the rate constants for the hydrolysis of the given compound and of the standard compound, respectively, and $\rho_\mathrm{A}{}^{*}\sigma^{*}$ measures the polar effect, E_r the resonance effect, and E_s the steric effect. Except for esters in which there is conjugation with the ester's carbonyl group, the resonance effect is assumed to be negligible. Polar effects on acid-catalyzed ester hydrolysis are known to be quite small. The two ρ values listed by Jaffé are 0.106 and 0.144. Therefore polar factors are also neglected. Values of E_s thus calculated from rates of acid-catalyzed ester hydrolyses and esterifications (in which polar effects are also known to be small) are listed in Table 12-1. These values relate to the structure of the acyl part of the ester. A few E_s values for the alkyl part of esters are available.[40] The values of E_s listed in Table 12-1 are applicable only in the case of hydrolysis by the carbonyl-addition mechanism.

The relative insensitivity of acidic esterification and hydrolysis reac-

[40] R. W. Taft, Jr., *J. Am. Chem. Soc.*, **74**, 3120 (1952).

tions to polar effects is the result of a compensation of two opposing influences. Electron-donating groups assist the initial protonation of the carbalkoxy group, but they inhibit the subsequent nucleophilic attack by water. Stated alternately, the transition-state grouping has about the same electron-withdrawing power as the original carbalkoxy group. The loss in electron-withdrawing power brought about by the transformation of the carbon-oxygen double bond largely to a saturated entity has been compensated by the positive charge the grouping has acquired.

TABLE 12-1. STERIC SUBSTITUENT CONSTANTS, E_s, FOR R IN RCO_2R' AT 25°[a]

Substituent	E_s	Substituent	E_s
H	+1.24	$(CH_3)_2CHCH_2$	−0.93
CH_3	0.00	CF_3	−1.16
Cyclobutyl	−0.06	$(CH_3)_3C$	−1.54
C_2H_5	−0.07	$(C_6H_5)_2CH$	−1.76
CH_3OCH_2	−0.19	Br_2CH	−1.86
$ClCH_2$	−0.24	CBr_3	−2.43
$BrCH_2$	−0.27	$(C_2H_5)_3C$	−3.8
ICH_2	−0.37	$(CH_3)_3C$	
$n\text{-}C_3H_7$	−0.36	$\quad\quad\mid$	
$n\text{-}C_8H_{17}$	−0.33	$CH_3\!\!-\!\!C\!\!-$	
$C_6H_5CH_2$	−0.38	$\quad\quad\mid$	−4.0
$(CH_3)_2CH$	−0.47	CH_2	
Cyclopentyl	−0.51	$\quad\quad\mid$	
Cyclohexyl	−0.79	$(CH_3)_3C$	

[a] R. W. Taft, Jr., in M. S. Newman (ed.), "Steric Effects in Organic Chemistry," p. 598, John Wiley & Sons, Inc., New York, 1956.

12-5. Formation and Hydrolysis of Amides of Carboxylic Acids.
12-5a. Ammonolysis and Aminolysis of Esters. Betts and Hammett found that the ammonolysis of methyl phenylacetate in methanol is catalyzed by base and slowed by acid (even by ammonium chloride, which cannot produce any net transformation of ammonia to ammonium ions).[41] Similar observations, sometimes in the form of reports that the reaction was of higher than first-order in ammonia or amine, were made by other workers.[42–44] Recent investigations by Bunnett and Davis[45] and by Jencks and Carriuolo[46] have permitted these observations to be explained plausibly in mechanistic terms. The formation of $N\text{-}n$-butylformamide

[41] R. L. Betts and L. P. Hammett, *J. Am. Chem. Soc.,* **59,** 1568 (1937).

[42] R. Baltzly, I. M. Berger, and A. A. Rothstein, *J. Am. Chem. Soc.,* **72,** 4149 (1950).

[43] P. J. Hawkins and D. S. Tarbell, *J. Am. Chem. Soc.,* **75,** 2982 (1953).

[44] W. H. Watanabe and L. R. DeFonso, *J. Am. Chem. Soc.,* **78,** 4542 (1956).

[45] J. F. Bunnett and G. T. Davis, *J. Am. Chem. Soc.,* **82,** 665 (1960).

[46] W. P. Jencks and J. Carriuolo, *J. Am. Chem. Soc.,* **82,** 675 (1960).

from ethyl formate and n-butylamine in ethanol solution[45] was found to obey the following kinetic equation, in the absence of any other reagents:

$$v = (k[\text{BuNH}_2]^{3/2} + k'[\text{BuNH}_2]^2)[\text{HCO}_2\text{Et}] \qquad (12\text{-}2)$$

The first term in this equation appears to be due to a reaction between butylamine and ethyl formate catalyzed by ethoxide ions, and, indeed, the reaction is found to be strongly catalyzed by sodium ethoxide with the observed increase in reaction rate being directly proportional to the ethoxide-ion concentration. In an ethanol solution containing butylamine as the only added acid or base the ethoxide-ion concentration will be equal to the square root of the butylamine concentration.

$$n\text{-}\text{C}_4\text{H}_9\text{NH}_2 + \text{EtOH} \rightleftharpoons n\text{-}\text{C}_4\text{H}_9\text{NH}_3^+ + \text{EtO}^- \qquad (12\text{-}3)$$

$$K_{\text{BuNH}_2} = \frac{[\text{BuNH}_3^+][\text{EtO}^-]}{[\text{BuNH}_2]}$$

Since

$$[\text{BuNH}_3^+] = [\text{EtO}^-]$$

$$[\text{EtO}^-] = \sqrt{K_{\text{BuNH}_2}[\text{BuNH}_2]}$$

Such a term in the kinetic equation would result from the operation of the following mechanism:

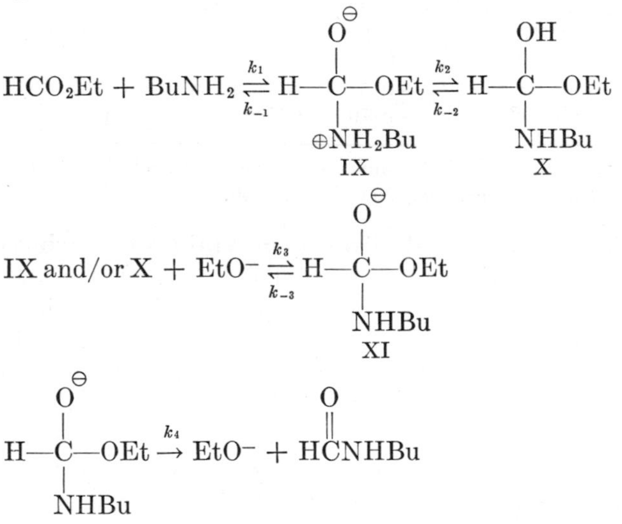

with either of the last two steps rate-controlling. The second term in kinetic Eq. (12-2), relating to that part of the reaction whose rate is proportional to the square of the amine concentration, can be explained in terms of mechanism (12-4) in two possible ways. The rate-controlling step may be the deprotonation of IX and/or X, and the second butylamine molecule may be responsible for some of this deprotonation.

Although the solvent should also be capable of deprotonating IX and/or X to give, in the kinetic equation, a term proportional merely to the first power of the amine concentration, it seems quite possible that such a term would be too small to detect.   Nevertheless, because of the absence of such a term from the kinetic equation, Bunnett and Davis favored an alternate mechanism in which the rate-controlling step is the acid-catalyzed removal of ethoxide from intermediate XI.   That part of the reaction whose rate is proportional to the square of the butylamine concentration is attributed to the action of the butylammonium ion on XI. Bunnett and Davis pointed out how changes in the structures of the reactants could change the reaction mechanism in ester aminolysis.

Jencks and Carriuolo studied a somewhat different case, the reaction of phenyl acetate with a variety of amines in aqueous solution.   They pointed out certain difficulties (which do not seem insuperable, however) in explaining their observations in terms of mechanism (12-4) and suggested an alternate reaction path, in which the second molecule of amine exerts its catalyzing action not by its influence on the fate of a reaction intermediate like IX, X, or XI, but by facilitating the original nucleophilic attack on the ester, as shown in the following transition state:

$$
\begin{array}{ccccc}
 & & & & -\delta \\
 & & \mathrm{H} & \mathrm{O} & \\
 +\delta & +\delta\;| & & \| & \\
\mathrm{RNH_2}\text{------}\mathrm{H}\text{----}\mathrm{N}\text{----}\mathrm{C}\text{---}\mathrm{OR} \\
 & & | & | & \\
 & & \mathrm{R} & \mathrm{R} &
\end{array}
$$

12-5b. Isomerization of Ammonium Cyanate.   Wöhler's classic study was the first investigation of the transformation of ammonium cyanate to urea.[47]   The reaction has since been the object of a number of kinetic studies.[48]   Walker and Hambly found that the reaction was second-order, first-order in ammonium ions and first-order in cyanate ions and concluded that the reaction was between the two ions.[49]

$$\mathrm{NH_4^+ + NCO^- \rightarrow (H_2N)_2CO}$$

Chattaway pointed out that because of the equilibrium

$$\mathrm{NH_4^+ + NCO^- \rightleftharpoons NH_3 + HNCO}$$

which lies very largely to the left, the kinetic equation is equally in

[47] F. Wöhler, *Pogg. Ann.*, **3**, 177 (1825); **12**, 253 (1828).

[48] A more detailed discussion of the reaction mechanism has been given by Frost and Pearson, *op. cit.*, chap. 12B.

[49] J. Walker and F. J. Hambly, *J. Chem. Soc.*, **67**, 746 (1895).

agreement with a mechanism in which ammonia reacts with cyanic acid (or its tautomer, isocyanic acid)

$$NH_3 + HNCO \xrightarrow{k} (H_2N)_2CO$$

since this mechanism predicts the rate expression

$$v = \frac{kK_W}{K_{NH_3}K_{HNCO}} [NH_4^+][NCO^-] \tag{12-5}$$

where K_W is the autoprotolysis constant of water, the solvent, K_{NH_3} is the ionization constant of ammonia, and K_{HNCO} that of cyanic acid.[50]

Inspection of the Brønsted-Bjerrum-Christiansen equation [Eq. (3-7)] shows that the reaction between two oppositely charged ions like the ammonium ion and the cyanate ion should show a large negative ionic-strength effect whose size (in dilute solution) is predicted quantitatively. On the other hand, the specific rate constant for reaction between two neutral species like ammonia and isocyanic acid should be relatively little affected by the ionic strength. The isomerization of ammonium cyanate to urea has been found to have a considerable negative ionic-strength effect, whose magnitude is within a factor of two of that predicted theoretically, and the reaction has for this reason been said to occur directly between the ions.[51] However, Weil and Morris have pointed out that although the ionic strength will have little effect on the value of k in Eq. (12-5), it will affect the equilibrium constants in this equation because they are for equilibria between ions and neutral molecules.[52] In fact, these workers show that both mechanisms make the same prediction of the effect of ionic strength on rate.

It has therefore not been possible to distinguish between the two proposed mechanisms by use of kinetics, but there are certainly some very good arguments of other types that make the reaction between ammonia and isocyanic (or cyanic) acid preferable.[53] The carbon atom of cyanic acid (or its tautomer, isocyanic acid) must be relatively electron-deficient and hence susceptible to attack by the unshared electron pair of ammonia.

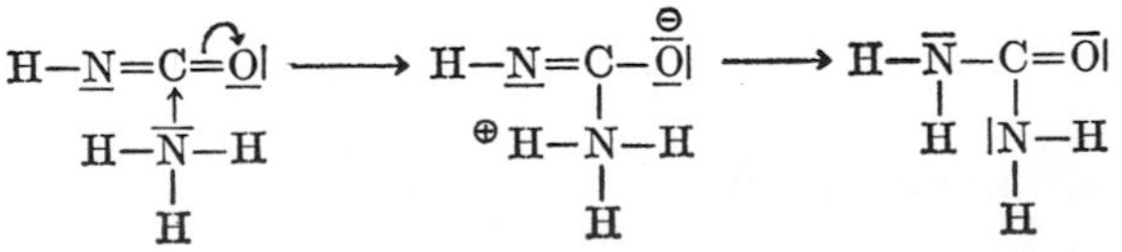

The ease of this reaction is shown by the high reactivity of ammonia and

[50] F. D. Chattaway, *J. Chem. Soc.*, **101**, 170 (1912).

[51] J. C. Warner and F. B. Stitt, *J. Am. Chem. Soc.*, **55**, 4807 (1933).

[52] I. Weil and J. C. Morris, *J. Am. Chem. Soc.*, **71**, 1664 (1949).

[53] Cf. T. M. Lowry, *Trans. Faraday Soc.*, **30**, 375 (1934).

primary and secondary amines (and, to a smaller extent, weaker nucleo-philic reagents such as water and alcohols) toward alkyl and aryl iso-cyanates. The carbon atom of the cyanate anion must be much less electrophilic, and the ammonium ion has *no* unshared electrons. The only reasonable reaction one could postulate between these two ions is the proton transfer to give ammonia and cyanic (or isocyanic) acid, which we have already mentioned.

12-5c. *Hydrolysis of Amides.* The hydrolysis of amides, like that of esters, is speeded by both acids and bases, and the reaction is usually second-order, first-order in amide and first-order in acid or base.[54] Bender and Ginger found that the base-catalyzed O^{18} exchange of benz-amide was about five times as fast as the hydrolysis reaction at 109°.[55] Thus, apparently the reaction proceeds by the carbonyl-addition mecha-nism with the reversible formation of an intermediate (XII) in which the carbonyl oxygen atom has become equivalent to the oxygen atom from the attacking hydroxide ion.

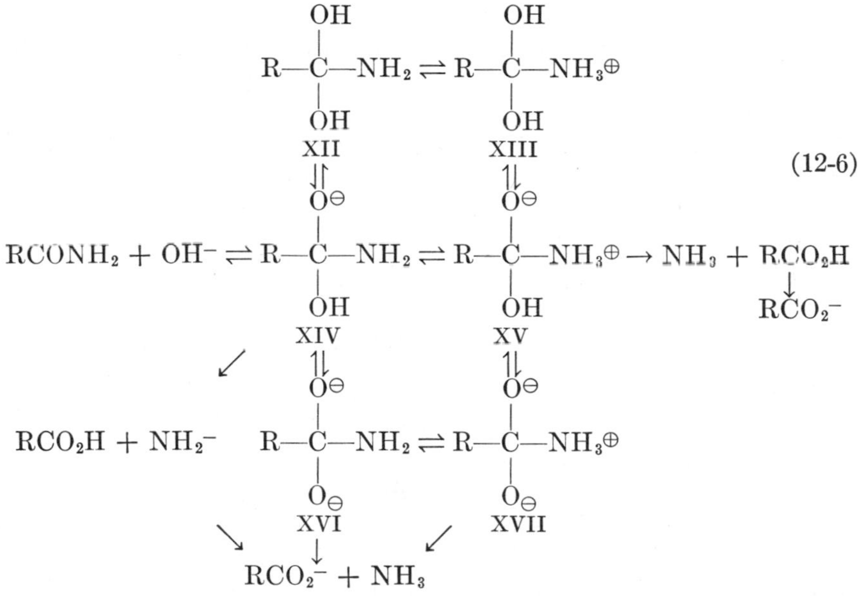

$$(12\text{-}6)$$

It is not presently clear which of the various possible intermediates shown actually undergoes the carbon-nitrogen bond cleavage that leads to the final products. Biechler and Taft found that with the *N*-methyl-anilides of trifluoroacetic and, to a smaller extent, other acids, the rate of

[54] Cf. E. E. Reid, *Am. Chem. J.*, **21**, 284 (1899); **24**, 397 (1900).
[55] M. L. Bender and R. D. Ginger, *J. Am. Chem. Soc.*, **77**, 348 (1955).

basic hydrolysis follows the equation[56]

$$v = (k_1[\text{OH}^-] + k_2[\text{OH}^-]^2) \, [\text{amide}] \qquad (12\text{-}7)$$

The part of the reaction that is second-order in hydroxide ion very probably involves the decomposition of an intermediate of the type of XVI. Such a course for the reaction is particularly probable for N-methyltrifluoroacetanilide because of the electron-withdrawing power of the fluorine atoms and the phenyl group.

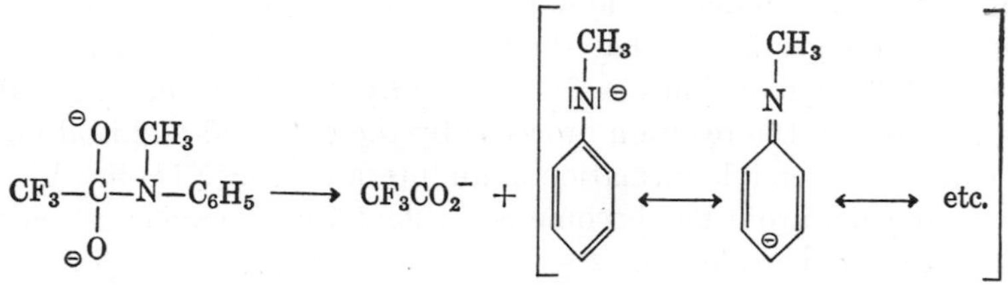

The acidic hydrolysis of benzamide is not accompanied by any detectable amount of O^{18} exchange of the reactant.[55]   Apparently the active intermediate (perhaps XIII) is transformed to products much more rapidly than it reverts to reactants.

The hydrolysis of amides may be catalyzed by agents other than acids and bases in a manner analogous to that already described for ester hydrolysis (Sec. 12-3).[29]   The mechanisms of such reactions are of interest in relation to learning more about the manner in which such naturally occurring amides as proteins are hydrolyzed and synthesized by the action of enzymes.

Bender, Chow, and Chloupek described convincing evidence that the hydrolysis of phthalamic acid is subject to internal catalysis and involves the intermediate formation of phthalic anhydride.[57]   The rate of hydrolysis of this compound is proportional to the concentration of the undissociated acid (and hence of any of its tautomers) between pH 1.0 and 5.0.   Furthermore, at pH 3.0, the hydrolysis rate is about 10^5 times as fast as that of benzamide (whose hydrolysis rate is proportional to the concentrations of benzamide and hydrogen ion in this pH range).   In view of the fact that p-carboxybenzamide hydrolyzes somewhat slower than benzamide, it seems very unlikely that the effect of the o-carboxy group in phthalamic acid is simply a polar substituent effect.   Nor can the effect of the o-carboxy group be explained very plausibly as a simple steric effect; o-nitrobenzamide hydrolyzes only one-tenth as fast as benzamide.   The great reactivity of phthalamic acid can be explained

[56] S. S. Biechler and R. W. Taft, Jr., *J. Am. Chem. Soc.*, **79**, 4927 (1957).
[57] M. L. Bender, Y.-L. Chow, and F. Chloupek, *J. Am. Chem. Soc.*, **80**, 5380 (1958).

in terms of nucleophilic attack by the carboxylate anion group on the protonated form of the carboxamide group.

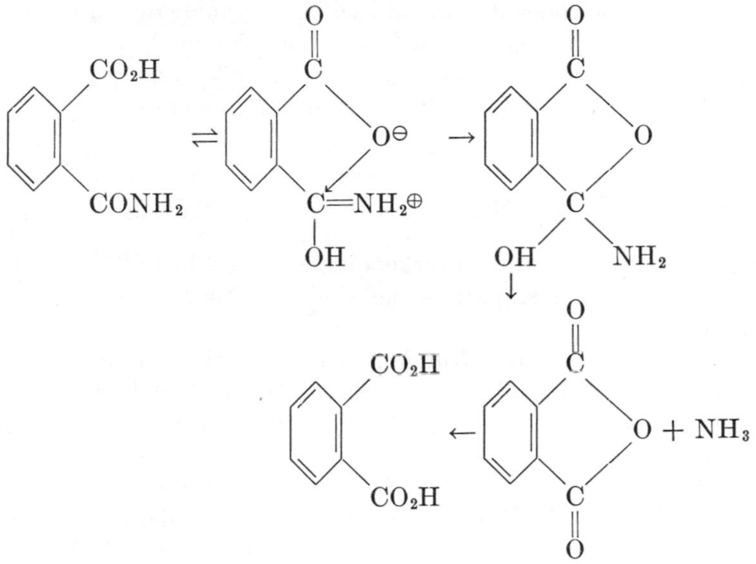

Double-labeling experiments using C^{13} and O^{18} were also employed to provide evidence for the intermediacy of phthalic anhydride.[57]

The rate of hydrolysis of amides in very strongly acidic solutions has been found to reach a maximum for acid of a certain strength (about $6 N$ in the hydrolysis of formamide in hydrochloric acid[58]) and then to decrease with increasing strength of the acid.[58,59] The approach of the rate to a maximum may be explained qualitatively by the fact that the amides are transformed essentially completely to their conjugate acids in the strongly acidic solution, so that further acidification serves no useful purpose. Augmenting this effect and tending to reduce the reaction rate is the decrease in the activity of water that accompanies the increasing acidity of the solvent. There may be more complications than these two, however.

12-6. Hydrolysis of Other Carboxylic Acid Derivatives.   12-6a. *Hydrolysis of Nitriles.* Nitriles may be hydrolyzed by either strong acids or strong bases. The hydrolysis of the simplest of the nitriles, hydrogen cyanide, in strongly acidic solutions has been studied carefully. Krieble and McNally showed that the reaction rate increases more rapidly than the acid concentration for hydrochloric, hydrobromic, and sulfuric acids

[58] V. K. Krieble and K. A. Holst, *J. Am. Chem. Soc.*, **60**, 2976 (1938).
[59] T. W. J. Taylor, *J. Chem. Soc.*, 2741 (1930).

and to an extent that varies with the nature of the acid.[60]   Thus at a concentration of 5.5 M in sulfuric acid $k = 0.02$, in hydrobromic acid $k = 0.20$, and in hydrochloric acid $k = 1.25$.[61]   The rate constants are independent of the concentration of hydrogen cyanide over a wide range. From these results it seems possible that the anion of the catalyzing acid is also involved in the reaction, so that the hydrolysis in hydrochloric acid may involve formimino chloride as an intermediate.

$$\text{HCN} \xrightarrow{\text{HCl}} \text{H}-\overset{\overset{\textstyle \text{Cl}}{\textstyle |}}{\text{C}}=\text{NH} \xrightarrow{\text{H}_2\text{O}} \text{H}-\text{CONH}_2 \rightarrow \text{HCO}_2\text{H} + \text{NH}_4^+$$

Rabinovitch, Winkler, and coworkers have also studied nitrile hydrolysis in strong acids,[62] and Kilpatrick has similarly studied the hydrolysis of cyanamide.[63]

12-6b. *Hydrolysis of Acid Halides.*   Hudson and coworkers found that under certain conditions the hydrolytic reactivity of acyl halides may be increased by either electron-withdrawing or electron-donating groups[64] in a manner rather similar to that described for certain saturated halides in Sec. 7-3b.   They further substantiate the proposal that the S_N1 character of the reaction is greatest with the acid halides having strongly electron-donating substituents by showing that it is these halides whose hydrolysis rates are most sensitive to the ion-solvating power of the medium.[65]   They also show that although the hydrolysis of acid chlorides is not catalyzed by acids, that of acid fluorides is.[66]

According to a preliminary report, the relative nucleophilicities of various species toward $ClCO_2Et$ are quite different from their nucleophilicities in S_N2 attack on saturated carbon.[67]   Attack by first-row atoms is the fastest, and it is not always the most basic nucleophile that reacts fastest.   This is illustrated by the observed order $Me_2C{=}NO^- >$ $OH^- > C_6H_5O^- > NO_2^- > N_3^- > F^- > Br^-, I^-, SCN^-, OCN^-, ClO_4^-,$ and NO_3^-.

12-6c. *Hydrolysis of Acid Anhydrides.*   As compounds of the type RCOX, acid anhydrides occupy a position between acid halides (in which X has a strong tendency to be lost as the anion) and esters (in

[60] V. K. Krieble and J. G. McNally, *J. Am. Chem. Soc.*, **51**, 3368 (1929).

[61] V. K. Krieble and A. L. Peiker, *J. Am. Chem. Soc.*, **55**, 2326 (1933).

[62] B. S. Rabinovitch and C. A. Winkler, *Can. J. Research*, **20B**, 221 (1942), and earlier references quoted.

[63] M. L. Kilpatrick, *J. Am. Chem. Soc.*, **69**, 40 (1947).

[64] R. F. Hudson and J. E. Wardill, *J. Chem. Soc.*, 1729 (1950); D. A. Brown and R. F. Hudson, *J. Chem. Soc.*, 3352 (1953).

[65] B. L. Archer and R. F. Hudson, *J. Chem. Soc.*, 3259 (1950); D. A. Brown and R. F. Hudson, *J. Chem. Soc.*, 883 (1953).

[66] C. W. L. Bevan and R. F. Hudson, *J. Chem. Soc.*, 2187 (1953).

[67] M. Green and R. F. Hudson, *Proc. Chem. Soc.*, 149 (1959).

which X usually requires assistance for displacement).   Acids have only a slight tendency to catalyze the hydrolysis of anhydrides in aqueous solution,[68] but in acetic acid solutions, where the existence of much more powerful proton donors is possible, acids are effective catalysts.[69]

Gold and coworkers studied the hydrolysis of acetic and several other aliphatic anhydrides.[70]   Among other effects, they studied the catalysis of the reaction by tertiary amines.   The catalysis appears to be subject to steric hindrance by bulky groups on the amines and is thus not dependent simply upon the basicity of the amines.   This suggests a nucleophilic attack at the carbonyl carbon atom.

$$CH_3\overset{O}{\overset{\|}{C}}O\overset{O}{\overset{\|}{C}}CH_3 + R_3N \rightarrow CH_3\overset{O}{\overset{\|}{C}}\overset{\oplus}{N}R_3 + CH_3CO_2^-$$
$$\downarrow H_2O$$
$$CH_3CO_2^- + R_3NH^+$$

Such a mechanism cannot explain the small but unmistakable catalytic action of acetate ion,[71] however, since nucleophilic attack by acetate ion would be recognizable only by isotopic-labeling experiments.   The acetate ion probably removes a proton from water, perhaps after and perhaps as it attacks the anhydride, e.g.,

$$CH_3\overset{O}{\overset{\|}{C}}-O-\overset{O}{\overset{\|}{C}}CH_3 + H_2O \rightleftharpoons CH_3\overset{\overset{\ominus}{O}}{\overset{|}{C}}-O-\overset{O}{\overset{\|}{C}}CH_3$$
$$\underset{\oplus OH_2}{}$$
$$\downarrow AcO^-$$

$$CH_3\overset{O}{\overset{\|}{C}} + CH_3CO_2^- \leftarrow CH_3\overset{\overset{\ominus}{O}}{\overset{|}{C}}-O-\overset{O}{\overset{\|}{C}}CH_3$$
$$\underset{OH}{} \qquad\qquad \underset{OH}{}$$

Berliner and Altschul studied the hydrolysis of substituted benzoic anhydrides.[72]   The reaction rate in 75 per cent dioxane–25 per cent water solution was increased by electron-withdrawing substituents, since these increase the tendency of water molecules to attack the carbonyl carbon atoms.   The data are in satisfactory agreement with the Hammett equation despite the fact that the entropies of activation vary quite widely for the different compounds studied.   This variation is correlated with the σ constants, of course.

[68] S. C. J. Olivier and G. Berger, *Rec. trav. chim.*, **46**, 609 (1927).
[69] T. Yvernault, *Compt. rend.*, **233**, 411 (1951); **235**, 167 (1952).
[70] V. Gold and E. G. Jefferson, *J. Chem. Soc.*, 1409, 1416 (1953), and earlier references.
[71] M. Kilpatrick, Jr., *J. Am. Chem. Soc.*, **50**, 2891 (1928); V. Gold, *Trans. Faraday Soc.*, **44**, 506 (1948).
[72] E. Berliner and L. H. Altschul, *J. Am. Chem. Soc.*, **74**, 4110 (1952).

12-7. Hydrolysis of Derivatives of Noncarboxylic Acids.   12-7a. *Hydrolysis of Phosphoric Acid Derivatives.*   The rate of hydrolysis of trimethyl phosphate is independent of the acidity of the solution between pH 7 and 3 M perchloric acid.[73]   In more basic and more acidic solutions faster reactions involving the base or acid become important.   It has been shown by O^{18} studies that the hydrolysis in basic solution involves P—O bond cleavage and the first-order hydrolysis involves C—O bond cleavage.[74]   Apparently the ester is attacked by hydroxide ions largely at phosphorus and by water largely at carbon,

$$
\begin{array}{ccc}
\text{O} & \text{HO}\quad \text{O}^{\ominus} & \text{HO} \\
\parallel & \diagdown\!\diagup & \mid \\
\text{CH}_3\text{O}\!-\!\text{P}\!-\!\text{OCH}_3 \xrightarrow{\text{OH}^-} \text{CH}_3\text{O}\!-\!\text{P}\!-\!\text{OCH}_3 \rightarrow \text{CH}_3\text{O}\!-\!\text{P}\!=\!\text{O} \;+\; \text{CH}_3\text{O}^- \\
\mid & \mid & \mid \\
\text{OCH}_3 & \text{OCH}_3 & \text{OCH}_3 \\
\downarrow\text{H}_2\text{O} & & \downarrow \\
\text{O} & & (\text{CH}_3\text{O})_2\text{PO}_2^- \;+\; \text{CH}_3\text{OH}
\end{array}
$$

$$
\begin{array}{c}
\text{O} \\
\parallel \\
\text{CH}_3\text{O}\!-\!\text{P}\!-\!\text{OCH}_3 \;+\; \text{CH}_3\text{OH}_2{}^+ \\
\mid \\
\text{O} \\
\ominus \qquad \downarrow \\
\\
(\text{CH}_3\text{O})_2\text{PO}_2\text{H} \;+\; \text{CH}_3\text{OH}
\end{array}
$$

In the transition state for nucleophilic attack at the phosphorus-oxygen double bond, like that for attack at the carbon-oxygen double bond of an ester, the attacking nucleophile has probably formed a new bond to a greater extent than in the transition state for S_N2 attack at a methyl group.   Hence the greater basicity of the hydroxide ion causes its rate of attack at phorphorus to be larger relative to its rate of attack at methyl than is the case for water (cf. β-lactones, Sec. 12-2b).

The study of the hydrolysis of the mono- and diesters of phosphoric acid is complicated by the number of different extents of protonation available for these species.   In Fig. 12-1 is a plot of log k for the first-order hydrolyses of monomethyl phosphate and dimethyl phosphate at various pH's.[75–77]   From this figure it may be seen that the anion of

[73] P. W. C. Barnard, C. A. Bunton, D. R. Llewellyn, K. G. Oldham, B. L. Silver, and C. A. Vernon, *Chem. & Ind.* (*London*), 760 (1955).

[74] E. Blumenthal and J. B. M. Herbert, *Trans. Faraday Soc.*, **41**, 611 (1945).

[75] J. Kumamoto, J. R. Cox, Jr., and F. H. Westheimer, *J. Am. Chem. Soc.*, **78**, 4858 (1956).

[76] C. A. Bunton, D. R. Llewellyn, K. G. Oldham, and C. A. Vernon, *J. Chem. Soc.*, 3574 (1958).

[77] C. A. Bunton, M. M. Mhala, K. G. Oldham, and C. A. Vernon, *J. Chem. Soc.*, 3293 (1960).

dimethyl phosphate undergoes second-order nucleophilic attack by hydroxide ions in strongly basic solution[75] (since K_W for water is 10^{-12} at 100°, a pH of 12 at this temperature corresponds to 1 M hydroxide ions). This attack, shown by O^{18} studies[77] to involve C—O bond cleavage, must be an S_N2 displacement at a methyl group. The slow hydrolysis in weakly basic and neutral solutions has not been studied, but in weakly acidic solution the rate of hydrolysis is proportional to the concentration

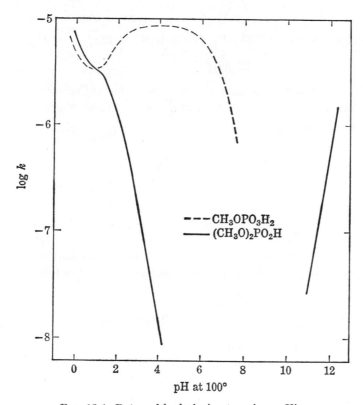

FIG. 12-1. Rates of hydrolysis at various pH's.

of undissociated dimethyl phosphate. With increasing acidity the hydrolysis rate starts to level off around pH 1.0 (because of the transformation of essentially all the dimethyl phosphate, an acid of pK 1.6, to the undissociated form), but then an acid-catalyzed reaction enters the picture. Both the "uncatalyzed" and the acid-catalyzed hydrolysis of undissociated dimethyl phosphate involve largely C—O bond cleavage but also 10 to 25 per cent P—O bond cleavage. The C—O cleavages are no doubt due to S_N2 attack by water on the methyl groups of dimethyl phosphate and its conjugate acid, whereas the P—O cleavages are probably due to attack on phosphorus.   M. C. Bailly[78] and Desjobert[79] discovered that the hydrolysis of several monoesters of phosphoric acid shows a rate maximum around pH 4.   This is the pH at which the maximum fraction of such esters are present as the monoanion.   The reaction, which involves P—O bond cleavage, is believed to proceed via the reactive intermediate PO_3^-.   This may arise from a zwitterionic tautomer of the monoanion as shown below,

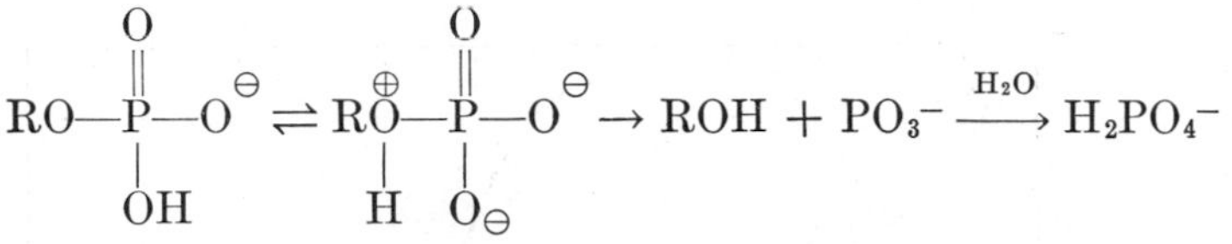

$$RO-\overset{\overset{O}{\|}}{\underset{\underset{OH}{|}}{P}}-O^{\ominus} \rightleftharpoons RO-\overset{\overset{O}{\|}}{\underset{\underset{\underset{O\ominus}{|}}{H}}{\overset{\oplus}{P}}}-O^{\ominus} \rightarrow ROH + PO_3^- \xrightarrow{H_2O} H_2PO_4^-$$

but Butcher and Westheimer prefer a somewhat different mechanism in which two proton transfers occur in the following manner:[80]

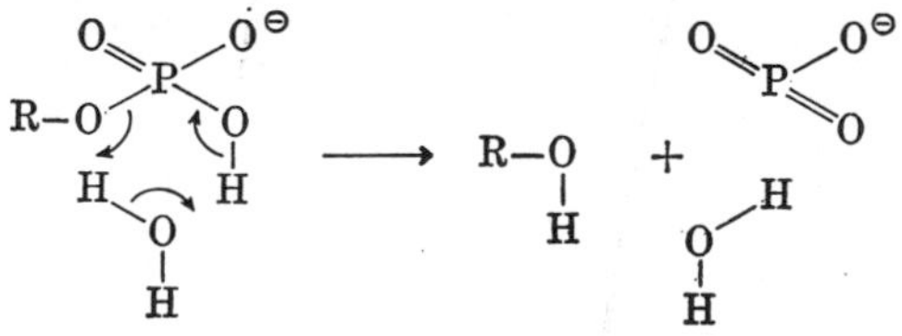

Neighboring-group participation can be of importance in the hydrolysis of phosphates.   O. Bailly and Gaumé noted that methyl 1-glyceryl phosphate undergoes a reversible base-catalyzed isomerization to methyl 2-glyceryl phosphate.[81]   A reaction mechanism like the following

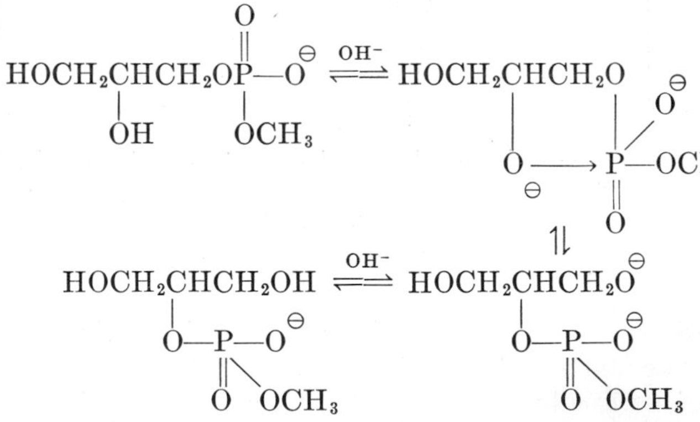

can, with only slight modification, explain why methyl β-hydroxyethyl phosphate undergoes basic hydrolysis more readily than methyl β-

---

[78] M. C. Bailly, *Bull. soc. chim. France,* **9,** 314, 340, 405, 421 (1942).

[79] A. Desjobert, *Compt. rend.,* **224,** 575 (1947); *Bull. soc. chim. France,* 809 (1947).

[80] W. W. Butcher and F. H. Westheimer, *J. Am. Chem. Soc.,* **77,** 2420 (1955).

[81] O. Bailly and J. Gaumé, *Compt. rend.,* **199,** 793 (1934); *Bull. soc. chim. France,* **2,** 354 (1935); **3,** 1396 (1936).

methoxyethyl phosphate,[81] and also why ribonucleic acids are so much more susceptible to basic hydrolysis than are deoxyribonucleic acids[82] (which, unlike ribonucleic acids, do not have a hydroxy group on the carbon atom adjacent to the one to which the phosphate group is attached). In these cases an ethylene phosphate is formed as a reaction intermediate. Kumamoto, Cox, and Westheimer showed that ethylene phosphate itself is a tremendously reactive diester of phosphoric acid, undergoing both acid- and base-catalyzed hydrolysis around 10^7 times as fast as its acyclic analog, dimethyl phosphate.[75,83] Studies using O^{18} showed that the hydrolysis of ethylene phosphate involves P—O bond cleavage.[84] The great reactivity of this compound may result from the release of a considerable amount of internal ring strain.   From comparisons of heats of hydrolysis, Cox, Wall, and Westheimer estimated this strain to amount to about 8 kcal/mole.[85]

12-7b. *Hydrolysis of Derivatives of Sulfur Acids.*   Since sulfonic acids are much stronger than carboxylic acids, reaction with alkyl-oxygen fission proceeds much more rapidly with alkyl sulfonates than with the corresponding carboxylic acid esters.   Cleavage at the sulfur-oxygen bond, however, occurs quite slowly.   Since the sulfur atom already bears a considerable positive charge, formation of the $RSO_2^{\oplus}$ ion is difficult.   Nucleophilic attack on sulfur is hindered, probably largely sterically but perhaps also by the energy required for the sulfur atom to expand its octet.   For these reasons the most common reactions of esters of sulfonic acids are nucleophilic displacements, by the S_N1 and S_N2 mechanisms, at the alkyl carbon atom.   Studies on a number of reactions of this type were described in Chap. 6.

It is found, however, that if the reactivity at the alkyl carbon atom is decreased sufficiently, nucleophilic attack at sulfur may be observed. Bunton and Frei have demonstrated this fact with phenyl *p*-toluenesulfonate.[86]   The alkaline hydrolysis of this compound in H_2O^{18} solution introduced O^{18} into the *p*-toluenesulfonate but not the phenolate.

$$C_7H_7SO_2{:}OC_6H_5 + O^{18}\overset{\ominus}{H} \rightarrow C_7H_7SO_2O^{18}H + C_6H_5\overset{\ominus}{O}$$
$$C_7H_7SO_2O^{18}H \rightarrow C_7H_7SO_2\overset{\ominus}{O^{18}}$$

Burwell has found a somewhat different situation in the hydrolysis of *sec*-butyl hydrogen sulfate.[87]   The sodium salt undergoes a first-order

[82] G. Schmidt and S. J. Thannhauser, *J. Biol. Chem.*, **161**, 83 (1945).

[83] J. R. Cox, Jr., Ph. D. thesis, Harvard University, 1958.

[84] P. C. Haake and F. H. Westheimer, *J. Am. Chem. Soc.*, **83**, 1102 (1961).

[85] J. R. Cox, Jr., R. E. Wall, and F. H. Westheimer, *Chem. & Ind. (London)*, 929 (1959).

[86] C. A. Bunton and Y. F. Frei, *J. Chem. Soc.*, 1872 (1951).

[87] R. L. Burwell, Jr., and H. E. Holmquist, *J. Am. Chem. Soc.*, **70**, 878 (1948); R. L. Burwell, Jr., *J. Am. Chem. Soc.*, **74**, 1462 (1952).

reaction, presumably involving S_N2 attack by water on carbon, to give alcohol with inverted configuration. In acidic solution, however, the alcohol has retained configuration, although it is somewhat racemized. Under these conditions, then, the reaction apparently involves mostly sulfur-oxygen bond cleavage.

A number of sulfite esters have been found to be much more reactive than the corresponding sulfates, even in basic solution where the over-all free energy of reaction must be more favorable for the sulfates, in view of the greater strength of the acids in which sulfur is in the $+6$ valence state.[88] These reactivity data provide strong evidence that the hydrolysis of sulfites proceeds by S—O bond cleavage, and this point has been proved by O^{18} studies.[88]

12-7c. Hydrolysis of Derivatives of Other Inorganic Acids. Lucas and Hammett pointed out, in their study of the hydrolysis of t-butyl and benzyl nitrates, that the similarity of these reactions to those of the corresponding halides reveals a probability that the reactions involve carbon-oxygen bond fission.[89] This mechanism is further supported by Baker and Easty's study of the $E1$, $E2$, S_N1, and S_N2 mechanisms in the solvolysis of t-butyl, isopropyl, ethyl, and methyl nitrates[90] and is confirmed by Cristol, Franzus, and Shadan's observation that the neutral hydrolysis of optically active 2-octyl nitrate gives 2-octanol with inversion of configuration.[91] Although the nucleophilic attack of water occurs largely at carbon in neutral solution, in alkaline solution it appears that hydroxide ion attacks nitrogen to a considerable extent, since 2-octyl nitrate gives 2-octanol with retained configuration though partially racemized.[91]

By stereochemical studies on 2-octyl nitrite as well as by other means, Allen showed that the hydrolysis involves nitrogen-oxygen bond cleavage in acidic, basic, and neutral solution.[92]

Anbar, Dostrovsky, Samuel, and Yoffe used O^{18} to study the hydrolysis of a considerable number of organic esters of inorganic oxyacids.[93]

PROBLEMS

1. The hydrolysis of alkyl benzoates in 99.9 per cent sulfuric acid (in which these esters are strong bases) is a first-order reaction whose rate constant is independent of the concentration of water present and of the initial concentration of ester. The

[88] C. A. Bunton, P. B. D. de la Mare, J. G. Tillett, et al., *J. Chem. Soc.*, 4751, 4754, 4761, 4766 (1958); 1766 (1959); 37 (1960).

[89] G. R. Lucas and L. P. Hammett, *J. Am. Chem. Soc.*, **64,** 1928 (1942).

[90] J. W. Baker and D. M. Easty, *J. Chem. Soc.*, 1193, 1208 (1952); see also Sec. 8-2a.

[91] S. J. Cristol, B. Franzus, and A. Shadan, *J. Am. Chem. Soc.*, **77,** 2512 (1955).

[92] A. D. Allen, *J. Chem. Soc.*, 1968 (1954); cf. Ref. 91.

[93] M. Anbar, I. Dostrovsky, D. Samuel, and A. D. Yoffe, *J. Chem. Soc.*, 3603 (1954).

following relative rates of hydrolysis have been observed: $C_6H_5CO_2CH_3 > C_6H_5-CO_2C_2H_5 < C_6H_5CO_2CH(CH_3)_2 < C_6H_5CO_2C(CH_3)_3$. The following Hammett ρ constants have been observed: for the acid-catalyzed hydrolysis of ethyl benzoates in 60 per cent acetone (second-order rate constants), ρ is $+0.1$; in 99.9 per cent sulfuric acid, the ρ's for the hydrolysis of methyl and isopropyl benzoates are -3.2 and approximately $+2.9$, respectively, and for the hydrolysis of ethyl benzoates the plot of log k's vs. σ's shows a minimum in the log k values around a σ of 0.7. Tell what hydrolysis mechanism is probably operating for each of the cases described above and give your reasons. Explain how changes in reactant structure and reaction conditions bring about any changes in reaction mechanism that you suggest.

2. The dipole moments of γ-butyrolactone and δ-valerolactone are around 4 D, whereas those of the acetates, propionates, butyrates, etc., of normal primary alcohols are around 1.8 D. The alkaline hydrolyses of γ-butyrolactone and δ-valerolactone are, respectively, about 400 and 13,000 times as fast as the hydrolyses of the "corresponding" acyclic esters. Suggest explanations for these differences.

3. Ester hydrolysis may be slowed sterically in two ways. (a) *Some* of the rotational conformers that existed in the reactant may be too unstable, because of steric interactions, to exist to an appreciable extent in the transition state. (b) Even the most stable of the various rotational conformers may be sterically strained.

How could one use values of E_s determined at various temperatures to learn something about the relative importance of factors (a) and (b)?

4. List the following compounds in the order of their decreasing reactivity in alkaline hydrolysis to yield a diester of phosphoric acid: $(C_2H_5O)_3PO$, $(p\text{-}O_2NC_6H_4O)_3PO$, $(C_2H_5O)_2(C_6H_5O)PO$, $(C_2H_5O)_2(p\text{-}O_2NC_6H_4O)PO$. Give your reasons.

Chapter 13

SOME CARBON-CARBON BOND FORMATION AND CLEAVAGE REACTIONS OF CARBOXYLIC ACID DERIVATIVES

13-1. Decarboxylation and Carboxylation.[1] In order to separate carbon dioxide from a carboxylic acid, RCO_2H, it is necessary to remove both the hydrogen atom and the R group. The removal of hydrogen is usually accomplished by having the reaction proceed through a form containing the carboxylate anion group, or at least the carboxyl group hydrogen-bonded to a base. When the hydrogen atom is thus removed *without* its bonding electron pair, it is necessary to remove R *with* its bonding pair. It would therefore be expected that acids with properly situated, strongly electron-withdrawing groups should be decarboxylated with relative ease. The reaction mechanism

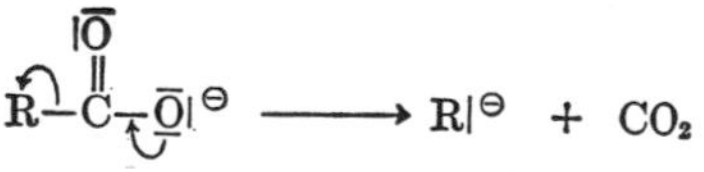

$$R \overset{\overline{|\underline{O}|}}{\underset{}{\overset{\|}{C}}} \overline{\underline{O}}|^{\ominus} \longrightarrow R|^{\ominus} + CO_2$$

resembles several others we have discussed, e.g., the reverse aldol condensation (Sec. 11-3c).

13-1a. *The Carbanion Mechanism for Decarboxylation.* From the generalization above it would be expected that carboxylate anions should lose carbon dioxide easily if they may thereby form relatively stable carbanions. Pedersen described some of the most convincing evidence for this reaction mechanism in his study of the decarboxylation of nitroacetic acid and α-nitroisobutyric acid.[2] Both these processes are first-order reactions of the anions of the acids. The addition of bromine has no effect on the rate of decomposition of the α-nitroisobutyrate anion, although it completely changes the nature of the product, from 2-nitropropane to 2-bromo-2-nitropropane. This must be due to the reaction

---

[1] For a review of polar mechanisms for decarboxylation, see B. R. Brown, *Quart. Revs.* (*London*), **5,** 131 (1951).

[2] K. J. Pedersen, *J. Phys. Chem.*, **38,** 559 (1934).

of bromine with the anion of 2-nitropropane,

$$O_2NC(CH_3)_2CO_2{}^- \rightarrow [O_2NC(CH_3)_2]^- + CO_2$$
$$[O_2NC(CH_3)_2]^- + Br_2 \rightarrow O_2NC(CH_3)_2Br + Br^-$$

since 2-nitropropane cannot be brominated under the reaction conditions.

Similarly the rate of decarboxylation of dibromomalonic acid is proportional to the concentration of the singly charged anion, HO_2CCBr_2-$CO_2{}^-$, and the product may be changed from dibromo- to tribromoacetic acid by addition of bromine to the reaction mixture.[3]   Again, bromine has no effect on the reaction rate and is incapable of brominating dibromoacetic acid under the reaction conditions.

Several other acids capable of yielding fairly stable carbanions have been found to decarboxylate by first-order reactions of their anions. These include phenylpropiolic acid,[4] 2,4,6-trinitrobenzoic acid,[5] and several trihaloacetic acids.[5-6]   The importance of carbanion stability in the latter series of reactions has been attested by the close approach to a linear relationship between log k's for decarboxylation of trihaloacetate ions and log k's for the formation of the same trihalomethyl anions from deuteriohaloforms and base.[7]   It is perhaps surprising to note the stability of the 2,4,6-trinitrophenyl anion, evidenced by the relative ease of decomposition of the related carboxylate anion,

but this is further supported by the observation of deuterium exchange of trinitrobenzene in alkaline ethanol solution[8] and by the fact that the formation of phenyl anions is known to be greatly facilitated by electron-withdrawing ortho substituents.[9]

13-1b. *Decarboxylation of β-Keto Acids.* If the R group of $RCO_2{}^-$ contains a sufficiently basic functional group, it may accept a proton and greatly increase its electron-withdrawing power, thus facilitating the

[3] J. Muus, *J. Phys. Chem.*, **39**, 343 (1935); **40**, 121 (1936).

[4] R. A. Fairclough, *J. Chem. Soc.*, 1186 (1938).

[5] F. H. Verhoek, *J. Am. Chem. Soc.*, **56**, 571 (1934); **61**, 186 (1939).

[6] I. Auerbach, F. H. Verhoek, and A. L. Henne, *J. Am. Chem. Soc.*, **72**, 299 (1950).

[7] J. Hine, N. W. Burske, M. Hine, and P. B. Langford, *J. Am. Chem. Soc.*, **79**, 1406 (1957).

[8] M. S. Kharasch, W. G. Brown, and J. McNab, *J. Org. Chem.*, **2**, 36 (1937); cf., however, J. A. A. Ketelaar, A. Bier, and H. T. Vlaar, *Rec. trav. chim.*, **73**, 37 (1954).

[9] G. E. Hall, R. Piccolini, and J. D. Roberts, *J. Am. Chem. Soc.*, **77**, 4540 (1955).

loss of carbon dioxide.   This fact appears to be important in the decarboxylation of β-keto acids.

The kinetic equation for the decarboxylation of acetoacetic acid in aqueous solution has the form[10]

$$v = k[CH_3COCH_2CO_2H] + k'[CH_3COCH_2CO_2^-]$$

That part of the reaction due to the acetoacetate anion very probably proceeds by the carbanion mechanism.   In a study of the reaction as a whole, Pedersen has used α,α-dimethylacetoacetic acid to avoid the type of keto-enol tautomerism that could complicate the study of acetoacetic acid itself.[11]   From the identical form of the observed kinetic equation and the fact that the k's were of the same general order of magnitude, it seems very likely that the reactions involve the same (keto) form with both acids.   Subsequently Pedersen suggested that the reaction proceeds through a dipolar-ion tautomer whose concentration is proportional to that of the undissociated acid.[2]

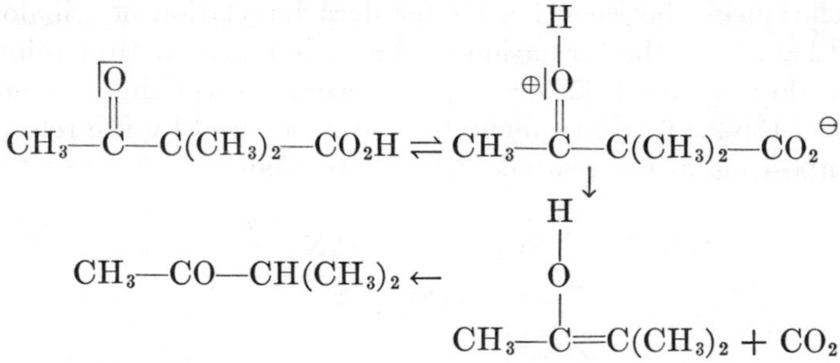

$$CH_3-\overset{\overset{\displaystyle |\overline{O}}{\|}}{C}-C(CH_3)_2-CO_2H \rightleftharpoons CH_3-\overset{\overset{\displaystyle H}{|}}{\underset{}{\overset{\oplus|O}{\|}}}C-C(CH_3)_2-CO_2^{\ominus}$$

$$CH_3-CO-CH(CH_3)_2 \leftarrow \quad \overset{\overset{\displaystyle H}{|}}{\underset{\displaystyle CH_3-\overset{\overset{|}{O}}{C}=C(CH_3)_2 + CO_2}{O}}$$

Excellent evidence that the enol and not the keto form of methyl isopropyl ketone is the initial product of the reaction is found in the observation that in the presence of iodine or bromine the iodo or bromo ketone is formed, although the iodine (or bromine) has no effect on the reaction rate and is incapable of halogenating the ketone significantly under the reaction conditions.[12]   Westheimer and Jones pointed out that the fraction of the acid present as the dipolar ion should decrease with the dielectric constant of the solvent.[13]   Since they find considerable changes in solvent to have little effect on the reaction rate, they state that the dipolar ion is an unlikely intermediate.   It would seem, however, that although the relative concentration of the dipolar ion should decrease with the ion-solvating power of the medium, its specific rate constant for reaction

[10] E. M. P. Widmark, *Acta Med. Scand.*, **53**, 393 (1920); *Chem. Abstr.*, **15**, 2763 (1921).

[11] K. J. Pedersen, *J. Am. Chem. Soc.*, **51**, 2098 (1929).

[12] *Ibid.*, **58**, 240 (1936).

[13] F. H. Westheimer and W. A. Jones, *J. Am. Chem. Soc.*, **63**, 3283 (1941).

should increase, because this reaction involves charge destruction in the transition state. Actually, the coulombic and hydrogen-bonding interactions between the unlike charges should give the dipolar ion a cyclic

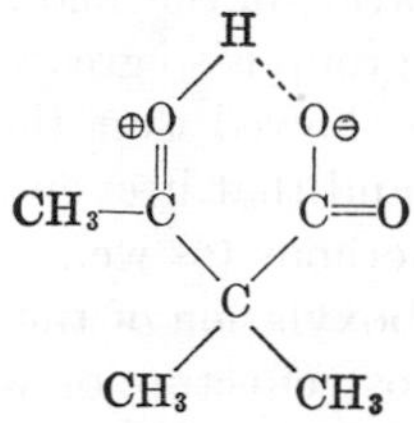

$$
\begin{array}{c}
\text{H} \\
\diagup \quad \cdots \\
\oplus\text{O} \qquad \text{O}\ominus \\
\| \qquad\qquad | \\
CH_3\!-\!C \qquad C\!=\!O \\
\diagdown \quad\diagup \\
C \\
\diagup \quad \diagdown \\
CH_3 \qquad CH_3
\end{array}
$$

configuration not differing greatly from that of the chelated intermediate suggested by Westheimer and Jones.

The decarboxylation of β-keto acids is not subject to general base catalysis but is specifically catalyzed by primary amines.[14] This strongly suggests an intermediate of the following type[1,12] (cf. Sec. 11-3c).

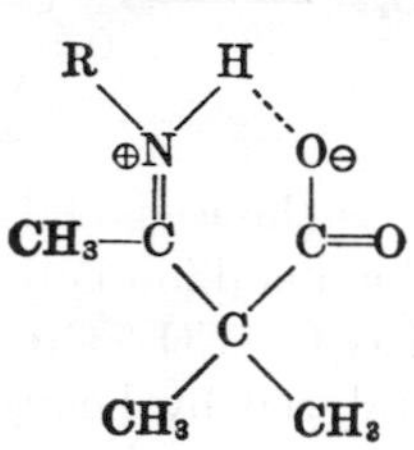

$$
\begin{array}{c}
\text{R} \qquad \text{H} \\
\diagdown \diagup \quad \cdots \\
\oplus\text{N} \qquad \text{O}\ominus \\
\| \qquad\qquad | \\
CH_3\!-\!C \qquad C\!=\!O \\
\diagdown \quad\diagup \\
C \\
\diagup \quad \diagdown \\
CH_3 \qquad CH_3
\end{array}
$$

The suggestions that the decarboxylations of β-keto acids yield enols and enolate anions are also supported by the fact that ketopinic acid,

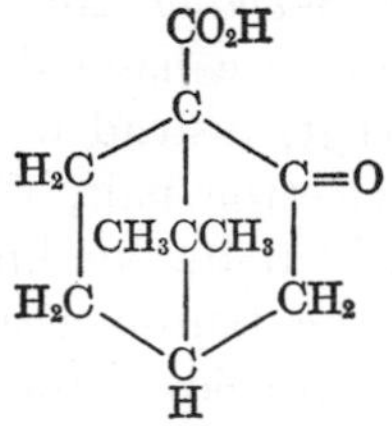

$$
\begin{array}{c}
CO_2H \\
| \\
C \\
H_2C \diagup \ | \ \diagdown C\!=\!O \\
| \ \ CH_3CCH_3 \ | \\
H_2C \diagdown \ | \ \diagup CH_2 \\
C \\
| \\
H
\end{array}
$$

which should be incapable of yielding a stable enol or enolate anion (Bredt's rule, see Sec. 10-1b), is remarkably resistant to decarboxylation.[15]

The decarboxylation of malonic acid probably has a mechanism like that of acetoacetic acid, since both the acid and the monoanion contribute to the reaction.[4]

Steinberger and Westheimer made a particularly careful study of the

[14] K. J. Pedersen, *J. Am. Chem. Soc.*, **60**, 595 (1938).

[15] Cf. J. Bredt, *Ann. Acad. Sci. Fennicae*, **29A**, no. 2 (1927); *Chem. Abstr.*, **22**, 1152 (1928).

decarboxylation of dimethyloxaloacetic acid.[16] They found that both the monoanion (presumably only one of the two possible forms) and dianion take part in the reaction but find no term in the kinetic equation proportional to the concentration of the acid itself. This may be because the basicity of the carbonyl group is so greatly decreased by the adjacent carboxyl group. They also showed that the enol (or enolate anion) is the initial reaction product and that it even has sufficient stability to be detected by spectroscopic methods (as well as by reaction with halogen). It was found that the decarboxylation of the acid is catalyzed by several heavy-metal cations, the most effective of which (and the one actually present in the enzymatic decarboxylation of oxaloacetic acid) is the manganous ion. A plot of the effect of these metal ions vs. pH shows that they coordinate with the dianion. The rate-controlling step of the reaction evidently has the mechanism

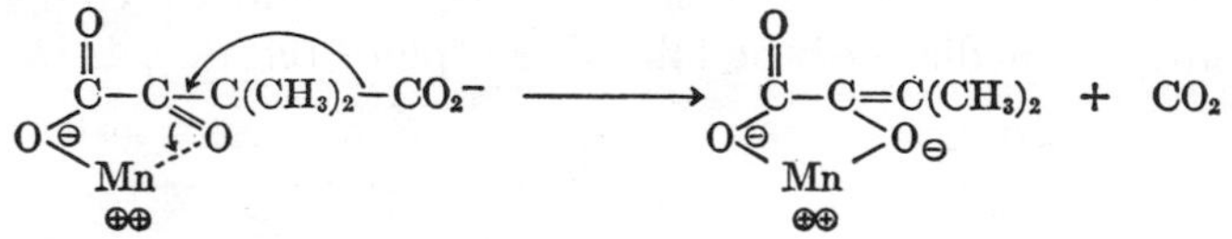

Part of the evidence that it is the carboxylate ion group shown that is involved in chelation with the metal ion is the fact that the decarboxylation of the monoester $C_2H_5O_2C$—CO—$C(CH_3)_2CO_2H$, which was also studied carefully, is not catalyzed by heavy-metal ions, and neither is that of acetoacetic acid.

Pedersen studied the metal-ion-catalyzed decarboxylation of oxaloacetic acid[17] and Prue that of acetonedicarboxylic acid.[18]

13-1c. *Decarboxylation of α-Imino Acids and Related Compounds.* The decarboxylation reactions we have described have all resembled acid-base reactions except that, instead of the removal of a hydrogen atom without its bonding electron pair, there has been the loss of a carboxylate anion group without its bonding electron pair. From this fact and by analogy with the tautomerism of methyleneazomethines (Sec. 10-2b) it seems logical that acids of the type of I should decarboxyl-

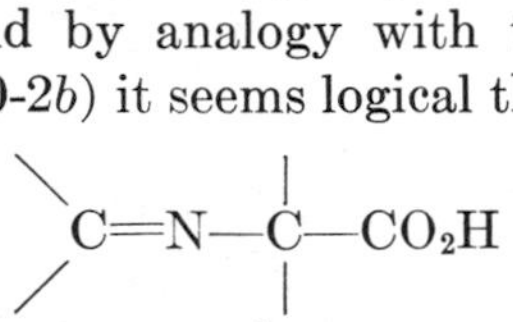

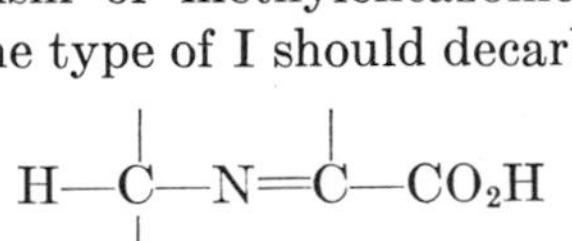

ate fairly readily and that those of the type of II could also, after tautomerizing to I. In fact, the enzymatic decarboxylation of α-amino and α-keto acids is related to the interconversion of such compounds via

[16] R. Steinberger and F. H. Westheimer, *J. Am. Chem. Soc.*, **73**, 429 (1951).

[17] K. J. Pedersen, *Acta Chem. Scand.*, **6**, 285 (1952).

[18] J. E. Prue, *J. Chem. Soc.*, 2331 (1952).

intermediates like I and II.   Snell and coworkers studied the biological activity of pyridoxal (III—vitamin B_6), whose presence is required for the activity of most decarboxylases and transaminases.[19]   For all these enzymes certain metal ions act as catalysts, and in most cases mixtures of pyridoxal and metal salts show considerable catalytic activity (although not so much activity as when the enzyme is present).   The reactions of amino acids appear to involve an initial condensation with pyridoxal followed by chelation with the metal ion.   The intermediate IV, formed

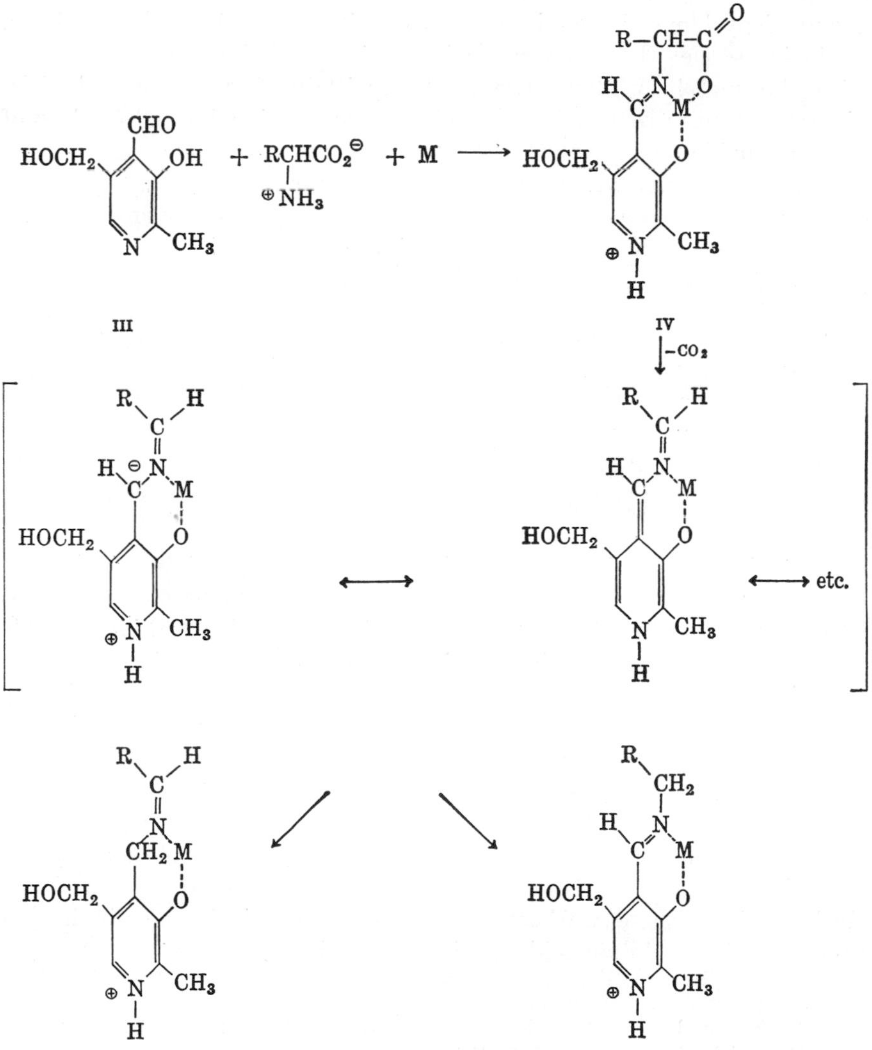

III   + $RCHCO_2^{\ominus}$ ($\overset{\oplus}{N}H_3$)   +   M $\longrightarrow$ IV

$\downarrow -CO_2$

[19] E. E. Snell et al., *J. Am. Chem. Soc.*, **76**, 637, 639, 644, 648, 653 (1954).

thereby, may lose carbon dioxide to give a resonance-stabilized intermediate that may be protonated in either of two ways, one of which yields a compound that can be hydrolyzed to an amine and pyridoxal and the other a compound that can be hydrolyzed to an aldehyde and pyridoxamine (which has vitamin B_6 activity equal to that of pyridoxal). If, instead of carbon dioxide, a proton is removed from the α-carbon of the amino acid part of IV, the resonance-stabilized species formed can be protonated to give a compound that can be hydrolyzed to an α-keto acid and pyridoxamine. According to this reaction mechanism, the aldehyde group of pyridoxal is required to react with the amino acid to give a Schiff base, the o-hydroxy group is required to take part in chelating with the metal ion, and the electron-withdrawing protonated ring-nitrogen atom is required to stabilize the carbanion formed by loss of carbon dioxide or a proton. The hydroxymethyl group appears to be useless in the nonenzymatic reaction but is believed to furnish the point of attachment to the enzyme in the reaction in vivo. Strong support for this explanation of the function of the various substituent groups in pyridoxal comes from the observation that in the presence of certain metal salts 4-nitrosalicylaldehyde and, to a smaller extent, 6-nitrosalicylaldehyde are effective catalysts for several of the same transamination reactions that are known to be catalyzed by pyridoxal. Salicylaldehyde, 4-nitrobenzaldehyde, and 3,5-dinitrosalicylaldehyde show little or no catalytic activity under the same conditions.

Picolinic acid, quinaldinic acid, and related compounds bear a certain resemblance to acids I and II. However, they cannot tautomerize in the manner I and II can, and unlike the anion of I, they cannot lose carbon dioxide to give a carbanion with two essentially equivalent structures; nor can they react by the concerted addition of a proton at one carbon atom and loss of carbon dioxide at another to give a stable product. Nevertheless, acids of this type can be decarboxylated without excessive difficulty. Hammick and coworkers described good evidence that these reactions involve some type of α-pyridyl (or α-quinolyl) carbanion. They observed that the decarboxylations in the presence of aldehydes and ketones produce α-pyridyl- (and α-quinolyl-) carbinols.[20] By showing that the decarboxylation of quinaldinic acid is first-order, that the rate may be slowed by the addition of either acid or base, and that the dipolar ion, 1-methylquinolinium-2-carboxylate, reacts relatively rapidly, these workers demonstrated that it is the dipolar-ion tautomer of quinaldinic acid that is the reactive form.[21]

[20] P. Dyson and D. L. Hammick, *J. Chem. Soc.*, 1724 (1937); M. R. F. Ashworth, R. P. Daffern, and D. L. Hammick, *J. Chem. Soc.*, 809 (1939); cf. B. R. Brown and D. L. Hammick, *J. Chem. Soc.*, 173 (1949).

[21] B. R. Brown and D. L. Hammick, *J. Chem. Soc.*, 659 (1949).

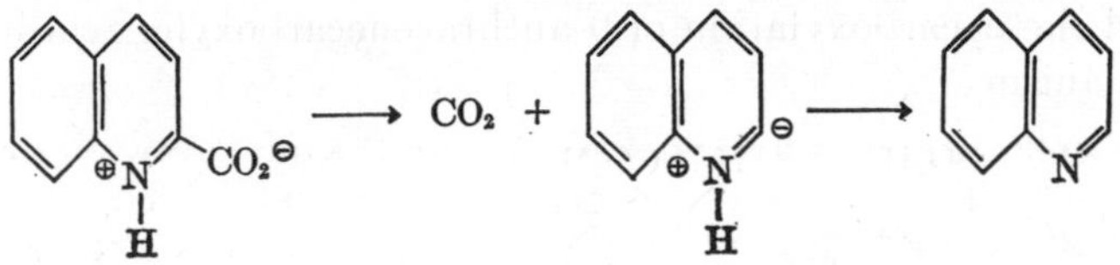

13-1d. The Carbonium-ion Mechanism for Decarboxylation. The formation of a carbonium ion at the β-carbon atom of a carboxylic acid (or better still, a carboxylate anion) might well be expected to lead to decarboxylation by a mechanism of the type

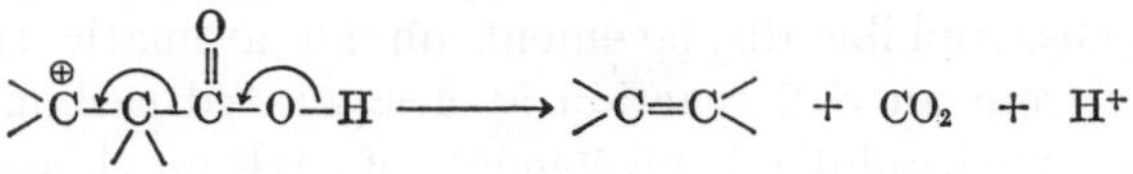

or

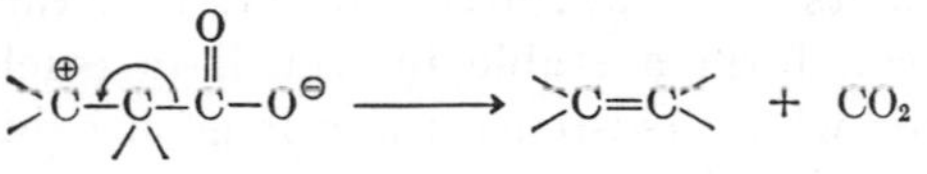

The work of Grovenstein and Lee on certain decarboxylations of this kind[22] has been described under elimination reactions (Sec. 8-2c). These workers pointed out that the reaction may proceed via an intermediate carbonium ion in some cases, whereas in others the loss of carbon dioxide may be simultaneous with the removal from the β-carbon atom of a group with its bonding electron pair. Johnson and Heinz suggested a carbonium-ion mechanism for the decarboxylation of cinnamic acids by acidic catalysts.[23]

For β,γ-unsaturated acids a decarboxylation mechanism rather similar to that we have described for β-keto acids has been suggested.[24]

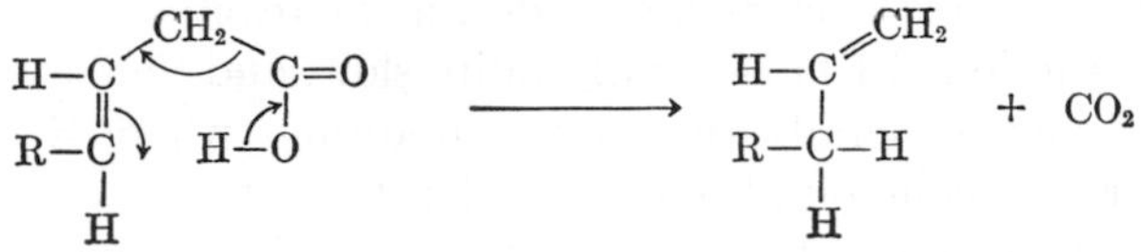

Both these reactions are of the carbonium-ion type in so far as the proton transfer precedes the loss of carbon dioxide and are of the four-center type in so far as the reaction is entirely concerted. There need be no sharp line between the two classes of reactions.

There are a number of aromatic substitution–type reactions in which carboxy groups on an aromatic ring are replaced by hydrogen atoms in the presence of an acid catalyst. They can be regarded as an example of carbonium-ion decarboxylations. Schenkel and Schenkel-Rudin have

[22] E. Grovenstein, Jr., and D. E. Lee, J. Am. Chem. Soc., **75**, 2639 (1953).
[23] W. S. Johnson and W. E. Heinz, J. Am. Chem. Soc., **71**, 2913 (1949).
[24] R. T. Arnold, O. C. Elmer, and R. M. Dodson, J. Am. Chem. Soc. **72**, 4359 (1950).

discussed the decarboxylation of 9-anthracenecarboxylic acid in terms of the mechanism

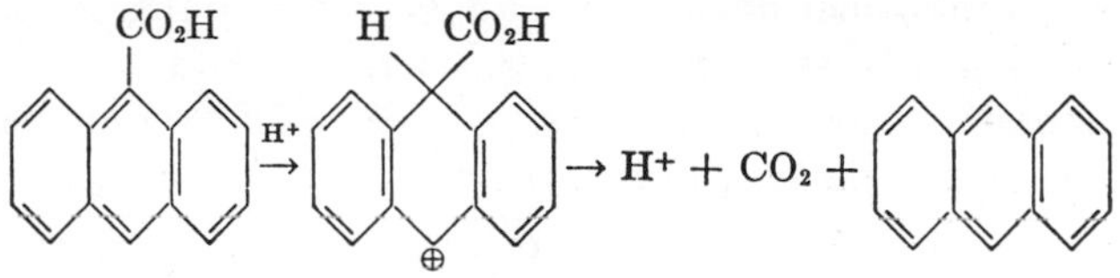

which they refer to as the S_E2 mechanism by analogy to the S_N2 reaction mechanism.[25] It should be noted that this mechanism for the bimolecular electrophilic displacement on an aromatic ring (see Sec. 16-2g) differs from an S_E2 reaction at a saturated carbon atom in the same way that nucleophilic displacements of acyl, vinyl, and aryl halides differ from displacements at a saturated carbon; viz., the reactions at unsaturated carbon can form a stable intermediate (such as the carbonium ion above) for whose existence there is good evidence in some cases, e.g., ester hydrolysis, whereas the reaction at a saturated carbon is best interpreted as proceeding by a concerted one-step process.

Brown, Hammick, Scholefield, and Elliott studied the acid-catalyzed decarboxylation of some o- and p-hydroxybenzoic acids, and Schubert and Gardner studied the reaction of mesitoic acid and 2,4,6-trihydroxybenzoic acid.[26]

13-1e. *Carboxylation.* It appears that the equilibrium

$$RH + CO_2 \rightleftharpoons RCO_2H \tag{13-1}$$

is often closely enough balanced at ordinary temperatures and concentrations that the reaction can be driven to either side (e.g., by removing or adding high pressures of carbon dioxide), if the reaction rate is sufficient. The rate of reaction (13-1) is usually quite slow, and therefore carboxylation reactions are generally run under conditions where R exists as the carbanion or as a compound with considerable carbanion character. In some cases, such as the Kolbe-Schmitt synthesis of salicylic acid, it is sufficient to use a basic catalyst that will transform some of the RH to its anion, but often the base only reacts with the carbon dioxide. The synthesis of carboxylic acids from organometallic compounds

$$RMgX + CO_2 \rightleftharpoons RCO_2MgX$$

is made feasible by the carbanion character of the organic reactant and

[25] H. Schenkel and M. Schenkel-Rudin, *Helv. Chim. Acta*, **31**, 514 (1948).

[26] B. R. Brown, D. L. Hammick, and A. J. B. Scholefield, *J. Chem. Soc.*, 778 (1950); B. R. Brown, W. W. Elliott, and D. L. Hammick, *J. Chem. Soc.*, 1384 (1951); W. M. Schubert, *J. Am. Chem. Soc.*, **71**, 2639 (1949); W. M. Schubert and J. D. Gardner, *J. Am. Chem. Soc.*, **75**, 1401 (1953).

also by the fact that the equilibrium is more favorable in the cases that, and to the extent that, RCO_2H is more acidic than RH.

Stiles and Finkbeiner demonstrated that the equilibrium in carboxylation reactions may be shifted well to the right by chelation of the product. Thus, magnesium methyl carbonate is capable of transforming primary nitro compounds to the magnesium salts of their carboxylation products.[27]

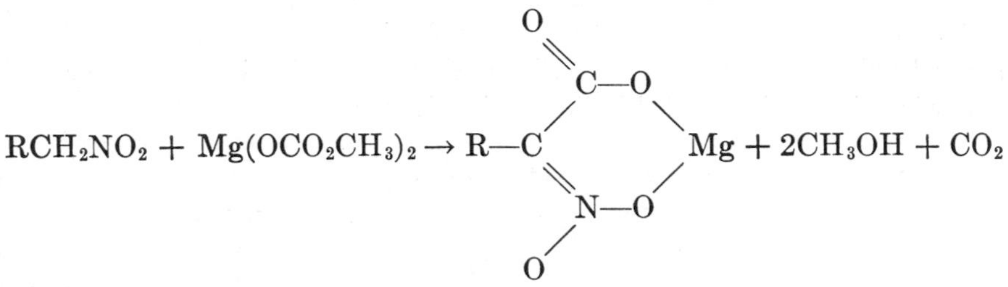

$$RCH_2NO_2 + Mg(OCO_2CH_3)_2 \rightarrow R-C \underset{N-O}{\overset{C-O}{\Big\langle}} Mg + 2CH_3OH + CO_2$$

Similarly, β-keto acids can be made from magnesium methyl carbonate and ketones of the type $RCOCH_2R'$.[28]

13-2. Decarbonylation and Carbonylation. By decarbonylation we mean the loss of carbon monoxide from a molecule. If the R group in RCHO can form a stable enough anion, the compound can be decarbonylated by the base-catalyzed mechanism

$$RCHO + B^{\ominus} \rightarrow BH + RCO^{\ominus} \rightarrow CO + R^{\ominus}$$

$$R^{\ominus} + BH \rightarrow RH + B^{\ominus}$$

and if R in RCO_2H can form a sufficiently stable cation, the acid-catalyzed mechanism

$$RCO_2H + H^+ \rightarrow RCOOH_2^{\oplus} \rightarrow H_2O + RCO^{\oplus}$$

$$RCO^{\oplus} \rightarrow CO + R^{\oplus} \xrightarrow{H_2O} ROH + H^+ \quad \text{etc.}$$

can occur. Examples of decarbonylations and their reverse, carbonylations, by both mechanisms are known. As an example of the base-catalyzed reaction, it is known that alkali-metal alkoxides catalyze the decomposition of formate esters into carbon monoxide and alcohols (the reverse reaction takes place as well, of course).[29] The acid-catalyzed decarbonylation reaction is general for α-hydroxy acids, since the "car-

---

[27] M. Stiles and H. L. Finkbeiner, *J. Am. Chem. Soc.*, **81**, 505 (1959).

[28] M. Stiles, *J. Am. Chem. Soc.*, **81**, 2598 (1959).

[29] F. Adickes and G. Schäfer, *Ber.*, **65B**, 950 (1932); J. A. Christiansen and J. C. Gjaldbaek, *Kgl. Danske Videnskab. Selskab., Mat.-fys. Medd.*, **20**, no. 3 (1942); *Chem. Abstr.*, **38**, 3898⁶ (1944).

bonium ions" formed are simply the conjugate acids of aldehydes and ketones. The reaction also occurs with formic acid, benzoylformic acid, and compounds of the type RCO_2H, where R is a tertiary alkyl radical. Although the mechanisms of these reactions probably have the general character of that shown above, there may be differences in some of the mechanistic details, since the reaction rates are variously affected by changes in the acidity function, H_0.[30]

As one of a large number of examples of acid-catalyzed carbonylations, several polyhalomethanes react with carbon monoxide in the presence of a Lewis acid to give acetyl halide derivatives.[31]

$$CCl_4 + AlCl_3 \rightarrow AlCl_4^{\ominus} + CCl_3^{\oplus}$$
$$CCl_3^{\oplus} + CO \rightarrow Cl_3CCO^{\oplus} \xrightarrow{AlCl_4^{\ominus}} Cl_3CCOCl$$

Pincock, Grigat, and Bartlett showed that in the preparation of the decalin-9-carboxylic acids via carbonylation of the corresponding carbonium ion, the trans acid is the kinetically controlled product and the cis acid the thermodynamically controlled product.[32] Meinwald and coworkers showed that the rearrangement of α-cinenic acid to geronic acid involves decarbonylation and recarbonylation.[33]

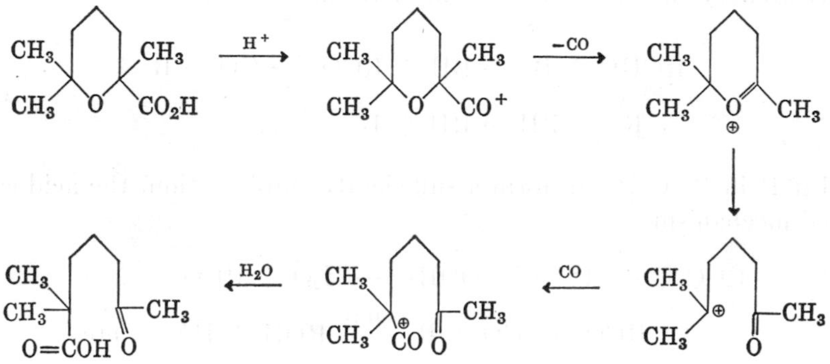

13-3. The Acetoacetic Ester Condensation and Related Reactions.

13-3a. *The Acetoacetic Ester Condensation.* A carbon-carbon bond-forming condensation between two molecules of an ester to give a β-keto ester is called an acetoacetic ester, or Claisen, condensation. The latter term is also applied to related reactions. Hauser and Renfrow

[30] L. P. Hammett, *Chem. Revs.*, **16**, 67 (1935); W. W. Elliott and D. L. Hammick, *J. Chem. Soc.*, 3402 (1951), and references cited therein.

[31] C. W. Theobald, U.S. Patent 2,378,048; *Chem. Abstr.*, **39**, 4085⁶ (1945).

[32] R. E. Pincock, E. Grigat, and P. D. Bartlett, *J. Am. Chem. Soc.*, **81**, 6332 (1959).

[33] J. Meinwald, H. C. Hwang, D. Christman, and A. P. Wolf, *J. Am. Chem. Soc.*, **82**, 483 (1960).

appear to be the first to have suggested the currently accepted reaction mechanism in terms of modern physical organic chemical concepts.[34] This mechanism, rather similar to that of the aldol condensation, involves the reaction of the basic catalyst with one molecule of ester to yield a carbanion, which adds to the carbonyl group of another ester molecule, displacing an alkoxide ion.

(1) $EtO^{\ominus} + CH_3CO_2Et \rightleftharpoons EtOH + \overset{\ominus}{C}H_2CO_2Et$

(2) $CH_3\overset{\overset{O}{\|}}{C} + \overset{\ominus}{C}H_2CO_2Et \rightleftharpoons CH_3\overset{\overset{\overset{\ominus}{O}}{|}}{C}CH_2CO_2Et$
$\quad\quad |$
$\quad\quad OEt \quad\quad\quad\quad\quad\quad\quad\quad\quad OEt$

$\rightleftharpoons CH_3\overset{\overset{O}{\|}}{C}CH_2CO_2Et + EtO^{\ominus}$

(3) $CH_3COCH_2CO_2Et + EtO^{\ominus} \rightleftharpoons CH_3CO\overset{\ominus}{C}HCO_2Et + EtOH$

The last step, in which the β-keto ester is transformed into its conjugate base, is of the greatest importance for the practical success of the reaction. The equilibrium constant for the formation of the β-keto ester itself, as shown in the first two steps of the mechanism, has the form

$$K = \frac{[CH_3COCH_2CO_2Et][EtOH]}{[CH_3CO_2Et]^2}$$

and the equilibrium constant for the whole reaction (all three steps) is

$$K' = \frac{[CH_3CO\overset{\ominus}{C}HCO_2Et][EtOH]^2}{[CH_3CO_2Et]^2[EtO^{\ominus}]}$$

It may thus be seen that

$$K' = KK_3$$

where
$$K_3 = \frac{[CH_3CO\overset{\ominus}{C}HCO_2Et][EtOH]}{[CH_3COCH_2CO_2Et][EtO^{\ominus}]}$$

K_3, the equilibrium constant for step 3, must be at least 10^5, since acetoacetic ester, whose pK_a is about 10.7 in water,[35] is probably at least 10^5 times as strong an acid as ethanol under these reaction conditions. That is, the equilibrium constant for the reaction is made much more favorable by the acidity of the β-keto ester, which permits most of the ester to be changed to a form (its conjugate base) incapable of direct cleavage to

[34] C. R. Hauser and W. B. Renfrow, Jr., J. Am. Chem. Soc., **59**, 1823 (1937).
[35] R. G. Pearson and R. L. Dillon, J. Am. Chem. Soc., **75**, 2439 (1953).

starting material. In fact, esters that have only one α-hydrogen atom do not undergo the acetoacetic ester condensation in the presence of sodium alkoxide catalysts, because the β-keto ester formed would not be sufficiently acidic to be transformed to its conjugate base to any significant extent. The equilibrium in such reactions, however, may be forced to the right by the use of more strongly basic catalysts. Triphenylmethylsodium, mesitylmagnesium bromide, sodium amide, and diisopropylaminomagnesium bromide are among the bases that have been found capable of bringing about the formation of β-keto esters that contain no α-hydrogen atoms.[34,36]

13-3b. *Alkaline Cleavage of β-Diketones and Related Compounds.* In general, compounds of the type RCOX may be easily cleaved by alkali if $X^\ominus$ is a fairly stable anion (cf. the alkaline hydrolysis of esters). It is therefore not unexpected that β-diketones are fairly readily cleaved by bases. Pearson and Sandy studied the cleavage of acetylacetone by sodium ethoxide in ethanol.[37] They made the observation, surprising at first glance, that the reaction is of the first order, its rate being proportional to the concentration of the reactant present in the smaller concentration. This observation is quite rational, however, in view of the following reaction mechanism:

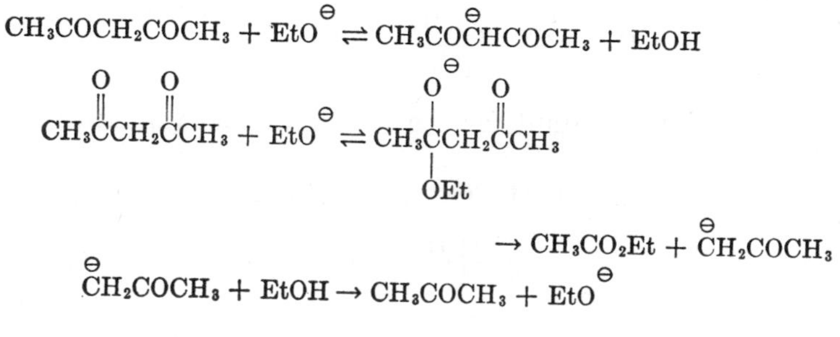

$$CH_3COCH_2COCH_3 + EtO^\ominus \rightleftharpoons CH_3CO\overset{\ominus}{C}HCOCH_3 + EtOH$$

$$\underset{\substack{\displaystyle\| \\ O}}{CH_3C}\underset{\substack{\displaystyle\| \\ O}}{CH_2C}CH_3 + EtO^\ominus \rightleftharpoons CH_3\underset{\substack{| \\ OEt}}{\overset{O^\ominus}{C}}CH_2\overset{O}{\underset{\|}{C}}CH_3$$

$$\rightarrow CH_3CO_2Et + \overset{\ominus}{C}H_2COCH_3$$

$$\overset{\ominus}{C}H_2COCH_3 + EtOH \rightarrow CH_3COCH_3 + EtO^\ominus$$

If we assume that the rate-controlling step is the decomposition of the acetylacetone–ethoxide-ion adduct and that all the other steps are relatively fast, we can derive the kinetic equation

$$v = k[CH_3COCH_2COCH_3][EtO^-]$$

However, if the acid-base equilibrium of the first equation is established rapidly,

$$[CH_3COCH_2COCH_3][EtO^-] = K[CH_3CO\overset{\ominus}{C}HCOCH_3]$$

Therefore
$$v = kK[CH_3CO\overset{\ominus}{C}HCOCH_3] \qquad (13\text{-}2)$$

[36] C. R. Hauser et al., *J. Am. Chem. Soc.*, **68**, 2647 (1946); **69**, 2649 (1947); **71**, 1350 (1949).

[37] R. G. Pearson and A. C. Sandy, *J. Am. Chem. Soc.*, **73**, 931 (1951).

Acetylacetone is an acid of such a strength that the concentration of its anion present in a basic solution will be essentially equal to the concentration of sodium ethoxide or of acetylacetone added, depending on which is the limiting reagent. The mechanism given, then, explains the observation of first-order kinetics. The alkaline ethanolysis of phenacylpyridinium ions follows a similar kinetic equation and has the same type of mechanism.[37]

The alkaline hydrolysis of acetylacetone is a more complicated reaction, the kinetic equation having the form of (13-2) with an additional term involving hydroxide ion.[38]

$$v = k[CH_3CO\overset{\ominus}{C}HCOCH_3] + k'[CH_3CO\overset{\ominus}{C}HCOCH_3][OH^-]$$

The first term corresponds to cleavage by a mechanism like that followed in alcoholic solution. As Pearson and Mayerle point out, the second term, involving an additional hydroxide ion, is probably due to cleavage by the mechanism

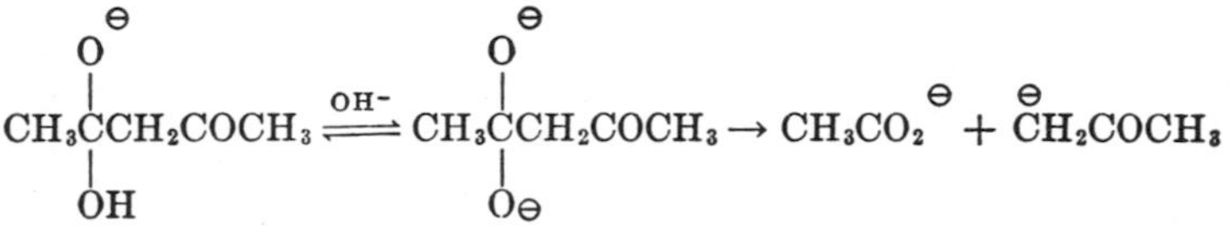

$$CH_3\underset{OH}{\overset{\overset{\ominus}{O}}{\underset{|}{C}}}CH_2COCH_3 \overset{OH^-}{\rightleftharpoons} CH_3\underset{O\ominus}{\overset{\overset{\ominus}{O}}{\underset{|}{C}}}CH_2COCH_3 \rightarrow CH_3\overset{\ominus}{C}O_2 + \overset{\ominus}{C}H_2COCH_3$$

Although the doubly charged intermediate must be present at a much smaller concentration than the singly charged one, its specific rate constant for cleavage is probably much larger. This situation is somewhat similar to that found in the Cannizzaro reaction (Sec. 11-4a).

Dimethylacetylacetone cannot be "protected" by transformation to its conjugate base and is therefore cleaved much more rapidly than the unsubstituted compound or its monomethyl derivative.

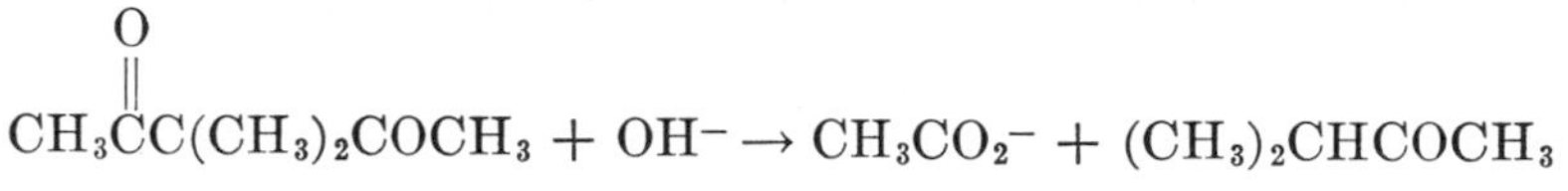

$$CH_3\overset{\overset{O}{||}}{C}C(CH_3)_2COCH_3 + OH^- \rightarrow CH_3CO_2^- + (CH_3)_2CHCOCH_3$$

The kinetics in this case are second-order, first-order in hydroxide ion and first-order in ketone.[38]

The study of the basic cleavage of chloral hydrate carried out by Gustafsson and Johanson showed that this part of the haloform reaction probably has a mechanism of the type described for the cleavage of β-diketones.[39] This mechanism should be modified to agree with the observation of general base catalysis,[39] and, indeed, it is possible that the reactions studied by Pearson and coworkers are also subject to general base catalysis.

[38] R. G. Pearson and E. A. Mayerle, *J. Am. Chem. Soc.*, **73**, 926 (1951).
[39] C. Gustafsson and M. Johanson, *Acta Chem. Scand.*, **2**, 42 (1948).

PROBLEMS

1. Suggest a reasonable mechanism to explain why potassium cyanide is a catalyst in the transformation of $RCOCO_2H$ to $RCHO$.

2. Write a plausible mechanism for the reaction of threonine ($CH_3CHOHCHNH_2$-CO_2H) with pyridoxal and aluminum ions. The products include considerable acetaldehyde and glycine and less α-ketobutyric acid and ammonia.

3. Using equilibrium constants, as in the explanation in Sec. 13-3a, explain how the mixed Claisen condensation of ethyl benzoate with ethyl isobutyrate is driven to the right by the use of triphenylmethylsodium.

4. Suggest a reasonable mechanism for the alkaline cleavage (general base-catalyzed) of chloral hydrate to chloroform and formate ions.

Chapter 14

REARRANGEMENTS DUE
TO ELECTRON-DEFICIENT CARBON

Several types of rearrangement reactions have already been described under various other headings. The Stevens rearrangement has been mentioned as an example of an S_Ni reaction (Sec. 6-3); allylic rearrangements by the S_N1 and S_N2' mechanisms were discussed in Sec. 6-5; and the Favorskii rearrangement in Sec. 10-3c. Other carbon-skeleton rearrangements will be discussed here.

14-1. Carbonium-ion Rearrangements. Of the very large number of carbonium-ion rearrangements we shall discuss only a few, most of them chosen because of the care with which they have been studied. It is hoped, nevertheless, that certain generalizations pointed out in this and subsequent sections will lead to an understanding of the nature of most of the rearrangements not mentioned specifically.

14-1a. Evidence that Carbonium Ions Are Intermediates. Meerwein and van Emster obtained some of the first strong evidence that carbonium ions may be intermediates in carbon-skeleton rearrangements in their study of the rearrangement of camphene hydrochloride to isobornyl chloride.[1] They observed that the reaction rate increased with the nature of the solvent in roughly the same order that had been found for the ability of solvents to ionize triphenylmethyl chloride. They further established that compounds such as $HgCl_2$, $FeCl_3$, $SnCl_4$, etc., capable of coordinating with a chloride ion, were excellent catalysts for the reaction. It was later shown that the rearrangement of the chloride was slower than that of the bromide or arylsulfonate but faster than that of the trichloroacetate or *m*-nitrobenzoate.[2] This order of reactivity is understandable, since the ease of ionization should increase with the stability of the anion being formed.

The probability of a carbonium-ion mechanism for rearrangement reactions was emphasized (in somewhat different language) by Whitmore

[1] H. Meerwein and K. van Emster, *Ber.*, **55**, 2500 (1922).

[2] H. Meerwein, O. Hammel, A. Serini, and J. Vorster, *Ann.*, **453**, 16 (1927).

and coworkers in a number of subsequent papers. It was further supported by Dostrovsky and Hughes, who showed that the second-order reaction of neopentyl bromide with sodium ethoxide gave the normal product, ethyl neopentyl ether, but that the unimolecular solvolyses and the reaction with silver ion gave rearranged products.[3]

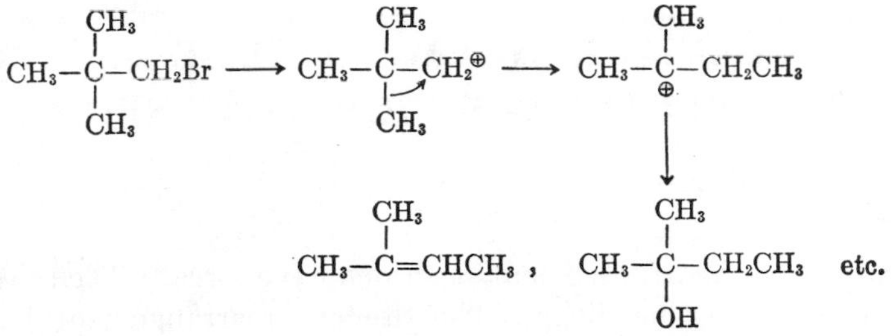

$$CH_3-\underset{\underset{CH_3}{|}}{\overset{\overset{CH_3}{|}}{C}}-CH_2Br \longrightarrow CH_3-\underset{\underset{CH_3}{|}}{\overset{\overset{CH_3}{|}}{C}}-CH_2^{\oplus} \longrightarrow CH_3-\underset{\oplus}{\overset{\overset{CH_3}{|}}{C}}-CH_2CH_3$$

$$CH_3-\underset{\overset{CH_3}{|}}{C}=CHCH_3 \; , \quad CH_3-\underset{\underset{OH}{|}}{\overset{\overset{CH_3}{|}}{C}}-CH_2CH_3 \quad \text{etc.}$$

Although carbonium-ion rearrangements were originally written as simply involving the shift of an alkyl group or hydrogen atom from an adjacent atom to the electron-deficient carbon atom (as shown in the case of neopentyl above), it was subsequently suggested, first by Nevell, de Salas, and Wilson,[4] that there may be formed a carbonium ion of intermediate structure.

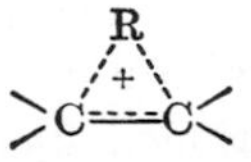

$$>C\overset{\overset{\displaystyle R}{\overset{\diagup+\diagdown}{}}}{=\!=\!=}C<$$

A configuration like this would be expected to be formed at some time as R migrates from one carbon atom to the next. The uniqueness of the present suggestion is that this configuration corresponds to an energy minimum.

14-1b. *Migration of β-Aryl Groups.* There is good evidence that neighboring aromatic radicals may participate in displacement reactions in a manner analogous to that described for RS—, R_2N—, halogen, etc., groups (Sec. 6-4). Cram's investigations of 3-phenyl-2-butyl and of 3-phenyl-2-pentyl and 2-phenyl-3-pentyl compounds are of interest in this regard.[5] The four stereoisomers of 3-phenyl-2-butanol were separated. These compounds will be called IA and IB (one *dl* pair) and IIA and IIB (the other *dl* pair). Evidence based on the structures of the olefins obtained by the Chugaev decomposition of these alcohols suggests that *dl* pair I has the threo structure shown and that II is the erythro

[3] I. Dostrovsky and E. D. Hughes, *J. Chem. Soc.*, 157, 164, 166, 169, 171 (1946).

[4] T. P. Nevell, E. de Salas, and C. L. Wilson, *J. Chem. Soc.*, 1188 (1939).

[5] D. J. Cram, *J. Am. Chem. Soc.*, **71**, 3863, 3875, 3883 (1949).

racemate. Furthermore, the fact that the p-toluenesulfonate (tosylate) of IA reacts with potassium acetate in absolute ethanol (conditions thought likely to cause an S_N2 replacement of tosylate by acetate) to yield the acetate of IIA implies that in IA and IIA the configuration around the carbon atom to which the phenyl group is attached is identical. This evidence for the configurations of these compounds is considerably strengthened by the fact that these configurations are those which must be used in the most reasonable mechanism for acetolysis of the tosylates. The acetolysis of these tosylates was found to proceed as follows: an active threo form (IA) yields racemic threo acetate, whereas an active erythro form (IIA) yields the acetate of the same active erythro alcohol (IIA).

$$\text{IA tosylate} \xrightarrow{\text{HOAc}} \text{IA and IB acetates} \qquad \text{(racemic I acetate)}$$
$$\text{IIA tosylate} \xrightarrow{\text{HOAc}} \text{IIA acetate}$$

As described in the discussion of bromonium ions (Sec. 6-4e), in an S_N2 reaction a IA compound would yield a IIA derivative, whereas by an S_Ni mechanism, an active (IA) product would be formed, rather than the racemic product observed. The results of an ordinary S_N1 reaction would be the same as the S_N2 or S_Ni noted above, or a combination, depending on which side of the carbonium ion the entering group is assumed to attack. The reaction is evidently not an entirely concerted one like that suggested[6] for the related rearrangement of neopentyl halides

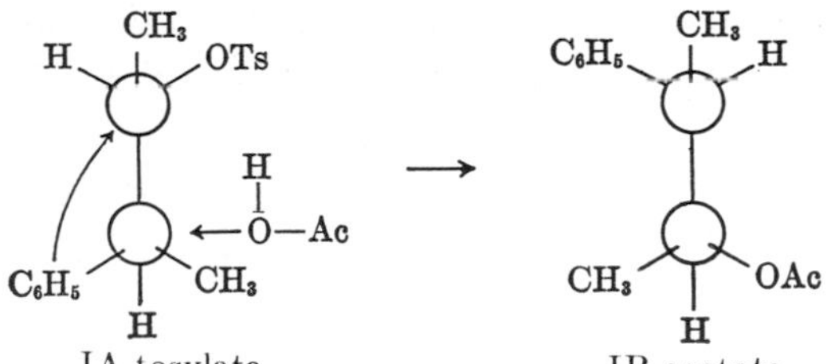

IA tosylate   IB acetate

since this should lead to the formation of only IB acetate. Therefore the most suitable mechanism appears to be that suggested by Cram, involving the participation of the neighboring phenyl group. With the threo IA, the "phenonium ion" formed has a plane of symmetry and hence can yield only racemic products, since it is equally likely to give the acetate of IA or IB.

[6] C. G. Swain, *J. Am. Chem. Soc.*, **70**, 1126 (1948).

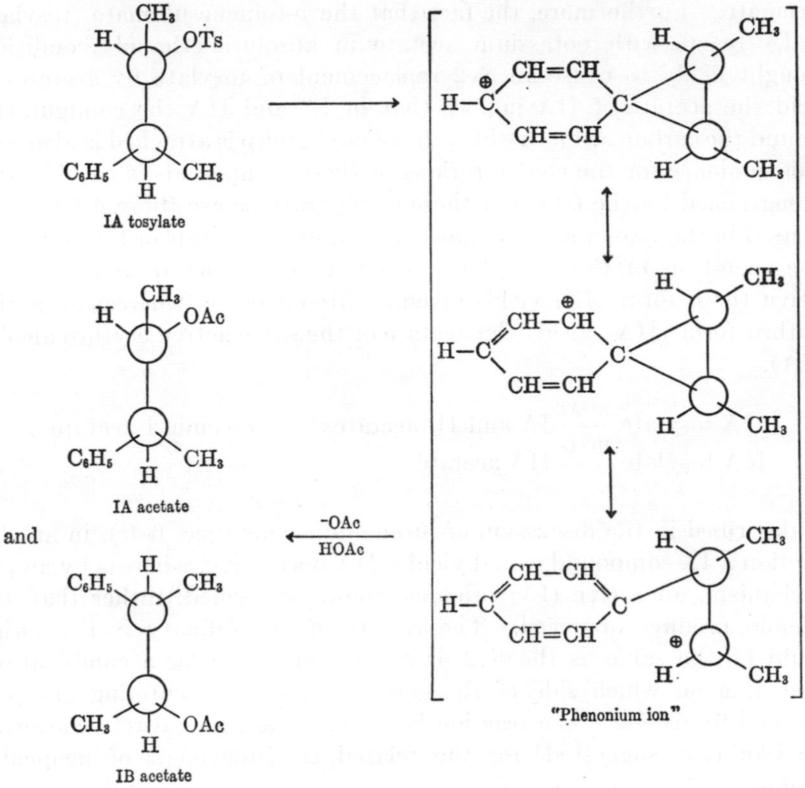

CH₃
H   OTs

C₆H₅   CH₃
H
IA tosylate

CH₃
H   OAc

C₆H₅   CH₃
H
IA acetate

and

C₆H₅   CH₃
H

CH₃   OAc
H
IB acetate

$\xleftarrow[\text{HOAc}]{^-\text{OAc}}$

H   CH₃
CH=CH
H—C⊕   C
CH=CH
H   CH₃

H   CH₃
CH—CH
H—C   C
CH=CH
H   CH₃

H   CH₃
CH—CH
H—C   C
CH=CH
H⊕   CH₃

"Phenonium ion"

In the case of the erythro compound, IIA, the intermediate phenonium ion is optically active and gives the acetate of IIA by reaction at either carbon atom.

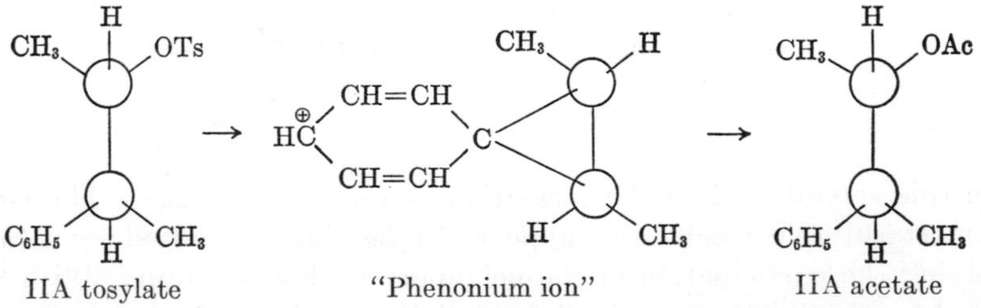

H
CH₃   OTs

C₆H₅   CH₃
H
IIA tosylate

→

CH₃   H
CH=CH
HC⊕   C
CH=CH
H   CH₃
"Phenonium ion"

→

H
CH₃   OAc

C₆H₅   CH₃
H
IIA acetate

It is highly unlikely that the acetate of I formed in the acetolysis of IA tosylate is inactive due to racemization after formation. It has also been shown that although the tosylate does racemize somewhat as it reacts, the extent of this racemization is not nearly enough to account for the formation of the racemic product.[7] When the acetolysis of IA tosylate was studied kinetically both by measuring the rate of loss of optical activity and by titrating the p-toluenesulfonic acid formed, it was found that both methods yielded good first-order rate constants but that the polarimetric rate constants were about five times as large as those determined titrimetrically.[8] This observation suggests that IA forms racemic ion pairs, which revert to tosylate 80 per cent of the time and form acetate only 20 per cent of the time. This suggestion is similar to that described in Sec. 6-5c for α,α-dimethylallyl chloride and is supported by further evidence of the type mentioned for the allyl halide.

In the acetolysis of certain related compounds, such as 2-phenyl-1-propyl tosylate[9,10] and 2,2,2-triphenylethyl tosylate,[11] neighboring-group participation by phenyl is manifest in the form of added driving force for the reaction.[12] The solvolysis of the 3-phenyl-2-butyl tosylates is only about 60 per cent as fast as that of 2-butyl tosylate,[11] but it can be argued that even this rate represents some augmentation in rate over that expected for simple carbonium-ion formation, since an electron-withdrawing β-phenyl substituent might be expected to decrease the S_N1 reactivity of 2-butyl tosylate by as much as a factor of ten.[15]

It is often possible to obtain evidence for various types of intermediates by studying related compounds containing structural features that would be expected to stabilize the intermediates. Evidence for the formation of carbonium ions, carbanions, and free radicals is strengthened by the fact that when enough stabilizing structural changes are made, one may find intermediates existing in concentrations large enough for direct detection and, in some cases, one may even find isolable species of the type in question. So it is with phenonium ions. A p-methoxy group has a marked effect, increasing the acetolysis rate of 3-phenyl-2-butyl

[7] D. J. Cram, *J. Am. Chem. Soc.*, **74**, 2129 (1952).

[8] S. Winstein and K. Schreiber, *J. Am. Chem. Soc.*, **74**, 2165 (1952).

[9] S. Winstein and H. Marshall, *J. Am. Chem. Soc.*, **74**, 1120 (1952).

[10] S. Winstein and K. C. Schreiber, *J. Am. Chem. Soc.*, **74**, 2171 (1952).

[11] S. Winstein, B. K. Morse, E. Grunwald, K. C. Schreiber, and J. Corse, *J. Am. Chem. Soc.*, **74**, 1113 (1952).

[12] For such rate enhancement due to participation, Ingold and coworkers use the term *synartetic acceleration*,[13] and Winstein and coworkers use the term *anchimeric acceleration*.[14]

[13] F. Brown, E. D. Hughes, C. K. Ingold, and J. F. Smith, *Nature*, **168**, 65 (1951).

[14] S. Winstein, C. R. Lindegren, H. Marshall, and L. L. Ingraham, *J. Am. Chem. Soc.*, **75**, 147 (1953).

[15] In making this comparison, it does not seem desirable to use the rate of ion-pair formation by 3-phenyl-2-butyl tosylate as determined by the rate of its racemization[7,8] because the rate of ion-pair formation by 2-butyl tosylate is unknown.

tosylate by about eightyfold.[16] With the p-O^- substituent the intermediate (III) can be detected spectrally and isolated.[17]

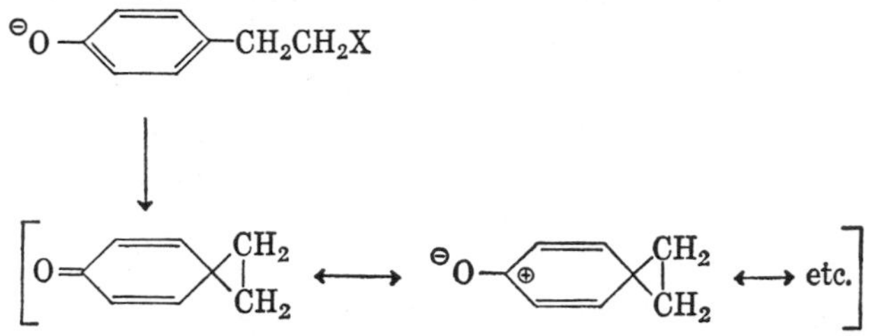

Although the term "phenonium ion" must be stretched to be applied to III, the isolation of III does add credibility to the suggestion that less stable phenonium ions are also formed as reaction intermediates.

Although the solvolysis mechanism for I and II involving carbon-skeleton rearrangement is certainly reasonable, it is not evident in the structure of the reaction products. This is not so, however, in the case of the 3-phenyl-2-pentyl tosylates, which yield 2-phenyl-3-pentyl acetates as well as unrearranged products. Studies of the solvolysis of these compounds and the related 2-phenyl-3-pentyl tosylates further support the rearrangement mechanisms we have discussed.[5]

14-1c. *Neighboring Vinyl Groups.* There is evidence that neighboring vinyl (and substituted vinyl) groups may behave in a manner analogous to that described for phenyl groups. Shoppee suggested that the double bond in cholesteryl halides and related compounds may display neighboring-group participation in replacement reactions.[18] Kinetic studies have yielded evidence that a resonance-stabilized carbonium ion of the type (IV) shown below is an intermediate in the reactions of certain cholesteryl compounds.[19-21] A rapid tautomeric equilibrium between two isomeric carbonium ions might be suggested to explain this rearrangement, but this would not explain the fact that cholesteryl tosylate solvolyzes about 100 times as rapidly as does cyclohexyl tosylate. The proposed mecha-

[16] S. Winstein, M. Brown, K. C. Schreiber, and A. H. Schlesinger, *J. Am. Chem. Soc.*, **74**, 1140 (1952).

[17] S. Winstein and R. Baird, *J. Am. Chem. Soc.*, **79**, 756, 4238 (1957).

[18] C. W. Shoppee, *J. Chem. Soc.*, 1147 (1946).

[19] S. Winstein et al., *J. Am. Chem. Soc.*, **70**, 838 (1948); **81**, 4399 (1959).

[20] R. G. Pearson, L. C. King, et al., *J. Am. Chem. Soc.*, **70**, 3479 (1948); **73**, 4149 (1951); **74**, 6238 (1952).

[21] C. W. Shoppee et al., *J. Chem. Soc.*, 3361 (1952); 540 (1953); 679, 686 (1955).

nism involving ionization directly to the resonance-stabilized IV does explain this fact, although perhaps not uniquely.

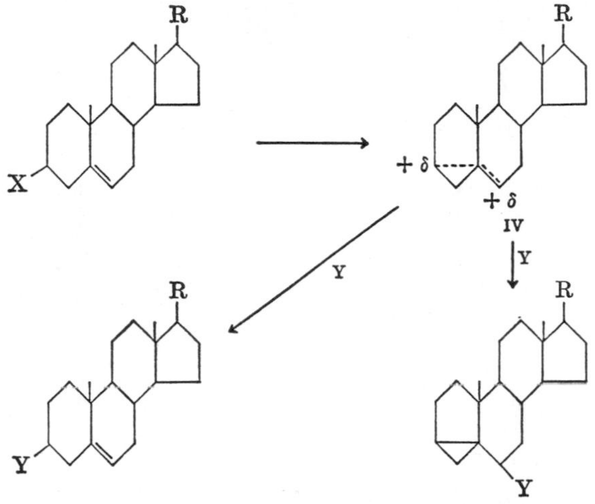

Similar participation of a double bond has been used to explain the solvolyses of dehydronorbornyl compounds that yield nortricyclyl derivatives.[22,23]

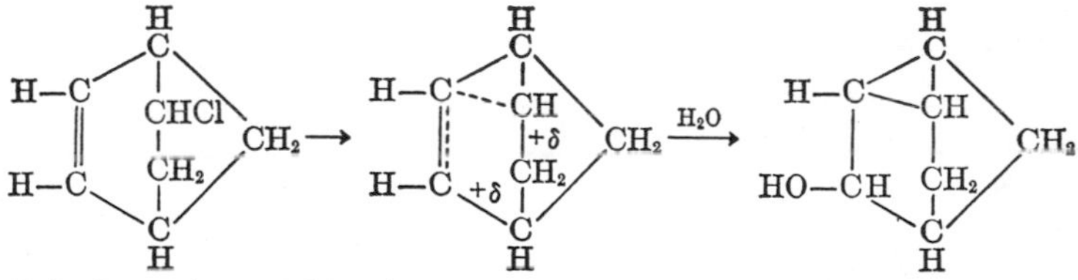

Dehydronorbornylchloride   3-Hydroxynortricylene

14-1d. *Migration of Saturated Alkyl Groups.*   The first strong evidence for the participation of a saturated hydrocarbon radical in a nucleophilic displacement reaction on carbon appears to be due to Winstein and Trifan. These workers showed that optically active *exo*-norbornyl *p*-bromobenzenesulfonate, upon acetolysis, yields racemic *exo*-norbornyl acetate.[24] The racemization was found to proceed more rapidly than the solvolysis, indicating that the initial product of reaction is an ion pair that may recombine or dissociate (see Sec. 6-5c).

[22] J. D. Roberts et al., *J. Am. Chem. Soc.*, **72**, 3116, 3329 (1950); **76**, 4623 (1954); **77**, 3034 (1955).

[23] S. Winstein, H. M. Walborsky, and K. Schreiber, *J. Am. Chem. Soc.*, **72**, 5795 (1950).

[24] S. Winstein and D. S. Trifan, *J. Am. Chem. Soc.*, **71**, 2953 (1949); **74**, 1154 (1952).

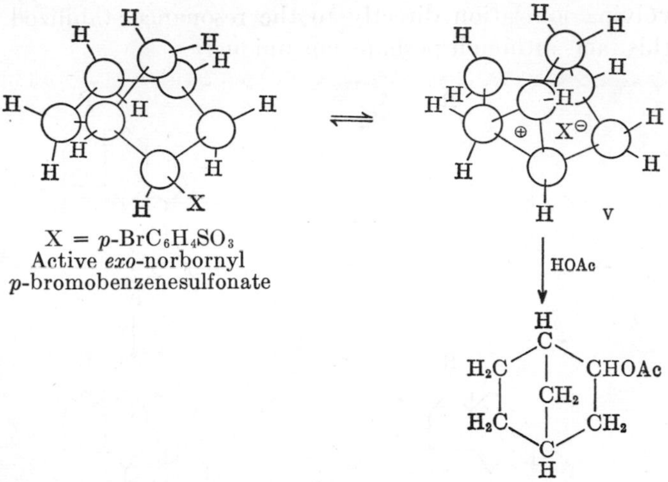

X = p-BrC$_6$H$_4$SO$_3$
Active *exo*-norbornyl
p-bromobenzenesulfonate

Racemic *exo*-norbornyl acetate

Any explanation postulating a rapid tautomeric equilibrium between two enantiomorphic norbornyl carbonium ions does not explain the fact that *exo*-norbornyl p-bromobenzenesulfonate solvolyzes about 350 times as rapidly as does the endo isomer. This fact is explained by neighboring-group participation to form a carbonium ion with structure V, since with the exo compound the neighboring group may attack from

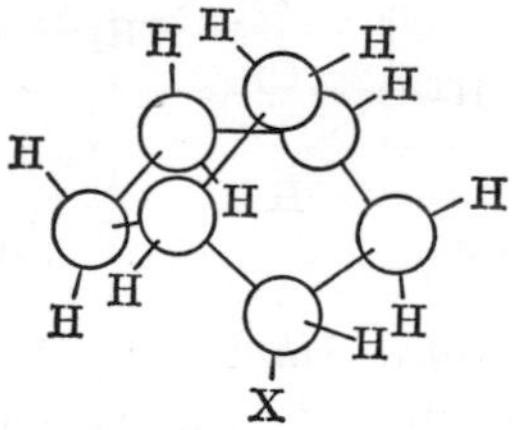

Optically active *endo*-norbornyl p-bromobenzenesulfonate

the rear while the p-bromobenzenesulfonate ion is leaving the front. Thus in the case of the exo compound the more stable bridged carbonium ion (V) is being formed in the transition state, whereas with the endo compound it appears that an ordinary carbonium ion is formed first and that this rearranges to carbonium ion V (which does appear to be an intermediate in the solvolysis of the endo compound also, because racemic *exo*-acetate is formed in the acetolysis of optically active *endo*-p-bromobenzenesulfonate). Roberts and Lee reported studies using C^{14} that show that the carbon-skeleton rearrangement is more drastic than would be expected from the intermediacy of carbonium ion V alone, and it therefore appears that V is not the only carbonium-ion intermediate in the reaction.[25]

The carbonium-ion-type reactions of a number of cyclobutyl and cyclopropylcarbinyl compounds give about the same mixture of products and appear to involve nonclassical carbonium ions.[26]

14-1e. *Types of Rearrangement Mechanisms.* Even in the simplest example of a rearrangement by the carbonium-ion mechanism we could suggest as many as four steps

$$
\begin{array}{ccccccccc}
 & R & & & R & & & R & & & R & & & R \\
 & | & | & & | & | & & | & & & | & | & & | & | \\
-C & - & C- & \longrightarrow & -C & - & C- & \longrightarrow & -C & = & C- & \longrightarrow & -C & - & C- & \longrightarrow & -C & - & C- \\
 & | & | & & | & \oplus & & | & | & & \oplus & | & & | & | \\
 & X & & & & & & & & & & & & & Y \\
\end{array}
$$

$$\text{VI} \qquad\qquad \text{VII} \qquad\qquad \text{VIII}$$

involving an unrearranged carbonium ion (VI), an intermediate carbonium ion (VII), and a rearranged carbonium ion (VIII). There are cases, however, in which the experimental evidence suggests that some of the steps have become fused, i.e., in which one or more of the three possible intermediate carbonium ions are no longer a true intermediate. Considering the existence or nonexistence of three different carbonium ions, there will be $2^3 = 8$ different kinds of cases. More work is needed to learn the relation between the structure of the reactants, reaction conditions, etc., and the type of case that will appear. However, it seems that some reactions may be classified as being of a certain type and furthermore that not all are of the same type. Thus in the case of the solvolysis of *exo*-norbornyl *p*-bromobenzenesulfonate the reactant is transformed directly to an ion of type VII (as part of an ion pair), and this ion goes directly to the products, with ions of type VI and VIII never appearing as true intermediates. The solvolysis of the corresponding endo isomer, on the other hand, appears to involve the initial formation of a carbonium ion like VI, which may then rearrange to the bridged ion VII; VII then yields the observed exo product directly. In some rearrangements there is no need to postulate any species like VII as a true reaction intermediate. Winstein and Morse, for example, showed that optically active α-phenylneopentyl chloride and *p*-toluenesulfonate solvolyze to give completely racemic rearranged products and largely racemic, but partly inverted, unrearranged products.[27] Although a number of the other types of cases seem to have been observed, there

[25] J. D. Roberts and C. C. Lee, *J. Am. Chem. Soc.*, **73**, 5009 (1951); J. D. Roberts, C. C. Lee, and W. H. Saunders, Jr., *J. Am. Chem. Soc.*, **76**, 4501 (1954).

[26] C. G. Bergstrom and S. Siegel, *J. Am. Chem. Soc.*, **74**, 145 (1952); J. D. Roberts et al., *J. Am. Chem. Soc.*, **73**, 2509, 3542 (1951); **81**, 4390 (1959).

[27] S. Winstein and B. K. Morse, *J. Am. Chem. Soc.*, **74**, 1133 (1952).

appears to be no good evidence for the suggested[6] completely concerted mechanism, with no carbonium-ion intermediates.

Useful semiquantitative information about whether certain carbonium-ion formation reactions involve the initial formation of the unrearranged carbonium ion like VI or a bridged ion like VII may be obtained by application of the generalizations concerning the effect of structure on reactivity in neighboring-group participation reactions described in Sec. 7-3c. We shall take, as a point of reference, the acetolysis of the 3-phenyl-2-butyl tosylates, a reaction whose rate showed little driving force due to neighboring-group participation, being only about five times as fast as that expected for simple carbonium-ion formation. According to Winstein and Grunwald, a β-methyl group increases the rate of neighboring-group participation (direct formation of VII) by about twentyfold and has little effect on the rate of formation of the ordinary carbonium ion (VI).[28] Therefore the introduction of a β-methyl group into our reference compounds should yield compounds that acetolyze about 20 times as fast, via direct formation of a phenonium ion. Experimentally it is found that 3-methyl-3-phenyl-2-butyl tosylate acetolyzes about 60 times as fast as do the reference compounds.[11,29] The acetolysis of 1-phenyl-2-propyl tosylate, which has one less β-methyl substituent than our reference compounds, appears, as might be expected, to be a borderline case with most of the reaction not involving phenyl participation.[16]

14-1f. *The Pinacol and Related Rearrangements.*[30] Acid-catalyzed transformation to an aldehyde or ketone is a rather general reaction for 1,2-diols. When both hydroxy groups are tertiary, the reaction occurs particularly easily, the intermediate carbonium ion(s) being relatively stable, and is called the pinacol rearrangement. For the simplest such rearrangement, that of pinacol itself, we may write the mechanism

[28] S. Winstein, E. Grunwald, et al., *J. Am. Chem. Soc.*, **70**, 812, 816, 821, 828 (1948).

[29] This comparison involves the assumption that in the present case *p*-bromobenzenesulfonates acetolyze three times as fast as tosylates, just as they are known to do in other cases.

[30] For a detailed discussion of the mechanism of pinacol rearrangements, see C. J. Collins, *Quart. Revs. (London)*, **14**, 357 (1960).

Apparently the reaction does proceed via the intermediate carbonium ion IX; at least, this seems the simplest explanation of the observation that in dilute sulfuric acid pinacol undergoes oxygen exchange with the solvent almost three times as fast as it rearranges.[31] Furthermore, the rate of oxygen exchange plus rearrangement by pinacol is reasonable for the formation of IX; it is about one-fiftieth the rate of acid-catalyzed oxygen exchange of t-butyl alcohol, and, indeed, an electron-withdrawing β-hydroxy group would be expected to decrease the rate of carbonium-ion formation by about this factor.[32] Although the reaction scheme given shows two ways in which the carbonium ion IX reacts, there are at least two other possible modes of reaction for IX that might seem possible: namely, participation of β-hydroxy to give an epoxide, and loss of a β-hydrogen to give 1,1,2-trimethylallyl alcohol. The possibility of epoxide formation does not seem to have been investigated thoroughly in the case of pinacol, but Gebhart and Adams have shown that in the rearrangement of benzopinacol (tetraphenylethanediol) in acetic acid solution about 80 per cent of the reaction involves the intermediate formation of the epoxide.[34] Since the pinacol rearrangement gives pinacolone in almost quantitative yield, any 1,1,2-trimethylallyl alcohol formed must be reprotonated to IX. Apparently, in concentrated sulfuric acid, no trimethylallyl alcohol is formed; Kursanov and Parnes found that the pinacolone formed in cold D_2SO_4 contained deuterium in the methyl (methyl ketones exchange with D_2SO_4) but not the t-butyl group.[35] This latter observation also shows reversal of the pinacol rearrangement is relatively slow. Under stronger conditions, however, such reversal has been shown both by carbon- and by deuterium-labeling of the methyl group of pinacolone.[36]

$$\overset{*}{C}H_3COC(CH_3)_3 \xrightarrow{H_2SO_4} CH_3CO\overset{\overset{\displaystyle \overset{*}{C}H_3}{|}}{C}(CH_3)_2$$

[31] C. A. Bunton, T. Hadwick, D. R. Llewellyn, and Y. Pocker, *J. Chem. Soc.*, 403 (1958).

[32] The effect of the β-hydroxy group (neglecting the effect of the β-methyl groups) on log k is about half of the effect that a β-chlorine has on the rate of carbonium-ion formation from t-butyl chloride.[33] The value of σ^* for the hydroxymethyl group is about half of that for the chloromethyl group.

[33] H. C. Brown, M. S. Kharasch, and T. H. Chao, *J. Am. Chem. Soc.*, **62,** 3435 (1940).

[34] H. J. Gebhart, Jr., and K. H. Adams, *J. Am. Chem. Soc.*, **76,** 3925 (1954).

[35] D. N. Kursanov and Z. N. Parnes, *Zhur. Obshchei. Khim.*, **27,** 668 (1957); *Chem. Abstr.*, **51,** 16285h (1957).

[36] T. S. Rothrock and A. Fry, *J. Am. Chem. Soc.*, **80,** 4349 (1958); Z. N. Parnes, S. V. Vitt, and D. N. Kursanov, *Zhur. Obshchei. Khim.*, **28,** 410 (1958); *Chem. Abstr.*, **52,** 13618f (1958).

In many pinacol and related rearrangements reversal and other subsequent reactions of initially formed products appear to occur much more readily and have been shown in some cases to be responsible for "anomalous migration aptitudes," variations in products with reaction conditions, etc.[30]

Winstein and Ingraham have discussed the mechanism of pinacolic rearrangements in terms of the theory of reactivity in neighboring-group participation (Sec. 7-3c).[28,37] A number of observations may be correlated in these terms. In the semipinacolic deamination of optically active 1,1-diphenyl-2-amino-1-propanol, for example, Bernstein and Whitmore observed that rearrangement was accompanied by inversion of configuration at the asymmetric carbon atom.[38]

$$
\begin{array}{ccc}
& \underset{\displaystyle |}{C_6H_5} & \\
C_6H_5-\underset{\displaystyle \underset{HO}{|}}{C}-\underset{\displaystyle \underset{NH_2}{|}}{C}HCH_3 & \xrightarrow{HNO_2} & C_6H_5-\underset{\displaystyle \underset{HO}{|}}{\overset{\displaystyle \overset{C_6H_5}{|}}{C}}-\underset{\displaystyle N_2^{\oplus}}{C}HCH_3
\end{array}
$$

$$
\begin{array}{ccc}
\underset{\displaystyle \underset{O}{\|}}{C}_6H_5-C-CHCH_3 & \longleftarrow & C_6H_5-\overset{\displaystyle \overset{C_6H_5}{|}}{\underset{\displaystyle \underset{HO}{|}}{\overset{\oplus}{C}}}-CHCH_3
\end{array}
$$

This observation suggests but does not establish participation by a β-phenyl group. Such participation is probably a welcome alternative to the formation of a secondary carbonium ion with three electron-withdrawing β substituents.

The nature of the rearrangement product formed when there is more than one β-R group that may migrate has been studied considerably. In many cases it is necessary to learn which carbon atom is going to lose a group with its bonding electron pair, but in the case of the pinacols formed by the bimolecular reduction of unsymmetrical ketones it is only necessary to learn which of two possible groups migrates. Bachmann and coworkers made a number of observations like the following

$$
\begin{array}{c}
\underset{\displaystyle \underset{HO}{|}}{p\text{-}MeOC_6H_4}\ \ \underset{\displaystyle \underset{OH}{|}}{C_6H_4OMe\text{-}p} \\
C_6H_5-C-C-C_6H_5 \xrightarrow{H^+}
\end{array}
\begin{array}{c}
O\ \ C_6H_5 \\
98.6\%\ C_6H_5-\overset{\displaystyle \|}{C}-\overset{\displaystyle |}{C}-(C_6H_4OMe\text{-}p)_2 \\
+ \\
O\ \ C_6H_4OMe\text{-}p \\
1.4\%\ p\text{-}MeOC_6H_4-\overset{\displaystyle \|}{C}-\overset{\displaystyle |}{C}-(C_6H_5)_2
\end{array}
$$

and derived from them the following order of migration aptitudes: p-methoxyphenyl > p-tolyl > p-biphenyl > m-tolyl > p-fluorophenyl >

[37] S. Winstein and L. L. Ingraham, *J. Am. Chem. Soc.*, **77**, 1738 (1955).

[38] H. I. Bernstein and F. C. Whitmore, *J. Am. Chem. Soc.*, **61**, 1324 (1939).

phenyl > p-chlorophenyl > m-methoxyphenyl > o-tolyl and o- and m-chlorophenyl.[39] This order and the fact that aryl groups usually migrate more readily than methyl groups can be explained in terms of the stability of the intermediate phenonium (or other) ion. The migration aptitudes of the ortho-substituted groups are believed to be reduced by steric hindrance.

Although Bachmann and coworkers gave little attention to whether the diols they used were meso or dl, it may have made little difference since it is possible that the rearrangement of benzopinacols proceeds via relatively long-lived benzhydryl-type carbonium ions. In other cases, however, different diastereomers may yield different products. This point has been very effectively established by Curtin and Pollak in the related pinacolic deamination reaction.[40] With the amino alcohols studied in this reaction, there are two racemates (dl pairs). It was found that in the reaction of the α racemate of 1,2-diphenyl-1-p-chlorophenyl-2-aminoethanol with nitrous acid the phenyl group migrates, whereas with the β racemate the p-chlorophenyl group does.

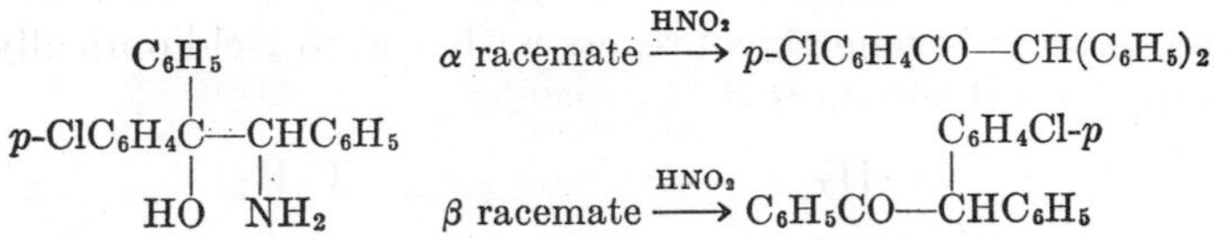

$$C_6H_5$$

$$p\text{-ClC}_6H_4\overset{|}{\underset{\underset{\text{HO}}{|}}{C}}\text{—}\overset{|}{\underset{\underset{\text{NH}_2}{|}}{C}}HC_6H_5$$

$$\alpha \text{ racemate} \xrightarrow{\text{HNO}_2} p\text{-ClC}_6H_4CO\text{—CH}(C_6H_5)_2$$

$$C_6H_4Cl\text{-}p$$

$$\beta \text{ racemate} \xrightarrow{\text{HNO}_2} C_6H_5CO\text{—}\overset{|}{C}HC_6H_5$$

Pollak and Curtin point out that the most stable rotational conformer for each of the racemates will be the one that places the larger (aryl) groups attached to the hydroxylated carbon atom on either side of the smallest group (a hydrogen atom) on the amino carbon atom. In one racemate this will place the phenyl group trans to the amino group—ready to migrate when the amino group is changed to a diazonium-ion group and then lost as nitrogen. In the most stable conformer of the other racemate it is the p-chlorophenyl group that is trans to the amino group.

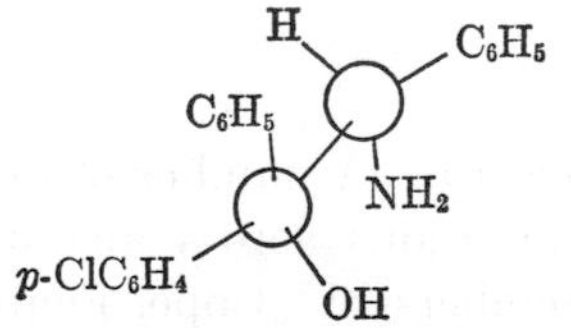

α Racemate β Racemate

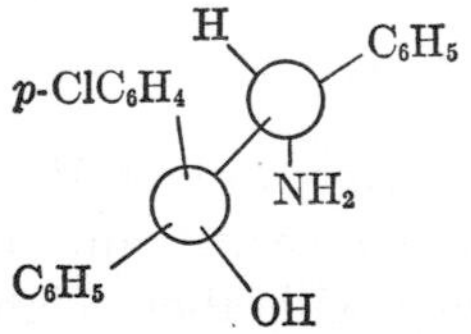

[39] W. E. Bachmann and F. H. Moser, $J.\ Am.\ Chem.\ Soc.$, **54**, 1124 (1932).

[40] D. Y. Curtin and P. I. Pollak, $J.\ Am.\ Chem.\ Soc.$, **73**, 992 (1951); cf. P. I. Pollak and D. Y. Curtin, $J.\ Am.\ Chem.\ Soc.$, **72**, 961 (1950); D. Y. Curtin, E. E. Harris, and P. I. Pollak, $J.\ Am.\ Chem.\ Soc.$, **73**, 3453 (1951).

The configurations assigned these two diastereomeric racemates on the basis of this interpretation were subsequently confirmed by independent means. This explanation, of course, depends on the assumption that the intermediate that rearranges (the diazonium ion in this case) does so at a rate that is rapid compared with the rate of interconversion of the different rotational conformations of a given species. This assumption is plausible in the present case, but it may not be in certain others, and when it is not, the extent of migration of the two groups will depend only on the stability of the two possible transition states for aryl-group migration.[41]

14-2. Some Rearrangements Involving Migration of Hydrogen.
14-2a. 1,2 Migrations of Hydrogen. Hydrogen atoms differ from most substituents in the ease with which they may be removed without their bonding electron pairs. For this reason rearrangements involving migration of hydrogen should proceed with relative ease by an elimination-readdition mechanism. Nevertheless, rearrangement by an internal mechanism has been found to be the preferred reaction path in a number of cases. Mislow and Siegel, for example, showed that optically active 1-phenyl-1-*o*-tolylethylene glycol reacts with acid to yield optically active phenyl-*o*-tolylacetaldehyde.[42]

$$
\begin{array}{ccc}
\text{C}_7\text{H}_7 & & \text{C}_7\text{H}_7 \\
| & & | \\
(+)\ \text{C}_6\text{H}_5\text{---C---CH}_2\text{OH} & \rightarrow & (+)\ \text{C}_6\text{H}_5\text{---CHCHO} \\
| & & \\
\text{OH} & &
\end{array}
$$

Ley and Vernon showed that the rearrangement of 1,2-dimethoxy-2-methylpropane to 1,1-dimethoxy-2-methylpropane in deuteriomethanol solution involved largely, if not entirely, internal migration of hydrogen,[43] and Collins and coworkers showed that 2-deuterio-1,1,2-triphenylethylene glycol yields deuteriated phenyl benzhydryl ketone.[44]

$$
\begin{array}{ccc}
(\text{C}_6\text{H}_5)_2\text{C---CDC}_6\text{H}_5 \xrightarrow{\text{H}^+} & (\text{C}_6\text{H}_5)_2\overset{\oplus}{\text{C}}\text{---CDC}_6\text{H}_5 \rightarrow & (\text{C}_6\text{H}_5)_2\text{CDCOC}_6\text{H}_5 \\
| \quad | & | & + \\
\text{HO} \quad \text{OH} & \text{OH} & \text{H}^+
\end{array}
$$

14-2b. Transannular Migrations of Hydrogen. A number of investigations have revealed unusually facile interactions across medium-sized rings (rings with about eight to twelve members).[45] Cope, Fenton, and

[41] D. Y. Curtin and M. C. Crew, *J. Am. Chem. Soc.*, **77**, 356 (1955).

[42] K. Mislow and M. Siegel, *J. Am. Chem. Soc.*, **74**, 1060 (1952).

[43] J. B. Ley and C. A. Vernon, *J. Chem. Soc.*, 2987 (1957).

[44] C. J. Collins et al., *J. Am. Chem. Soc.*, **81**, 460 (1959).

[45] For recent reviews (in English) of two aspects of this work, see (*a*) V. Prelog, *Bull. soc. chim. France*, 1433 (1960); (*b*) J. Sicher et al., *Bull. soc. chim. France*, 1438 (1960).

Spencer,[46] for example, found that the reaction of *cis*-cyclooctene with hydrogen peroxide in formic acid or the formolysis of *cis*-cyclooctene oxide gives about as much 1,4-cyclooctanediol (later shown to be cis[47]) as *trans*-1,2-cyclooctanediol, the expected product.

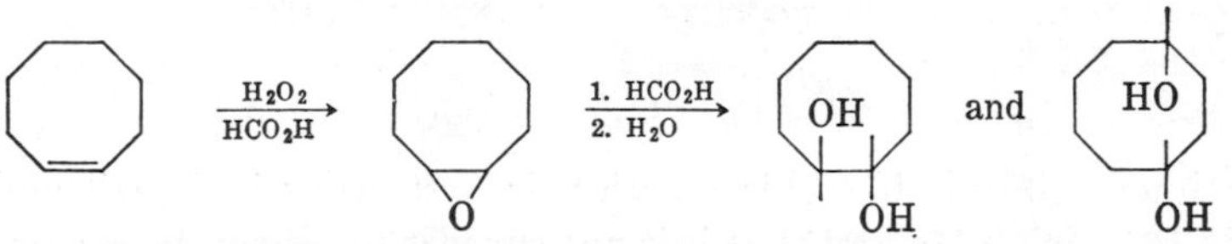

Urech and Prelog studied the acetolysis of cyclodecyl *p*-toluenesulfonate labeled with C^{14} (half at the substituted carbon and half adjacent to it).[48] They found that the major product, cyclodecene, had C^{14} scattered throughout the ring. The distribution, shown below for the *trans*-cyclodecene, proved that the rearrangement could not have proceeded by a series of 1,2-hydrogen shifts.

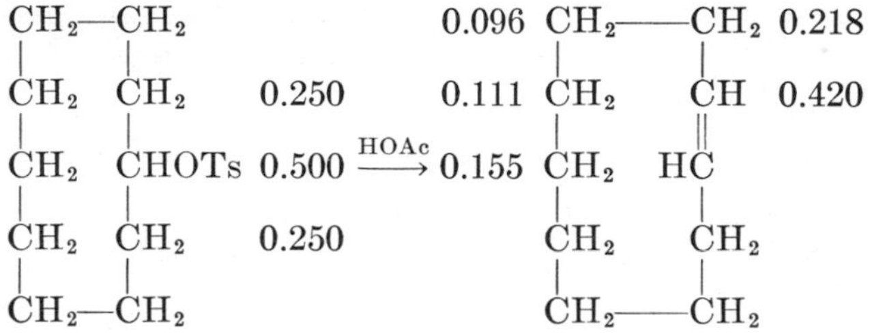

Fraction of C^{14} at various positions

The rearrangement is believed to come at least partly from 1,5- and perhaps 1,6-hydride transfers. Such internal hydride transfers seem particularly likely since X-ray determinations of the structure of cyclodecane derivatives show that some of the hydrogen atoms are quite near those five carbons away.[45a, 49]

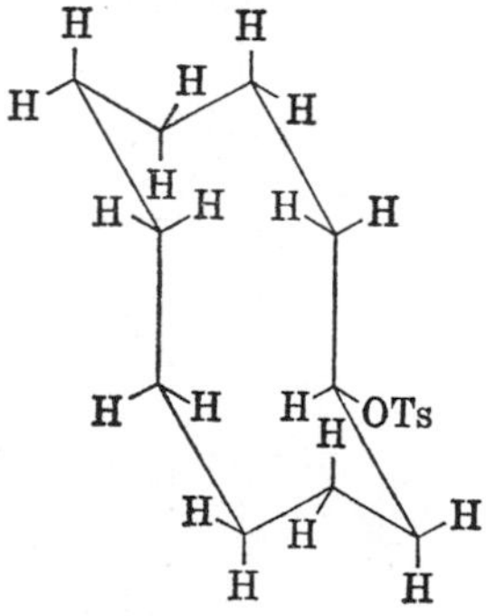

[46] A. C. Cope, S. W. Fenton, and C. F. Spencer, *J. Am. Chem. Soc.*, **74**, 5884 (1952).
[47] A. C. Cope and B. C. Anderson, *J. Am. Chem. Soc.*, **79**, 3892 (1957).
[48] H. J. Urech and V. Prelog, *Helv. Chim. Acta*, **40**, 477 (1957).
[49] J. D. Dunitz and V. Prelog, *Angew. Chem.*, **72**, 896 (1960).

14-3. The Benzilic Acid Rearrangement. The benzilic acid rearrangement is the base-catalyzed rearrangement of α-diketones to the salts of α-hydroxy acids, the most familiar example being the reaction of benzil.

$$C_6H_5COCOC_6H_5 + OH^- \rightarrow (C_6H_5)_2\overset{\displaystyle OH}{\underset{\displaystyle |}{C}}-CO_2^-$$

Westheimer showed that this reaction is first-order in benzil and first-order in hydroxide ion and that it is not subject to *general* base catalysis.[50] Roberts and Urey found that benzil undergoes base-catalyzed O^{18} exchange at a rate that is much faster than its rearrangement.[51] These workers point out that their observations are consistent with the assumption that benzil reversibly and relatively rapidly adds a hydroxide ion to give an intermediate that undergoes a rate-controlling rearrangement.

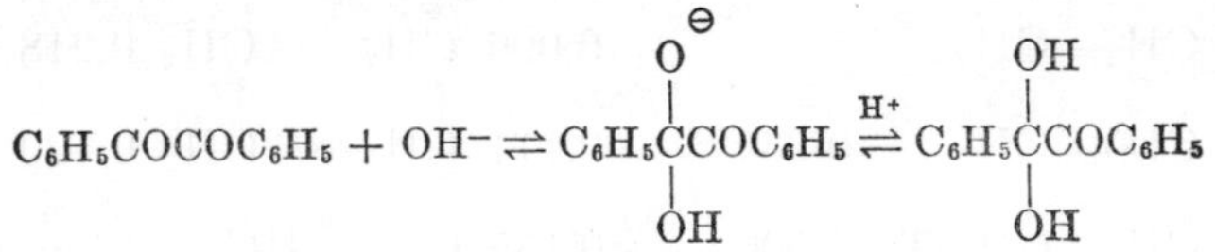

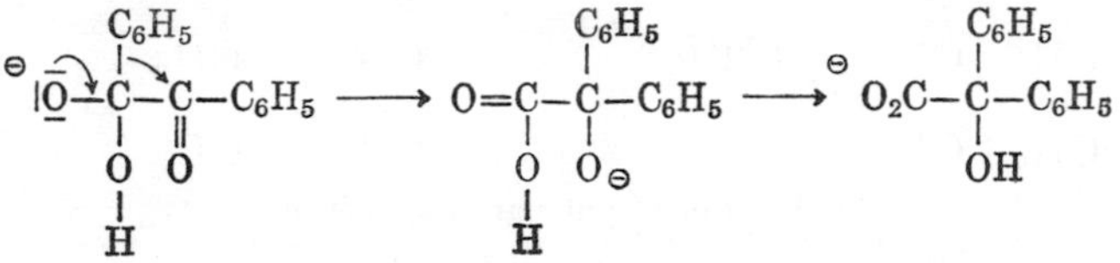

$$C_6H_5COCOC_6H_5 + OH^- \rightleftharpoons C_6H_5\overset{\displaystyle \overset{\ominus}{O}}{\underset{\displaystyle \underset{\displaystyle OH}{|}}{C}}COC_6H_5 \overset{H^+}{\rightleftharpoons} C_6H_5\overset{\displaystyle OH}{\underset{\displaystyle \underset{\displaystyle OH}{|}}{C}}COC_6H_5$$

$$\overset{\ominus}{|O}-\overset{\displaystyle \overset{C_6H_5}{|}}{\underset{\displaystyle \underset{H}{\overset{|}{O}}}{C}}-\overset{\displaystyle \underset{\displaystyle \overset{||}{O}}{C}}{}-C_6H_5 \longrightarrow O{=}\overset{\displaystyle \overset{C_6H_5}{|}}{\underset{\displaystyle \underset{H}{\overset{|}{O}}}{C}}-\overset{\displaystyle \underset{\displaystyle \overset{|}{O_\ominus}}{C}}{}-C_6H_5 \longrightarrow O_2C-\overset{\displaystyle \overset{C_6H_5}{|}}{\underset{\displaystyle \underset{OH}{|}}{C}}-C_6H_5$$

To the extent to which the carbonyl carbon atom to which the phenyl group migrates has a positive charge, this rearrangement may be seen to resemble the carbonium-ion rearrangements described earlier in this chapter. As a part of the driving force for the reaction, the phenyl group may be seen to be "pushed" from the carbon atom from which it migrates by the strong tendency of the negatively charged oxygen atom to share one of its unshared electron pairs. The loss of a proton by the carboxyl group and its gain by the alkoxy group, shown above as rapid steps that follow the rate-controlling migration of the phenyl group, could have been written as occurring simultaneously with the phenyl migration. Such a concerted mechanism, however, is rendered implausible by the observation that the hydroxide-ion catalyst may be replaced by an alkoxide ion. Thus Doering and Urban found that benzil reacts with sodium methoxide or potassium t-butoxide to yield the methyl or t-butyl ester of benzilic acid.[52] Furthermore, the reaction is about twice as fast in $66\frac{2}{3}$ per cent dioxane–$33\frac{1}{3}$ per cent D_2O as in $66\frac{2}{3}$ per cent dioxane–

[50] F. H. Westheimer, *J. Am. Chem. Soc.*, **58**, 2209 (1936).

[51] I. Roberts and H. C. Urey, *J. Am. Chem. Soc.*, **60**, 880 (1938).

[52] W. v. E. Doering and R. S. Urban, *J. Am. Chem. Soc.*, **78**, 5938 (1956).

$33\frac{1}{3}$ per cent H_2O.[53]   This solvent kinetic isotope effect, due to the greater basicity of deuteroxide ions compared with hydroxide ions (cf. Sec. 5-3c), is large for base-catalyzed reactions, and it therefore seems unlikely that it carries concealed within it any significant compensating primary kinetic isotope effect of the type that would be expected to slow a reaction involving a hydrogen transfer in the rate-controlling step.

PROBLEMS

1. The *p*-toluenesulfonate of one of the alcohols below undergoes acetolysis about 10^{11} times as fast as does the corresponding saturated *p*-toluenesulfonate and yields the acetate of the starting alcohol. The *p*-toluenesulfonate of the other alcohol acetolyzes about 10^4 times as fast as its saturated analog and gives an acetoxybicyclo [3.2.0]-2-heptene. Give the structures of the two compounds and explain the relative reactivities.

2. The following reaction is catalyzed by both acids and bases. Suggest a reasonable mechanism for each type of catalysis.

$$\begin{matrix} & R' & & & R' \\ & | & & & | \\ R-&C&-CO-R'' \rightarrow R-CO-&C&-R'' \\ & | & & & | \\ & OH & & & OH \end{matrix}$$

3. In the presence of sulfuric acid, di-*t*-butyl ketone gives 3,3,4,4-tetramethyl-2-pentanone and a little methyl isopropyl ketone. When the reactant was labeled in the carbonyl group with C^{14}, the major product was also found to be so labeled (the location of C^{14} in the methyl isopropyl ketone was not studied).

$$\begin{matrix} & CH_3 & & CH_3 & & & & CH_3 \\ & | & & | & & H_2SO_4 & & | \\ CH_3-&C&-C^*O-&C&-CH_3 & \xrightarrow{} & CH_3-C^*O-&C&-C(CH_3)_3 \\ & | & & | & & & & | \\ & CH_3 & & CH_3 & & & & CH_3 \end{matrix}$$

Suggest a reasonable mechanism for the formation of each of the two products.

4. When diphenyl-*p*-tolyl acetaldehyde rearranges in the presence of acid, about four times as much as ketone A is formed as of ketone B.

$$\begin{matrix} p\text{-}MeC_6H_4 & & & & p\text{-}MeC_6H_4 \\ | & & & & | \\ (C_6H_5)_2C-CHO \rightarrow (C_6H_5)_2CHCOC_6H_4Me\text{-}p & \text{and} & C_6H_5CHCOC_6H_5 \\ & A & & & B \end{matrix}$$

One mechanism that may be written for this rearrangement seems plausible except that it requires the assumption that a phenyl group migrates more readily than a *p*-tolyl group. Another reasonable mechanism may be written that is in accord with the usual relative migrating aptitudes (*p*-tolyl > phenyl). Experiments involving isotopic labeling have shown that the latter mechanism is correct. Write the two mechanisms and describe the isotopic-labeling experiments that could be used to distinguish between them.

[53] J. Hine and H. W. Haworth, *J. Am. Chem. Soc.*, **80**, 2274 (1958).

Chapter 15

REARRANGEMENTS DUE
TO ELECTRON-DEFICIENT NITROGEN
AND OXYGEN ATOMS

15-1. Rearrangements Due to Electron-deficient Nitrogen Atoms. *15-1a. The Hofmann Reaction of Amides.* Nitrogen compounds may rearrange just as their carbon analogs do when a group attached to the nitrogen atom is removed with its pair of bonding electrons. This similarity of the Hofmann and related reactions to many carbon-skeleton rearrangements was pointed out by Whitmore in the initial paper of his important series "The Common Basis of Intramolecular Rearrangements."[1]

The Hofmann degradation of amides by the action of halogen and alkali almost certainly involves the initial formation of a N-haloamide, since these intermediates have been isolated in a number of cases and found to undergo the rearrangement in the presence of alkali. The removal of a halide anion is facilitated by transformation of the N-haloamide to its conjugate base (compare the basic hydrolysis of chloroform, Sec. 24-1a). However, as Hauser and Kantor point out, it is not likely that there is a true intermediate in which a nitrogen atom has only a sextet of electrons.[2] It is more probable that the migration of the R group occurs simultaneously with the loss of the halide ion. This concerted mechanism

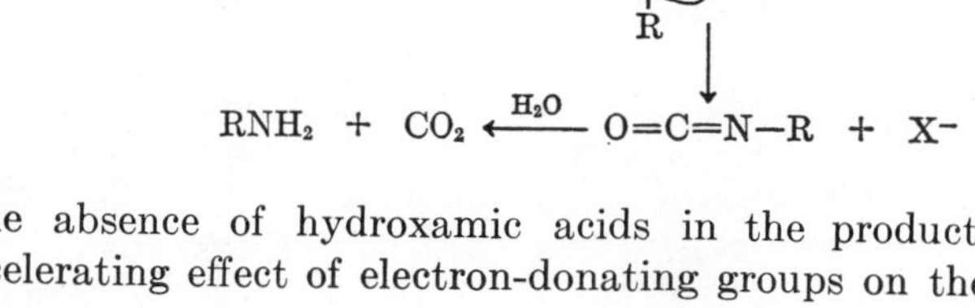

$$RCONH_2 \xrightarrow[OH^-]{X_2} R-\overset{\overset{\displaystyle O}{\|}}{C}-\overset{\overset{\displaystyle H}{|}}{N}-X \xrightarrow{OH^-} \overset{\overset{\displaystyle O}{\|}}{\underset{\underset{\displaystyle R}{|}}{C}}-\overset{\ominus}{\underset{..}{N}}-\underset{..}{\overset{..}{X}}|$$

$$RNH_2 + CO_2 \xleftarrow{H_2O} O=C=N-R + X^-$$

explains the absence of hydroxamic acids in the product[2] and the marked accelerating effect of electron-donating groups on the reaction;

[1] F. C. Whitmore, *J. Am. Chem. Soc.*, **54**, 3274 (1932).
[2] C. R. Hauser and S. W. Kantor, *J. Am. Chem. Soc.*, **72**, 4284 (1950).

the ρ (-2.5)[3] for this removal of a halide ion separated by two atoms from the aromatic ring is considerably nearer that found for the solvolysis of 2-aryl-2-methyl-1-propyl p-bromobenzenesulfonates (-3.0), in which aryl participation is believed to occur,[4] than it is to the ρ (-1.1) for the solvolysis of 1-aryl-2-methyl-2-propyl chlorides, in which there appears to be no neighboring-group participation.[5]

A number of workers have shown that in the Hofmann reaction, as well as in the related Lossen, Curtius, and Schmidt reactions, the R group migrates with complete retention of configuration. For example, Noyes found that β-camphoramic acid (known to be cis from its relation to the anhydride-forming camphoric acid, prepared by the oxidation of camphor) undergoes the Hofmann reaction to give an amino acid whose

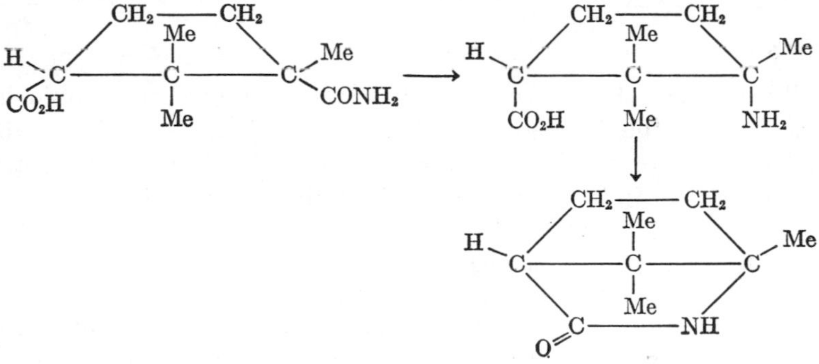

cis configuration is shown by its ability to form a lactam.[6] Wallis and Nagel showed that active α-benzylpropionamide gives optically pure α-benzylethylamine.[7]

The Lossen rearrangement of acyl derivatives of hydroxamic acids is very similar to the Hofmann reaction.

Renfrow and Hauser[8] found that electron-donating groups in R increase the reactivity ($\rho = -2.597$ for R = substituted phenyl[9]), while in R' they decrease the reactivity ($\rho = +0.865$ for R' = substituted phenyl[9]).

[3] C. R. Hauser and W. B. Renfrow, Jr., *J. Am. Chem. Soc.*, **59**, 121 (1937).

[4] R. Heck and S. Winstein, *J. Am. Chem. Soc.*, **79**, 3432 (1957).

[5] A. Landis and C. A. VanderWerf, *J. Am. Chem. Soc.*, **80**, 5277 (1958).

[6] W. A. Noyes and R. S. Potter, *J. Am. Chem. Soc.*, **34**, 1067 (1912); **37**, 189 (1915).

[7] E. S. Wallis and S. C. Nagel, *J. Am. Chem. Soc.*, **53**, 2787 (1931).

[8] W. B. Renfrow, Jr. and C. R. Hauser, *J. Am. Chem. Soc.*, **59**, 2308 (1937).

[9] L. P. Hammett, "Physical Organic Chemistry," pp. 190–191, McGraw-Hill Book Company, Inc., New York, 1940.

This would be expected since electron-donating groups in R should increase its migration aptitude, whereas in R′ they should decrease the stability of the ion R′CO$_2^-$.

15-1b. The Curtius Reaction. The Curtius reaction is often carried out by decomposing the acyl azide in an inert solvent so that the isocyanate, which is usually produced only as an intermediate in the Hofmann, Lossen, and Schmidt reactions, may in this case be isolated. The reaction may be written as involving a preliminary loss of nitrogen to give an intermediate (I), in which the R group migrates from carbon to the nitrogen atom with only six electrons in its outer shell.

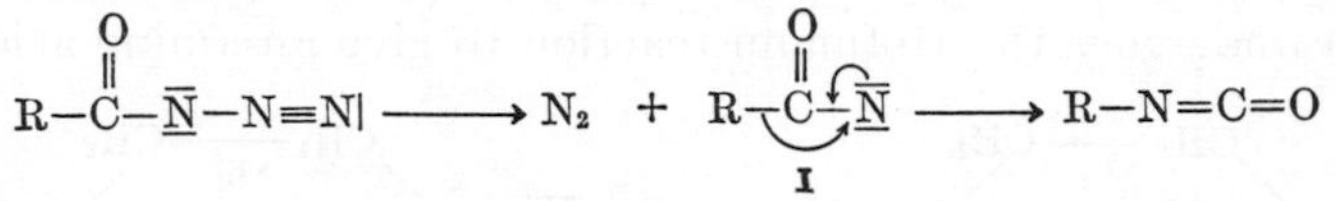

$$R-\overset{\overset{O}{\|}}{C}-\overline{N}-N\equiv N| \longrightarrow N_2 + R-\overset{\overset{O}{\|}}{C}-\overline{N} \longrightarrow R-N=C=O$$
$$I$$

Actually, though, the intermediate I is the same one that would have been written in the Hofmann reaction if the migration of R had been suggested to occur subsequent to, rather than simultaneously with, the loss of halogen. It is about equally improbable as an intermediate in the Curtius reaction.[2] That is, the loss of nitrogen and the migration of R probably occur simultaneously.

After a consideration of the mechanism of the Schmidt reaction of carboxylic acids, Newman and Gildenhorn made and verified the prediction that the decomposition of acyl azides should be subject to acid catalysis.[10] Perhaps the increased positive charge on nitrogen created by protonation increases the tendency of the R group to migrate with its bonding electron pair, displacing nitrogen. This explanation is in agreement with Yukawa and Tsuno's observation that the acid-catalyzed Curtius reaction is speeded by electron-donating substituents ($\rho = -1.1$).[11]

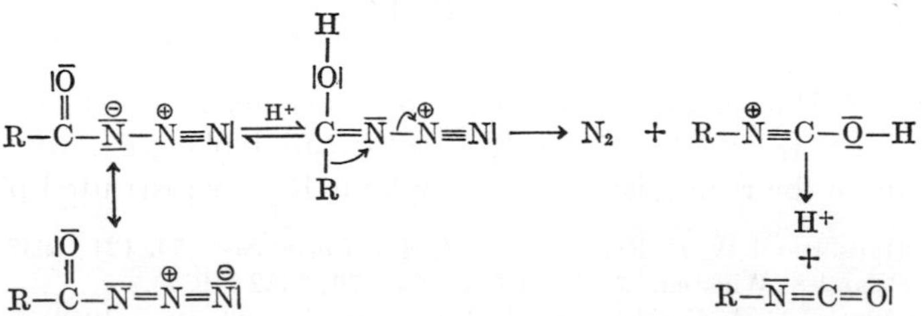

$$R-\overset{\overset{|\overline{O}}{\|}}{C}-\overset{\ominus}{\overline{N}}-\overset{\oplus}{N}\equiv N| \underset{}{\overset{H^+}{\rightleftharpoons}} \overset{\overset{H}{|}}{\underset{R}{C}}=\overline{N}-\overset{\oplus}{N}\equiv N| \longrightarrow N_2 + R-\overset{\oplus}{N}=C-\overline{O}-H$$

$$R-\overset{\overset{|\overline{O}}{\|}}{C}-\overset{\oplus}{\overline{N}}=\overset{\ominus}{N}=\overline{N|} \qquad\qquad R-\overline{N}=C=\overline{O|}$$

$$H^+$$
$$+$$

[10] M. S. Newman and H. L. Gildenhorn, *J. Am. Chem. Soc.*, **70**, 317 (1948); R. A. Coleman, M. S. Newman, and A. B. Garrett, *J. Am. Chem. Soc.*, **76**, 4534 (1954).
[11] Y. Yukawa and Y. Tsuno, *J. Am. Chem. Soc.*, **81**, 2007 (1959).

Newman and Gildenhorn suggest that the azide is protonated at nitrogen to give

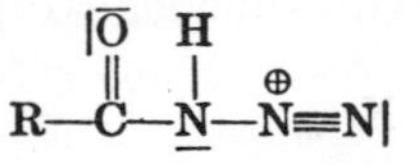

$$\overset{\displaystyle |\overline{O}}{\underset{\displaystyle}{\|}} \quad \overset{\displaystyle H}{\underset{\displaystyle}{|}}$$
$$R-C-N-\overset{\oplus}{N}\equiv N|$$

Although this possibility has not been disproved, the alternative interpretation is more closely analogous to the mechanism of the Schmidt reaction of ketones given in the next section.

15-1c. *The Schmidt Reaction.*   The Schmidt reaction is usually considered to include the transformation of acids to amines and of ketones to amides by the action of hydrazoic acid and a strong acid catalyst, as well as certain related reactions.   Smith proposed the most plausible mechanism for the reaction of ketones.[12]   In this mechanism the conjugate acid of the ketone coordinates with a molecule of hydrazoic acid to give II, which is dehydrated to III; III then rearranges in a manner analogous to the Beckmann rearrangement.

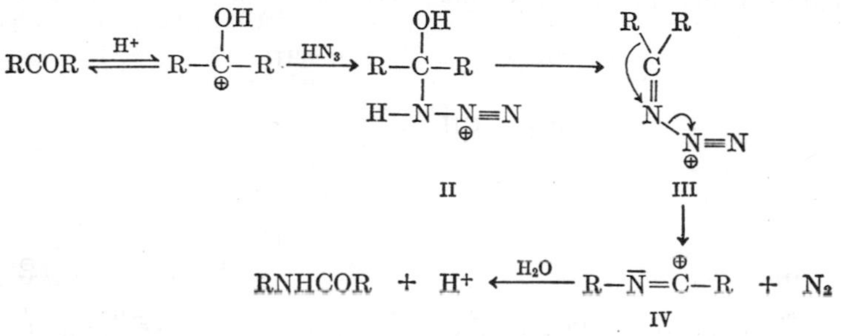

$$RCOR \underset{}{\overset{H^+}{\rightleftharpoons}} R-\underset{\oplus}{\overset{\overset{\displaystyle OH}{|}}{C}}-R \xrightarrow{HN_3} R-\underset{\underset{\displaystyle H-N-\underset{\oplus}{N}\equiv N}{|}}{\overset{\overset{\displaystyle OH}{|}}{C}}-R \longrightarrow$$

II

III

$$RNHCOR + H^+ \xleftarrow{H_2O} R-\overline{N}=\overset{\oplus}{C}-R + N_2$$

IV

Although II could be written as rearranging directly to the conjugate acid of the amide, there are several reasons for believing that this does

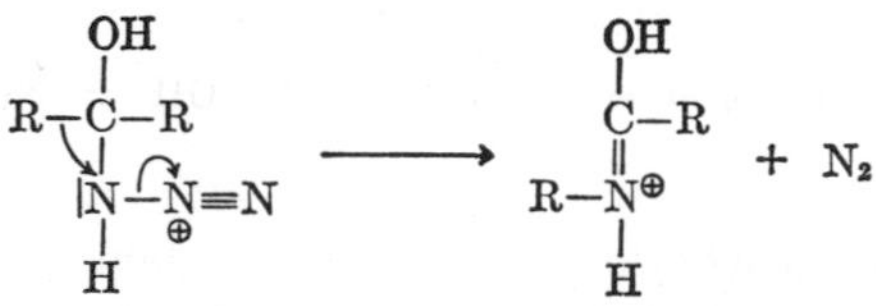

$$R-\underset{\underset{\displaystyle H}{|}}{\overset{\overset{\displaystyle OH}{|}}{\underset{\displaystyle |N-\underset{\oplus}{N}\equiv N}{C}}}-R \longrightarrow \underset{\underset{\displaystyle H}{|}}{\overset{\overset{\displaystyle OH}{|}}{\underset{\displaystyle R-\overset{\oplus}{N}}{C}}}-R + N_2$$

not occur.   One reason is that intermediate III explains why alkyl azides usually may not be substituted for hydrazoic acid in this reaction.[12,13] The most convincing reason is the fact that if excess hydrazoic acid is used, tetrazoles are formed, although it has been shown that these could

[12] P. A. S. Smith, *J. Am. Chem. Soc.*, **70**, 320 (1948).
[13] L. H. Briggs, G. C. De Ath, and S. R. Ellis, *J. Chem. Soc.*, 61 (1942); J. H. Boyer and J. Hamer, *J. Am. Chem. Soc.*, **77**, 951 (1955).

not have been formed by the subsequent reaction of the amides that are the initial products.[14] This fact is explained by the intermediate IV, which may combine with hydrazoic acid to yield a tetrazole.

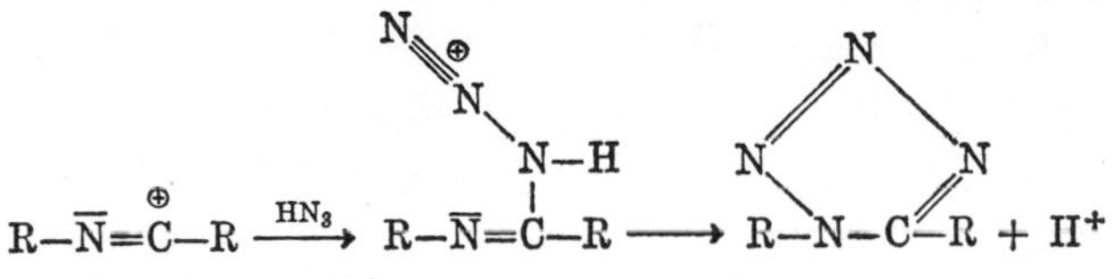

Under the conditions used for the reaction an amide would be expected to be transformed to its conjugate acid but not to an ion like IV.

Newman and Gildenhorn found that 2,4,6-trimethylbenzoic acid undergoes the Schmidt reaction in sulfuric acid readily at 0°, although benzoic acid requires a reaction temperature of 25° or more. It is certainly reasonable to ascribe the reactivity of the trimethyl compound to the ease with which it forms an oxocarbonium ion (see Sec. 12-1c). The reaction mechanism then appears to be

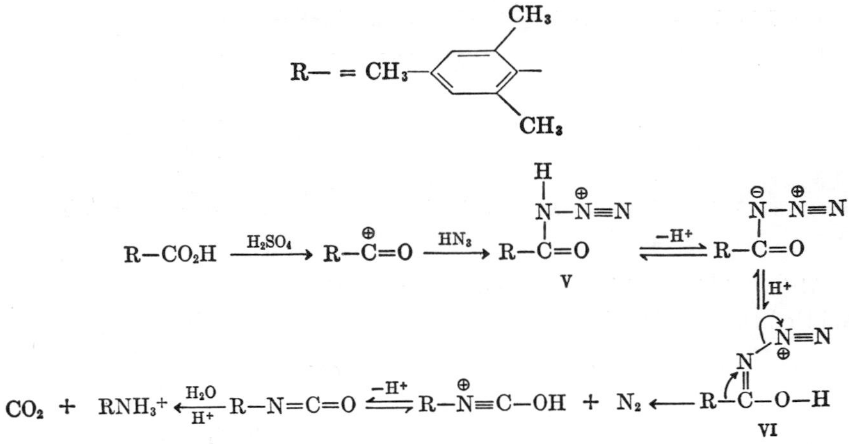

It is difficult to tell whether the reaction of benzoic acid goes through some very small amount of the oxocarbonium ion or whether it involves the less reactive but more plentiful conjugate acid $C_6H_5C(OH)_2^+$.

15-1d. *The Beckmann Rearrangement.* Some of the first evidence that it is trans R groups that migrate most readily in rearrangement reactions was obtained in studies of the Beckmann rearrangement. This fact became obvious from a great many previously published data when

[14] K. F. Schmidt, *Ber.*, **57**, 707 (1924); *Chem. Abstr.*, **19**, 3248 (1925); M. A. Spielman and F. L. Austin, *J. Am. Chem. Soc.*, **59**, 2658 (1937).

Meisenheimer and others finally correctly established the configuration of oximes.[15] The mechanism of the Beckmann rearrangement involves the removal of the hydroxyl group with its bonding electron pair and the simultaneous migration of the R group trans to it. This mechanism explains why the reaction is brought about by conditions favoring the removal of the hydroxyl group. Acid catalysts are effective, as are acyl halides. The latter often give intermediate esters that can be isolated and studied kinetically. Kuhara and coworkers pointed out the increase in rearrangement rate with increase in the stability of the anion X^- in

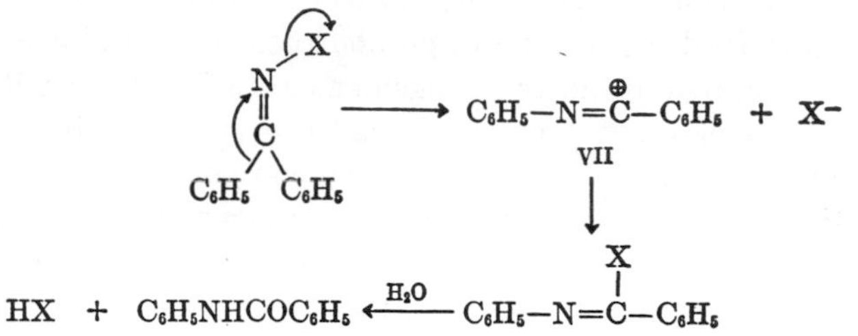

the series[16] $X^- = C_6H_5SO_3^-$, $ClCH_2CO_2^-$, $C_6H_5CO_2^-$, $CH_3CO_2^-$. Chapman found a correlation of the reaction rate (where $X^- =$ picrate anion) with the ion-solvating power of the medium.[17] It seems likely, especially in the case of the poorer ion-solvating media studied, that completely dissociated ions are never formed, i.e., that VII is an ion pair (by analogy to some of the related carbon-skeleton rearrangements described in Sec. 14-1).

Pearson, Baxter, and Martin observed the activating influence of electron-donating substituents in the Beckmann rearrangement of acetophenones ($\rho = -2.0$) and also pointed out the desirability of using for certain substituents another group of σ's, which later became the σ^+ values.[18]

Pearson, Carter, and Greer found that hydrazones, upon diazotization, undergo a rearrangement analogous to the Beckmann.[19] The first inter-

---

[15] J. Meisenheimer, *Ber.*, **54**, 3206 (1921); J. Meisenheimer, P. Zimmermann, and U. v. Kummer, *Ann.*, **446**, 205 (1926); O. L. Brady and G. Bishop, *J. Chem. Soc.*, **127**, 1357 (1925).

[16] M. Kuhara, K. Matsumiya, and N. Matsunami, *Mem. Coll. Sci. Kyoto Imp. Univ.*, **1**, 105 (1914); M. Kuhara and H. Watanabe, *Mem. Coll. Sci. Kyoto Imp. Univ.*, 1(9), 349 (1916); *Chem. Abstr.*, **9**, 1613 (1915); **11**, 579 (1917).

[17] A. W. Chapman and C. C. Howis, *J. Chem. Soc.*, 806 (1933); A. W. Chapman, *J. Chem. Soc.*, 1550 (1934).

[18] D. E. Pearson, J. F. Baxter, and J. C. Martin, *J. Org. Chem.*, **17**, 1511 (1952).

[19] D. E. Pearson and C. M. Greer, *J. Am. Chem. Soc.*, **71**, 1895 (1949); D. E. Pearson, K. N. Carter, and C. M. Greer, *J. Am. Chem. Soc.*, **75**, 5905 (1953).

mediate in this reaction is presumably the same as in the Schmidt reaction of ketones.

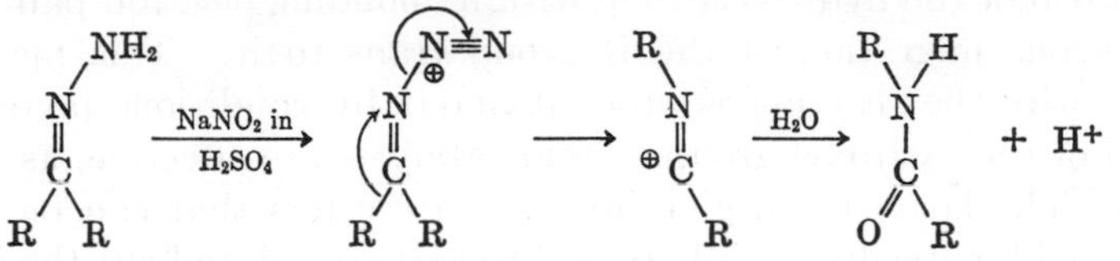

15-2. Rearrangements Due to Electron-deficient Oxygen Atoms.[20]

15-2a. Rearrangements of Esters of Peracids. Criegee studied the rearrangement of the 9-decalyl esters of perbenzoic acid and *p*-nitroperbenzoic acid.[21] The facts that the rearrangement rate increases with the ion-solvating ability of the solvent and that the *p*-nitroperbenzoate reacts much more rapidly than the unsubstituted compound show that this is a polar reaction quite analogous to the rearrangements due to electron-deficient carbon and nitrogen atoms. Criegee's work was subsequently extended by Bartlett and Kice[22] and by Goering and Olson.[23] Both groups of workers showed that the rearrangement does not involve the formation of any significant fraction of free ions (in methanol solution) but presumably proceeds through an ion-pair intermediate instead.

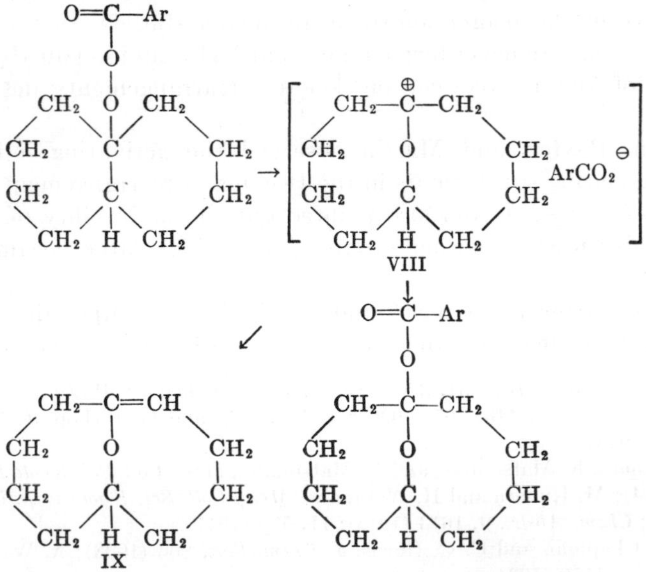

[20] For a review of this subject, see J. E. Leffler, *Chem. Revs.*, **45**, 385 (1949).
[21] R. Criegee, *Ann.*, **560**, 127 (1948).
[22] P. D. Bartlett and J. L. Kice, *J. Am. Chem. Soc.*, **75**, 5591 (1953).
[23] Harlan L. Goering and A. C. Olson, *J. Am. Chem. Soc.*, **75**, 5853 (1953).

Bartlett and Kice isolated the by-product IX. It seems unlikely that the reaction involves the intermediate formation of an ion containing an oxygen with only six electrons in its outer shell, since such an intermediate (of the type $RO^{\oplus}$) should be much less stable than even a primary carbonium ion. It therefore seems more probable that the rearrangement to the cation VIII occurs simultaneously with the removal of the $ArCO_2^-$ group. Denney showed that the RCO_2^- never becomes entirely free by observing that when the starting perester is labeled in the carbonyl group with O^{18}, all the O^{18} in the product was in the carbonyl group.[24]

15-2b. *Reaction of Ketones with Peracids.* The reaction of ketones with peracids to give esters probably has a mechanism of the type suggested by Criegee[21] and extended by Friess,[25] in which there is a rate-controlling addition of the peracid to the carbonyl group followed by a rapid rearrangement.

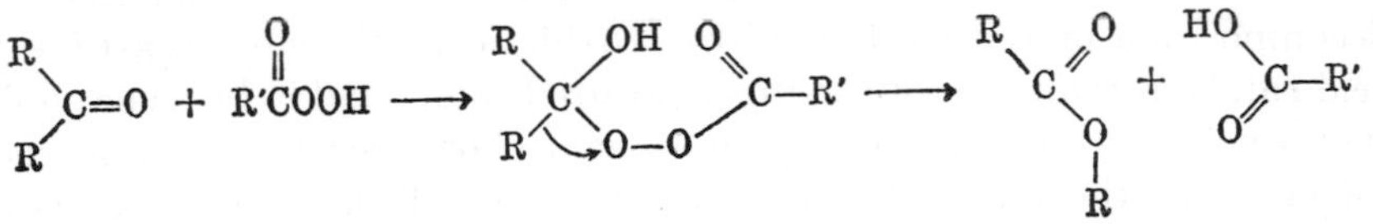

Doering and Dorfman ruled out two alternative mechanisms by using O^{18} to show that the carbonyl oxygen atom of the reacting ketone becomes the carbonyl oxygen atom of the ester.[26]

The Dakin reaction, in which the formyl groups of *o*- and *p*-hydroxy- and -aminobenzaldehydes are replaced by a hydroxy group by the action of hydrogen peroxide in alkaline solution, probably has a similar mechanism.

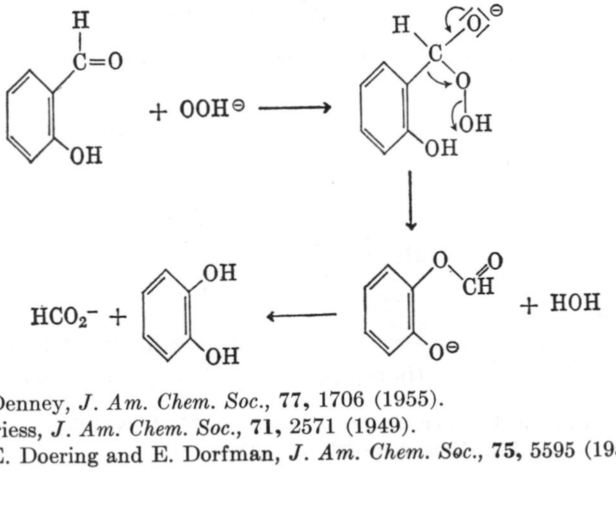

[24] D. B. Denney, *J. Am. Chem. Soc.*, **77**, 1706 (1955).
[25] S. L. Friess, *J. Am. Chem. Soc.*, **71**, 2571 (1949).
[26] W. v. E. Doering and E. Dorfman, *J. Am. Chem. Soc.*, **75**, 5595 (1953).

Mislow and Brenner found that in the reaction of optically active methyl α-phenylethyl ketone with peracetic acid,

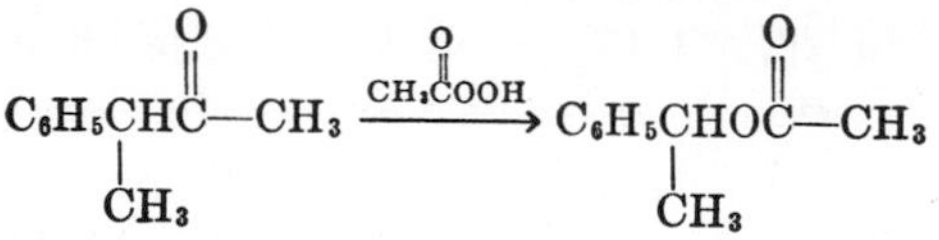

$$C_6H_5CH\overset{\overset{\displaystyle O}{\|}}{C}-CH_3 \xrightarrow{CH_3COOH} C_6H_5CHO\overset{\overset{\displaystyle O}{\|}}{C}-CH_3$$
$$\underset{CH_3}{|} \qquad\qquad\qquad \underset{CH_3}{|}$$

the α-phenylethyl group migrates with complete retention of its stereo-chemical configuration.[27] This, of course, would be expected from analogy with the Hofmann, Curtius, Schmidt, and other reactions.

15-2c. *Rearrangements and Decompositions of Peroxides.* Leffler found that although p-methoxy-p'-nitrobenzoyl peroxide, like most acyl per-oxides (Sec. 21-1a), decomposes by a free-radical mechanism in nonpolar solvents, the reaction rate is much faster in better ion-solvating media, suggesting the incursion of a polar decomposition.[28] The fact that the decomposition of the methoxynitroperoxide is vastly more sensitive to acid catalysis than that of the unsubstituted compound in benzene shows that a polar decomposition may be brought about even in a hydrocarbon solvent. In the relatively polar solvent thionyl chloride the reaction yields a rearrangement product—a mixed ester-anhydride of carbonic acid.

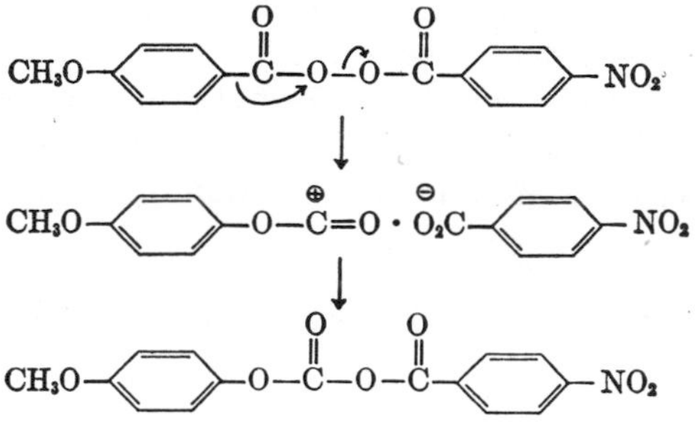

Although an intermediate cation of bridged structure

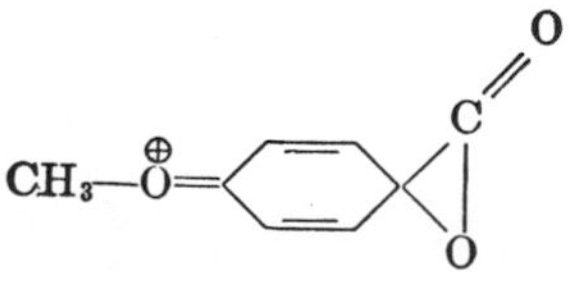

may be formed, and analogous intermediates in other rearrangements described in this chapter, there appears to be no compelling evidence for

[27] K. Mislow and J. Brenner, *J. Am. Chem. Soc.*, **75**, 2318 (1953).
[28] J. E. Leffler, *J. Am. Chem. Soc.*, **72**, 67 (1950).

such an intermediate, although it provides a good explanation for the effect of the p-methoxy group.

Bartlett and Leffler presented evidence that both free-radical and polar reaction paths may be used in the decomposition of phenylacetyl peroxide.[29]

One of the commercial methods for preparing phenol, the acid-catalyzed decomposition of cumene hydroperoxide,[30] probably has a mechanism

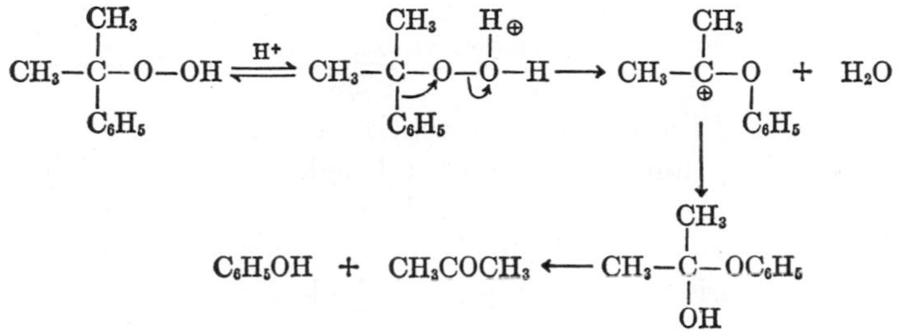

$$
\underset{\underset{C_6H_5}{|}}{\overset{\overset{CH_3}{|}}{CH_3-C-O-OH}} \overset{H^+}{\rightleftharpoons} \underset{\underset{C_6H_5}{|}}{\overset{\overset{CH_3}{|}}{CH_3-C-O-\overset{H\oplus}{O}-H}} \longrightarrow \underset{\underset{C_6H_5}{|}}{\overset{\overset{CH_3}{|}}{CH_3-C-\overset{\oplus}{O}}} + H_2O
$$

$$
C_6H_5OH + CH_3COCH_3 \longleftarrow \underset{\underset{OH}{|}}{\overset{\overset{CH_3}{|}}{CH_3-C-OC_6H_5}}
$$

as Seubold and Vaughan, who found the reaction to be subject to specific acid catalysis, have suggested.[31]

PROBLEM

1. The reaction of 1-phenyl-1-p-anisylethylene with sodium azide in 90 per cent sulfuric acid gives a mixture of two amines and two ketones.   Write a mechanism for the reaction including the structural formulas of the products and a statement as to which are formed in greater yields.

[29] P. D. Bartlett and J. E. Leffler, *J. Am. Chem. Soc.*, **72**, 3030 (1950).
[30] H. Hock and S. Lang, *Ber.*, **77B**, 257 (1944).
[31] F. H. Seubold, Jr., and W. E. Vaughan, *J. Am. Chem. Soc.*, **75**, 3790 (1953).

Chapter 16

ELECTROPHILIC AROMATIC SUBSTITUTION[1]

Because most of the common aromatic substitution reactions—nitration, halogenation, sulfonation, and the Friedel-Crafts reaction—involve the attack of electrophilic species on the aromatic ring, we shall discuss *electrophilic* aromatic substitutions before the nucleophilic or free-radical varieties. Since orientation is a part of the larger problem of reactivity and since reactivity is best discussed in terms of the reaction mechanism, we shall begin our discussion by considering the mechanisms of these reactions.

16-1. Mechanism of Aromatic Nitration.[2] 16-1a. *Nitration in Sulfuric Acid Solution.* Nitration has probably received the most careful study, from a mechanistic viewpoint, of any aromatic substitution reaction. The exact nature of the mechanism varies considerably with the type of nitrating reagent used and with the character of the compound being nitrated. An example that is, in certain respects, relatively simple, which has received considerable attention, is the nitration of nitrobenzene by nitric acid in sulfuric acid solution. Martinsen studied this reaction kinetically in 1905 and found agreement with the rate equation

$$v = k[C_6H_5NO_2][HNO_3]$$

where the bracketed expressions refer to the *formal* concentrations of the reagents without regard to the nature and relative concentrations of the various forms in which they may exist in sulfuric acid solution.[3] Second-order kinetics of this type were subsequently found by several other groups of workers in studies of the nitration of relatively unreactive

---

[1] For a more detailed treatment of aromatic substitution with emphasis on the reactions of hydrocarbons, see K. L. Nelson and H. C. Brown, in "The Chemistry of Petroleum Hydrocarbons," vol. III, chap. 56, Reinhold Publishing Corporation, New York, 1955.

[2] For a detailed treatment of nitration and halogenation, see P. B. D. de la Mare and J. H. Ridd, "Aromatic Substitution, Nitration and Halogenation," Butterworth & Co. (Publishers), Ltd., London, 1959.

[3] H. Martinsen, *Z. physik. Chem.*, **50**, 385 (1905); **59**, 605 (1907).

aromatic compounds in sulfuric acid solution.[4] As Westheimer and Kharasch[5] and also Bennett, Brand, and Williams[6] pointed out, previous cyroscopic measurements by Hantzsch[7] show that the number of particles formed from a nitric acid molecule in sulfuric acid solution approaches four, suggesting ionization by the equation

$$HNO_3 + 2H_2SO_4 \rightarrow NO_2^+ + H_3O^+ + 2HSO_4^- \qquad (16\text{-}1)$$

The latter investigators note other work, including spectral data and their own electrolytic demonstration that nitrogen is in a cationic form, in support of this suggestion that sulfuric acid produces nitronium ions (NO_2^+) from nitric acid. They also point out that these nitronium ions are very probably the actual nitrating species in the reactions studied. The activating and orienting influences of substituents, the catalytic effect of acids, and other factors reveal the electrophilic character of the reagent that attacks the aromatic ring. The nitronium ion should be the most electrophilic nitrating agent possible, and in pure sulfuric acid it appears to be the principal form in which nitric acid exists. It therefore seems likely that it is the effective nitrating agent. Westheimer and Kharasch presented a more direct argument that this is the case. They noted that the second-order rate constant for the nitration of nitrobenzene increases by about 3,000-fold as the concentration of sulfuric acid is increased from 80 to 90 per cent. This increase in rate may be attributed to an increase in the fraction of the nitric acid present as nitronium ions due to the shift of the equilibrium (16-1) to the right. However, it might also be suggested that the nitration is due to the nitric acidium ion, $H_2ONO_2^+$, whose relative concentration should also increase with the acidity of the medium.

$$HONO_2 + H_2SO_4 \rightleftharpoons H_2\overset{\oplus}{O}NO_2 + HSO_4^{\ominus} \qquad (16\text{-}2)$$

It was found possible to distinguish between these two alternatives by use of two indicators.[5] One, anthraquinone, ionizes in sulfuric acid simply by the addition of a proton, as suggested in Eq. (16-2) for nitric acid.

$$C_{14}H_8O_2 + H_2SO_4 \rightleftharpoons C_{14}H_8O_2H^{\oplus} + HSO_4^{\ominus} \qquad (16\text{-}3)$$

The other, tris-(p-nitrophenyl)methanol ionizes in a manner like that shown for nitric acid in Eq. (16-1).

$$(p\text{-}O_2NC_6H_4)_3COH + 2H_2SO_4 \rightleftharpoons (O_2NC_6H_4)_3C^{\oplus} + H_3O^{\oplus} + 2HSO_4^{\ominus}$$

[4] A. Klemenc and R. Schöller, *Z. anorg. u. allgem. Chem.*, **141,** 231 (1924); K. Lauer and R. Oda, *J. prakt. Chem.*, **144,** 176 (1936); *Ber.*, **69,** 1061 (1936).

[5] F. H. Westheimer and M. S. Kharasch, *J. Am. Chem. Soc.*, **68,** 1871 (1946).

[6] G. M. Bennett, J. C. D. Brand, and G. Williams, *J. Chem. Soc.*, 869, 875 (1946).

[7] A. Hantzsch, *Z. physik. Chem.*, **61,** 257 (1907); **65,** 41 (1908).

The effect of sulfuric acid concentration on the positions of the equilibria shown was determined for the two indicators.   It was then seen that the rate constant for nitration varied with the sulfuric acid concentration in very nearly the same way as did the fraction of tris-(p-nitrophenyl)-methanol present as the carbonium ion and not at all like the fraction of anthraquinone present as its conjugate acid (this fraction changed by less than a factor of ten between 80 and 90 per cent sulfuric acid).   In acidity-function terms (Sec. 2-3d) log k was found to be much more nearly proportional to J_0 than to H_0.

Hughes, Ingold, and coworkers made the evidence for the nitronium ion overwhelming: by more accurate cryoscopic measurements in sulfuric acid solution;[8] by isolation of nitronium perchlorate, nitronium disulfate, and other nitronium salts;[9] and by spectral studies of nitronium salts and of solutions of nitric acid and related compounds.[10]

16-1b. *Nitration in Nitromethane and Acetic Acid Solutions.*   A series of investigations that added greatly to our knowledge of nitration and of aromatic substitution reactions in general was begun by Benford and Ingold with a study of nitration in nitromethane solution.[11]   Some such solvent capable of dissolving both nitric acid and relatively non-polar organic compounds is necessary to study the nitration of aromatic hydrocarbons, whose rate of nitration in sulfuric acid solution is often controlled by the rate of solution therein.   Since the water formed in the reaction enters into certain equilibria involving nitric acid, the effect of this complication was minimized by using the nitric acid in large excess (about 5 M) over the aromatic compound (about 0.1 M).   Under these conditions it was found that the nitration of benzene was a *zero-order reaction,*

$$v = k$$

the rate simply remaining constant until all the benzene had reacted. The value of the rate constant did depend upon the concentration of nitric acid used, but this did not change significantly during the course of a given run.   The nitrations of toluene and ethylbenzene were also found to be zero-order and to have the same rate constant as benzene. This independence of the concentration and even the nature of the aromatic compound suggests that the reactions all have the same rate-controlling step(s) and that this step(s) does not involve the aromatic compound.   Hughes, Ingold, and Reed found the same zero-order

[8] R. J. Gillespie, J. Graham, E. D. Hughes, C. K. Ingold, and E. R. A. Peeling, *J. Chem. Soc.,* 2504 (1950).

[9] D. R. Goddard, E. D. Hughes, and C. K. Ingold, *J. Chem. Soc.,* 2559 (1950).

[10] C. K. Ingold, D. J. Millen, and H. G. Poole, *J. Chem. Soc.,* 2576 (1950); D. J. Millen, *J. Chem. Soc.,* 2589, 2600, 2606 (1950).

[11] G. A. Benford and C. K. Ingold, *J. Chem. Soc.,* 929 (1938).

kinetics for the nitration of toluene, p-xylene, mesitylene, ethylbenzene, and (under some conditions) benzene in acetic acid solution, and they showed that the rate of reaction in all cases is controlled by the rate of formation of nitronium ions as shown in the scheme below.[12]

$$HNO_3 + HNO_3 \rightleftharpoons H_2\overset{\oplus}{O}NO_2 + NO_3^{\ominus}$$
$$H_2\overset{\oplus}{O}NO_2 \rightleftharpoons H_2O + NO_2^{\oplus}$$
$$NO_2^{\oplus} + ArH \rightarrow ArNO_2 + \overset{\oplus}{H} \tag{16-4}$$

Nitronium-ion formation will be rate-controlling only if the nitronium ions react with the aromatic compound more rapidly than they recombine with water. This fact permits a test of the proposed mechanism. With the less reactive halobenzenes the reaction was between zero- and first-order in aromatic compound and with the still less reactive compounds, o-, m-, and p-dichlorobenzene, 1,2,4-trichlorobenzene, and ethyl benzoate, the nitration in acetic acid solution was definitely first-order. This would be predicted from the steady-state approximation if the nitronium ions recombine with water much more rapidly than they attack the aromatic reactant. The first-order rate constants obtained *were*, of course, dependent on the nature of the aromatic compound.

The mechanism (16-4) was subjected to a large number of other experimental tests, all of which supported it. Sulfuric acid was found to increase and potassium nitrate to decrease the rates of both the zero- and first-order nitrations without appreciably changing their kinetic form. The effects of these two reagents are on the rapid and reversible first step. These effects are quantitatively as well as qualitatively in agreement with theory. Small amounts of water have relatively little effect on the rate of zero-order nitration, but larger amounts considerably slow the reaction and may change a zero-order reaction to a first-order one (by making the second step reversible).

Although the nitronium ion itself was found to be the active nitrating reagent in the examples described, we cannot exclude the possibility that nitronium-ion donors such as $H_2\overset{\oplus}{O}NO_2$, $AcONO_2$, O_2NONO_2, $HONO_2$, etc., are responsible for nitration in some other cases.

16-1c. *Mechanism of the Attack of the Nitrating Agent on the Aromatic Ring.* On the basis of the evidence described so far, the replacement of aromatic H^+ by NO_2^+ could be written as a concerted process with the introduction of the nitro group and the removal of a proton (by a base) occurring as a single step.

$$O_2\overset{\oplus}{N} + Ar\!-\!H + B \rightarrow ArNO_2 + B\overset{\oplus}{H} \tag{16-5}$$

[12] E. D. Hughes, C. K. Ingold, and R. I. Reed, *J. Chem. Soc.*, 2400 (1950).

It seems much more likely, though, that a definite reaction intermediate is formed, as Pfeiffer and Wizinger suggested.[13]   Such an intermediate may be formed rapidly and reversibly (16-6), or it may be formed in the rate-controlling step and then rapidly deprotonated (16-7).

$$\text{Ar—H} + \overset{\oplus}{NO_2} \underset{k_{-1}}{\overset{k_1}{\rightleftharpoons}} \overset{\oplus}{Ar} \overset{\displaystyle NO_2}{\underset{\displaystyle H}{<}}$$

$$\overset{\oplus}{Ar} \overset{\displaystyle NO_2}{\underset{\displaystyle H}{<}} + B \overset{k_2}{\rightarrow} Ar\text{—}NO_2 + \overset{\oplus}{BH}$$

$$k_{-1} \gg k_2 \tag{16-6}$$

or $$k_2 \gg k_{-1} \tag{16-7}$$

One point of evidence in favor of the two-step mechanism comes from studies of hydrogen kinetic isotope effects.   As Melander pointed out, since both mechanisms (16-5) and (16-6) involve a hydrogen transfer in the rate-controlling step, the reaction will show a primary kinetic isotope effect if either of these mechanisms is operative, whereas if (16-7) is the correct mechanism, there will be only a secondary kinetic isotope effect.[14] In the nitration of tritiated samples of benzene, nitrobenzene, toluene, bromobenzene, and naphthalene Melander found that tritium atoms were replaced to about the same extent, and therefore at about the same rate, as protium atoms.[14]   Other workers obtained similar results using deuterium.[15,16]   In few cases are the uncertainties in Melander's experiments small enough to prove that the kinetic isotope effect k_H/k_T is smaller than 1.25.   Nevertheless, since there seem to be very few, if any, primary tritium or deuterium kinetic isotope effects (other than for certain reactions that may be diffusion-controlled) that have been shown to be this small, it seems very probable that the observed isotope effect is not a primary one and hence that (16-7) is the correct mechanism for the attack of nitronium ions on aromatic rings.

Additional strong evidence for the intermediate in electrophilic aromatic substitution, as for many other intermediates, comes from direct observations and isolation experiments in cases where the intermediates are particularly stable.   McCaulay and Lien showed that the methylbenzenes behave as bases in liquid hydrogen fluoride containing boron

[13] P. Pfeiffer and R. Wizinger, *Ann.*, **461**, 132 (1928).

[14] L. Melander, *Nature*, **163**, 599 (1949); *Acta Chem. Scand.*, **3**, 95 (1949); *Arkiv Kemi*, **2**, 211 (1951); cf. U. Berglund-Larsson and L. Melander, *Arkiv Kemi*, **6**, 219 (1953).

[15] W. M. Lauer and W. E. Noland, *J. Am. Chem. Soc.*, **75**, 3689 (1953).

[16] T. G. Bonner, F. Bowyer, and G. Williams, *J. Chem. Soc.*, 2650 (1953).

trifluoride, and they determined basicity constants.[17]   Kilpatrick and Luborsky obtained values (Table 16-1) in reasonable agreement with these by making conductimetric measurements in hydrogen fluoride without added boron trifluoride.[18]

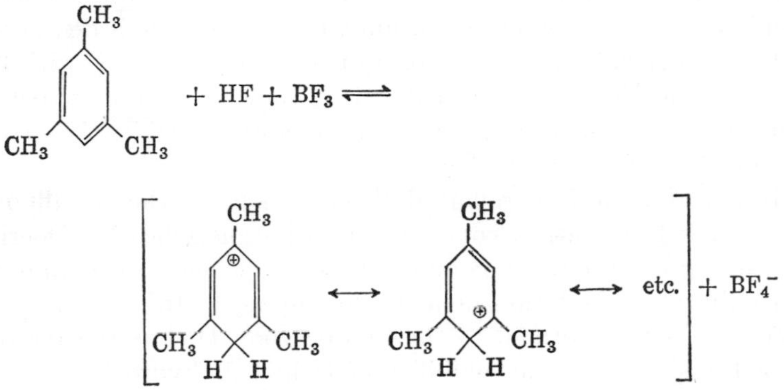

$$\tag{16-8}$$

The cation formed in this case, corresponding to the intermediate in electrophilic substitution of hydrogen for hydrogen (or its isotopes), has

TABLE 16-1. EQUILIBRIUM IN π- AND σ-COMPLEX FORMATION AND RATES OF HALOGENATION OF METHYLBENZENES[a]

Benzene	π complex[b]	σ complex[c]	Halogenation rate[d]
Unsubstituted..........	0.61	0.09	0.0004
Methyl...............	0.92	0.63	0.24
1,4-Dimethyl..........	1.00	1.00	1.00
1,2-Dimethyl..........	1.13	1.1	2.1
1,3-Dimethyl..........	1.26	26	204
1,2,4-Trimethyl........	1.36	63	600
1,2,3-Trimethyl........	1.46	69	660
1,2,4,5-Tetramethyl.....	....	140	1,100
1,2,3,4-Tetramethyl.....	1.63	400	4,400
1,3,5-Trimethyl........	1.59	13,000	75,000
1,2,3,5-Tetramethyl.....	1.67	16,000	170,000
Pentamethyl..........	....	29,000	360,000
Hexamethyl...........	....	97,000	

[a] All relative to p-xylene.
[b] Equilibrium constants for π-complex formation with HCl.[19]
[c] Basicity constants in hydrogen fluoride.[18]
[d] Rates of halogenation in acetic acid solution.[20]

[17] D. A. McCaulay and A. P. Lien, *J. Am. Chem. Soc.*, **73**, 2013 (1951).
[18] M. Kilpatrick and F. E. Luborsky, *J. Am. Chem. Soc.*, **75**, 577 (1953).
[19] H. C. Brown and J. D. Brady, *J. Am. Chem. Soc.*, **74**, 3570 (1952).
[20] H. C. Brown and L. M. Stock, *J. Am. Chem. Soc.*, **79**, 1421 (1957).

been called a σ *complex* by Brown and Brady in order to distinguish it from the more weakly bonded π *complexes*, in which a reagent coordinates with the π electrons of the aromatic ring without forming a strong σ bond to any one particular atom.[19] The σ complexes differ from π complexes in that they are electrical conductors,[18] are colored, are formed more slowly, and give carbon-bound deuterium when made from a deuterium halide.[1,19] The correlation of rates of aromatic halogenation[20] with the stabilities of σ complexes seen in Table 16-1 shows that the transition state in electrophilic aromatic substitution probably resembles the σ complex (but not the π complex) rather closely.

Olah, Kuhn, and Pavlath reported the isolation (as the tetrafluoroborates) of several aromatic σ-complex derivatives as solids.[21] Doering and coworkers prepared the σ complex I by the exhaustive aluminum chloride–catalyzed methylation of methylbenzenes.[22] In aqueous solution this cation is transformed to the hydrocarbon II, but the relative stability of I is shown by the solubility of II in 4 M hydrochloric acid, where it gives I.

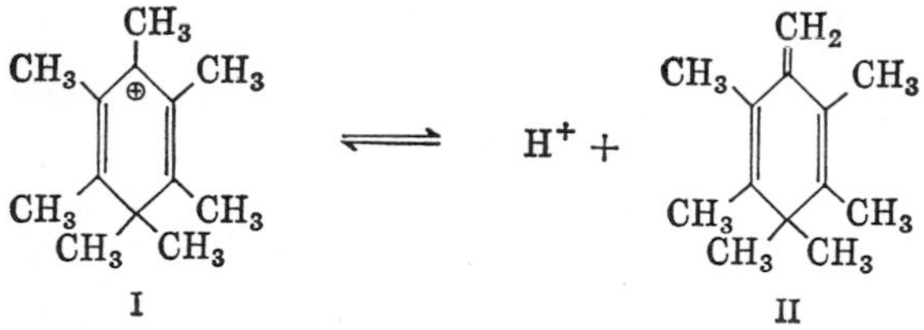

16-1d. *Nitrations Involving Prior Nitrosation.* A number of aromatic nitration reactions have been found to be slowed by nitrous acid.[12] This appears to be partly due to the action of nitrous acid as a base

$$HNO_2 + 2H_2SO_4 \rightleftharpoons NO^+ + H_3O^+ + 2HSO_4^-$$

and partly to its ability to transform nitric acid to dinitrogen tetroxide.

$$HNO_2 + HNO_3 \rightarrow N_2O_4 + H_2O$$

In certain nitration reactions oxidation of the aromatic compound is a significant side reaction. Since this oxidation transforms some of the nitric acid to nitrous acid and oxides of nitrogen, many such nitrations are greatly slowed by this side reaction. The nitration of aromatic amines and phenols, however, has been found to be catalyzed by nitrous acid, so that nitrous acid–producing side reactions cause autocatalysis.[3] The mechanisms of reactions of this type have been studied carefully by

[21] G. A. Olah, S. J. Kuhn, and A. Pavlath, *Nature*, **178,** 693 (1956); *J. Am. Chem. Soc.*, **80,** 6535, 6541 (1958).

[22] W. v. E. Doering et al., *Tetrahedron*, **4,** 178 (1958).

Hughes, Ingold, and coworkers,[23] who showed that the nitrous acid catalysis is due to the occurrence of a nitrosation reaction to yield a nitroso compound that is subsequently oxidized to the nitro compound usually observed as the product. The nitrosation is due both to the action of the nitrosonium ion, NO^+, and to dinitrogen tetroxide, which, although less active, is present in much greater concentrations.[23,24] Even the more reactive nitrosonium ion is not nearly so strong an electrophilic reagent as the nitronium ion. For this reason it is capable of attacking only the most reactive aromatic compounds, such as amines, phenols, phenol ethers, etc., and thus it is only these compounds whose nitrations are catalyzed by nitrous acid. For the same reason, the nitrosonium ion is capable of existence in much more weakly acidic solutions than is the nitronium ion. In relatively dilute acid solutions, where nitronium ions have very little tendency to be formed, practically all the nitration involves nitrosation and oxidation. On the other hand, in strongly acidic solutions the nitration of even amines and phenols may be due to nitronium ions, and under these conditions nitrous acid, which interferes with nitronium-ion formation, is a negative rather than a positive catalyst.[23]

16-2. Mechanisms of Other Electrophilic Aromatic Substitution Reactions. *16-2a. Hydrogen Exchange.* Since Lowry-Brønsted acids are electrophilic reagents, they would be expected to be capable of performing an aromatic substitution reaction by donating a proton to the ring to replace one already there. This symmetrical substitution reaction will be observable only by the use of deuterium or tritium tracers.

Ingold, Raisin, and Wilson showed that benzene could be deuteriated by shaking with deuterium sulfate.[25] They further observed that the rate of deuterium exchange decreases as the strength of the catalyzing acid decreases but may be greatly increased by the presence of electron-donating groups on the aromatic ring. Thus, although aqueous hydrochloric acid is essentially incapable of catalyzing the deuterium exchange of benzene at room temperature, it is quite effective with anisole and dimethylaniline, even though the latter compound must be changed largely to its unreactive salt under the reaction conditions. The nuclear deuterium exchange of phenol, unlike that of most aromatic compounds, is base-catalyzed. This is due to the transformation of the —OH group

[23] C. A. Bunton, E. D. Hughes, C. K. Ingold, D. I. H. Jacobs, M. H. Jones, G. J. Minkoff, and R. I. Reed, *J. Chem. Soc.*, 2628 (1950); J. Glazer, E. D. Hughes, C. K. Ingold, A. T. James, G. T. Jones, and E. Roberts, *J. Chem. Soc.*, 2657 (1950). See also R. M. Schramm and F. H. Westheimer, *J. Am. Chem. Soc.*, **70**, 1782 (1948).

[24] E. L. Blackall, E. D. Hughes, and C. K. Ingold, *J. Chem. Soc.*, 28 (1952).

[25] C. K. Ingold, C. G. Raisin, and C. L. Wilson, *Nature*, **134**, 734 (1934); *J. Chem. Soc.*, 1637 (1936).

to the much more strongly electron-donating —O⁻ group. The ortho and para hydrogen atoms of the phenoxide anion are thus reactive enough to be replaced by the action of the weak acid, water. We know that it is the positions activated toward ordinary electrophilic aromatic substitution that are active in these acid-catalyzed deuterium exchanges, not only because phenol, aniline, anisole, dimethylaniline, etc., exchange three carbon-bound hydrogen atoms fairly readily and any others much more difficultly (if at all), but also because deuterium has been shown to enter only the ortho and para positions of aniline and phenol.[26] This was established by Best and Wilson, who showed that deuteriated samples of these compounds lost all their deuterium when transformed to their 2,4,6-tribromo derivatives.

On the basis of H_0-rate correlations it has been suggested that aromatic hydrogen exchange proceeds by the rapid reversible formation of an H⁺-aromatic π complex, which then undergoes a slower isomerization to a σ complex.[27] In view of the uncertainties that attend the use of the H_0 function, this evidence must be regarded as quite weak, and, in fact, Kresge and Chiang[28] have shown that such a mechanism is not applicable to 1,3,5-trimethoxybenzene, and Colapietro and Long[29] have made a similar observation for azulene. Both groups of workers found aromatic hydrogen exchange to be subject to general acid catalysis, in agreement with the simple mechanism

$$\text{ArH}' + \text{HA} \rightleftharpoons \text{ArH}'\text{H}^+ + \text{A}^- \rightleftharpoons \text{ArH} + \text{H}'\text{A}$$

and in disagreement with any mechanism involving a rate-limiting interconversion of isomeric π and σ complexes.

16-2b. Sulfonation. A study of the mechanism of aromatic sulfonation reactions reveals certain complications not found in some of the other electrophilic aromatic substitution reactions. For example, Melander showed that in the sulfonation of tritium-labeled benzene and bromobenzene by oleum the tritium atoms are replaced considerably more slowly than protium.[14] This observation reveals the improbability of a mechanism like (16-7), in which a rate-controlling attack of the substituting agent on the aromatic ring is followed by a relatively rapid loss of a hydrogen cation to restore aromaticity to the ring. Since there appears to be no particular evidence in favor of a concerted mechanism like (16-5), we shall assume by analogy with other aromatic substitutions that the reaction proceeds by a two-step mechanism but that in the case of sulfonation the intermediate reverts to reactant at a rate at least com-

[26] A. P. Best and C. L. Wilson, *J. Chem. Soc.*, 28 (1938).

[27] V. Gold and D. P. N. Satchell, *J. Chem. Soc.*, 3609, 3619, 3622 (1955); 1635 (1956).

[28] A. J. Kresge and Y. Chiang, *J. Am. Chem. Soc.*, **81**, 5509 (1959); **83**, 2877 (1961).

[29] J. Colapietro and F. A. Long, *Chem. & Ind. (London)*, 1056 (1960).

parable to its transformation to product. After all, the reversals of electrophilic aromatic substitutions are also electrophilic aromatic substitutions. Hence if the two-step mechanism is general, sometimes the first step will be rate-limiting and sometimes the second step will be.

The studies of Brand and coworkers on sulfonations in concentrated and fuming sulfuric acid are in agreement with a mechanism in which the intermediate $H-Ar-SO_3H^+$ is formed reversibly and donates a proton to the solvent in the rate-limiting step of the reaction.[30,31] One manner in which this could occur is the following:

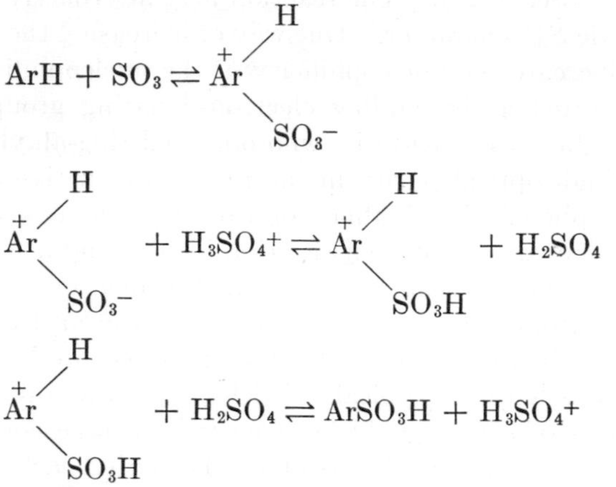

$$ArH + SO_3 \rightleftharpoons \overset{+}{Ar}\!\!\begin{array}{c} H \\ \diagup \\ \diagdown \\ SO_3^- \end{array}$$

$$\overset{+}{Ar}\!\!\begin{array}{c} H \\ \diagup \\ \diagdown \\ SO_3^- \end{array} + H_3SO_4^+ \rightleftharpoons \overset{+}{Ar}\!\!\begin{array}{c} H \\ \diagup \\ \diagdown \\ SO_3H \end{array} + H_2SO_4 \qquad (16\text{-}9)$$

$$\overset{+}{Ar}\!\!\begin{array}{c} H \\ \diagup \\ \diagdown \\ SO_3H \end{array} + H_2SO_4 \rightleftharpoons ArSO_3H + H_3SO_4^+$$

The function of the proton added in the second step of the reaction is to increase the acidity of the hydrogen being replaced. Hinshelwood and coworkers found the sulfonation of *p*-nitrotoluene, chlorobenzene, *m*-dichlorobenzene, bromobenzene, *α*-nitronaphthalene, *p*-nitroanisole, and benzene by sulfur trioxide in nitrobenzene solution to be, in all cases, first-order in aromatic compound and second-order in sulfur trioxide.[32] Perhaps a second molecule of sulfur trioxide assumes the role played by the proton in the second step of mechanism (16-9), or perhaps S_2O_6 is the electrophilic attacking reagent. It is not likely that the second sulfur trioxide acts purely as a base in view of the greater basicity of the more abundant solvent.

16-2c. *Friedel-Crafts Alkylation.* Carbonium ions appear to be relatively effective at attacking aromatic nuclei to introduce alkyl groups. This fact is, no doubt, the basis of many Friedel-Crafts alkylation reactions involving alkyl halides, olefins, alcohols, esters, and other reactants capable of forming carbonium ions. A carbonium-ion mechanism gives

[30] J. C. D. Brand, *J. Chem. Soc.*, 997, 1004 (1950).
[31] J. C. D. Brand et al., *J. Chem. Soc.*, 3922 (1952); 3844 (1959).
[32] C. N. Hinshelwood et al., *J. Chem. Soc.*, 1372 (1939); 469, 649 (1944).

an explanation for the many rearrangements that have been found to accompany the reaction. Although carbonium ions may be reaction intermediates, there is evidence that they need not always be so. The reaction of d-sec-butyl alcohol with benzene in the presence of boron fluoride, hydrogen fluoride, phosphoric acid, or sulfuric acid gives sec-butylbenzene which is largely (> 99 per cent) racemized but which definitely has *some* optical activity.[33] If the reaction is regarded as a nucleophilic substitution on carbon, it may be considered an S_N1 reaction in which racemization is not quite complete because of the operation of a shielding effect (Sec. 6-2b). The reaction may alternately be considered S_N1 with a little S_N2 character. One way of increasing the S_N2 character would be to increase the nucleophilicity of the nucleophilic reagent (the aromatic reactant) as by adding electron-donating groups. This has been done by Hart and Eleuterio, who obtained ring-alkylated products of relatively high optical purity in the reaction of active α-phenylethyl chloride with phenols in alkaline solutions.[34] These workers demonstrated the S_N2 character of their reaction by showing it to be first-order in each reactant and to lead to inversion in configuration at the asymmetric carbon atom and also by showing that the aryl ether is not an intermediate in the formation of the ring-alkylated product. Of course, this example was hardly a typical Friedel-Crafts reaction.

Aromatic methylations would be expected to have considerable S_N2 character in view of the relative instability of methyl carbonium ions and the large S_N2 reactivity of methyl halides. Brown and Jungk showed that the methylation of toluene cannot be an S_N1 reaction. About 50 per cent more m-xylene is obtained when methyl bromide is used than when methyl iodide is used, under the same conditions.[35] The orientation obtained on attack by a carbonium ion on an aromatic ring should be independent of the source of the carbonium ion, under a given set of conditions.

Hart, Spliethoff, and Eleuterio described evidence for aromatic substitution by the S_Ni mechanism.[36] They found that the para-alkylated product of the reaction of phenol or of 2,6-xylenol with optically active α-phenylethyl chloride had undergone inversion of configuration at the asymmetric carbon atom, presumably because of an S_N2-type reaction. The products of the ortho alkylations of phenol, p-cresol, and p-chlorophenol all had the same configuration as the starting α-phenylethyl chloride. This retention of configuration that occurs only for substitu-

[33] C. C. Price and M. Lund, *J. Am. Chem. Soc.*, **62**, 3105 (1940); R. L. Burwell, Jr., and S. Archer, *J. Am. Chem. Soc.*, **64**, 1032 (1942).

[34] H. Hart and H. S. Eleuterio, *J. Am. Chem. Soc.*, **76**, 516, 519 (1954).

[35] H. C. Brown and H. Jungk, *J. Am. Chem. Soc.*, **77**, 5584 (1955).

[36] H. Hart, W. L. Spliethoff, and H. S. Eleuterio, *J. Am. Chem. Soc.*, **76**, 4547 (1954).

tion ortho to the hydroxy group suggests an $S_{N}i$-type mechanism in which the hydroxy group solvates the chloride ion as it is displaced.

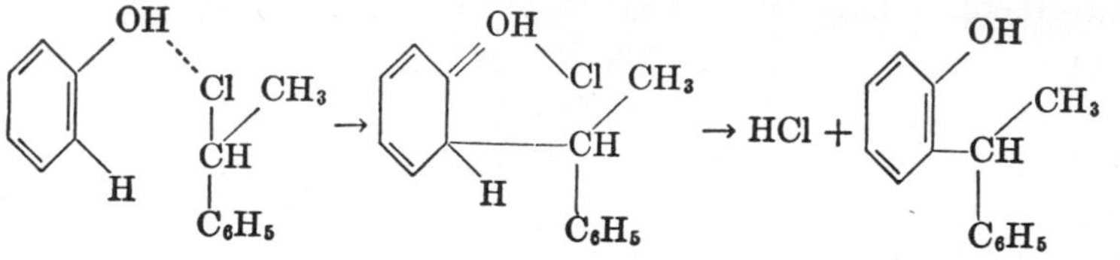

The kinetics of a more representative Friedel-Crafts alkylation have been investigated by Ulich and Heyne, who studied the condensation of benzene with n-propyl chloride, using a gallium trichloride catalyst in carbon disulfide solution.[37] They report that the reaction rate is proportional to the concentration of benzene and to that of the gallium chloride–propyl chloride complex,

$$v = k[C_6H_6][C_3H_7Cl \cdot GaCl_3]$$

or, alternately, third-order

$$v = k'[C_6H_6][C_3H_7Cl][GaCl_3]$$

since $[C_3H_7Cl][GaCl_3] = K[C_3H_7Cl \cdot GaCl_3]$

Brown and Grayson reported analogous third-order kinetics for the condensations of several benzyl chlorides with aromatic compounds in the presence of aluminum chloride in nitrobenzene solution.[38] There are conductimetric[39] and spectroscopic[40] data, as well as dipole-moment determinations[41] and radioactive-halogen-exchange[42] and vapor pressure–composition phase studies,[43] that show that several Friedel-Crafts catalysts may coordinate with the halogen atom of alkyl halides and in some cases thus promote their ionization. It therefore seems most reasonable to explain the observed third-order kinetics by assuming a reversible and relatively rapid reaction between the alkyl halide and the Friedel-Crafts catalyst to give a small concentration of an intermediate, which then

[37] H. Ulich and G. Heyne, *Z. Elektrochem.*, **41**, 509 (1935).

[38] H. C. Brown and M. Grayson, *J. Am. Chem. Soc.*, **75**, 6285 (1953); for reports that the order of the reaction with respect to aluminum chloride is about 0.5 for the reactions of secondary and tertiary chlorides in nitrobenzene and as much as 2.0 in other solvents, see N. N. Lebedev, *J. Gen. Chem. USSR (Eng. Transl.)*, **24**, 673 (1954); **28**, 1211 (1958).

[39] P. Walden, *Ber.*, **35**, 2018 (1902).

[40] V. V. Korshak and N. N. Lebedev, *Zhur. Obshchei Khim.*, **18**, 1766 (1948); F. Fairbrother and B. Wright, *J. Chem. Soc.*, 1058 (1949).

[41] F. Fairbrother, *Trans. Faraday Soc.*, **37**, 763 (1941); *J. Chem. Soc.*, 503 (1945).

[42] F. Fairbrother, *J. Chem. Soc.*, 503 (1937).

[43] H. C. Brown, L. P. Eddy, and R. Wong, *J. Am. Chem. Soc.*, **75**, 6275 (1953); H. C. Brown and W. J. Wallace, *J. Am. Chem. Soc.*, **75**, 6279 (1953).

reacts with the aromatic compound in the rate-controlling step. Depending upon the exact nature of all the constituents of the reaction mixture, this intermediate may be an alkyl halide–catalyst complex, which undergoes an S_N2-type attack by the aromatic compound, or it may be a carbonium ion (perhaps as part of an ion pair), which subsequently replaces a proton on the aromatic ring. Furthermore, there will be a gradual and continuous transition between the S_N1 and S_N2 extremes, as described in Secs. 6-2c and 7-3b.

Brown and Grayson suggested that their reactions of benzyl chlorides are of the S_N2 type and stated that their data appear to eliminate the possibility of a carbonium-ion mechanism. Thus the mechanism

$$\text{ArCH}_2\text{Cl} + \text{AlCl}_3 \underset{k_{-1}}{\overset{k_1}{\rightleftharpoons}} \text{ArCH}_2^{\oplus} + \text{AlCl}_4^{\ominus}$$

$$\text{ArCH}_2^{\oplus} + \text{Ar}'\text{H} \overset{k_2}{\rightarrow} \text{ArCH}_2\text{Ar}'\text{H}^{\oplus}$$

$$\text{ArCH}_2\text{Ar}'\text{H}^{\oplus} + \text{AlCl}_4^{\ominus} \overset{k_3}{\rightarrow} \text{ArCH}_2\text{Ar}' + \text{HCl} + \text{AlCl}_3$$

$$k_{-1} \gg k_2; \; k_3 \gg k_2$$

involving a rate-controlling attack of the carbonium ion (or ion pair) on the aromatic ring is said to be unacceptable because it predicts that p-nitrobenzyl chloride should react more rapidly than benzyl chloride, since it will form a more reactive carbonium ion. It is certainly true that k_2 should be larger for the p-nitro compound, but it is the fraction $k_1 k_2/k_{-1}$ upon which the rate depends, and the rest of this fraction, k_1/k_{-1} (the equilibrium constant for the formation of the carbonium ion), must be much larger for the unsubstituted compound. In fact, as Nenitzesco, Tzitzeica, and Ioan[44] pointed out, since the side-chain carbon atom of the benzyl chloride must be considerably more positive in the transition state of the rate-controlling step of the reaction, a p-nitro group would be expected to decrease the reactivity, exactly as found.

It is also stated that reaction by the ionization mechanism should be powerfully affected by changes in the ionizing properties of the medium, whereas the observed reaction rate is merely cut in two by replacement of 60 per cent of the nitrobenzene solvent by methylcyclohexane.[38] However, this change in solvent probably affects the reaction rate in more ways than one. Since nitrobenzene forms a complex with aluminum chloride, a shift to a hydrocarbon solvent should increase the activity of the catalyst. In fact, as Brown and Grayson note, the reaction appears to go faster in benzene than in nitrobenzene solution.

As Brown and Grayson further note, the formation of n- (rather than iso-) propylbenzene and neopentyl- (rather than t-amyl-) benzene in the

[44] C. D. Nenitzesco, S. Tzitzeica, and V. Ioan, *Bull. soc. chim. France*, 1272, 1279 (1955).

reactions of the corresponding alcohols with benzene in the presence of aluminum chloride suggests that these reactions proceed by the S_N2 mechanism.   It appears, however, that Friedel-Crafts alkylations rarely, if ever, acquire enough S_N2 character to cause the *making* of the new carbon-carbon bond to have more effect on the reactivity than the *breaking* of the carbon-halogen (or other) bond.   Thus methyl halides undergo the reaction less readily than their ordinary primary alkyl counterparts[45] even though both may react by an S_N2 type of mechanism.   Despite this fact, the continuous addition of electron-donating groups does not result in an interminable increase in reactivity, because sooner or later the point will be reached where a carbonium ion may be formed with the greatest of ease but, having been formed, will be too stable to react further.   Thus, benzene may be alkylated with methyl chloride, or even more easily with benzyl chloride or benzhydryl chloride, but triphenylmethyl chloride fails to react.   This is not due to the impossibility of the existence of the product, since tetraphenylmethane is a quite stable compound, boiling at 431° without decomposition.[46]   Nor is it due to steric hindrance, since triphenylmethyl chloride and the more reactive aromatic compound phenol yield *p*-hydroxytetraphenylmethane with no Friedel-Crafts catalysts (except the anion-solvating reagents phenol and hydrogen chloride).[47]

16-2*d*. *Friedel-Crafts Acylation.*   For the Friedel-Crafts acylation reaction we may postulate either a carbonyl addition mechanism

$$
\begin{array}{ccc}
\ominus MX_3 & \ominus MX_3 & \\
| & | & \\
O & \oplus O & \\
| & \| & \\
RCOX + MX_3 \rightleftharpoons R\!-\!\underset{\oplus}{C}\!-\!X \leftrightarrow R\!-\!C\!-\!X
\end{array}
$$

$$
\begin{array}{ccc}
\ominus MX_3 & \ominus MX_3 & \downarrow \text{ArH} \\
| & | & \\
O & \oplus O & \ominus OMX_3 \\
| & \| & | \\
R\!-\!\underset{\oplus}{C}\!-\!Ar \leftrightarrow R\!-\!C\!-\!Ar + HX \leftarrow R\!-\!C\!-\!X \\
& & | \\
& & \oplus ArH
\end{array}
$$

or an acyl carbonium-ion mechanism

$$
RCOX + MX_3 \rightleftharpoons MX_4^{\ominus} + R\overset{\oplus}{C}O
$$

$$
\begin{array}{cc}
O\!\cdot\! MX_3 & \downarrow \text{ArH} \\
\| & \\
R\!-\!C\!-\!Ar + HX \overset{MX_4^{\ominus}}{\longleftarrow} RCOArH^{\oplus}
\end{array}
$$

[45] C. R. Smoot and H. C. Brown, *J. Am. Chem. Soc.*, **78**, 6249 (1956).
[46] F. Ullmann and A. Münzhuber, *Ber.*, **36**, 404 (1903).
[47] H. Hart and F. A. Cassis, *J. Am. Chem. Soc.*, **76**, 1634 (1954).

Just as in Friedel-Crafts alkylation, the prevalence of rearrangements in the reaction of primary halides and other facts suggest that the reaction tends to have more S_N1 character than most nucleophilic substitutions so in the present case does the acyl carbonium-ion mechanism appear to be more common than in most acid halide and ester hydrolysis reactions. This is probably due to the fact that most aromatic reactants are not very highly nucleophilic.

Burton and Praill found considerable evidence that at least some Friedel-Crafts acetylation reactions are due to the "acetylium" ion $CH_3CO^\oplus$.[48] These workers showed that solutions of sulfuric or perchloric acid in acetic anhydride are capable of acetylating anisole, although similar solutions in acetic acid are inactive. This fact could be explained by assuming the acetylating agent is either the acetylium ion or the conjugate acid of acetic anhydride, either of which should be more reactive than the conjugate acid of acetic acid. Burton and Praill also found that a powerful acetylating agent could also be prepared by the reaction of acetyl chloride with silver perchlorate in nitromethane or acetic anhydride solution, and they postulated that this agent is the salt, acetylium perchlorate.[48] It is reasonable that this compound should be ionized in such solvents, since Seel has shown conductimetrically that CH_3COBF_4, $CH_3COSbCl_6$, and $C_6H_5COSbCl_6$ are ionized in sulfur dioxide solution.[49,5]

Jensen and Brown[51] showed that the benzoylation of toluene in benzoyl chloride solution gave essentially the same product distribution with six different metallic chloride catalysts, even though the activity of the catalysts varied over a range of about 10^6. This suggests that toluene is undergoing attack by the same species, namely, a benzoyl cation, in each case.

16-2e. *Halogenation.*[2] From our discussions of the mechanisms of other aromatic substitution reactions we should expect X^+ (where X is halogen) to be the most active halogenating agent possible and compounds of the type X—Y, where Y may be removed with the bonding electron pair, also to be capable of attacking aromatic rings. The activity of reagents of the type X—Y should increase with the electron-withdrawing power of Y. In a number of studies these expectations have been borne out (in so far as they have been investigated).

Soper and Smith studied the chlorination of phenol by hypochlorous

[48] H. Burton and P. F. G. Praill, *J. Chem. Soc.*, 1203, 2034 (1950); 522, 529, 72 (1951); 755 (1952); 827, 837 (1953).

[49] F. Seel, *Z. anorg. u. allgem. Chem.*, **250**, 331 (1943); **252**, 24 (1943).

[50] F. Seel and H. Bauer, *Z. Naturforsch.*, **2b**, 397 (1947).

[51] F. R. Jensen and H. C. Brown, *J. Am. Chem. Soc.*, **80**, 3039 (1958).

acid and found that the rate data agreed with either of the equations[52]

$$v = k[OCl^-][C_6H_5OH]$$
or
$$v = k'[HOCl][C_6H_5O^-] \qquad (16\text{-}10)$$

which are kinetically indistinguishable, since

$$[OCl^-][C_6H_5OH] = \frac{K_a^{HOCl}}{K_a^{C_6H_5OH}} [HOCl][C_6H_5O^-] \qquad (16\text{-}11)$$

We may calculate from the ionization constants of hypochlorous acid and phenol that the concentration term, $[OCl^-][C_6H_5OH]$, for hypochlorite-ion chlorination will be about 100 times as large as $[HOCl]$ $[C_6H_5O^-]$, the term for hypochlorous acid chlorination. However, since hypochlorous acid should be a tremendously more powerful chlorinating agent than hypochlorite ion and since the phenoxide ion should be vastly more reactive than phenol, it is very likely that k' is, at least, millions of times larger than k. Hence the chlorination must be almost entirely due to hypochlorous acid. However, the reaction is specifically catalyzed by hydrochloric acid, which transforms hypochlorous acid to chlorine. This shows that chlorine is a more reactive halogenating agent than hypochlorous acid.

Analogously, Francis showed that hypobromous acid is a less effective aromatic brominating agent than is bromine.[53]

Shilov and Kanyaev found that the hypobromous acid bromination of sodium m-anisolesulfonate is catalyzed by acids.[54]

$$v = k[ArH][HOBr][H^+]$$

The brominating agent here could be either H_2OBr^+ or Br^+. Wilson and Soper and Derbyshire and Waters obtained similar results with other aromatic reactants.[55]

The rates of iodination of phenol and of aniline in aqueous solution in the presence of iodide ions have been found to be inversely proportional to the square of the iodide-ion concentration.[56-58] Under the conditions,

[52] F. G. Soper and G. F. Smith, *J. Chem. Soc.*, 1582 (1926).
[53] A. W. Francis, *J. Am. Chem. Soc.*, **47**, 2340 (1925).
[54] E. Shilov and N. Kanyaev, *Compt. rend. acad. sci. U.R.S.S.*, **24**, 890 (1939); *Chem. Abstr.*, **34**, 4062 (1940).
[55] W. J. Wilson and F. G. Soper, *J. Chem. Soc.*, 3376 (1949); D. H. Derbyshire and W. A. Waters, *J. Chem. Soc.*, 564, 574 (1950).
[56] F. G. Soper and G. F. Smith, *J. Chem. Soc.*, 2757 (1927).
[57] B. S. Painter and F. G. Soper, *J. Chem. Soc.*, 342 (1947).
[58] E. Berliner, *J. Am. Chem. Soc.*, **72**, 4003 (1950); **73**, 4307 (1951).

most of the material titratable as iodine was in the form of the triiodide ion.

$$I_3^- \rightleftharpoons I_2 + I^-$$

The reaction therefore cannot involve a rate-controlling attack of iodine on the aromatic reactant, since this would give a rate proportional simply to the reciprocal of the iodide-ion concentration. A rate-controlling attack by hypoiodous acid, the hypoiodous acidium ion, or the iodine cation would be in agreement with the observed dependence of the rate on the iodide-ion concentration.

$$I_3^- + H_2O \rightleftharpoons HOI + H^+ + 2I^-$$
$$I_3^- + H_2O \rightleftharpoons H_2OI^+ + 2I^-$$
$$I_3^- \rightleftharpoons I^+ + 2I^-$$

However, as Grovenstein and Henderson pointed out, the observed data may equally well be explained by a reversible reaction of iodine with the aromatic species followed by a rate-limiting loss of the hydrogen atom being replaced.[59]

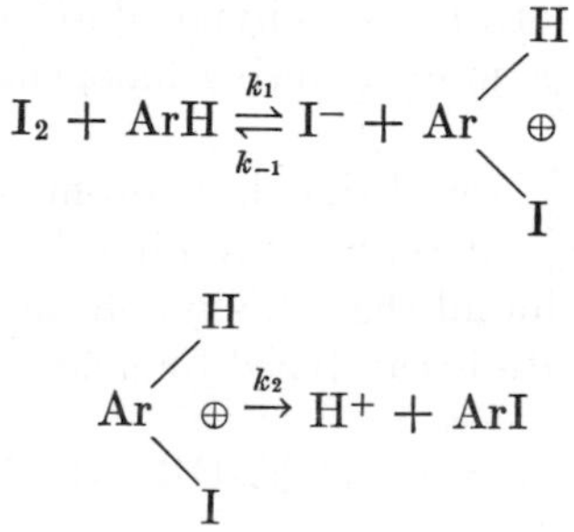

$$I_2 + ArH \underset{k_{-1}}{\overset{k_1}{\rightleftharpoons}} I^- + Ar\overset{H}{\underset{I}{\diagdown}}{}^{\oplus}$$

$$Ar\overset{H}{\underset{I}{\diagdown}}{}^{\oplus} \xrightarrow{k_2} H^+ + ArI$$

(16-12)

Mechanism (16-12) has been shown to be the correct alternative by the observation of large deuterium kinetic isotope effects in the iodination of phenol,[60] aniline,[61] N-methylaniline,[61] N,N-dimethylaniline,[61] sodium m-aminobenzoate,[61] p-nitrophenol,[62] and anisole,[62] but, surprisingly, not for the m-dimethylaminobenzenesulfonate anion.[61] The primary kinetic isotope effect characteristic of k_2 will be reflected fully in the experimental rate only if the term $k_{-1}[I^-]$ is considerably larger than k_2. Therefore Grovenstein and Aprahamian's observation that the kinetic isotope effect found in the iodination of p-nitrophenol decreases at very low iodide-ion concentrations[62] is added evidence for mechanism (16-12). Because

[59] E. Grovenstein, Jr., and U. V. Henderson, Jr., *J. Am. Chem. Soc.*, **78**, 569 (1956).

[60] E. Grovenstein, Jr., and D. C. Kilby, *J. Am. Chem. Soc.*, **79**, 2972 (1957).

[61] E. A. Shilov and F. Weinstein, *Nature*, **182**, 1300 (1958); *Doklady Akad. Nauk S.S.S.R.*, **123**, 93 (1958).

[62] E. Grovenstein, Jr., and N. S. Aprahamian, *J. Am. Chem. Soc.*, **84**, 212 (1962).

of the greater susceptibility of organic iodides to nucleophilic attack on halogen and the greater ability of iodide ions to attack halogen substituents (cf. Sec. 8-2b) $k_{-1}[I^-]$ would be expected to be larger compared with k_2 for iodine than the analogous terms would be for bromine and chlorine. This offers an explanation for the fact that the rate-limiting step in bromination and chlorination is more likely to be the attack on the aromatic ring (rather than the subsequent deprotonation) than in the case of iodination.   Melander, for example, found no significant tritium kinetic isotope effect in the iodine-catalyzed bromination of toluene,[14] and Berliner and Schueller found only a 14 per cent, and therefore probably secondary, deuterium kinetic isotope effect in the bromination of biphenyl.[63]   From the form of the kinetic equation obtained by Grovenstein and Henderson in the bromination of 2,6-dibromophenol it appears that deprotonation of the intermediate is rate-limiting at high bromide-ion concentrations but not otherwise.[59]   Deuterium kinetic isotope effects $(k_H/k_D \sim 2)$ have been observed in the bromination of 2-naphthol-6,8-disulfonic acid and of dimethylaniline.[64]

De la Mare, Hughes, Ketley, and Vernon obtained evidence for the intermediacy of the chloronium ion (Cl^+) in certain aromatic chlorinations that is analogous in some respects to the evidence for the intermediacy of the nitronium ion in certain aromatic nitrations that we have already discussed.[65,66]   Working with considerable concentrations of silver perchlorate (to keep the chloride concentration low and thus prevent chlorination by molecular chlorine) and acid (to minimize competing chlorination by Cl_2O), these workers reported that the chlorination of anisole in aqueous solution obeyed the following kinetic equation, the reaction rate being independent of the anisole concentration over the range 0.004 to 0.01 M.

$$v = k[HOCl] + k'[HOCl][H^+] \qquad (16\text{-}13)$$

where $k = 7.5 \times 10^{-3}$ min^{-1} and $k' = 7.1 \times 10^{-2}$ liters mole^{-1} min^{-1} at 25°.   The reaction is said to proceed via a rate-controlling formation of the chloronium ion, the first term representing the spontaneous and the second the acid-catalyzed removal of OH from hypochlorous acid.   The values of k and k' show that, compared with other polar dehydroxylations to give highly reactive intermediates, this reaction is phenomenally insensitive to acid catalysis.   There are other observations that are

---

[63] E. Berliner and K. E. Schueller, *Chem. & Ind. (London)*, 1444 (1960).

[64] H. Zollinger, *Experienta*, **12**, 165 (1956); P. G. Farrell and S. F. Mason, *Nature*, **183**, 250 (1959).

[65] P. B. D. de la Mare, E. D. Hughes, and C. A. Vernon, *Research (London)*, **3**, 192, 242 (1950).

[66] P. B. D. de la Mare, A. D. Ketley, and C. A. Vernon, *J. Chem. Soc.*, 1290 (1954)

difficult to explain in terms of the chloronium-ion (or any other) mechanism. The chlorination of phenol at concentrations around 0.003 M, for example, follows Eq. (16-13) with the same value of k but with about *twice* as large a value of k'. Many observations, on the other hand, are fit neatly and almost uniquely by the chloronium-ion mechanism. If the aromatic reactant is not sufficiently active, the chloronium ion will commonly combine with water, giving a slower reaction whose rate is dependent on the concentration of the aromatic reactant. This provides a reasonable interpretation of the observation of Derbyshire and Waters that the chlorination of the α-toluenesulfonate ion is relatively slow and follows the rate equation[67]

$$v = k[H^+][HOCl][C_6H_5CH_2SO_3^-]$$

and also for de la Mare, Ketley, and Vernon's observation that the chlorination of methyl p-tolyl ether, in the concentration range 0.04 to 0.10 M, is a first-order reaction (at a given acidity) whose rate is independent of the ether concentration and equal to the rate of chlorination of dilute solutions of phenol or anisole under the same conditions, whereas at lower ether concentrations the reaction is slower. Apparently phenol and anisole at higher concentrations are sufficiently reactive to undergo attack by the less reactive agent $ClOH_2^+$. At least this seems the most reasonable explanation for the third term that must be added to Eq. (16-13) to cover the rate of chlorination of these compounds at concentrations above 0.01 M, where

$$v = k[HOCl] + k'[H^+][HOCl] + k''[H^+][HOCl][ArH] \quad (16\text{-}14)$$

Bell and Gelles made estimates of

$$K_{X^+} = \frac{[X^+][X^-]}{[X_2]}$$

of 10^{-40}, 10^{-50}, and 10^{-60} for I, Br, and Cl, respectively, and of

$$K_{XOH_2^+} = \frac{[XOH_2^+][X^-]}{[X_2]}$$

of 10^{-10}, 10^{-20}, and 10^{-30} for I, Br, and Cl, respectively, all in aqueous solution.[68] From these estimates it might seem unlikely that any halogenation due to X^+ (and perhaps even $ClOH_2^+$) had ever really been observed. The uncertainties in such estimates are probably enormous, however, as can be best illustrated as follows for $K_{XOH_2^+}$. It seems reasonable to assume that XOH_2^+ will be more acidic than H_3O^+ by

[67] D. H. Derbyshire and W. A. Waters, *J. Chem. Soc.*, 73 (1951).
[68] R. P. Bell and E. Gelles, *J. Chem. Soc.*, 2734 (1951).

about the same factor that HOX is more acidic than H_2O.[69]   From this assumption and the known values for the hydrolysis constants of the halogens in aqueous solution,[70]

$$K_{HOX} = \frac{[H^+][X^-][HOX]}{[X_2]}$$

it is possible to calculate values of $K_{XOH_2^+}$ of about 10^{-20} and 10^{-15} for I and Cl.   This value of $K_{ClOH_2^+}$ and the related value of $K_{IOH_2^+}/K_{ClOH_2^+}$ differ from Bell and Gelles's estimates by factors of 10^{15} and 10^{25}, respectively, but seem at least as reliable as theirs.

In sum, it is believed that additional investigations should be made on the subject of when and whether X^+ cations act as intermediates in aromatic substitution.[71]

16-2f. *Diazo Coupling*.[72]   Wistar and Bartlett made a careful kinetic study of diazo coupling with phenols and aromatic amines and thereby determined the composition of the transition state in the rate-limiting step of the reaction.[73]   Variations in the location of various hydrogen atoms and in the presence or absence of molecules of solvent permit the following three interpretations of the results they obtained with anilines. The rate-limiting step may be (1) reaction of an anilinium ion with a diazo hydroxide, (2) reaction of an aniline with a diazonium cation, or (3) deprotonation (by solvent) of the adduct formed between an aniline and a diazonium cation.   Wistar and Bartlett pointed out that since diazo coupling occurs only with strongly activated aromatic compounds and gives ortho-para orientation with aromatic amines, the first alternative is implausible.   Certainly, the more electrophilic diazonium ion should be much more active than a diazo hydroxide in attacking aromatic nuclei.   It is probable that the rate-limiting step may be either 2 or 3, depending on the nature of the reactants and reaction conditions.   At least, this was found, by Zollinger and coworkers, to be the case for diazo couplings with phenols.   The rate of coupling of 2-deuterio-1-naphthol-4-sulfonic acid with 2-methoxybenzenediazonium ions was within 3 per cent of the rate for the corresponding protium compound; 2-naphthol-6,8-disulfonic acid, however, coupled with 4-chlorobenzenediazonium ions

---

[69] For the acidity of HOX's, see E. A. Shilov, *J. Am. Chem. Soc.*, **60**, 490 (1938) and M.-L. Josien and G. Sourisseau, *Bull. soc. chim. France*, 255 (1950).

[70] G. Zimmerman and F. C. Strong, *J. Am. Chem. Soc.*, **79**, 2063 (1957); W. C. Bray and E. L. Connolly, *J. Am. Chem. Soc.*, **33**, 1485 (1911).

[71] For interesting direct evidence that the iodide cation, I^+, is formed when iodine is dissolved in fuming sulfuric acid, see M. C. R. Symons, *J. Chem. Soc.*, 387, 2187 (1957); J. Arotsky, H. C. Mishra, and M. C. R. Symons, *J. Chem. Soc.*, 12 (1961).

[72] For a more detailed discussion of this subject, see H. Zollinger, "Chemie der Azofarbstoffe," chap. 9, Birkhäuser Verlag, Basel, 1958.

[73] R. Wistar and P. D. Bartlett, *J. Am. Chem. Soc.*, **63**, 413 (1941).

about 6.5 times as rapidly as did its 1-deuterio derivative.[74]   Apparently, in the first case the attack on the aromatic reactant is rate-limiting (kinetic studies show that such attack is ordinarily on the phenoxide ion[72,73]), and in the second case the deprotonation step is.   This interpretation is also supported by the effect of the concentration and nature of added bases on the rate, isotope effect, and position of coupling; and by the effect of changing the diazonium ion used.[74,75]

16-2g. *Replacement of Groups Other than Hydrogen.*   Although all the electrophilic aromatic substitution reactions discussed thus far have involved the replacement of hydrogen, there is no reason why all such reactions should.   There are several reasons why it is usually hydrogen that is replaced.   Hydrogen is the most common substituent and offers little steric hindrance to an attacking group.   Another important factor is that in most intermediates of the type H—Ar$^+$—X the ease of proton-transfer reactions usually makes it much easier to lose H$^+$ than X$^+$.

It would be expected, then, that the ease of replacement of a group X will depend on the stability of the species that is formed when X is expelled without its bonding electron pair.   Since the ease with which halogens can be removed without their bonding electron pairs (cf. Sec. 8-2b) varies in the order I > Br > Cl > F, the same order of ease of electrophilic displacement would be expected, and is found.   Iodination reactions are quite subject to reversal due to the replacement of iodine by hydrogen by the action of the electrophilic reagent hydrogen iodide. Gold and Whittaker pointed out that the effect of substituents on the reaction

$$ArI + HI \rightarrow ArH + I_2$$

is the same as in a typical aromatic substitution reaction.[76]   Aromatic brominations are not ordinarily sufficiently reversible to produce directly measurable amounts of bromine, but the occurrence of reactions like the following[77]

$$C_6H_5Br \xrightarrow[HBr]{AlBr_3} C_6H_6 \text{ and } C_6H_4Br_2\text{'s}$$

can be explained on this basis (and others).   As in the examples quoted, in most of the electrophilic aromatic substitutions in which hydrogen is not the group displaced, it is the displacing group.   Thus Benkeser,[78]

[74] H. Zollinger, *Helv. Chim. Acta,* **38,** 1597, 1617, 1623 (1955).

[75] H. Zollinger et al., *Helv. Chim. Acta,* **40,** 1955 (1957); **41,** 1816, 2274 (1958).

[76] V. Gold and M. Whittaker, *J. Chem. Soc.,* 1184 (1951).

[77] F. Fairbrother and N. Scott, *Chem. & Ind. (London),* 998 (1953).

[78] R. A. Benkeser et al., *J. Am. Chem. Soc.,* **75,** 4528 (1953); **76,** 6353 (1954); **80,** 2279, 2283, 5289 (1958); **82,** 4881 (1960).

Eaborn,[79] and their coworkers studied the acid-catalyzed replacement of various —SiR_3 groups, and Schubert and coworkers[80] studied the replacement of various acyl groups by hydrogen. The substitution of hydrogen for —HgR,[81,82] —SO_3H,[83] —PO_3H_2,[84] and —CO_2H[85] has also been investigated. Eaborn and Pande observed the following relative rates of cleavage of $C_6H_5XEt_3$ by acid: $C_6H_5PbEt_3 > C_6H_5SnEt_3 > C_6H_5GeEt_3 > C_6H_5SiEt_3$.[86] Electrophilic aromatic substitutions in which hydrogen does not figure include the replacement of —$B(OH)_2$,[87] —SiR_3,[88,89] —SO_3^-,[90] and —CO_2^-[59] by bromine and of various groups by —NO_2.[91]

16-2h. *The Benzidine Rearrangement.* The benzidine rearrangement

has proved of great interest to theoretical organic chemists. A number of mechanisms and modifications of mechanisms have been proposed. Most of these are in agreement with the fact that the rearrangement is *intramolecular.* Jacobson pointed out that no rearrangement of an unsymmetrical hydrazobenzene ArNHNHAr′ had ever been found to give any of either of the symmetrical benzidines, $H_2NArArNH_2$ or $H_2NAr′Ar′NH_2$.[92] Wheland and Schwartz used a radioactive-carbon-tracer technique to show that not more than 0.3 per cent of the symmetrical product, *o*-tolidine, was formed in the rearrangement of 2-methyl-2′-ethoxyhydrazobenzene.[93] These data show that the reaction is

[79] C. Eaborn et al., *J. Chem. Soc.*, 3148 (1953); 4858 (1956); 2299, 2303, 3031, 3034, 3640 (1959).

[80] W. M. Schubert et al., *J. Am. Chem. Soc.*, **74**, 1829 (1952); **76**, 1 (1954); **78**, 64 (1956).

[81] M. S. Kharasch et al., *J. Org. Chem.*, **3**, 347, 405, 409 (1938).

[82] A. H. Corwin et al., *J. Am. Chem. Soc.*, **69**, 1004 (1947); **77**, 6280 (1955).

[83] A. A. Spryskov et al., *J. Gen. Chem. USSR (Eng. Transl.)* **20**, 1083 (1950); **21**, 1649 (1951); **22**, 1911 (1952); **24**, 1777 (1954); **28**, 1693 (1958).

[84] A. Viout and P. Rumpf, *Bull. soc. chim. France*, 768 (1957).

[85] Sec. 13-1*d* and W. M. Schubert et al., *J. Am. Chem. Soc.*, **71**, 2639 (1949); **75**, 1401 (1953); **76**, 9 (1954).

[86] C. Eaborn and K. C. Pande, *J. Chem. Soc.*, 1566 (1960).

[87] H. G. Kuivila et al., *J. Am. Chem. Soc.*, **73**, 4629 (1951); **74**, 5068 (1952); **76**, 2675, 2679 (1954); **77**, 4834 (1955).

[88] R. A. Benkeser and A. Torkelson, *J. Am. Chem. Soc.*, **76**, 1252 (1954).

[89] C. Eaborn et al., *J. Chem. Soc.*, 4449 (1957); 179 (1960).

[90] L. G. Cannell, *J. Am. Chem. Soc.*, **79**, 2932 (1957).

[91] D. V. Nightingale, *Chem. Revs.*, **40**, 117 (1947).

[92] P. Jacobson, *Ann.*, **428**, 76 (1922).

[93] G. W. Wheland and J. R. Schwartz, *J. Chem. Phys.*, **17**, 425 (1949); cf. G. J. Bloink and K. H. Pausacker, *J. Chem. Soc.*, 950 (1950).

probably not a cleavage to two fragments of the same type (such as an ArNH· and an Ar'NH· radical) followed by their recombination in such a way as to yield the final product. It does not, however, rule out a mechanism in which cleavage yields two different types of fragments (such as $ArNH^+$ and $Ar'NH^-$), since if an unsymmetrical hydrazobenzene always cleaves in the same way, the combination of different types of fragments could never lead to the formation of a symmetrical benzidine.

Smith, Schwartz, and Wheland,[94] improving on an approach used earlier by Ingold and Kidd,[95] obtained evidence of a more general nature that the hydrazobenzene never cleaves into two independent fragments. They studied the rearrangement of a mixture of 2-methylhydrazobenzene labeled in the methyl group with C^{14} (III) and 2,2'-dimethylhydrazobenzene (IV).

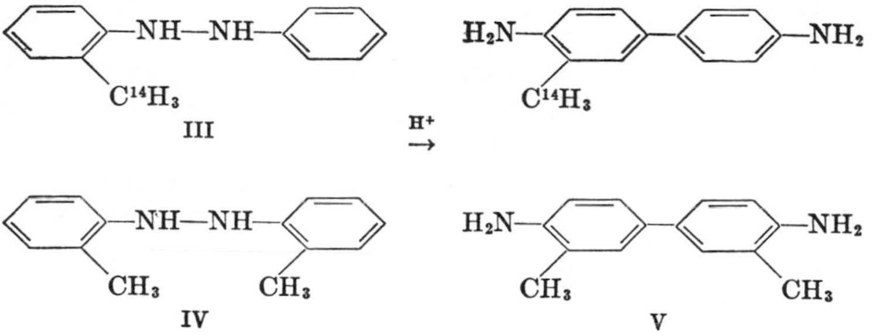

The o-tolidine (V) produced contained no excess C^{14} (over that present in ordinary carbon), and it was concluded that not more than 0.03 per cent of a mixed product could have been formed. Such a test for the intermolecular character of a reaction is valid, of course, only if both reactants undergo a considerable part of their reaction at the same time; that is, their reaction rates must be similar. The threefold difference in reaction rates known to exist between III and IV is small enough to make the test valid.

Hammond and Shine showed that the reaction is third-order (first-order in hydrazobenzene and second-order in hydrogen ions),[96] and others have verified this finding.[97] This kinetic order and the observation of general acid catalysis[98] may mean that the doubly protonated form of hydrazobenzene rearranges as it is formed or immediately

[94] D. H. Smith, J. R. Schwartz, and G. W. Wheland, *J. Am. Chem. Soc.*, **74**, 2282 (1952).

[95] C. K. Ingold and H. V. Kidd, *J. Chem. Soc.*, 984 (1933).

[96] G. S. Hammond and H. J. Shine, *J. Am. Chem. Soc.*, **72**, 220 (1950).

[97] R. B. Carlin, R. G. Nelb, and R. C. Odioso, *J. Am. Chem. Soc.*, **73**, 1002 (1951); L. J. Croce and J. D. Gettler, *J. Am. Chem. Soc.*, **75**, 874 (1953); cf. V. O. Lukashevich, *Doklady Akad. Nauk S.S.S.R.*, **133**, 115 (1960).

[98] M. D. Cohen and G. S. Hammond, *J. Am. Chem. Soc.*, **75**, 880 (1953).

thereafter,

$$C_6H_5NHNHC_6H_5 + H^+ \underset{k_{-1}}{\overset{k_1}{\rightleftharpoons}} C_6H_5\overset{\oplus}{N}H_2NHC_6H_5$$

$$C_6H_5\overset{\oplus}{N}H_2NHC_6H_5 + HA \underset{k_{-2}}{\overset{k_2}{\rightleftharpoons}} C_6H_5\overset{\oplus}{N}H_2\overset{\oplus}{N}H_2C_6H_5$$

$$C_6H_5\overset{\oplus}{N}H_2\overset{\oplus}{N}H_2C_6H_5 \overset{k_3}{\rightarrow} H_2NC_6H_4C_6H_4NH_2 + 2H^+$$

$$k_{-1} \gg k_2; \ k_3 \gg k_{-2}$$

although mechanisms of the other types used to explain general acid catalysis (Sec. 5-2) could also be devised for this case. The rearrangement step of the reaction may involve a simultaneous cleavage of the

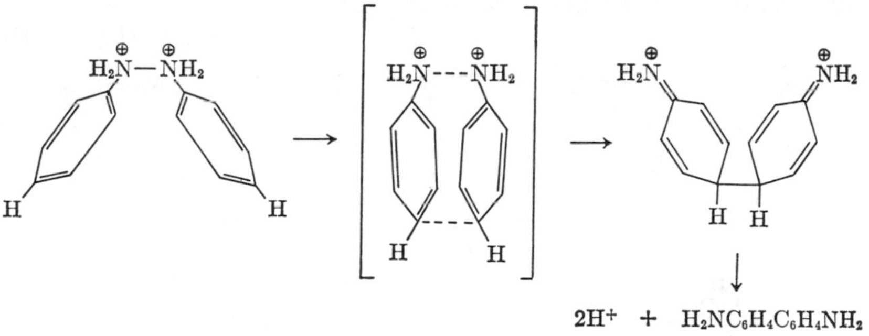

$$2H^+ \ + \ H_2NC_6H_4C_6H_4NH_2$$

nitrogen-nitrogen bond, which has been weakened by the electrostatic repulsion between two like charges, and the formation of a new carbon-carbon bond.[98,99] The mechanism of the formation of the by-products in the benzidine rearrangement has also been considered.[97,99]

16-2i. *Other Aromatic Rearrangements.*[100] The benzidine rearrangement is one of a number of rearrangements of the type

$$\underset{C_6H_5\overset{|}{N}-X}{\overset{R}{}} \overset{H^+}{\rightarrow} o\text{- and } p\text{-}XC_6H_4NHR$$

which have a considerable similarity in over-all result but which may differ widely in reaction mechanism. Although the benzidine rearrangement has been well established as an intramolecular process, there are other examples that have equally well been shown to be intermolecular in that the X that becomes attached to the ring in the product is not necessarily the one attached to the side chain in the reactant.

[99] Cf. R. Robinson, *J. Chem. Soc.*, 220 (1941); E. D. Hughes and C. K. Ingold, *J. Chem. Soc.*, 608 (1941); 1638 (1950); D. L. Hammick and S. F. Mason, *J. Chem. Soc.*, 638 (1946); M. J. S. Dewar, *J. Chem. Soc.*, 777 (1946).

[100] For a review of this topic, see E. D. Hughes and C. K. Ingold, *Quart. Revs. (London)*, **6**, 34 (1952).

One of the earliest and best established of these intermolecular rearrangements is the acid-catalyzed transformation of N-chloroacetanilide to o- and p-chloroacetanilide. Orton and coworkers pointed out that the reaction was catalyzed specifically by hydrochloric acid and succeeded in isolating both acetanilide and chlorine from the reaction mixture.[101] They also demonstrated that under a given set of conditions the hydrochloric acid–catalyzed reaction yielded the same ratio of o- to p-chloroacetanilide as did the reaction of acetanilide with chlorine.[102] These data strongly suggest that the "rearrangement" consists of the reaction of N-chloroacetanilide with hydrochloric acid to give acetanilide and chlorine.

$$CH_3CONClC_6H_5 + HCl \rightleftharpoons CH_3CONHC_6H_5 + Cl_2$$
$$\rightarrow CH_3CONHC_6H_4Cl + HCl$$

Heller, Hughes, and Ingold suggested an entirely different sort of mechanism for the formally similar acid-catalyzed rearrangement of phenylhydroxylamine to o- and p-aminophenol.[103] They proposed the intermediate formation of a resonance-stabilized carbonium ion capable of combining with nucleophilic reagents in more than one way.

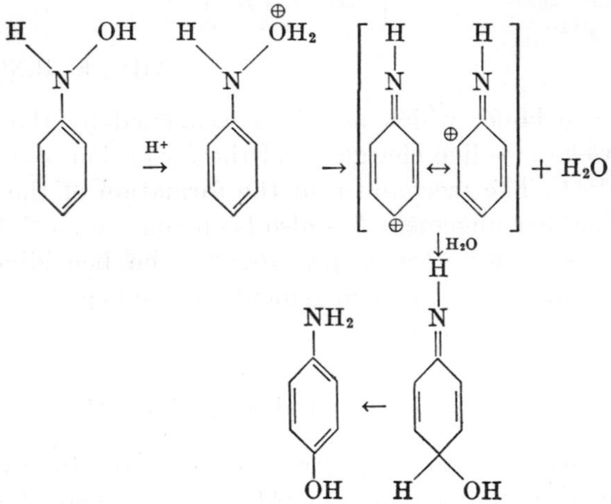

The reaction products may vary in the presence of other nucleophilic reagents. Thus in methanol and ethanol solutions the reaction yields methyl and ethyl ethers of o- and p-aminophenol. In the presence of phenol some p-$HOC_6H_4C_6H_4NH_2$-p was formed. A kinetic study showed the reaction rate to be proportional to the concentration of the conjugate acid of phenylhydroxylamine. In view of the report that the rate

[101] K. J. P. Orton and W. J. Jones, *J. Chem. Soc.*, **95**, 1456 (1909); K. J. P. Orton and H. King, *J. Chem. Soc.*, **99**, 1185 (1911).
[102] K. J. P. Orton and A. E. Bradfield, *J. Chem. Soc.*, 986 (1927).
[103] H. E. Heller, E. D. Hughes, and C. K. Ingold, *Nature*, **168**, 909 (1951).

of the hydrochloric acid–catalyzed reaction, in which considerable o- and p-chloroaniline are formed, is independent of the chloride concentration,[103] it seems that the reaction probably proceeds via a discrete intermediate like the resonance-stabilized cation depicted and is not a concerted S_N2'-type process. The matter is being investigated more thoroughly in this regard, however.

16-3. Reactivity and Orientation in Electrophilic Aromatic Substitution. 16-3a. *Basis for Electronic Effects in Orientation and Reactivity.* Modern concepts of physical organic chemistry offer a particularly satisfying qualitative explanation for the observed data on reactivity and orientation in electrophilic aromatic substitution reactions. This may be partly because aromatic substitution was the most widely used proving ground for many of these concepts when they were being developed by the English school of organic chemists during the 1920's. In accordance with a common practice, we shall discuss these concepts in terms of resonance instead of the T, M, E, etc., effects (Sec. 2-4e) generally used by the English workers. This should in no way obscure our debt to these workers.[104]

For *irreversible* reactions the orientation of substitution depends only on the relative reactivity of the various positions on the ring. Some of the complications introduced by reversibility, as in sulfonation and the Friedel-Crafts reaction, are considered in Sec. 16-3g. According to transition-state theory the reactivity depends on the free-energy difference between the reactants and the transition state. We have already described evidence for the formation of an intermediate like that shown below.

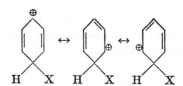

The transition state for the reaction as a whole will be either the transition state leading to the formation of the intermediate from reactant or the one leading from the intermediate to product. In either case, the transition state will differ from the reactant largely by looking much more like the intermediate. Therefore it is useful, if not rigorous, to discuss orientation and reactivity in terms of the stability of the intermediate (compared with that of the reactants). Since electron donation by the inductive effect should stabilize the positively charged ring of the inter-

[104] J. Allan, A. E. Oxford, R. Robinson, and J. C. Smith, *J. Chem. Soc.*, 401 (1926); C. K. Ingold and E. H. Ingold, *J. Chem. Soc.*, 1310 (1926); C. K. Ingold, *Ann. Repts. on Progr. Chem.* (*Chem. Soc. London*), **23**, 129 (1926); *Rec. trav. chim.*, **48**, 797 (1929); *Chem. Revs.*, **15**, 225 (1934).

mediate and electron withdrawal should destabilize it, we should expect groups capable of inductive electron supply to increase reactivity and inductive electron-withdrawing groups to decrease reactivity in electrophilic aromatic substitutions. Of the common substituents only alkyl groups may be very definitely claimed to have an electron-donating inductive effect. Therefore the inductive effect of most groups would be expected to tend to decrease reactivity. With most groups, however, resonance effects also appear to be quite important. Thus for all groups in which the atom attached directly to the ring has an unshared electron pair (and does not have a formal positive charge) a fourth particularly stable structure (VI) may contribute to the resonance-hybrid intermediate.

VI

This contributing structure (VI) has one more covalent bond than any of the others. A similar structure may be written for the intermediate for ortho (but not meta) substitution. This resonance factor causes an activation of the ortho and para positions that opposes the deactivation produced by the inductive effect. In the case of the halogen substituents the inductive effect appears to be more powerful than the resonance effect. That is, the inductive effect deactivates all positions, and although the resonance effect decreases this deactivation for the ortho and para positions, it does not eliminate it. The halogens therefore bring about deactivation with ortho-para orientation. With alkoxy and amino groups the inductive effect is smaller, and because of the greater basicity of oxygen and nitrogen the resonance effect is larger. These two types of groups cause activation with ortho-para orientation.

Orientation in aromatic substitution reactions has often been discussed in terms of the electron densities at the various positions of aromatic rings. The positions of high electron density are then said to be reactive in electrophilic substitution and those of low electron density reactive in nucleophilic substitution. However, this procedure is not rigorous from the viewpoint of transition-state theory, and although it usually gives the right answer (and is simpler to apply), there are several cases known in which orientation is incorrectly predicted. As Jaffé pointed out, the 2 and 4 positions of pyridine N-oxide and azobenzene, for example, are more reactive than the 3 position toward both electrophilic and nucleophilic substitution.[105] Thus in one of the two types of reactions the

[105] H. H. Jaffé, *J. Am. Chem. Soc.*, **76**, 3527 (1954).

reactivity is not dependent on the electron density alone.   There will be a tendency toward this type of behavior in the case of any compound with a substituent group capable of both tautomeric electron donation and withdrawal.

In terms of electron densities the deactivating and meta-orienting influence of such groups as nitro, carboxy, acyl, cyano, sulfonic acid, etc., is attributed to electron withdrawal from the ring as a whole by the inductive effect and in addition from the ortho and para positions by a resonance effect.

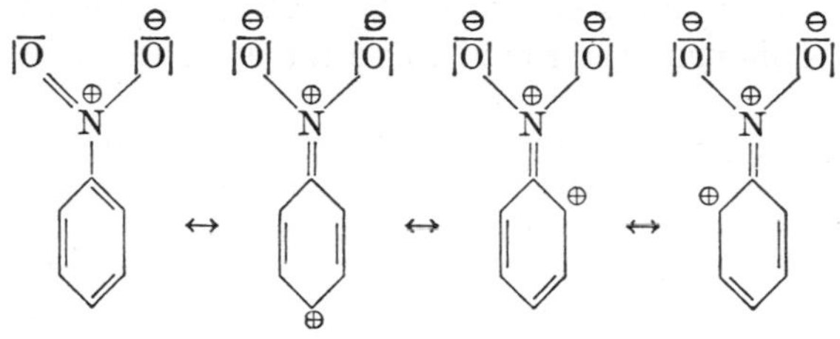

VII   VIII   IX

In terms of our approximation of transition-state theory, however, neglecting no-bond structures and structures with like charges on adjacent atoms, there are no contributing structures of the type of VII, VIII, and IX for the intermediate for para substitution of nitrobenzene and only one for ortho.   In fact, except for the internal resonance of the nitro group, only three contributing structures may be written for the ortho intermediate, and for the para intermediate only the following two:[106]

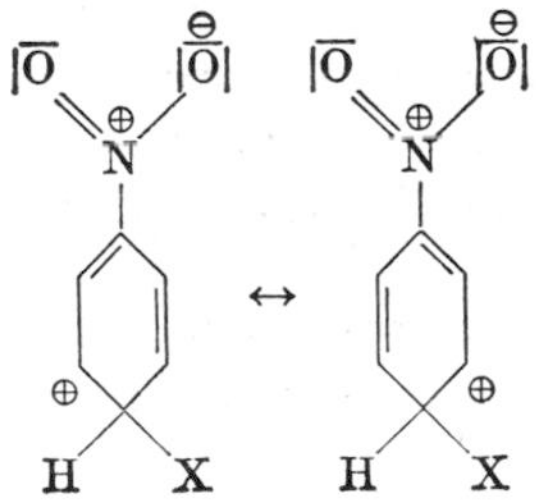

Five structures contribute to the intermediate for meta substitution.

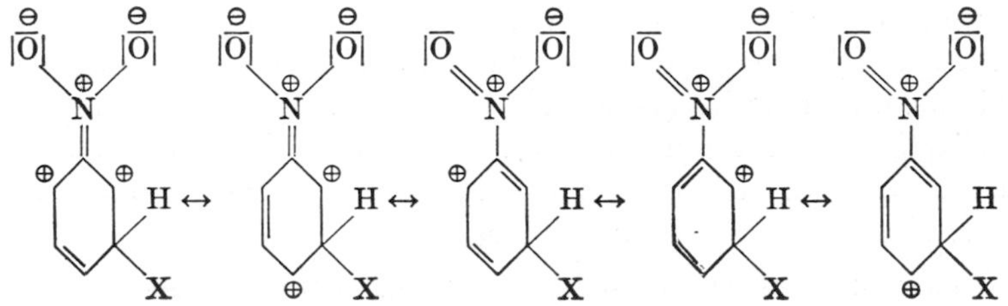

[106] Cf. G. E. K. Branch and M. Calvin, "The Theory of Organic Chemistry," p. 477, Prentice-Hall, Inc., Englewood Cliffs, N. J., 1941.

The foregoing explains why the nitro group (and analogously, cyano, acyl, carboxyl, and related groups) deactivates all positions and why this deactivation is greatest for the para position and least for the meta position.

16-3b. *Hammett-equation Correlations for Electrophilic Aromatic Substitutions.* Since resonance-electron-donating substituents should be capable of strong resonance interaction with the reaction center in the transition state of electrophilic aromatic substitution reactions, it would be expected that the use of σ^+ values for such substituents should lead to considerably better Hammett-equation correlations than would the use of σ values. This has been found to be the case, and in Table 16-2 are

TABLE 16-2. HAMMETT ρ CONSTANTS OF ELECTROPHILIC AROMATIC SUBSTITUTION REACTIONS[a]

Reaction	ρ
ArH + Br$_2$, in HOAc at 25°	−12.14
ArH + Cl$_2$, in HOAc at 25°	− 8.06
ArH + HNO$_3$, in CH$_3$NO$_2$ or Ac$_2$O at 0° or 25°	− 6.22
ArH + HOBr + HClO$_4$, in 50% dioxane at 25°	− 5.78
ArSiMe$_3$ + HClO$_4$ → ArH, in 72% MeOH at 50°	− 4.32
ArSiMe$_3$ + Br$_2$ → ArBr, in HOAc at 25°	− 6.04
ArB(OH)$_2$ + Br$_2$ → ArBr, in 20% HOAc at 25°	− 4.44
ArH + Hg(OAc)$_2$ → ArHgOAc, in HOAc at 25°	− 4.10[b]

[a] From Ref. 107 unless stated otherwise.
[b] H. C. Brown and M. Dubeck, *J. Am. Chem. Soc.*, **82**, 1939 (1960).

listed a number of data on such correlations, largely from the extensive investigations of Brown and coworkers in this field.[107] To correct for the fact that there are six equivalent positions available for substitution in benzene, two equivalent positions in the meta substitution of C$_6$H$_5$X, etc., we use partial rate factors (m_f and p_f) instead of k/k_0 values.

$$m_f = 3k/k_0$$
$$p_f = 6k/k_0 \tag{16-15}$$

The large differences in the ρ's in Table 16-2 are not solely due to differences in the nature of the intermediates formed in the respective reactions. They are probably due *more* to differences in how far along the reaction path the transition states lie. In general, the greater the energy content of the immediate reactant(s) relative to that of the immediate product(s), the earlier in the reaction the transition state will occur (cf. Fig. 5-3). The closer the transition state lies to the intermediate, in which the aromatic ring bears almost a full positive charge, the larger will be ρ.

[107] Summarized by H. C. Brown and Y. Okamoto, *J. Am. Chem. Soc.*, **80**, 4979 (1958).

Thus bromination by Br_2 and by HOBr and $HClO_4$ probably involves the rate-controlling formation of the intermediate

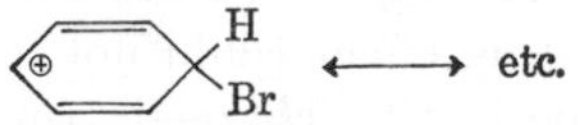

In the latter case, however, the attack on the aromatic ring is probably performed by the highly energetic Br^+ cation.   The transition state should come earlier in this reaction, and accordingly, ρ is found to be smaller (Table 16-2).   The effect of solvent on ρ is probably also important but has been little studied.

Brown and Nelson discovered a linear relation between the values of $\log p_f$ for various aromatic substitution reactions of toluene and the corresponding $\log (p_f/m_f)$ values.[108]   A more fundamental basis for such a relationship was demonstrated by de la Mare[109] and by McGary, Okamoto, and Brown,[110] who pointed out that from the Hammett equation

$$\log p_f = \rho^r \sigma_p{}^x \tag{16-16}$$

and

$$\log m_f = \rho^r \sigma_m{}^x \tag{16-17}$$

where the superscript r refers to a given reaction and x refers to a given substituent.   Therefore

$$\log (p_f/m_f) = \rho^r (\sigma_p{}^x - \sigma_m{}^x) \tag{16-18}$$

Dividing Eq. (16-16) by (16-18) and rearranging

$$\log p_f = \left(\frac{\sigma_p{}^x}{\sigma_p{}^x - \sigma_m{}^x} \right) \log \frac{p_f}{m_f} \tag{16-19}$$

According to this relation, a plot of $\log p_f$ for a number of substitution reactions of a given ArX (in the case of Brown and Nelson, X was CH_3) vs. the corresponding $\log (p_f/m_f)$ values should give a straight line of slope $\sigma_p{}^x/(\sigma_p{}^x - \sigma_m{}^x)$ passing through the origin.   The data available on toluene follow Eq. (16-19) more closely than they do the Hammett equation (with σ or σ^+).[111]   The agreement obtained with t-butylbenzene[111] seems satisfactory, but there are indications that significant deviations may be found for halogen[112] and phenyl substituents.[113]   Additional data

[108] H. C. Brown and K. L. Nelson, *J. Am. Chem. Soc.*, **75**, 6292 (1953).

[109] P. B. D. de la Mare, *J. Chem. Soc.*, 4450 (1954).

[110] C. W. McGary, Jr., Y. Okamoto, and H. C. Brown, *J. Am. Chem. Soc.*, **77**, 3037 (1955).

[111] L. M. Stock and H. C. Brown, *J. Am. Chem. Soc.*, **81**, 3323, 5621 (1959).

[112] K. L. Nelson, *J. Org. Chem.*, **21**, 145 (1956).

[113] Ref. 2, p. 243.

on these and other substituents are needed to tell whether Eq. (16-19) is, in general, more closely followed than the Hammett equation, from which it may be derived.

One group whose orienting effect could not have been predicted from its σ constants is the positively charged trimethylammonium group. Roberts, Clement, and Drysdale pointed out that this group has a large positive σ constant as both a meta and para substituent but that the meta σ constant is larger, showing that electrons are removed more strongly from the meta position than from the para.[114] From the existence of a Hammett-equation correlation we should expect the meta position to be more highly deactivated than the para, so that the group should give para orientation. It is reported, however, to be a meta-directing group.[115-117] Roberts and coworkers pointed out that the observed meta orientation may be explained in a manner similar to that suggested by Pfeiffer and Wizinger.[118] The transition state for meta substitution will be described by structures X, XI, and XII and that for para substitution by XIII, XIV, and XV (among others).

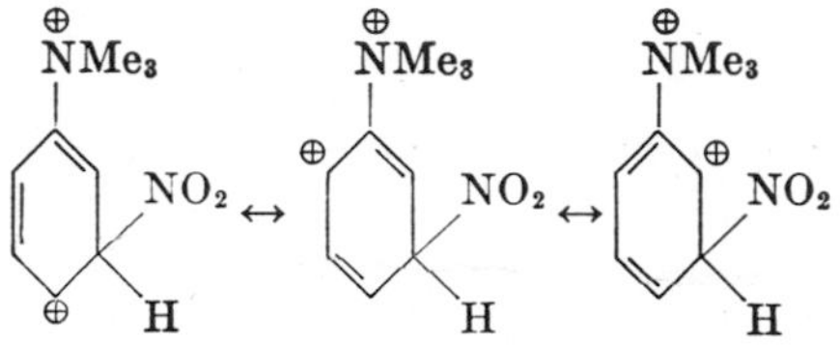

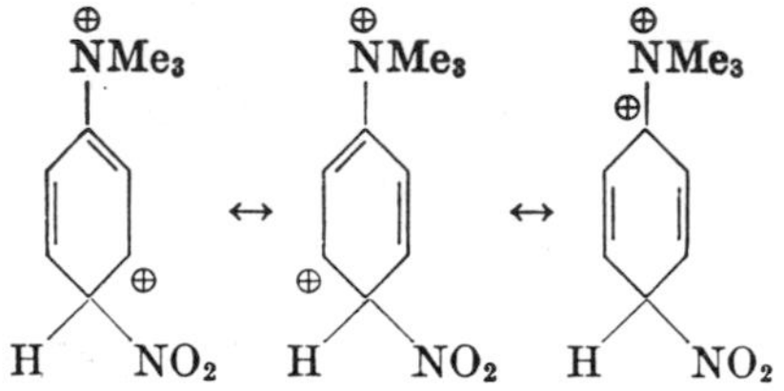

[114] J. D. Roberts, R. A. Clement, and J. J. Drysdale, *J. Am. Chem. Soc.*, **73**, 2181 (1951).

[115] D. Vorländer and E. Siebert, *Ber.*, **52B**, 283 (1919).

[116] A. N. Nesmeyanov, T. P. Tolstaya, L. S. Isaeva, and A. V. Grib, *Doklady Akad. Nauk S.S.S.R.*, **133**, 602 (1960); *Chem. Abstr.*, **54**, 24529i (1960).

[117] J. C. D. Brand and A. Rutherford, *J. Chem. Soc.*, 3927 (1952). It is of interest to note that if one makes the assumption that the solubility of any $Me_3\overset{+}{N}C_6H_4SO_3^-$ is a linear function of the total amount of $Me_3\overset{+}{N}C_6H_4SO_3^-$'s present in the solution (this is at least as plausible as the assumption actually made), the percentages of ortho, meta, and para substitution may be calculated to be 10, 62, and 28 instead of 8, 78, and 14, respectively.

[118] P. Pfeiffer and R. Wizinger, *Ann.*, **461**, 132 (1928).

However, structure XV, having like charges on adjacent atoms, will have such a high energy content as to decrease greatly its contribution to the total structure of the transition state for para substitution. Therefore the transition state for meta substitution will be more stable. This type of argument may be used to explain the fact that, except for the anomalous case of the triphenyloxonium ion,[116] meta substitution always occurs when the atom directly attached to the ring bears a formal positive charge. This generalization has alternately been explained[104,119] by the suggestion that the inductive effect removes electrons more effectively from the ortho and para positions than from the meta position due to the contributions of structures like

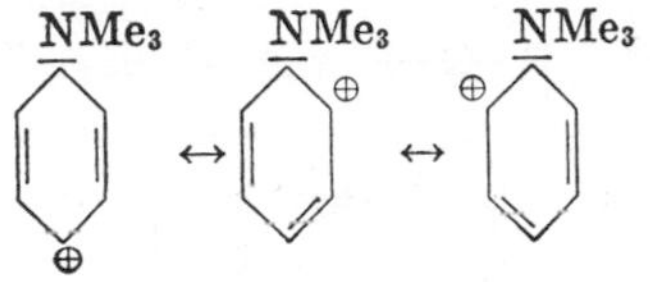

This explanation requires a new interpretation of the σ constants of the trimethylammonium group in terms of an electrostatic effect operating directly through space.[119]

Other compounds that appear to have a smaller electron density in the ortho and para positions than in the meta position but that nevertheless undergo substitution (less readily than benzene, however) in the ortho and para positions are $C_6H_5CH{=}CHCO_2H$, $C_6H_5CH{=}CHNO_2$, and $C_6H_5CH{=}CHSO_2Cl$.[120] This suggests that although structure XVI contributes less to the transition state for para substitution than do the other structures shown, its contribution is sufficient to cause this transition state to be more stable than that for meta substitution, for which no structure like XVI may be written.[120]

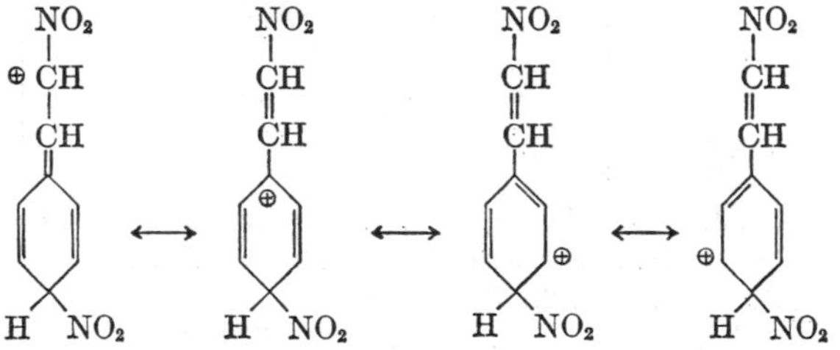

XVI

16-3c. *Effect of Alkyl and Substituted Alkyl Groups.* The activating and ortho-para directing influence of unsubstituted alkyl groups is probably due to both hyperconjugation and the inductive effect. It appears

[119] C. K. Ingold, "Structure and Mechanism in Organic Chemistry," sec. 19b, Cornell University Press, Ithaca, N.Y., 1953.

[120] F. G. Bordwell and K. Rohde, *J. Am. Chem. Soc.*, **70**, 1191 (1948).

that the relative importance of these two factors may vary from reaction to reaction. Thus the methyl group, for which hyperconjugation should be greater, usually, but not always, activates the para position more strongly than does the t-butyl group, which should have a larger inductive effect.[107] Solvation effects may also be important.

When halogen atoms are substituents on a saturated aliphatic side chain, they cannot donate electrons by a resonance effect because they are not conjugated (for this purpose) with the ring. Their electron-withdrawing inductive effect may still operate, however, resulting in a kind of hyperconjugation in the case of benzyl-type halides.

It is resonance of this sort that has been said to cause the p-CF$_3$ σ constant to have a larger positive value than that for m-CF$_3$.[121] The tendency of groups capable of strong electron withdrawal by the inductive effect to cause meta substitution increases with the number and electron-withdrawing power of these groups and decreases with their distance from the aromatic ring. This is shown by the data in Table 16-3.

TABLE 16-3. PERCENTAGE OF META SUBSTITUTION IN THE NITRATION OF
VARIOUS NEGATIVELY SUBSTITUTED ALKYLBENZENES

Compound	Meta, %	Compound	Meta, %
$C_6H_5CH_3$	3^a	$C_6H_5CH_2CH_2NO_2$	13^d
$C_6H_5CH_2Cl$	14^b	$C_6H_5CH_2NMe_3^{\oplus}$	88^e
$C_6H_5CHCl_2$	34^a	$C_6H_5CH_2CH_2NMe_3^{\oplus}$	19^e
$C_6H_5CCl_3$	64^a	$C_6H_5(CH_2)_3NMe_3^{\oplus}$	5^f
$C_6H_5CH_2NO_2$	67^c		

[a] A. F. Holleman, J. Vermeulen, and W. J. de Mooy, *Rec. trav. chim.*, **33**, 1 (1914).
[b] C. K. Ingold and F. R. Shaw, *J. Chem. Soc.*, 575 (1949).
[c] J. W. Baker, *J. Chem. Soc.*, 2257 (1929).
[d] J. W. Baker and I. S. Wilson, *J. Chem. Soc.*, 842 (1927).
[e] F. R. Goss, W. Hanhart, and C. K. Ingold, *J. Chem. Soc.*, 250 (1927).
[f] C. K. Ingold and I. S. Wilson, *J. Chem. Soc.*, 810 (1927).

16-3d. *Orientation and Reactivity in Nonbenzenoid Aromatic Rings.*
Orientation and reactivity in the electrophilic aromatic substitution reactions of nonbenzenoid compounds may be treated in a manner similar to that we have used for benzene derivatives. Thus for furan (and

[121] J. D. Roberts, R. L. Webb, and E. A. McElhill, *J. Am. Chem. Soc.*, **72**, 408 (1950).

analogously for pyrrole and thiophene) a larger number of relatively stable contributing structures may be written for the intermediate for alpha substitution

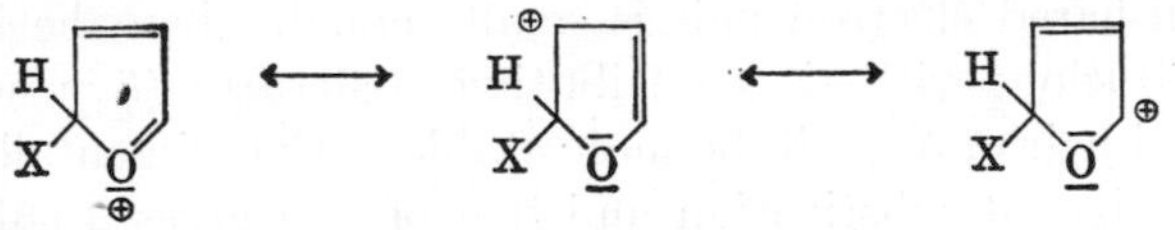

than for the intermediate for beta substitution.

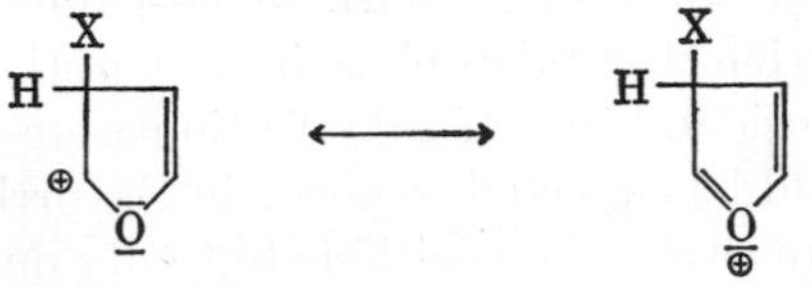

Since these intermediates, and hence the transition states leading to them, appear more stable than that of benzene, activated alpha substitution would be expected and is found.

For the intermediate in the alpha substitution of naphthalene, the positive charge may be distributed between two different atoms without disturbing the benzenoid resonance of the nonreacting ring,

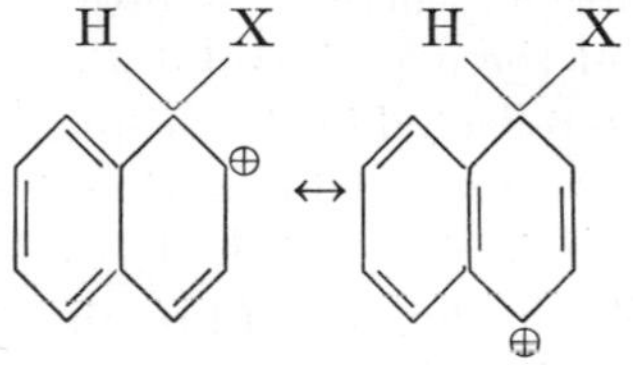

and a total of seven relatively stable contributing structures may be written. For beta substitution there are only six relatively stable contributing structures and only one atom on which the positive charge may be placed without disturbing the benzene ring.

The electrophilic aromatic substitution reactions of indole and benzofuran may be represented as follows:

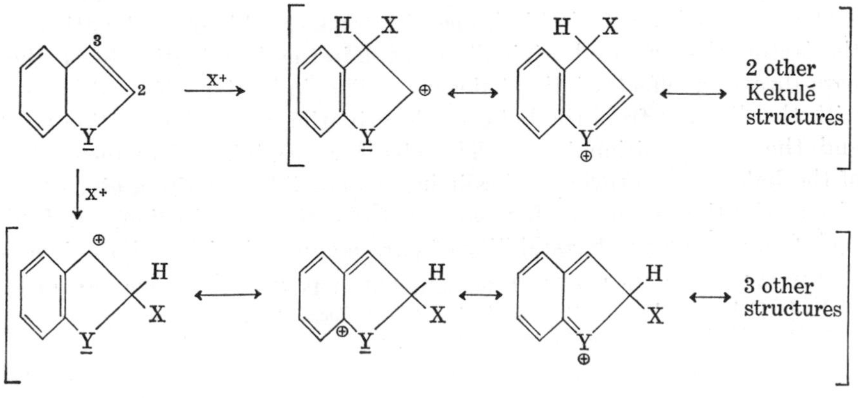

If the ability of Y to share its unshared electron pair is sufficient, the intermediate for 3 substitution, in which there are two contributing structures involving such sharing, will be more stable.   If the ability of Y to share its unshared electron pair is small enough, then the intermediate for 2 substitution, with six contributing structures (five not involving the unshared pair of Y), will be more stable.   Experimentally it is found that indole gives 3 substitution and benzofuran gives 2 substitution in most electrophilic substitution reactions.

16-3e. *The Ortho-Para Ratio.* With ortho-para directing groups the fraction of substitution that takes place in each of the two types of active positions varies considerably.   If substitution were completely random, twice as much would be expected to occur in the ortho position as in the para, since there are two ortho positions but only one para.   Actually it appears that there is usually less than twice as much ortho as para substitution.   One reason for the general preference for para substitution is steric.   The nitration of toluene gives 56.5 per cent ortho, 3.5 per cent meta, and 40 per cent para substitution, whereas *t*-butylbenzene gives 12 per cent ortho, 8.5 per cent meta, and 79.5 per cent para substitution. Thus the ratio of ortho to para substitution is almost 10 times as large for toluene as for *t*-butylbenzene, whereas the ratios of meta to para substitution hardly differ beyond the experimental error.[122]   Steric inhibition of ortho substitution is also reflected in the decreases of the ortho-para ratio that have been found to accompany increasing size of the new group being introduced into the aromatic ring.   Thus Holleman pointed out that the ortho-para ratio is smaller in the case of bromination than in chlorination and that it is still smaller in sulfonation.[123]

There are certain data that make it obvious, however, that the ortho-para ratio is influenced by electronic as well as steric factors.   Holleman found that the nitration of fluorobenzene gives 12 per cent ortho substitution, chlorobenzene 30 per cent, bromobenzene 38 per cent, and iodobenzene 41 per cent (none of these halobenzenes give an appreciable amount of meta substitution).[123]   Although this order is the reverse of that expected from steric hindrance, it seems capable of explanation by the factors that we used originally to explain the deactivating but ortho-para-directing effect of the halogens.   The inductive effect decreases with the distance from the halogen, deactivating the ortho position most and the para position least.   The electron-supplying resonance effect of the halogens, however, makes it just as possible to write a given type of contributing structure for para as for ortho substitution.   Indeed, such facts as the greater stability of para compared with ortho quinones

[122] H. Cohn, E. D. Hughes, M. H. Jones, and M. G. Peeling, *Nature*, **169**, 291 (1952); cf. K. L. Nelson and H. C. Brown, *J. Am. Chem. Soc.*, **73**, 5605 (1951).

[123] A. F. Holleman, *Chem. Revs.*, **1**, 218 (1925).

have been used as evidence for the innately greater stability of a structure like XVII compared with one like XVIII.

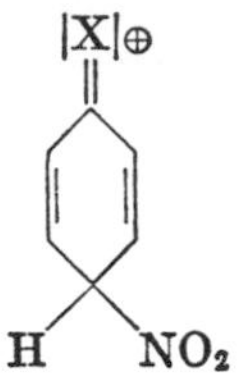

XVII 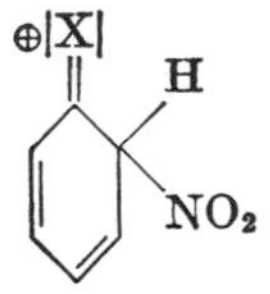   XVIII

Since the inductive effect decreases with distance and the resonance effect does not, we should expect the deactivation of the ortho position relative to the para to increase with increasing electronegativity of the halogen.

The high ortho-para ratio obtained with groups capable of withdrawing electrons by resonance has been explained in Sec. 16-3a.

In certain reactions ortho substitution occurs in high yield, apparently via a mechanism in which the attacking reagent coordinates with the substituent already on the ring.[124]

16-3f. *The Effect of Several Substituents.*   As a first approximation it is useful to assume that the effects of substituents are additive, so that in the compound

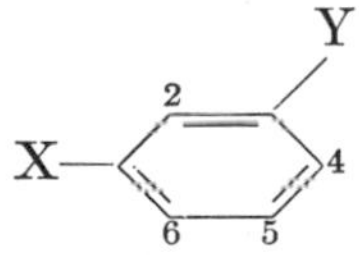

XIX

the partial rate factors for substitution in the 2, 4, 5, and 6 positions would be $o_f{}^x o_f{}^y$, $p_f{}^x o_f{}^y$, $m_f{}^x m_f{}^y$, and $o_f{}^x p_f{}^y$, where the o_f's, m_f's, and p_f's are those obtained from data on C_6H_5X and C_6H_5Y.   Although there are not many reliable data available to check this generalization, there appear to be a number of significant deviations from it.   Some of these deviations seem explicable.   In o-XC_6H_4Y one should consider the possibility of steric inhibition of resonance.   In o- and p-XC_6H_4Y resonance interactions between X and Y may give the molecule extra stabilization.   In a reaction like 2 substitution in XIX steric hindrance may be greater than the sum of the hindrance found for ortho substitution in the individual compounds C_6H_5X and C_6H_5Y.   In other cases careful examination of the stability of the intermediate can give clues as to the reason for the observed orientation.   Thus, in the case of m-XC_6H_4Y (XIX), where X is a resonance electron-donating and Y a -withdrawing group, substitu-

---

[124] Cf. W. Seaman and J. R. Johnson, *J. Am. Chem. Soc.*, **53,** 711 (1931).

tion often tends to occur at the 2 position, between the two groups.[125] In such a case there are more resonance structures (neglecting structures with like charges on adjacent atoms) involving the positive charge on X that contribute to the intermediate for 2 substitution than for substitution in any other position.

16-3g. Effect of Reversibility on Orientation in Aromatic Substitution. Many sulfonation, Friedel-Crafts, iodination, and other aromatic substitution reactions are appreciably reversible under the conditions under which they are usually run. When this occurs, the ordinary orientation rules we have described will not necessarily be followed. These rules are for determining the *kinetically controlled product*, i.e., the one formed most rapidly. However, if reversal is fast enough, the various products will form in amounts proportional to their stabilities. That is, we shall obtain most of the *thermodynamically controlled product*. Since the reverse of an electrophilic aromatic substitution is simply another electrophilic aromatic substitution, the product formed most rapidly is almost always the one that also reverts to starting material most rapidly. Therefore it is impossible to predict the thermodynamically controlled product simply from a knowledge of the kinetically controlled one.

The data on the sulfonation of naphthalene may be explained in these terms. It is well known that the reaction at about 80° yields practically pure α-naphthalenesulfonic acid, whereas at 160° the β isomer is the predominant product. Studies of the reaction[126,127] show that it makes no difference whether the α sulfonation occurring below 100° is reversible or not, since the rate of β sulfonation is negligible below about 110°. Above this temperature the sulfonation of naphthalene yields largely the β isomer if the reaction is allowed to proceed to equilibrium (rapidly attained at 160°). The greater thermodynamic stability of this isomer is probably at least partly due to the less sterically hindered nature of the β position. The rearrangement of α- to β-naphthalenesulfonic acid in H_2SO_4 that is labeled with S^{35} has been shown to be somewhat faster

[125] Cf. C. K. Ingold, in E. D. Rodd (ed.), "Chemistry of Carbon Compounds," vol. IIIA, p. 47, Elsevier Publishing Company, Amsterdam, 1954.

[126] R. Lantz, *Compt. rend.*, **201**, 149 (1935); *Bull. soc. chim. France*, [5], **2**, 2092 (1935).

[127] A. A. Spryskov, *Zhur. Obshchei Khim.*, **14**, 833 (1944); **16**, 1057, 2126 (1946); **17**, 591, 1309 (1947); *Chem. Abstr.*, **40**, 1821 (1946); **41**, 2720 (1947); **42**, 894, 1921 (1948); **43**, 471 (1949).

than the appearance of S^{35} in the β-naphthalenesulfonic acid.[128]   It was suggested that part of the rearrangement is internal, but it was recognized that the active sulfonating species ejected from the α position may attack the β position faster than it equilibrates with the solvent.

McCaulay and Lien obtained stronger evidence for an internal mechanism for rearrangement in the acid-catalyzed equilibration of the xylenes.[129]   They found that in the initial stages of the isomerization of o-xylene considerable meta but very little para isomer was formed.   If the reaction had involved toluene formation and remethylation, p-xylene initially should have been formed more rapidly than the meta isomer. The reaction mechanism is, therefore, probably of the following type:

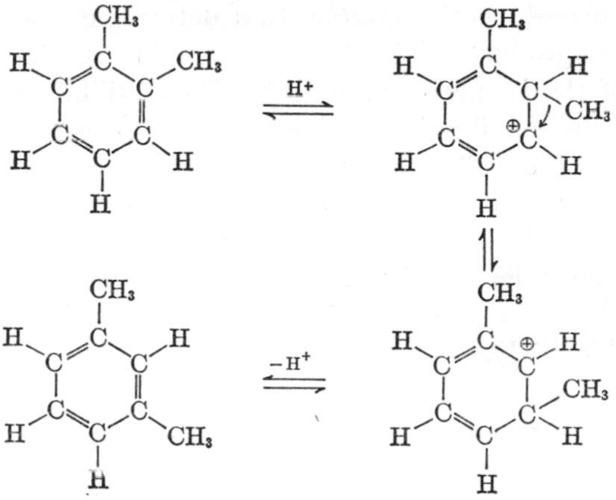

A similar mechanism may operate for the Jacobsen rearrangement. Rearrangements of alkylbenzenes probably do occur by the dealkylation-realkylation mechanism in the case of alkyl groups, like t-butyl or even isopropyl,[130] that form relatively stable cations.

By application of the Hammett equation it is possible to calculate the relative stabilities of various meta and para isomers, and also equilibrium constants for disproportionation reactions.[131]   Thus for the equilibration of meta and para isomers in a case where there are no resonance interactions between X and Y,

$$K_{mp} = \frac{[p\text{-}XC_6H_4Y]}{[m\text{-}XC_6H_4Y]}$$

$$\log 2K_{mp} = \tau_m\sigma_{m-X}\sigma_{m-Y} - \tau_p\sigma_{p-X}\sigma_{p-Y}$$

[128] S. E. Shnol, Ya. K. Syrkin, V. I. Yakerson, and L. A. Blyumenfeld, *Doklady Akad. Nauk S.S.S.R.*, **101**, 1075 (1955); F. M. Vainshtein and E. A. Shilov, *J. Gen. Chem U.S.S.R.*, (*Eng. Transl.*), **27**, 2616 (1957).

[129] D. A. McCaulay and A. P. Lien, *J. Am. Chem. Soc.*, **74**, 6246 (1952).

[130] R. H. Allen, T. Alfrey, Jr., and L. D. Yats, *J. Am. Chem. Soc.*, **81**, 42 (1959).

[131] J. Hine, *J. Am. Chem. Soc.*, **82**, 4877 (1960).

where the "2" is made necessary by the fact that there are two positions meta but only one para to a given substituent, and the τ's are those used in Eq. (4-7). Analogously, for a disproportionation reaction,

$$K_d = \frac{[C_6H_6][p\text{-}C_6H_4X_2]}{[C_6H_5X]^2}$$

$$\log 12K_d = -\tau_p(\sigma_{p-x})^2$$

Since in the cases under consideration changes are being made in the atom attached directly to the aromatic ring, the experimental data may not be predicted as accurately as in most Hammett-equation correlations.

We should remember that it is the relative stability of the various isomers in the *actual reaction mixture* that determines the thermodynamically controlled product. This has been emphasized in McCaulay and Lien's study of the isomeric xylenes.[129] The equilibrium mixture in the presence of hydrogen fluoride and a catalytic concentration of boron fluoride contains about 60 per cent of the meta isomer and about 20 per cent of each of the others. In the presence of a large excess of boron fluoride, however, the equilibrium mixture was more than 97 per cent meta. Since the xylenes exist as their salts in the presence of excess boron trifluoride, the equilibrium is shifted toward the more basic meta isomer (cf. Table 16-1).

PROBLEMS

1. Suggest a reasonable mechanism for each of the following reactions:

(a) $C_6H_6 + CH_2O + HCl \xrightarrow{\text{ZnCl}_2} C_6H_5CH_2Cl + H_2O$

(b) $Al(OC_6H_5)_3 + C_2H_4 \xrightarrow[\text{2. H}_2\text{O}]{\text{1. heat}}$ o-ethylphenol and some 2,6-diethylphenol

(c) Azoxybenzene $\xrightarrow{\text{H}^+}$ $p\text{-HOC}_6H_4N{=}NC_6H_5$

2. Suggest an explanation for each of the following observations:

(a) The bromination of 1,3,5-tri-t-butylbenzene was found to show a large tritium kinetic-isotope effect under conditions where benzene, toluene, mesitylene, etc., did not.

(b) The diazo coupling reactions of N-methyl-N-t-butylaniline are much slower than those of N,N-dimethylaniline.

3. It is found in a certain aromatic iodination that the over-all deuterium kinetic isotope effect (k_H/k_D) is 5.32 at an iodide-ion concentration of 2.50×10^{-3} M and 1.95 at an iodide-ion concentration of 1.00×10^{-5} M. If it is assumed that the only kinetic-isotope effect is on the rate of loss of the hydrogen being replaced (i.e., there are no secondary isotope effects), what is the deuterium kinetic-isotope effect on the individual step in which the hydrogen is lost? What are the relative rates at which the reaction intermediate is transformed back to reactants and on to products? State any assumptions that are made in your calculations.

4. In which position(s) of each of the following would you expect electrophilic aromatic substitution to occur the most rapidly?   Give your reasons.

(a) $(C_6H_5)_3C^+$

(b) $C_6H_5N_3$

(c) $C_6H_5PO(OEt)_2$

(d) $C_6H_5SF_5$

(e) Anthracene

(f) $p\text{-}HOC_6H_4F$

(g) C_6H_5NO

(h) $C_6H_5B(OH)_2$

(i) $C_6H_5SiMe_3$

(j) $C_6H_5ClO_3$

(k) $(C_6H_5)_2Cl^+$

(l) $p\text{-}CH_3C_6H_4C(CH_3)_3$

(m) 2-Cyano-6-methylphenol

(n) $C_6H_5C(NO_2)_3$

Chapter 17

NUCLEOPHILIC SUBSTITUTION AT AROMATIC AND VINYL CARBON[1]

17-1. Aromatic Diazonium Salts.   Because of the great stability of elemental nitrogen, its formation in a reaction may add tremendously to the driving force for the reaction.   It is for this reason, no doubt, that aromatic diazonium salts undergo nucleophilic displacement reactions so much more easily than aromatic halides.   Before we discuss the reactions of these compounds, though, we shall discuss the mechanism of their formation by the diazotization of amines, since this topic has not yet been considered.

17-1a. Mechanism of Diazotization of Amines.   Taylor showed that the reaction of primary amines with nitrous acid follows the rate equation[2]

$$v = k[RNH_2][HNO_2]^2$$

where the bracketed expressions refer to the actual (rather than formal) concentrations of the enclosed species.   This observation was confirmed by Schmid and Muhr[3] and most reasonably interpreted by Hammett, who suggested that the amine performs a nucleophilic attack on nitrogen trioxide present in equilibrium with the nitrous acid.[4]   Such a nucleophilic attack could lead to the conjugate acid of a nitrosamine and, via further proton transfers, to the diazo hydroxide, which reacts with acid to give the diazonium ion (17-1).   This mechanism involving the intermediate formation of nitrogen trioxide was subsequently established by

---

[1] For reviews of this subject, see J. F. Bunnett and R. E. Zahler, *Chem. Revs.*, **49,** 273 (1951); J. Miller, *Rev. Pure Appl. Chem.*, **1,** 171 (1951); J. F. Bunnett, *Quart. Revs.*, (*London*), **12,** 1 (1958).

[2] T. W. J. Taylor, *J. Chem. Soc.*, 1099, 1897 (1928).

[3] H. Schmid and G. Muhr, *Ber.*, **70,** 421 (1937).

[4] L. P. Hammett, "Physical Organic Chemistry," p. 294, McGraw-Hill Book Company, Inc., New York, 1940.

Hughes, Ingold, and Ridd, who studied the reaction in considerably less acidic solution.[5] The change in pH so decreased the nitrous acid concentration and increased the amine concentration as to make the first rather than the second step of mechanism (17-1) rate-controlling; i.e., the kinetic equation had the form

$$v = k[HNO_2]^2$$

Bunton, Llewellyn, and Stedman showed that the O^{18} exchange of nitrous acid under similar conditions also follows this rate equation with the same value of k.[6] This shows that the process whose rate was being studied was indeed the formation of nitrogen trioxide, and it also shows that this nitrogen trioxide formation involves the attack of nitrite ion on nitrous acidium ion and not the combination of nitrite ions with nitrosonium ions (NO^+).

$$HNO_2 \rightleftharpoons H^+ + NO_2^-$$
$$H^+ + HNO_2 \rightleftharpoons H_2NO_2^+$$
$$H_2NO_2^+ + NO_2^- \rightleftharpoons N_2O_3 + H_2O$$

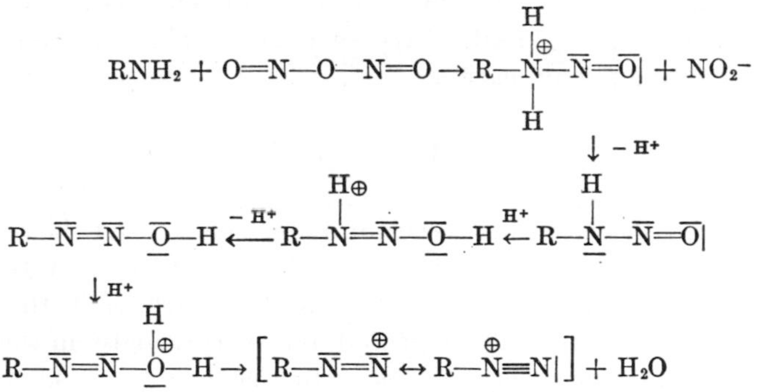

(17-1)

In acidic solutions containing enough chloride, bromide, or iodide ions the kinetic equation has an additional term involving the halide ion.

$$v = k[RNH_2][HNO_2]^2 + k'[RNH_2][HNO_2][H^+][X^-]$$

Apparently, this added term in the kinetic equation is due to the action of nitrosyl halides in addition to nitrogen trioxide.

[5] E. D. Hughes, C. K. Ingold, and J. H. Ridd, *Nature*, **166,** 642 (1950); *J. Chem. Soc.*, 58, 65, 70, 77, 82, 88 (1958).
[6] C. A. Bunton, D. R. Llewellyn, and G. Stedman, *Nature*, **175,** 83 (1955); *J. Chem. Soc.*, 568 (1959).

17-1b. Decomposition of Diazonium Salts. The transformations of a number of aromatic diazonium salts to phenols have been found to be first-order in aqueous solution, the rate being independent of the nature and concentrations of the anion in dilute solutions.[7-10] Waters[11] suggested that the reaction proceeds by the S_N1 mechanism, although he pointed out that a nucleophilic attack by water may be occurring. The fact that the decomposition of benzenediazonium chloride proceeds only about 20 per cent faster in 12 N hydrochloric acid (where 60 per cent of the reaction product is chlorobenzene) than in dilute solution[12] has been quoted in support of the S_N1 mechanism. In this concentrated a solution, however, we cannot discount the possibility that the accelerating effect of an additional nucleophilic attack by chloride ions may be partially offset by a solvent effect, decreasing the rate of nucleophilic attack by water. The fact that *p*-nitrobenzenediazonium chloride hydrolyzes more slowly than the unsubstituted compound supports the carbonium-ion mechanism for the latter, at least, since the *p*-nitro group should greatly increase the rate of nucleophilic attack by water, as it does the rate of other aromatic nucleophilic displacements. The decompositions of benzenediazonium fluoroborate in the presence of such weakly nucleophilic species as chlorobenzene, diphenyl ether, etc., to yield the corresponding onium salts[13] also seem best explained as reactions of the highly reactive phenyl cation.

$$C_6H_5N_2{}^+ \rightarrow N_2 + C_6H_5{}^+ \xrightarrow{\ C_6H_5Cl\ } (C_6H_5)_2Cl^+$$

Assuming an S_N1 mechanism, it is certainly reasonable that the reactivity should be decreased, as it has been found to be,[12] by the nitro and other electron-withdrawing groups. Electron-donor groups in the meta position increase the reactivity as expected, but, surprisingly, in the para position they decrease the reactivity.[12,14] Hughes, however, has pointed out that this is consistent with the S_N1 mechanism, since groups capable of supplying electrons by resonance from the para position should stabilize

[7] H. Euler, *Ann.*, **325**, 292 (1902).

[8] H. A. H. Pray, *J. Phys. Chem.*, **30**, 1417, 1477 (1926).

[9] E. A. Moelwyn-Hughes and P. Johnson, *Trans. Faraday Soc.*, **36**, 948 (1940).

[10] D. F. DeTar and A. R. Ballentine, *J. Am. Chem. Soc.*, **78**, 3916 (1956).

[11] W. A. Waters, *J. Chem. Soc.*, 266 (1942).

[12] M. L. Crossley, R. H. Kienle, and C. H. Benbrook, *J. Am. Chem. Soc.*, **62**, 1400 (1940).

[13] A. N. Nesmeyanov and T. P. Tolstaya, *Doklady Akad. Nauk S.S.S.R.*, **105**, 94 (1955); *Chem. Abstr.*, **50**, 11266f (1956); A. N. Nesmeyanov, L. G. Makarova, and T. P. Tolstaya, *Tetrahedron*, **1**, 145 (1957).

[14] E. S. Lewis and E. B. Miller, *J. Am. Chem. Soc.*, **75**, 429 (1953).

the diazonium ion,[1]

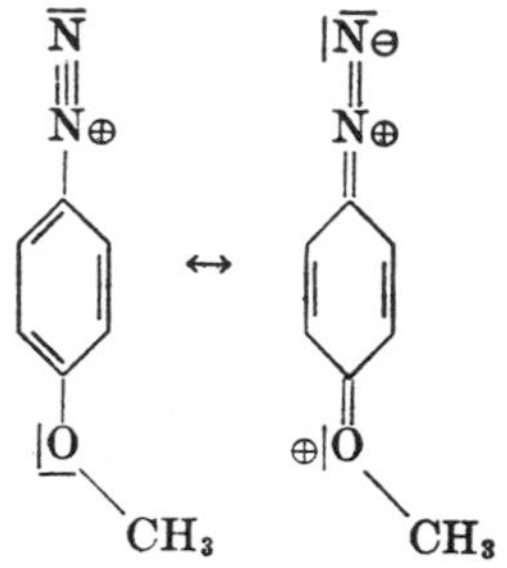

even though such resonance stabilization is impossible for the carbonium ion.

Lewis and Hinds have shown that with the p-nitro compound reactivity toward bimolecular nucleophilic attack is sufficiently developed to be observable.[15] They found the increase in the rate of decomposition of the diazonium ion in the presence of bromide ion to be proportional to the bromide-ion concentration, as was the ratio of p-nitrobromobenzene to p-nitrophenol in the product. They also found the reaction to be very sensitive to catalysis by copper, its effect being noticeable at a concentration of 10^{-5} M. This catalysis is probably related to the Sandmeyer reaction, but there appears to be no generally accepted explanation for it.

17-2. Bimolecular Nucleophilic Displacements on Aromatic Rings. 17-2a. *Mechanism of Aromatic Bimolecular Nucleophilic Displacement.* Although the aromatic bimolecular nucleophilic displacement reaction

$$Y| + Ar—X \rightarrow Ar—Y + X|$$

bears a considerable formal resemblance to the S_N2 mechanism for substitution at saturated carbon atoms, in the present case, as in ester hydrolysis (Sec. 12-1b), the carbon atom undergoing nucleophilic attack is never bonded to more than four other atoms at once. This greatly facilitates the stability of a discrete reaction intermediate in which both the entering and departing groups are covalently bound to carbon. In certain cases such an intermediate may be observed directly. For example, a number of aromatic polynitro compounds are known to form adducts with metal alkoxides, cyanides, and amines. Meisenheimer showed that the adduct formed from potassium ethoxide and trinitroanisole and that from potassium methoxide and trinitrophenetole are identical, the same mixture of trinitroanisole and trinitrophenetole being

[15] E. S. Lewis and W. H. Hinds, *J. Am. Chem. Soc.*, **74**, 304 (1952).

obtained from each on acidification.[16] This constitutes evidence that the adduct has structure I.

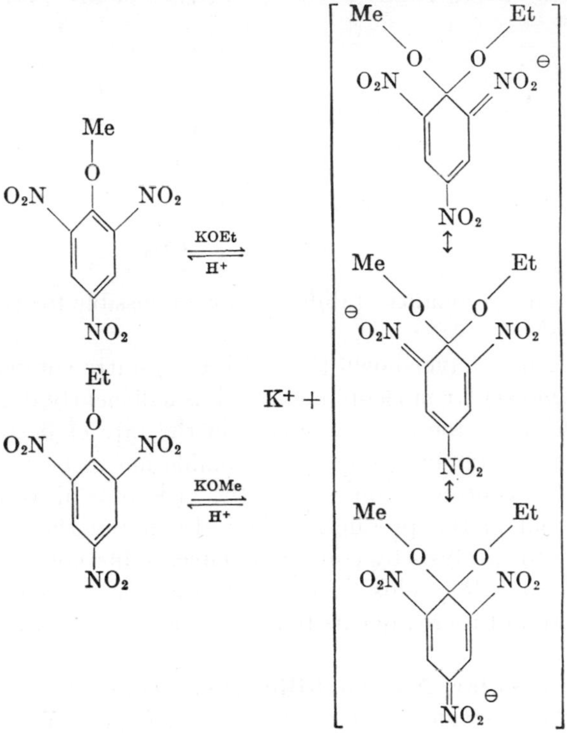

Farr, Bard, and Wheland showed that solution of *m*-dinitrobenzene in liquid ammonia to give a purple solution capable of conducting an electric current is probably due to the formation of the anion II.[17]

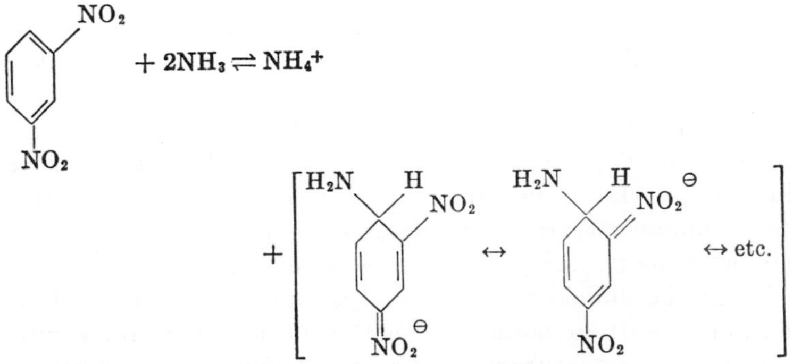

[16] J. Meisenheimer, *Ann.*, **323**, 205 (1902).

[17] J. D. Farr, C. C. Bard, and G. W. Wheland, *J. Am. Chem. Soc.*, **71**, 2013 (1949).

Bolton, Miller, and Parker found that in dimethylformamide solution p-fluoronitrobenzene combines with an equivalent of azide ions without the liberation of any fluoride ions.[18] These observations and the spectrum of the resulting solution show that the following anion was formed:

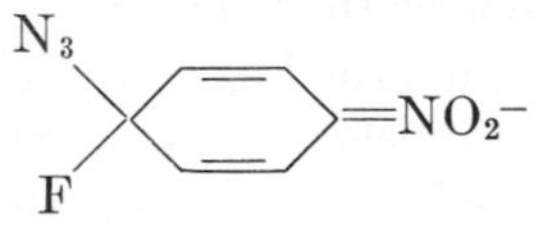

   Bunnett and Randall made a kinetic study of the reaction of N-methylaniline with 2,4-dinitrofluorobenzene that seems to rule out the possibility of a simple one-step nucleophilic displacement mechanism.[19] They found that the reaction rate was dependent on the concentrations of amine and aromatic fluoride but also that the reaction was subject to general base catalysis. The reaction rate in ethanol containing potassium acetate increased linearly with the potassium acetate concentration. This was explained by the following mechanism

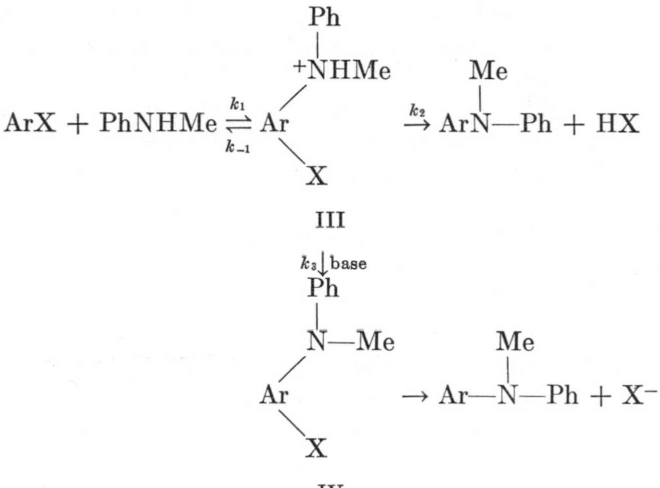

in which the function of the base is to dehydrohalogenate the reactive intermediate III, probably via the deprotonated intermediate IV. Although the observation of potassium acetate catalysis could be explained by the transformation of a small amount of N-methylaniline to its more reactive conjugate base in a rapid equilibrium process (a rate-controlling reaction of acetate ion with N-methylaniline would give an over-all reaction rate independent of the aromatic halide concentration instead of first-order in aromatic halide as observed), this explanation

[18] R. Bolton, J. Miller, and A. J. Parker, *Chem. & Ind.* (*London*), 1026 (1960).
[19] J. F. Bunnett and J. J. Randall, *J. Am. Chem. Soc.*, **80**, 6020 (1958).

was disposed of by the observation that the potassium acetate–catalyzed reaction was not slowed by the addition of acetic acid. The reaction was also shown to be catalyzed by hydroxide ion in aqueous dioxane in a manner that was said to become somewhat less than linear at higher hydroxide-ion concentrations as the reaction step governed by k_3 becomes comparable in rate to that governed by k_{-1}. This point does not appear to have been as firmly established by the experimental results, whose interpretation is complicated by the competing direct reaction of the hydroxide catalyst, a reaction that in some instances proceeded more than 100 times as fast as the reaction whose catalysis was being investigated. The analogous reactions of 2,4-dinitrochlorobenzene and 2,4-dinitrobromobenzene are not subject to general base catalysis, presumably because the intermediates in these cases decompose to product much more rapidly than they revert to reactant. The decomposition of these intermediates is said to proceed by the step governed by k_2, whose mechanism is not given explicitly but is implied to involve the initial loss of X as an anion (the proton transfer governed by k_3 would not be expected to be much faster than in the case of the fluoride, although a concerted dehydrohalogenation would be). One unusual (but not impossible) aspect of the reaction mechanism should be noted. The loss of a proton from the fairly highly acidic ammonium ion III, even in the presence of appreciable concentrations of acetate or hydroxide ion, is said to be slower than the cleavage of a carbon-nitrogen bond. The studies by Ross and coworkers of the reactions of 2,4-dinitrochlorobenzene with amines also support the two-step mechanism.[20]

17-2b. *Effect of the Structure of the Aryl Group on Reactivity.* By analogy with the examples in the previous section it seems probable that there is a true intermediate (though usually not nearly so stable as I or II) in most bimolecular nucleophilic aromatic substitutions. Structural changes that stabilize this intermediate will practically always stabilize the transition state leading to it and hence increase the reactivity. Since the nucleophilic reagent uses one of its own electron pairs to form the new bond to carbon, one of the electron pairs of the aromatic ring is thereby set free, so that groups capable of accepting electrons increase the reactivity.

Berliner and Monack studied the solvolysis of a number of 4-substituted 2-nitrobromobenzenes in piperidine solution, with substituents ranging from halogen through alkyl and alkoxyl to amino (the 4-nitro compound reacted too rapidly to measure).[21] They obtained a fairly satisfactory Hammett-equation plot with $\rho = +4.95$, showing the great

[20] S. D. Ross et al., *J. Am. Chem. Soc.*, **79**, 6547 (1957); **80**, 5319 (1958); **81**, 2113, 5336 (1959).

[21] E. Berliner and L. C. Monack, *J. Am. Chem. Soc.*, **74**, 1574 (1952).

extent to which the reaction is aided by electron withdrawal and slowed by electron donation. Using the nitro, methylsulfonyl, acetyl, chloro, and positively charged trimethylammonio groups as substituents, Bunnett and coworkers found $\rho = +3.9$ for the reaction of sodium methoxide with 4-substituted 2-nitrochlorobenzenes.[22]

As expected, reasonable agreement with the Hammett equation was obtained only when σ^- values were used for groups capable of resonance-electron withdrawal. Miller and coworkers have included discussions of the reasons for remaining deviations from the Hammett equation in their extensive investigations of nucleophilic aromatic substitution.[23]

The diazonium cation group ($-N_2^{\oplus}$) appears to be even more effective than the nitro group at activating o- and p-halogen toward nucleophilic replacement.[23a] Reactions of this type are probably run more often by accident than by design, since many of the known examples were discovered when a reaction series involving the diazotization of an aromatic amine yielded an unexpected product. For example, when 1-nitro-2-naphthylamine is diazotized in hydrochloric acid solution, the nitro group is replaced by a chlorine atom.[24]

Halogen atoms in the α and γ positions of pyridine rings are also activated to nucleophilic attack.

From comparisons of the reactivity of 2-chloro-5-nitropyridine and 2,4-dinitrochlorobenzene toward various nucleophilic reagents, the activation of these positions appears to be less than that due to an o- or p-nitro group.[25]

[22] J. F. Bunnett et al., *J. Am. Chem. Soc.*, **75**, 642 (1953).

[23] J. Miller et al., (*a*) *J. Chem. Soc.*, 750 (1956); *Australian J. Chem.*, **11**, 302 (1958); (*b*) *Australian J. Chem.*, **11**, 297 (1958); (*c*) *ibid.*, **9**, 61, 74, 299, 382 (1956); *J. Chem. Soc.*, 2926, 2929 (1955); 2329 (1956), and references cited therein.

[24] N. N. Vorozhtsov, V. V. Kozlov, and I. S. Travkin, *Zhur. Obshchei Khim.*, **9**, 522 (1939); *Chem. Abstr.*, **34**, 410 (1940).

[25] A. Mangini and B. Frenguelli, *Gazz. chim. ital.*, **69**, 86 (1939); R. R. Bishop, E. A. S. Cavell, and N. B. Chapman, *J. Chem. Soc.*, 437 (1952).

Banks found that the nucleophilic substitution reactions of halides of the type of α- and γ-halopyridines are acid-catalyzed.[26]   As he pointed out, this behavior might be expected since the protonation (or alkylation[23b]) of a heteronitrogen atom should increase its electron-withdrawing power.

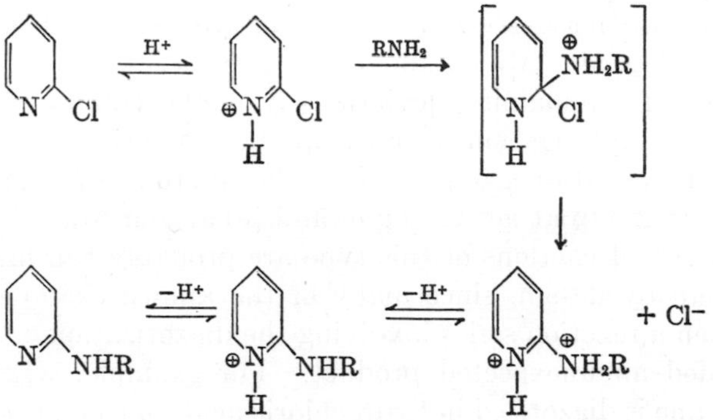

17-2c. Effect of the Nature of the Displaced Group on Reactivity.   The effect of the nature of X on the rate of nucleophilic substitution reactions of Ar—X is rather different from the effect in the analogous aliphatic substitution reactions.   As in the aliphatic cases, however, the relative reactivity of various X's may depend upon the exact nature of the reactions.   Some of the observed variations in reactivity that accompany changes in the nature of X appear rational in view of the mechanism we have described for the reaction.

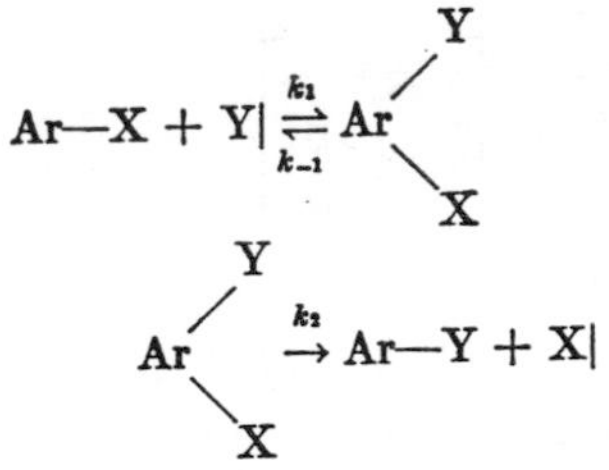

$$V$$

If k_{-1} is negligible in comparison to k_2, the first step of the reaction is rate-controlling; otherwise, the second step is.   Other things being equal, we should expect the intermediate V to lose preferentially whichever of X or Y represents the more stable (less basic) species.   In most of the aromatic nucleophilic displacements that have been studied kinetically, Y was a primary amine or alkoxide or hydroxide ion and X a halogen

[26] C. K. Banks, *J. Am. Chem. Soc.*, **66**, 1127 (1944).

atom. It therefore seems likely that in these cases the first step of the reaction is rate-controlling. In reactions of this type the C—X bond is not being broken to any great extent in the transition state, although there is interference with the resonance that gives this bond a small amount of double-bond character. For this reason the great strength of the carbon-fluorine bond, which is probably responsible for a great deal of the unreactivity of saturated fluorides, is no longer such an important factor. Consequently the high electronegativity of fluorine, which appears to form considerably stronger bonds to sp^3 carbon than to sp^2 carbon, becomes relatively more important. This may explain the data of Bevan, who found that the reactivities of the various p-nitro-halobenzenes toward sodium ethoxide were in the ratio F:Cl:Br:I::3,100:13.6:11.8:1 at 90.8°.[27] Although it seems that fluorine is usually the most reactive halogen, Hammond and Parks showed that it is possible to change the reaction conditions so as to obtain the reactivity sequence ArBr > ArCl > ArF.[28] This was done by changes designed to increase the extent to which the carbon-halogen bond is broken in the transition state. The variation in the relative ease of nucleophilic displacement of groups that may accompany a change in nucleophilic reagent was shown quite strikingly by Loudon and Robson.[29] These workers found that by use of the proper nucleophilic reagent, any one of the three substituents of 3-chloro-4-(p-toluenesulfonyl)nitrobenzene could be preferentially displaced.

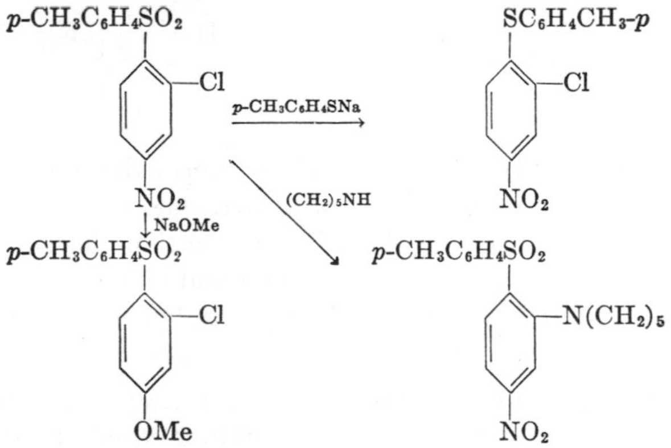

Despite the existence of cases of the type described above, the following order of mobilities listed by Bunnett and Zahler[1] has considerable utility:

[27] C. W. L. Bevan, *J. Chem. Soc.*, 2340 (1951).
[28] G. S. Hammond and L. R. Parks, *J. Am. Chem. Soc.*, **77**, 340 (1955).
[29] J. D. Loudon and T. D. Robson, *J. Chem. Soc.*, 242 (1937); cf. J. D. Loudon and N. Shulman, *J. Chem. Soc.*, 722 (1941).

$$-F > -NO_2 > -Cl \sim -Br \sim -I > -OSO_2R > -NR_3^+ > -OAr$$
$$> -OR > -SAr \sim -SR > -SO_2R > -NR_2 > -H.$$

The Chichibabin method of amination of pyridine and related compounds[30] is an example of a nucleophilic displacement of hydrogen (as a hydride ion).

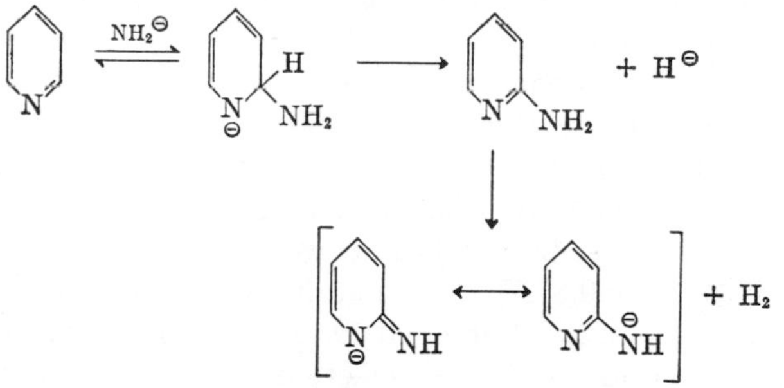

17-3. The Elimination-Addition Mechanism for Nucleophilic Aromatic Substitution.[31]
17-3a. *The Mechanism of Benzyne Formation.* A large number of instances have been recorded in which a strongly basic reagent reacted with an aromatic halide to replace the halogen atom by hydrogen and to introduce the attacking base into the ortho position. In an early example, Haeussermann showed that each of the three isomeric dichlorobenzenes reacts with potassium diphenylamide to give the same tetraphenylphenylenediamine (among other products).[32]

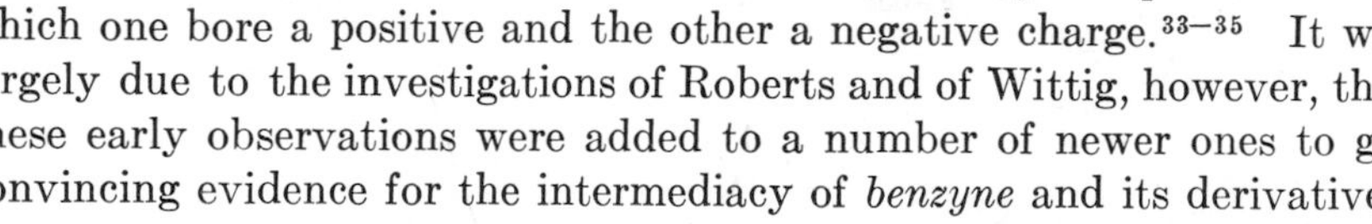

$$o\text{-}, m\text{-}, \text{ or } p\text{-}C_6H_4Cl_2 + (C_6H_5)_2NK \rightarrow$$

There were several early suggestions of an intermediate aromatic species in which two adjacent carbon atoms were attached by a triple bond or in which one bore a positive and the other a negative charge.[33–35] It was largely due to the investigations of Roberts and of Wittig, however, that these early observations were added to a number of newer ones to get convincing evidence for the intermediacy of *benzyne* and its derivatives.

[30] A. E. Chichibabin and O. A. Seide, *Zhur. Russ. Fiz.-Khim. Obshchestva*, **46,** 1216 (1914); *Chem. Abstr.*, **9,** 1901 (1915); M. T. Leffler, "Organic Reactions," vol. I, chap. 4., John Wiley & Sons, Inc., New York, 1942.

[31] For a review of this topic, see R. Huisgen and J. Sauer, *Angew. Chem.*, **72,** 91, 294 (1960).

[32] C. Haeussermann, *Ber.*, **33,** 939 (1900); **34,** 38 (1901).

[33] R. Stoermer and B. Kahlert, *Ber.*, **35,** 1633 (1902).

[34] G. Wittig, *Naturwissenschaften*, **30,** 696 (1942).

[35] A. A. Morton, J. B. Davidson, and B. L. Hakan, *J. Am. Chem. Soc.*, **64,** 2242 (1942).

Chlorobenzene labeled at the chlorine-bearing carbon atom with C^{14} was found to react with potassium amide in liquid ammonia to give aniline, in which only half of the original C^{14} was still at the carbon atom bearing the functional group.[36] It was further shown that o-deuteriochlorobenzene reacts slower than its protium analog, that the reactants and products are not isomerized under the reaction conditions, and that halides (such as bromomesitylene and bromodurene) having no o-hydrogen do not react.[36,37] It was pointed out that the reaction therefore probably consists in a rate-controlling dehydrohalogenation to give an electrically neutral "benzyne" intermediate to which ammonia may add in two possible ways.

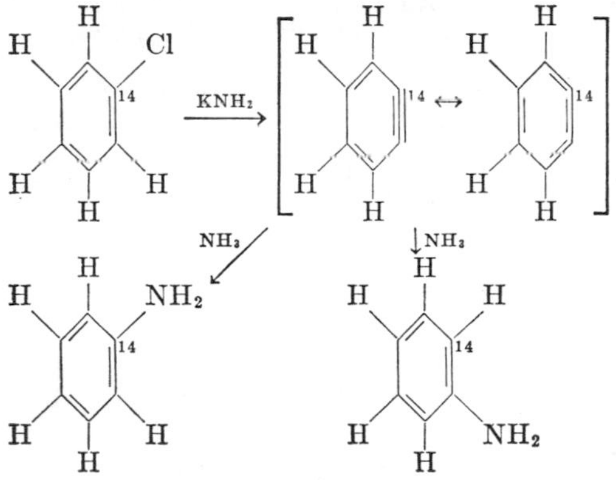

Wittig and Pohmer added to the evidence for this intermediate, which they called "dehydrobenzene," by capturing it via a Diels-Alder addition to furan.[38]

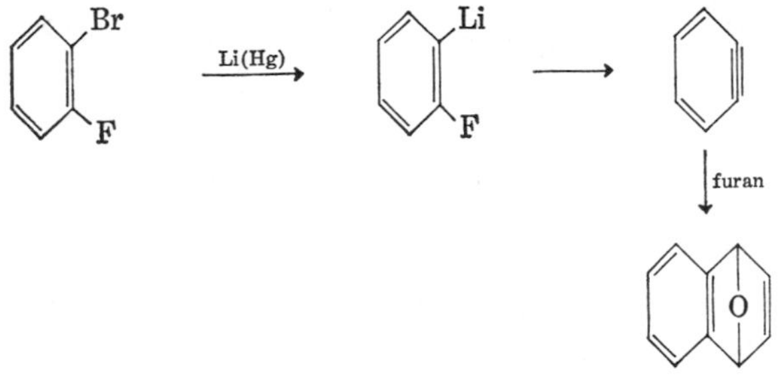

[36] J. D. Roberts, H. E. Simmons, Jr., L. A. Carlsmith, and C. W. Vaughan, *J. Am. Chem. Soc.*, **75**, 3290 (1953).

[37] J. D. Roberts, D. A. Semenow, H. E. Simmons, Jr., and L. A. Carlsmith, *J. Am. Chem. Soc.*, **78**, 601 (1956); J. D. Roberts, C. W. Vaughan, L. A. Carlsmith, and D. A. Semenow, *J. Am. Chem. Soc.*, **78**, 611 (1956).

[38] G. Wittig and L. Pohmer, *Angew. Chem.*, **67**, 348 (1955); *Chem. Ber.*, **89**, 1334 (1956).

In liquid ammonia containing potassium amide, o-deuteriofluorobenzene exchanges deuterium much more rapidly than it forms aniline, showing that in this case benzyne is formed by the carbanion mechanism for elimination (cf. Sec. 8-1a).[39] With chlorobenzene, under the same conditions, the intermediate carbanion appears to revert to reactant and go on to product at comparable rates; and with bromobenzene it is not possible to tell whether the dehydrohalogenation is concerted or whether there is formed an intermediate carbanion that almost invariably loses a bromide ion to give benzyne.[37] In no case, however, does there appear to be good evidence for the formation of a benzyne by a concerted elimination.

17-3b. Orientation and Yield in Reactions Involving Benzynes. In the formation of a benzyne by removal of HX from a given ArX, the removal of either the H or the X can be rate-limiting. In either case, it seems probable that in the rate-limiting step the transition state has considerable carbanion character. Since the addition of HY to the intermediate benzyne to give ArY is simply the reverse of benzyne formation, it is not surprising that the transition states in such reactions usually seem to have considerable carbanion character also. Rates of benzyne formation are therefore expected to depend greatly upon the basicity of the nucleophilic reagent and the stability of the intermediate carbanion. The ease of displacement of X with its bonding electron pair and steric hindrance are two other factors of importance.

The relative stability of various aryl anions usually depends largely on the inductive effect of substituent groups. This was shown, for example, by Hall, Piccolini, and Roberts's observation of the following relative rates of deuterium exchange of various monosubstituted deuteriobenzenes with potassium amide in liquid ammonia solution: o-F > o-CF$_3$ > o-OCH$_3$ ∼ m-CF$_3$ ∼ p-CF$_3$ > m-F > p-F > m-OCH$_3$ > unsubstituted > p-OCH$_3$.[39] Although chlorobenzene and bromobenzene react readily with potassium amide in liquid ammonia to give aniline, fluorobenzene, which gives the requisite intermediate carbanion readily under these conditions, reacts only very slowly at −40° because of the difficulty of displacing the fluoride ion, especially in such a poorly hydrogen-bonding solvent. This unreactivity of aromatic fluorides toward benzyne formation and their considerable reactivity in bimolecular nucleophilic aromatic substitution is reflected in Huisgen and coworkers' observation that the 1- and 2-chloro, -bromo, and -iodo derivatives of naphthalene all react with lithium piperidide in piperidine practically entirely by the benzyne mechanism, but 1- and 2-fluoronaphthalene react, to a considerable extent, by the bimolecular mechanism.[40] This conclusion follows from the fact that a mixture of about 31 per cent N-1-naphthylpiperidine

[39] G. E. Hall, R. Piccolini, and J. D. Roberts, *J. Am. Chem. Soc.*, **77**, 4540 (1955).

[40] R. Huisgen, J. Sauer, W. Mack, and I. Ziegler, *Chem. Ber.*, **92**, 441 (1959).

and 69 per cent N-2-naphthylpiperidine is obtained from all the halides except the fluorides, both of which yield considerably larger amounts of unrearranged product (under some conditions). These observations also show that the reaction of the 2-naphthyl chloride, bromide, and iodide proceeded very largely via 1,2-naphthyne just as did that of the 1-naphthyl halides.

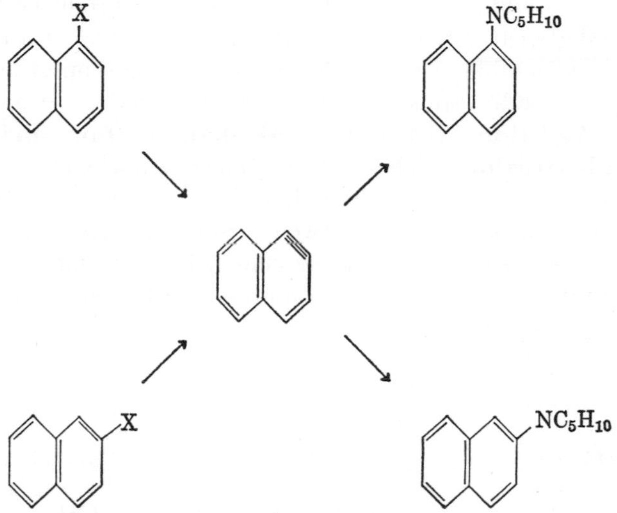

Nucleophilic reagents usually add to benzynes so as to produce the more stable intermediate carbanion. This is shown by the data of Roberts and coworkers (Table 17-1) and by other data on the reactions

TABLE 17-1. PRODUCTS OF THE REACTIONS OF NaNH$_2$ OR KNH$_2$ WITH RC$_6$H$_4$X IN LIQUID AMMONIA[37]

R	X	Yield, %	Orientation of product, %		
			Ortho	Meta	Para
o-OCH$_3$	Br	33	. . .	100	
m-OCH$_3$	Br	59	. . .	100	
p-OCH$_3$	Br	31	. . .	49	51
o-CF$_3$	Cl	28	. . .	100	
m-CF$_3$	Cl	16	. . .	100	
p-CF$_3$	Cl	25	. . .	50	50
p-F	Br	30	. . .	20	80
o-CH$_3$	Cl	66	45	55	
o-CH$_3$	Br	64	48	52	
m-CH$_3$	Br	61	22	56	22
m-CH$_3$	Cl	66	40	52	8
p-CH$_3$	Cl	35	. . .	62	38

of aryl halides with sodium and potassium amides. Thus, *o*- and *m*-bromoanisole, *o*- and *m*-chlorobenzotrifluoride, and *o*-chloro- and *o*-iodoanisole all react with sodamide to give the *m*-amino derivatives in essentially pure form.[37,41,42]   All these reactions apparently involve an intermediate 2,3-benzyne; the *o*-halides because they must, and the *m*-halides because the *o*-hydrogen is so much more acidic than the *p*-hydrogen.   The addition to the 2,3-benzyne then produces the more stable intermediate carbanion with a full negative charge, or a transition state with a considerable negative charge, on the 2-carbon atom.   With the weakly polar methyl substituent, mixtures of products are obtained.

17-4. The Addition-Elimination Mechanism for Nucleophilic Aromatic Substitution.   There are a number of nucleophilic substitution reactions of aromatic compounds known that appear to proceed via nonaromatic intermediates in which two or more of the ring carbon atoms have become sp^3-hybridized.   Phloroglucinol, for example, reacts with cold aqueous ammonia to give 5-aminoresorcinol and 3,5-diaminophenol,[43] presumably by way of cyclohexane-1,3,5-trione, for whose presence there is considerable evidence.

In ketonization of a phenol, loss of resonance stabilization of the aromatic ring is compensated (to varying degrees in different cases) by the greater

[41] H. Gilman and S. Avakian, *J. Am. Chem. Soc.*, **67**, 349 (1945).

[42] R. A. Benkeser and R. G. Severson, *J. Am. Chem. Soc.*, **71**, 3838 (1949).

[43] J. Pollack, *Monatsh.*, **14**, 401 (1893).

stability of ordinary ketones compared with the corresponding enols (cf. Sec. 10-2d).[44]   With polyhydric phenols the gain in stability due to ketonization is much larger, and the loss of aromatic stabilization is about the same.

Naphthols ketonize more easily than simple phenols because the stabilization lost is just the *difference* between the resonance stabilization of a naphthalene ring and that of a benzene ring.   Rieche and Seeboth have described convincing evidence that in the Bucherer reaction (the bisulfite-ion-catalyzed interconversion of naphthols and naphthylamines) the ketonic form of the naphthol and the imine form of the naphthylamine are stabilized by the addition of bisulfite ion to a double bond in the same ring.[45]

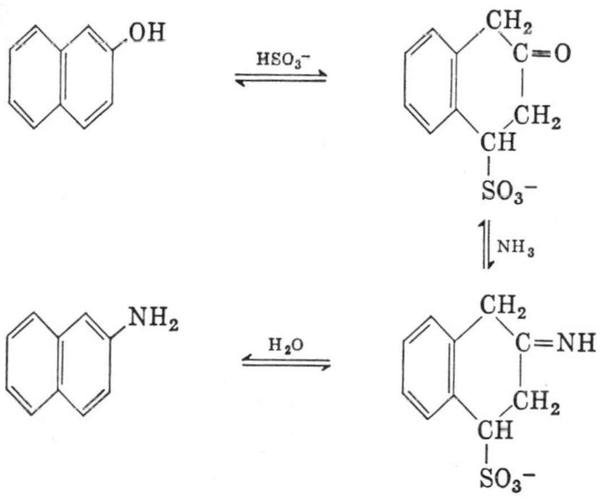

17-5. Nucleophilic Substitution at Vinyl Carbon.

Most of the mechanisms conceivable for nucleophilic substitution at vinyl carbon are analogous to those we have discussed previously for aromatic compounds. The elimination-addition mechanism operates not only for such reactants as 1-halocyclohexenes and 1-halocyclopentenes, where the intermediate acetylene is quite reactive,[46–48] but also for acyclic compounds such as certain haloethylenes.   Truce and coworkers found that the reaction of

[44] For a detailed discussion of this point, see G. W. Wheland, "Advanced Organic Chemistry," secs. 14-4 and 14-8, John Wiley & Sons, Inc., New York, 1960.

[45] A. Rieche and H. Seeboth, *Ann.*, **638**, 43, 57, 66, 76, 81, 92, 101 (1960).

[46] A. E. Favorsky et al., *Ann.*, **390**, 122 (1912); *J. Gen. Chem. U.S.S.R.*, **6**, 720 (1936); *Chem. Abstr.*, **30**, 6337⁷ (1936).

[47] G. Wittig et al., *Ber.*, **77**, 306 (1944); *Angew. Chem.*, **72**, 324 (1960).

[48] J. D. Roberts et al., *Tetrahedron*, **1**, 343 (1957); *J. Am. Chem. Soc.*, **82**, 4750 (1960).

cis-1,2-dichloroethylene with the sodium salts of thiophenols is powerfully catalyzed by sodium ethoxide and proceeds much more rapidly than the reaction of the more difficultly dehydrohalogenated *trans*-dichloroethylene.[49] Furthermore, chloroacetylene, the first intermediate in the proposed reaction scheme below, was shown to give VII under the reaction conditions and some VI under milder conditions.

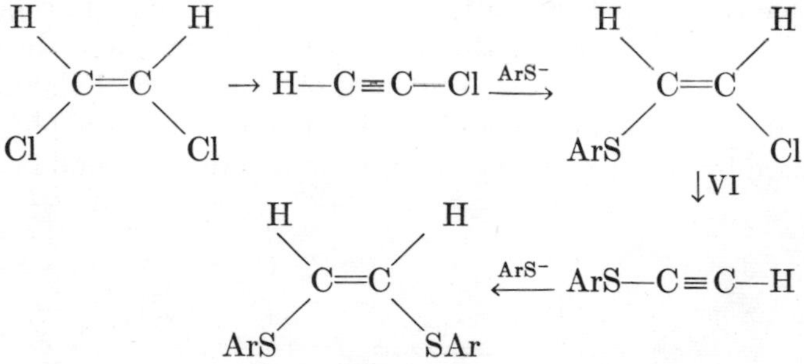

1,1-Dichloroethylene, on the other hand, reacts by the addition-elimination mechanism to give the same final product via the intermediacy of VIII and IX, both of which can be isolated under the proper conditions.[50]

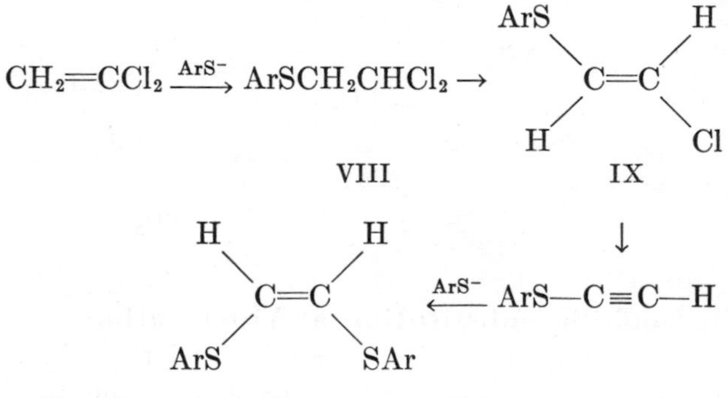

Reactions like those studied by Truce and coworkers, involving trans elimination and trans addition, result in over-all nucleophilic substitution with retention of geometric configuration. Miller and Yonan found that the reactions of *cis*- and *trans*-1-*p*-nitrophenyl-2-bromoethylene with iodide ion give largely retention of geometric configuration,[51] and similar observations have been made for the reactions of ethyl β-chloro-*cis*- and -*trans*-crotonate with a number of nucleophilic reagents.[52] Both sets of

[49] W. E. Truce et al., *J. Am. Chem. Soc.*, **78**, 2743 (1956).

[50] W. E. Truce and M. M. Boudakian, *J. Am. Chem. Soc.*, **78**, 2748 (1956).

[51] S. I. Miller and P. K. Yonan, *J. Am. Chem. Soc.*, **79**, 5931 (1957).

[52] D. E. Jones, R. O. Morris, C. A. Vernon, and R. F. M. White, *J. Chem. Soc.*, 2349 (1960).

reactions were discussed in terms of a mechanism resembling the mechanism for bimolecular nucleophilic aromatic substitution, e.g.,

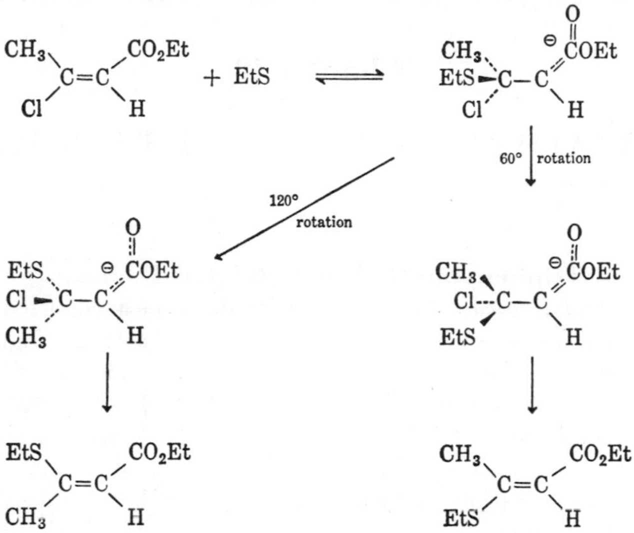

PROBLEMS

1. The reaction of 1-fluoronaphthalene with excess phenyllithium followed by carbonation yields a mixture of A and B, two isomeric acids with the molecular formula $C_{17}H_{12}O_2$. The analogous reaction of 2-fluoronaphthalene yields A, B, and a third isomer, C. In each case it is A that is formed in the largest yield. Write structural formulas for A, B, and C and give mechanisms for their formation.

2. Give reasonable mechanistic explanations for each of the following observations:

(*a*) The reaction of *p*-chloronitrobenzene with potassium cyanide at 160° gives nitrogen and *m*-chlorobenzoic acid in about 50 per cent yield. The hydrolyses of *m*-chlorobenzonitrile and *m*-chlorobenzamide in the presence of the same concentration of potassium cyanide and at the same temperature are slower reactions. The addition of N^{15}-labeled ammonia to the reaction solution has no effect on the isotope content of the nitrogen produced in the reaction. The use of *p*-chloronitrobenzene containing 9 atom per cent N^{15} causes the nitrogen formed to contain 9 per cent N_2^{29}. When the reaction is carried out in water enriched in H_2O^{18}, the carboxy oxygen atoms are found to contain half as much (atom per cent) O^{18} as the water.

(*b*) At 340° the reaction of *p*-bromotoluene with 4 *M* aqueous sodium acetate gives 33 per cent *p*-cresol (and no other cresols), whereas 4 *M* aqueous sodium hydroxide gives a 55 per cent yield of a mixture of 45 per cent *p*-cresol and 55 per cent *m*-cresol.

Chapter 18

RELATIVELY STABLE FREE RADICALS

18-1. The Triphenylmethyl and Related Radicals. 18-1a. *The Triphenylmethyl Radical.* The first molecule to be recognized as a free organic radical was synthesized by Gomberg.[1] In attempting to prepare hexaphenylethane, he treated triphenylmethyl chloride with finely divided silver in benzene solution. When the resultant white crystalline solid was found to be triphenylmethyl peroxide, the experiment was repeated in the absence of air. There then resulted a yellow solution, which upon evaporation yielded a white crystalline solid. The solid was originally thought to be triphenylmethyl on the basis of its great reactivity in solution. It combined very rapidly with chlorine, bromine, and iodine to form the corresponding triphenylmethyl halides, with nitric oxide to yield triphenylnitrosomethane, $(C_6H_5)_3CNO$, and with oxygen to yield the peroxide. It was later shown, however, that the white solid was hexaphenylethane, which in solution dissociates reversibly into the yellow free triphenylmethyl radical. This equilibrium and the dissociation of other hexaarylethanes have been studied most often by colorimetric measurements,[2] by molecular-weight determinations,[3] and by magnetic-susceptibility measurements[4] (free radicals tend to be paramagnetic, while nonradicals are diamagnetic). Each of the methods has its weaknesses. The colorimetric method is often rendered inaccurate by side reactions that yield colored products. The molecular-weight method often depends upon a small difference between two large numbers, one of which cannot be determined very accurately. It is also somewhat restricted to specific temperatures, such as the melting point and boiling point of the solvent. The magnetic method was long thought to be

[1] M. Gomberg, *Ber.*, **33**, 3150 (1900); *J. Am. Chem. Soc.*, **22**, 757 (1900).

[2] J. Piccard, *Ann.*, **381**, 347 (1911); K. Ziegler and L. Ewald, *Ann.*, **473**, 163 (1929).

[3] M. Gomberg and L. H. Cone, *Ber.*, **37**, 2037 (1904); W. Schlenk, T. Weickel, and A. Herzenstein, *Ann.*, **372**, 1 (1910); M. Gomberg and C. S. Schoepfle, *J. Am. Chem. Soc.*, **39**, 1652 (1917); **41**, 1655 (1919).

[4] (a) N. W. Taylor, *J. Am. Chem. Soc.*, **48**, 854 (1926); (b) E. Müller, I. Müller-Rodloff, and W. Bunge, *Ann.*, **520**, 235 (1935); (c) M. F. Roy and C. S. Marvel, *J. Am. Chem. Soc.*, **59**, 2622 (1937).

relatively reliable, but it has since been shown that previously used corrections for the diamagnetic contributions to the susceptibility are very probably much too low and that it would be extremely difficult to make an accurate correction.[5] The most suitable method for determining free-radical concentrations, and hence dissociation constants for hexaarylethanes, is probably by use of electron-spin-resonance (ESR, or paramagnetic-resonance) spectral measurements,[6] but not very many measurements of this type have been made.[7] Independent of the method of measurement, there are a large number of experimental precautions that should be taken in studies of this sort if the results are to be dependable. In most of the reported investigations not all these precautions were taken.

The ease of dissociation of hexaarylethanes (compared, say, with that of ethane itself) may be described in terms of two factors. One of these, the resonance stabilization of the triarylmethyl radicals, results in a reluctance of these radicals to form a covalent bond to any atom or radical. The other, presumably a steric hindrance factor, results in an added reluctance of triarylmethyl radicals to form covalent bonds to each other (or other bulky groups). We shall discuss the factors by comparison of the triphenylmethyl radical with the unsubstituted methyl radical as a standard. Bent and Cuthbertson found the hydrogenolysis of hexaphenylethane to be exothermic by about 35 kcal/mole in solution.[8] When this thermochemical equation

$$(C_6H_5)_3CC(C_6H_5)_3 + H_2 \rightarrow 2(C_6H_5)_3CH \qquad \Delta H = -35 \text{ kcal}$$

is combined with those for the association of triphenylmethyl radicals and of hydrogen atoms,

$$2(C_6H_5)_3C\cdot \rightarrow (C_6H_5)_3CC(C_6H_5)_3 \qquad \Delta H = -11 \text{ kcal}$$
$$2H\cdot \rightarrow H_2 \qquad \Delta H = -103 \text{ kcal}$$

we get

$$2(C_6H_5)_3C\cdot + 2H\cdot \rightarrow 2(C_6H_5)_3CH \qquad \Delta H = -149 \text{ kcal}$$
$$(C_6H_5)_3C\cdot + H\cdot \rightarrow (C_6H_5)_3CH \qquad \Delta H = -74.5 \text{ kcal}$$

Thus the tertiary carbon-hydrogen bond–dissociation energy is found to be 74.5 kcal/mole, or 27.5 kcal smaller than the value (102 kcal[9]) for

[5] P. W. Selwood and R. M. Dobres, *J. Am. Chem. Soc.*, **72**, 3860 (1950); T. L. Chu and S. I. Weissman, *J. Am. Chem. Soc.*, **73**, 4462 (1951).

[6] D. J. E. Ingram, "Free Radicals as Studied by Electron Spin Resonance," Butterworth & Co. (Publishers), Ltd., London, 1958.

[7] Cf. F. C. Adam and S. I. Weissman, *J. Am. Chem. Soc.*, **80**, 2057 (1958).

[8] H. E. Bent and G. R. Cuthbertson, *J. Am. Chem. Soc.*, **58**, 170 (1936).

[9] H. C. Andersen and G. B. Kistiakowsky, *J. Chem. Phys.*, **11**, 10 (1943); G. B. Kistiakowsky and E. R. Van Artsdalen, *J. Chem. Phys.*, **12**, 469 (1944).

the formation of our standard methyl radical from methane. The heat of dimerization of triphenylmethyl radicals is about 72 kcal less than that of methyl radicals (83 kcal). Although a large fraction of this 72 kcal can be attributed to the 55-kcal resonance stabilization of the two triphenylmethyl radicals involved, the remaining 17-kcal steric effect is also of considerable importance. In fact, data on the heat of reaction of hexaphenylethane with oxygen lead to a larger estimate of the steric effect and smaller estimate of the resonance effect.[10]

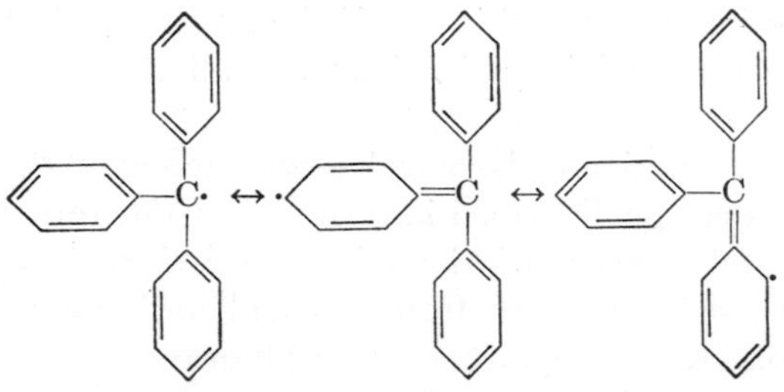

The resonance stabilization of triphenylmethyl would probably be much greater if coplanarity of the molecule were not prevented by steric interference between the *o*-hydrogen atoms of the phenyl groups.

Further support for a steric contribution to the ease of dissociation of hexaarylethanes is found in the observation that while methyl substituents in any position increase the extent of dissociation, those in the ortho position do so best.[11] Also the report that the carbon-carbon single bond is 1.58 ± 0.03 A in hexamethylethane rather than the usual 1.54 A suggests that it is stretched by steric repulsions.[12]

Ziegler, Orth, and Weber studied the rate of dissociation of hexaphenylethane by measuring its rate of combination with nitric oxide.[13] With a sufficiently high concentration of nitric oxide the reaction rate is independent of its concentration and simply first-order in hexaphenylethane. The rate-controlling step, then, must be dissociation to triphenylmethyl radicals, which combine with nitric oxide much more rapidly than they dimerize.

$$(C_6H_5)_3CC(C_6H_5)_3 \rightarrow 2(C_6H_5)_3C\cdot$$
$$(C_6H_5)_3C\cdot + NO \rightarrow (C_6H_5)_3C\text{—NO}$$

[10] H. E. Bent, G. R. Cuthbertson, M. Dorfman, and R. E. Leary, *J. Am. Chem. Soc.*, **58**, 165 (1936).

[11] C. S. Marvel, M. B. Mueller, C. M. Himel, and J. F. Kaplan, *J. Am. Chem. Soc.*, **61**, 2771 (1939); C. S. Marvel, J. F. Kaplan, and C. M. Himel, *J. Am. Chem. Soc.*, **63**, 1892 (1941).

[12] S. H. Bauer and J. Y. Beach, *J. Am. Chem. Soc.*, **64**, 1142 (1942).

[13] K. Ziegler, P. Orth, and K. Weber, *Ann.*, **504**, 131 (1933).

The rate constants thus found are in satisfactory agreement with ones obtained by somewhat similar methods involving the reaction of hexaphenylethane with iodine and with oxygen.[14] The reaction rate was measured in each of the following solvents: CCl_4, $CHCl_3$, $C_2H_4Br_2$, $C_6H_5NH_2$, C_2H_5OH, $C_6H_5CH_3$, $C_6H_5NO_2$, $C_6H_5N(CH_3)_2$, ClC_2H_4OH, $CH_2(CO_2C_2H_5)_2$, $NCCH_2CO_2C_2H_5$, o-$HOC_6H_4CO_2CH_3$, C_5H_5N, CH_3CN, and CS_2.[13, 14] None of the rate constants determined differed from a mean value by as much as a factor of two. This relative independence of the nature of the solvent is found in most free-radical reactions except those (a large number) in which the solvent enters directly into the reaction by being attacked by an intermediate radical. In a few radical reactions, however, in which the radical(s) appears to form a complex with the solvent (cf. Sec. 22-1d), there are large solvent effects. The rate constants for a typical polar reaction would vary by a factor of millions or more in a range of solvents like that listed above. The nature of the solvent has also been found to have little effect on the magnitude of the *equilibrium* constant for dissociation.[15]

Dyachkovskii, Bubnov, and Shilov used paramagnetic-resonance measurements to determine rate and equilibrium constants for dimerization of triphenylmethyl radicals.[16] They found the process to require activation, the heat of activation for dimerization (7 kcal/mole) being equal to the difference between the heat of reaction (11 kcal/mole) and the heat of activation for dissociation (18 kcal/mole). Apparently, the central carbon atom of the radical, having all three valences in the same plane, must begin to assume a tetrahedral configuration in order to form the new carbon-carbon bond; and this deviation from planarity results in a loss of resonance stabilization.

18-1b. *Effect of Structure on the Dissociation of Hexaarylethanes and Related Compounds.* Marvel and coworkers used the magnetic method to determine the dissociation constants for a large number of hexaarylethanes and found that alkyl, cyclohexyl, methoxy, and phenyl substituents all increase the extent of dissociation.[11,17] Since all these groups are capable of electron donation, it may be somewhat surprising to learn that electron-withdrawing groups can also increase radical stability. Tris-(p-nitrophenyl)methyl, for example, exists as a radical in the solid state.[18] Nevertheless, the presence of either type of substituent may

[14] K. Ziegler, L. Ewald, and P. Orth, *Ann.*, **479**, 277 (1930); K. Ziegler, A. Seib, F. Knoevenagel, P. Herte, and F. Andrews, *Ann.*, **551**, 150 (1942).

[15] K. Ziegler and L. Ewald, *Ann.*, **473**, 163 (1929).

[16] F. S. Dyachkovskii, N. N. Bubnov, and A. E. Shilov, *Doklady Akad. Nauk S.S.S.R.*, **122**, 629 (1958); *Chem. Abstr.*, **54**, 23651g (1960); cf. Ref. 14.

[17] C. S. Marvel and coworkers, *J. Am. Chem. Soc.*, **59**, 2622 (1937); **61**, 2008, 2769 (1939); **62**, 1550 (1940); **66**, 415, 914 (1944).

[18] F. L. Allen and S. Sugden, *J. Chem. Soc.*, 440 (1936).

permit extra contributing structures to be written.   Those for *p*-alkyl groups show the operation of a type of hyperconjugation.

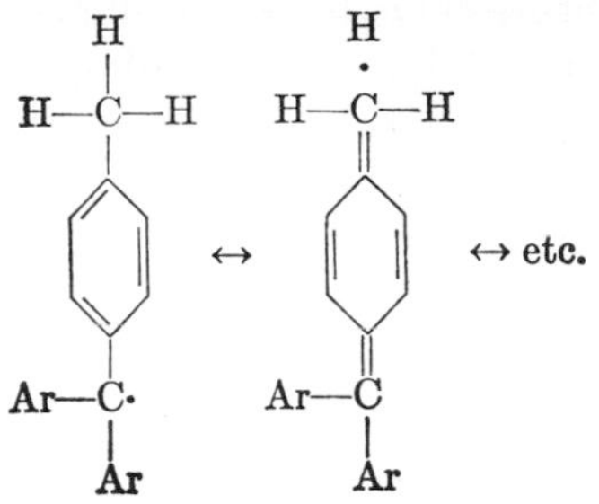

With *p*-nitro groups we may write structures of the type

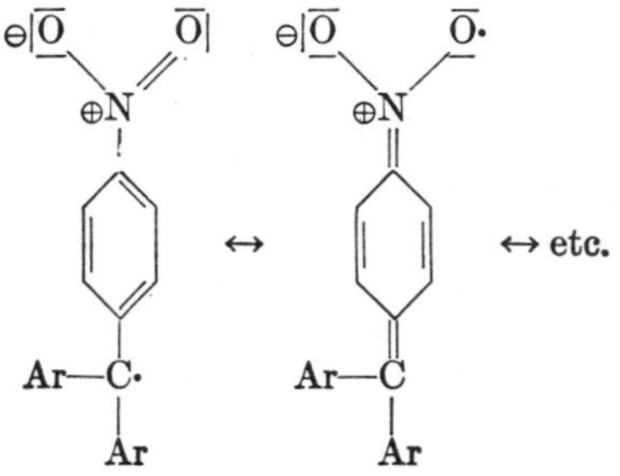

Dipole-dipole repulsions are also probably important in promoting dissociation of hexaarylethanes.

The effect of certain fused-ring systems on radical stability has also been studied.   The dimer of the 9-phenylfluoryl radical,

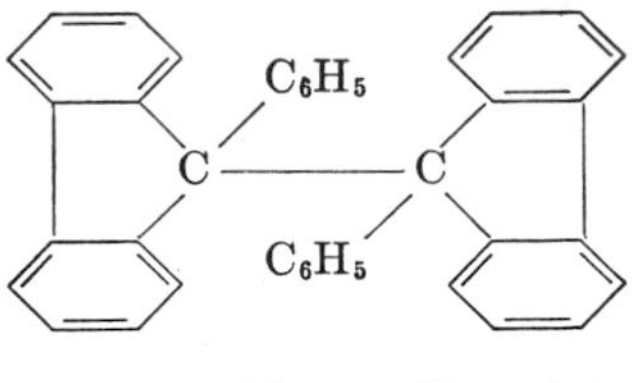

is colorless and hence presumably undissociated in solution at room temperature.   The brown color formed reversibly upon heating suggests that dissociation does occur at higher temperatures.   It has been stated that the decrease in dissociation (compared with hexaphenylethane) is due to a smaller amount of resonance energy in the radical.[19]   On the other hand, according to the measurements of Bent and Cline, who find that the heat of reaction with oxygen is about 20 kcal less than it is for hexaphenylethane, the decreased dissociation is due to an increase in the stability of the ethane rather than a decrease in the stability of the free

[19] L. C. Pauling and G. W. Wheland, *J. Chem. Phys.*, **1**, 362 (1933).

radical.[20]   In fact, according to these measurements the radical is probably *more* stable than the triphenylmethyl radical (it has the advantage of a coplanar structure of at least two of the benzene rings).

Unlike the fluoryl radical, the xanthyl radical appears to increase the extent of dissociation into free radicals.[21]

Pentaphenylcyclopentadienyl, in which the unpaired electron may be written on 20 different carbon atoms,

$$C_6H_5$$
$$|$$
$$C\cdot$$

$$C_6H_5\!-\!C \qquad\qquad C\!-\!C_6H_5$$
$$C_6H_5\!-\!C\!-\!C\!-\!C_6H_5$$

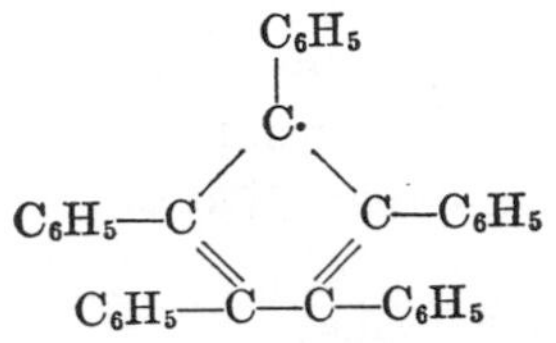

appears to exist entirely as the free radical in the solid form.[22]   Although steric hindrance in this molecule probably decreases the amount of resonance by preventing coplanarity, it probably aids dissociation by hampering dimerization.   Tris-(*p*-phenylphenyl)methyl is another radical that exists as such in the solid form and in solution.[4b,5]   The unpaired electron may be written on any of 19 different carbon atoms.

$$(p\text{-}C_6H_5C_6H_4)_2C = \!\!\!\bigcirc\!\!\!=\!\!\!\bigcirc\!\!\!\cdot \leftrightarrow \text{etc.}$$

The resonance stabilization due to distribution of the unpaired electron over only two benzene rings can be sufficient to permit the formation of a fairly stable free radical if there is enough steric hindrance to dimerization.   This appears to be the case with the pentaphenylethyl radical,

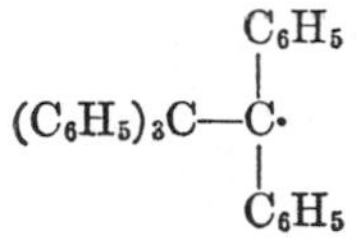

$$C_6H_5$$
$$|$$
$$(C_6H_5)_3C\!-\!C\cdot$$
$$|$$
$$C_6H_5$$

which exists (according to molecular-weight determinations) entirely as the free radical in benzene solution.[23]   This is probably due to a large extent to the bulk of the triphenylmethyl group.   Even a *t*-butyl group can be effective in this position.   Conant and Bigelow observed that tetraphenyldi-*t*-butylethane forms a yellow color reversibly when its benzene solution is heated to about 50°.[24]   That this represents

[20] H. E. Bent and J. E. Cline, *J. Am. Chem. Soc.*, **58**, 1624 (1936).

[21] J. B. Conant and coworkers, *J. Am. Chem. Soc.*, **47**, 572, 3068 (1925); **48**, 1743 (1926); **49**, 2080 (1927); **51**, 1925 (1929).

[22] E. Müller and I. Müller-Rodloff, *Ber.*, **69B**, 665 (1936).

[23] W. Schlenk and H. Mark, *Ber.*, **55B**, 2285 (1922).

[24] J. B. Conant and N. M. Bigelow, *J. Am. Chem. Soc.*, **50**, 2041 (1928).

dissociation to radicals is also shown by the rapid absorption of the theoretical amount of oxygen even at room temperature (where a slight yellow color is present).

There are a number of compounds that do not dissociate sufficiently under ordinary conditions to yield a directly measurable concentration of free radicals but that may be shown by somewhat less direct methods to dissociate under fairly mild conditions. Conant and Evans found that a number of 9,9'-dialkyl derivatives of bixanthyl react with oxygen at room temperature at a rate that is independent of the oxygen concentration and is first-order in the bixanthyl.[25] The rate-controlling step of the reaction is presumably a dissociation to two 9-alkylxanthyl radicals. Bachmann and Wiselogle observed that pentaarylethanes, upon heating, disproportionate to hexaarylethanes and tetraarylethanes, probably by the mechanism[26]

$$(C_6H_5)_3CCH(C_6H_5)_2 \rightleftharpoons (C_6H_5)_3C\cdot + (C_6H_5)_2CH\cdot$$
$$2(C_6H_5)_3C\cdot \rightleftharpoons (C_6H_5)_3CC(C_6H_5)_3$$
$$2(C_6H_5)_2CH\cdot \rightarrow (C_6H_5)_2CHCH(C_6H_5)_2$$

The rate of reaction has been determined for a number of derivatives, and the effect of structure on reactivity has been discussed.[26,27]

18-1c. *Diradicals of the Triphenylmethyl Type.* A considerable amount of research has been directed toward learning whether certain compounds have the diradical structure I or the quinoid structure II.

I and II cannot simply be two contributing structures for the same molecule, since I has two unpaired electrons and II has none (Sec. 1-1a, rule 5). Structure II is favored by having one more covalent bond than I. However, I has the advantage of the resonance stabilization of n more benzene rings. We would therefore expect an increase in n to favor I. This appears to be the case.

According to magnetic-susceptibility measurements, the compound for which $n = 1$ and the one (known as the Chichibabin hydrocarbon) for which $n = 2$ do not have structure I to any detectable extent,[28,29] but for $n = 3$ and 4 there is a detectable fraction of material with struc-

[25] J. B. Conant and M. W. Evans, *J. Am. Chem. Soc.*, **51**, 1925 (1929).

[26] W. E. Bachmann and F. Y. Wiselogle, *J. Org. Chem.*, **1**, 354 (1936).

[27] J. Coops, H. Galenkamp, J. Haantjes, H. L. Luirink, and W. T. Nauta, *Rec. trav. chim.*, **67**, 469 (1948).

[28] E. Müller, *Z. Elektrochem.*, **45**, 593 (1939).

[29] E. Müller and I. Müller-Rodloff. *Ann.*, **517**, 134 (1935).

ture I in equilibrium with the more stable II.[30] Schwab and Agliardi studied some of these compounds by a more sensitive and probably more reliable method based on the fact that free radicals catalyze the conversion of p- to o-hydrogen.[31] They found that not more than 0.2 per cent of the compound where $n = 1$ was present as a diradical but that about 10 per cent of the Chichibabin hydrocarbon ($n = 2$) was. By the method of paramagnetic resonance absorption the Chichibabin hydrocarbon has been found to be 4 to 5 per cent diradical in character.[32]

Diradicals may also be favored by making quinoid structures impossible, as in the Schlenk hydrocarbon,

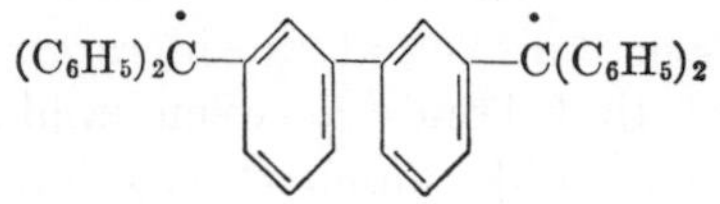

$$(C_6H_5)_2\overset{\cdot}{C}\text{——}\hspace{-0.3em}\langle \rangle\hspace{-0.3em}\langle \rangle\text{——}\overset{\cdot}{C}(C_6H_5)_2$$

whose diradical character is readily detectable even by magnetic-susceptibility measurements.[29]

The diradical structure may also be favored by steric interference with the coplanarity necessary for the maximum stability of the quinoid form.   Müller and Neuhoff showed this to be the case by synthesizing the compound

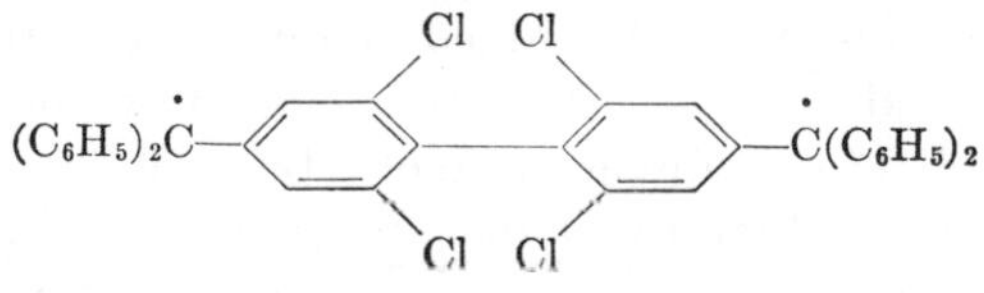

$$(C_6H_5)_2\overset{\cdot}{C}\text{——}\hspace{-0.3em}\langle \overset{Cl}{\underset{Cl}{}}\rangle\text{——}\langle \overset{Cl}{\underset{Cl}{}}\rangle\text{——}\overset{\cdot}{C}(C_6H_5)_2$$

and observing that it has enough diradical character to detect by magnetic measurements.[33]

18-2. Nitrogen, Oxygen, and Sulfur Free Radicals.   18-2a. *Radicals with the Unpaired Electron on Nitrogen.*   Wieland found that when the colorless solution of tetraphenylhydrazine in toluene is heated, a greenish-brown color is formed reversibly.[34] This is attributed to dissociation to the diphenylamino radical.

$$(C_6H_5)_2NN(C_6H_5)_2 \rightleftharpoons 2(C_6H_5)_2N\cdot$$

In agreement with this suggestion, Cain and Wiselogle found that at a

[30] E. Müller and H. Pfanz, *Ber.*, **74B**, 1051, 1075 (1941).

[31] G.-M. Schwab and N. Agliardi, *Ber.*, **73B**, 95 (1940).

[32] C. A. Hutchison, Jr., A. Kowalsky, R. C. Pastor, and G. W. Wheland, *J. Chem. Phys.*, **20**, 1485 (1952).

[33] E. Müller and H. Neuhoff, *Ber.*, **72B**, 2063 (1939); cf. E. Müller and E. Tietz, *Ber.*, **74B**, 807 (1941).

[34] H. Wieland, *Ann.*, **381**, 200 (1911); *Ber.*, **48**, 1078 (1915); "Die Hydrazine," pp. 71ff., Ferd. Enke Verlag, Stuttgart, 1913.

pressure of greater than 0.2 atm nitric oxide reacts with tetraphenyl-hydrazine to give diphenylnitrosoamine at a rate independent of the nitric oxide concentration and first-order in that of the hydrazine.[35] The rate-controlling step of this reaction is apparently a dissociation to diphenylamino radicals, which then react with nitric oxide more rapidly than they dimerize. It has also been found that tetraphenylhydrazine and hexaphenylethane, each of which is considerably dissociated at 90°, react at about this temperature to yield the tertiary amine expected from a coupling of unlike radicals.[34]

$$(C_6H_5)_3CC(C_6H_5)_3 + (C_6H_5)_2NN(C_6H_5)_2 \rightarrow 2(C_6H_5)_3CN(C_6H_5)_2 \quad (18\text{-}1)$$

This amine is not dissociated even at considerably higher temperatures. Analogous to the fact that bonds between unlike atoms tend to be stronger than those between like atoms (Sec. 1-4c), it is often found that unlike radicals couple to form stronger bonds than identical radicals. With regard to reaction (18-1) we may see from Table 1-4 that two typical carbon-nitrogen bonds should have a greater combined strength than one carbon-carbon and one nitrogen-nitrogen bond. This greater strength of the bond between unlike radicals may be found even when the bond joins identical (but differently substituted) atoms, as when both are carbon, and presumably has the same source as that for bonds between unlike atoms, viz., the added resonance contribution of ionic structures. Wieland reports that the electron-withdrawing p-nitro group decreases the ease of dissociation of tetraphenylhydrazine, whereas the p-methoxy and p-dimethylamino groups increase it.[34] Molecular-weight determinations show that tetra-(p-dimethylaminophenyl)hydra-zine is appreciably dissociated in benzene and nitrobenzene solutions even at room temperature.

Tetrazane derivatives are in general even more highly dissociated than the related hydrazines. The white crystalline hexaphenyltetrazane (III) gives blue solutions that disobey Beer's law, become more deeply colored at higher temperatures, and react with nitric oxide and with triphenylmethyl to yield the expected products.[36] This radical may be

$$(C_6H_5)_2N-\underset{\underset{C_6H_5}{\diagup}}{N}-\underset{\underset{C_6H_5}{\diagdown}}{N}-N(C_6H_5)_2 \rightleftharpoons 2(C_6H_5)_2N-\overset{\displaystyle\cdot}{N}C_6H_5$$

III

stabilized by electron-withdrawing groups on the lone phenyl group. The compound 1,1-diphenyl-2-picrylhydrazyl (IV), for example, exists entirely as a free radical, both in solution and in the solid state. This has been established by both molecular-weight determinations[37] and

[35] C. K. Cain and F. Y. Wiselogle, *J. Am. Chem. Soc.*, **62**, 1163 (1940).
[36] S. Goldschmidt, *Ber.*, **53**, 44 (1920).
[37] S. Goldschmidt and K. Renn, *Ber.*, **55**, 628 (1922).

magnetic-susceptibility studies.[4b,38] Some of the contributing structures below show how the electron-withdrawing power of the nitro group may aid in the stabilization of this radical.

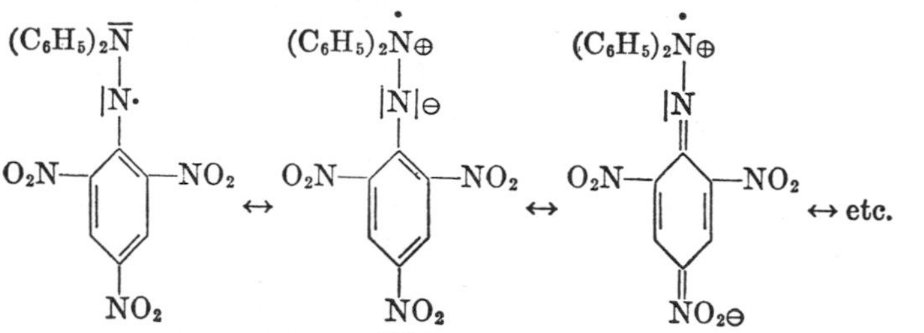

IV

Goldschmidt and coworkers found that the dissociation of the tetraphenyldibenzoyltetrazane (V) is increased by electron-donating substituents on the phenyl groups and decreased by electron-withdrawing substituents there.[39]

$$(C_6H_5)_2N—N—N—N(C_6H_5)_2 \rightleftharpoons 2(C_6H_5)_2N—\overset{\cdot}{N}COC_6H_5$$

V

Wilmarth and Schwartz showed that the equilibrium constants for the dissociation of such tetrazanes, substituted in the phenyl groups, follow the Hammett equation with ρ equal to -1.52. For the rates of dissociation ρ is -0.55, and for the recombination of the radicals ρ is $+0.97$.[40]

Trivalent nitrogen free radicals are also known. One type includes the so-called Wurster salts which are prepared by the oxidation of p-phenylenediamine derivatives at the proper pH. The formation of the resonance-stabilized radical shown below has been proved by study of the oxidation-reduction curves at various pH's, spectroscopic measurements, and magnetic-susceptibility determinations.[41] This species is,

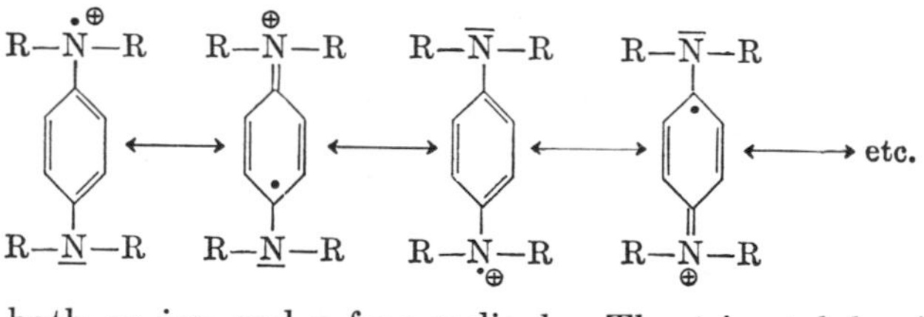

of course, both an ion and a free radical. The tri-p-tolylaminium ion,

$$(p\text{-}CH_3C_6H_4)_3\overset{\oplus}{N}\cdot \leftrightarrow \text{etc.}$$

[38] (a) H. Katz, Z. Physik., **87**, 238 (1933); (b) J. Turkevich and P. W. Selwood, J. Am. Chem. Soc., **63**, 1077 (1941).

[39] S. Goldschmidt, A. Wolf, E. Wolffhardt, I. Drimmer, and S. Nathan, Ann., **437**, 194 (1924); S. Goldschmidt and J. Bader, Ann., **473**, 137 (1929).

[40] W. K. Wilmarth and N. Schwartz, J. Am. Chem. Soc., **77**, 4543, 4551 (1955).

[41] L. Michaelis, Chem. Revs., **16**, 243 (1935).

which may be prepared by the oxidation of tri-p-tolyamine, is also in this category.[38a,42]   The oxidation of N,N-diphenylhydroxylamine yields

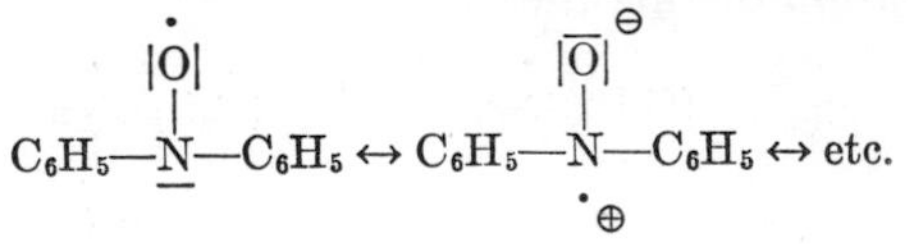

$$\overset{\displaystyle \cdot}{\underset{\displaystyle |}{|O|}} \qquad \overset{\displaystyle \ominus}{\underset{\displaystyle |}{\overline{|O|}}}$$

$$C_6H_5-\underline{N}-C_6H_5 \leftrightarrow C_6H_5-\underline{N}-C_6H_5 \leftrightarrow \text{etc.}$$
$$\underset{\displaystyle \cdot\oplus}{}$$

a radical[4a,38a,43] that is perhaps as much an oxygen as a nitrogen free radical.

18-2b. Oxygen and Sulfur Free Radicals.   Bis-(9-ethoxy-10-phenanthryl) peroxide has been shown by magnetic methods to undergo a light-catalyzed dissociation to the 9-ethoxy-10-phenanthroxy radical.[44]

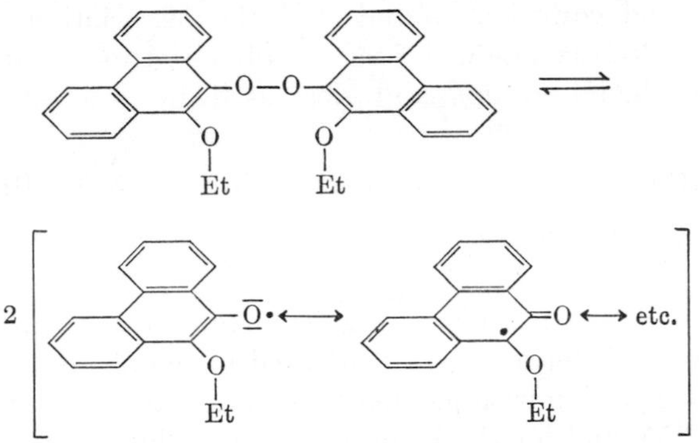

The semiquinones are somewhat similar to Würster salts except that the radicals are anions rather than cations.   They are intermediates in the alkaline oxidation of hydroquinones to quinones and have been investigated thoroughly by Michaelis.[41,45]

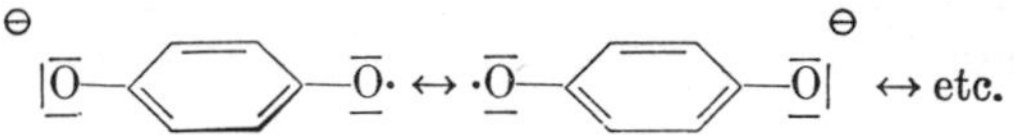

Cook and coworkers found that several 4-substituted derivatives of the 2,6-di-t-butylphenoxy radical can be prepared as relatively stable species in solution.   They pointed out that in order to be stable such radicals must have large enough ortho- and para-substituent groups to interfere with dimerization.   If such ortho and para substituents have benzyl-type hydrogen atoms, however, side reactions occur.[46]

[42] P. Rumpf and F. Trombe, *Compt. rend.*, **206**, 671 (1938); *J. chim. phys.*, **35**, 110 (1938).

[43] L. Cambi, *Gazz. chim. ital.*, **63**, 579 (1933).

[44] H. G. Cutforth and P. W. Selwood, *J. Am. Chem. Soc.*, **70**, 278 (1948).

[45] L. Michaelis and M. P. Schubert, *Chem. Revs.*, **22**, 437 (1938).

[46] C. D. Cook et al., *J. Org. Chem.*, **18**, 261 (1953); **24**, 1356 (1959); *J. Am. Chem. Soc.*, **75**, 6242 (1953); **78**, 2002 (1956); **81**, 1176 (1959); cf. E. Müller et al., *Ber.*, **88**, 601, 1819 (1955); **90**, 1530, 2660 (1957); **91**, 2682 (1958); **92**, 474 (1959).

Schlenk and Weickel found that the blue solution formed by the reaction of benzophenone with sodium in ether reacted rapidly with iodine or oxygen to regenerate the ketone.[47] From this and other evidence they suggested that the metal ketyl, VI, a salt whose anion is also a radical, is formed.   Magnetic measurements have shown that an equilibrium is established between the metal ketyl and its dimer, a metal pinacolate.[48]

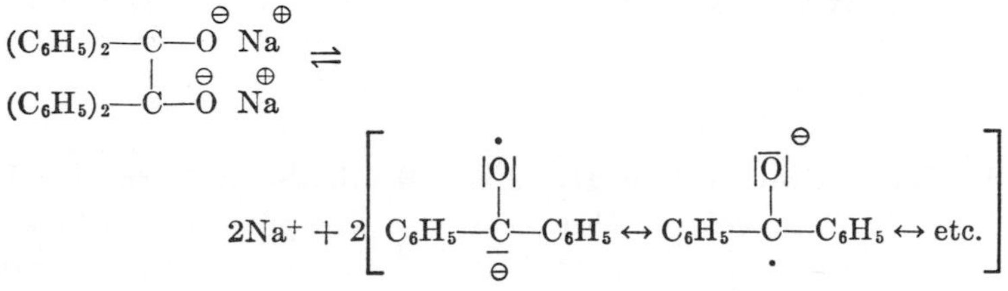

Cutforth and Selwood showed by means of magnetic methods that bis-(2-benzothiazolyl) disulfide dissociates reversibly to form measurable concentrations of free radicals at 100° and above.[44]

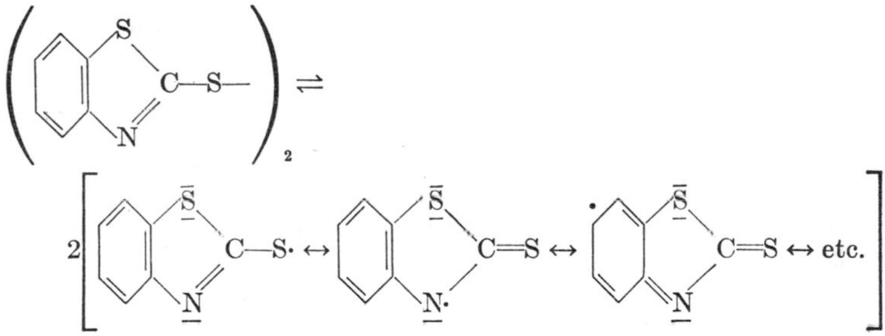

PROBLEMS

1. Suggest an explanation for each of the following observations:

(*a*) When triphenylmethylsodium reacts with triphenylmethyl chloride, the concentration of radicals (measured by ESR) initially becomes quite high but it rather quickly decreases to a lower level, where it remains essentially constant for a relatively long time.

(*b*) The oxidation conditions that transform 2,4,6-tri-*t*-butylphenol to the stable 2,4,6-tri-*t*-butylphenoxy radical transform 2,6-di-*t*-butyl-4-methylphenol to 1,2-bis-(3,5-di-*t*-butyl-4-hydroxyphenyl)ethane.

(*c*) Solutions of hexaphenylethane in sunlight slowly give triphenylmethane and 9,9′-diphenyl-9,9′-difluorenyl.

[47] W. Schlenk and T. Weickel, *Ber.*, **44**, 1182 (1911).

[48] S. Sugden, *Trans. Faraday Soc.*, **30**, 18 (1934); R. N. Doescher and G. W. Wheland, *J. Am. Chem. Soc.*, **56**, 2011 (1934).

Chapter 19

THE FORMATION OF REACTIVE FREE RADICALS

19-1. Formation of Reactive Free Radicals in Pyrolysis Reactions. *19-1a. Pyrolysis of Organometallic Compounds.* Ordinary aliphatic free radicals, such as methyl, ethyl, etc., are much too reactive to be produced and kept at the fairly high concentrations required for their study by the methods used for triphenylmethyl and related radicals. Accordingly, it has been necessary to adopt other techniques.

The first work to yield good evidence for the formation of radicals of this type was that of Paneth and Hofeditz,[1] who used a modification of a method by which atomic hydrogen had been previously studied.[2]   A

$Pb(CH_3)_4$ in H_2 $\longrightarrow$ B A $\longrightarrow$ To liquid air trap and pump

FIG. 19-1. Schematic diagram of apparatus used by Paneth and Hofeditz to study tetramethyllead decomposition.

stream of an inert gas, such as hydrogen or nitrogen, containing some tetramethyllead vapor was allowed to flow through a glass tube.   If the tube was heated strongly at any point A (Fig. 19-1), a lead mirror was deposited on the wall of the tube at this point.   Thus the heat had caused the decomposition of the tetramethyllead to yield metallic lead as one product.   Evidence as to the nature of another product was obtained by subsequently heating the tube at a point B upstream from the lead mirror deposited at A.   When this was done, a new mirror was deposited at B and the old mirror at A was removed.   It seemed reasonable that it was methyl radicals formed by the decomposition of tetramethyllead at B that combined with and removed the lead at A.   This hypothesis was confirmed when it was shown that the product of the removal of lead at A could be caught in a liquid air trap and identified as tetramethyllead.   It was further shown that removal of bismuth,

[1] F. A. Paneth and W. Hofeditz, *Ber.*, **62**, 1335 (1929).

[2] K. F. Bonhoeffer, *Z. physik. Chem.*, **113**, 199 (1924).

zinc, and antimony mirrors could be brought about with the formation of their methyl derivatives.   Mirrors were not removed by the carrier gases, hydrogen, nitrogen, etc., nor by possible decomposition products such as methane, ethane, ethylene, and acetylene.   Under given conditions, the rate of disappearance of the metal at A was found to decrease as the distance between A and B increased.   This showed that the methyl radicals disappeared rather rapidly after they were formed. When hydrogen was used as the carrier gas, most of the methyl radicals appeared as methane, but when the carrier was nitrogen or helium, the principal product was ethane.

By the same technique tetraethyllead was shown to yield ethyl radicals.[3]   Tetrabenzyltin similarly yielded benzyl radicals, but the gaseous products of the decomposition of tetra-n-propyllead and tetraisobutyllead removed zinc and antimony mirrors to form dimethylzinc and $(CH_3)_2Sb$—$Sb(CH_3)_2$, respectively.[4]   Apparently if any n-propyl or isobutyl radicals were formed, they decomposed very rapidly.

19-1b. *Pyrolysis of Other Types of Organic Compounds.*   The formation of free radicals upon heating is not a special characteristic of certain types of compounds.   It appears that essentially all organic compounds will decompose to give radicals if heated to a high enough temperature. Rice, Johnston, and Evering secured the first good evidence that this is true.[5]   They found that a large number of different organic compounds, including hydrocarbons, alcohols, aldehydes, ketones, ethers, and acids, removed metallic mirrors following pyrolysis at 800 to 1000° in an apparatus of the type used by Paneth.   No organometallic compounds containing other than methyl or ethyl groups could be detected in the products of reaction with metallic mirrors.   It is suggested that at the pyrolysis temperature higher radicals decompose to lower radicals and olefins, e.g.,[6]

$$CH_3CH_2CH_2\cdot \rightarrow CH_3\cdot + CH_2{=}CH_2$$
$$(CH_3)_2CHCH_2\cdot \rightarrow CH_3\cdot + CH_3CH{=}CH_2$$

Although most compounds will give radicals when heated to high enough temperatures, it does not follow that *all* high-temperature vapor-phase reactions are free-radical in character.   Some have been shown to be reactions of the multicenter type (Chap. 25).   The mechanisms of high-temperature vapor-phase pyrolysis reactions have been studied with great care by a number of capable investigators.   Steacie has written

[3] F. A. Paneth and W. Lautsch, *Ber.*, **64B**, 2702 (1931).

[4] F. A. Paneth and W. Lautsch, *J. Chem. Soc.*, 380 (1935).

[5] F. O. Rice, W. R. Johnston, and B. L. Evering, *J. Am. Chem. Soc.*, **54**, 3529 (1932).

[6] F. O. Rice, *J. Am. Chem. Soc.*, **53**, 1959 (1931).

an authoritative and comprehensive monograph on this work.[7]   In later sections we shall discuss the mechanisms of the free-radical decompositions of a number of classes of compounds in more detail, particularly for certain compounds that decompose in solution at temperatures below 200° and are therefore useful as initiators in common free-radical reactions.

19-2. Other Methods of Forming Free Radicals.   **19-2a. Photolysis.**[8]   By removing metallic mirrors with the radicals formed, Pearson[9] verified earlier suggestions that the photolysis of acetone involves the intermediate formation of free radicals.   Analysis of the organometallic compounds formed in mirror removals shows that photolysis produces methyl radicals from acetone, ethyl radicals from diethyl ketone, and both phenyl and methyl radicals from acetophenone.[10]   Since biacetyl has been found in the photolysis-reaction mixture from acetone, it appears likely that acetyl radicals are also an intermediate in the reaction.[11,12] From the sharp decrease in the yield of biacetyl that accompanies an increase in the reaction temperature, it seems that the acetyl radicals are easily decarbonylated.   Methane, ethane, carbon monoxide, biacetyl, methyl ethyl ketone, and acetonyl acetone have been found as reaction products under conditions where only a small fraction of the acetone used was allowed to decompose.   The relative amounts of these products formed depend on the reaction temperature, the acetone concentration, and other variables.   Noyes and Dorfman considered these and other data in summarizing the evidence for the following reaction mechanism.[13]

$$CH_3COCH_3 \xrightarrow{h\nu} CH_3\cdot + CH_3CO\cdot$$
$$CH_3\cdot + CH_3CO\cdot \rightarrow CH_3COCH_3$$
$$CH_3CO\cdot \rightarrow CH_3\cdot + CO$$
$$2CH_3\cdot \rightarrow CH_3CH_3$$
$$2CH_3CO\cdot \rightarrow CH_3COCOCH_3 \qquad (19\text{-}1)$$
$$CH_3\cdot + CH_3COCH_3 \rightarrow CH_4 + CH_3COCH_2\cdot$$
$$CH_3\cdot + CH_3COCH_2\cdot \rightarrow CH_3COCH_2CH_3$$
$$2CH_3COCH_2\cdot \rightarrow CH_3COCH_2CH_2COCH_3$$
$$CH_3CO\cdot + CH_3COCH_2\cdot \rightarrow CH_3COCH_2COCH_3$$

[7] E. W. R. Steacie, "Atomic and Free Radical Reactions," 2d ed., Reinhold Publishing Corporation, New York, 1954.

[8] For reviews of certain topics in this area, see J. N. Pitts, Jr., H. E. Gunning, N. Davidson, and J. R. McNesby, *J. Chem. Educ.*, **34**, 112, 121, 126, 130 (1957).

[9] T. G. Pearson, *J. Chem. Soc.*, 1718 (1934).

[10] T. G. Pearson and R. H. Purcell, *J. Chem. Soc.*, 1151 (1935); H. H. Glazebrook and T. G. Pearson, *J. Chem. Soc.*, 589 (1939).

[11] M. Barak and D. W. G. Style, *Nature*, **135**, 307 (1935).

[12] R. Spence and W. Wild, *Nature*, **138**, 206 (1936); *J. Chem. Soc.*, 352 (1937).

[13] W. A. Noyes, Jr., and L. M. Dorfman, *J. Chem. Phys.*, **16**, 788 (1948).

The situation would become even more complicated if the reaction were allowed to proceed long enough for a significant amount of attack by the various free radicals on the products shown.

Under certain conditions the principal reaction products are carbon monoxide and ethane. It has been suggested that the reaction involves a direct rearrangement of the acetone into these two products without the intermediate formation of radicals. Dorfman and Noyes, however, showed that such a mechanism cannot contribute significantly to the total reaction, since the ratio of methane to ethane formed increases as the intensity of catalyzing light is decreased, and at sufficiently low intensities considerably more methane than ethane is produced.[14] The direct mechanism, which offers no explanation for these data, is apparently not of great importance, if it occurs at all. Mechanism (19-1) provides a good explanation of the data, since ethane is formed by a reaction second-order in radicals and methane by one first-order in radicals, and since the radical concentrations should decrease with decreasing light intensity. The mechanism has met other tests that Noyes and Dorfman describe.[13]

The photolysis of ketones with γ-hydrogen atoms ordinarily follows quite another path. An olefin and a smaller ketone molecule are formed. Norrish and Appleyard, for example, showed that methyl n-butyl ketone yields acetone and propylene.[15] Davis and Noyes suggested that this reaction involves the decomposition of the excited state of the ketone via a cyclic transition state giving the enol form of acetone, which subsequently ketonizes.[16]

$$
\begin{array}{c}
\text{O} \quad\quad \text{H---CH---CH}_3 \\
\parallel \quad\quad\quad\quad | \\
\text{CH}_3\text{---C} \quad\quad\quad \text{CH}_2 \\
\diagdown \quad\quad \diagup \\
\text{CH}_2
\end{array}
\;\xrightarrow{h\nu}\;
\begin{array}{c}
\text{OH} \\
| \\
\text{CH}_3\text{---C} \\
\diagdown \\
\text{CH}_2
\end{array}
\;+\;
\begin{array}{c}
\text{CH---CH}_3 \\
\parallel \\
\text{CH}_2
\end{array}
\qquad (19\text{-}2)
$$

This mechanism is supported by Ausloos and Murad's observation that $CD_3COCD_2CH_2CH_3$ yields pentadeuterioacetone[17] and by Blacet and Miller's observation that cyclohexanone, cyclopentanone, and cyclobutanone, for which the cyclic transition state of mechanism (19-2) would be excessively strained, photolyze by a mechanism similar to (19-1).[18]

[14] L. M. Dorfman and W. A. Noyes, Jr., *J. Chem. Phys.*, **16,** 557 (1948).
[15] R. G. W. Norrish and M. E. S. Appleyard, *J. Chem. Soc.*, 874 (1934).
[16] W. Davis, Jr., and W. A. Noyes, Jr., *J. Am. Chem. Soc.*, **69,** 2153 (1947).
[17] P. Ausloos and E. Murad, *J. Am. Chem. Soc.*, **80,** 5929 (1958).
[18] F. E. Blacet and A. Miller, *J. Am. Chem. Soc.*, **79,** 4327 (1957).

$$
\begin{array}{ccc}
\underset{\displaystyle\text{CH}_2\text{—}\text{CH}_2}{\overset{\displaystyle\text{CH}_2\underset{|}{}\text{CH}_2}{\text{CO}}} & \xrightarrow{h\nu} & \cdots
\end{array}
$$

CO
CH₂   CH₂ $\xrightarrow{h\nu}$ ·CO
CH₂   CH₂   CH₂   CH₂
    CH₂

·CO
CH₂   CH₂   →   ĊH₂   ĊH₂   →   ĊH₂   ĊH₂   + CO
CH₂   CH₂   CH₂   CH₂   CH₂   CH₂
    CH₂   CH₂   CH₂

CH₃   CH₂   CH₂——————CH₂
CH₂   CH   CH₂   CH₂
    CH₂   CH₂

Franck and Rabinowitsch pointed out that the photoexcitation of a small molecule may be less likely to result in dissociation in solution than in the gas phase because in the former case there is a greater chance that the excited molecule will be deactivated by collision before it dissociates.[19] The *Franck-Rabinowitsch principle* also explains why the gas-phase pairing of reactive radicals is often relatively inefficient. It might be thought that the large amount of energy given off by the pairing of electrons would make the pairing of radicals occur with great ease. This is often so (allowing for the fact that the concentration of reactive radicals can never be very large), but in the case of atoms and some very simple radicals it is just this energy that proves the reaction's undoing. The energy remains in the molecule and is usually not lost readily as a quantum of light. In a complex molecule the energy is distributed over the whole molecule, adding to the energy of a number of internal vibrations and rotations. It is very improbable that all the energy will be concentrated in one bond so as to cause its rupture before the molecule is energetically equilibrated with the rest of the system by collisions with other molecules and the walls of the container. In a diatomic molecule, however, the energy must be transferred by collision *very* quickly after the molecule is formed or it will dissociate again. For this reason most atom- and some simple radical-pairing reactions in the gas phase occur only at walls or at termolecular collisions where the third colliding molecule may absorb some of the energy given off.

19-2b. *Reaction of Organic Halides with Sodium Vapor.*[20] Von Hartel and Polanyi studied the vapor-phase reaction between sodium and certain organic halides by a technique that has been used a great deal subsequently. At elevated temperatures and reduced pressures they let

[19] J. Franck and E. Rabinowitsch, *Trans. Faraday Soc.*, **30**, 120 (1934).

[20] Much of the work on this subject has been summarized by E. Warhurst, *Quart. Revs. (London)*, **5**, 44 (1951).

sodium vapor in an inert carrier gas, such as hydrogen or nitrogen, flow through a nozzle into a reaction vessel containing an organic halide and carrier gas.[21]   The reaction appears to involve radical formation.   With hydrogen as a carrier gas methyl halides yield methane.   The best evidence for radical formation in this reaction is due to Horn, Polanyi, and Style, who used a reactor like that shown in Fig. 19-2, in which the effluent gases from the Na-RX reaction chamber, A, flowed into a second

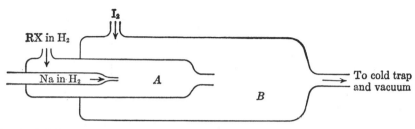

FIG. 19-2. Schematic diagram of apparatus for reaction of organic halides with sodium in the vapor phase.

vessel, B, containing iodine vapor.[22]   Under these conditions the reaction of methyl chloride yielded some methyl iodide.   The reaction scheme must have been

$$CH_3Cl + Na \rightarrow NaCl + CH_3 \cdot \xrightarrow{I_2} CH_3I$$

since methyl chloride does not react with iodine under the conditions used and since the methyl iodide cannot be formed from a sodium compound because all the sodium is used up in reaction vessel A, where it is deposited on the wall as sodium chloride.   No sodium salts are deposited in vessel B or subsequently.   Similar evidence has been obtained for the formation of the ethyl[22] and phenyl[23] radicals.

Von Hartel, Meer, and Polanyi determined the rate of reaction of a number of organic halides with sodium vapor.[21,24]   This was done by using a sodium lamp to illuminate the transparent apparatus at the point where the sodium vapor flowed into an excess of organic halide. Since the sodium vapor was visible under the sodium lamp, it was possible to estimate how far the average sodium atom penetrated the halide vapor before reaction.   From this distance the collision yield (the average number of collisions required to bring about reaction) was calculated.

[21] H. von Hartel and M. Polanyi, *Z. physik. Chem.*, **11B,** 97 (1930).

[22] E. Horn, M. Polanyi, and D. W. G. Style, *Z. physik. Chem.*, **23B,** 291 (1933); *Trans. Faraday Soc.*, **30,** 189 (1934).

[23] E. Horn and M. Polanyi, *Z. physik. Chem.*, **25B,** 151 (1934).

[24] H. von Hartel, N. Meer, and M. Polanyi, *Z. physik. Chem.*, **19B,** 139 (1932).

Some of the data obtained by these workers and by Warhurst[20] are shown in Table 19-1.

TABLE 19-1. REACTIVITIES OF ORGANIC HALIDES TOWARD SODIUM VAPOR
AT 275°C[20,21,24]

Organic halide	Collision yield[a]	Organic halide	Collision yield[a]
CH_3F	$>10^6$	$CHCl_3$	100
CH_3Cl	10,000	CCl_4	25
CH_3Br	50	$CH_2{=}CHCl$	11,000
CH_3I	1	$CH_2{=}CHCH_2Cl$	250
C_2H_5Cl	7,000	$C_6H_5CH_2Cl$	$\sim$1
$n\text{-}C_3H_7Cl$	4,400	CH_3OCH_2Cl	2,780
$n\text{-}C_4H_9Cl$	3,300	C_2H_5OCOCl	1,220
$n\text{-}C_5H_{11}Cl$	2,200	CH_3COCl	20
$i\text{-}C_3H_7Cl$	3,300	$C_2H_5OCOCH_2Cl$	36
$t\text{-}C_4H_9Cl$	1,500	CH_3COCH_2Cl	5
CH_2Cl_2	900	$HOCH_2CH_2Cl$	980

[a] Average number of collisions required to bring about reaction.

From the results shown it appears that radical stability is of importance in determining the reactivity but that polar effects (see Sec. 20-1c) are also present, electron-withdrawing groups increasing the reactivity toward the electron-donating sodium atom. No satisfying explanation has been advanced for the continued increase in reactivity with increasing length of the aliphatic chain.

19-2c. *Electrolysis.* It seems likely that a number of electrolysis reactions involve the formation of free radicals. Clusius and coworkers described some of the evidence that the first step of the Kolbe electrolysis of salts of carboxylic acids is the loss of an electron by the carboxylate anion.[25,26] The observed products in the case of sodium propionate, for example, are explained on the basis of the following reactions:

$$CH_3CH_2CO_2^{\ominus} \xrightarrow{-e} CH_3CH_2CO_2\cdot$$
$$CH_3CH_2CO_2\cdot \rightarrow CH_3CH_2\cdot + CO_2$$
$$2CH_3CH_2\cdot \rightarrow CH_2{=}CH_2 + CH_3CH_3 \tag{19-3}$$
$$2CH_3CH_2\cdot \rightarrow CH_3CH_2CH_2CH_3$$
$$CH_3CH_2CO_2\cdot + CH_3CH_2\cdot \rightarrow CH_3CH_2CO_2CH_2CH_3 \quad \text{etc.}$$

Such reactions as the methylation of trinitrotoluene, which takes place

[25] P. Hölemann and K. Clusius, *Z. physik. Chem.*, **35B**, 261 (1937); W. Schanzer and K. Clusius, *Z. physik. Chem.*, **190A**, 241 (1941); **192A**, 273 (1943).

[26] Cf. C. L. Wilson and W. T. Lippincott, *J. Am. Chem. Soc.*, **78**, 4290 (1956).

in the presence of electrolyzing sodium acetate,[27] probably involve the action of a methyl radical formed by decarboxylation of an acetoxy radical.

Free radicals also appear to be produced by the electrolysis of solutions of certain organometallic compounds. The electrolysis of solutions of ethylsodium in diethylzinc yields, at the anode, mostly ethane and ethylene but also some methane, propane, and butane.[28] The first two compounds are probably formed by disproportionation and the latter by dimerization of ethyl radicals, as shown in the third and fourth equations of scheme (19-3). No explanation for methane and propane formation is apparent. Additional evidence for the formation of ethyl radicals is provided by the fact that tetraethyllead is formed if a lead anode is used. In view of the ability of zinc to increase its coordination number beyond two, it seems likely that in diethylzinc solution ethylsodium may exist as Na^+ and $Zn (C_2H_5)_4^-$ ions.

Evans and coworkers explained the results of Grignard electrolysis experiments in terms of a free-radical mechanism.[29]

There is also evidence for the formation of free radicals in the bombardment of alkyl halides with neutrons,[30] alpha particles,[31] and X rays.[31]

19-3. Trapped Reactive Radicals. Lewis and coworkers found evidence that the ultraviolet irradiation of certain aromatic molecules trapped in "EPA" glass (a mixture of ether, isopentane, and ethyl alcohol cooled to liquid nitrogen temperatures) transforms these molecules to free radicals, which, because of their immobility, are unable to dimerize or disproportionate.[32] Related procedures enabling direct measurements to be made on trapped atoms, radicals, and highly reactive molecules were developed by Rice and coworkers and others.[33,34] Jen has reviewed the ESR studies that have been made on trapped methyl radicals.[35] After gamma irradiation at 4°K methane shows the ESR spectrum charac-

[27] L. F. Fieser, R. C. Clapp, and W. H. Daudt, *J. Am. Chem. Soc.*, **64**, 2052 (1942).

[28] F. Hein, E. Petzchner, K. Wagler, and F. A. Segitz, *Z. anorg. u. allgem. Chem.*, **141**, 161 (1924).

[29] W. V. Evans and coworkers, *J. Am. Chem. Soc.*, **56**, 654 (1934); **58**, 720, 2284 (1936); **61**, 898 (1939); **62**, 534 (1940); **63**, 2574 (1941).

[30] E. Glückauf and J. W. J. Fay, *J. Chem. Soc.*, 390 (1936); C. S. Lu and S. Sugden, *J. Chem. Soc.*, 1273 (1939).

[31] H. Eyring, J. O. Hirschfelder, and H. S. Taylor, *J. Chem. Phys.*, **4**, 479 (1936); C. B. Allsopp, *Trans. Faraday Soc.*, **40**, 79 (1944); J. Weiss, *Nature*, **153**, 748 (1944).

[32] G. N. Lewis et al., *J. Am. Chem. Soc.*, **64**, 2801 (1942); **65**, 520, 2419, 2424 (1943); **66**, 1579, 2100 (1944).

[33] F. O. Rice et al., *J. Am. Chem. Soc.*, **73**, 5529 (1951); **75**, 548 (1953); **77**, 291 (1955); **81**, 1856 (1959).

[34] A. M. Bass and H. P. Broida (eds.), "Formation and Trapping of Free Radicals," Academic Press, Inc., New York, 1960.

[35] C. K. Jen, chap. 7, sec. V*A*, in Ref. 34.

teristic of small but equal amounts of methyl radicals and hydrogen atoms.[36,37] ESR absorption bands with the same shape as those attributed to the methyl radicals were also found in the frozen products of the exposure of gaseous methane to electric discharge.[38] The formation of deuterium- and C^{13}-labeled methyl radicals by similar methods gave products whose ESR spectra differed in the expected manner.

19-4. Bond-dissociation Energies. As pointed out in Sec. 1-4c, a bond-dissociation energy is the energy required for the homolytic cleavage of a given bond in a molecule

$$R\text{—}X \rightarrow R\cdot + X\cdot \tag{19-4}$$

and, for polyatomic molecules, is not the same as the average bond energy. The determination of the energy of reaction (19-4) for the same X but different R's gives the relative stabilities of the different R's as

TABLE 19-2. DISSOCIATION ENERGIES (D) OF R—X IN KILOCALORIES PER MOLE[a]

Bond	D	Bond	D
CH_3—H	101	CH_3S—H	89
C_2H_5—H	98	CH_3—SH	74
$(CH_3)_2CH$—H	89?	CH_3S—SCH_3	73
$(CH_3)_3C$—H	85?	CH_3—Br	68
$C_6H_5CH_2$—H	78	$BrCH_2$—Br	63
CH_3—CH_3	83	Br_2CH—Br	56
HO—H	118	Br_3C—Br	49
OC=O	127	F_3C—Br	64
HO—OH	54	Br_3C—H	93
H_2N—H	102	CH_3—CHO	75
H_2N—NH_2	54	CH_3CO—$COCH_3$	60
HS—H	95	CH_3—NH_2	79

[a] From the tables of Steacie,[7] pp. 95–98.

radicals, relative to the stability of the RX's. Except for varying amounts of steric, resonance, and polar interactions between R and X (or Y) these relative stabilities of R·'s will be the same when referred to species in which the R is bonded to some other group (as in RY). For this reason the accompanying collection of bond-dissociation energies (Table 19-2) is of considerable value in discussions of free-radical reactions. The uncertainties are on the order of several kilocalories per mole for most of the values listed, but the data are sufficient to show the

[36] B. Smaller and M. S. Matheson, *J. Chem. Phys.*, **28**, 1169 (1958).

[37] L. A. Wall, D. W. Brown, and R. E. Florin, *J. Phys. Chem.*, **63**, 1762 (1959).

[38] C. K. Jen, S. N. Foner, E. L. Cochran, and V. A. Bowers, *Phys. Rev.*, **112**, 1169 (1958).

greater stability of tertiary, benzyl, α-halo, and certain other types of radicals. Methods for the determination of bond-dissociation energies have been discussed critically by Steacie,[7] Cottrell,[39] and others.

PROBLEMS

1. The photolysis of methyl acetate at 29° was carried out so that less than 5 per cent of the methyl acetate was used up. The products found were biacetyl, acetone, methanol, dimethyl ether, ethane, methane, carbon monoxide, carbon dioxide, ketene, formaldehyde, methyl methoxyacetate, and methyl propionate. Suggest a reasonable mechanism for the reaction.

2. According to Table 19-2, the dissociation energy of the CH_3—H bond is 16 kcal larger than that of the $(CH_3)_3C$—H bond. Would you expect the dissociation energy of the CH_3—Cl bond to be more than 16 kcal larger, almost exactly 16 kcal larger, or less than 16 kcal larger than that of the $(CH_3)_3C$—Cl bond? Why?

[39] T. L. Cottrell, "The Strengths of Chemical Bonds," Butterworth & Co. (Publishers), Ltd., London, 1954.

Chapter 20

ADDITION OF FREE RADICALS TO OLEFINS

20-1. Vinyl Polymerization.[1] *20-1a. Mechanism of Vinyl Polymerization.* Since the addition of a carbonium ion, carbanion, or free radical to an olefin should yield another intermediate of the same type, it is not surprising that vinyl polymerization may be catalyzed by either acids, bases, or free radicals.

$$R^{\oplus} + CH_2{=}CHX \rightarrow RCH_2\overset{\oplus}{C}HX \xrightarrow{CH_2{=}CHX} RCH_2CHXCH_2\overset{\oplus}{C}HX \rightarrow etc.$$

$$R^{\ominus} + CH_2{=}CHX \rightarrow RCH_2\overset{\ominus}{C}HX \xrightarrow{CH_2{=}CHX} RCH_2CHXCH_2\overset{\ominus}{C}HX \rightarrow etc.$$

$$R\cdot + CH_2{=}CHX \rightarrow RCH_2\overset{\cdot}{C}HX \xrightarrow{CH_2{=}CHX} RCH_2CHXCH_2\overset{\cdot}{C}HX \rightarrow etc.$$

Of these three possible methods of polymerization the free-radical type is by far the most important.

There is good evidence that certain vinyl polymerizations proceed by a free-radical mechanism. Schulz and Wittig[2] found that tetraphenylsuccinonitrile, known to be about 1 per cent dissociated into free radicals, initiates the polymerization of styrene. The kinetics of the polymerization could be explained readily by a free-radical mechanism. Vinyl polymerizations have been found to be brought about by benzoyl peroxide,[3] triphenylmethylazobenzene,[4] *N*-nitrosoacylarylamines,[5] and several aliphatic azo compounds.[6] There is other evidence that all these compounds decompose readily by a free-radical mechanism. Schulz showed that in the tetraphenylsuccinonitrile-initiated polymerization of styrene and methyl methacrylate[7] and the triphenylmethylazobenzene-initiated polymerization of styrene[4,7] approximately one polymer chain

---

[1] For a more complete discussion of this topic, see C. Walling, "Free Radicals in Solution," chaps. 3, 4, and 5, John Wiley & Sons, Inc., New York, 1957.

[2] G. V. Schulz and G. Wittig, *Naturwissenschaften*, **27**, 387 (1939).

[3] G. V. Schulz and E. Husemann, *Z. physik. Chem.*, **39B**, 246 (1938).

[4] G. V. Schulz, *Naturwissenschaften*, **27**, 659 (1939).

[5] A. T. Blomquist, J. R. Johnson, and H. J. Sykes, *J. Am. Chem. Soc.*, **65**, 2446 (1943).

[6] F. M. Lewis and M. S. Matheson, *J. Am. Chem. Soc.*, **71**, 747 (1949).

[7] G. V. Schulz, *Z. Elektrochem.*, **47**, 265 (1941).

was formed per molecule of initiator used. The actual participation of the initiator in the reaction was shown by the detection of chlorine in the polystyrene formed by chloroacetyl peroxide[8] and p-chlorobenzoyl peroxide initiation,[9] and of bromine in the polystyrene formed using p-bromobenzoyl peroxide,[8,9] 3,4,5-tribromobenzoyl peroxide,[10] and m-bromobenzoyl peroxide.[11] Fraenkel, Hirshon, and Walling even identified free radicals formed during vinyl polymerization directly by electron-spin-resonance spectroscopy.[12] A number of structural studies on polymers[13] verified the mechanistically probable hypothesis that vinyl monomers undergo, at least predominantly, head-to-tail polymerization as shown above rather than head-to-head and tail-to-tail polymerization to give

$$\cdots CH_2CHX\text{---}CHXCH_2\text{---}CH_2CHX\text{---}CHXCH_2 \cdots$$

Probably the principal reason why radicals almost always add to CH_2= CHX to form the new radical on the substituted carbon atom is that practically all X groups stabilize radicals better than hydrogen does.

Free-radical polymerizations have mechanisms of the following general type:

$$I \rightarrow R\cdot \qquad (20\text{-}1)$$

$$R\cdot + CH_2\text{=}CHX \rightarrow RCH_2\dot{C}HX$$
$$RCH_2\dot{C}HX + CH_2\text{=}CHX \rightarrow R(CH_2CHX)_2\cdot \qquad (20\text{-}2)$$
$$R(CH_2CHX)_n\cdot + CH_2\text{=}CHX \rightarrow R(CH_2CHX)\cdot_{n+1}$$

$$R(CH_2CHX)_n\cdot + R\cdot \rightarrow R(CH_2CHX)_nR$$
$$R(CH_2CHX)_n\cdot + R(CH_2CHX)_m\cdot \rightarrow R(CH_2CHX)_n(CHXCH_2)_mR$$
$$\downarrow \qquad\qquad\qquad (20\text{-}3)$$
$$R(CH_2CHX)_nH + R(\dot{C}H_2CHX)_{m-1}CH\text{=}CHX$$
$$R(CH_2CHX)_n\cdot + HS \rightarrow R(CH_2CHX)_nH + S\cdot$$

The reaction starts by what is known as the *initiation step* (20-1) in which the *initiator*, I, in some way yields a free radical (R·). In the *propagation*, or *chain-carrying*, steps (20-2) the free radical adds to an olefinic double bond to yield a new radical, which then adds to another

---

[8] C. C. Price, R. W. Kell, and E. Krebs, *J. Am. Chem. Soc.*, **64**, 1103 (1942).

[9] P. D. Bartlett and S. G. Cohen, *J. Am. Chem. Soc.*, **65**, 543 (1943).

[10] C. C. Price and B. E. Tate, *J. Am. Chem. Soc.*, **65**, 517 (1943).

[11] H. F. Pfann, D. J. Salley, and H. Mark, *J. Am. Chem. Soc.*, **66**, 983 (1944).

[12] G. K. Fraenkel, J. M. Hirshon, and C. Walling, *J. Am. Chem. Soc.*, **76**, 3606 (1954).

[13] C. S. Marvel and E. C. Horning, in H. Gilman, "Organic Chemistry," 2d ed., vol. 1, chap. 8, John Wiley & Sons, Inc., New York, 1943, and references given therein; cf. C. S. Marvel, E. D. Weil, L. B. Wakefield, and C. W. Fairbanks, *J. Am. Chem. Soc.*, **75**, 2326 (1953).

molecule of olefin, etc., so that the polymer chain may become hundreds of monomer units long. This process may be interrupted by any of several types of *termination steps* (20-3). In the initiation step a reactive radical is created; in each propagation step one radical is used up and one is formed with no net change; whereas in the termination step one or more reactive radicals are destroyed. Termination may occur by the pairing of two radicals or by a disproportionation reaction between two radicals. In addition, there may be present in the reaction mixture an *inhibitor*, HS, that is very reactive toward free radicals. This inhibitor may react with a *growing-chain radical* to yield a radical, S·, that is relatively incapable of adding to a monomer molecule, and that thereby inhibits polymerization. Since the propagation step may occur hundreds of times to every occurrence of an initiation, termination, or inhibition step, the free-radical chain mechanism gives a very satisfactory explanation of how a few molecules of initiator may cause the polymerization of a large number of molecules of monomer, and similarly how a small amount of inhibitor may prevent the polymerization of a large amount of monomer. The action of inhibitor also explains the existence of an *induction period*, a period at the beginning of the reaction when no polymerization occurs, due to inhibition. Since the inhibitor is being used up during this period, the reaction begins to pick up speed as the inhibitor supply becomes exhausted.

Most of the italicized terms in the preceding paragraph are used generally not only in discussions of polymerization but in all types of chain-reaction mechanisms.

Many of the effects of changes in reactants and reaction conditions on the properties of the polymer produced may be explained in terms of a mechanism of the type just described. For example, an increase in the concentration of initiator causes a decrease in the average molecular weight of the resultant polymer. The increase in initiator concentration causes an increase in the number of growing-chain radicals. Since the rate of the polymerization (propagation) reaction is first-order in these radicals, there is an increase in polymerization rate. However, since the termination reaction is second-order in these radicals, its rate increases faster, and thus the average length of the polymer chains is shorter.[3]

The chain length, and hence molecular weight, of polymers formed by a free-radical mechanism may also be influenced by a process known as *chain transfer*. This reaction involves the abstraction, by the radical end of a growing-chain polymer, of an atom from another molecule to terminate the polymer chain but to create a new radical capable of bringing about further polymerization. The net result is that there may be a decrease in the average molecular weight of the polymer but no necessary decrease in the rate of polymerization. Chain-transfer agents (mercaptans are often used) are frequently introduced deliberately in order to control the molecular weight of the polymer and are sometimes called *regulators*. However, chain transfer may occur to many of the solvents in which polymerization has been studied, to unchanged monomer, or to previously formed polymer chains. In the latter case, of course, no decrease in average molecular weight occurs, since the process terminates one chain but adds to the length of another.

20-1b. *Absolute Rate Constants for Vinyl Polymerization.* Let us consider the mechanism of the benzoyl peroxide–catalyzed polymerization of vinyl acetate.[14] The benzoyl peroxide (P) acts as an initiator by decomposing to yield two free radicals (B) (Sec. 21-1a). We shall assume that essentially all these radicals add to vinyl acetate monomer molecules (M) to form a growing-chain radical (R), that the rate constant for the addition of the growing-chain radical to monomer is independent of the chain length, and that termination is due to a bimolecular reaction between growing-chain radicals.

$$P \xrightarrow{k_1} 2B$$
$$B + M \to R$$
$$R + M \xrightarrow{k_2} R$$
$$R + R \xrightarrow{k_3} R \quad R \qquad \text{or disproportionation products}$$

Since vinyl acetate forms a polymer with fairly long chains, we may describe its polymerization rate, as measured by the rate of disappearance of monomer, by the equation

$$-\frac{dM}{dt} = k_2 RM \tag{20-4}$$

ignoring the relatively small amount of monomer used up by reaction with the initiating radical B. Since the growing-chain radical should be quite reactive, we may use the steady-state treatment (Sec. 3-1a), in which we equate its rate of formation to its rate of disappearance. Having postulated that every initiating radical forms a growing-chain radical, we may say

$$k_1 P = k_3 R^2 \qquad \text{or} \qquad R = \sqrt{\frac{k_1 P}{k_3}}$$

Substituting in (20-4),

$$-\frac{dM}{dt} = k_2 \sqrt{\frac{k_1}{k_3}} MP^{1/2} \tag{20-5}$$

[14] K. Nozaki and P. D. Bartlett, *J. Am. Chem. Soc.*, **68**, 2377 (1946).

Nozaki and Bartlett[14] found that the reaction is first-order in vinyl acetate and one-half-order in benzoyl peroxide, as required by Eq. (20-5) and the mechanism leading to it. The rate constant they obtain for the reaction is equal to $k_2 \sqrt{k_1/k_3}$ in terms of the rate constants for individual steps. From separate experiments on benzoyl peroxide the value of k_1 may be determined. From this a value of $k_2/\sqrt{k_3}$ may be determined, but by none of the ordinary methods of measuring polymerization rates has it been possible to obtain a rate constant (like k_2 or k_3) governing a reaction of a reactive growing-chain radical.

However, by use of a special technique known as the *rotating-sector method*, Melville determined the average life of intermediate radicals in the vapor-phase polymerization of methyl methacrylate.[15] It appears that this method was first applied accurately to a liquid-phase polymerization by Bartlett and Swain,[16] who studied the radical-induced polymerization of vinyl acetate. When the rotating-sector method is employed, an initiator is used whose decomposition to radicals is light-catalyzed. The reaction vessel is illuminated by a beam of the catalyzing light. This beam is required to pass through the area swept by a rotating opaque disk from which a sector has been cut. Thus by rotation of the disk the reaction mixture may be subjected to alternate periods of light and dark of known duration. Since the rate of decomposition of the initiator is directly proportional to the light intensity, we see from Eq. (20-5) that the polymerization rate will be proportional to the square root of the light intensity.

Let us consider the case where one-fourth of the opaque disk has been cut out, so that the dark periods are three times as long as the light periods. When the disk is rotated slowly so that the periods of light and dark are relatively long, the concentration of growing-chain radicals (and therefore the rate, which is proportional to this concentration) will quickly increase at the beginning of a light period until the steady state is reached, where the rate of termination has become equal to the rate of initiation. At the beginning of a dark period the rate will quickly fall to zero. This situation is depicted in the plot of rate (or concentration of growing chains) vs. time shown in Fig. 20-1a. When the disk is rotated more rapidly (Fig. 20-1b), the light period is too short for the reaction rate to approach the steady state closely and the dark period is too short for the rate to fall very close to zero. As the disk-rotation

[15] H. W. Melville, *Proc. Roy. Soc. (London)*, **163A,** 511 (1937).

[16] P. D. Bartlett and C. G. Swain, *J. Am. Chem. Soc.*, **67,** 2273 (1945); **68,** 2381 (1946); cf. G. M. Burnett and H. W. Melville, *Nature*, **156,** 661 (1945); *Proc. Roy. Soc. (London)*, **189A,** 456 (1947); G. M. Burnett, H. W. Melville, and L. Valentine, *Trans. Faraday Soc.*, **45,** 960 (1949); M. S. Matheson, E. E. Auer, E. B. Bevilacqua, and E. J. Hart, *J. Am. Chem. Soc.*, **71,** 2610 (1949); H. Kwart, H. S. Broadbent, and P. D. Bartlett, *J. Am. Chem. Soc.*, **72,** 1060 (1950).

rate continues to increase, we begin (Fig. 20-1c) to approach the situation in which the light is uninterrupted but only one-fourth as intense. Under these conditions the reaction will proceed one-half as fast as during the light periods with slow disk rotation.   Thus, from comparison of Figs. 20-1a and c, we see that as the rate of disk rotation is increased,

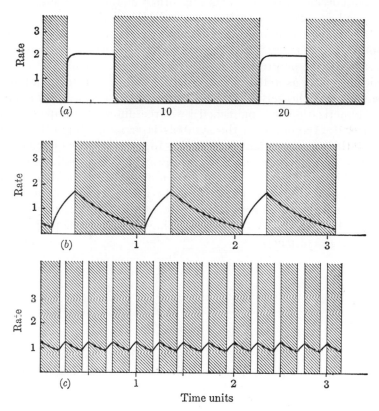

FIG. 20-1. Rate of polymerization plotted against time for various speeds of the rotating disk.   Shaded areas represent dark periods; unshaded areas represent periods of illumination.

the average polymerization rate over a full cycle of light and dark will double.   Flory had shown that the rate of approach to the steady state is directly related to the ratio k_3/k_2.[17]   Therefore k_3/k_2 may be determined from the rate of disk rotation at which the polymerization rate is the mean between the rate it approaches at slow rotation and that which it approaches at rapid rotation.   From this value of k_3/k_2 and the value of $k_2/\sqrt{k_3}$, determined from ordinary methods of following

[17] P. J. Flory, *J. Am. Chem. Soc.*, **59**, 241 (1937).

polymerization rates, the individual values of k_2 and k_3 are obtained. The value of k_2 appears to be about 10^3 liters mole^{-1} sec^{-1} and that of k_3 about 3×10^7 liters mole^{-1} sec^{-1} for vinyl acetate at 25°.[16]

The rotating-sector method has been used to determine the absolute rate constants and steady-state concentration of growing-chain radicals for other polymerizations. This and other methods for obtaining such data have been reviewed by Burnett and Melville.[18]

20-1c. *Copolymerization*.[19] The polymerization of a mixture of two or more monomers, giving a polymer chain in which both monomers become units, is of great interest from the standpoint of theoretical organic chemistry as well as being of considerable commercial importance.

A satisfactory theoretical basis for comparing the behavior of monomers in copolymerization was published by three different groups of workers in 1944.[20-22] Expressed in the symbols later agreed upon by all three groups,[23] this treatment may be stated as follows: In the copolymerization of the two monomers M_1 and M_2, the growing-chain radicals can be of only two kinds, those ending in an M_1 unit (designated $M_1\cdot$) and those ending in an M_2 unit ($M_2\cdot$). If we make the reasonable approximation that the reactivity of the growing-chain radical depends only upon the nature of the radical end, the polymer chain is seen to be propagated by only four types of reactions.

$$
\begin{aligned}
M_1\cdot + M_1 &\xrightarrow{k_{11}} M_1\cdot \\
M_1\cdot + M_2 &\xrightarrow{k_{12}} M_2\cdot \\
M_2\cdot + M_1 &\xrightarrow{k_{21}} M_1\cdot \\
M_2\cdot + M_2 &\xrightarrow{k_{22}} M_2\cdot
\end{aligned}
\tag{20-6}
$$

In the formation of polymers of relatively high molecular weight the amount of monomer consumed in initiation may be neglected. Thus the rate of disappearance of M_1 and M_2 is

$$
-\frac{d[M_1]}{dt} = k_{11}[M_1\cdot][M_1] + k_{21}[M_2\cdot][M_1]
$$

$$
-\frac{d[M_2]}{dt} = k_{12}[M_1\cdot][M_2] + k_{22}[M_2\cdot][M_2]
$$

---

[18] G. M. Burnett, *Quart. Revs. (London)*, **4**, 292 (1950); G. M. Burnett and H. W. Melville, *Chem. Revs.*, **54**, 225 (1954).

[19] For comprehensive and authoritative reviews of this topic, see T. Alfrey, J. J. Bohrer, and H. Mark, "Copolymerization," Interscience Publishers, Inc., New York, 1952, and Ref. 1, chap. 4.

[20] T. Alfrey and G. Goldfinger, *J. Chem. Phys.*, **12**, 205 (1944).

[21] F. R. Mayo and F. M. Lewis, *J. Am. Chem. Soc.*, **66**, 1594 (1944).

[22] F. T. Wall, *J. Am. Chem. Soc.*, **66**, 2050 (1944).

[23] T. Alfrey, F. R. Mayo, and F. T. Wall, *J. Polymer Sci.*, **1**, 581 (1946).

and the relative rate of disappearance is

$$\frac{d[M_1]}{d[M_2]} = \frac{k_{11}[M_1\cdot][M_1] + k_{21}[M_2\cdot][M_1]}{k_{12}[M_1\cdot][M_2] + k_{22}[M_2\cdot][M_2]} \tag{20-7}$$

We may assume that a steady state is soon reached at which the rate of conversion of $M_1\cdot$ radicals to $M_2\cdot$ radicals is equal to the rate of the reverse reaction.

$$k_{12}[M_1\cdot][M_2] = k_{21}[M_2\cdot][M_1] \tag{20-8}$$

Substituting the value of $[M_1\cdot]$ from (20-8) into (20-7) and multiplying numerator and denominator by $[M_2]/k_{21}[M_2\cdot]$, we obtain

$$\frac{d[M_1]}{d[M_2]} = \frac{[M_1]}{[M_2]} \frac{(k_{11}/k_{12})[M_1] + [M_2]}{(k_{22}/k_{21})[M_2] + [M_1]}$$

If we define r_1 as k_{11}/k_{12} and r_2 as k_{22}/k_{21}, this becomes

$$\frac{d[M_1]}{d[M_2]} = \frac{[M_1]}{[M_2]} \frac{r_1[M_1] + [M_2]}{r_2[M_2] + [M_1]} \tag{20-9}$$

The constants r_1 and r_2 are referred to as *monomer reactivity ratios*, since each is the ratio of the rate constant for the combination of the given type of radical with the corresponding monomer to its rate constant for combination with the other monomer.

Equation (20-9), which describes the relative rate at which two monomers enter the polymer chain in terms of their concentrations and of their monomer reactivity ratios, may be used in the differential form shown when the relative concentration of the monomers does not change greatly. The integration of (20-9) yields a more generally applicable (but considerably more complex) expression. The use of the integrated expression to obtain values for r_1 and r_2 is described by Mayo and Lewis[21] for the copolymerization of styrene and methyl methacrylate. The equations fit the experimental results quite satisfactorily, and Alfrey, Bohrer, and Mark have tabulated monomer reactivity ratios for 275 different pairs of monomers.[19] Equations developed to treat the copolymerization of three[24] and of n monomers[25] have also been found to give satisfactory results.[25,26]

In the case of the copolymerization of styrene and methyl methacrylate, Mayo and Lewis[21] obtained the values

$$r_1 = 0.50 \pm 0.02 \qquad r_2 = 0.50 \pm 0.02$$

These values mean that a growing-chain radical ending with a styrene

---

[24] T. Alfrey and G. Goldfinger, *J. Chem. Phys.*, **12**, 322 (1944).

[25] C. Walling and E. R. Briggs, *J. Am. Chem. Soc.*, **67**, 1774 (1945).

[26] E. C. Chapin, G. E. Ham, and R. G. Fordyce, *J. Am. Chem. Soc.*, **70**, 538 (1948).

unit has twice the tendency to combine with methyl methacrylate molecules that it has to combine with styrene. The methyl methacrylate–type radicals, on the other hand, show a preference for combination with styrene. Thus there will be a considerable tendency for the units in the copolymer chain to alternate between styrene and methyl methacrylate (the *alternating effect*). It seems reasonable to expect that the rate at which radicals combine with a given monomer to form a new radical would depend upon the stability of the radical being formed. For the values of r_1 and r_2 found, however, it is obvious that this is, at least, not the *only* factor of importance. If it were, both types of radicals would prefer to react with the same monomer, and this preference would be of the same magnitude for both; i.e., it would be expected that $k_{11}/k_{12} = k_{21}/k_{22}$ or $r_1r_2 = 1$. Actually r_1r_2 is usually considerably less than unity and is never significantly more. In other words, the tendency of radical $M_1\cdot$ to react with M_1 as compared with its tendency to react with M_2 is never more and usually less than would have been predicted from the relative reactivities of M_1 and M_2 toward radical $M_2\cdot$. The accumulation of a large number of monomer reactivity ratios led to the explanation of these facts by the development of a theory of polar effects on free-radical reactions, which has proved very useful in the interpretation of data on other radical reactions. Mayo and Walling pointed out a few of the areas of applicability.[19]

The importance of polar factors was first discussed by Price,[27] who pointed out that the magnitude of the alternating effect in copolymerization increases with an increasing difference in the electron-donating ability of the substituents on the two monomers. Alfrey and Price described the copolymerization behavior of olefins in terms of two factors: Q, a measure of the stability of the radical formed when the olefin adds to a growing polymer chain, and e, a measure of the sum of the electron-withdrawing power of the substituents on the double bond.[28] Each olefin has its characteristic value of Q, the general monomer reactivity factor, and e, the polar factor. From these two values for two olefins there may be calculated, by the equations of Alfrey and Price, values to be expected for r_1 and r_2 in copolymerization. The agreement of these r values with those determined experimentally is only fair. At least part of the deviation from experimental values is no doubt due to the complete neglect of steric factors. From a tabulation of the Q and e values that have been obtained[19] it appears that radical-stabilizing ability varies approximately as follows: $p\text{-Me}_2\text{NC}_6\text{H}_4 \sim p\text{-O}_2\text{NC}_6\text{H}_4 > \text{C}_6\text{H}_5 > \text{CN} \sim \text{CO}_2\text{Me} > \text{Cl} \sim \text{Me} > \text{H}$, while as electron withdrawers,

[27] C. C. Price, *J. Polymer Sci.*, **1**, 83 (1946).

[28] T. Alfrey and C. C. Price, *J. Polymer Sci.*, **2**, 101 (1947); C. C. Price, *J. Polymer Sci.*, **3**, 772 (1948).

$CN > CO_2Me > p\text{-}O_2NC_6H_4 > Cl > H > Me > C_6H_5 > p\text{-}Me_2NC_6H_4.$
Instead of attributing the polar effect to a coulombic attraction between two dipoles, as Alfrey and Price do, Walling and Mayo[29] suggest that it is due to stabilization of the transition state by resonance structures involving electron donation and acceptance. For example, in the transition state of the addition of a methyl acrylate molecule to a growing polymer chain ending in a styrene unit, the following resonance structure may be considered:

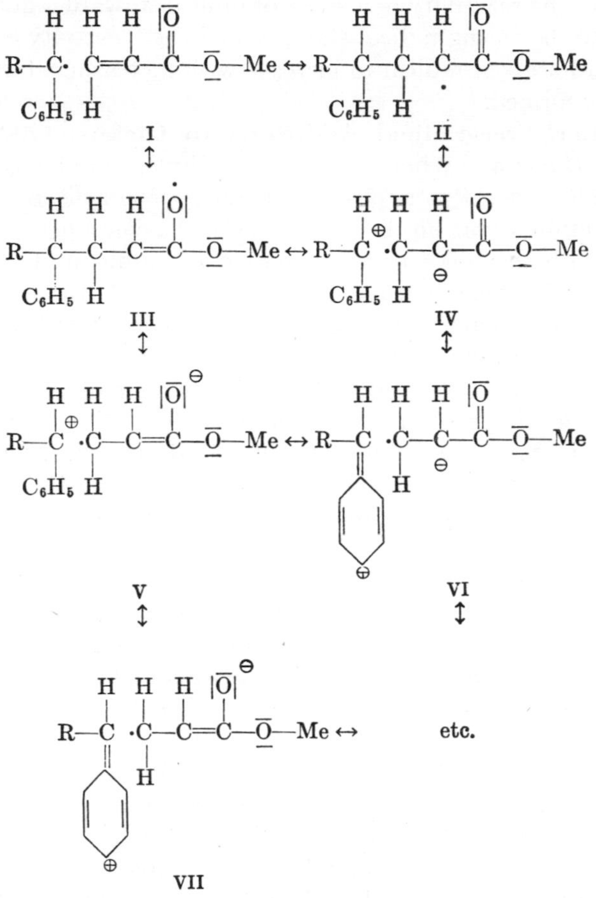

$$
\begin{array}{ccc}
& \text{H H H } |\bar{O} & \\
\text{R—C· C=C—C—O—Me} & \leftrightarrow & \text{R—C—C—C—C—O—Me} \\
\text{C}_6\text{H}_5 \text{ H} & & \text{C}_6\text{H}_5 \text{ H} \quad \cdot \\
\text{I} & & \text{II}
\end{array}
$$

$$
\begin{array}{ccc}
\text{H H H } |O| & & \text{H H H } |\bar{O} \\
\text{R—C—C—C=C—O—Me} & \leftrightarrow & \text{R—C } \overset{\oplus}{\cdot}\text{C—C—C—O—Me} \\
\text{C}_6\text{H}_5 \text{ H} & & \text{C}_6\text{H}_5 \text{ H} \quad \ominus \\
\text{III} & & \text{IV}
\end{array}
$$

$$
\begin{array}{ccc}
\text{H H H } |\bar{O}|^{\ominus} & & \text{H H H } |\bar{O} \\
\text{R—}\overset{\oplus}{\text{C}} \cdot\text{C—C=C—O—Me} & \leftrightarrow & \text{R—C } \cdot\text{C—C—C—O—Me} \\
\text{C}_6\text{H}_5 \text{ H} & & \text{H} \quad \ominus \\
\text{V} & & \text{VI}
\end{array}
$$

$$
\begin{array}{cc}
\text{H H H } |\bar{O}|^{\ominus} & \\
\text{R—C } \cdot\text{C—C=C—O—Me} \leftrightarrow & \text{etc.} \\
\text{H} & \\
\text{VII} &
\end{array}
$$

Structures like I, II, and IV may be written for the addition of any olefin to any free radical, and those of the type of III and V could explain the reactivity of methyl acrylate toward the addition of free radicals in general. Structures like VI could contribute in the addition of styrene-type radicals to any olefin. It is structures like VII, and similar ones,

[29] C. Walling and F. R. Mayo, *J. Polymer Sci.*, **3**, 895 (1948).

that explain the particularly great reactivity of electron-donor radicals toward electron-acceptor olefins. If we define an electron-donor radical as one that could form an especially stable cation upon loss of its unpaired electron and an electron-acceptor radical as one that forms an especially stable anion upon gaining an electron, we may say that added reactivity will be found at any time a donor radical adds to an olefin to form an acceptor radical or vice versa. This added reactivity is in addition to that which would be expected from radical-stability considerations. We should expect the reactivity of a series of radicals toward a given olefin to decrease with increasing radical stability and the reactivity of a series of olefins toward a given radical to increase with increasing stability of the radical being formed.

20-2. Other Free-radical Additions to Olefins. *20-2a. Radical Additions of Halogen.* When bromine and ethylene are mixed in the gas phase in a glass vessel, it appears that much, if not all, of the ensuing addition reaction occurs on the glass walls of the container, since coating these walls with paraffin decreases the reaction rate more than tenfold.[30] An even larger difference is found in the reaction of chlorine and ethylene.[31] The reaction goes faster when the container is coated with cetyl alcohol than when it is paraffin-coated and still faster when coated with stearic acid. From this fact, together with the fact that small amounts of water vapor catalyze the reaction in glass (on which the water could be adsorbed) but not in a paraffin-coated vessel, it appears most likely that this surface reaction has a polar mechanism. In investigating reaction rates and mechanisms, it is important to keep the possibility of surface reactions in mind.

However, despite the frequency of polar addition, there is good evidence that it is often possible to bring about the addition of halogen to olefinic double bonds by a free-radical chain reaction. For example, the addition of chlorine to tetrachloroethylene in the vapor phase[32] and in carbon tetrachloride solution[33] has been found to be catalyzed by light (with high quantum yields) and strongly inhibited by oxygen. Bromine and cinnamic acid in carbon tetrachloride solution combine very rapidly, even in the dark, if oxygen is excluded. In the presence of oxygen the reaction is very slow in the dark but occurs readily in the light.[34] The addition of chlorine to benzene[35] and of bromine to phenanthrene[36] are also light-catalyzed, and the latter is slowed by certain inhibitors. These data

[30] R. G. W. Norrish, *J. Chem. Soc.*, **123**, 3006 (1923).

[31] R. G. W. Norrish and G. G. Jones, *J. Chem. Soc.*, 55 (1926).

[32] R. G. Dickinson and J. L. Carrico, *J. Am. Chem. Soc.*, **56**, 1473 (1934).

[33] R. G. Dickinson and J. A. Leermakers, *J. Am. Chem. Soc.*, **54**, 3852, 4648 (1932).

[34] W. H. Bauer and F. Daniels, *J. Am. Chem. Soc.*, **56**, 2014 (1934).

[35] H. P. Smith, W. A. Noyes, Jr., and E. J. Hart, *J. Am. Chem. Soc.*, **55**, 4444 (1933).

[36] C. C. Price, *J. Am. Chem. Soc.*, **58**, 1835 (1936).

may be explained by a mechanism of the type

$$Cl_2 \xrightarrow{h\nu} 2Cl\cdot$$
$$Cl\cdot + Cl_2C{=}CCl_2 \rightarrow Cl_3C{-}\overset{\cdot}{C}Cl_2$$
$$Cl_3C{-}\overset{\cdot}{C}Cl_2 + Cl_2 \rightarrow Cl_3C{-}CCl_3 + Cl\cdot$$

That oxygen functions as an inhibitor by combining with the chain-propagating pentachloroethyl radicals is evidenced by the production of trichloroacetyl chloride and phosgene in the oxygen-inhibited chlorination of tetrachloroethylene.[33]

Miller and coworkers obtained strong evidence that fluorine reacts with olefins by a free-radical mechanism without the necessity of an initiator.[37] A mixture of oxygen and fluorine was found to react with tetrachloroethylene to give the oxygenated products $Cl_2CFCOCl$, Cl_3CCOCl, and $COCl_2$ in high yield in the dark at 0°. The thermal cleavage of fluorine molecules (of bond energy 37 kcal) would be too slow to lead to the observed reaction. It seems more plausible that the reaction is initiated by the formation of a fluorine atom and a haloalkyl radical from a molecule of olefin and one of fluorine.

$$Cl_2C{=}CCl_2 + F{-}F \rightarrow Cl_2\overset{\cdot}{C}CCl_2F + F\cdot$$

It should be noted that the new C—F bond created in this process has a large enough bond energy (102 kcal) to compensate for the cleavage of the F—F bond (37 kcal) and the transformation of a double bond to a single bond ($\sim$ 63 kcal). With any of the other halogens such a process would be endothermic by at least 20 kcal.

20-2b. Radical Additions of Hydrogen Halides. The addition of hydrogen bromide to unsymmetrical olefins is a reaction that is of particular interest because it was used extensively in testing older theories of physical organic chemistry. Many anomalous results obtained earlier were explained by the observation of Kharasch and Mayo that, although the addition of hydrogen bromide to allyl bromide under "ordinary" conditions occurs rapidly to yield largely 1,3-dibromopropane, the reaction of the carefully purified reactants in the absence of air occurs slowly and yields almost entirely 1,2-dibromopropane.[38] That the predominant formation of the 1,3-dibromide is due to the presence of peroxides was shown by the following facts: On standing in the presence of air in either the light or dark, allyl bromide becomes contaminated with peroxides. This peroxide-containing allyl bromide reacts rapidly and yields largely 1,3-dibromide in either the presence or absence of air. Even carefully purified allyl bromide in the absence of air may be made

[37] W. T. Miller, Jr., et al., *J. Am. Chem. Soc.*, **78**, 2793 (1956); **79**, 3084 (1957).
[38] M. S. Kharasch and F. R. Mayo, *J. Am. Chem. Soc.*, **55**, 2468 (1933).

to yield largely the 1,3-dibromide if a little benzoyl peroxide is added. It seems likely that hydrogen bromide does add to allyl bromide by a polar mechanism to give mostly 1,2-dibromopropane. Although the presence of peroxides has no effect upon this reaction, the peroxides serve as initiators for a rapid competing free-radical chain reaction that produces largely 1,3-dibromopropane. In agreement with this explanation are the facts that the formation of 1,3-dibromide is largely eliminated by inhibitors such as hydroquinone, diphenylamine, etc., and that the formation of the 1,3-dibromide is catalyzed by light.[38] The mechanism suggested for the free-radical part of the reaction is as follows:[39,40]

$$\text{Peroxide} \rightarrow \text{R·}$$
$$\text{R·} + \text{HBr} \rightarrow \text{RH} + \text{Br·}$$
$$\text{Br·} + \text{CH}_2{=}\text{CHCH}_2\text{Br} \rightarrow \text{BrCH}_2\dot{\text{C}}\text{HCH}_2\text{Br} \qquad (20\text{-}10)$$
$$\text{BrCH}_2\dot{\text{C}}\text{HCH}_2\text{Br} + \text{HBr} \rightarrow \text{BrCH}_2\text{CH}_2\text{CH}_2\text{Br} + \text{Br·} \quad (20\text{-}11)$$

The nature of the product is determined by the chain-propagating steps (20-10) and (20-11). In step (20-10) the bromine atom would be expected to add to the end rather than the middle carbon atom, since a secondary radical is more stable than a primary one. Hydrogen bromide is unique among the hydrogen halides in its tendency to undergo reversal in the mode of its addition to olefins in the presence of peroxides. This fact may be rationalized in terms of the bond energies listed in Table 1-4.

First we must argue that if the free-radical mechanism is to be effective, both of the propagation steps must be rapid, since if they are not, the intermediate radicals will tend to accumulate and termination will be facilitated. The activation energy for these steps must be at least equal to the energy of reaction, and therefore it would be best if both steps were exothermic. This is the case with hydrogen bromide. In step (20-10) the difference in energy between a double bond and a single bond (about 63 kcal) is offset by the energy of the new carbon-bromine bond (65 kcal) that is formed. In (20-11) an 87-kcal hydrogen-bromine bond is broken, but a 98-kcal carbon-hydrogen bond is formed. Hydrogen fluoride does not react because its bond (energy about 135 kcal) is too strong to be easily broken by a radical. The failure with hydrogen iodide is apparently due to the unreactivity of the iodine atom, which does not form a strong enough bond to carbon to compensate for breaking a double bond. The situation with regard to hydrogen chloride is rather delicately balanced. Raley, Rust, and Vaughan showed that hydrogen chloride may be added to ethylene by a free-radical chain

[39] M. S. Kharasch, H. Engelmann, and F. R. Mayo, *J. Org. Chem.*, **2,** 288 (1937).
[40] D. H. Hey and W. A. Waters, *Chem. Revs.*, **21,** 202 (1937).

mechanism.[41]   However, in this case the reaction analogous to (20-11) involves the attack of a reactive *primary* radical.   The more stable secondary radicals are evidently not reactive enough, since propylene is relatively unreactive toward free-radical addition of hydrogen chloride.[41] This rationalization would be more convincing, of course, if it were based on bond-dissociation energies rather than just average bond energies. Unfortunately, the necessary data of the former type are less readily available.

Even with hydrogen bromide the presence of peroxides does not always result in a reversal in the direction of addition.   In some cases the polar addition proceeds so rapidly that the free-radical reaction cannot be made to compete with it.   In other cases each of the two mechanisms would be expected to give the same product.   Thus methyl methacrylate would be expected to yield a primary bromide by the free-radical as well as a polar mechanism.

$$
\begin{array}{ccc}
\overset{\displaystyle Me}{\underset{\displaystyle |}{}} & & \overset{\displaystyle Me}{\underset{\displaystyle |}{}} \\
CH_2\!\!=\!\!C\!\!-\!\!CO_2Me \xrightarrow{\ Br\cdot\ } BrCH_2\!\!-\!\!\underset{\bullet}{C}\!\!-\!\!CO_2Me \xrightarrow{\ HBr\ } \\
\end{array}
$$

$$
\overset{\displaystyle Me}{\underset{\displaystyle |}{}}
$$
$$
BrCH_2\!\!-\!\!CH\!\!-\!\!CO_2Me + Br\cdot
$$

This reaction was found to yield the product shown,[42] but we do not know which mechanism is operative.

20-2c. Miscellaneous Free-radical Additions to Olefins.   There are a wide variety of organic compounds that have been found to add to olefins by a free-radical mechanism.   A large number of these reactions were discovered by Kharasch and coworkers.   Many of the additions proceed well only with terminal olefins.   A number of polyhalomethanes have been found to react in the presence of a diacyl peroxide or ultraviolet light.[43]   The mechanism, illustrated for the case of carbon tetrachloride, is

$$Initiator \rightarrow R\cdot$$
$$R\cdot + CCl_4 \rightarrow RCl + \cdot CCl_3$$
$$R\!-\!CH\!\!=\!\!CH_2 + \cdot CCl_3 \rightarrow R\!-\!\overset{\bullet}{C}H\!-\!CH_2CCl_3$$
$$R\!-\!\overset{\bullet}{C}H\!-\!CH_2CCl_3 + CCl_4 \rightarrow R\!-\!CHCl\!-\!CH_2CCl_3 + \cdot CCl_3$$

Other types of compounds probably add by an analogous mechanism,

[41] J. H. Raley, F. F. Rust, and W. E. Vaughan, *J. Am. Chem. Soc.*, **70**, 2767 (1948).

[42] C. C. Price and E. C. Coyner, *J. Am. Chem. Soc.*, **62**, 1306 (1940).

[43] M. S. Kharasch, E. V. Jensen, and W. H. Urry, *J. Am. Chem. Soc.*, **69**, 1100 (1947); M. S. Kharasch, O. Reinmuth, and W. H. Urry, *J. Am. Chem. Soc.*, **69**, 1105 (1947).

aldehydes, primary alcohols, and amines yielding ketones,[44] secondary alcohols,[45] and larger amines,[46] respectively.

$$RCHO + R'CH\!=\!CH_2 \rightarrow R'CH_2CH_2COR$$
$$RCH_2OH + R'CH\!=\!CH_2 \rightarrow R'CH_2CH_2CHOHR$$
$$RCH_2NH_2 + R'CH\!=\!CH_2 \rightarrow R'CH_2CH_2CHNH_2R$$

PROBLEMS

1. In the presence of azo-bis-isobutyronitrile, an initiator that decomposes readily to yield free radicals, methyl mercaptan adds to cis-2-butene to give methyl 2-butyl sulfide. When the reaction was stopped after only a small amount of sulfide had been formed a large amount of trans-2-butene was found in the reaction mixture. Under the same conditions, in the absence of methyl mercaptan, azo-bis-isobutyronitrile and cis-2-butene yield no trans-2-butene. Write a reasonable reaction mechanism for the formation of the sulfide and the trans olefin.

2. In the copolymerization of vinyl chloride (M_1) and vinyl acetate (M_2), r_1 and r_2 are 1.7 and 0.23, respectively. From the data in Sec. 20-1b, estimate the absolute values of k_{11}, k_{12}, k_{21}, and k_{22} in this copolymerization. Tell what assumptions, if any, are made in your estimate and assess the relative reliability for the four estimates.

3. For a large number of substituent groups (X's) where the compounds $CH_2\!=\!CHX$ and $CH_2\!=\!CX_2$ readily form high polymers, the compounds $XCH\!=\!CHX$, $XCH\!=\!CX_2$, and $X_2C\!=\!CX_2$ do not. The substituent fluorine seems to be an exception to this generalization. Suggest an explanation for this behavior.

4. The presence of allyl hydrogen atoms in the monomer (as in propylene) seems to interfere with the ease of formation of high polymers by a free-radical mechanism. Suggest an explanation for this fact.

5. Pairs of compounds are known that give copolymers of high molecular weight, although neither compound alone is capable of forming a high polymer. Suggest structural features in the reactants that might lead to such behavior and tell why.

6. Express the over-all activation energy for the polymerization of vinyl acetate in terms of the activation energies of the individual steps.

[44] M. S. Kharasch, W. H. Urry, and B. M. Kuderna, *J. Org. Chem.*, **14**, 248 (1949).

[45] W. H. Urry, F. W. Stacey, O. O. Juveland, and C. H. McDonnell, *J. Am. Chem. Soc.*, **75**, 250 (1953); W. H. Urry, F. W. Stacey, E. S. Huyser, and O. O. Juveland, *J. Am. Chem. Soc.*, **76**, 450 (1954).

[46] W. H. Urry, O. O. Juveland, and F. W. Stacey, *J. Am. Chem. Soc.*, **74**, 6155 (1952).

Chapter 21

RADICAL DECOMPOSITION OF PEROXIDES AND AZO AND DIAZONIUM COMPOUNDS

21-1. Decomposition of Peroxides. 21-1*a. Mechanism of the Decomposition of Benzoyl Peroxide.* There is good evidence that the decomposition of diacyl peroxides may proceed by a free-radical mechanism. Walker and Wild showed that the decomposition of acetyl peroxide is catalyzed by light.[1] The decomposition of benzoyl peroxide is slowed by inhibitors and catalyzed by known radicals.[2] The decomposition of benzoyl peroxide in nitrobenzene solution brings about considerable para phenylation of the nitrobenzene, although substitution by electrophilic reagents gives almost entirely meta substitution.[3] Acyl peroxides are common catalysts for a large number of reactions for which there is much good evidence for a free-radical mechanism.

Since benzoyl peroxide has been investigated more thoroughly than any other acyl peroxide, its decomposition will be discussed in some detail. The reaction does not consist simply of a first-order decomposition of the peroxide. In addition, to an extent that varies widely with the solvent, it is a chain reaction in which intermediate radicals attack benzoyl peroxide molecules to produce further decomposition. This is shown by the observations of Nozaki and Bartlett[2] that the reaction is inhibited by reagents such as oxygen, hydroquinone, and picric acid and accelerated by free radicals such as triphenylmethyl and diphenyl-amino. Further evidence is the fact that in many solvents the kinetic order of the reaction is greater than one. The fact that the reaction rate varies widely with the solvent suggests that the solvent is participating in the rate-controlling step of the reaction. The order of reactivity in various solvents (see Table 21-1) would be utterly inexplicable for a polar reaction, and it would not be reasonable to expect such large variation in the rate of cleavage of an uncharged molecule to neutral radicals. However, it can be rationalized fairly well by the assumption that the radicals

[1] O. J. Walker and G. L. E. Wild, *J. Chem. Soc.*, 1132 (1937).
[2] K. Nozaki and P. D. Bartlett, *J. Am. Chem. Soc.*, **68**, 1686 (1946).
[3] D. H. Hey, *J. Chem. Soc.*, 1966 (1934).

439

originally formed from the peroxide decomposition react with the various solvents to form radicals that differ widely in their ability to induce the further decomposition of benzoyl peroxide.[2]

Information about the "spontaneous," as distinguished from the "induced," decomposition of benzoyl peroxide was obtained by Hammond and Soffer.[4] These workers used the very effective inhibitor iodine in a solvent (carbon tetrachloride) that does not react with the initial product of inhibition, benzoyl hypoiodite. Under anhydrous

TABLE 21-1. PERCENTAGE DECOMPOSITION OF BENZOYL PEROXIDE IN VARIOUS SOLVENTS[a] AT 79.8°[2,5]

Solvent	Time, min	De-compn., %	Solvent	Time, min	De-compn., %
Carbon tetrachloride....	60	13.0	t-Butylbenzene.........	60	28.5
Cyclohexene...........	60	14.0	Acetic anhydride.......	60	48.5
Anisole...............	60	14.0	Cyclohexane...........	60	51.0
Methyl benzoate........	60	14.5	Ethyl acetate..........	60	53.5
Chloroform............	60	14.5	Acetic acid............	60	59.3
Nitrobenzene..........	60	15.5	Dioxane...............	60	82.4
Benzene..............	60	15.5	Diethyl ether..........	10	75.2
Toluene..............	60	17.4	Ethyl alcohol..........	10	81.8
Chlorobenzene.........	60	18.0	m-Cresol..............	10	87.6
Styrene...............	60	19.0	Isopropyl alcohol.......	10	95.1
Ethyl iodide...........	60	23.4	2,4,6-Trimethylphenol..	10	98.8[b]
Acetone..............	60	28.5			

[a] Aniline, triethylamine, dimethylaniline, and n-butylamine all reacted explosively.
[b] At 60.0°.

conditions iodobenzene from the decomposition of the hypoiodite was isolated as the reaction product in more than 80 per cent yield. However, when a separate layer of water, which is known to hydrolyze benzoyl hypoiodite rapidly, was added to the reaction mixture, the product was benzoic acid in almost quantitative yield. That the reaction studied was only the "spontaneous" decomposition seems assured because no further decrease in reaction rate was found when the inhibitor concentration was more than doubled. The benzoic acid could not have been formed from water and benzoyl peroxide or some intermediate ordinarily present in the carbon tetrachloride decomposition of the peroxide since no benzoic acid was found among the products of the decomposition of benzoyl peroxide in carbon tetrachloride with no iodine but an excess of water present. It was also shown that the water had no effect on the rate-controlling step of the reaction in which benzoic acid is produced

[4] G. S. Hammond and L. M. Soffer, *J. Am. Chem. Soc.*, **72**, 4711 (1950).

quantitatively, because the reaction rate was identical to that in the absence of water. Since the rate-controlling step of the spontaneous decomposition thus appears to produce intact benzoate groups quantitatively, it must be simply a fission of the peroxide into two benzoate radicals.

$$C_6H_5COO-OCOC_6H_5 \rightarrow 2C_6H_5COO\cdot$$

The mechanism of the "induced" decomposition of benzoyl peroxide is also of interest, since this is the principal reaction path in those solvents in which decomposition is rapid and even in the "slower" solvents if the solution is fairly concentrated. Bartlett and Nozaki studied the decomposition of benzoyl peroxide in n-butyl ether solution.[5] The relatively high rate of the reaction and the marked inhibition by oxygen show that the reaction is largely induced. In a 50 per cent solution of vinyl acetate in n-butyl ether containing 1 per cent benzoyl peroxide, there is more decomposition of benzoyl peroxide but less polymerization of vinyl acetate than in a similar solution in cyclohexane. This shows, among other things, that the additional decomposition of benzoyl peroxide (induced decomposition) does not cause any additional polymerization and hence that it probably does not involve the formation of additional radicals. This is in agreement with an induced decomposition in which only one radical is created for every radical destroyed. The principal products of the decomposition in n-butyl ether solution are benzoic acid (almost 1 mole per mole of peroxide decomposed) and high-boiling material that is probably largely α-butoxybutyl benzoate (by analogy with the work of Cass,[6] who isolated more than 0.8 mole of α-ethoxyethyl benzoate per mole of benzoyl peroxide in a similar peroxide decomposition in diethyl ether solution). The nature of the product shows that it is probably the α-butoxybutyl radical that is responsible for the induced decomposition of the peroxide and hence that the mechanism of the major portion[7] of the reaction is

$$C_6H_5COO-OCOC_6H_5 \xrightarrow{k_1} 2C_6H_5COO\cdot$$
$$C_6H_5COO\cdot + C_3H_7CH_2OC_4H_9 \xrightarrow{k_2} C_6H_5CO_2H + C_3H_7\overset{\cdot}{C}HOC_4H_9$$
$$C_3H_7\overset{\cdot}{C}HOC_4H_9 + C_6H_5COO-OCOC_6H_5$$

$$\xrightarrow{k_3} C_3H_7\overset{\displaystyle OCOC_6H_5}{\underset{\displaystyle OC_4H_9}{\overset{\diagup}{\underset{\diagdown}{CH}}}} + C_6H_5COO\cdot$$

---

[5] P. D. Bartlett and K. Nozaki, *J. Am. Chem. Soc.*, **69**, 2299 (1947).

[6] W. E. Cass, *J. Am. Chem. Soc.*, **69**, 500 (1947).

[7] The decomposition must also consist to a minor extent of a simultaneous carbon dioxide-producing reaction since some carbon dioxide is obtained.

Some possible termination reactions are:

1. Recombination of benzoate radicals to give benzoyl peroxide.

$$2C_6H_5COO\cdot \xrightarrow{k_4} C_6H_5COO\!-\!OCOC_6H_5$$

2. Combination of benzoate radicals to give phenyl benzoate or some other molecule not likely to reenter the reaction.

$$2C_6H_5COO\cdot \xrightarrow{k_5} C_6H_5COOC_6H_5 + CO_2$$

3. Crossed termination between a benzoate and an α-butoxybutyl radical to give the principal reaction product (or other nonradicals).

$$C_6H_5COO\cdot + C_3H_7\overset{\cdot}{C}HOC_4H_9 \xrightarrow{k_6} C_3H_7\underset{OC_4H_9}{\overset{OCOC_6H_5}{CH}}$$

4. Dimerization or disproportionation of the α-butoxybutyl radicals.

$$2C_3H_7\overset{\cdot}{C}HOC_4H_9 \xrightarrow{k_7} C_4H_9O\underset{\underset{C_3H_7}{|}}{CH}\!-\!\underset{\underset{C_3H_7}{|}}{CH}OC_4H_9$$

In the case of each of these possibilities the kinetic order of the reaction may be ascertained by assuming that the given possibility is the only termination step and by applying the steady-state assumption (Sec. 3-1a) to the resultant mechanism.

Using the symbols

$$P = [C_6H_5COO\!-\!OCOC_6H_5] \qquad\qquad SH = n\text{-}C_4H_9OC_4H_9\text{-}n$$

$$R = [C_6H_5COO\cdot] \qquad\qquad\qquad S = [n\text{-}C_3H_7\overset{\cdot}{C}HOC_4H_9\text{-}n]$$

and assuming termination mechanism 3, crossed termination, we get the reaction mechanism

$$P \xrightarrow{k_1} 2R$$

$$R + SH \xrightarrow{k_2} RH + S$$

$$S + P \xrightarrow{k_3} R\!-\!S + R$$

$$R + S \xrightarrow{k_6} R\!-\!S$$

Therefore

$$-\frac{dP}{dt} = k_1P + k_3PS \tag{21-1}$$

From the steady-state assumption (and treating k_2 as a first-order rate constant, since the solvent concentration remains essentially constant),

$$\frac{dR}{dt} = 2k_1P - k_2R + k_3PS - k_6RS = 0 \qquad (21\text{-}2)$$

and

$$\frac{dS}{dt} = k_2R - k_3PS - k_6RS = 0 \qquad (21\text{-}3)$$

From Eq. (21-3)

$$R = \frac{k_3PS}{k_2 - k_6S}$$

Substitution in (21-2) and solution of the quadratic gives

$$S = \frac{k_c}{k_3}$$

where

$$k_c = \frac{-k_1k_6 + \sqrt{k_1{}^2k_6{}^2 + 4k_1k_2k_3k_6}}{2k_6}$$

Therefore,

$$-\frac{dP}{dt} = k_1P + k_cP = (k_1 + k_c)P$$

Thus, if mechanism 3 is the only mechanism for termination, the entire reaction, induced as well as spontaneous, will be first-order.

Termination mechanism 1 leads to the kinetic equation

$$-\frac{dP}{dt} = k_2 \sqrt{\frac{k_1}{k_4}} P^{1/2}$$

in which the entire reaction is one-half-order in peroxide.

From 2, the equation

$$-\frac{dP}{dt} = k_1P + k_2 \sqrt{\frac{k_1}{k_5}} P^{1/2}$$

is obtained, in which the induced part of the reaction is one-half-order.
Mechanism 4 yields the relation

$$-\frac{dP}{dt} = k_1P + k_3 \sqrt{\frac{k_1}{k_7}} P^{3/2}$$

in which the induced part of the reaction is three-halves-order with respect to peroxide.   Bartlett and Nozaki found the reaction to be first-order.[5]   Since the reaction is apparently very largely induced and since only termination mechanism 3 yields first-order kinetics for the induced part of the reaction, they concluded that termination probably occurs, at least largely, by reaction between the two unlike radicals.   They also point out that there is considerable precedent for such a tendency of unlike radicals to prefer reaction with each other to dimerization (cf. Sec. 18-2a).   In the present case it is noteworthy that the α-butoxybutyl radical would be a strongly electron-donating radical, whereas the benzoate radical should be electron-accepting.

It appears that in alcoholic solutions the benzoate radicals attack the α-hydrogens to form radicals that induce the further decomposition of benzoyl peroxide.[5] With phenols, phenoxy radicals are thought to be produced.[5] There are several possible ways in which amines may be attacked.[5,8]

In the solvents in which the decomposition of benzoyl peroxide is relatively slow (see Table 21-1), the reaction mechanism appears to be somewhat different. In some cases there are simultaneous first- and three-halves-order reactions, and in others simultaneous first- and second-order reactions. In these solvents a considerable portion of the induced reaction is due to the radicals originally produced from the benzoyl peroxide, although some induced decomposition due to radicals from the solvent still occurs.[2]

21-1b. Effect of Structure on the Rate of Decomposition of Diaroyl Peroxides. Swain, Stockmayer, and Clarke investigated the effect of structure on the rate of the *spontaneous* thermal decomposition of meta- and para-substituted benzoyl peroxides in dioxane solution.[9] They used 3,4-dichlorostyrene to inhibit the *induced* decomposition of the peroxide. By neglecting the relative stability of the radicals formed and attributing the influence of substituents on rate to a polar factor, fair agreement with experimental results was obtained.

A benzoyl peroxide molecule may be considered to be two dipoles joined at their negative ends.

$$
\begin{array}{ccc}
 & \overset{\text{O}}{\underset{\|}{}} & \overset{\text{O}}{\underset{\|}{}} \\
\text{C}_6\text{H}_5-\text{C}-\text{O}-\text{O}-\text{C}-\text{C}_6\text{H}_5 \\
\underset{+\delta}{\vdash} \quad \underset{-\delta}{\longrightarrow} \quad \underset{-\delta}{\longleftarrow} \quad \underset{+\delta}{\dashv}
\end{array}
$$

If the reaction is thought of as due to this repulsion, the rate would be expected to be proportional to the magnitude of the dipoles. Therefore electron-donating groups, which would increase the size of the dipoles, should increase the reaction rate. In general, this is found to be the case, whereas electron-withdrawing groups slow the reaction. In fact, the data agree fairly well with the Hammett equation, where ρ has the value -0.38 for this reaction and σ is the sum of the σ (or σ^+) values for the substituents (e.g., since $\sigma^+ = -0.778$ for the p-methoxy group, this value is used for p-methoxybenzoyl peroxide and the value -1.556 is used for p,p'-dimethoxybenzoyl peroxide). The experimental

[8] J. E. Leffler, *J. Am. Chem. Soc.*, **72**, 3809 (1950); C. E. Boozer and G. S. Hammond, *J. Am. Chem. Soc.*, **76**, 3861 (1954).

[9] C. G. Swain, W. H. Stockmayer, and J. T. Clarke, *J. Am. Chem. Soc.*, **72**, 5426 (1950).

values of the rate constants are compared with those predicted from the Hammett equation in Table 21-2.

Blomquist and Buselli obtained similar results; in addition, they studied five ortho-substituted derivatives and determined entropies and energies of activation for the reactions.[10] They found that p,p'-dinitrobenzoyl peroxide (and also the m,m' isomer) decomposes much more rapidly than would be expected from the Hammett equation; they

TABLE 21-2. RATES OF SPONTANEOUS DECOMPOSITION OF
SUBSTITUTED BENZOYL PEROXIDES IN DIOXANE AT 80°[9]

| Substituents | σ | 10^3k | | Deviation, %[a] |
		Predicted	Actual	
p,p'-Di-MeO	−1.556	7.38	7.06	4
p-MeO	−0.778	4.31	4.54	5
p,p'-Di-Me	−0.622	3.87	3.68	5
p,p'-Di-t-Bu	−0.512	3.55	3.65	3
m,m'-Di-MeO	+0.230	2.15	3.45	60
m-MeO	+0.115	2.33	2.89	24
p-Me, m'-Br	+0.080	2.38	2.66	12
m,m'-Di-Me	−0.138	2.77	2.64	5
None	0.000	2.52	2.52	0
p,p'-Di-Cl	+0.228	2.15	2.17	1
p,p'-Di-Br	+0.300	2.05	1.94	5
m-CN	+0.560	1.71	1.64	4
m,m'-Di-Cl	+0.746	1.51	1.58	5
m,m'-Di-Br	+0.782	1.47	1.54	5
p,p'-Di-CN	+1.32	1.01	1.22	21
m,m'-Di-CN	+1.12	1.16	1.02	12
Median.........	....	....	....	5

[a] Percentage by which the actual deviated from the predicted.

pointed out that this may be due to the nitro group's being such a strong electron-withdrawing group that it reverses the direction of the dipole. There is a possibility, however, that the large rate of the nitro compounds is due to the difficulty in suppressing the *induced* part of the decomposition.   The effect of substituents on the rate of the induced decomposition is the opposite of that on the spontaneous decomposition.   Since electron-donor radicals are most active at inducing decomposition, it is not surprising that electron-withdrawing groups in the peroxide molecule should aid the reaction.[9]

21-1c. *Decomposition of Dialkyl Peroxides.* The decomposition of di-t-butyl peroxide has been studied in some detail. The reaction is

[10] A. T. Blomquist and A. J. Buselli, *J. Am. Chem. Soc.*, **73**, 3883 (1951).

catalyzed by ultraviolet radiation.[11]   It was studied kinetically in the vapor phase at temperatures around 150° by Raley, Rust, and Vaughan.[12]   These workers showed that packing the reaction vessel with enough glass rodding to change the surface-volume ratio by more than elevenfold had no detectable effect on the rate of reaction of the peroxide and hence that the rate-controlling step must occur in the gas phase.   The reaction appeared to be entirely spontaneous, with no noticeable amount of induced chain decomposition, since it followed good first-order kinetics and was not slowed by such inhibitors as oxygen, nitric oxide, and propylene.   When run in a vessel packed with glass wool, the reaction products are entirely acetone and ethane and are believed to be formed by the mechanism.[13]

$$(CH_3)_3COOC(CH_3)_3 \rightarrow 2(CH_3)_3CO\cdot$$
$$(CH_3)_3CO\cdot \rightarrow (CH_3)_2CO + CH_3\cdot$$
$$2CH_3\cdot \rightarrow CH_3CH_3$$

Apparently, the dimerization of the methyl radicals is at least partially a surface reaction, for when the reaction is run in a large unpacked vessel, some methane and methyl ethyl and higher ketones are formed, presumably because the methyl radicals now enter into reactions other than dimerization.[12]

$$CH_3\cdot + CH_3COCH_3 \rightarrow CH_4 + CH_3COCH_2\cdot$$
$$CH_3COCH_2\cdot + CH_3\cdot \rightarrow CH_3COCH_2CH_3 \qquad \text{etc.}$$

The rate-controlling step in the decomposition of di-t-butyl peroxide in cumene, t-butylbenzene, or tri-n-butylamine solution is apparently the same as in the vapor-phase reaction, since the reaction proceeds at the same rate (within experimental error) in all three solvents, and only about 30 per cent slower in the gas phase.[14]   The decomposition of benzoyl peroxide, a compound sensitive to induced decomposition, is slow in t-butylbenzene solution at 80° but explosive in amine (e.g., aniline and triethylamine) solutions at room temperature.[2]   Although the formation of t-butoxy radicals occurs at about the same rate in these different solvents, their fate, which is settled after the rate-controlling step of the reaction, varies considerably.   At 125° in tri-n-butylamine, where the donor element nitrogen makes the α-hydrogen atoms very susceptible to attack by an acceptor alkoxy radical, about 95 per cent

[11] (a) E. R. Bell, F. F. Rust, and W. E. Vaughan, *J. Am. Chem. Soc.*, **72,** 337 (1950); (b) L. M. Dorfman and Z. W. Salsburg, *J. Am. Chem. Soc.*, **73,** 255 (1951).

[12] J. H. Raley, F. F. Rust, and W. E. Vaughan, *J. Am. Chem. Soc.*, **70,** 88 (1948).

[13] N. A. Milas and D. M. Surgenor, *J. Am. Chem. Soc.*, **68,** 205 (1946).

[14] J. H. Raley, F. F. Rust, and W. E. Vaughan, *J. Am. Chem. Soc.*, **70,** 1336 (1948).

of the t-butoxy radicals form t-butyl alcohol.[15]   At the same temperature in cumene, where a benzyl-type radical, $C_6H_5\dot{C}(CH_3)_2$, may be formed, about 80 per cent of the t-butoxy radicals form alcohol, and in t-butylbenzene, where the resultant radical is not so stabilized by resonance, only 37 per cent form t-butyl alcohol. As previously stated, in the vapor phase, where the butoxy radical collides with other molecules only relatively rarely, no t-butyl alcohol was found. In cumene and t-butylbenzene it is noted that as the reaction temperature is increased, the yield of t-butyl alcohol decreases and that of acetone and methane increases. This shows that the activation energy for the decomposition of the radical is higher than for its abstraction of a hydrogen atom from solvent. Under certain conditions the reaction may become partly induced. This appears to be the case with the decomposition of the pure liquid. In the first place, the reaction is about three times as fast as would be expected from the rate constants in the other solvents studied.[11a]   Also, from the nature of the products it appears certain that di-t-butyl peroxide molecules have been attacked by intermediate radicals. This would certainly be expected to cause chain decomposition. Under these conditions isobutylene oxide becomes a principal reaction product. The following mechanism explains these facts.

$$(CH_3)_3COOC(CH_3)_3 \rightarrow 2(CH_3)_3CO\cdot$$

$$(CH_3)_3CO\cdot \rightarrow CH_3COCH_3 + CH_3\cdot$$

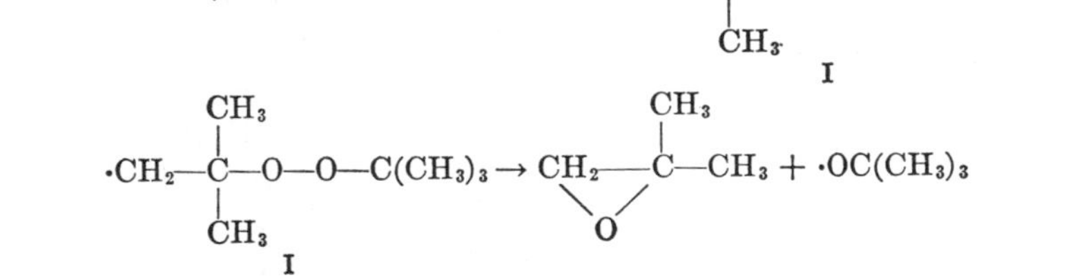

$$(CH_3)_3CO\cdot + (CH_3)_3COOC(CH_3)_3 \rightarrow (CH_3)_3COH + \underset{\underset{CH_3}{|}}{\overset{\overset{CH_3}{|}}{\cdot CH_2C}}-O-O-C(CH_3)_3 \quad \text{I}$$

$$CH_3\cdot + (CH_3)_3COOC(CH_3)_3 \rightarrow CH_4 + \underset{\underset{CH_3}{|}}{\overset{\overset{CH_3}{|}}{\cdot CH_2C}}-O-O-C(CH_3)_3 \quad \text{I}$$

$$\underset{\underset{CH_3}{|}}{\overset{\overset{CH_3}{|}}{\cdot CH_2-C}}-O-O-C(CH_3)_3 \rightarrow CH_2\underset{O}{\overset{\overset{CH_3}{|}}{\diagdown C \diagup}}-CH_3 + \cdot OC(CH_3)_3$$

I

By a similar mechanism, small amounts of hydrogen chloride can cause induced chain decomposition. The addition of about 30 mm of HCl causes the initial rate of the vapor-phase decomposition of 180 mm

[15] This fact, incidentally, shows that the initial, rate-controlling cleavage is indeed into two butoxy radicals rather than directly to $CH_3\cdot + CH_3COCH_3 + (CH_3)_3CO\cdot$.

of di-t-butyl peroxide at about 140° to become at least 10 times as fast.[16] This may be attributed to chain induction by chlorine atoms.[16]

$$(CH_3)_3CO\cdot + HCl \rightarrow (CH_3)_3COH + Cl\cdot$$

$$Cl\cdot + (CH_3)_3COOC(CH_3)_3 \rightarrow HCl + \cdot CH_2\overset{\overset{\displaystyle CH_3}{|}}{\underset{\underset{\displaystyle CH_3}{|}}{C}}OOC(CH_3)_3$$

$$\text{I}$$

$$\cdot CH_2\overset{\overset{\displaystyle CH_3}{|}}{\underset{\underset{\displaystyle CH_3}{|}}{C}}OOC(CH_3)_3 \rightarrow \underset{\underset{\displaystyle O}{\diagdown\diagup}}{CH_2}\!-\!\overset{\overset{\displaystyle CH_3}{|}}{C}\!-\!CH_3 + \cdot OC(CH_3)_3$$

$$\text{I}$$

$$\underset{\underset{\displaystyle O}{\diagdown\diagup}}{CH_2}\!-\!C(CH_3)_2 + HCl \rightarrow ClCH_2\underset{\underset{\displaystyle OH}{|}}{C}(CH_3)_2$$

Here the isobutylene oxide is not isolated as such but as the chlorohydrin. Evidence that the radical (I) may have a separate existence, rather than decomposing at the same time it is formed, may be seen in the fact that di-t-butyl peroxide may be photochlorinated to $ClCH_2C(CH_3)_2$-$OOC(CH_3)_3$ at about 35°.[16] The radical (I) must have been an intermediate in this reaction. A study of the photolysis of di-t-butyl peroxide[11b] has given added evidence for certain of the conclusions drawn from thermal-decomposition data.

Studies on the decomposition of di-t-amyl peroxide suggest that the mechanism is very similar to that for di-t-butyl peroxide.[12] As might be expected, the decomposition of the t-amyloxy radical yields acetone and an ethyl radical rather than the less stable methyl radical (and methyl ethyl ketone). In general, it appears that the rate at which an alkoxy radical loses an alkyl radical (or hydrogen atom) depends upon the stability of the alkyl radical and the carbonyl compound formed. Hence the ease of decomposition (compared to the ease of abstraction of a hydrogen atom from solvent) of a series of alkoxy radicals was found to vary thus:[17] CH_3—$C(CH_3)_2O\cdot \sim (CH_3)_2CH$—$CH_2O\cdot > CH_3$—$CH(CH_3)O\cdot > n\text{-}C_3H_7$—$CH_2O\cdot > CH_3$—$CH_2O\cdot > CH_3O\cdot$.

21-1d. Decomposition of Alkyl Hydroperoxides. In contrast to di-t-alkyl peroxides, t-alkyl hydroperoxides appear to be quite sensitive to induced chain decomposition. The decomposition of t-butyl hydroperoxide is strongly accelerated by that of 2,2'-azo-bis-isobutyronitrile, $Me_2C(CN)$—$N{=}N$—$(CN)CMe_2$,[18] a compound known to give free radicals (Sec. 21-2b)

[16] J. H. Raley, F. F. Rust, and W. E. Vaughan, *J. Am. Chem. Soc.*, **70**, 2767 (1948).

[17] F. F. Rust, F. H. Seubold, Jr., and W. E. Vaughan, *J. Am. Chem. Soc.*, **72**, 338 (1950). For similar data, see N. A. Milas and L. H. Perry, *J. Am. Chem. Soc.*, **68**, 1938 (1946).

[18] V. Stannett and R. B. Mesrobian, *J. Am. Chem. Soc.*, **72**, 4125 (1950).

and also by that of di-t-butyl peroxide.[19]   The mechanism of the latter induced decomposition appears to involve an attack on the hydroxylic hydrogen atom.[19]

$$(CH_3)_3COOC(CH_3)_3 \rightarrow 2(CH_3)_3CO\cdot$$
$$(CH_3)_3CO\cdot \rightarrow (CH_3)_2CO + CH_3\cdot$$
$$CH_3\cdot + (CH_3)_3COOH \rightarrow CH_4 + (CH_3)_3COO\cdot$$
$$(CH_3)_3COO\cdot + CH_3\cdot \rightarrow (CH_3)_3COOCH_3$$
$$(CH_3)_3COOCH_3 \rightarrow (CH_3)_3CO\cdot + CH_3O\cdot \qquad \text{etc.}$$

Bateman and Hughes described evidence for a bimolecular initiation reaction in the decomposition of cyclohexenyl hydroperoxide.[20]

$$2C_6H_9OOH \rightarrow H_2O + C_6H_9O\cdot + C_6H_9OO\cdot$$

21-1e. The Concerted Cleavage of Several Bonds in Peroxide Decompositions.   Bartlett and coworkers described convincing evidence that the initial step in the decomposition of certain peroxides consists of the simultaneous cleavage of several bonds and the formation of one or more stable molecules as well as two radicals.   Such a mechanism would explain the relatively great reactivity of phenylacetyl peroxide, but the complications observed in this reaction, which is subject to induced chain decomposition and also acid catalysis, weaken the argument somewhat.[21] In subsequent studies of the decomposition of t-butyl esters of various peracids, enormous differences in reactivity were observed and seen to be inexplicable in terms of a simple O—O bond fission.[22]   From the fact that the reaction rates increased with the stability, as a radical, of the R group in these $RCO_2OC(CH_3)_3$ compounds, it appears that R is being liberated as a radical in the rate-controlling step of the reaction.

$$R\!-\!\overset{\displaystyle O}{\overset{\|}{C}}\!-\!O\!-\!OC(CH_3)_3 \rightarrow R\cdot + CO_2 + \cdot OC(CH_3)_3$$

In this mechanism the stability of the carbon dioxide molecule being formed adds to the driving force.

From the kinetic data listed in Table 21-3 it may be seen that the entropy of activation tends to be lower in those reactions in which resonance-stabilized radicals, such as benzyl and benzhydryl, are formed. The extent of resonance stabilization of these radicals depends on how nearly coplanar their carbon skeletons are.   For this reason such radicals

[19] F. H. Seubold, Jr., F. F. Rust, and W. E. Vaughan, *J. Am. Chem. Soc.*, **73,** 18 (1951).

[20] L. Bateman and H. Hughes, *J. Chem. Soc.*, 4594 (1952).

[21] P. D. Bartlett and J. E. Leffler, *J. Am. Chem. Soc.*, **72,** 3030 (1950).

[22] P. D. Bartlett et al., *J. Am. Chem. Soc.*, **80,** 1398 (1958); **82,** 1753, 1756, 1762, 1769, (1960).

are formed only from certain conformers of the appropriate peresters. The perester in which R is 1-phenylallyl is more reactive than the one in which R is 3-phenylallyl, not because of any difference in the stability of the radicals, which are identical, but at least partly because the *reactant*, in the case of the 3-phenylallyl compound, is stabilized by conjugation between the double bond and the aromatic ring.

TABLE 21-3. DECOMPOSITION RATES OF PERESTERS $RCO_2OC(CH_3)_3$ AT $60°$[22]

R	Half-life, min	$\Delta H^{\ddagger}$, kcal	$\Delta S^{\ddagger}$, cal/degree
Methyl................	500,000	38	17
Phenyl................	30,000	33.5	7.8
Benzyl................	1,700	28.1	2.2
Trichloromethyl........	970	30.3	9.4
t-Butyl................	300	30.0	11.1
3-Phenylallyl	100	23.5	−5.9
Benzhydryl............	26	24.3	−1.0
2-Phenyl-2-propyl.......	12	26.1	5.8
1,1-Diphenylethyl.......	6	24.7	3.3
1-Phenylallyl..........	4	23.0	−1.1

21-2. Free-radical Decompositions to Yield Nitrogen. 21-2a. *Decomposition of Azomethane and Related Compounds.* In experiments of the Paneth type (Sec. 19-1a) Leermakers showed that the products of the decomposition of azomethane at 475° are capable of removing metallic mirrors.[23] This shows that free radicals are formed in the reaction but does not rule out the possibility that part of the reaction involves a rearrangement directly to ethane and nitrogen, the principal reaction products. However, Davis, Jahn, and Burton showed that in the closely related photolysis of azomethane no significant fraction of the reaction occurs by such a mechanism.[24] They found that the addition of nitric oxide to the reaction mixture prevented the formation of any gaseous alkanes. Nitric oxide, itself a free radical, is often a very effective reagent at "capturing" free alkyl radicals formed as reaction intermediates. From the quantum yield of unity,[24] unaffected by nitric oxide, and other facts it may be seen that the reaction does not consist to any appreciable extent of an induced chain decomposition.

The reaction seems to involve the transformation of azomethane into two methyl radicals and a nitrogen molecule; however this trans-

[23] J. A. Leermakers, *J. Am. Chem. Soc.*, **55**, 3499 (1933).
[24] T. W. Davis, F. P. Jahn, and M. Burton, *J. Am. Chem. Soc.*, **60**, 10 (1938).

formation may be depicted as a concerted mechanism[23]

$$CH_3—N{=}N—CH_3 \rightarrow 2CH_3\cdot + N_2$$
$$2CH_3\cdot \rightarrow C_2H_6 \tag{21-4}$$

or a stepwise process in which the $CH_3—N{=}N\cdot$ radical has an independent existence.[25]

$$CH_3—N{=}N—CH_3 \rightarrow CH_3—N{=}N\cdot + CH_3\cdot$$
$$CH_3—N{=}N\cdot \rightarrow CH_3\cdot + N_2 \tag{21-5}$$
$$2CH_3\cdot \rightarrow C_2H_6$$

No experiments have been reported in which the $CH_3—N{=}N\cdot$ radical has been "captured," but this cannot be construed as good evidence against mechanism (21-5), since it does not appear that any work has been carried out with this purpose and since the radical may just be intrinsically difficult to capture.   Ramsperger made a sound argument for mechanism (21-4) on the basis of relative reactivities.[26]   This argument may be restated in terms of the much larger reactivity effects quoted by Overberger and DiGiulio.[27]   The reactivity of $C_6H_5CH(CH_3)$-$N{=}NCH(CH_3)_2$ is much greater than that of $(CH_3)_2CHN{=}NCH(CH_3)_2$, as would be expected (because of the greater stability of the α-phenylethyl radical compared with the isopropyl radical) from either mechanism. However, the concerted mechanism (21-4) offers a good explanation for why $C_6H_5CH(CH_3)N{=}NCH(CH_3)C_6H_5$, in which two α-phenylethyl radicals are formed, is about 40 times as reactive as $C_6H_5CH(CH_3)N{=}NCH(CH_3)_2$.   From mechanism (21-5) no increase in reactivity larger than the twofold statistical effect would be expected.

21-2b. *Decomposition of Azo Nitriles.*   The azo nitriles comprise a class of azo compounds whose decomposition has been studied relatively thoroughly, partly because of the commercial use of some of them as initiators in free-radical polymerization.   Lewis and Matheson showed that 2-azo-bis-isobutyronitrile decomposes at very nearly the same rate in a wide variety of solvents.[28]   The reaction is cleanly first-order,

[25] M. Page, H. O. Pritchard, and A. F. Trotman-Dickenson, *J. Chem. Soc.*, 3878 (1953).

[26] H. C. Ramsperger, *J. Am. Chem. Soc.*, **49,** 912, 1495 (1927); **50,** 714 (1928); **51,** 2134 (1929); cf. S. G. Cohen and C. H. Wang, *J. Am. Chem. Soc.*, **77,** 2457, 3628 (1955).

[27] C. G. Overberger and A. V. DiGiulio, *J. Am. Chem. Soc.*, **81,** 2154 (1959).

[28] F. M. Lewis and M. S. Matheson, *J. Am. Chem. Soc.*, **71,** 747 (1949); cf. Ref. 29 and K. Ziegler, W. Deparade, and W. Meye, *Ann.*, **567,** 141 (1950); C. E. H. Bawn and S. F. Mellish, *Trans. Faraday Soc.*, **47,** 1216 (1951).

[29] C. G. Overberger et al., *J. Am. Chem. Soc.*, **71,** 2661 (1949); **73,** 2618, 4880 (1951); **75,** 2078 (1953); **76,** 2722, 6185 (1954).

and the rate is practically unaffected by the addition of an inhibitor. The reaction, then, like that of other azo compounds, has as its rate-controlling step a simple cleavage without a significant amount of induced chain decomposition.

TABLE 21-4. RATE CONSTANTS, SEC^{-1}, FOR THE DECOMPOSITION OF AZO NITRILES IN TOLUENE AT 80°[29]

I. Compounds of the type

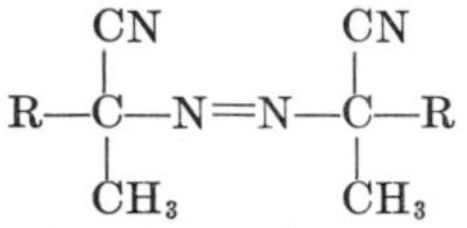

$$\underset{\underset{CH_3}{|}}{\overset{\overset{CN}{|}}{R-C}}-N=N-\underset{\underset{CH_3}{|}}{\overset{\overset{CN}{|}}{C-R}}$$

R	10^4k	R	10^4k
Methyl.........	1.66	Benzyl...............	1.16
Ethyl...........	0.87	p-Chlorobenzyl........	0.88
n-Propyl........	1.70	p-Nitrobenzyl........	1.00
Isopropyl........	1.02	Cyclopropyl[a]..........	25
n-Butyl.........	1.58		33
Isobutyl[a]........	7.1	Cyclobutyl[a]..........	1.51
	10		1.51
t-Butyl[a]........	0.77	Cyclopentyl[a].........	1.30
	1.09		1.31
Neopentyl[a]......	136	Cyclohexyl...........	2.27
	158		

II. Alicyclic azo nitriles of the type $(CH_2)_n$

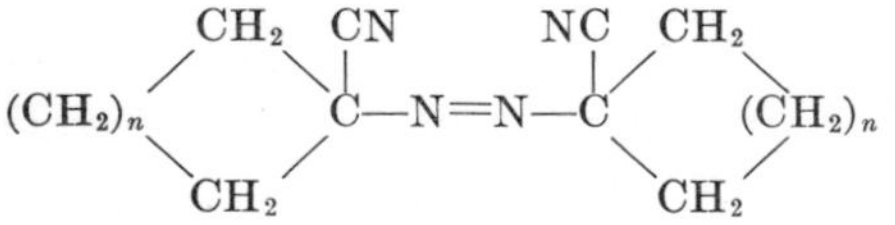

$$(CH_2)_n \overset{\overset{\overset{CH_2}{\diagdown}}{}}{\underset{\underset{CH_2}{\diagup}}{}} \overset{CN}{\underset{}{C}}-N=N-\overset{NC}{\underset{}{C}} \overset{\overset{CH_2}{\diagdown}}{\underset{\underset{CH_2}{\diagup}}{}} (CH_2)_n$$

Alicyclic ring	10^4k	Alicyclic ring	10^4k
Cyclobutyl..........	0.0017	Cycloheptyl........	12.2
Cyclopentyl.........	0.726	Cyclooctyl..........	83.5
Cyclohexyl..........	0.063	Cyclodecyl.........	18.4

[a] Both dl and meso isomers studied. It is not known which isomer has which structure.

Overberger and coworkers have studied the effect of structure on reactivity in these reactions.[29] Their results are summarized in Table 21-4. It was not known whether the various compounds studied had the meso or the dl configuration, but the cases where both diastereomers were studied show both to have about the same reactivity. It may be seen that most simple alkyl groups have about the same effect on reactivity. The reaction-accelerating influence of isobutyl and neopentyl groups is attributed to steric effects.   The reactivity of the cyclopropyl compound may be rationalized in terms of the resemblance of the three-membered ring to a double bond.   The radical formed upon decomposition of the azo compound is then seen to resemble an allyl radical.[29] The relative ease with which free radicals are formed on alicyclic rings of various sizes was also studied (Table 21-4, II) and was found to resemble the relative ease of formation of carbonium ions on alicyclic rings (Sec. 7-3a).   The reasons for this variation in reactivity are probably also similar.

21-2c. *Decomposition of Aromatic Diazonium Compounds.*   We have already described some of the evidence that aromatic diazonium compounds may decompose by a polar mechanism under certain conditions when the aromatic diazonium cation is present (Sec. 17-1b).   There is also good evidence for decomposition by a free-radical mechanism under some conditions where the diazonium compound is present in a covalent state.   We do not imply, of course, that diazonium cations cannot react by a radical mechanism nor the covalent compounds by a polar one.

Some of the first evidence for the free-radical mechanism was described by Grieve and Hey, who found that the decomposition of N-nitroso-acetanilide in several aromatic solvents led to the substitution of phenyl radicals into the aromatic ring.[30]   Since the decomposition in nitrobenzene solution yields predominantly o- and p-nitrobiphenyl, it is obvious that the reaction is not an ordinary electrophilic aromatic substitution.   These workers suggested that the reaction goes through phenyl diazoacetate, which decomposes to give nitrogen, acetoxy radicals, and phenyl radicals, which bring about the aromatic phenylation.

$$C_6H_5-\overset{\overset{\displaystyle NO}{|}}{N}-Ac \rightarrow C_6H_5-N{=}N-OAc \rightarrow C_6H_5\cdot + N_2 + AcO\cdot$$
$$\downarrow {C_6H_5NO_2}$$
$$\text{mostly } o\text{- and } p\text{-}C_6H_5C_6H_4NO_2$$

The mechanism of the aromatic substitution reaction will be discussed in Sec. 22-3a.   Grieve and Hey also found the first-order reaction rate constants to be very little affected by the nature of the solvent over a range (viz., carbon tetrachloride, benzene, and nitrobenzene) in which the ion-solvating abilities should vary quite widely.   Assuming the rate-controlling step of the reaction to be the decomposition of the diazoacetate, this insensitivity to solvent is very reasonable for a free-radical decomposition but not for a decomposition to ions.   However, Huisgen and Horeld showed that the rate of decomposition of nitrosoacetanilide in benzene, as measured by the rate of formation of nitrogen, is the same as the rate of formation of phenylazo-β-naphthol in the presence of

[30] W. S. M. Grieve and D. H. Hey, *J. Chem. Soc.*, 1797 (1934).

β-naphthol, suggesting that the two reactions have a common rate-controlling step.[31]

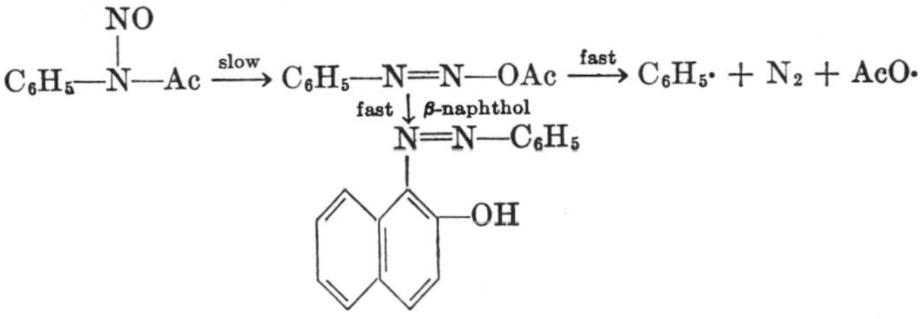

$$\underset{\substack{|\\ C_6H_5\text{—N—Ac}}}{\overset{NO}{}} \xrightarrow{\text{slow}} C_6H_5\text{—N}=\text{N—OAc} \xrightarrow{\text{fast}} C_6H_5\cdot + N_2 + AcO\cdot$$

fast ↓ β-naphthol

$$\underset{\substack{|\\ \text{OH}}}{\overset{N=N\text{—}C_6H_5}{}}$$

Considerable other evidence supports this explanation, which shows that comparison of the rates of decomposition in various solvents gives no evidence for or against the free-radical mechanism.[32] Nevertheless, there are many other data in support of the free-radical mechanism. DeTar made strong arguments based on the fact that in several cases relatively small changes in the reaction conditions bring about a drastic change in the course of the reaction. In each case he pointed out how reasonable it is that the change in conditions should have brought about a change in reaction mechanism and how the change in mechanism should have changed the course of reaction in the manner found. Thus he found that the decomposition of N-nitrosoacetanilide in methanol at 25° gives 25 to 30 per cent benzene and 5 to 10 per cent anisole.[33] The addition of sodium acetate increases the decomposition rate by about twentyfold and the yield of benzene to 40 to 45 per cent, and it decreases the yield of anisole to about 2 per cent. In the presence of 0.04 M sulfuric acid the yield of anisole increases to 55 to 75 per cent, and that of benzene drops to 10 per cent or less. It therefore appears that the nitrosoacetanilide rearranges to phenyl diazoacetate, which, in methanol, ionizes even faster than it decomposes. There is thus a small amount of the reactive covalent acetate in equilibrium with a larger amount of the more stable salt. The addition of acetate ions drives this equilibrium to the left, whereas sulfuric acid changes the acetate ions to acetic acid and drives the equilibrium to the right.

$$C_6H_5\text{—N}=\text{N—OAc} \rightleftharpoons C_6H_5\text{—N}_2^+ + OAc^-$$

---

[31] R. Huisgen and G. Horeld, *Ann.*, **562**, 137 (1949).

[32] R. Huisgen, *Ann.*, **573**, 163 (1951); D. H. Hey, J. Stuart-Webb, and G. H. Williams, *J. Chem. Soc.*, 4657 (1952).

[33] D. F. DeTar, *J. Am. Chem. Soc.*, **73**, 1446 (1951); cf. D. F. DeTar and M. N. Turetzky, *J. Am. Chem. Soc.*, **77**, 1745 (1955); **78**, 3925, 3928 (1956).

The phenyl radical abstracts a hydrogen atom from methanol, preferring the more weakly bonded one attached to carbon, and the phenyl cation coordinates with the unshared electron pair of the oxygen atom. Huisgen and Nakaten made observations similar to these.[34]

PROBLEMS

1. In the free-radical decomposition of a series of $ArCH_2CO_2OC(CH_3)_3$'s a better Hammett-equation correlation was obtained by use of σ^+ values than ordinary σ values. Suggest an explanation for this fact and tell whether ρ is positive or negative (cf. Sec. 20-1c).

2. Describe experiments designed to show whether the induced decomposition of benzoyl peroxide in ether is due to attack on the carbonyl or the peroxy oxygen atom of the peroxide. Point out what check experiments, if any, would be necessary to assure the validity of the results.

3. Suggest a reasonable mechanism for the decomposition of 1 mole of di-*t*-butyl peroxide in 10 moles of 1,1-diethoxybutane. The yields of products are as follows (in moles):

CH_4	0.32
C_2H_6	~1.1
$n\text{-}C_4H_{10}$	0.09
MeCHO	2.86
Me_2CO	0.32
t-BuOH	~1.7
n-PrCHO	0.54
n-BuOEt	0.61
$n\text{-}PrCO_2Et$	0.92
4,5-Diethoxyoctane	0.88

4. The reaction

$$ArN_2^+ + H_3PO_2 + H_2O \rightarrow ArH + H_3PO_3 + H^+ + N_2$$

carried out in acidic aqueous solution, is inhibited by quinones and catalyzed by ferrous sulfate (among other reagents). Suggest a plausible reaction mechanism.

[34] R. Huisgen and H. Nakaten, *Ann.*, **573**, 181 (1951).

Chapter 22

SOME REACTIONS INVOLVING
RADICAL DISPLACEMENTS

22-1. Free-radical Halogenation. *22-1a. Mechanism of Radical Halogenations.* The chlorination and bromination of saturated hydrocarbons have long been known to be light-catalyzed. This fact shows that these reactions are very probably free-radical in nature, and the high quantum yields observed show that they are chain reactions of considerable chain length.

Vaughan and Rust showed that although chlorine and ethane alone react at a negligible rate at 120°, the reaction rapidly goes to completion if 0.002 mole per cent of tetraethyllead, which decomposes to give ethyl radicals at this temperature, is added.[1] They also found that hexaphenylethane is an effective catalyst of chlorinations in the liquid phase at lower temperatures. In the case of $Pb(C_2H_5)_4$ the reaction mechanism is presumably

$$Pb(C_2H_5)_4 \rightarrow Pb + 4C_2H_5 \cdot$$
$$C_2H_5 \cdot + Cl_2 \rightarrow C_2H_5Cl + Cl \cdot$$
$$Cl \cdot + C_2H_6 \rightarrow HCl + C_2H_5 \cdot$$

A number of mechanisms of termination are possible. Some involve "wall reactions" and collisions with a third body to absorb the energy given off. Oxygen is a powerful inhibitor for the reaction, probably because of its great ability to combine with ethyl radicals to form C_2H_5—O—O·, a radical much less capable of removing hydrogen atoms from ethane than is Cl·.

From the nature of the products, we know that the chlorine atom attacks the hydrocarbon molecule to break a carbon-hydrogen bond rather than a carbon-carbon bond, although the latter type of bond is weaker. This is probably due partly to the possibility of forming the hydrogen-chlorine bond (stronger than carbon-chlorine) and partly to

[1] W E. Vaughan and F. F. Rust, *J. Org. Chem.*, **5,** 449 (1940).

the fact that the carbon atoms are shielded, being surrounded entirely by hydrogen atoms.

The bromination of saturated hydrocarbons is rather similar to the chlorination, except that bromine atoms are less reactive than chlorine atoms. Iodine atoms are still less reactive, and for this reason the iodination of alkanes is not a synthetically useful reaction.

Miller and coworkers obtained evidence that fluorine molecules may react with organic compounds to yield free radicals directly without the requirement of any added initiator.[2]

$$R—H + F—F \rightarrow R\cdot + H—F + F\cdot$$

In this reaction the energy of the H—F bond being formed is essentially equal to the sum of the energies of the C—H and F—F bonds being broken. An analogous reaction for any other halogen would be endothermic by at least 50 kcal/mole.

Steacie has discussed the kinetics and mechanisms of a number of vapor-phase halogenation reactions.[3]

22-1b. *Halogenation with Reagents Other than Elemental Halogen.* In addition to the elemental halogens, a number of reagents may be used in free-radical halogenation reactions. One of the most widely used of these reagents is sulfuryl chloride. The extensive use of this compound stems from the studies of Kharasch and Brown, who found that in the presence of a benzoyl peroxide catalyst it is capable of replacing the hydrogen atoms attached to saturated carbon atoms in a wide variety of organic compounds.[4]

Ziegler and coworkers showed that N-bromosuccinimide is a particularly suitable reagent for the replacement of allylic hydrogen by bromine.[5] The fact that the reaction is catalyzed by light[6] and benzoyl peroxide[7] and slowed by picric acid and other inhibitors[8] points to a free-radical mechanism. There is disagreement, however, as to whether the N-bromosuccinimide acts simply to furnish very low concentrations of elemental bromine[8] or whether it takes a more direct part in the reaction, as shown

[2] W. T. Miller, Jr., et al., *J. Am. Chem. Soc.*, **78,** 2793, 4992 (1956).

[3] E. W. R. Steacie, "Atomic and Free Radical Reactions," 2d ed., pp. 657–747, Reinhold Publishing Corporation, New York, 1954.

[4] M. S. Kharasch and H. C. Brown, *J. Am. Chem. Soc.*, **61,** 2142, 3432 (1939); **62,** 925 (1940).

[5] K. Ziegler, A. Späth, E. Schaaf, W. Schumann, and E. Winkelmann, *Ann.*, **551,** 80 (1942).

[6] C. Meystre, L. Ehmann, R. Neher, and K. Miescher, *Helv. Chim. Acta*, **28,** 1252 (1945).

[7] H. Schmid and P. Karrer, *Helv. Chim. Acta*, **29,** 573 (1946).

[8] J. Adam, P. A. Gosselain, and P. Goldfinger, *Nature*, **171,** 704 (1953); *Bull. soc. chim. Belges*, **65,** 525 (1956); B. P. McGrath and J. M. Tedder, *Proc. Chem. Soc.*, 80 (1961).

in the following mechanism:[9]

$$\text{R·} + \underset{CH_2}{\overset{CH_2}{\diagdown}}\!\!\!\!\underset{CO}{\overset{CO}{>}}\!\!N\!-\!Br \rightarrow \underset{CH_2}{\overset{CH_2}{\diagdown}}\!\!\!\!\underset{CO}{\overset{CO}{>}}\!\!N\!· + RBr$$

$$\underset{CH_2}{\overset{CH_2}{\diagdown}}\!\!\!\!\underset{CO}{\overset{CO}{>}}\!\!N\!· + -\overset{H}{\underset{|}{C}}\!-\!C\!=\!C- \rightarrow \underset{CH_2}{\overset{CH_2}{\diagdown}}\!\!\!\!\underset{CO}{\overset{CO}{>}}\!\!NH + -\overset{·}{C}\!-\!C\!=\!C-$$

$$-\overset{·}{C}\!-\!C\!=\!C- + \underset{CH_2}{\overset{CH_2}{\diagdown}}\!\!\!\!\underset{CO}{\overset{CO}{>}}\!\!N\!-\!Br \rightarrow -\overset{Br}{\underset{|}{C}}\!-\!C\!=\!C- + \underset{CH_2}{\overset{CH_2}{\diagdown}}\!\!\!\!\underset{CO}{\overset{CO}{>}}\!\!N\!·$$

22-1c. Orientation in Radical Halogenations. Unless the intermediate radical rearranges (cf. Sec. 23-2), the nature of the product of a free-radical halogenation depends only on which hydrogen atom of the organic reactant is removed by the attacking halogen atom. As Mayo and Walling have pointed out, apparently both radical stability and polar factors are important here.[10] There will be a tendency for the most stable possible radical to be formed, but there will also be a tendency for the electron-withdrawing halogen atom to attack a point of high electron density and avoid a point of low electron density.

The existence of polar factors is perhaps most convincingly illustrated by certain correlations that have been obtained between Hammett's σ constants and reactivity in radical halogenations. In a study of the photobromination of several substituted toluenes Kooyman, van Helden, and Bickel obtained data that gave a reasonable fit to the Hammett equation, with a ρ value of -1.05.[11] Walling and Miller found that ρ is -0.76 in the photochlorination of toluenes.[12] With aliphatic hydrocarbons both polar and radical-stability effects tend to favor the order of reactivity, tertiary > secondary > primary. This order was observed by Hass, McBee, and Weber, who studied the thermal chlorination of propane, butane, isobutane, pentane, and isopentane.[13] By assuming that the relative reactivities of primary, secondary, and tertiary hydrogen

[9] G. F. Bloomfield, *J. Chem. Soc.*, 114 (1944); H. J. Dauben, Jr., and L. L. McCoy, *J. Am. Chem. Soc.*, **81**, 4863 (1959).

[10] (a) F. R. Mayo and C. Walling, *Chem. Revs.*, **46**, 269 (1950); cf. (b) A. B. Ash and H. C. Brown, *Record Chem. Progr. (Kresge-Hooker Sci. Lib.)*, **9**, 81 (1948).

[11] E. C. Kooyman, R. van Helden, and A. F. Bickel, *Koninkl. Ned. Akad. Wetenschap. Proc.*, **56B**, 75 (1953).

[12] C. Walling and B. Miller, *J. Am. Chem. Soc.*, **79**, 4181 (1957).

[13] H. B. Hass, E. T. McBee, and P. Weber, *Ind. Eng. Chem.*, **28**, 333 (1936).

atoms are 1.00 to 3.25 to 4.43, these workers were able to calculate the composition of the monochloride mixtures formed at 300° with the agreement seen in Table 22-1.   The magnitude of these differences in ease

TABLE 22-1. CALCULATED AND EXPERIMENTAL COMPOSITION OF PRODUCTS OF MONOCHLORINATION OF FIVE LOWER HYDROCARBONS AT 300°[13]

Hydrocarbon	Monochlorides, %							
	Calculated				Experimental			
	1-Cl	2-Cl	3-Cl	4-Cl	1-Cl	2-Cl	3-Cl	4-Cl
Propane..............	48	52	....	...	48	52		
Butane...............	32	68	....	...	32	68		
Isobutane............	67	33	....	...	67	33		
2-Methylbutane........	30	22	33	15	33.5	22	28	16.5
Pentane..............	23.5	51	25.5	...	23.8	48.8	27.4	

of substitution varies with the reactivity of the halogen atom in the expected manner.   Substitution becomes more nearly random at higher temperatures, where many chlorine atoms have the ability to remove almost every hydrogen atom with which they collide.[13]   Bromine atoms appear to have a stronger tendency to attack tertiary hydrogen than do the more reactive chlorine atoms.   Roberts and Coraor, for example, noted that the photobromination of isopentane yields almost entirely *t*-amyl bromide, whereas the chlorination reaction under the same conditions gave only a little of the tertiary chloride mixed with a large amount of other monochlorides.[14]

A chlorine atom would be expected to facilitate further chlorination on the same carbon atom by stabilization of the radical formed, but to inhibit reaction at this carbon atom (and to a smaller extent at adjacent carbons) by its electron-withdrawing inductive effect.   Apparently, the polar effect is more important than the radical-stabilization effect.   In most of the data that have been reported on free-radical chlorinations it appears that chlorine (compared with hydrogen) decreases the extent of chlorination on the same (α-) and adjacent (β-) carbon atoms and to a smaller extent on the γ-carbon atom.   Thus Tishchenko reported that the photochlorination of *n*-butyl chloride at 35 to 40° gives substitution on the various carbon atoms to the extent shown below.[15]

$$\overset{\delta}{C}H_3 - \overset{\gamma}{C}H_2 - \overset{\beta}{C}H_2 - \overset{\alpha}{C}H_2 - Cl$$
$$25\% \quad 50\% \quad 17\% \quad 3\%$$

[14] J. D. Roberts and G. R. Coraor, *J. Am. Chem. Soc.*, **74**, 3586 (1952).

[15] D. V. Tishchenko, *Zhur. Obshchei Khim.*, **7**, 658 (1937); *Chem. Abstr.*, **31**, 5755 (1937); cf. M. S. Kharasch and H. C. Brown, *J. Am. Chem. Soc.*, **61**, 2142 (1939).

According to Rust and Vaughan, as the reaction temperature is increased the deactivating influence of chlorine on the α-carbon atom is decreased whereas the deactivation of the β-carbon atom is increased.[16] At 380°, they reported, the dichlorides from the chlorination of n-butyl chloride contain about 22 per cent of the 1,1 isomer; 53 per cent 1,3; 25 per cent 1,4; and only a negligible fraction of the 1,2-dichloride.   Ash and Brown, however, point out that the decrease in the yield of the 1,2 isomer may be due to the decomposition of the intermediate radical at the high temperature used.[10b]

$$C_2H_5CH_2CH_2Cl + Cl\cdot \rightarrow HCl + C_2H_5\overset{\cdot}{C}HCH_2Cl \rightarrow C_2H_5CH{=}CH_2 + Cl\cdot$$

The interplay of radial stability and polar factors may be seen in many of the data of Henne and coworkers on the chlorination of aliphatic fluoride derivatives.   The following results were obtained from chlorinations in the presence of sunlight.[17]

$$F_3CCH_2CH_2CH_3 \rightarrow 44\%\ F_3CCH_2\overset{|}{\underset{Cl}{C}}HCH_3,\ 56\%\ F_3CCH_2CH_2CH_2Cl$$

$F_3CCH_2CH_3 \rightarrow F_3CCH_2CH_2Cl$ (only monochloride isolated)

$F_2CHCH_3 \rightarrow 70\%\ F_2CClCH_3,\ 6\%\ F_2CClCH_2Cl$, but no F_2CHCH_2Cl

$F_3CH \rightarrow F_3CCl$ (very slowly)

$F_3CCH_2CF_3 \rightarrow$ no reaction

Some of the observed data are difficult to explain, however.   For example, although CH_3CF_3 is chlorinated with difficulty, as expected, the only product isolated was Cl_3CCF_3, showing that $ClCH_2CF_3$ is more reactive toward chlorination than is CH_3CF_3.[17]   This tendency of the hydrogen atoms next to a $-CF_3$ group to be deactivated but to be completely replaced once chlorination has started has been observed with a number of other compounds.   Perhaps the inductive effect of the fluorine atoms is so great that the inductive effect added by the chlorine is relatively small.   In this way the radical-stabilizing ability of chlorine becomes the more important factor.   Another unexpected result is that the chlorination of $CH_3CH_2CCl_3$ yields more $CH_3CHClCCl_3$ than $ClCH_2CH_2CCl_3$.[17]

There are a number of possible explanations for these anomalous reactions that have not received careful mechanistic study.   A related anomaly that seems to have a reasonable explanation appears in the chlorination of diethyl ether at room temperature, where α-hydrogen is replaced first, then the three β-hydrogens successively, and only then is the α'-hydrogen attacked.   This is apparently due to the loss of

[16] F. F. Rust and W. E. Vaughan, *J. Org. Chem.*, **6**, 479 (1941).

[17] A. L. Henne et al., *J. Am. Chem. Soc.*, **64**, 1157 (1942); **67**, 1197, 1906 (1945).

hydrogen chloride by the α-chloro ether under the reaction conditions,

$$\text{Et}_2\text{O} \xrightarrow{\text{Cl}_2} \underset{\underset{\text{Cl}}{|}}{\text{CH}_3\text{CHOEt}} \xrightarrow{-\text{HCl}} \text{CH}_2{=}\text{CHOEt} \xrightarrow{\text{Cl}_2} \underset{\underset{\text{Cl}\ \ \text{Cl}}{|\ \ |}}{\text{CH}_2\text{CHOEt}}$$

$$\downarrow -\text{HCl}$$

$$\underset{\underset{\text{Cl}}{|}}{\text{Cl}_3\text{CCHOEt}} \xleftarrow{\text{etc.}} \underset{\underset{\text{Cl}}{|}}{\text{Cl}_2\text{CHCHOEt}} \xleftarrow{\text{Cl}_2} \underset{\underset{\text{Cl}}{|}}{\text{CH}{=}\text{CHOEt}}$$

since chlorination at -25 to $-30°$ yields the α,α'-dichloride.[18]

In the free-radical halogenation of a n-alkylbenzene such as n-propylbenzene there are seen to be three types of hydrogens that could be removed by the attack of a halogen atom. These are the aromatic hydrogens, the benzyl hydrogens, and the others, which we shall call aliphatic hydrogens. Only by removal of a benzyl hydrogen may a resonance-stabilized radical be formed. It is therefore not surprising that the free-radical halogenation of n-alkylbenzenes ordinarily yields benzyl-type halides. Evidently the slight electron-withdrawing power of the aromatic ring has a smaller effect than radical stabilization. For the removal of an aromatic hydrogen this polar factor should be stronger, and there is evidence that the phenyl radical is less stable than ordinary aliphatic radicals. As an illustration of the fact that aliphatic-type hydrogens are replaced in preference to aromatic ones, the chlorination of t-butylbenzene (which has no benzyl hydrogens) to 1-chloro-2-methyl-2-phenylpropane[4,19] may be mentioned (cf. Sec. 23-2b). The difficulty of replacing an aromatic hydrogen atom is seen in the radical chlorination of benzene, which yields the addition product benzene hexachloride at ordinary temperatures and gives substitution only at considerably elevated temperatures.

By letting a small amount of chlorine react with an excess of a hydrocarbon mixture and analyzing the organic chloride mixture formed, Brown and Russell showed that the hydrogen atoms in cyclohexane are about four times as reactive as the side-chain hydrogens in toluene toward photochlorination at $80°$.[20] This is a surprising result, even though the hydrogen atoms in cyclohexane are secondary and those in toluene are primary. Judging from the relative rates of decomposition of azo nitriles (Table 21-3, II), the stability of a cyclohexyl radical should be comparable to that of a secondary aliphatic radical. By using deuterium-labeled hydrocarbons, it was shown that the situation was not being complicated by the attack of hydrocarbon radicals on hydrocarbon molecules.[20]

[18] G. E. Hall and F. M. Ubertini, *J. Org. Chem.*, **15**, 715 (1950).

[19] W. E. Truce, E. T. McBee, and C. C. Alfieri, *J. Am. Chem. Soc.*, **71**, 752 (1949).

[20] H. C. Brown and G. A. Russell, *J. Am. Chem. Soc.*, **74**, 3995 (1952).

The polar effect on radical halogenation is perhaps found most strikingly in the reactions of compounds with such strongly electron-withdrawing substituents as the carboxyl group and its derivatives. In the radical chlorination of aliphatic acids and their derivatives the α-hydrogen atoms appear to be relatively inactive.[4] In the chlorination of isobutyryl chloride with sulfuryl chloride in the presence of benzoyl peroxide, 80 per cent β-chloro- and 20 per cent α-chloroisobutyryl chloride are formed.[4] It appears that neutral or electron-donating radicals preferentially remove α-hydrogen atoms because of the greater stability of the radicals formed. At least, this is the interpretation given[10] the observation of Kharasch and Gladstone that the decompositions of acyl peroxides in aliphatic acid solutions yield succinic acid derivatives.[21]

$$(CH_3CO_2)_2 \rightarrow 2CH_3\cdot + 2CO_2$$

$$CH_3\cdot + (CH_3)_2CHCO_2H \rightarrow CH_4 + (CH_3)_2\overset{\cdot}{C}CO_2H$$

$$2(CH_3)_2\overset{\cdot}{C}CO_2H \rightarrow \begin{array}{c} (CH_3)_2C-CO_2H \\ | \\ (CH_3)_2C-CO_2H \end{array}$$

It has been suggested that the reaction above involves the formation of radicals on the β-carbon and that these then remove an α-hydrogen atom from another molecule to give the radical on the α-carbon.[10b] Price and Morita, however, showed that this does not occur to any large extent, if it occurs at all. They studied the decomposition of acetyl peroxide in α-deuterioisobutyryl chloride and found considerable quantities of deuteriomethane in the gases produced.[22] Their data show that methyl radicals attack α-hydrogen about 12 times as rapidly as they attack β-hydrogen atoms. The orientation observed in the radical chlorination of acids and their derivatives makes it appear likely that such halogenation methods as the Hell-Volhard-Zelinski, which give preferential alpha substitution, proceed by a different mechanism, probably one analogous to that of the acid-catalyzed halogenation of ketones.

Ash and Brown discussed the effect of structure on reactivity and orientation in radical chlorinations and concluded that various groups have the following net effect on the ease of replacement of a hydrogen atom attached to the same carbon atom: $C_6H_5 > CH_3 > H > AcO > ClCH_2 > Cl_2CH > Cl_3Si > CO_2H > Cl > COCl > Cl_3CCO_2 > CCl_3 > CF_3$.[10b,23]

22-1d. *Solvent Effects on Radical Halogenation.* The suggestion of Russell and Brown that the reactive chlorinating species in free-radical

[21] M. S. Kharasch and M. T. Gladstone, *J. Am. Chem. Soc.,* **65,** 15 (1943).
[22] C. C. Price and H. Morita, *J. Am. Chem. Soc.,* **75,** 3686 (1953).
[23] A. B. Ash and H. C. Brown, *J. Am. Chem. Soc.,* **77,** 4019 (1955).

chlorinations in aromatic solvents may be a π complex[24] was investigated and extended by Russell.[25] The ratio of the reactivity of the tertiary hydrogen atoms of 2,3-dimethylbutane to that of the primary hydrogens, which has the value 3.7 for chlorinations in the pure hydrocarbon solvent at 55°, remains in the range 3.5 to 4.8 for a considerable variety of added solvents (CCl_4, CH_3NO_2, CH_3OAc, C_2HCl_3, $SiCl_4$, t-BuOH, EtCN, and cyclohexene) all at a concentration of 4 M. For a number of aromatic compounds and a few others [CS_2, $SOCl_2$, S_2Cl_2, $HCON(CH_3)_2$, (n-C_4-$H_9)_2O$, and dioxane], higher values ranging up to 30 and above were obtained under the same conditions. These solvents are believed to form complexes with the chlorine atom that greatly decrease its reactivity. Thus the immediate reactants in the attack of Cl· on RH are stabilized so that the transition state is caused to lie nearer the products HCl and R· (cf. Fig. 5-3). Since the transition states have more resemblance to the R· products, the difference in stability of the transition states will be nearer the difference in stability between a primary and tertiary radical (about 13 kcal [Table 19-2], corresponding to a k_{tert}/k_{prim} value of $\sim 10^9$). The complexes formed with most aromatic compounds seem more reasonably depicted as π complexes than as σ complexes. The strength of the complexes formed during chlorination, as measured by k_{tert}/k_{prim} values (increasing in the order $C_6H_5NO_2 < C_6H_5COCl \sim C_6H_5CF_3 < C_6H_5CO_2CH_3 \sim C_6H_5Cl \sim C_6H_5F < C_6H_6 < C_6H_5CH_3 < C_6H_5OCH_3 < C_6H_5I$), seems more closely related to the stabilities of π complexes formed with HCl than to the rates of formation of radical-aromatic σ complexes by attack of methyl[26] or phenyl radicals[27] (cf. Sec. 22-3b). In the case of a few compounds containing iodine and sulfur atoms the k_{tert}/k_{prim} values are much higher than would be expected for an aromatic π complex, and the complex may instead involve the formation of a chlorine-iodine (or sulfur) bond, e.g.,

$$C_6H_5-\overset{\diagup\cdot}{\underline{I}}-\overline{Cl} \quad \longleftrightarrow \quad C_6H_5-\underline{I}-\overline{Cl}\backslash$$

Indeed, when iodobenzene was used, the formation of iodobenzene dichloride ($C_6H_5ICl_2$) accompanied the hydrocarbon-chlorination reaction. This polar interaction that leads to complex formation might be expected to be particularly important for chlorine atoms, which have the highest electron affinity (88.2 kcal/mole) of any common radicals. From the electron affinities of fluorine (83.5), bromine (81.6), and

[24] G. A. Russell and H. C. Brown, *J. Am. Chem. Soc.*, **77**, 4031 (1955).

[25] G. A. Russell, *J. Am. Chem. Soc.*, **80**, 4987 (1958).

[26] W. J. Heilman, A. Rembaum, and M. Szwarc, *J. Chem. Soc.*, 1127 (1957).

[27] D. R. Augood and G. H. Williams, *Chem. Revs.*, **57**, 123 (1957).

iodine (74.6) atoms, and of HOO· (70), hydroxyl (50), triphenylmethyl (48), phenoxy (27), and methyl (25) radicals, such solvent effects might be expected to be of decreasing importance for the various radicals in the order listed.[24] Russell described evidence for complex formation by other radicals.

22-2. Autooxidation.   *22-2a. Mechanism of Autooxidation Reactions.* The autooxidation of benzaldehyde, like that of many other compounds, has been found to be a light-catalyzed reaction.   From this fact a free-radical mechanism appears probable, and since Bäckström found quantum yields on the order of 10,000 for the reaction, a chain mechanism is likely.[28] The light-catalyzed autooxidation reaction was found to be markedly inhibited by small concentrations of diphenylamine, phenol, and anthracene.[28] Ziegler and Ewald showed that hexaphenylethane, known to dissociate to free radicals, catalyzes the autooxidation of aldehydes.[29] This direct evidence for the intermediacy of free radicals and the chemical course of the reaction (hydrogen is replaced by the hydroperoxy group) suggest the following mechanism for initiation and chain propagation:

$$(C_6H_5)_3CC(C_6H_5)_3 \rightarrow (C_6H_5)_3C\cdot$$
$$(C_6H_5)_3C\cdot + O_2 \rightarrow (C_6H_5)_3COO\cdot$$
$$(C_6H_5)_3COO\cdot + C_6H_5CHO \rightarrow (C_6H_5)_3COOH + C_6H_5CO\cdot$$
$$C_6H_5CO\cdot + O_2 \rightarrow C_6H_5COO_2\cdot$$
$$C_6H_5COO_2\cdot + C_6H_5CHO \rightarrow C_6H_5COO_2H + C_6H_5CO\cdot$$

The oxygen molecule has two unpaired electrons and may thus be regarded as a sort of diradical.   It is not a sufficiently reactive radical to attack most organic molecules under ordinary conditions, but it is very facile at combining with free radicals.   The hydroperoxy radicals formed by the addition of oxygen to organic radicals are much more reactive than is oxygen, but they are only moderately reactive compared with most organic radicals.

The determination of the nature of the termination step or steps requires a more careful and quantitative study of the reaction.   Such a study appears to have been carried out first in connection with the autooxidation of olefins and has been reviewed by Bolland[30] and by Bateman.[31] If we assume that all radicals formed from the initiator attack either oxygen or the hydrocarbon and that termination is due to a reac-

[28] H. L. J. Bäckström, *J. Am. Chem. Soc.*, **49**, 1460 (1927).

[29] K. Ziegler and L. Ewald, *Ann.*, **504**, 162 (1933).

[30] J. L. Bolland, *Quart. Revs. (London)*, **3**, 1 (1949).

[31] L. Bateman, *Quart. Revs. (London)*, **8**, 147 (1954).

tion between two chain-carrying radicals, we may write the mechanism

$$\text{Initiator} \rightarrow \text{radicals (R}\cdot \text{ and/or RO}_2\cdot) \qquad \text{rate} = r_1$$

$$\text{R}\cdot + \text{O}_2 \xrightarrow{k_2} \text{RO}_2\cdot \tag{22-1}$$

$$\text{RO}_2\cdot + \text{RH} \xrightarrow{k_3} \text{R}\cdot + \text{RO}_2\text{H} \tag{22-2}$$

$$2\text{R}\cdot \xrightarrow{k_4} \left. \begin{array}{l} \\ \end{array} \right\} \tag{22-3}$$

$$\text{R}\cdot + \text{RO}_2\cdot \xrightarrow{k_5} \left\{ \begin{array}{l} \text{products not} \\ \text{further entering} \\ \text{the reaction} \end{array} \right. \tag{22-4}$$

$$2\text{RO}_2\cdot \xrightarrow{k_6} \left. \begin{array}{l} \\ \end{array} \right. \tag{22-5}$$

If the chain length is long, so that the number of radicals produced in propagation steps is very large compared with those produced by initiation,

$$k_2[\text{R}\cdot][\text{O}_2] = k_3[\text{RO}_2\cdot][\text{RH}] \tag{22-6}$$

From the steady-state treatment the rate of formation of radicals is equal to their rate of disappearance

$$r_1 = k_4[\text{R}\cdot]^2 + 2k_5[\text{R}\cdot][\text{RO}_2\cdot] + k_6[\text{RO}_2\cdot]^2 \tag{22-7}$$

A value for $[\text{RO}_2\cdot]$ from Eq. (22-6) may be substituted in (22-7) to give

$$[\text{R}\cdot] = \frac{r_1^{1/2}[\text{RH}]}{(k_4[\text{RH}]^2 + 2k_2k_3^{-1}k_5[\text{O}_2][\text{RH}] + k_2^2k_3^{-2}k_6[\text{O}_2]^2)^{1/2}} \tag{22-8}$$

The rate of the over-all reaction is measured by the uptake of oxygen and may be taken to be equal to the rate of reaction (22-1). Substitution of Eq. (22-8) into the rate equation for (22-1) gives

$$v = \frac{r_1^{1/2}[\text{RH}][\text{O}_2]}{(k_2^{-2}k_4[\text{RH}]^2 + 2k_2^{-1}k_3^{-1}k_5[\text{O}_2][\text{RH}] + k_3^{-2}k_6[\text{O}_2]^2)^{1/2}} \tag{22-9}$$

In agreement with this equation, the reaction rate has, in a number of cases, been shown to be proportional to the square root of the initiation rate. Thus the rate of the benzoyl peroxide–induced reaction is proportional to the square root of the initiator concentration,[32] and the light-catalyzed reaction has a rate proportional to the square root of the light intensity.[33] Since k_2 is much larger than k_3, as expected, at fairly high oxygen concentrations, $\text{RO}_2\cdot$ radicals tend to reach a concentration much higher than that of $\text{R}\cdot$ radicals. Under these conditions termination occurs essentially entirely by mechanism (22-5), and the first two terms

[32] J. L. Bolland, *Proc. Roy. Soc.* (*London*), **186A**, 218 (1946); *Trans. Faraday Soc.*, **44**, 669 (1948).

[33] L. Bateman, *Trans. Faraday Soc.*, **42**, 266 (1946); L. Bateman and G. Gee, *Proc. Roy. Soc.* (*London*), **195A**, 376, 391 (1948).

in the denominator of Eq. (22-9) may be neglected compared with the third. Thus at high oxygen pressures the rate equation approaches the form

$$v = k_3 k_6^{-\frac{1}{2}} r_1^{\frac{1}{2}} [RH] \qquad (22\text{-}10)$$

For analogous reasons, at low oxygen pressures the reaction may become first-order in oxygen.

$$v = k_2 k_4^{-\frac{1}{2}} r_1^{\frac{1}{2}} [O_2]$$

The intermediate cases are also known, of course.

In a study of the autooxidation of n-decanal Cooper and Melville suggested that initiation may take place by the attack of oxygen on the aldehyde.[34] A similar initiation mechanism has been suggested for the vapor-phase oxidation of ethers.[35]

22-2b. *Reactivity in Autooxidations.* The reactivity of organic compounds in autooxidation reactions is probably most commonly controlled by the rate of attack of radicals on the compound.

$$RO_2\cdot + RH \rightarrow RO_2H + R\cdot$$

The reactivity in this step of the reaction is rather similar to that in free-radical halogenations, being increased by increasing stability of the radical being formed and by the presence of electron-donating substituents. Due to the relatively low reactivity of $RO_2\cdot$ radicals, many autooxidation reactions show considerable selectivity.

Because of the greater reactivity of tertiary hydrogen atoms and the greater stability of tertiary hydroperoxides, it is feasible to prepare hydroperoxides from saturated hydrocarbons only in the case of the tertiary compounds. Thus, 9-decalyl hydroperoxide may be prepared by the autooxidation of decalin, but only in poor yield.[36] The reaction is greatly facilitated when the hydrogen atoms replaced are of the benzyl type, cumene and tetralin giving good yields of hydroperoxides.[37]

The relative reactivity of olefins toward autooxidation has been studied rather carefully. Due to the intermediacy of resonance-stabilized allylic free radicals, these reactions may involve double-bond migrations. Indeed, the fact that the autooxidation of ethyl linoleate yields hydroperoxides with conjugated double bonds[38] has been quoted

[34] H. R. Cooper and H. W. Melville, *J. Chem. Soc.*, 1984 (1951).

[35] T. A. Eastwood and C. Hinshelwood, *J. Chem. Soc.*, 733 (1952).

[36] A. C. Cope and G. Holzman, *J. Am. Chem. Soc.*, **72**, 3062 (1950).

[37] M. Hartmann and M. Seiberth, *Helv. Chim. Acta*, **15**, 1390 (1932); H. Hock and S. Lang, *Ber.*, **77B**, 257 (1944).

[38] J. L. Bolland and H. P. Koch, *J. Chem. Soc.*, 445 (1945); cf. E. H. Farmer, H. P. Koch, and D. A. Sutton, *J. Chem. Soc.*, 541 (1943).

as evidence that organic free radicals are intermediates in autooxidation processes.

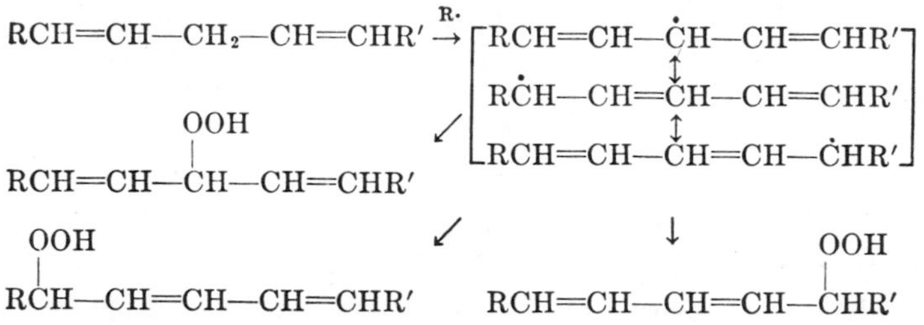

$$RCH{=}CH{-}CH_2{-}CH{=}CHR' \overset{R\cdot}{\rightarrow} \begin{bmatrix} RCH{=}CH{-}\overset{\cdot}{C}H{-}CH{=}CHR' \\ R\overset{\cdot}{C}H{-}CH{=}\overset{\updownarrow}{C}H{-}CH{=}CHR' \\ RCH{=}CH{-}\overset{\updownarrow}{C}H{=}CH{-}\overset{\cdot}{C}HR' \end{bmatrix}$$

$$\underset{\substack{| \\ RCH{=}CH{-}CH{-}CH{=}CHR'}}{OOH} \qquad \swarrow \qquad \downarrow \qquad \underset{\substack{| \\ RCH{=}CH{-}CH{=}CH{-}CHR'}}{OOH}$$

$$\underset{\substack{| \\ R\overset{\cdot}{C}H{-}CH{=}CH{-}CH{=}CHR'}}{OOH} \qquad \swarrow$$

where R $= n\text{-}C_5H_{11}$; R' $= (CH_2)_7CO_2Et$

Bolland pointed out some correlations between olefin structure and reactivity in the high-oxygen-pressure region where the kinetic Eq. (22-10) is obeyed and the reactivity is controlled by step (22-2).[39] Taking propylene

$$\underset{\gamma \qquad \beta \qquad \alpha}{CH_3{-}CH{=}CH_2}$$

as a reference compound, he stated that at 45° (1) replacement of n hydrogen atoms at α or γ by alkyl groups increases the reactivity 3.3^n-fold, (2) replacement of an α-hydrogen by phenyl increases the reactivity 23-fold, and (3) replacement of a γ-hydrogen by a 1-alkenyl group increases the reactivity 107-fold.

It should be pointed out that, although we have discussed only the abstraction of hydrogen atoms from olefins, attacking peroxy radicals may also add to the double bonds, often with interesting results.[30,31]

Ethers are particularly sensitive to autooxidation, as might be expected, since the reaction involves attack by an electron-withdrawing radical. Diisopropyl ether has long been known to oxidize more easily than diethyl ether, and Eastwood and Hinshelwood found that dimethyl ether is particularly resistant to oxidation.[40] The ease of oxidation of aldehydes is probably due to the stability of the acyl radical. Walling and McElhill showed that electron-donating substituents increase the reactivity of benzaldehydes toward a given perbenzoate radical.[41] It was found that n-butyraldehyde was several times as reactive as benzaldehyde, suggesting that the polar effect in this case is more important than the radical-stability effect.

[39] J. L. Bolland, *Trans. Faraday Soc.*, **46,** 358 (1950).
[40] T. A. Eastwood and C. Hinshelwood, *J. Chem. Soc.*, 733 (1952).
[41] C. Walling and E. A. McElhill, *J. Am. Chem. Soc.*, **73,** 2927 (1951).

22-3. Free-radical Aromatic Substitution.[42,43]   *22-3a. Mechanism of Free-radical Aromatic Substitution.*   It is likely that certain aromatic substitution reactions, such as high-temperature vapor-phase halogenations, proceed by a mechanism quite analogous to that of most free-radical aliphatic substitutions with such chain-propagating steps as

$$Cl\cdot + C_6H_6 \rightarrow HCl + C_6H_5\cdot$$
$$C_6H_5\cdot + Cl_2 \rightarrow C_6H_5Cl + Cl\cdot$$

For certain free-radical substitution reactions occurring in solution, however, there are reasons for considering other types of reaction mechanisms.   The decomposition of phenyldiazoacetate (Sec. 21-2c), of benzoyl peroxide, and of several other compounds in aromatic solvents leads to the phenylation of the aromatic rings.   The mechanism of these reactions was first considered carefully by Grieve and Hey, who suggested that a phenyl radical was simply displacing a hydrogen atom from the aromatic ring.[44]

$$C_6H_5\cdot + ArH \rightarrow C_6H_5{-}Ar + H\cdot \tag{22-11}$$

However, as DeTar and Sagmanli pointed out, no good evidence (such as the formation of certain reduction products) for the intermediacy of hydrogen atoms appears to have been found.[45]   Furthermore, it does not seem likely that such a one-step displacement of a hydrogen atom would be a rapid enough reaction to compete with the possible competing reactions (such as dimerization) which must be very rapid, since the phenyl radical is known from other studies to be quite reactive.   The reaction simply involves the formation of a carbon-carbon bond while breaking a carbon-hydrogen bond, and it appears that carbon-hydrogen bonds are invariably much stronger than the corresponding carbon-carbon bonds.   Therefore it seems that reaction (22-11) should have too high an activation energy to occur fast enough to explain the observed data.   It could also be suggested that the aromatic substitution is part of an induced chain decomposition reaction,[45] in which radicals formed from the solvent attack the source of phenyl radicals.

$$C_6H_5{-}N{=}N{-}OAc \rightarrow C_6H_5\cdot + N_2 + AcO\cdot$$
$$C_6H_5\cdot + ArH \rightarrow C_6H_5H + Ar\cdot$$
$$Ar\cdot + C_6H_5{-}N{=}N{-}OAc \rightarrow Ar{-}C_6H_5 + N_2 + AcO\cdot$$

If this is the reaction mechanism, then the unsubstituted phenyl radical must be capable of bringing about induced decomposition in this way,

[42] O. C. Dermer and M. T. Edmison, *Chem. Revs.*, **57,** 77 (1957).

[43] G. H. Williams, "Homolytic Aromatic Substitution," Pergamon Press, New York, 1960.

[44] W. S. M. Grieve and D. H. Hey, *J. Chem. Soc.*, 1797 (1934).

[45] D. F. DeTar and S. V. Sagmanli, *J. Am. Chem. Soc.*, **72,** 965 (1950).

since benzene is among the aromatic compounds that can be phenylated by the methods described. Yet biphenyl appears never to have been isolated as a by-product in the phenylation of other aromatic compounds.

The most probable mechanism seems to be a two-stage version of (22-11). The phenyl radical adds to the aromatic ring to give a radical from which a hydrogen atom may then be abstracted.

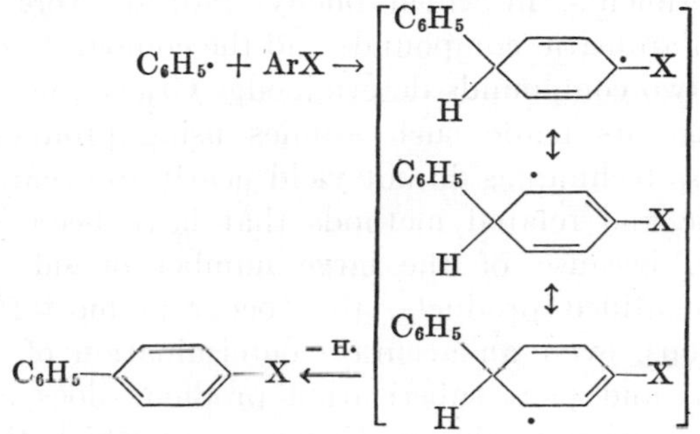

$$C_6H_5\cdot + ArX \rightarrow \text{(radical intermediate)} \qquad (22\text{-}12)$$

DeTar and Long supplied strong evidence for the intermediacy of the above radical by isolation of a dimerization and a disproportionation product from the intermediate radical in the phenylation of benzene.[46]

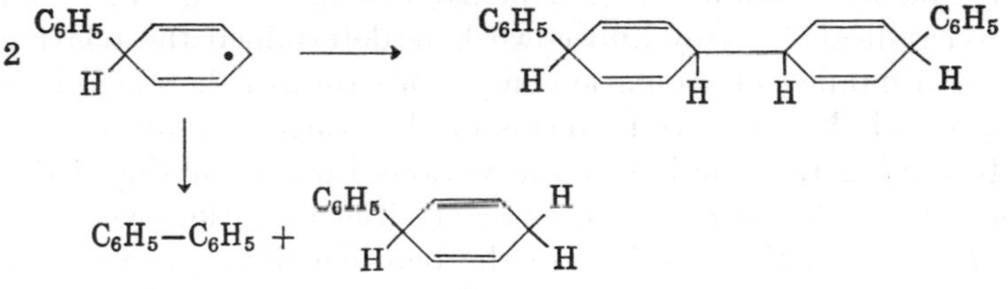

$$C_6H_5\text{—}C_6H_5 + \text{(I)}$$

I

Although the phenylcyclohexadiene (I) can be isolated by working under nitrogen, it is oxidized to biaryl by air in conventional methods of carrying out the reaction.

Additional evidence for the phenylcyclohexadienyl-radical mechanism comes from hydrogen kinetic-isotope-effect studies. Several groups of workers have shown that when deuterium- or tritium-labeled aromatics are phenylated, the unreacted material remaining after a considerable fraction has reacted has the same isotopic content as the original reactant.[47,48] This shows that the attack of radicals on the aromatic ring is probably not appreciably reversible.

[46] D. F. DeTar and R. A. J. Long, *J. Am. Chem. Soc.*, **80**, 4742 (1958); cf. R. O. C. Norman and W. A. Waters, *J. Chem. Soc.*, 167 (1958).

[47] R. J. Convery and C. C. Price, *J. Am. Chem. Soc.*, **80**, 4101 (1958); C. Shih, D. H. Hey, and G. H. Williams, *J. Chem. Soc.*, 1871 (1959).

[48] E. L. Eliel, S. Meyerson, Z. Welvart, and S. H. Wilen, *J. Am. Chem. Soc.*, **82**, 2936 (1960).

22-3b. Reactivity and Orientation in Free-radical Aromatic Substitution.
The relative reactivities of the various positions on a given aromatic
ring toward free-radical aromatic substitution may be studied, of course,
by determining the product ratios in the radical substitution of the
appropriate aromatic compound. In addition, the relative reactivities
of different aromatic rings have been determined in a number of cases by
"competition" experiments, in which phenyl radicals were generated
in a mixture of two aromatic compounds and the comparative extent of
phenylation of the two compounds determined. Grieve and Hey[44] and
other early investigators made such studies using product-isolation
techniques, but these techniques do not yield nearly so accurate results
as the spectroscopic and related methods that have been used more
recently. However, because of the large number of side reactions,
many yielding unidentified products, that occur in most free-radical
aromatic substitutions, even an accurate determination of the yields
of the ortho, meta, and para substitution products does not give a
very reliable determination of the relative rates at which the original
radical attacked the various positions of the aromatic ring.

Hey, Nechvatal, and Robinson showed that essentially the same
results were obtained whether *N*-nitrosoacetanilide, phenylazotriphenyl-
methane, phenyl azohydroxide, or benzoyl peroxide is used as the source
of phenyl radicals.[49] Hey and coworkers determined the relative reactivities of a number of aromatic compounds toward the phenyl radical[50]
and expressed their results in terms of the *reactivity relative to benzene*.
By determining the fraction of the various isomers produced they also
obtained the *partial rate factors*, or reactivities of various positions relative to any one of the positions in the benzene ring. The collection of
such data (due to a number of workers) listed in Augood and Williams'
review[43] is given in Table 22-2. Also listed are the relative reactivities
of some of the aromatic species toward methyl radicals. These *methyl
affinities*, due to Szwarc and coworkers, are measures of the *total* rate at
which the aromatic compound reacts with methyl radicals, not just the
rate at which ring hydrogens are replaced by methyl groups.[26,51]

Since it appears that practically any α substituent stabilizes radicals
better than hydrogen does, we should expect 4 substitution and (barring
excessive steric effects) 2 substitution, in which the carbon atom to which
the substituent is attached is given radical character during the reaction,
to be faster than 3 substitution. In addition to this type of stabilization
of the intermediate in radical substitution, illustrated by contributing

---

[49] D. H. Hey, A. Nechvatal, and T. S. Robinson, *J. Chem. Soc.*, 2892 (1951); cf. D. F.
DeTar and H. J. Scheifele, Jr., *J. Am. Chem. Soc.*, **73**, 1442 (1951).

[50] D. H. Hey et al., *J. Chem. Soc.*, 2094 (1952); 3412 (1953); 794 (1954).

[51] M. Levy and M. Szwarc, *J. Am. Chem. Soc.*, **77**, 1949 (1955).

TABLE 22-2. RELATIVE REACTIVITIES IN RADICAL ATTACK ON
VARIOUS AROMATIC COMPOUNDS

Compound	Partial rate factors in phenylation			Relative reactivity[43] (phenylation)	Relative methyl affinity[26]
	2 Substitution	3 Substitution	4 Substitution		
C_6H_6	1.00	1.00	1.00	1.00	1.00
C_6H_5F	2.2	1.5	1.2	1.4	2.2
C_6H_5Cl	2.6	1.1	1.2	1.4	4.2
C_6H_5Br	2.1	1.8	1.8	1.8	3.6
C_6H_5I	2.1	1.7	1.8	1.8	
C_6H_5Me	3.5	1.0	1.4	1.7	
$C_6H_5CH_2Me$	2.0	1.0	1.4	1.2	
$C_6H_5CHMe_2$	0.8	1.1	1.4	0.9	
$C_6H_5CMe_3$	0.6	1.3	1.4	0.9	
$C_6H_5C_6H_5$	2.9	1.4	3.4	4.0	5
$C_6H_5NO_2$	6.9	1.2	7.7	4.0	
C_6H_5CN	6.5	1.1	6.5	3.6	12.2
$C_6H_5SO_3CH_3$	2.4	1.5	1.3	1.5	
$C_6H_5CF_3$	0.6	1.2	2.4	1.0	
$C_6H_5SiMe_3$	1.0	1.4	1.5	1.1	
C_5H_5N	1.8	0.9	0.9	1.0	3
$C_{10}H_8$	5.5	30[a]	....	24	22

[a] 1 Substitution.

structure II for the 2 substitution of nitrobenzene, it may be pointed out
that 2 substitution gives less interference with resonance interaction
between the substituent and the aromatic ring. Thus, as shown below,
for the intermediate radical in the 2 substitution of nitrobenzene three
contributing structures involving resonance interaction between the
nitro group and the ring may be written, whereas only two such structures may be written for the intermediate for 4 substitution. Such
activation of the 2 position is opposed by steric hindrance to a varying
extent.

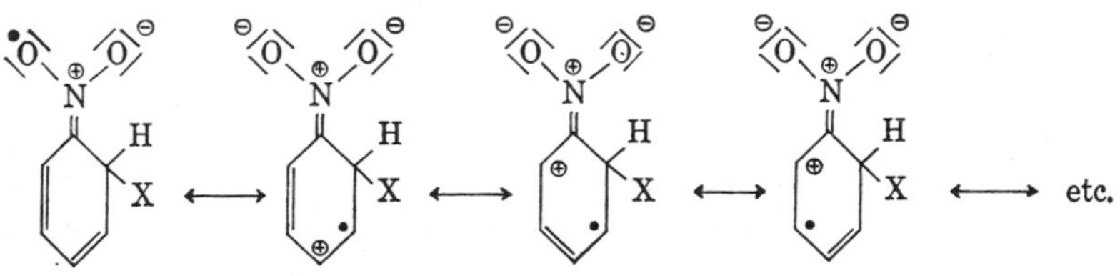

II

The data in Table 22-2 tend to agree with these theoretical generalizations. Some of the disagreements are probably due to experimental

error, but there are also, no doubt, other theoretical factors that should be considered.

PROBLEMS

1. The decomposition of $(C_6H_5)_2CHN\!=\!NCH(C_6H_5)_2$ in the presence of $(C_6H_5)_2$-$C^{14}H_2$ gives $(C_6H_5)_2CHCH(C_6H_5)_2$ containing essentially no C^{14}. When a small amount of thiophenol is present in the reaction mixture, the $(C_6H_5)_2CHCH(C_6H_5)_2$ formed (in somewhat diminished yield) contains a considerable amount of C^{14}. The presence of benzyl mercaptan or of 2,4,6-trimethylthiophenol, however, does not cause C^{14} to appear in the $(C_6H_5)_2CHCH(C_6H_5)_2$ formed. Explain these results.

2. The free-radical chlorination of 2,3-dimethylbutane by elemental chlorine at 55° gives about 38 per cent of the tertiary and 62 per cent of the primary monochloride. Under the same conditions, sulfuryl chloride gives about 63 per cent tertiary and 37 per cent primary product. Give a qualitative explanation for these results.

3. Suggest a mechanism for the following reaction:

$$(RO)_2PHO + CCl_4 + 2NH_3 \xrightarrow[\text{light}]{\text{air}} (RO)_2PONH_2 + CHCl_3 + NH_4Cl$$

Chapter 23

STEREOCHEMISTRY AND REARRANGEMENTS
OF FREE RADICALS

23-1. Stereochemistry of Free Radicals. *23-1a. Optical Activity of Free Radicals.* Brown, Kharasch, and Chao carried out an important investigation that has added greatly to our knowledge of the stereochemistry of free radicals.[1] They studied the free-radical chlorination of optically active 2-methylbutyl chloride and found that the 1,2-dichloro-2-methylbutane formed was racemic. This shows that the intermediate 1-chloro-2-methyl-2-butyl radicals racemized. This observation is also good evidence that the attack of a chlorine atom on the alkyl chloride does yield hydrogen chloride and an alkyl radical

$$
\underset{\underset{\text{CH}_3}{|}}{\text{CH}_3\text{CH}_2\text{CHCH}_2\text{Cl}} + \text{Cl}\cdot \rightarrow \text{HCl} + \underset{\underset{\text{CH}_3}{|}}{\text{CH}_3\text{CH}_2\overset{\cdot}{\text{C}}\text{CH}_2\text{Cl}}
$$

rather than the dichloride and a hydrogen atom

$$
\underset{\underset{\text{CH}_3}{|}}{\text{CH}_3\text{CH}_2\text{CHCH}_2\text{Cl}} + \text{Cl}\cdot \rightarrow \underset{\underset{\underset{\text{Cl}}{|}}{\overset{\text{CH}_3}{|}}}{\text{CH}_3\text{CH}_2\text{C}\text{CH}_2\text{Cl}} + \text{H}\cdot
$$

since it would be a very improbable coincidence that the chlorine atom should attack the asymmetric carbon atom from the front and from the back at exactly the same rate.

There have been a number of other observations of the racemization of free-radical intermediates. In the *t*-butyl peroxide–induced decarbonylation[2] of optically active methylethylisobutylacetaldehyde Doering and coworkers found the 2,4-dimethylhexane produced to be racemic.[3]

---

[1] H. C. Brown, M. S. Kharasch, and T. H. Chao, *J. Am. Chem. Soc.*, **62,** 3435 (1940).

[2] S. Winstein and F. H. Seubold, Jr., *J. Am. Chem. Soc.*, **69,** 2916 (1947).

[3] W. v. E. Doering, M. Farber, M. Sprecher, and K. B. Wiberg, *J. Am. Chem. Soc.*, **74,** 3000 (1952).

$$
\underset{\overset{|}{\text{Et}}}{\overset{\overset{\text{Me}}{|}}{i\text{-Bu}-\text{C}}}-\text{CHO} + t\text{-BuO}\cdot \rightarrow t\text{-BuOH} + \underset{\overset{|}{\text{Et}}}{\overset{\overset{\text{Me}}{|}}{i\text{-Bu}-\text{C}}}-\overset{\cdot}{\text{C}}{=}\text{O}
$$

$$\downarrow$$

$$
\underset{\overset{|}{\text{Et}}}{\overset{\overset{\text{Me}}{|}}{i\text{-Bu}-\text{CH}}} + \underset{\overset{|}{\text{Et}}}{\overset{\overset{\text{Me}}{|}}{i\text{-Bu}-\text{C}}}-\overset{\cdot}{\text{C}}{=}\text{O} \xleftarrow[\underset{\text{Et}}{i\text{-Bu}-\text{C}-\text{CHO}}]{\overset{\text{Me}}{}} \underset{\overset{|}{\text{Et}}}{\overset{\overset{\text{Me}}{|}}{i\text{-Bu}-\text{C}\cdot}}
$$

Gruver and Calvert showed that the 2-butyl iodide formed in the photolysis of active 2-methylbutanal was optically inactive[4]

$$
\underset{}{\overset{\overset{\text{Me}}{|}}{\text{EtCHCHO}}} \overset{h\nu}{\rightarrow} \cdot\text{CHO} + \underset{}{\overset{\overset{\text{Me}}{|}}{\text{EtCH}\cdot}} \overset{\text{I}_2}{\rightarrow} \underset{}{\overset{\overset{\text{Me}}{|}}{\text{EtCHI}}}
$$

and Denney and Beach obtained racemic α-phenylethyl chloride from the decomposition of 1,1-dimethyl-2-phenylpropyl hypochlorite, in which the chain-propagation steps are probably as follows:[5]

$$
\underset{}{\overset{\overset{\text{Me}}{|}}{\text{C}_6\text{H}_5\text{CH}}}-\underset{\overset{|}{\text{Me}}}{\overset{\overset{\text{Me}}{|}}{\text{CO}\cdot}} \rightarrow \text{Me}_2\text{CO} + \underset{}{\overset{\overset{\text{Me}}{|}}{\text{C}_6\text{H}_5\text{CH}\cdot}}
$$

$$
\underset{}{\overset{\overset{\text{Me}}{|}}{\text{C}_6\text{H}_5\text{CH}\cdot}} + \underset{\overset{|}{\text{Me}}}{\overset{\overset{\text{Me}}{|}}{\text{C}_6\text{H}_5\text{CH}}}-\text{COCl} \rightarrow \underset{}{\overset{\overset{\text{Me}}{|}}{\text{C}_6\text{H}_5\text{CHCl}}} + \underset{\overset{|}{\text{Me}}}{\overset{\overset{\text{Me}}{|}}{\text{C}_6\text{H}_5\text{CH}}}-\text{CO}\cdot
$$

Kharasch, Kuderna, and Nudenberg found that the decomposition of optically active methylethylacetyl peroxide yields 2-butyl methylethylacetate that may be hydrolyzed to optically active 2-butanol with the same configuration about its asymmetric carbon atom as the original peroxide had.[6] DeTar and Weis analogously obtained 1-phenyl-2-propanol with 75 per cent retention of configuration from the ester produced in the decomposition of β-phenylisobutyryl peroxide in carbon tetrachloride solution.[7] This result was not attributed to any ability of the intermediate radicals to maintain their configuration but to the fact that the radicals whose combination yields the ester are formed

[4] J. T. Gruver and J. G. Calvert, *J. Am. Chem. Soc.*, **80**, 3524 (1958).

[5] D. B. Denney and W. F. Beach, *J. Org. Chem.*, **24**, 109 (1959).

[6] M. S. Kharasch, J. Kuderna, and W. Nudenberg, *J. Org. Chem.*, **19**, 1283 (1954).

[7] D. F. DeTar and C. Weis, *J. Am. Chem. Soc.*, **79**, 3045 (1957).

"facing" each other in a solvent cage; they combine faster than the 1-phenyl-2-propyl radical turns around to present its other face to the carboxy radical.

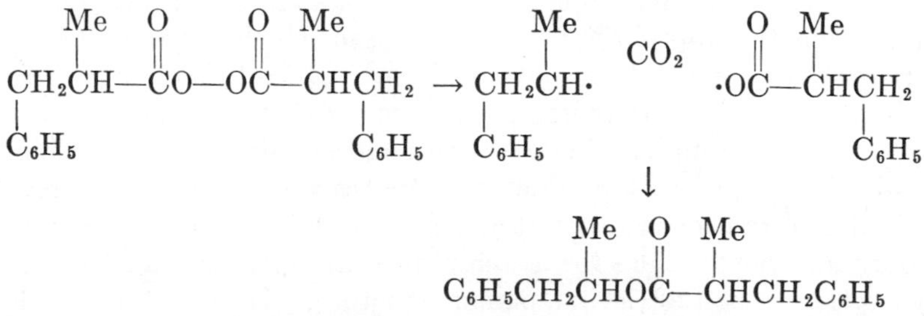

$$\underset{\overset{|}{C_6H_5}}{\overset{\overset{Me}{|}}{CH_2CH}}—\overset{\overset{O}{||}}{CO}—\overset{\overset{O}{||}}{OC}—\underset{\overset{|}{C_6H_5}}{\overset{\overset{Me}{|}}{CHCH_2}} \rightarrow \underset{\overset{|}{C_6H_5}}{\overset{\overset{Me}{|}}{CH_2CH\cdot}} \quad \overset{CO_2}{} \quad \cdot\overset{\overset{O}{||}}{OC}—\underset{\overset{|}{C_6H_5}}{\overset{\overset{Me}{|}}{CHCH_2}}$$

$$\downarrow$$

$$C_6H_5CH_2\overset{\overset{Me}{|}}{CH}\overset{\overset{O}{||}}{OC}—\overset{\overset{Me}{|}}{CH}CH_2C_6H_5$$

This conclusion is supported by evidence that the ester is not formed by induced decomposition of the peroxide. Furthermore, the 1-phenyl-2-propyl chloride formed when the 1-phenyl-2-propyl radicals abstract a chlorine atom from carbon tetrachloride is racemic. Brenner and Mislow found that the 2-butyl 2-methylbutanoate formed in the electrolysis of optically active 2-methylbutanoic acid, where the two combining radicals need not have any particular preferential orientation with respect to each other, gave racemic 2-butanol on hydrolysis.[8]

23-1b. *Bridgehead Radicals.*   The previous section describes good evidence that free radicals racemize readily; there are two possible explanations for this racemization. On one hand, it may be that radicals are most stable when in a planar configuration, as carbonium ions appear to be (cf. Sec. 7-3a). On the other hand, the radicals may prefer a pyramidal configuration, but, as in the case of amines, there may be a very rapid equilibrium between the two enantiomorphic pyramidal forms.

Some information relevant to this point has been obtained in studies of the formation and reactions of radicals at the bridgeheads of bicyclic ring systems.

Kharasch, Engelmann, and Urry showed that the peroxide of 1-apocamphanecarboxylic acid

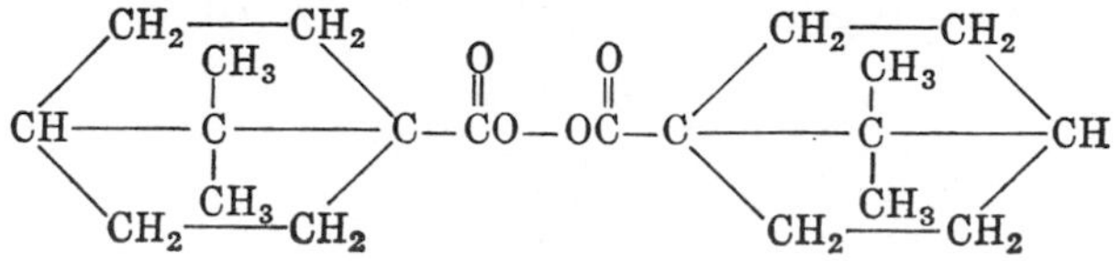

[8] J. Brenner and K. Mislow, *J. Org. Chem.*, **21**, 1312 (1956).

decomposes much more slowly than analogous acyclic diacyl peroxides and, in carbon tetrachloride solution, yields 36 per cent 1-chloroapo-camphane.[9]   Both observations show that the 1-apocamphyl radical is less stable than tertiary aliphatic radicals, which are formed relatively rapidly from the appropriate peroxides[10] and which are not sufficiently reactive to abstract chlorine atoms from carbon tetrachloride to any appreciable extent.[11]   It therefore seems clear that the greater ease with which a t-butyl group can increase the bond angles (about $109.5°$ in the nonradical reactant and products) of the carbon atom that bears the unpaired electron leads to greater radical stability.   Apparently, then, the optimum bond angles for a free radical are greater than $109.5°$, and they *might* even be $120°$, as in a planar structure.   Even stronger evidence concerning radical structure comes from the work of Herzberg and Shoosmith, who deduced, from the spectra of the methyl radical and its deuterio derivatives (obtained in flash photolysis experiments), that the methyl radical is planar (or very nearly so) with a C—H bond distance of about 1.08 A.[12]

Kooyman and Vegter studied the halogenation of norbornane (bicy-clo[2.2.1]heptane) using a variety of halogenating agents.[13]   In all cases attack on norbornane gave 2-norbornyl radicals almost exclusively. The bridgehead radicals are not formed, probably because of the small C—C—C bond angles at this position.   The reason why so little substitu-tion takes place at the 7 position is less clear.   The ratio of *exo*- to *endo*-2-norbornyl halide formed

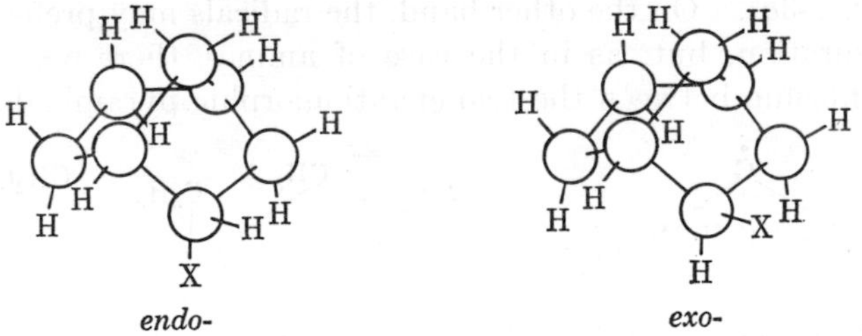

endo-   exo-

2-Norbornyl halides

must depend on the manner in which the 2-norbornyl radical attacks the halogenating agent.   In all cases the major product is exo, presumably

---

[9] M. S. Kharasch, F. Engelmann, and W. H. Urry, *J. Am. Chem. Soc.*, **65**, 2428 (1943).

[10] Cf. P. D. Bartlett and R. R. Hiatt, *J. Am. Chem. Soc.*, **80**, 1398 (1958).

[11] M. S. Kharasch, S. S. Kane, and H. C. Brown, *J. Am. Chem. Soc.*, **64**, 1621 (1942).

[12] G. Herzberg and J. Shoosmith, *Can. J. Phys.*, **34**, 523 (1956); G. Herzberg, *Proc. Roy. Soc. (London)*, **A262**, 291 (1961).

[13] E. C. Kooyman and G. C. Vegter, *Tetrahedron*, **4**, 382 (1958).

because there is more space available near the 1-carbon bridge than near the 2-carbon bridge.   The yield of exo product is greater (about 95 per cent) with such bulky reagents as sulfuryl chloride, phosphorus pentachloride, and carbon tetrachloride than it is with chlorine and bromine (which give about 70 per cent exo).

The chlorination of bicyclo[2.2.2]octane yields about as much of the 1-chloride as of the 2-chloride.[14]

1-Chlorobicyclo[2.2.2]octane   2-Chlorobicyclo[2.2.2]octane

In this case the strain present in the bridgehead radical being formed is a smaller factor than the other effects that ordinarily favor tertiary chlorination relative to secondary chlorination. Since the transition state in the attack of chlorine atoms on hydrocarbons comes relatively early in the reaction, there is not nearly so much strain in the transition state as in the bridgehead radical.

Although Blickenstaff and Hass reported a large amount of bridgehead substitution in the vapor-phase nitration of bicyclo[2.2.1]heptane,[15] Smith (working at a lower temperature) observed none.[16]

23-1c. *Stereochemistry of Radical Additions.*   In a number of cases the free-radical addition of a given reagent has yielded the same product or mixture of products from each of a pair of cis-trans isomeric olefins. Skell and Woodworth, for example, found that the photochemical addition of bromotrichloromethane to either *cis-* or *trans-*2-butene yields the same mixture of diastereomeric 2-bromo-3-trichloromethylbutanes, under conditions where the reacting olefins do not isomerize.[17]   Similarly both cis and trans olefins were reported to give the same copolymer in the case of vinyl acetate and the 1,2-dichloroethylenes[18] and of sulfur dioxide and the 2-butenes.[19]   Apparently, the establishment of equilibrium between

[14] A. F. Bickel, J. Knotnerus, E. C. Kooyman, and G. C. Vegter, *Tetrahedron*, **9**, 230 (1960).

[15] R. T. Blickenstaff and H. B. Hass, *J. Am. Chem. Soc.*, **68**, 1431 (1946).

[16] G. W. Smith, *J. Am. Chem. Soc.*, **81**, 6319 (1959).

[17] P. S. Skell and R. C. Woodworth, *J. Am. Chem. Soc.*, **77**, 4638 (1955).

[18] F. R. Mayo and K. E. Wilzbach, *J. Am. Chem. Soc.*, **71**, 1124 (1949).

[19] P. S. Skell, R. C. Woodworth, and J. H. McNamara, *J. Am. Chem. Soc.*, **79**, 1253 (1957).

the various conformers of the intermediate radical (I) is rapid compared with the rate of subsequent reaction of the radical.

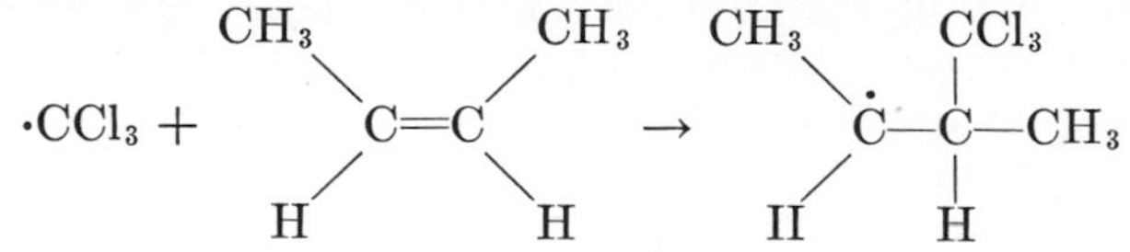

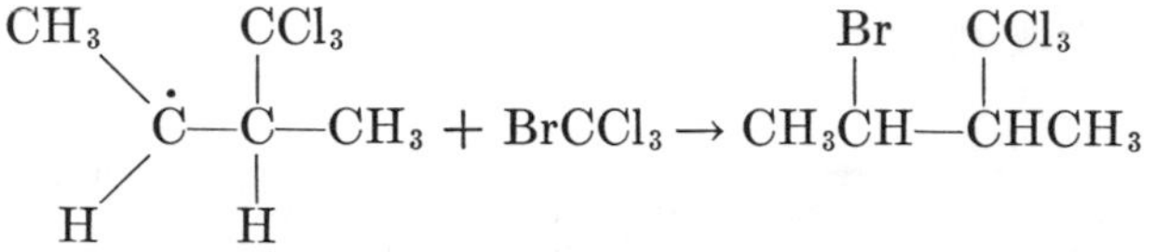

In the addition of hydrogen bromide, however, a different result may be obtained. This result was foreshadowed by the observations of Goering and coworkers that the radical additions of hydrogen bromide to 1-methyl-, 1-bromo-, and 1-chlorocyclohexene are all cleanly trans reactions (yielding the appropriate *cis*-1-X-2-bromocyclohexanes).[20] Subsequently, Goering and Larsen found that the 2-bromo-2-butenes add hydrogen bromide in liquid hydrogen bromide solution at $-80°$ by a stereospecifically trans process, the cis olefin giving *meso*-2,3-dibromobutane and the trans olefin giving *dl*-dibromide.[21] Skell and Allen obtained analogous results in the addition of deuterium bromide to the 2-butenes.[22] In these cases the intermediate radicals abstract a hydrogen atom from hydrogen bromide faster than they attain conformational equilibrium.

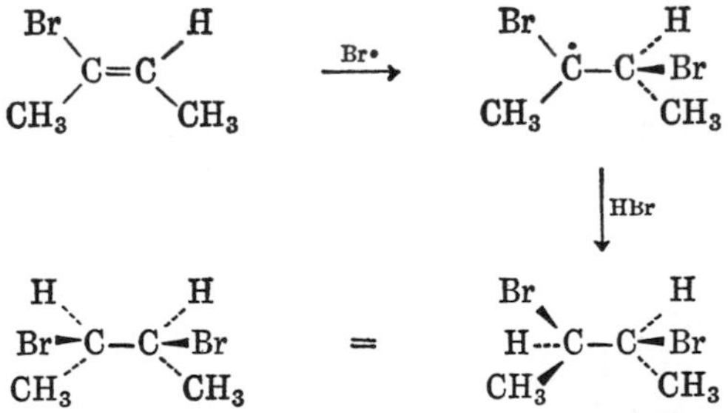

This hydrogen abstraction occurs at the side of the radical least hindered by the bulky bromine atom attached to the adjacent carbon. In the vapor phase, where the intermediate radical collides with hydrogen bromide molecules much less frequently, and therefore abstracts hydrogen

[20] H. L. Goering, P. I. Abell, and B. F. Aycock, *J. Am. Chem. Soc.*, **74**, 3588 (1952); H. L. Goering and L. L. Sims, *J. Am. Chem. Soc.*, **77**, 3465 (1955).

[21] H. L. Goering and D. W. Larsen, *J. Am. Chem. Soc.*, **79**, 2653 (1957); **81**, 5937 (1959).

[22] P. S. Skell and R. G. Allen, *J. Am. Chem. Soc.*, **81**, 5383 (1959).

atoms from them much more slowly, cis and trans isomers yield about the same mixture of products.[21]

This explanation for the trans character of radical additions by liquid hydrogen bromide was supported by Skell and Allen's observation that methyl mercaptan, which ordinarily gives a mixture of cis and trans addition via a radical mechanism, adds to 2-butene stereospecifically trans in liquid hydrogen bromide solution.[23]   The reaction mechanism (for the deuterio compounds, which were used to make the stereochemical course of the reaction observable) is presumably

$$CH_3S\cdot + CH_3CH{=}CHCH_3 \rightarrow CH_3CH{-}\overset{\cdot}{C}HCH_3$$
$$\underset{CH_3\overset{|}{S}}{}$$

$$CH_3CH{-}\overset{\cdot}{C}HCH_3 + DBr \rightarrow Br\cdot + CH_3CH{-}CHDCH_3$$
$$\underset{CH_3\overset{|}{S}}{} \qquad\qquad \underset{CH_3\overset{|}{S}}{}$$

$$Br\cdot + CH_3SD \rightarrow DBr + CH_3S\cdot$$

The free-radical addition of silicochloroform to several acetylenes[24] and of hydrogen bromide to propyne at $-70°$[25] has been found to go trans.

23-2. Rearrangements of Free Radicals.   *23-2a. Rearrangements of Diradicals.*   The isomerization of cyclopropane to propylene appears to proceed via the reversible formation of trimethylene radicals, in which, in the rate-controlling step, a secondary hydrogen atom shifts to the adjacent carbon.   At least, as Rabinovitch, Schlag, and Wiberg pointed out, this seems the simplest way to explain the concomitant cis-trans isomerization of 1,2-dideutericocyclopropane that is observed.[26,27]

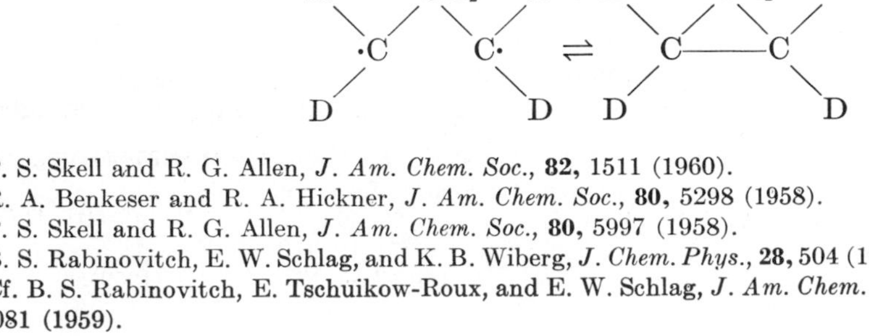

[23] P. S. Skell and R. G. Allen, *J. Am. Chem. Soc.*, **82**, 1511 (1960).

[24] R. A. Benkeser and R. A. Hickner, *J. Am. Chem. Soc.*, **80**, 5298 (1958).

[25] P. S. Skell and R. G. Allen, *J. Am. Chem. Soc.*, **80**, 5997 (1958).

[26] B. S. Rabinovitch, E. W. Schlag, and K. B. Wiberg, *J. Chem. Phys.*, **28**, 504 (1958).

[27] Cf. B. S. Rabinovitch, E. Tschuikow-Roux, and E. W. Schlag, *J. Am. Chem. Soc.*, **81**, 1081 (1959).

The cleavage of 1,1-dimethylcyclopropane would be expected to yield much more of the more stable tertiary-primary diradical (II) than of the primary-primary diradical (III). Therefore the principal products of the pyrolysis of 1,1-dimethylcyclopropane would be expected to be 2-methyl-2-butene and 3-methyl-1-butene. Flowers and Frey found that each of these olefins is formed in about 50 per cent yield and that only about 1 per cent 2-methyl-1-butene is formed.[28]

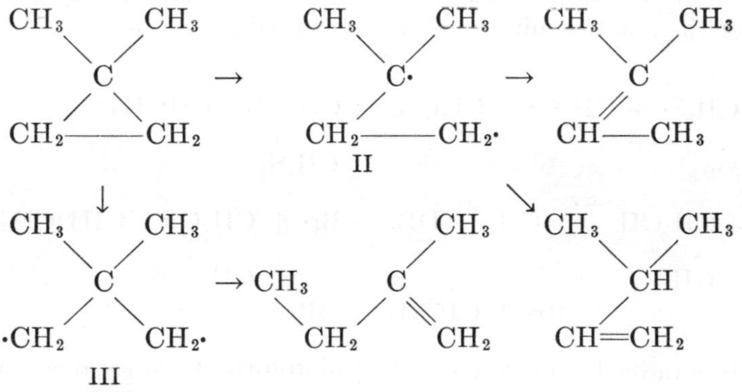

That the radical III would indeed give rise to 2-methyl-1-butene is supported by Walker and Wood's isolation of this olefin as the product of the Kolbe electrolysis of potassium β,β-dimethylglutarate.[29]

23-2b. Rearrangements of Monoradicals. The migration of a group in a trimethylene radical permits the pairing of two electrons and the formation of a new bond. This driving force for rearrangement is absent in monoradicals. Qualitatively there are some of the same driving forces for the rearrangement of ordinary radicals as for the rearrangement of carbonium ions. Many primary radicals produced as reaction intermediates could give more stable tertiary radicals by the migration of a β substitutent to the α-carbon atom. Quantitatively, however, the difference in stability between a primary and a tertiary radical is not so great as between a primary and tertiary carbonium ion. This follows from the fact that the ionization potentials (energy required to remove an electron) of radicals decrease in the order methyl > primary > secondary > tertiary.[30] The relative extents to which radical and carbonium-ion rearrangements occur do not depend solely on the energy change that accompanies rearrangement. The *rate* of rearrangement

---

[28] M. C. Flowers and H. M. Frey, *J. Chem. Soc.*, 3953 (1959); 2758 (1960). These workers give a somewhat different explanation for their results.

[29] J. Walker and J. K. Wood, *J. Chem. Soc.*, **89**, 598 (1906); cf. L. Vanzetti, *Atti accad. nazl. Lincei. Rend. Classe sci. fis. mat. e nat.*, [5], **13**, 112 (1904).

[30] F. P. Lossing and J. B. deSousa, *J. Am. Chem. Soc.*, **81**, 281 (1959).

and the rate of competing reactions (that is, the average lifetime of the intermediates) are also important.   In very few cases do we have reliable information about the average lifetimes of corresponding radicals and carbonium ions.   Nevertheless, whatever the cause, radical rearrangements have been found to be distinctly less common than carbonium-ion rearrangements.

Several workers generated substituted norbornyl radicals as intermediates in radical additions to norbornenes.[31–33]   Unlike the corresponding cations (Sec. 14-1d), however, these radicals did not rearrange.

A number of radical rearrangements involving migration of a β-aryl substituent have been investigated.   In an early example of this type Urry and Kharasch treated neophyl chloride (2-methyl-2-phenyl-1-chloropropane) with cobaltous chloride and phenylmagnesium bromide. This would be expected from earlier work to lead to the neophyl radical.[34] Among the reaction products were several, including 15 per cent isobutylbenzene, 9 per cent 2-methyl-3-phenyl-1-propene, and 4 per cent β,β-dimethylstyrene, that showed carbon-skeleton rearrangement.   Winstein and Seubold generated neophyl radicals by the t-butyl peroxide–induced decarbonylation of β-phenylisovaleraldehyde.[2]   The reaction product consisted of approximately equal amounts of the rearranged product, isobutylbenzene, and the unrearranged product t-butylbenzene.

$$
\underset{\underset{\displaystyle \text{Me}}{|}}{\overset{\overset{\displaystyle \text{Me}}{|}}{C_6H_5C}}CH_2CHO \xrightarrow{R\cdot} \underset{\underset{\displaystyle \text{Me}}{|}}{\overset{\overset{\displaystyle \text{Me}}{|}}{C_6H_5C}}CH_2\dot{C}O \rightarrow \underset{\underset{\displaystyle \text{Me}}{|}}{\overset{\overset{\displaystyle \text{Me}}{|}}{C_6H_5C}}CH_2\cdot + CO
$$

$$\downarrow \text{RCHO}$$

$$
\underset{\underset{\displaystyle \text{Me}}{|}}{\overset{\overset{\displaystyle \text{Me}}{|}}{\text{CH}}}CH_2C_6H_5 \xleftarrow{\text{RCHO}} \underset{\underset{\displaystyle \text{Me}}{|}}{\overset{\overset{\displaystyle \text{Me}}{|}}{\cdot C}}CH_2C_6H_5 \qquad C_6H_5\dot{C}Me_3 + R\dot{C}O
$$

Seubold showed that the extent of rearrangement in this reaction increases with the average lifetime of the intermediate radicals.[35]   He found 57 per cent rearrangement when the reaction was carried out in pure aldehyde (6.4 M) and 80 per cent rearrangement when a 1.0 M solution of aldehyde in chlorobenzene was used.   The increase in extent of rearrangement with dilution of the aldehyde is due to the fact that in the more dilute solution the neophyl radical collides with aldehyde molecules less fre-

[31] M. S. Kharasch and H. N. Friedlander, *J. Org. Chem.*, **14**, 239 (1949).

[32] S. J. Cristol and G. D. Brindell, *J. Am. Chem. Soc.*, **76**, 5699 (1954).

[33] J. A. Berson and W. M. Jones, *J. Am. Chem. Soc.*, **78**, 6045 (1956).

[34] W. H. Urry and M. S. Kharasch, *J. Am. Chem. Soc.*, **66**, 1438 (1944).

[35] F. H. Seubold, Jr., *J. Am. Chem. Soc.*, **75**, 2532 (1953).

quently.   Since the hydrogen-abstraction reaction is thus slowed, the rearrangement is relatively favored.   Seubold estimated, from studies at several temperatures, that the activation energy for rearrangement is about 8 kcal.   The rearrangement probably proceeds via the intermediate bridged radical IV.

At least, this seems the most plausible explanation for the migration of phenyl rather than methyl, whose migration would lead to a more stable new radical.   The results obtained show that IV is not the initial radical produced but rather that it is formed from intermediate unrearranged neophyl radicals.

The importance of the lifetime of neophyl radicals in determining whether rearrangement occurs has also been demonstrated in studies of the chlorination of t-butylbenzene.   The free-radical chlorination of t-butylbenzene in solution must involve intermediate neophyl radicals, but only neophyl chloride is formed.[36]   In the vapor phase, however, particularly at low partial pressures of chlorine, considerable rearrangement occurs.[37]   The observation that the vapor-phase side-chain nitration of t-butylbenzene yields increasing amounts of rearranged products as the pressure is lowered was used by Duffin, Hughes, and Ingold as evidence for the free-radical nature of the reaction.[38]

Urry and Nicolaides found that p-methylneophyl radicals, produced either by the reaction of the chloride with a Grignard reagent in the presence of cobaltous chloride or by the radical-induced decarbonylation of β-p-tolylisovaleraldehyde, rearranged to essentially the same extent as neophyl radicals do under the same conditions.[39]

Curtin, Hurwitz, and Kauer generated a number of related radicals by the decarbonylation of the appropriate aldehydes.[40]   The β,β,β-triphenyl-

[36] M. S. Kharasch and H. C. Brown, *J. Am. Chem. Soc.*, **61**, 2142 (1939); W. E. Truce, E. T. McBee, and C. C. Alfieri, *J. Am. Chem. Soc.*, **71**, 752 (1949).

[37] J. D. Backhurst, E. D. Hughes, and C. K. Ingold, *J. Chem. Soc.*, 2742 (1959).

[38] H. C. Duffin, E. D. Hughes, and C. K. Ingold, *J. Chem. Soc.*, 2734 (1959).

[39] W. H. Urry and N. Nicolaides, *J. Am. Chem. Soc.*, **74**, 5163 (1952).

[40] D. Y. Curtin and M. J. Hurwitz, *J. Am. Chem. Soc.*, **74**, 5381 (1952); D. Y. Curtin and J. C. Kauer, *J. Org. Chem.*, **25**, 880 (1960).

ethyl (V), α-methyl-β,β,β-triphenylethyl (VI), and β,β-diphenylpropyl (VII) radicals were found to give almost complete rearrangement.

$$(C_6H_5)_3CCH_2 \cdot \qquad (C_6H_5)_3\overset{\cdot}{C}CHCH_3 \qquad (C_6H_5)_2\overset{\overset{\displaystyle CH_3}{|}}{C}CH_2 \cdot$$

$$\text{V} \qquad\qquad\qquad \text{VI} \qquad\qquad\qquad \text{VII}$$

The β-phenyl-β-p-anisylethyl radical did not rearrange at all under the conditions used.

Slaugh showed that in the $C_6H_5C^{14}H_2CH_2 \cdot$ radical the β-phenyl group migrates to an extent that may be decreased by the addition of thiophenol, an effective hydrogen donor whose presence decreases the average lifetime of the intermediate radicals.[41]

Bartlett and Cotman showed that the pyrolysis of p-nitrotriphenylmethyl hydroperoxide gives more of the products of migration of the p-nitrophenyl group (e.g., p-nitrophenol) than of migration of the phenyl groups.[42] The contrast of this result with that (preferential phenyl migration) obtained in the acid-catalyzed decomposition reaction and in certain related processes suggested the use of the relative migration aptitudes of phenyl and p-nitrophenyl as a means of distinguishing polar from free-radical mechanisms. In other studies of various triarylmethoxy radicals Kharasch and coworkers found p-phenylphenyl and α-naphthyl groups to migrate about six times as fast as phenyl; p-tolyl migrated at about the same rate as phenyl.[43]

PROBLEMS

1. The decomposition of 2-octanesulfonyl chloride to 2-octyl chloride and sulfur dioxide can be induced by small amounts of benzoyl peroxide and is also found to be catalyzed by ultraviolet light. Suggest a mechanism for the reaction and predict the stereochemistry of the product when optically active reactant is used.

2. Establishment of equilibrium between the cis and trans isomers of decahydronaphthalene can be greatly speeded by the addition of small amounts of di-t-butyl peroxide to the reaction mixture. Explain this observation.

3. Tell whether you would expect the β-phenylethyl or the β-phenylpropyl radical to rearrange more rapidly under a given set of conditions. Give your reasons.

4. At 450° benzyl bromide (2 mole per cent) induces the isomerization of isopropylbenzene. In one run 47 per cent unchanged reactant, 43 per cent n-propylbenzene, and 10 per cent by-products were observed. Suggest a reasonable reaction mechanism.

[41] L. H. Slaugh, *J. Am. Chem. Soc.*, **81**, 2262 (1959).

[42] P. D. Bartlett and J. D. Cotman, Jr., *J. Am. Chem. Soc.*, **72**, 3095 (1950).

[43] M. S. Kharasch, A. C. Poshkus, A. Fono, and W. Nudenberg, *J. Org. Chem.*, **16**, 1458 (1951).

Chapter 24

METHYLENES[1]

One of the most rapidly growing parts of the field of physical organic chemistry is the study of reactions involving intermediates containing divalent carbon. Such intermediates may be formed from homolytic *and* heterolytic reactions, and both methods of formation will be considered here. Doering, one of the most active workers in this field, has suggested the term *carbenes* for divalent-carbon compounds,[2] and this term has been widely used. In this chapter, however, we shall name such compounds as derivatives of "methylene," as they are named in *Chemical Abstracts*. Certain reactions of one compound that can be considered a methylene, namely, carbon monoxide, have already been discussed in Sec. 13-2.

24-1. Dihalomethylenes. *24-1a. Mechanism of the Basic Hydrolysis of Chloroform.* In one of the earliest suggestions of a methylene reaction intermediate that still seems plausible, Geuther suggested that the basic hydrolysis of chloroform involves the intermediacy of dichloromethylene.[3] The following detailed mechanism for the reaction is supported by Horiuti and Sakamoto's observation that in the presence of heavy water chloroform undergoes base-catalyzed deuterium exchange at a rate that is much faster than its hydrolysis.[4]

$$CHCl_3 + OH^- \rightleftharpoons H_2O + CCl_3^-$$
$$CCl_3^- \rightarrow Cl^- + CCl_2 \qquad (24\text{-}1)$$
$$CCl_2 \xrightarrow[\text{several steps}]{OH^-,\ H_2O} CO \text{ and } HCO_2^-$$

Data on the hydrolytic reactivity of chloroform and other halides of methane give strong added support for the above mechanism. In reactions with such weakly basic nucleophilic reagents as water, ethanol, and piperidine the reactivities of the chlorides of methane stand in the

---

[1] For a more detailed treatment of this subject, see I. L. Knunyants, N. P. Gambaryan, and E. M. Rokhlin, *Uspekhi Khim.*, **27**, 1361 (1958).

[2] W. v. E. Doering and L. H. Knox, *J. Am. Chem. Soc.*, **78**, 4947 (1956).

[3] A. Geuther, *Ann.*, **123**, 121 (1862).

[4] Y. Sakamoto, *J. Chem. Soc. Japan*, **57**, 1169 (1936); J. Horiuti and Y. Sakamoto, *Bull. Chem. Soc. Japan*, **11**, 627 (1936); *Chem. Abstr.*, **31**, 931⁴ (1937).

order $CH_3Cl \gg CH_2Cl_2 > CHCl_3 < CCl_4$.[5] The decrease in reactivity observed with the first three members of the series is attributable to the decrease in S_N2 reactivity brought about by α-chloro substituents (Sec. 7-3b). The increased reactivity of carbon tetrachloride may be due to the incursion of the S_N1 mechanism, in which α-chlorines increase reactivity (Sec. 7-3a). With strongly basic nucleophilic reagents such as alkoxide and hydroxide ions the reactivity sequence is $CH_3Cl \gg CH_2Cl_2 \ll CHCl_3 \gg CCl_4$.[5,6] The greatly enhanced relative reactivity of chloroform observed cannot be explained by any combination of the S_N1 and S_N2 mechanisms. In the dichloromethylene mechanism, however, the concentration of the required intermediate trichloromethyl anion is proportional to the strength and concentration of the base used.

Probably the strongest evidence for mechanism (24-1) comes from experiments in which the intermediate dichloromethylene is captured in various ways. Thus, for example, although chloroform is relatively inert to the action of sodium thiophenoxide alone, it reacts rapidly in the presence of hydroxide ions, giving triphenyl orthothioformate.[6] Although thiophenoxide ions are sufficiently nucleophilic to combine effectively with dichloromethylene, the much more basic hydroxide ions are required to transform chloroform to the requisite intermediate trichloromethyl anions. Chloride, bromide, and iodide ions are also capable of capturing dichloromethylene.[7] At a salt concentration of 0.16 M, where none of the weakly nucleophilic salts sodium perchlorate, sodium nitrate, and sodium fluoride has any definitely detectable effect on the reaction rate, sodium chloride decreases the rate of the basic hydrolysis of chloroform by about 15 per cent. This is due to a mass-law effect (cf. Sec. 6-2a); the chloride ions transform dichloromethylene to trichloromethyl anion and thus to chloroform. The reaction rate, as measured by the rate of disappearance of alkali, is decreased even more by the addition of sodium bromide or sodium iodide. With bromide and iodide, however, the calculated second-order rate constants increase as the reaction proceeds. This would be expected since the calculated k's are rate constants not for the disappearance of chloroform, but for the disappearance of total haloform. In capturing dichloromethylene, bromide and iodide ions prevent the hydrolysis of chloroform by transforming it to another haloform.

$$CCl_2 + Cl^- \xrightarrow{k_{Cl}} CCl_3^- \rightarrow CHCl_3$$

$$CCl_2 + Br^- \xrightarrow{k_{Br}} CCl_2Br^- \rightarrow CHCl_2Br$$

$$CCl_2 + I^- \xrightarrow{k_I} CCl_2I^- \rightarrow CHCl_2I$$

[5] P. Petrenko-Kritschenko and V. Opotsky, *Ber.*, **59B**, 2131 (1926).

[6] J. Hine, *J. Am. Chem. Soc.*, **72**, 2438 (1950).

[7] J. Hine and A. M. Dowell, Jr., *J. Am. Chem. Soc.*, **76**, 2688 (1954).

This interpretation of the rate data was supported by the isolation of dichloroiodomethane from the reaction of chloroform with iodide ions in alkaline solution (none is formed in neutral solution). It was further found that the relative magnitudes of k_{H_2O}, k_{Cl}, k_{Br}, and k_I are just those that would be expected from the relative nucleophilicities of these ions in other organic reactions (cf. Table 7-4). Although dichloromethylene can be regarded as both a carbanion (having an unshared electron pair on carbon) and carbonium ion (having a carbon atom with only six electrons in its outer shell), it is obviously reacting as an electrophilic reagent in combination with water and halide ions. Mechanism (24-1), as written, does not explain the rate maximum around pH 4 in a plot of rate vs. pH reported by Horiuti, Tanabe, and coworkers.[8] The reaction rate in the vicinity of this pH maximum appears to be abnormally sensitive to ionic strength and, in fact, the maximum disappears at ionic strengths as high as 0.2 M or more.[9,10] The alternative mechanism for chloroform hydrolysis suggested, to explain the rate maximum, involves the intermediate formation of an isomer of chloroform, which is then transformed to dichloromethylene.[8]

24-1b. *Relative Stabilities of Dihalomethylenes.* The basic hydrolysis of a number of other haloforms has been studied and in all cases evidence for the intermediacy of dihalomethylenes was obtained. The following generalized mechanism, involving the reversible formation of an intermediate trihalomethyl anion, operates for most of these haloforms.

$$CHXYZ + OH^- \underset{k_{-1}}{\overset{k_1}{\rightleftharpoons}} CXYZ^- + H_2O$$

I

Data on the rates of hydrolysis of these compounds, listed in Table 24-1 with the rates of carbanion formation, can be explained in terms of three factors: (1) the acidity of the haloforms, (2) the ease with which the halogen that is lost as an anion (Z in the mechanism above) performs this role, and (3) the ability of the two halogens (X and Y) that are left behind

[8] J. Horiuti, K. Tanabe, and K. Tanaka, *J. Research Inst. Catalysis, Hokkaido Univ.*, **3**, 119, 147 (1955); *Chem. Abstr.*, **50**, 1428c,f (1956); J. Horiuti and M. Katayama, *J. Research Inst. Catalysis, Hokkaido Univ.*, **6**, 57 (1958).

[9] J. Hine and P. B. Langford, *J. Am. Chem. Soc.*, **80**, 6010 (1958); J. Hine, *J. Research Inst. Catalysis, Hokkaido Univ.*, **6**, 202 (1958).

[10] K. Tanabe and Y. Watanabe, *J. Research Inst. Catalysis, Hokkaido Univ.*, **7**, 79 (1959): **8**, 12 (1960); *J. chim. phys.*, 486 (1960).

to stabilize the dihalomethylene being formed.   The effect of structure on reactivity in haloform hydrolysis has been treated in terms of these three factors by use of a semiempirical linear free-energy relationship.[11]   The

TABLE 24-1. RATE CONSTANTS FOR BASE-CATALYZED CARBANION FORMATION AND HYDROLYSIS BY HALOFORMS[a]

Haloform	$10^5\,k$ in liters mole^{-1} sec^{-1} at 0° in water	
	Carbanion formation[b]	Hydrolysis
CHF_3	. . . . . .	<0.00001[c]
CHI_3	105,000	~0.001[d]
$CHClI_2$	. . . . . .	0.01
$CHCl_3$	820	0.06
$CHBrClI$	. . . . . .	0.12
$CHBr_3$	101,000	0.24
$CHCl_2I$	4,800	0.26
$CHBr_2Cl$	25,000	0.66
$CHCl_2F$	16	1.23
$CHBrCl_2$	5,100	1.49
$CHClF_2$	[e]	1.7
$CHFI_2$	8,800	15
$CHBrClF$	365	132
$CHBrF_2$	[e]	208
$CHBr_2F$	3,600	277
CHF_2I	[e]	960

[a] Data from Refs. 11, 12, and 13.

[b] Assuming (except for $CHCl_2F$ and $CHBrClF$, where k_H/k_D values of 1.76 and 1.74, respectively, have been determined) that k_H/k_D is 1.75.

[c] Assuming reactivity toward the t-amyloxide anion is at least as large as toward the hydroxide ion and that the activation energy is at least as large as for other haloforms.

[d] Assuming the reactivity relative to that of $CHClI_2$ is the same as at 50° in aqueous dioxane.

[e] These compounds give concerted α dehydrohalogenation rather than carbanion formation (cf. Sec. 24-1c).

effect of the various halogens on the acidity of the hydrogen atom was assumed to be measured adequately by the rates of base-catalyzed deuterium exchange of haloforms, which show that as α substituents halogens facilitate carbanion formation in the order $I \sim Br > Cl > F$.[12]   The difference in ease with which chlorine, bromine, and iodine are lost as

[11] J. Hine and S. J. Ehrenson, *J. Am. Chem. Soc.*, **80**, 824 (1958); J. Hine and F. P. Prosser, *J. Am. Chem. Soc.*, **80**, 4282 (1958).

[12] J. Hine, N. W. Burske, M. Hine, and P. B. Langford, *J. Am. Chem. Soc.*, **79**, 1406 (1957).

[13] J. Hine and P. B. Langford, *J. Am. Chem. Soc.*, **79**, 5497 (1957); J. Hine and A. D. Ketley, *J. Org. Chem.*, **25**, 606 (1960).

anions from the intermediate trihalomethyl anion does not appear to be very large, but from the very low reactivity of fluoroform it appears that fluorine is lost only with great difficulty. The most important factor in determining the relative reactivities of the various haloforms (except fluoroform) is the relative stability of the dihalomethylene being formed. The ability of the various halogens to stabilize dihalomethylenes appears to vary in the order $F \gg Cl > Br > I$, the difference in the effect of a fluorine and an iodine substituent corresponding to a rate effect of about 10^5-fold at room temperature. The variations in the methylene-stabilizing abilities of the various halogens are probably a function of relative abilities of the halogen to share one of their unshared electron pairs and thus stabilize the methylene by contribution of structures like I above.

24-1c. *Concerted Dehydrohalogenation of Haloforms.* Since fluoro substituents (compared with other halogens) destabilize trihalomethyl anions but stabilize methylenes, they greatly increase the probability of a trihalomethyl anion decomposing to dihalomethylene (rather than being reprotonated to haloform). The dibromofluoromethyl anion, for example, decomposes to dihalomethylene about 1 time in every 13 that it is formed in water at 0°, whereas the tribromomethyl anion decomposes less than 1 time in every 400,000 under the same conditions (cf. Table 24-1). From an extrapolation of these observations one might expect the bromodifluoromethyl anion to decompose every time that it is formed, so that for bromodifluoromethane the rate-limiting step in basic hydrolysis would be the initial carbanion formation. Actually, it is found that the effect of a second fluoro substituent is even more profound. Bromodifluoromethane, chlorodifluoromethane, and difluoroiodomethane all undergo alkaline hydrolysis at a rate considerably faster than they would even be expected (from data on other haloforms) to form carbanions under the given conditions.[13] Apparently, in these cases the trihalomethyl anion has become so unstable and the dihalomethylene so stable that the anion is bypassed and the dihalomethylene formed directly in a concerted alpha elimination.

$$\text{HO}^- + \text{H}-\overset{\overset{\text{F}}{|}}{\underset{\underset{\text{F}}{|}}{\text{C}}}-\text{Br} \rightarrow \text{HO---H---}\overset{\overset{\text{F}}{|}}{\underset{\underset{\text{F}}{|}}{\text{C}}}\text{---Br} \rightarrow \text{HOH} + \overset{\overset{\text{F}}{|}}{\underset{\underset{\text{F}}{|}}{\text{C}}} + \text{Br}^-$$

In agreement with this proposal is the fact that the basic hydrolysis of deuteriobromodifluoromethane, unlike that of all the other deuteriohaloforms that had been studied, is not accompanied by deuterium exchange of the reactant.

24-1d. *Addition of Dihalomethylenes to Olefins.* Doering and Hoffmann showed that when dichloro- and dibromomethylene are generated in the

presence of olefins, 1,1-dihalocyclopropane derivatives are obtained.[14]
When applied to cis and trans olefins, the addition was found to be
stereospecific.[15,16]

A mechanism proceeding through an intermediate zwitterion (II) or
diradical (III), which might isomerize by rotation around a carbon-carbon
single bond, therefore seems less likely than a one-step reaction in which
both new carbon-carbon bonds are formed simultaneously.

By competition experiments, in which dihalomethylenes were generated
in the presence of mixtures of various pairs of olefins, it was found that
electron-donating substituents increase the reactivity of the olefins.[15,17]
The resultant data (Table 24-2) confirm the electrophilic character of

TABLE 24-2. RELATIVE REACTIVITIES OF OLEFINS TOWARD CCl_2, CBr_2,
$\cdot CCl_3$, AND Br_2[15,17]

Olefin	$k_{olefin}/k_{isobutylene}$			
	CCl_2	CBr_2	$\cdot CCl_3$	Br_2
$(CH_3)_2C{=}C(CH_3)_2$	6.6	3.5	....	2.5
$(CH_3)_2C{=}CHCH_3$	2.9	3.2	0.17	1.9
$(CH_3)_2C{=}CH_2$	1.00	1.00	1.00	1.00
trans-Pentene-2	0.26			
$CH_2{=}CHOC_2H_5$	0.23			
cis-Pentene-2	0.20			
$CH_2{=}CH{-}CH{=}CH_2$	....	0.5	>40	
Cyclopentene	....	0.5	0.15	
Cyclohexene	0.12	0.4	0.045	
$C_6H_5CH{=}CH_2$	....	0.4	>20	0.59
$n\text{-}C_4H_9CH{=}CH_2$	0.023	0.07	0.19^a	0.36

^{a} $n\text{-}C_6H_{13}CH{=}CH_2$.

[14] W. v. E. Doering and A. K. Hoffmann, *J. Am. Chem. Soc.*, **76**, 6162 (1954).
[15] P. S. Skell and A. Y. Garner, *J. Am. Chem. Soc.*, **78**, 3409, 5430 (1956).
[16] W. v. E. Doering and P. M. LaFlamme, *J. Am. Chem. Soc.*, **78**, 5447 (1956).
[17] W. v. E. Doering and W. A. Henderson, Jr., *J. Am. Chem. Soc.*, **80**, 5274 (1958).

dihalomethylenes. Comparison of these results with those obtained in a known radical-addition reaction (by trichloromethyl radicals) shows such differences as to make the intermediacy of the diradical III implausible. The relative reactivities toward CX_2's are seen to stand in about the same order as toward the electrophilic reagent bromine, which reacts by forming bonds simultaneously to both carbon atoms of the olefin.

Parham and coworkers found that the adducts of dihalomethylenes and indene rearrange to 2-halonaphthalenes.[18]

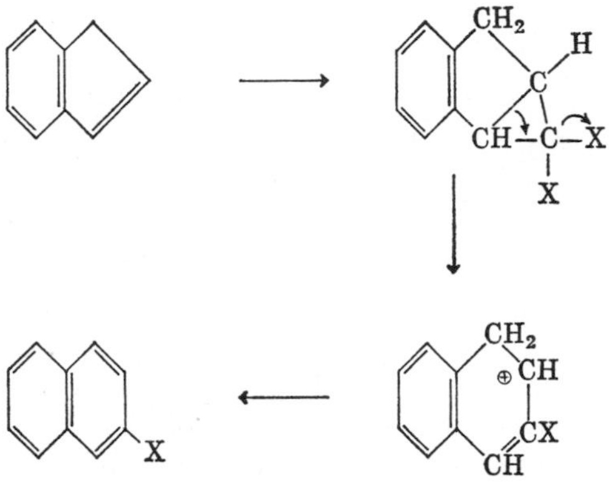

Very little,[19] if any,[20,21] 1,4 addition is reported in reactions with conjugated dienes.

24-1e. Other Methods for Forming Dihalomethylenes. In several instances evidence has been found for the decomposition of organic halides to dihalomethylenes in the vapor phase. Spectroscopic observations show that electrical-discharge decomposition[22,23] and pyrolysis[24] of gaseous fluorocarbons give rise to difluoromethylene, a singlet (non-radical) molecule with a nonlinear structure. Semeluk and Bernstein found the principal products of the pyrolysis of chloroform at temperatures around 500° to be hydrogen chloride and tetrachloroethylene.[25] In kinetic studies they showed that the decomposition is a homogeneous first-order reaction somewhat slowed, as the reaction proceeds, by the hydrogen chloride formed. Shilov and Sabirova observed a significant

[18] W. E. Parham and H. E. Reiff, *J. Am. Chem. Soc.*, **77**, 1177 (1955); W. E. Parham, H. E. Reiff, and P. Swartzentruber, *J. Am. Chem. Soc.*, **78**, 1437 (1956).

[19] M. Orchin and E. C. Herrick, *J. Org. Chem.*, **24**, 139 (1959).

[20] R. C. Woodworth and P. S. Skell, *J. Am. Chem. Soc.*, **79**, 2542 (1957).

[21] A. Ledwith and R. M. Bell, *Chem. & Ind. (London)*, 459 (1959).

[22] P. Venkateswarlu, *Phys. Rev.*, **77**, 676 (1950).

[23] R. K. Laird, E. B. Andrews, and R. F. Barrow, *Trans. Faraday Soc.*, **46**, 803 (1950).

[24] J. L. Margrave and K. Wieland, *J. Chem. Phys.*, **21**, 1552 (1953).

[25] G. P. Semeluk and R. B. Bernstein, *J. Am. Chem. Soc.*, **76**, 3793 (1954); **79**, 46 (1957).

deuterium kinetic isotope effect in reactions followed only to a small extent of completion.[26] They also found that the reaction is not significantly inhibited by toluene, usually an effective inhibitor for high-temperature free-radical chain reactions, and they suggested that the initial step is a direct transformation to dichloromethylene and hydrogen chloride.

In solution, almost any reaction that yields trihalomethyl anions can be used to generate dihalomethylenes. Some of these reactions, such as the decarboxylation of trihaloacetate ions[27] and the reaction of bases with hexachloroacetone[28] and esters of trichloroacetic acid,[29] are particularly useful in the preparation of 1,1-dihalocyclopropanes from olefins. Just as difluorochloromethane reacts with base to give difluoromethylene by a concerted mechanism rather than via an intermediate carbanion, the difluorochloroacetate anion also undergoes a concerted decomposition to yield difluoromethylene.[30]

24-1f. *Other Reactions of Dihalomethylenes.* Dichloromethylene was shown to be an intermediate in the Reimer-Tiemann reaction.[6,31,32]

It is probably also an intermediate in the formation of isocyanides from primary amines, the transformation of potassium pyrrole to 3-chloropyridine, and a number of other reactions of chloroform involving strong bases.[6] The reaction of dihalomethylene with alkoxide ions is of interest because of the large number of different products that may be obtained. Potassium isopropoxide reacts with chlorodifluoromethane to give iso-

[26] A. E. Shilov and R. D. Sabirova, *Doklady Akad. Nauk S.S.S.R.*, **114**, 1058 (1957); *Zhur. Fiz. Khim.*, **34**, 860 (1960).

[27] W. M. Wagner, *Proc. Chem. Soc.*, 229 (1959).

[28] P. K. Kadaba and J. O. Edwards, *J. Org. Chem.*, **25**, 1431 (1960); F. W. Grant and W. B. Cassie, *J. Org. Chem.*, **25**, 1433 (1960).

[29] W. E. Parham and E. E. Schweizer, *J. Org. Chem.*, **24**, 1733 (1959).

[30] J. Hine and D. C. Duffey, *J. Am. Chem. Soc.*, **81**, 1131 (1959).

[31] H. Wynberg, *J. Am. Chem. Soc.*, **76**, 4998 (1954); *Chem. Revs.*, **60**, 169 (1960).

[32] J. Hine and J. M. van der Veen, *J. Am. Chem. Soc.*, **81**, 6446 (1959); *J. Org. Chem.*, **26**, 1406 (1961).

propyl difluoromethyl ether and triisopropyl orthoformate.[33] The orthoformate cannot arise from further attack on the difluoromethyl ether, which is quite inert under the reaction conditions. All other reasonable reaction mechanisms involve the intermediate formation of isopropoxyfluoromethylene, e.g.,

$$CHClF_2 \xrightarrow{i\text{-PrOK}} CF_2 \qquad \rightarrow \qquad i\text{-PrOCHF}_2$$
$$\downarrow$$
$$i\text{-PrO}\text{—}C\text{—}F \xrightarrow[\text{steps}]{\text{several}} (i\text{-PrO})_3CH$$

With fluorine-free haloforms other alkoxyhalomethylenes are probably reaction intermediates. Skell and Starer pointed out that the prevalence of rearrangements in reactions involving these other alkoxyhalomethylenes seems to show that they lose halide ions easily to give cations such as IV.[34]

$$n\text{-}C_4H_9O^- + CBr_2 \rightarrow n\text{-}C_4H_9O\text{—}C\text{—}Br \xrightarrow{-Br^-} n\text{-}C_4H_9O\text{=}C^+$$
$$\text{IV}$$
$$\downarrow$$

cis and trans $CH_3CH\text{=}CHCH_3$, $CH_3CH\text{—}CH_2$, $\leftarrow n\text{-}C_4H_9^+ + CO$

and $C_2H_5CH\text{=}CH_2$ $\qquad\qquad CH_2$

The ease of formation and reactivity of IV can be rationalized in terms of the fact that it is isoelectronic with a diazonium cation, whose tendency to form carbonium ions is well known. The reactions of dichloro- and dibromomethylene with alkoxides ions yield methylene halides as well as olefins, ethers, and ortho esters.[34–36] Apparently, these dihalomethylenes are sufficiently electrophilic to be effective hydride-ion abstractors.

24-2. Methylene. *24-2a. Formation of Methylene from Diazomethane and Ketene.* The first good evidence for the formation of a methylene intermediate appears to be due to Staudinger and Kupfer,[37] who obtained ketene from the pyrolysis of diazomethane in the presence of carbon monoxide. The reaction probably took the following course

$$CH_2N_2 \rightarrow N_2 + CH_2 \xrightarrow{CO} CH_2\text{=}CO$$

but the possibility of direct reaction between carbon monoxide and diazo-

[33] J. Hine and K. Tanabe, *J. Am. Chem. Soc.*, **79**, 2654 (1957); **80**, 3002 (1958).

[34] P. S. Skell and I. Starer, *J. Am. Chem. Soc.*, **81**, 4117 (1959).

[35] J. Hine, A. D. Ketley, and K. Tanabe, *J. Am. Chem. Soc.*, **82**, 1398 (1960).

[36] J. Hine, E. L. Pollitzer, and H. Wagner, *J. Am. Chem. Soc.*, **75**, 5607 (1953).

[37] H. Staudinger and O. Kupfer, *Ber.*, **45**, 501 (1912).

methane was not ruled out. Rice and Glasebrook used the Paneth technique (Sec. 19-1a) to show that the pyrolysis of diazomethane leads to methylene.[38] It was found that tellurium, selenium, antimony, and arsenic mirrors are removed, but zinc, cadmium, lead, thallium, and bismuth mirrors are not. These results show that the mirror removal is not due to monovalent radicals like methyl and ethyl, which are known to remove any of the mirrors described. The methylene was further characterized by the identification of polytelluroformaldehyde, $(CH_2Te)_x$, as the product of removal of tellurium mirrors. Pearson, Purcell, and Saigh isolated polytelluroformaldehyde in similar experiments showing that methylene is produced in the photolysis of diazomethane and of ketene.[39]

Herzberg and Shoosmith were able to observe the spectra of CH_2, CD_2, and CHD in the vapor-phase flash photolysis of CH_2N_2, CD_2N_2, and $CHDN_2$.[40,41] The methylene is initially formed in a singlet state (having no unpaired electrons), at least some of which then rapidly changes to a triplet state (with two unpaired electrons).[42] The triplet form, which is linear and has C—H bond distances of 1.03 A, apparently has two sp bonding orbitals and an unpaired electron in each of the two remaining p orbitals. The singlet form, with C—H bond distances of about 1.12 A and an H—C—H bond angle of about 103°, may be regarded as having one empty p orbital and three other orbitals with both s and p character; two, with a large fraction of p character, are used for bonding, and the other, largely s, orbital contains two unshared electrons.

Norrish, Crone, and Saltmarsh found that the photolysis of ketene yields twice as much carbon monoxide as ethylene and suggested the mechanism[43]

$$CH_2{=}CO \rightarrow CH_2 + CO$$
$$CH_2 + CH_2{=}CO \rightarrow CH_2{=}CH_2 + CO$$

An alternative hypothesis, that the ethylene arises from the dimerization of methylene, was disproved by Kistiakowsky and Rosenberg, who found that the rate of carbon monoxide formation can be almost halved by the addition of enough ethylene (which competes with the ketene for methylene),[44] and by Strachan and Noyes, who found that, with pure ketene,

[38] F. O. Rice and A. L. Glasebrook, *J. Am. Chem. Soc.*, **56**, 2381 (1934).

[39] T. G. Pearson, R. H. Purcell, and G. S. Saigh, *J. Chem. Soc.*, 409 (1938).

[40] G. Herzberg and J. Shoosmith, *Nature*, **183**, 1801 (1959).

[41] Cf. G. W. Robinson and M. McCarty, Jr., *J. Am. Chem. Soc.*, **82**, 1859 (1960); T. D. Goldfarb and G. C. Pimentel, *J. Am. Chem. Soc.*, **82**, 1865 (1960).

[42] G. Herzberg, *Proc. Roy. Soc. (London)*, **A262**, 291 (1961).

[43] R. G. W. Norrish, H. G. Crone, and O. Saltmarsh, *J. Chem. Soc.*, 1533 (1933).

[44] G. B. Kistiakowsky and N. W. Rosenberg, *J. Am. Chem. Soc.*, **72**, 321 (1950).

the quantum yields of carbon monoxide and ethylene are approximately 2.0 and 1.0, respectively.[45]

24-2b. *Reactions of Methylene.* One of the most interesting reactions of methylene is the *insertion reaction*, discovered by Meerwein, Rathjen, and Werner, in which the methylene group is inserted between a hydrogen atom and the atom to which the hydrogen was attached. Thus diethyl ether yields ethyl n-propyl ether and ethyl isopropyl ether, and isopropyl alcohol yields isopropyl methyl ether and *sec-* and *tert-*butyl alcohol.[46]

$$CH_2N_2 \xrightarrow{h\nu} N_2 + CH_2 \xrightarrow{CH_3CHOHCH_3} (CH_3)_3COH,$$
$$(CH_3)_2CHOCH_3, \text{ and } CH_3CH_2CHOHCH_3$$

Doering and coworkers showed that the methylene initially formed from the photolysis of diazomethane is a tremendously reactive species giving insertion reactions almost at random. Reaction with n-pentane gives n-hexane, 2-methylpentane, and 3-methylpentane in the ratio 48:35:17 at $-75°$ and 49:34:17 at 15°.[47] Random insertion would give the ratio $50:33\frac{1}{3}:16\frac{2}{3}$. Addition to a carbon-carbon double bond, the *only* reaction of dihalomethylenes with olefins at ordinary temperatures, is also observed with methylene, but it does not compete particularly well with insertion. The mixture of 7-carbon products from the photolysis of diazomethane in cyclohexene at $-75°$ has the composition shown below.[47]

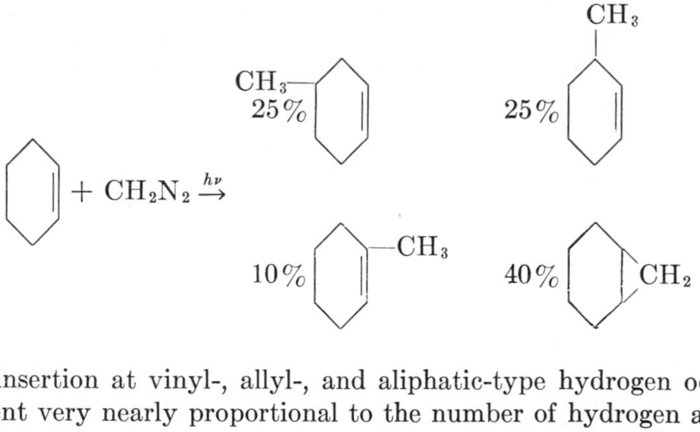

Thus, insertion at vinyl-, allyl-, and aliphatic-type hydrogen occurs to an extent very nearly proportional to the number of hydrogen atoms of each of these types present in the reactant molecule.

[45] A. N. Strachan and W. A. Noyes, Jr., *J. Am. Chem. Soc.*, **76**, 3258 (1954).

[46] H. Meerwein, H. Rathjen, and H. Werner, *Ber.*, **75**, 1610 (1942).

[47] W. v. E. Doering, R. G. Buttery, R. G. Laughlin, and N. Chaudhuri, *J. Am. Chem. Soc.*, **78**, 3224 (1956).

Frey and Kistiakowsky found a larger degree of selectivity in the gas-phase reaction of propane with methylene generated by the photolysis of ketene. About 37 per cent of the C_4H_{10} formed was isobutane, whereas random reaction would have given 25 per cent isobutane.[48] It was suggested that methylene formed from the photolysis of diazomethane contains more energy than that formed from ketene and that this is responsible for the more nearly random reactions observed by Doering and coworkers. Frey and Kistiakowsky further showed that even the methylene formed from ketene has more kinetic energy than the thermal equilibrium amount. In the presence of an inert gas such as argon or carbon dioxide, to which the initially formed energy-rich methylene can give up (by collision) some of its excess energy, the selectivity in reaction with propane increased, the isobutane comprising about 42 per cent of the C_4H_{10} in the presence of 64 mm of carbon dioxide.[48] Frey found that methylene from diazomethane reacts more nearly randomly than that from ketene in the vapor phase,[49] showing that the results of Doering and coworkers were not due solely to the fact that their studies were carried out in solution.[47]

The evidence of Herzberg and Shoosmith that methylene is usually formed as a singlet that then changes to a more stable triplet[40,42] was in disagreement with some, though not all, of the earlier theoretical discussions based on quantum-mechanical principles,[50-55] but it had been foreshadowed by certain observations on the reactions of methylenes. Skell and Woodworth found that the photolysis of a mixture of diazomethane and excess *cis*- or *trans*-2-butene gave *cis*- or *trans*-1,2-dimethylcyclopropane in a stereospecific reaction. By reasoning like that used in the case of the dihalomethylenes (Sec. 24-1*d*), they concluded that it is singlet methylene that adds under these conditions.[56] In the presence of an inert gas, however, it appears that this addition reaction is no longer stereospecific.[57] Anet, Bader, and Van der Auwera reported that in the

[48] H. M. Frey and G. B. Kistiakowsky, *J. Am. Chem. Soc.*, **79**, 6373 (1957).

[49] H. M. Frey, *J. Am. Chem. Soc.*, **80**, 5005 (1958).

[50] J. Lennard-Jones, *Trans. Faraday Soc.*, **30**, 70 (1934); J. Lennard-Jones and J. A. Pople, *Discussions Faraday Soc.*, **10**, 9 (1951).

[51] H. H. Voge, *J. Chem. Phys.*, **4**, 581 (1936).

[52] K. J. Laidler and E. J. Casey, *J. Chem. Phys.*, **17**, 213 (1949).

[53] G. H. Duffey, *J. Chem. Phys.*, **17**, 840 (1949).

[54] B. F. Gray, *J. Chem. Phys.*, **28**, 1252 (1958).

[55] Cf. G. A. Gallup, *J. Chem. Phys.*, **26**, 716 (1957); **28**, 1252 (1958).

[56] P. S. Skell and R. C. Woodworth, *J. Am. Chem. Soc.*, **78**, 4496 (1956); **81**, 3383 (1959).

[57] F. A. L. Anet, R. F. W. Bader, and A.-M. Van der Auwera, *J. Am. Chem. Soc.*, **82**, 3217 (1960).

presence of large amounts of nitrogen some of the initially formed singlet decays to the more stable triplet, which then adds as a diradical.

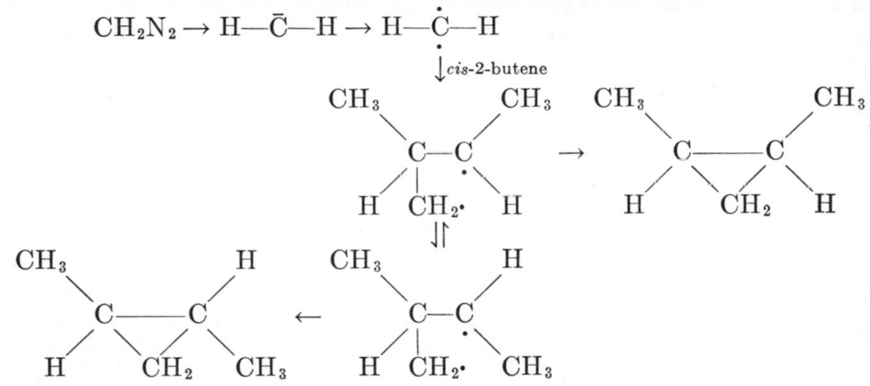

$$CH_2N_2 \rightarrow H-\bar{C}-H \rightarrow H-\overset{\cdot}{\underset{\cdot}{C}}-H$$

The possibility that initially energy-rich dimethylcyclopropane may have isomerized before deactivation was ruled out in similar experiments by Frey.[58]

Two different mechanisms have been found to operate simultaneously for the insertion reaction. Frey and Kistiakowsky[48] secured evidence that some of the methylation due to methylene involves the intermediate formation of methyl radicals, as follows:

$$R-H + CH_2 \rightarrow R\cdot + CH_3\cdot \rightarrow R-CH_3$$

The probability of this mechanism was revealed by the detection of ethane as a by-product in several gas-phase methylations. The ethane appears to be formed by the pairing of methyl radicals, since its formation is inhibited by oxygen. More evidence for this radical-combination mechanism comes from Frey's identification of n-hexane, 2-methylpentane, and 2,3-dimethylbutane among the products of the reaction of methylene with propane.[59] In the liquid phase, where R· and CH_3· are formed within a solvent cage, their combination is more probable than it is in the gas phase, where larger amounts of such products as R—R and ethane are observed.

Doering and Prinzbach obtained convincing evidence for direct methylation by a mechanism that does not involve intermediate radicals.[60] They found that the gas-phase photolysis of diazomethane in the presence of $(CH_3)_2C{=}C^{14}H_2$ gives 2-methyl-1-butene that has 92 per cent of its C^{14} in the unsaturated CH_2 group. This shows that only 16 per cent of the reaction proceeded via an intermediate methallyl radical (V), in which the C^{14} becomes one of two chemically equivalent

[58] H. M. Frey, J. Am. Chem. Soc., **82**, 5947 (1960).
[59] H. M. Frey, Proc. Chem. Soc., 318 (1959).
[60] W. v. E. Doering and H. Prinzbach, Tetrahedron, **6**, 24 (1959).

carbon atoms, and therefore 84 per cent of the methylation is a direct insertion.

$$CH_2 + CH_3 \overset{CH_3}{\underset{|}{\,\,\,}} C{=}C^{14}H_2 \xrightarrow{84\%} CH_3CH_2\overset{CH_3}{\underset{|}{\,\,\,}}C{=}C^{14}H_2$$

$$\downarrow 16\%$$

$$CH_3\cdot + \left[\cdot CH_2\overset{CH_3}{\underset{|}{\,\,\,}}C{=}C^{14}H_2 \leftrightarrow CH_2{=}\overset{CH_3}{\underset{|}{\,\,\,}}C{-}C^{14}H_2\cdot \right] \rightarrow CH_2{=}\overset{CH_3}{\underset{|}{\,\,\,}}C{-}C^{14}H_2CH_3$$

$$\mathbf{V}$$

Other products are formed, of course, besides those shown in the equation above.

Richardson, Simmons, and Dvoretzky suggested that the direct-insertion reaction is due to singlet methylene and the radical-forming process is due to triplet methylene.[61]   This suggestion is in agreement with Frey and Kistiakowsky's observation that the presence of inert gas increases the yield of ethane in the reaction of methylene with propane,[48] but more work is needed before we can be sure to what extents the various reactions of methylene are due to the singlet and triplet forms.

24-3. Other Reactions Involving Methylenes. *24-3a. Unsaturated Methylenes.*   Since in all the stable compounds of divalent carbon that are known (carbon monoxide and isocyanides) the divalent carbon is attached by a double bond, it is not surprising that divalent carbon intermediates of the type $R_2C{=}C$ have been sought.   The reaction of 1,1-diaryl-2-haloethylenes with strong bases to give diarylacetylenes[62-64] can be written as proceeding by such an intermediate, but Bothner-By and also Curtin and coworkers ruled out this mechanism by showing that the aryl group trans to the departing halogen migrates preferentially.[65-68]

$$\underset{C_6H_5}{\overset{p\text{-}BrC_6H_4}{\diagdown}}C^{14}{=}C\underset{Br}{\overset{H}{\diagup}} \xrightarrow{t\text{-}BuO^-} \underset{C_6H_5}{\overset{p\text{-}BrC_6H_4}{\diagdown}}C^{14}{=}\overset{\ominus}{C}\underset{Br}{}$$

$$\downarrow$$

$$C_6H_5C^{14}{\equiv}CC_6H_4Br\text{-}p$$

[61] D. B. Richardson, M. C. Simmons, and I. Dvoretzky, *J. Am. Chem. Soc.*, **82,** 5001 (1960); **83,** 1934 (1961).

[62] P. Fritsch, *Ann.*, **279,** 319 (1894).

[63] W. P. Buttenberg, *Ann.*, **279,** 327 (1894).

[64] H. Wiechell, *Ann.*, **279,** 337 (1894).

[65] A. A. Bothner-By, *J. Am. Chem. Soc.*, **77,** 3293 (1955).

[66] D. Y. Curtin, E. W. Flynn, R. F. Nystrom, and W. H. Richardson, *Chem. & Ind.* (*London*), 1453 (1957).

[67] D. Y. Curtin, E. W. Flynn, and R. F. Nystrom, *J. Am. Chem. Soc.*, **80,** 4599 (1958).

[68] D. Y. Curtin and E. W. Flynn, *J. Am. Chem. Soc.*, **81,** 4714 (1959).

The reaction is therefore analogous to the Beckmann rearrangement (Sec. 15-1d).

In the case of 9-bromomethylenefluorene (VI), where alpha elimination and migration of one of the β substituents would give a highly unstable nonlinear acetylene (9,10-anthracyne), the reaction takes a different course. As Hauser and Lednicer[69] and also Curtin and coworkers showed,[66,70] the principal product of the reaction of VI with strong bases is 1,4-dibiphenylcnebutatriene (VII).

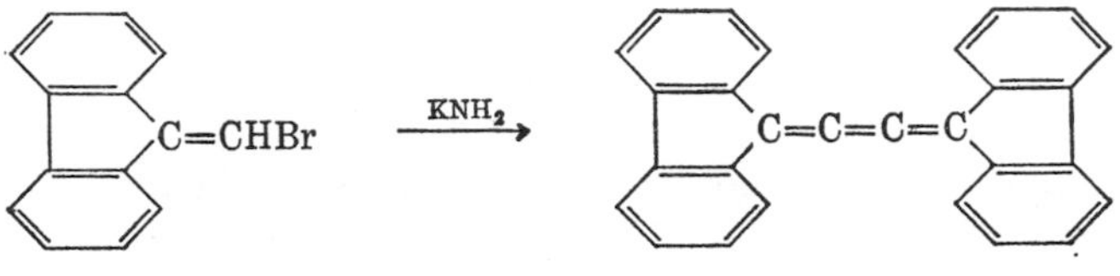

This reaction may proceed by alpha elimination to give an intermediate unsaturated methylene, but the possibility of an alkylation-dehydrohalogenation mechanism (cf. Sec. 24-3b) has not been entirely ruled out.

Hennion and Maloney observed that the solvolysis of 3-chloro-3-methyl-1-butyne (VIII) is greatly speeded by alkali and suggested the unsaturated methylene IX as an intermediate in the reaction.[71]

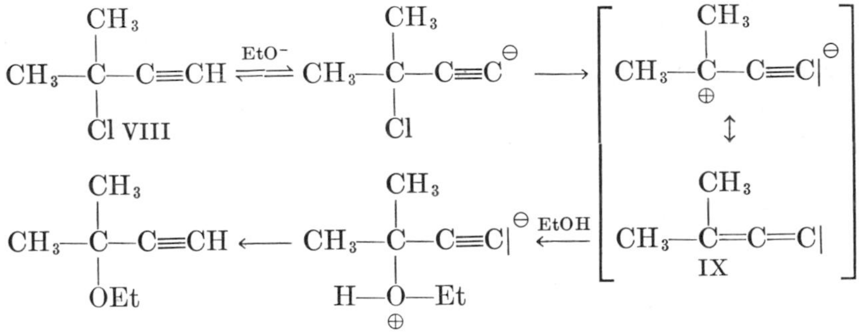

Hennion and Nelson pointed out that the insensitivity to base of compounds like $(CH_3)_2CClC{\equiv}CCH_3$, in which the acidic acetylenic hydrogen atom has been replaced by an alkyl group,[72] gives strong added support for the intermediacy of IX.[73] Hartzler reported the capture of IX by several olefins.[74]

[69] C. R. Hauser and D. Lednicer, *J. Org. Chem.*, **22**, 1248 (1957).

[70] D. Y. Curtin and W. H. Richardson, *J. Am. Chem. Soc.*, **81**, 4719 (1959).

[71] G. F. Hennion and D. E. Maloney, *J. Am. Chem. Soc.*, **73**, 4735 (1951).

[72] A. Burawoy and E. Spinner, *J. Chem. Soc.*, 3752 (1954).

[73] G. F. Hennion and K. W. Nelson, *J. Am. Chem. Soc.*, **79**, 2142 (1957).

[74] H. D. Hartzler, *J. Am. Chem. Soc.*, **81**, 2024 (1959); **83**, 4990, 4997 (1961).

24-3b. Formation of Methylenes by Alpha Eliminations.   Simmons and Smith found that the reaction of a zinc-copper couple with methylene iodide in diethyl ether gives a solution that appears to contain iodomethylzinc iodide.[75]   The filtered solution was essentially free of copper; it reacted with water to give methyl iodide and with iodine to give methylene iodide.   The fact that this homogeneous solution reacts readily with cyclohexene to give bicyclo[4.1.0]heptane (X) shows that the reaction very probably does not involve free methylene as an intermediate.

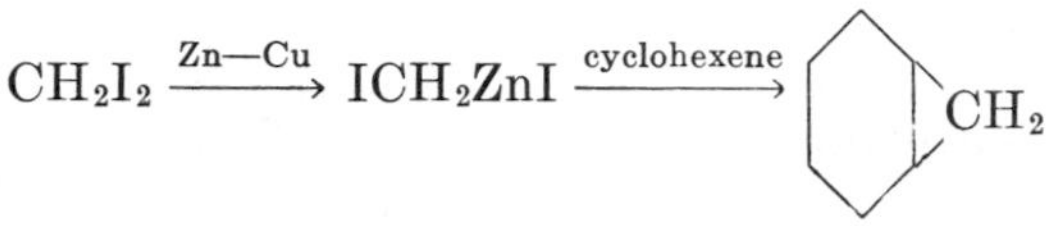

$$CH_2I_2 \xrightarrow{\text{Zn—Cu}} ICH_2ZnI \xrightarrow{\text{cyclohexene}}$$

X

Further evidence against the intermediacy of methylene is found in the absence of insertion products in the numerous reactions with olefins (to give cyclopropane derivatives) that were carried out.   After pointing out the improbability of certain alternative mechanisms, Simmons and Smith suggested that the reactant may be considered to have some bonding between the zinc and the carbon-bound iodine and therefore to be a complex of methylene and zinc iodide.   The reaction with olefins may then consist of a one-step displacement of zinc iodide from this complex via attack by the olefin.

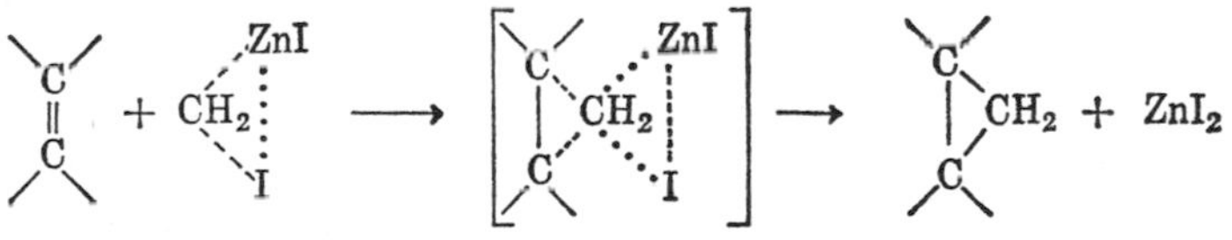

This mechanism is supported by the observed stereospecificity of the addition, but detailed structural studies of the reactant would be desirable.   In any event the formation of cyclopropanes by use of methylene iodide and zinc, a reaction that might at first seem to involve an α dehalogenation to give methylene, appears not to proceed via a methylene at all.   Similar conclusions have been reached for certain reactions that appear to involve α dehydrohalogenation.

The transformation of benzyl halides to stilbene derivatives by the action of strong bases[76,77] has been said to proceed via the formation and

[75] H. E. Simmons and R. D. Smith, *J. Am. Chem. Soc.*, **80**, 5323 (1958); **81**, 4256 (1959).

[76] C. A. Bischoff, *Ber.*, **21**, 2072 (1888).

[77] M. S. Kharasch et al., *J. Am. Chem. Soc.*, **61**, 2318 (1939); **62**, 2035 (1940); **65**, 11 (1943); **66**, 1276 (1944).

dimerization of an arylmethylene,[78-80] but Hahn[81] and Kleucker[82] pointed out that the reaction might consist of carbanion formation, alkylation, and dehydrohalogenation.

$$ArCH_2Cl \xrightarrow{B^-} Ar\overset{\ominus}{C}HCl \xrightarrow{ArCH_2Cl} ArCHCH_2Ar$$
$$\underset{Cl}{\overset{|}{}} \ XI$$
$$_{B^-}\downarrow$$
$$ArCH{=}CHAr$$

This explanation was strongly supported by Hauser and coworkers' isolation of the intermediate chloride XI in the reaction of several benzyl chlorides with potassium amide in liquid ammonia.[83]

Friedman and Berger found that the reaction of methyl chloride with phenylsodium in the presence of cyclohexene gives, after carbonation, 41 per cent toluene, 19 per cent benzene, 3 per cent bicyclo[4.1.0]heptane, 3 per cent benzoic acid, 2 per cent ethylbenzene, some ethylene, and small amounts of n-propylbenzene, isopropyl benzene, phenylacetic acid, and ethane.[84]   The formation of a bicyclo[4.1.0]heptane suggests that methylene may be an intermediate in this reaction, but the observation of Simmons and Smith[75] showed that the transformation of olefins to cyclopropane is not convincing evidence for the intermediacy of a methylene.   Perhaps chloromethylsodium ($ClCH_2Na$) is an intermediate in this reaction just as iodomethylzinc iodide appears to be in the Simmons-Smith reaction.   It is true that chloromethylsodium should have a greater tendency to lose metal halide to give methylene, but the absence of methylene-insertion reactions in the present case weakens the argument that methylene is an intermediate.   Use of *cis*-2-butene in the reaction showed that the cyclopropane formation is stereospecific.

Friedman and Berger also reported extensive α dehydrohalogenation in the reaction of other primary chlorides with phenylsodium.   The formation of 1,1-dimethylcyclopropane from neopentyl chloride and organosodium compounds[85] was shown to be an alpha rather than a gamma

[78] A. Michael, *J. Am. Chem. Soc.*, **42**, 820 (1920).

[79] E. Bergmann and J. Hervey, *Ber.*, **62B**, 893 (1929).

[80] C. K. Ingold and J. A. Jessop, *J. Chem. Soc.*, 709 (1930).

[81] G. Hahn, *Ber.*, **62B**, 2485 (1929).

[82] E. Kleucker, *Ber.*, **62B**, 2587 (1929).

[83] C. R. Hauser et al., *J. Am. Chem. Soc.*, **78**, 1653 (1956); **79**, 397 (1957).

[84] L. Friedman and J. G. Berger, *J. Am. Chem. Soc.*, **82**, 5758 (1960); **83**, 492, 500 (1961).   Cf. W. Kirmse and W. v. E. Doering, *Tetrahedron*, **11**, 266 (1960).

[85] F. C. Whitmore et al., *J. Am. Chem. Soc.*, **61**, 1616 (1939); **63**, 124, 2633 (1941); **64**, 1783 (1942).

elimination.   The 1,1-dideuterio compound was found to give a mono-deuteriocyclopropane.[84]

$$
\begin{array}{ccc}
\underset{\mathrm{CH_3}}{\overset{\mathrm{CH_3}}{\diagdown}}\mathrm{C}\underset{\mathrm{CH_3}}{\overset{\mathrm{CD_2Cl}}{\diagup}}
& \xrightarrow{\mathrm{C_6H_5Na}} &
\underset{\mathrm{CH_3}}{\overset{\mathrm{CH_3}}{\diagdown}}\mathrm{C}\underset{\mathrm{CH_3}}{\overset{\overset{\mathrm{Na}}{\mid}}{\mathrm{CDCl}}}
& \rightarrow &
\underset{\mathrm{CH_3}}{\overset{\mathrm{CH_3}}{\diagdown}}\mathrm{C}\underset{\mathrm{CH_3}}{\overset{\mathrm{C{-}D}}{\diagup}}
\end{array}
$$

XII

$$
\mathrm{CH_3\underset{\overset{\mid}{CH_3}}{C}{=}CDCH_3}
\qquad \swarrow \qquad \downarrow \qquad
\underset{\mathrm{CH_3}}{\overset{\mathrm{CH_3}}{\diagdown}}\mathrm{C}\underset{\mathrm{CH_2}}{\overset{\mathrm{CHD}}{\diagup}}
$$

In this case the intermediacy of t-butylmethylene (XII) is supported by the fact that the alkaline decomposition of $(CH_3)_3CCH{=}NNHSO_2C_7H_7$ in diethylcarbitol, a reaction believed to proceed via a methylene intermediate (see Sec. 24-3c), yields 1,1-dimethylcyclopropane and 2-methyl-2-butene in about the same ratio.

Using the weaker base potassium amide, Hauser and coworkers found some alpha elimination with rearrangement but no cyclopropane formation.   For example, the reaction of n-octyl bromide labeled with 1.84 atoms of deuterium per molecule at the β-carbon atom yielded 1-octene with about 1.05 deuterium atoms per molecule.[86]   Apparently about 10 per cent of the reaction consisted of alpha elimination.

$$
n\text{-}C_6H_{13}CD_2CH_2Br + NH_2^- \xrightarrow{\alpha} n\text{-}C_6H_{13}CD_2\overset{\ominus}{C}HBr
$$
$$
{\scriptstyle\beta}\downarrow \qquad\qquad\qquad\qquad \downarrow
$$
$$
n\text{-}C_6H_{13}CD{=}CH_2 \qquad\qquad n\text{-}C_6H_{13}CD{=}CHD
$$

The alpha elimination need not involve a methylene intermediate, however, since hydrogen migration may accompany bromide loss.

Closs and Closs obtained derivatives of chlorocyclopropane by the action of butyllithium on methylene chloride in the presence of olefins.[87] Although such observations may not constitute conclusive evidence that a methylene was formed, the known tendency of α-halogen substituents to facilitate methylene formation makes the intermediacy of chloro-methylene particularly plausible.   Using competition experiments, Closs and Schwartz found that the reactivity of olefins toward chloromethylene

[86] S. M. Luck, D. G. Hill, A. T. Stewart, Jr., and C. R. Hauser, *J. Am. Chem. Soc.*, **81**, 2784 (1959).

[87] G. L. Closs and L. E. Closs, *J. Am. Chem. Soc.*, **81**, 4996 (1959); **82**, 5723 (1960).

is increased by electron-donating substituents but that changes in olefin structure affect the reactivity toward chloromethylene less than they affect the reactivity toward dichloromethylene (Table 24-2). That is, as would be expected, chloromethylene is a less selective reagent than is the more stable dichloromethylene.[88]

The suggestion that the reaction of dichloromethyl methyl ether with potassium isopropoxide leads to methoxychloromethylene is supported by the observation of a significant deuterium kinetic isotope effect (which also shows that the reaction does not involve initial rapid reversible carbanion formation).[89]

$$i\text{-PrO}^- + CDCl_2OCH_3 \rightarrow Cl\text{—}C\text{—}OCH_3 + i\text{-PrOD} + Cl^-$$

24-3c. Other Methods of Forming Methylenes. Several of the methods for forming methylene may also be used for the formation of substituted methylenes. Staudinger and coworkers suggested methylene intermediates to explain many of the reactions of diazo compounds.[37,90,91] The photolysis of diphenyldiazomethane leads to the formation of the corresponding ketazine (XIII).[90,91] The intermediate diphenylmethylene apparently attacks unchanged diphenyldiazomethane.

$$(C_6H_5)_2CN_2 \xrightarrow{h\nu} N_2 + (C_6H_5)_2C \xrightarrow{(C_6H_5)_2CN_2} (C_6H_5)_2C\text{=}NN\text{=}C(C_6H_5)_2$$
$$\text{XIII}$$

In contrast, the photolysis of diazoethane yields ethylene and 2-butene.[f]

$$CH_3CHN_2 \rightarrow CH_3CH \xrightarrow{CH_3CHN_2} CH_3CH\text{=}CHCH_3 + N_2$$
$$\downarrow$$
$$CH_2\text{=}CH_2$$

In the photolysis of methylketene, where the same organic products are obtained, the increase in 2-butene yield that accompanies an increase in the pressure of methylketene used shows that the ethylene formed does result from the rearrangement of intermediate ethylidene and not from a one-step reaction.[93] Skell and coworkers showed that the additions of diphenylmethylene[94] and propargylene ($HC\equiv C\text{—}C\text{—}H$)[95] to olefins are

[88] G. L. Closs and G. M. Schwartz, *J. Am. Chem. Soc.*, **82**, 5729 (1960).

[89] J. Hine, R. J. Rosscup, and O. C. Duffey, *J. Am. Chem. Soc.*, **82**, 6120 (1960).

[90] H. Staudinger et al., *Ber.*, **44**, 2194, 2197 (1911); **49**, 1884, 1897, 1923, 1928, 1951, 1961, 1969, 2522 (1916).

[91] For a review of the reactions of diazo compounds and their mechanisms, see R. Huisgen, *Angew. Chem.*, **67**, 439 (1955).

[92] R. K. Brinton and D. H. Volman, *J. Chem. Phys.*, **19**, 1394 (1951).

[93] G. B. Kistiakowsky and B. H. Mahan, *J. Chem. Phys.*, **24**, 922 (1956).

[94] R. M. Etter, H. S. Skovronek, and P. S. Skell, *J. Am. Chem. Soc.*, **81**, 1008 (1959).

[95] P. S. Skell and J. Klebe, *J. Am. Chem. Soc.*, **82**, 247 (1960).

not stereospecific, suggesting that these methylenes are triplets, at least by the time they react with olefins.

Friedman and Shechter showed that the basic decomposition of p-toluenesulfonylhydrazones, a reaction discovered by Bamford and Stevens,[96] may be used as a general method for the formation of methylenes, since the intermediate diazo compounds usually decompose under the reaction conditions.[97]

$$
\begin{array}{ccc}
\underset{\displaystyle \text{R}}{\overset{\displaystyle \text{R}}{|}} & & \underset{\displaystyle \text{R}}{\overset{\displaystyle \text{R}}{|}} \\
\text{R}-\text{C}=\text{NNHSO}_2\text{C}_7\text{H}_7 & \xrightarrow{\text{base}} & \text{R}-\text{C}=\text{N}\overset{\ominus}{\text{N}}\text{SO}_2\text{C}_7\text{H}_7 \\
& & \downarrow \\
\underset{\displaystyle \text{R}}{\overset{\displaystyle \text{R}}{|}} & & \underset{\displaystyle \text{R}}{\overset{\displaystyle \text{R}}{|}} \\
\text{R}-\text{CH}-\overset{\oplus}{\text{N}}\equiv\text{N} & \xleftarrow{\text{ROH}} & \text{R}-\text{C}=\text{N}=\text{N} \\
\downarrow & & \downarrow \\
\underset{\displaystyle \text{R}}{\overset{\displaystyle \text{R}}{|}} & & \\
\text{R}-\overset{}{\text{CH}}\oplus & & \text{R}-\text{C}-\text{R}
\end{array}
$$

Powell and Whiting found that the rate-limiting step of the reaction is the first-order decomposition of the anion of the tosylhydrazone.[98]   In ethylene glycol and certain other solvents that are considerably more strongly acidic than water, it appears that the intermediate diazo compound frequently abstracts a proton from the solvent to give an aliphatic diazonium ion, which may lose nitrogen to give a carbonium ion.   In more weakly acidic solvents, such as acetamide and ethers, carbonium-ion formation does not compete significantly with methylene formation.

PROBLEMS

1. The photolysis of ketene in the presence of a large excess of cyclobutane yields propylene and ethylene in essentially equal amounts and also methylcyclobutane. The ratio of the yield of propylene (or ethylene) to methylcyclobutane decreases with increasing pressure.   Using 3130 A radiation at 60°, the ratio decreases from 8.9 at 3.43 mm pressure to 0.081 at 726 mm pressure.   Suggest an explanation for these observations.

2. The reaction of methylene chloride with potassium t-butoxide in the presence of benzene gives (in poor yield) tropilium t-butoxide.   Suggest a mechanism for this reaction.

3. The photolysis of ethane gives ethylene and hydrogen.   Photolysis of CH_3CD_3 and of a mixture of C_2H_6 and C_2D_6 yield H_2 and D_2 but very little HD.   Suggest a mechanism for this reaction.

[96] W. R. Bamford and T. S. Stevens, *J. Chem. Soc.*, 4735 (1952).

[97] L. Friedman and H. Shechter, *J. Am. Chem. Soc.*, **81,** 5512 (1959); **82,** 1002 (1960).

[98] J. W. Powell and M. C. Whiting, *Tetrahedron*, **7,** 305 (1959).

4. Compounds of the type shown below (3-nitroso-2-oxazolidones) decompose in the presence of alkali, giving off nitrogen.

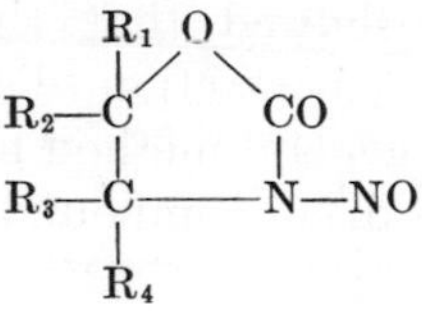

$$
\begin{array}{c}
\text{R}_1\ \ \text{O} \\
\text{R}_2-\text{C}\diagdown \\
\ \ \ \ \ \ \ \ \ \ \text{CO} \\
\text{R}_3-\text{C}-\!\!\!-\!\!\!-\text{N}-\text{NO} \\
\text{R}_4
\end{array}
$$

The following table lists some of the results obtained:

R_1	R_2	R_3	R_4	Product(s)	Yield, %
H	H	H	H	$HC{\equiv}CH$	12
CH_3	CH_3	H	H	$(CH_3)_2CHCHO$	80
C_6H_5	C_6H_5	H	H	$C_6H_5C{\equiv}CC_6H_5$	100
C_6H_5	CH_3	H	H	$C_6H_5C{\equiv}CCH_3$	74
				$C_6H_5CH_2COCH_3$	16
H	H	CH_3	CH_3	$(CH_3)_2CHCHO$	38
				$CH_2{=}C(CH_3)CH_2OH$	54
H	H	C_6H_5	H	$C_6H_5CHOHCH_2OH$	77
				$C_6H_5COCH_3$	12

Suggest at least two reasonable reaction mechanisms and describe experiments that might be used to distinguish between them.

5. Suggest methods that might be used to generate and establish the intermediacy of dialkoxymethylenes.

6. What does the catalytic action of finely divided silver (or copper or platinum) tell about the mechanism of the Wolff rearrangement?

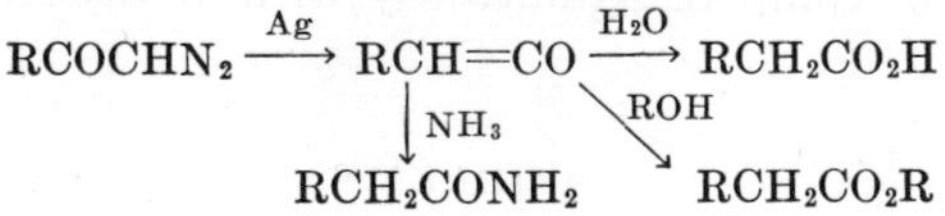

$$
RCOCHN_2 \xrightarrow{\ Ag\ } RCH{=}CO \xrightarrow{\ H_2O\ } RCH_2CO_2H
$$

$$
\Big\downarrow NH_3 \qquad \searrow ROH
$$

$$
RCH_2CONH_2 \qquad\quad RCH_2CO_2R
$$

Chapter 25

MULTICENTER-TYPE REACTIONS

According to the definition stated in Sec. 3-2c, multicenter-type reactions are those in which the atoms in the reactant(s) simply change their configuration to that of the product(s) without electron pairing or unpairing and without the formation or destruction of ions. There are usually three or four key atoms, each of which, in the transition state, is breaking an old single bond and/or forming a new single bond to an atom to which it was not previously bonded. Some three-center-type reactions were discussed in Chap. 24; the current chapter is devoted to other multicenter-type reactions. From the definition we should expect the following characteristics of reactions of this type: Their rates should not usually be greatly affected by the nature of the solvent, and in most cases they should proceed at a similar rate in the vapor phase and in solution. They should not be induced by initiators nor slowed by inhibitors nor have any of the other characteristics of chain reactions. They should not *require* acid or base catalysis, although cases are known that are *subject* to such catalysis.

25-1. Some Four-center-type Rearrangements. *25-1a. Cope Rearrangements.* Some excellent examples of four-center-type organic reactions are found in some of the rearrangements that were discovered and studied most extensively by Cope and coworkers. Malononitriles, cyanoacetic esters, and malonic esters, all containing both an allyl and a vinyl substituent on the α-carbon atom, quite generally undergo a rearrangement reaction upon heating to give α,β-unsaturated compounds.[1]

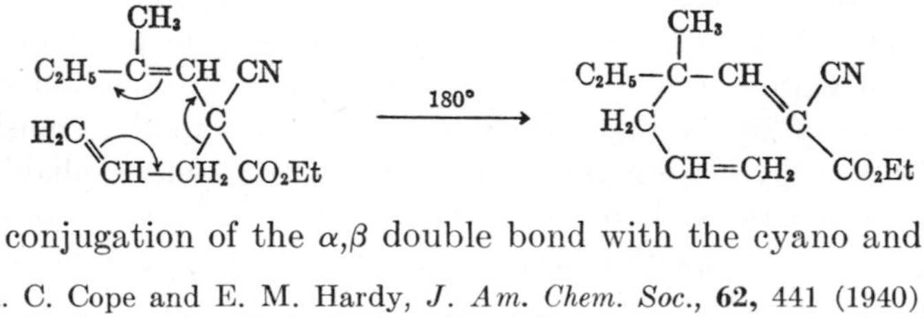

$$
\begin{array}{c}
\text{CH}_3 \\
\text{C}_2\text{H}_5-\text{C}=\text{CH} \quad \text{CN} \\
\text{H}_2\text{C} \qquad \text{C} \\
\text{CH}-\text{CH}_2 \quad \text{CO}_2\text{Et}
\end{array}
\xrightarrow{180°}
\begin{array}{c}
\text{CH}_3 \\
\text{C}_2\text{H}_5-\text{C}-\text{CH} \quad \text{CN} \\
\text{H}_2\text{C} \qquad \text{C} \\
\text{CH}=\text{CH}_2 \quad \text{CO}_2\text{Et}
\end{array}
\qquad (25\text{-}1)
$$

The conjugation of the α,β double bond with the cyano and carbethoxy

---

[1] A. C. Cope and E. M. Hardy, *J. Am. Chem. Soc.*, **62**, 441 (1940); A. C. Cope, K. E. Hoyle, and D. Heyl, *J. Am. Chem. Soc.*, **63**, 1843 (1941); A. C. Cope, C. M. Hofmann, and E. M. Hardy, *J. Am. Chem. Soc.*, **63**, 1852 (1941).

groups is a driving force for the reaction. The mechanism above was suggested by analogy with the Claisen rearrangement. In agreement with the requirements of this mechanism, the reaction shown and a number of related ones were found to be kinetically first-order.[1] As also required, it was shown that the reaction is *intramolecular*. This was done by rearranging a mixture of ethyl (1-methyl-1-hexenyl)allyl-cyanoacetate (I) and diethyl isopropenylcrotylmalonate (II).

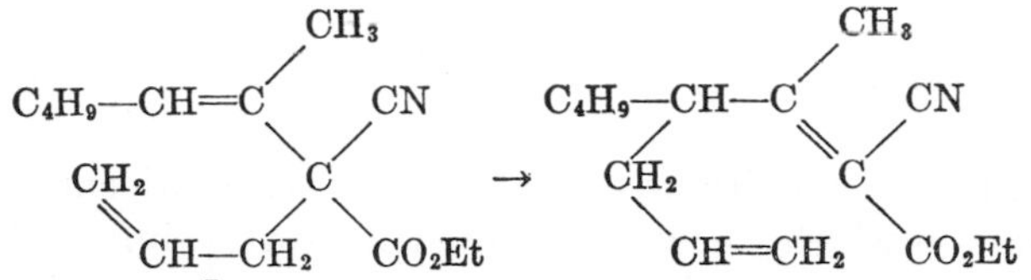

$$C_4H_9-CH=C\begin{smallmatrix}CH_3\end{smallmatrix}\quad CN$$

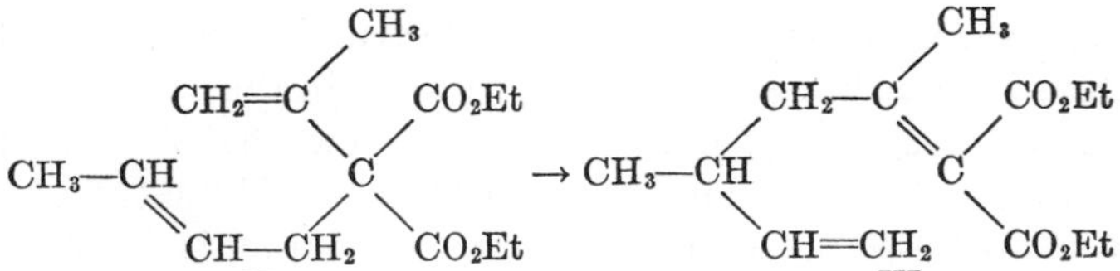

The reactions cleanly follow the equations above, showing that although both allyl and crotyl groups are migrating, evidently they do not become free, since no mixed products are formed.[1] The two groups are indeed migrating at the same time, since separate experiments on pure samples of I and II showed that their rearrangement rates are comparable. The rearrangement of II (which contains a crotyl group) to III (in which this has become an α-methylallyl group) is additional evidence for the mechanism (25-1), which requires that an allyl group "turn around" as it migrates.

Foster, Cope, and Daniels[2] studied the kinetics of the rearrangements of two of the substituted malononitriles and one cyanoacetic ester at several temperatures and found that the entropy of activation was about -12 e.u. They pointed out that this considerable negative entropy of activation is in agreement with the cyclic mechanism proposed, since the cyclic transition state will have lost several of the degrees of freedom of the reactant. Although there are other ways in which freedom of rotation around various bonds, etc., could be lost in a transition state, such correlations of the configuration of the reactant in the transition state with the entropy of activation are certainly much more reliable for four-center-type reactions than for reactions involving ions, because the entropy effects associated with ionic solvation may be large and not easily predictable.

[2] E. G. Foster, A. C. Cope, and F. Daniels, *J. Am. Chem. Soc.*, **69**, 1893 (1947).

Cope and coworkers studied the effect of structure on reactivity in these rearrangements and, among other things, found that the malononitriles rearranged faster than the corresponding cyanoacetic esters, which in turn were more reactive than the malonic esters.[1] This might be correlated with the greater electron-withdrawing power of the cyano group (compared to carbethoxy), or it might be that the cyano group better enters into conjugation with the α,β double bond being formed in the transition state. The latter explanation appears to be more probable (or more important). This follows from the fact that the phenyl group, which may conjugate with double bonds as effectively as cyano or carbethoxy but which is hardly comparable as an electron-withdrawing group, appears to be about as effective as a cyano or carbethoxy group at bringing about the Cope rearrangement. Thus Levy and Cope found that 3-phenyl-1,5-hexadiene rearranges to 1-phenyl-1,5-hexadiene smoothly at 177°.[3]

$$CH_2{=}CH \qquad\qquad CH_2{-}CH$$
$$CH_2 \qquad\quad CHC_6H_5 \xrightarrow{177°} CH_2 \qquad\quad CHC_6H_5$$
$$CH{-}CH_2 \qquad\qquad CH{=}CH_2$$

A methyl group was found to be definitely less effective than a phenyl group. 3-Methyl-1,5-hexadiene had to be heated to about 300° to rearrange, and even so the reaction was found to go only to about 95 per cent completion,[1] whereas all the other rearrangements described were complete within the experimental error.

Rearrangements of amine oxides,[4] sulfones,[5] and sulfinates[5] have also been investigated.

25-1b. The Claisen Rearrangement. Claisen discovered that the allyl ethers of phenols rearrange cleanly at temperatures around 200°.[6] The product was found to be an o-allylphenol when an unsubstituted ortho position was present and a p-allylphenol when both ortho positions were blocked. The reaction is first-order and does not require a catalyst. The rearrangements of crotyl phenyl ethers give o-α-methylallylphenols. This does not prove that the reaction mechanism requires the allyl group to become inverted during the reaction. One could make the rationalization that an intermediate resonance-stabilized carbonium ion or radical preferred to recombine at its reactive secondary carbon atom. This alternative explanation, however, is rendered untenable by the observation

[3] H. Levy and A. C. Cope, *J. Am. Chem. Soc.,* **66,** 1684 (1944).

[4] A. C. Cope and P. H. Towle, *J. Am. Chem. Soc.,* **71,** 3423 (1949).

[5] A. C. Cope, D. E. Morrison, and L. Field, *J. Am. Chem. Soc.,* **72,** 59 (1950).

[6] For references to the earlier work described, see D. S. Tarbell, The Claisen Rearrangement, in R. Adams, "Organic Reactions," vol. II, chap. 1, John Wiley & Sons, Inc., New York, 1944.

that α-methylallyl phenyl ethers rearrange to o-crotylphenols. The rearrangement is intramolecular, since mixtures of ethers rearrange without the formation of any cross products. Thus the Claisen rearrangement is very similar to the Cope rearrangement, and it appears to have the following mechanism.

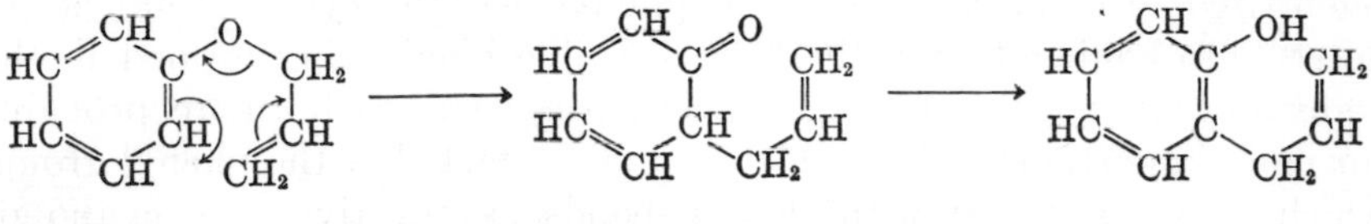

The second step, the enolization of a keto form of a phenol, may very well proceed by a polar mechanism, but since it is faster than the first step, it has not been studied kinetically.

The para Claisen rearrangement begins in the same manner as does the ortho rearrangement. The initially formed intermediate (IV) cannot enolize to a phenol, but it can undergo a second four-center-type rearrangement to give a product that can enolize.

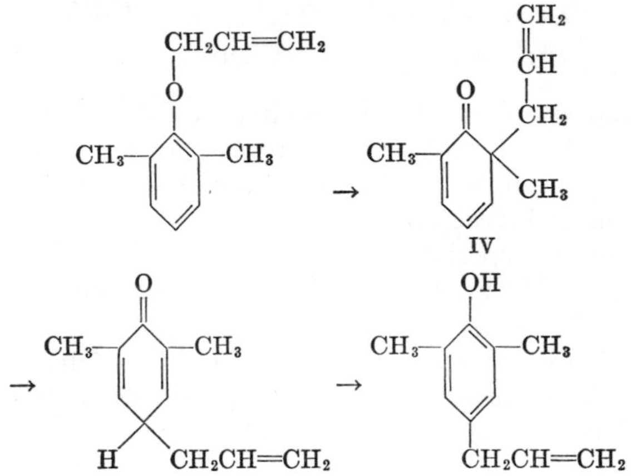

There is strong evidence for the above mechanism, which was suggested by Hurd and Pollack.[7] Rhoads, Raulins, and Reynolds showed that, contrary to an earlier report, allyl groups do migrate without net inversion (the second inversion reverses the first).[8,9] The intermediate IV was captured as its maleic anhydride adduct by Conroy and Firestone.[10]

[7] C. D. Hurd and M. A. Pollack, *J. Org. Chem.*, **3**, 550 (1939).

[8] S. J. Rhoads, R. Raulins, and R. D. Reynolds, *J. Am. Chem. Soc.*, **75**, 2531 (1953); **76**, 3456 (1954).

[9] J. P. Ryan and P. R. O'Connor, *J. Am. Chem. Soc.*, **74**, 5866 (1952); H. Schmid and K. Schmid, *Helv. Chim. Acta*, **36**, 489 (1953); E. N. Marvell, A. V. Logan, L. Friedman, and R. W. Ledeen, *J. Am. Chem. Soc.*, **76**, 1922 (1954).

[10] H. Conroy and R. A. Firestone, *J. Am. Chem. Soc.*, **75**, 2530 (1953); **78**, 2290 (1956).

Curtin and Crawford synthesized IV independently and showed that it rearranges relatively rapidly to give a mixture of allyl 2,6-dimethylphenyl ether and 4-allyl-2,6-dimethylphenol.[11]

The effect of structure and solvent on the rate of Claisen rearrangements was studied by several workers.[12–15] The reactivity was unexpectedly found to be increased by electron-donating groups on both the allyl group and the aromatic ring.[12] Illustrating the fact that no sharp line can be drawn between various categories of reaction mechanisms, the Claisen rearrangement seems clearly to have a certain amount of polar character. The rearrangement of p-cresyl allyl ether was found to proceed at the following relative rates in the solvents listed:[15] decalin, 1.0; benzonitrile, 1.6; benzyl alcohol, 6; ethylene glycol, 12; phenol, 29. There would be a much larger change in rate for a typical polar reaction, such as the reaction of methyl iodide with a tertiary amine, run in the solvents listed, however. The data given suggest the possibility of acid catalysis.

The Claisen rearrangement of allyl ethers is not limited to phenol ethers. Indeed, the first example that Claisen found was the rearrangement of the O-allyl ether of acetoacetic ester.[6] The reaction appears to be rather general for the allyl ethers of enols. Hurd and Pollack found that the reaction proceeds satisfactorily for the simplest possible case, allyl vinyl ether rearranging to allyl acetaldehyde.[16]

In a kinetic study Schuler and Murphy found this rearrangement to be a homogeneous first-order gas-phase reaction with an entropy of activation of -7.7 e.u.[17]

25-1c. Some Multicenter-type S_Ni Reactions. Although it is not entirely clear how broad an area of reaction mechanisms Hughes, Ingold, and coworkers intended to include when they suggested the term S_Ni for certain internal nucleophilic substitution reactions,[18] we shall include

[11] D. Y. Curtin and R. J. Crawford, *J. Am. Chem. Soc.*, **79**, 3156 (1957).

[12] W. N. White et al., *J. Am. Chem. Soc.*, **80**, 3271 (1958); *J. Org. Chem.*, **26**, 627 (1961).

[13] S. J. Rhoads and R. L. Crecelius, *J. Am. Chem. Soc.*, **77**, 5057 (1955).

[14] H. Schmid et al., *Helv. Chim. Acta*, **40**, 13 (1957); **41**, 657 (1958).

[15] H. L. Goering and R. R. Jacobson, *J. Am. Chem. Soc.*, **80**, 3277 (1958).

[16] C. D. Hurd and M. A. Pollack, *J. Am. Chem. Soc.*, **60**, 1905 (1938).

[17] F. W. Schuler and G. W. Murphy, *J. Am. Chem. Soc.*, **72**, 3155 (1950); cf. L. Stein and G. W. Murphy, *J. Am. Chem. Soc.*, **74**, 1041 (1952).

[18] W. A. Cowdrey, E. D. Hughes, C. K. Ingold, S. Masterman, and A. D. Scott, *J. Chem. Soc.*, 1252 (1937); cf. Sec. 6-3.

under this heading those reactions in which an atom with an unshared electron pair (after it is displaced, at least) is displaced from a carbon atom by another atom (with unshared electrons) that was attached to the carbon atom at which displacement occurs only through the atom displaced.

One reaction whose mechanism fits this definition is the decomposition of geminate diacyloxy compounds to aldehydes and anhydrides.

$$
\text{R}-\text{CH}
\begin{array}{c}
\overset{\displaystyle O}{\underset{\displaystyle\parallel}{}}\\[-2pt]
\text{O}-\text{C}-\text{R}'\\
\text{O}-\text{C}-\text{R}'\\[-2pt]
\underset{\displaystyle O}{\parallel}
\end{array}
\quad\longrightarrow\quad
\text{R}-\text{CHO} \;+\;
\begin{array}{c}
\overset{\displaystyle O}{\underset{\displaystyle\parallel}{}}\\[-2pt]
\text{C}-\text{R}'\\
\text{O}\\[-2pt]
\text{O}=\text{C}-\text{R}'
\end{array}
\tag{25-2}
$$

Coffin and coworkers studied the kinetics of a number of reactions of this type in the vapor phase and found all to be homogeneous first-order reactions.[19] The reaction in the liquid phase is subject to acid catalysis, but in the case of benzylidene diacetate and trichloroethylidene diacetate the liquid-phase reaction in the absence of catalysts has very nearly the same rate constant as the gas-phase reaction at the same temperature. It therefore appears likely that the uncatalyzed reaction has a multicenter-type mechanism like (25-2), whereas the acid-catalyzed reaction proceeds by a polar mechanism.   The reactivity of $(\text{R}'\text{CO}_2)_2\text{CHR}$ compounds has been found to be little affected by changing R' from methyl to ethyl to propyl.   Variations in R gave relative reactivities that varied in the following   order: $\text{C}_6\text{H}_5 \sim o\text{-ClC}_6\text{H}_4 \sim \alpha\text{-furyl} \sim \text{CH}_3\text{CH}{=}\text{CH} > n\text{-C}_6\text{H}_{13} \sim n\text{-C}_3\text{H}_7 \sim \text{CH}_3 \sim \text{CCl}_3 > (\text{AcO})_2\text{CH} > \text{H}$.   Thus,   in   general,   the reactivity increases with the stability of the carbonyl group of the aldehyde being formed in the transition state.   The position of the trichloromethyl and perhaps the diacetoxymethyl group seems anomalous, but steric factors are probably important in these cases.

25-2. Addition to Multiple Bonds. *25-2a. The Diels-Alder Reaction.*   A number of Diels-Alder additions of unsaturated compounds to conjugated dienes have been found to be simple second-order reactions that have no induction period and are unaffected by initiators and inhibitors.[20]   Usually the reactivity of the diene is increased by electron-donating groups, and that of the dienophile is increased by electron-withdrawing groups.[21]   Steric factors are also quite important.   The

[19] C. C. Coffin et al., *Can. J. Research,* **5,** 636 (1931); **6,** 417 (1932); **15B,** 229, 247, 254, 260 (1937); **18B,** 223, 410 (1940).

[20] A. Wassermann, *J. Chem. Soc.,* 828 (1935); G. A. Benford, H. Kaufmann, B. S. Khambata, and A. Wassermann, *J. Chem. Soc.,* 381 (1939), and other sources cited therein.

[21] For some of the many data on reactivity and the scope of the Diels-Alder reaction, see Ref. 6, Adams, *op. cit.,* vol. IV, chaps. 1 and 2, vol. V, chap. 3.

reactivity of the diene is greatly increased when, as in the case of cyclo-pentadiene, the diene is held in a conformation in which the double bonds are oriented cis to each other with respect to rotation around the single bond between them.[21] The reaction appears to involve the formation of a complex (perhaps a reactive intermediate) like V, which may, by a relatively small change in the positions of the atoms, be transformed into the product.[20]

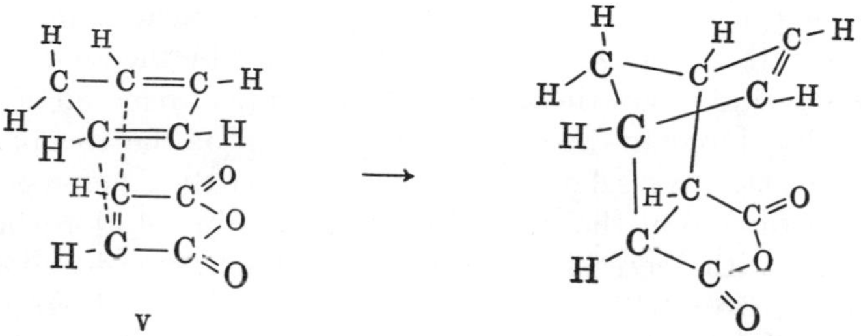

$$(25\text{-}3)$$

The simultaneous formation of both new carbon-carbon bonds explains the stereospecificity of the reaction. Additions to maleic and fumaric acids yield cis and trans dicarboxylic acids, respectively. Attractive interactions between the π-electron systems of the two reactant molecules are held responsible for the more rapid formation of the endo isomer in reactions like (25-3). To explain the nature of the products formed preferentially from the addition of unsymmetrical dienes to unsymmetrical dienophiles, it may be postulated that the formation of the two new bonds may not have proceeded to the same extent in the transition state.[22] Thus the transition state in the addition of 2-methoxybutadiene to acrolein is probably stabilized by the contribution of structures like those shown below, and therefore the methoxy and formyl groups are oriented 1,4 rather than 1,3 in the reaction product.

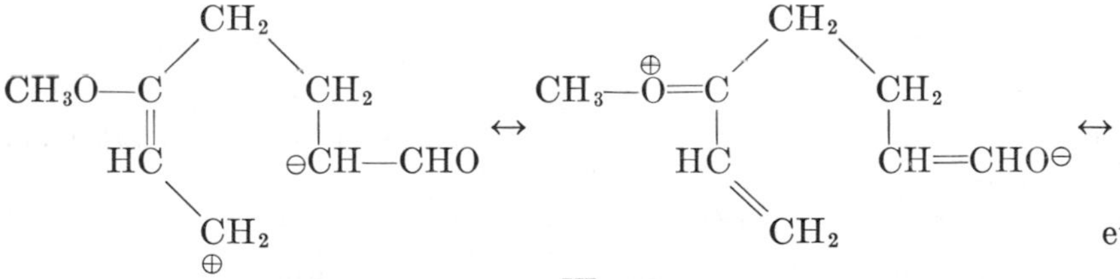

VI

An extreme version of this hypothesis is the suggestion that the reaction involves a zwitterionic species like VI, or an analogous diradical, as a

[22] Cf. R. B. Woodward and T. J. Katz, *Tetrahedron*, **5**, 70 (1959).

reaction intermediate.[23,24] The plausibility of such a suggestion varies with the stability that would be expected of the proposed intermediate. At the least, it seems certain that the difference in the extents to which the two new bonds are formed in the transition state will be much greater in some Diels-Alder reactions than in others.

Solvent and catalyst effects show that, although some Diels-Alder reactions have the characteristics of four-center-type reactions, in certain cases the transition state may have considerable polar character. The dimerization of cyclopentadiene is a homogeneous reaction in the gas phase,[25] and its rate in solution is not greatly affected by the ion-solvating power of the solvent.[26] On the other hand, the rate of reaction of cyclopentadiene with quinone in the vapor phase depends on the area of pyrex surface in the reaction vessel and is thus at least partly a heterogeneous reaction.[27] Furthermore, the rate of this reaction in solution shows a definite increase with increasing ion-solvating power.[28] Several Diels-Alder reactions have also been found to be subject to acid catalysis.[29] The acid catalysis is probably due to an increase in the electrophilicity of the dienophile resulting from its hydrogen-bonded association with the acid catalyst. Yates and Eaton observed that aluminum chloride, which is very effective at complexing with oxygen atoms, is a powerful catalyst for Diels-Alder additions to oxygenated dienophiles.[30] So it is seen that under certain conditions the reaction has much of the character of a polar reaction. As we have stated before, no sharp line may be drawn between the two types of reaction mechanisms.

There are a number of other addition reactions that might be included under the present heading with a broad enough definition of the Diels-Alder reaction. These include the addition of vinyl ethers to α,β-unsaturated carbonyl compounds and related reactions involving carbon-oxygen double bonds,[31]

[23] G. B. Kistiakowsky et al., *J. Am. Chem. Soc.*, **58**, 123 (1936); *J. Chem. Phys.*, **5**, 682 (1937); **7**, 725 (1939).

[24] C. Walling and J. Peisach, *J. Am. Chem. Soc.*, **80**, 5819 (1958).

[25] G. A. Benford and A. Wassermann, *J. Chem. Soc.*, 362 (1939).

[26] H. Kaufmann and A. Wassermann, *J. Chem. Soc.*, 870 (1939).

[27] A. Wassermann, *J. Chem. Soc.*, 1089 (1946).

[28] *Ibid.*, 623 (1942).

[29] *Ibid.*, 618 (1942); W. Rubin, H. Steiner, and A. Wassermann, *J. Chem. Soc.*, 3046 (1949).

[30] P. Yates and P. Eaton, *J. Am. Chem. Soc.*, **82**, 4436 (1960).

[31] R. I. Longley, Jr., and W. S. Emerson, *J. Am. Chem. Soc.*, **72**, 3079 (1950); C. W. Smith, D. G. Norton, and S. A. Ballard, *J. Am. Chem. Soc.*, **73**, 5267, 5270, 5273 (1951).

the addition of nitroso compounds to dienes,[32]

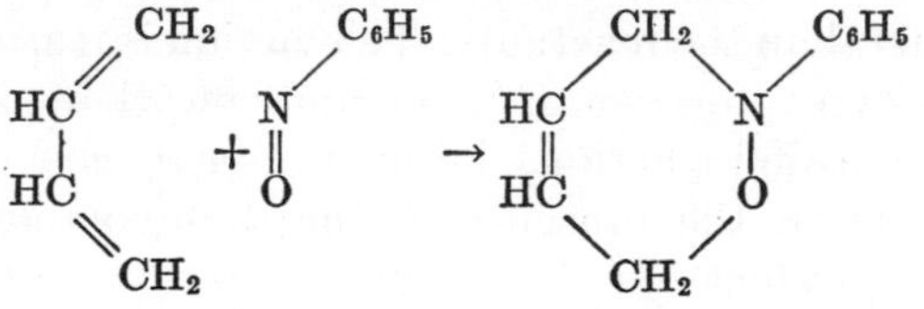

$$
\begin{array}{c}
\text{CH}_2 \qquad \text{C}_6\text{H}_5 \\
\text{HC} \qquad \text{N} \\
\text{HC} \qquad + \quad \text{O} \\
\text{CH}_2
\end{array}
\quad \rightarrow \quad
\begin{array}{c}
\text{CH}_2 \qquad \text{C}_6\text{H}_5 \\
\text{HC} \qquad \text{N} \\
\text{HC} \qquad \text{O} \\
\text{CH}_2
\end{array}
$$

and a number of reactions involving carbon-nitrogen double and triple bonds.[24]

25-2b. *1,3-Dipolar Addition Reactions.* Huisgen and coworkers pointed out that a number of different types of 1,3-addition reactions that have been known for some time may be described in terms of a single generalized reaction scheme.[33] Compounds for which we may write a reasonably stable contributing structure containing an atom with a positive charge and an electron sextet (atom *a*), and an atom with a negative charge and an unshared electron pair (atom *c*), separated by an atom with an unshared electron pair (atom *b*), will often add to the double bonds of other compounds.

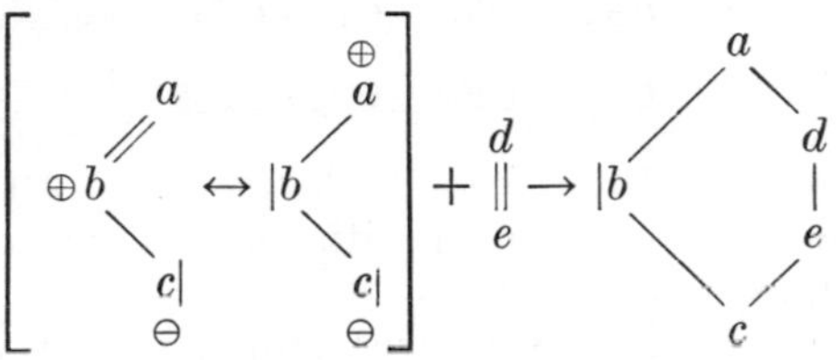

$$
\left[
\begin{array}{c}
a \qquad \overset{\oplus}{a} \\
\oplus b \qquad |b \\
c| \qquad c| \\
\ominus \qquad \ominus
\end{array}
\right]
+
\begin{array}{c}
d \\
\| \\
e
\end{array}
\rightarrow
|b
\begin{array}{c}
a \qquad d \\
\qquad | \\
c \qquad e
\end{array}
$$

The function of the pair of unshared electrons on *b* is to provide resonance stabilization of the reactant, as shown above, not to aid the addition reaction directly. A specific example of a 1,3-dipolar addition is the reaction of maleic anhydride with a nitrile oxide (only the most important contributing structure of which is shown below).

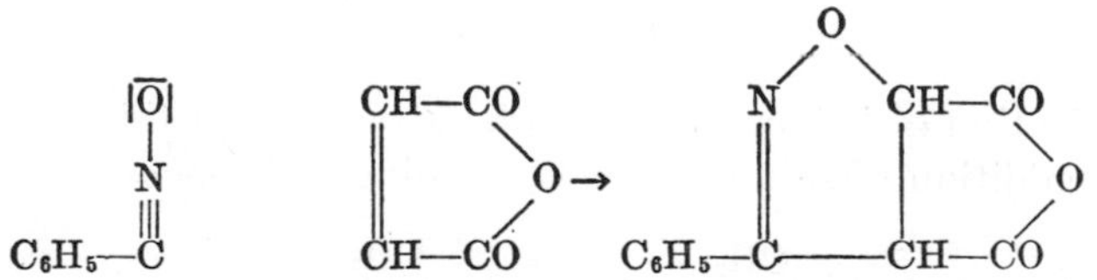

$$
\begin{array}{c}
|\overline{\text{O}}| \\
| \\
\text{N} \\
\| \\
\text{C}_6\text{H}_5\text{—C}
\end{array}
+
\begin{array}{c}
\text{CH—CO} \\
\| \qquad \qquad \text{O} \rightarrow \\
\text{CH—CO}
\end{array}
\begin{array}{c}
\text{O} \\
\text{N} \qquad \text{CH—CO} \\
\| \qquad \qquad \text{O} \\
\text{C}_6\text{H}_5\text{—C}\text{——}\text{CH—CO}
\end{array}
$$

In the case of the addition of phenyl azide to the carbon-carbon double bond of diethyl tetrahydro-3,6-methylenepyridazine-1,2-dicarboxylate, the reaction rate was found to be the same, within a factor of 2, in aceto-

---

[32] Y. A. Arbuzov and coworkers, *Doklady Akad. Nauk S.S.S.R.*, **60**, 993, 1173 (1948); **76**, 681 (1951); *Chem. Abstr.*, **42**, 7299h (1948); **43**, 650c (1949); **45**, 8535e (1951).

[33] Much of this work is summarized by R. Huisgen, *Proc. Chem. Soc.*, 357 (1961).

nitrile, ethanol, benzene, cyclohexane, and several other solvents. Since the transition state is therefore no more polar (actually it appears to be slightly less polar) than the reactants, the reaction is reasonably described as a multicenter-type process. Huisgen and coworkers discovered many new 1,3-dipolar addition reactions, in some cases by generating a reactive 1,3-dipolar system in the presence of the $d{=}e$ compound in order to obtain the cyclic product.

25-2c. Other Four-center-type Addition Reactions. A variety of addition reactions in which three-, four-, and five-membered rings are formed appear, with different degrees of certainty, to be of the multicenter type. The additions of methylenes to olefins (as well as the direct-insertion reactions of methylenes) can be called three-center-type reactions (cf. Chap. 24).

Although the 1,2 addition of one olefin to another is not so common a reaction as Diels-Alder addition, there are several types of compounds known that undergo this reaction. Lacher, Tompkin, and Park studied the kinetics of dimerization of tetrafluoroethylene and chlorotrifluoroethylene in the gas phase.[34]

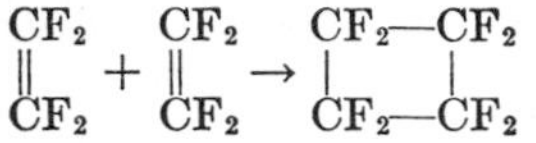

$$\begin{array}{ccc} CF_2 & CF_2 & CF_2-CF_2 \\ \| & + \| & \rightarrow \quad | \qquad | \\ CF_2 & CF_2 & CF_2-CF_2 \end{array}$$

The reactions were found to be second-order without an induction period and with no obvious demand for an initiator. Therefore it seems likely that they are of the four-center type.

Another 1,2 addition that may be of the four-center type is the addition of ketene to cyclopentadiene.[35]

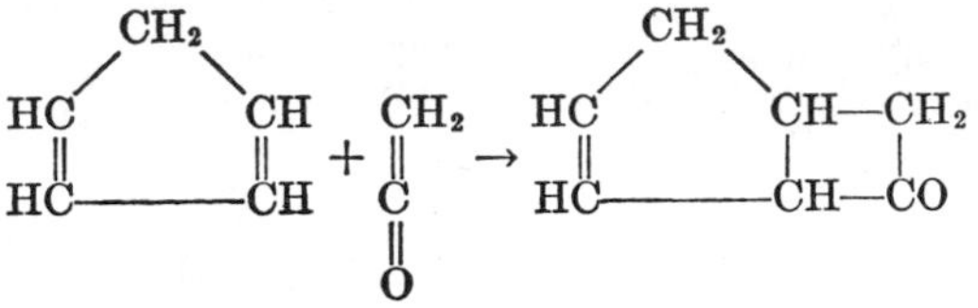

The dimerization of ketenes may also be a reaction of this type.

The 1,4 addition of sulfur dioxide to conjugated dienes,

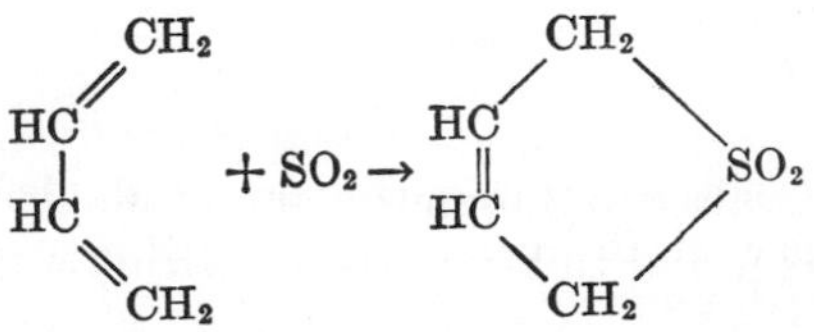

[34] J. R. Lacher, G. W. Tompkin, and J. D. Park, *J. Am. Chem. Soc.*, **74**, 1693 (1952).
[35] A. T. Blomquist and J. Kwiatek, *J. Am. Chem. Soc.*, **73**, 2098 (1951).

a three-center-type reaction that takes place when enough inhibitor is present to prevent the formation of a linear copolymer, has been found to be a smooth second-order reaction in at least three cases.[36]

25-3. Elimination Reactions.   *25-3a. The Chugaev Reaction.*   The Chugaev method of dehydrating an alcohol involves its transformation to a methyl xanthate, which upon pyrolysis yields methyl mercaptan, carbonyl sulfide, and the olefin.   Hückel, Tappe, and Legutke suggested that the reaction involves a cyclic transition state in which the β-hydrogen atom forms a bond to sulfur at the same time the α-carbon–oxygen bond is broken.[37]   Their mechanism was based to a considerable extent on the observation that the dehydration of menthols and α-decalols goes predominantly cis.   The mechanism was given a minor but reasonable modification by Stevens and Richmond, who suggested that it is the doubly bound sulfur atom that removes the β-hydrogen.[38]

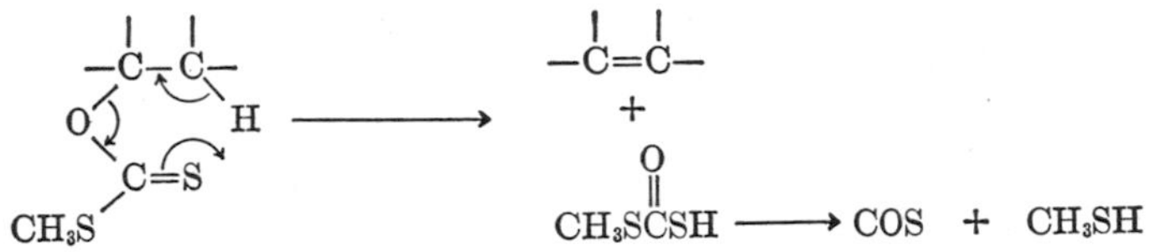

$$\tag{25-4}$$

These workers further observed that pinacolyl alcohol (3,3-dimethyl-2-butanol) gives the unrearranged olefin, *t*-butylethylene.   Barton noted that the reaction is probably of the four-center type,[39] and Alexander and Mudrak added to the evidence for preferentially cis elimination.[40] O'Connor and Nace studied the kinetics of several Chugaev reactions in the liquid phase.[41]   They found the reaction to be first-order and unaffected by several ordinary free-radical inhibitors.   The entropy of activation is negative.   These data support a cyclic multicenter-type mechanism such as (25-4), although, as O'Connor and Nace pointed out, there is no evidence that the electron pairs move as indicated by the arrows, or indeed that they move in pairs at all.   These workers also showed that when the methyl group attached to sulfur is replaced by a more strongly electron-withdrawing group, the reactivity of the xanthate increases.

---

[36] L. R. Drake, S. C. Stowe, and A. M. Partansky, *J. Am. Chem. Soc.*, **68**, 2521 (1946).

[37] W. Hückel, W. Tappe, and G. Legutke, *Ann.*, **543**, 191 (1940).

[38] P. G. Stevens and J. H. Richmond, *J. Am. Chem. Soc.*, **63**, 3132 (1941).

[39] D. H. R. Barton, *J. Chem. Soc.*, 2174 (1949).

[40] E. R. Alexander and A. Mudrak, *J. Am. Chem. Soc.*, **72**, 1810, 3194 (1950).

[41] G. L. O'Connor and H. R. Nace, *J. Am. Chem. Soc.*, **74**, 5454 (1952); **75**, 2118 (1953).

25-3b. *Other Multicenter-type Elimination Reactions.*[42] The preparation of olefins by the pyrolysis of esters bears a strong resemblance to the Chugaev reaction, and, indeed, Hurd and Blunck proposed a multicenter-type mechanism for the reaction before such a mechanism was suggested for the Chugaev reaction.[43]

Evidence for the four-center-type mechanism in this case includes preferred cis orientation,[40,44] first-order kinetics,[41,45] a negative entropy of activation,[41,45] no induction period,[45] and no effect of surface area or inhibitors on the reaction rate.[45]

Barton, Howlett, and coworkers studied the olefin-forming pyrolysis of alkyl chlorides in some detail. They found that the pyrolysis of ethylene dichloride (to give vinyl chloride) is, to a considerable extent, a radical chain reaction.[46] It is kinetically first-order, catalyzed by oxygen and chlorine and strongly inhibited by propylene. The pyrolysis reactions of ethyl chloride and ethylidene chloride, on the other hand, are not at all inhibited by propylene.[46] They are homogeneous first-order reactions in the gas phase and give excellent yields of the mono-dehydrochlorination products. These data suggest a four-center-type mechanism, and similar results have been obtained for *t*-butyl chloride,[47] isopropyl chloride,[48] propylene dichloride,[48] *n*-propyl chloride,[49] *n*-butyl chloride,[49] and 2,2-dichloropropane.[50] The fact that the pyrolysis of menthyl chloride is also homogeneous (in a reaction vessel with suitably coated walls), uninhibited (by propylene), and unimolecular in the vapor phase is of particular interest because the olefinic reaction products consist of about 25 per cent 2-menthene and 75 per cent 3-menthene.[51] The latter, major product must have been formed by a cis elimination,

[42] This subject has been reviewed by C. H. DePuy and R. W. King, *Chem. Revs.*, **60**, 431 (1960).

[43] C. D. Hurd and F. H. Blunck, *J. Am. Chem. Soc.*, **60**, 2419 (1938); cf. Ref. 39.

[44] R. T. Arnold, G. G. Smith, and R. M. Dodson, *J. Org. Chem.*, **15**, 1256 (1950); N. L. McNiven and J. Read, *J. Chem. Soc.*, 2067 (1952).

[45] D. H. R. Barton, A. J. Head, and R. J. Williams, *J. Chem. Soc.*, 1715 (1953); A. T. Blades, *Can. J. Chem.*, **32**, 366 (1954).

[46] D. H. R. Barton and K. E. Howlett, *J. Chem. Soc.*, 155, 165 (1949); 3695 (1952).

[47] D. H. R. Barton and P. F. Onyon, *Trans. Faraday Soc.*, **45**, 725 (1949).

[48] D. H. R. Barton and A. J. Head, *Trans. Faraday Soc.*, **46**, 114 (1950).

[49] D. H. R. Barton, A. J. Head, and R. J. Williams, *J. Chem. Soc.*, 2039 (1951).

[50] K. E. Howlett, *J. Chem. Soc.*, 945 (1953).

[51] D. H. R. Barton, A. J. Head, and R. J. Williams, *J. Chem. Soc.*, 453 (1952).

and the former, minor product may also have been formed by the same mechanism.   Each of the following chlorides is pyrolytically dehydrohalo-genated, at least partly by a free-radical chain reaction: ethylene dichlor-ide,[45] 1,1,1-trichloroethane,[52] *sym-* and *unsym-*tetrachloroethane,[53] 2,2'-di-chlorodiethyl ether,[49] 1,1-dichloropropane,[50] 1,4-dichlorobutane,[54] and 1,1,2-trichloroethane.[54]   It is pointed out that part of the difference between the two classes of compounds described is that many in the former group either react to give radical inhibitors or are themselves inhibitors.

The pyrolytic dehydrohalogenation of organic bromides bears consider-able resemblance to that of organic chlorides.   There is a homogeneous first-order reaction that can be isolated from the simultaneous free-radical chain reaction that sometimes occurs, by the addition of a suffi-cient amount of inhibitors (particularly olefins).   It is therefore reason-able to write for the reaction a four-center-type mechanism like (25-5).

$$
-\overset{|}{\underset{\underset{\text{H}}{|}}{\text{C}}}-\overset{|}{\underset{\underset{\text{Cl}}{|}}{\text{C}}}- \rightarrow -\overset{|}{\underset{\underset{\text{H}---\text{Cl}}{\vdots}}{\text{C}}}\overset{}{\underset{\vdots}{\text{-}}}\overset{|}{\underset{}{\text{C}}}- \rightarrow -\overset{|}{\underset{\underset{\text{H}-\text{Cl}}{}}{\text{C}}}=\overset{|}{\underset{}{\text{C}}}- \qquad (25\text{-}5)
$$

However, reactivity data show that there are significant differences in timing of the bond-breaking and bond-making processes between organic

TABLE 25-1. RELATIVE RATES OF PYROLYTIC OLEFIN FORMATION BY VARIOUS RX'S AT 400°[a]

X \ R	Et	*i*-Pr	*t*-Bu
Cl	1.00	178	19,500
Br	1.00	170	32,000
OAc	1.00	26	515
OCHO	1.00	20	720

[a] The reactivity of each RX is relative to that of the corresponding EtX.

bromides and chlorides, on one hand, and esters, on the other.[55]   Mac-coll[56] tabulated some of the data that illustrate this point (Table 25-1). The difference in reactivity between corresponding *t*-butyl and ethyl

[52] D. H. R. Barton and P. F. Onyon, *J. Am. Chem. Soc.*, **72**, 988 (1950).

[53] D. H. R. Barton and K. E. Howlett, *J. Chem. Soc.*, 2033 (1951).

[54] R. J. Williams, *J. Chem. Soc.*, 113 (1953).

[55] A. Maccoll, "Theoretical Organic Chemistry: Papers presented to the Kekulé Symposium," p. 230.   Butterworth & Co. (Publishers), Ltd., London, 1959.

[56] A. Maccoll, *J. Chem. Soc.*, 3398 (1958).

halides, amounting to about 13 kcal/mole in activation energy, is far too large to attribute to the difference in the stability of the olefins being formed (cf. Table 1-2). The over-all activation energies are too small for the reactions to involve simple homolysis of the C—X bonds; furthermore, the differences in reactivity are too large to attribute to differences in stability between primary, secondary, and tertiary radicals (Table 19-2), when it is realized that the stabilities of organic halides also follow the sequence primary < secondary < tertiary. The observed difference in halide reactivities is about the same (making allowance for the difference in reaction temperatures) as that found in S_N1 reactions (cf. Sec. 7-3c). For this reason, Maccoll and Thomas suggested that in the pyrolysis of halides the carbon-halogen bond has been broken to a much greater extent in the transition state than has the carbon-hydrogen bond, and therefore the halogen atom bears a considerable negative charge and the rest of the molecule a considerable positive charge.[57] An extreme form of this suggestion is the view that the reaction consists of a rate-controlling formation of an ion pair (not dissociated ions—the activation energy is much too small for dissociation to independent ions to be plausible), which then rapidly yields the olefin.[58] The observation of significant kinetic isotope effects ($k_H/k_D \sim 2$) in several cases,[55] however, shows that the carbon-hydrogen bond is probably being broken in the rate-controlling step. Thus, even if ionization to an ion pair does occur, it is probably not the rate-controlling step of the reaction.

The much smaller differences in reactivity observed with esters show that the transition state does not have any large amount of ionic character. This conclusion is supported by DePuy and Leary's observation that $p\text{-}CH_3OC_6H_4CHOAcCH_2CH_2C_6H_5$ loses acetic acid only about six times as fast as $C_6H_5CHOAcCH_2CH_2C_6H_5$ (in an S_N1 reaction a p-methoxy substituent would be expected to increase the reactivity by 10^3-fold or more).[59] It appears that in the transition state for ester pyrolysis the carbon-oxygen and carbon-hydrogen bonds have both been broken and the carbon-carbon double bond formed to a considerable extent. Therefore the reaction rate is considerably influenced by the stability of the olefin being formed. However, steric factors (due, for example, to increased eclipsing in the transition state), statistical factors (relating to the number of β-hydrogen atoms that can be lost with the acyloxy group), and polar factors are important, also.[60,61]

[57] A. Maccoll and P. J. Thomas, *Nature*, **176**, 392 (1955).

[58] C. K. Ingold, *Proc. Chem. Soc.*, 279 (1957).

[59] C. H. DePuy and R. E. Leary, *J. Am. Chem. Soc.*, **79**, 3705 (1957).

[60] D. H. Froemsdorf, C. H. Collins, G. S. Hammond, and C. H. DePuy, *J. Am. Chem. Soc.*, **81**, 643 (1959).

[61] C. H. DePuy, R. W. King, and D. H. Froemsdorf, *Tetrahedron*, **7**, 123 (1959).

PROBLEMS

1. Suggest experiments designed to show whether the thermal decomposition of cyclobutane to ethylene, a reaction known to be first-order in the gas phase and unaffected by the surface of the reaction vessel and by such free-radical inhibitors as propylene, toluene, and nitric oxide, is a one-step four-center-type reaction or a two-step reaction involving an intermediate diradical as shown below.

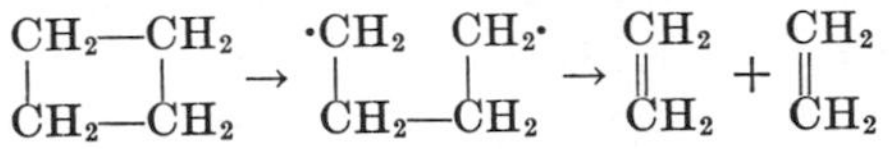

$$\begin{array}{cc} CH_2\!\!-\!\!CH_2 \\ | \quad\;\; | \\ CH_2\!\!-\!\!CH_2 \end{array} \rightarrow \begin{array}{cc} \cdot CH_2 \;\; CH_2\cdot \\ | \quad\;\; | \\ CH_2\!\!-\!\!CH_2 \end{array} \rightarrow \begin{array}{cc} CH_2 \\ \| \\ CH_2 \end{array} + \begin{array}{cc} CH_2 \\ \| \\ CH_2 \end{array}$$

What do the existing observations tell about the mechanism of the reaction?

2. Rationalize the observation that the pyrolysis of t-amyl acetate at 400° yields an olefin mixture containing 74 per cent 2-methyl-1-butene and 26 per cent 2-methyl-2-butene.

3. Suggest a mechanism for the reaction

$$C_2H_2 + C_2F_4 \xrightarrow[\text{Inhibitor}]{600° \; 1 \; atm} CF_2\!\!=\!\!CH\!\!-\!\!CH\!\!=\!\!CF_2$$

4. For several of the reactions of Sec. 25-1a, tell what fraction of the total number of conformers of the reactant should be capable of undergoing the reaction shown. Assuming equal stability for all the conformers, how large an increment will this difference between the number of conformers available to the reactant and the number available to the transition state contribute to the entropy of activation? Compare your answers with the experimentally observed entropies of activation.

5. Could the strong tendency of fluoroolefins to dimerize to give cyclobutane derivatives be explained in terms of an increase in the C—F bond energy accompanying the change of the sp^2 carbon to sp^3 hybridization, in which the carbon appears to be less electronegative? What is a reasonable estimate of the electronegativity difference between sp^2 and sp^3 carbon? How much would this affect the C—F bond energy? By how much would this factor cause the equilibrium constant for the dimerization of tetrafluoroethylene to differ from that for ethylene?

NAME INDEX

SUBJECT INDEX